Handbook of Aseptic Processing and Packaging

Nine years have passed since the second edition of the *Handbook of Aseptic Processing and Packaging* was published. Significant changes have taken place in several aseptic processing and packaging areas. These include aseptic filling of plant-based beverages for non-refrigerated shelf-stable formats for longer shelf life and sustainable packaging along with cost of environmental benefits to leverage savings on energy and carbon footprint. In addition, insight into safe processing of particulates using two- and three-dimensional thermal processing followed by prompt cooling is provided.

In the third edition, the editors have compiled contemporary topics with information synthesized from internationally recognized authorities in their fields. In addition to updated information, 12 new chapters have been added in this latest release with content on

- Design of the aseptic processing system and thermal processing
- Thermal process equipment and technology for heating and cooling
- Flow and residence time distribution (RTD) for homogeneous and heterogeneous fluids
- Thermal process and optimization of aseptic processing containing solid particulates
- Aseptic filling and packaging equipment for retail products and food service
- Design of facility, infrastructure, and utilities
- Cleaning and sanitization for aseptic processing and packaging operations
- Microbiology of aseptically processed and packaged products
- Risk-based analyses and methodologies
- Establishment of "validated state" for aseptic processing and packaging systems
- Quality and food safety management systems for aseptic and extended shelf life (ESL) manufacturing
- Computational and numerical models and simulations for aseptic processing
- Also, there are seven new appendices on original patents, examples of typical thermal process calculations, and particulate studies—single particle and multiple-type particles, and Food and Drug Administration (FDA) filing

The three editors and 22 contributors to this volume have more than 250 years of combined experience encompassing manufacturing, innovation in processing and packaging, R&D, quality assurance, and compliance. Their insight provides a comprehensive update on this rapidly developing leading-edge technology for the food processing industry.

The future of aseptic processing and packaging of foods and beverages will be driven by customer-facing convenience and taste, use of current and new premium clean label natural ingredients, use of multifactorial preservation or hurdle technology for maximizing product quality, and sustainable packaging with claims and messaging.

Handbook of Aseptic Processing and Packaging
Third Edition

Edited by
Jairus R. D. David
Pablo M. Coronel
Josip Simunovic

CRC Press
Taylor & Francis Group
Boca Raton London New York

CRC Press is an imprint of the
Taylor & Francis Group, an **informa** business

Third Edition published 2023
by CRC Press
6000 Broken Sound Parkway NW, Suite 300, Boca Raton, FL 33487-2742

and by CRC Press
4 Park Square, Milton Park, Abingdon, Oxon, OX14 4RN
CRC Press is an imprint of Taylor & Francis Group, LLC

First edition published by CRC Press 1996
Second edition published by CRC Press 2013

ISBN: 978-0-367-72480-1 (hbk)
ISBN: 978-0-367-74597-4 (pbk)
ISBN: 978-1-003-15865-3 (ebk)

DOI: 10.1201/9781003158653

Typeset in Kepler Std
by Deanta Global Publishing Services, Chennai, India

Contents

PART I Fundamentals and Frontiers, Framework for Regulations, and Marketing

PART II Science and Engineering Aspects of Aseptic Processing and Packaging Technologies

PART III   Risk-Based Analyses for Attaining Validated State for Production of Commercially Sterile Shelf-Stable Products and Guidance for Quality Assurance, Microbiological Food Safety, and Regulatory Compliance

PART IV   Frontiers and R&D Opportunities and Challenges

PART V   Appendices

Foreword

Aseptic processing and packaging is still very new to most of the food industry, while commercially catching fire outside the United States (US) over 50+ years ago. Traditional canning operations still make up the bulk of the world's ambient shelf-stable food supply. Although Dr. William McKinley Martin's (Dole) aseptic canning operation unit may well have been one of the first food aseptic systems in the US, Tetra Pak began cornering the commercial market of stable aseptic liquid food by introducing the web stock technology after success in the Nestlé research facility in Konolfingen, Switzerland, in the 1950s. Tetra Pak has since commercialized aseptic food packaging worldwide, providing high-quality fluid foods on every part of the planet. Although somewhat late to the game, the US cautiously dipped its toe into the aseptic water. During the late 1950s and early 1960s, Purdue University, led by Dr. Phil Nelson, and North Carolina State University (NC State), led by Dr. Bill Roberts, became the cornerstones for aseptic processing advances. At Purdue, focus was on one of the state's high-dollar commodity tomatoes, while at NCSU the focus was on dairy products. The first international aseptic conference was held in Raleigh, NC, in 1979 focusing on dairy products. This conference brought together over 300 aseptic experts from all over the world, including the highly regarded Dr. Harold Burton from Reading University in the UK. Many in the field believe his lifelong works earned him the place, certainly one of, if not the founding father of, modern food aseptic.

Tetra Pak had wanted to move into the large US market for decades. Their first attempt was not successful but after placing the TBA-3 filler in the pilot plant at NC State, prior to the international conference, industry migrated to test the novel operations. With the help of Bob Carlson at Cherry Burrell, both indirect and direct aseptic processing equipment was available. At Purdue, Scholle packaging provided bulk aseptic filling equipment and various indirect systems were donated for the vast tomato market.

Between Purdue and NC State, the industry was eventually populated with a highly trained aseptic work force. The Food and Drug Administration (FDA) came to NC State to be trained in this new technology they saw coming. Hydrogen peroxide (H_2O_2) as a food contact surface sterilant was at the time questioned by the FDA due to their concerns about residuals (H_2O_2 had been shown to be a carcinogenic agent). Data generated in the NC State pilot plant in the late 1970s allowed FDA to issue the ground-breaking decision published in February 1981 issued by the Federal Register that opened the commercial use of H_2O_2 as a product contact sterilant for flexible and semi-rigid packaging materials. However, the industry questioned the commercial use of low-acid aseptic, relative to the financial investment required.

In 1985, NC State sought to establish a National Science Foundation (NSF) Industry/University Cooperative Research Center in aseptic research (Center for Aseptic Processing and Packaging Studies-CAPPS) with a focus on industrially relevant research. With NSF seed funds, the industry bought into the program by way of financial memberships. Since the charter in 1987, the center has grown to three university sites, funded research in university labs throughout the country, sponsored international conferences, and basically became the "Johnny Appleseed" of aseptic technology for today's food industry. To add to this, FDA wanted certified operators in aseptic plants. Better Process Control Schools (BPCS) focused on aseptic were developed and are being taught throughout the country. NC State developed the first in-depth course coupled with the traditional canning focus of the BPCS. Rutgers University and UC-Davis followed. Purdue developed a course with more depth but did not couple it with the BPCS. Phil Nelson continued to develop his aseptic valuing technology which won him the World Food Prize in 2007, and which allowed large silos and cargo ships to store aseptically processed food products (tomato, banana puree, orange juice, and others) and then transfer the contents directly into consumer-sized units (perfecting bulk aseptic packaging and sterile transfer technology).

I have been fortunate to have come along to interact and witness the last 50 years of food aseptic advances. I have had the privilege to know, work with, and/or admire the brilliance of the three co-editors Dr. Jairus David, Dr. Pablo Coronel, and Dr. Josip Simunovic and all co-authors in this handbook. Collectively, they comprise the brain trust of dedicated work and advances in the aseptic food world. They have researched areas that needed more understanding. They have collectively gathered more industrial experiences than any group of aseptic experts throughout the world. Man has always been able to do long before he truly understood what actually was happening. This has never been truer than with aseptic. Processing and packaging aseptically were accomplished decades ago. True engineering, physical, chemical, and biological understanding followed. These authors have laid the building blocks for that fundamental understanding. Only with the basic knowledge can processes be safely improved and designed for regulatory review leading to "no objection" or more recently "no questions" rulings on process filings.

Processing products with particles during continuous flow baffled the aseptic community for many years. Technologies were developed using a variety of novel methods. Nestlé Carnation's Flash 18 system (US Trademark serial #72215292, 03/29/1965) used a pressure chamber (with operators inside) under increased pressure, 18 psi above ambient, to process particulate-containing food service products packaged in No. 10 cans. Alfa Laval developed its Jupiter system-batching, with most thermo-sensitive food items entering last. Rotohold made several attempts at physically controlled residence time during continuous flow. None of these attempts gained commercial traction. During 1995 and 1996, the National Center for Food Safety and Technology (NCFST) and the Center for Aseptic Processing and Packaging Studies (CAPPS) embarked on an ambitious undertaking to establish criteria to achieve an FDA no-objection letter for a particle-laden continuous flow aseptic process filing. This ground-breaking effort involving aseptic experts around the world provided technology of the current time to, in fact, lead to the first US FDA no-objection filing. However, once again the efforts needed and cost did not yield the industry investment hoped for. Not until technologies by Simunovic and others (between 1999 and 2019) with a variety of US patents, as described in this book, and several with rapid novel heating technologies, was that market truly opened to industry with regulatory concerns abated and numerous successful US FDA filing noted within.

Since the second edition, several updates and advances have happened. This third edition refocused the first edition's "A Food Industry Perspective." Added areas include fluid flow, residence time distribution, and hygienic design and example particulate filings. Since the second edition, dynamic growth in plant numbers and expansion of volume and product variety have occurred throughout the aseptic industry. Advances have been made in filler and sterilization technologies. New process modeling, simulation, and validation tools and techniques have been developed. As the process industry grows in use of aseptic technologies, the need for training will escalate. This handbook should prove to be an invaluable reference for all desiring a better understanding of the use of aseptic technologies.

Kenneth R. Swartzel, Ph.D.
William Neal Reynolds Distinguished Professor Emeritus
North Carolina State University at Raleigh
Raleigh, North Carolina

Preface

Aseptic processing and packaging is an attractive and challenging alternative compared to conventional methods of canning foods. Continuous sterilization of heat-sensitive pumpable fluids at ultra-high temperatures, followed by prompt cooling, results in a superior finished product. The final product can be filled into containers of varying compositions, of different shapes and with many appealing features. In addition, it is possible to aseptically fill and package either small portion packs or bulk containers such as bags, drums, totes, or tanks destined for remanufacturing.

This third edition of the book provides a comprehensive treatment of aseptic processing and packaging for people interested in the food and beverage processing industry. It is based primarily on the extensive experience of the 3 editors and 24 co-authors in processing, research and development, quality assurance, regulatory governance, marketing, and other areas related to aseptically processed and packaged foods and beverages.

There have been many dynamic changes that have occurred in the food industry since the publication of our older editions (David, Graves, and Carlson, 1996; David, Graves, and Szemplenski, 2013). Our objective was to assemble in one volume the large amount of information in science and engineering that has been published and to update the changes in food processing and packaging technologies. The new chapters that are featured in this new book include the following:

- Design of aseptic processing system and thermal processing (Chapter 4)
- Thermal process equipment and technology for heating and cooling (Chapter 5)
- Flow and residence time for homogeneous and heterogeneous fluids (Chapter 6)
- Thermal process optimization of aseptic processes containing solid particulates (Chapter 7)
- Aseptic filling and packaging equipment for retail products and food service (Chapter 8)
- Design of facility, infrastructure, and utilities (Chapter 11)
- Cleaning and sanitization (Chapter 12)
- Microbiology of aseptically processed and packaged products (Chapter 13)
- Risk-based analyses and methodologies (Chapter 14)
- Establishing "validated state" (Chapter 15)
- Quality and food safety management system (Chapter 16)
- Computational and numerical models and simulations for aseptic processing (Chapter 17)

The chapters are organized into five parts. It is important to note that most of the text book is written in a narrative form representing the respective authors' experience of the industry and its technology. We realize that there may be some duplication and overlap between chapters, but we think that readers can read and analyze specific chapters and obtain the information desired rather than having to read the entire book.

Part I Sets the Stage for Fundamentals and Framework for Regulations and Marketing and includes:

1) Aseptic Processing and Packaging: Fundamentals and Frontiers
2) US Federal Regulations for Aseptic Processing and Packaging of Food
3) The US Markets for Aseptically Processed and Packaged Products

Part II Covers in Detail the Science and Engineering Aspects of Aseptic Processing and Packaging and includes:

4) Design of Aseptic Processing System and Thermal Processing
5) Thermal Process Equipment and Technology for Heating and Cooling
6) Flow and Residence Time Distribution for Homogeneous and Heterogeneous Fluids
7) Thermal Process and Optimization of Aseptic Processes Containing Solid Particulates
8) Aseptic Filling and Packaging for Retail Products and Food Service
9) Aseptic Packaging Materials and Sterilants
10) Aseptic Bulk Packaging
11) Design of Facility, Infrastructure, and Utilities
12) Cleaning and Sanitization for Aseptic Processing Operations

Part III Encompasses Risk-Based Analyses for Attaining Validated State for Commercial Sterility and Regulatory Compliance:

13) Microbiology of Aseptically Processed and Packaged Products
14) Risk-Based Analyses and Methodologies
15) Establishing "Validated State" of Aseptic Processing and Packaging Systems
16) Quality and Food Safety Management System (QFSMS) for Aseptic and ESL Manufacturing Companies

Part IV covers Modeling and R&D Needs and Challenges:

17) Computational and Numerical Models and Simulations for Aseptic Processing
18) Frontiers and Research and Development: Challenges and Opportunities

Part V is Appendices Packed with Useful Information Supporting Book Chapters:

Appendix 1: United States History and Evolution
Appendix 2: Dr. William McKinley Martin—Father of Aseptic Canning
Appendix 3: Aseptic Filler Profiles
Appendix 4: Aseptic Contract Manufacturers in the United States
Appendix 5: Thermal Process Design Calculations for Aseptically Processed Fluids and Purees
Appendix 6: Particulate Study—Single Particle
Appendix 7: Particulate Study—Multiple-Type Particles
Appendix 8: Thermal Processing Methods

The organization of the book permits readers to selectively choose those sections in which they have the greatest interest. The sections written by the different authors reflect their personal styles and areas of expertise. The book provides a comprehensive update on this rapidly developing technology for the food processing industry.

Jairus R. D. David, Ph.D., CQE
JRD Food Technology Consulting, LLC.
Omaha, Nebraska

Pablo M. Coronel, Ph.D.
CRB Consulting Engineers
Raleigh, North Carolina

Josip Simunovic, Ph.D.
Department of Food, Bioprocessing and Nutrition Sciences
North Carolina State University
Raleigh, North Carolina

Note: References to commercial products and trade names are made with the understanding that no discrimination and/or no endorsement by the authors or the organizations that they are involved with are implied.

Acknowledgments

I would like to express my thanks to the following:

My wife, Shelley Zylstra-David for her loving encouragement of my work, and to our three children, Adriana, Brennan, and Blake for "daring" me to write "another book."

Dr. R. Larry Merson, Dr. R. Paul Singh, and Sherman J. Leonard for teaching me the rudiments of thermal processing and food safety during my graduate studies at the University of California at Davis.

"Merson's Herd": Mark Deniston, Elsa Fakhoury, Dr. Nick Stoforos, Dr. Bakri Hassan, Dr. Hiroshi Sawada, Dr. Rene Peralta, Elisa Pratt-Lowe, Dr. Kathleen Young-Perkins, and Teri Wolcott.

Ralph Graves (co-editor, first and second editions; 1996, 2013), V. R. (Bob) Carlson (co-editor, first edition; 1996), Tom Szemplenski (co-editor, second edition; 2013), Dr. Walter Dunkley, Dr. Kailash Purohit, and Dr. Dilip Chandarana—my mentors who guided my exodus from "ivory tower" to "plant floor" with practical insights on factory physics, quality control, commercial manufacturing, compliance, and production of saleable low-acid aseptic products.

Tony "Roy" Graves, Bob Machado, Jody Graves, Chuck Sanfilippo, Tony Barba, and Jack McGlashan (Real Fresh Aseptic, Inc., Visalia, California); Tracy Baker, Wayne Smalligan, Sherrie Harris, Jim Powell, Jay Burnett, Robert Weick, Arvie Shinavier, Jerry Stroven, Tammy Weaver, Jim Thomas, Dave Joslin, Barbara Ivens, Dr. Sandra Bartholomew, Dr. Al Bolles, and Dr. George Purvis (Gerber Baby Foods, Fremont, Michigan); Dr. Juan Menjivar, Dr. Shri Sharma, Ahmed Hussein, Dave Konst, and Dennis Peterson (Rich Products Corporation, Buffalo, New York).

My Core Team—Dr. Ferhan Ozadali (Reckitt Nutrition/Mead Johnson Nutrition; Purdue University), Dr. Dharmendra Mishra (Purdue University), Dr. Patnarin Benyathiar (Mahidol University, Thailand), Dr. Wilfredo Ocasio (Eurofins Microbiology Laboratories), and Dr. Nathan Anderson (US/FDA), and my co-editors Dr. Pablo Coronel and Dr. Josip Simunovic for their visionary leadership, expertise, passion, and teaming.

Sincere appreciation to 24 contributing authors, who, by giving freely of their expertise, have made this book possible. Many thanks are due to Stephen Zollo, Senior Editor, Taylor & Francis/CRC Press, Boca Raton, Florida, for his professionalism and unstinting support in bringing this edition of this book to publication. Our appreciation to Laura Piedrahita, Editorial Assistant, Life Sciences, Todd Perry, Production Editor, at Taylor & Francis/CRC Press, and Vijay Bose, Project Manager, Deanta Global, Chennai, India, for guidance and expediting the final stages of copy editing and printing.

Last but not least, my mentors—Al Wilder, Mike Klungness, Joaquin Pericas, Glen Weaver, Dr. Joseph Herskovic, and Dr. Richard McArdle.

Jairus R. D. David

I would like to thank my parents for encouraging my intellectual curiosity and showing me that the world is full of wonders. When the road can't be found, your wisdom keeps showing me ways to build new ones.

To my wife Ana Katalina who supports me physically and emotionally and reminds me that there is much more to life than just work. To my two beautiful children, being a parent has been the hardest work of them all, and I admit that I have learned from you many times more than I ever thought. All of you are the fresh air that keeps me going.

To all my past teachers, mentors, and colleagues, from whom I have learned so much and are an integral part of who I am today. I don't want to miss anyone by mentioning names but you know who you are.

To my fellow editors for the collaboration, knowledge, and encouragement to keep going.

And finally, to God, the giver of life for allowing me to wake up every day to serve Him and my fellow humans.

Pablo M. Coronel

This is for my beloved teachers, professors, mentors, and bosses who, in spite of all the evidence to the contrary, believed there was something worth encouraging and pursuing beyond the class clown persona, resistance and dislike of homework assignments, and perpetual daydreaming and aloofness. Your motivation, inspiration, challenges, threats, coaxing, bargaining, and skilled guidance steered me onto the path of commitment, dedication, and hard work to achieve things I never imagined possible.

To my friends, students, collaborators, and cofounders, thank you for your companionship and support in chasing our modest shared dream of revolutionizing the global food industry. I hope the journey was rewarding and brought to you as much passion, joy, and fulfillment as it did to me.

To my fellow editors Dr. Jairus David and Dr. Pablo M. Coronel, thank you for the opportunity to contribute to this book, for your creativity, patience, and understanding during a very difficult and complicated pandemic time. I am also very proud to be a part of the team of collaborating authors and hope that the readers will have a useful reference to turn to for information and hopefully inspiration.

Do not go where the path may lead, go instead where there is no path and leave a trail.

Ralph Waldo Emerson

Josip Simunovic

Editors

Jairus R. D. David, Ph.D., is a thought leader in the agro-food industry and an expert on food preservation science, food safety, and quality. His industrial experience spans family-owned small companies and large corporations—highly regulated baby foods, infant formula, consumer-packaged goods, and food service. Jairus has worked with the food industry in developing microbiology and thermal processing food safety objectives and compliance for canning and aseptic processing for over 38 years, balancing applied research with process and quality optimization, and launch of innovative products. He has developed natural and clean label ingredients systems, especially antimicrobial preservatives for designing minimally processed foods.

Jairus earned his Ph.D. in microbiology with emphasis on thermal processing from the University of California at Davis, under the tutelage of Dr. Larry Merson. His research focused on "Kinetics of Inactivation of Bacterial Spores at Ultra High Temperature in a Computer-Controlled Reactor." Jairus earned his M.Sc. (Food Technology) from UN/FAO Central Food Technological Research Institute, Mysore, India, and B.Sc. (Agriculture) from the University of Agricultural Sciences, Bangalore, India.

Jairus is a Certified Quality Manager (CQM) and Certified Quality Engineer (CQE), from the American Society for Quality (ASQ). He has participated in the leadership programs at the Kellogg School of Management, Center for Creative Leadership, Stephen Covey Leadership Forum, and the Massachusetts Institute of Technology.

A Fellow of the Institute of Food Technologists (IFT), he is the recipient of IFT's prestigious Industrial Scientist Award (2006). He is recognized for developing and influencing public health food safety policy on the use of honey in cereals and bakery products for the prevention of infant botulism in infants under 12 months of age. Currently, all honey and honey-containing food products in commerce carry a cautionary label "Do not feed honey to infants less than one year of age."

Jairus has co-authored several refereed papers, book chapters, and patents. He has also co-edited books on aseptic processing and packaging (first edition 1996, second edition 2013, Russian translation, 2014), Antimicrobials in Food (2021).

Jairus has served on the Industrial Board of Advisors at the Center for Food Safety (CFS), University of Georgia at Griffin, Georgia; and the Food Research Institute (FRI), University of Wisconsin at Madison, Wisconsin. Jairus is adjunct faculty at the Department of Food Science and Nutrition in Mississippi State University at Starkville, Mississippi; University of Nebraska at Lincoln, Nebraska; and Iowa State University at Ames, Iowa.

Jairus is currently Principal Consultant at JRD Food Technology Consulting, LLC., Omaha, Nebraska. More information can be found at WWW.JRDFoodTech.com. He is open for both domestic and international assignments, able and willing to do *pro bono* consulting projects in developing "Third-world" and "Fourth-world" countries, rural communities, and non-profit organizations.

Jairus and his wife of 30 years, Shelley, are proud parents of three millennial children.

Pablo M Coronel, Ph.D., is a thought leader in the food industry and authority on food safety and quality. Pablo has worked with the food industry in scaling up products from benchtop to industrial scale in several fields. Most of his work has been on developing advanced thermal processing and validation methods and technologies for the aseptic processing of heterogeneous foods. His industrial experience spans working for family-owned small companies and large corporations—highly regulated consumer-packaged goods and food service.

Pablo earned his Ph.D. in food science from North Carolina State University with research in the application of microwave heating to aseptic processing. He earned a Chemical Engineering degree from the Escuela Politécnica Nacional in Quito Ecuador.

Pablo is the recipient of IFT's prestigious Industrial Scientist Award (2015), as well as being part of two different IFT Industrial Achievement Award teams (2009 and 2015). He is recognized for developing continuous flow microwave processing in industrial settings.

Pablo is currently Director of Food Safety and thermal process authority in CRB Consulting Engineers, Kansas City, Missouri, and adjunct faculty at the Department of Food, Bioprocessing and Nutrition Sciences of North Carolina State University.

Pablo and his wife of 21 years, Ana Katalina, are proud parents of two children.

Josip Simunovic, Ph.D., is an educator, researcher, inventor, developer, and co-founder of food technology startups.

For over 30 years, his work has focused on development and commercialization of novel, advanced thermal and aseptic processing and packaging technologies, methods, and devices for their monitoring and safety validation, as well as production of new food and beverage products enabled by implementation of these technologies. Continuous flow microwave processing, an advanced thermal processing technology for pasteurization, sterilization, and aseptic processing of foods, was developed by his teams at NC State University, as well as the comprehensive system of tools for monitoring and validation of safety of aseptic foods containing solid particles. These developments resulted in the first FDA clearance of microwave-sterilized homogeneous low-acid food and the first FDA clearance of microwave-sterilized complex particulate food, and they have been recognized with IFT Industrial Achievement Awards in 2009 and 2015, IFT Food Expo Innovation Award in 2015, and Edison Innovation Awards in 2015 and 2016.

Josip has earned a Ph.D. in Food Science with a concentration in Food Process Engineering at North Carolina State University under the mentorship of Dr. Kenneth Ray Swartzel, his M.Sc. in Food Science and Technology at the University of Florida under the tutelage of Dr. John Peter Adams, and B.Sc. in Food Technology at Josip Juraj Strossmayer University in Osijek, Croatia.

Josip is a named inventor on 20 issued U.S. patents which provide the intellectual property foundation for several startup businesses. Josip is a Fellow of the Institute of Food Technologists, and has been recognized with the 2008 Food Engineering Award by the American Society of Agricultural and Biological Engineers, 2012 Marvin Tung Award from the Institute for Thermal Processing Specialists, and 2016 IFT Research and Development Award.

Josip is currently employed as a Research Professor with the Department of Food, Bioprocessing, and Nutrition Sciences at NC State University, and also serves as the Chief Science Officer of SinnovaTek Inc.

Contributors

Shirin J. Abd
Eurofins Microbiology Laboratories
Fresno, California

Nathan M. Anderson
Center for Food Safety and Applied
    Nutrition Office of Food Safety
U.S. Food and Drug Administration
Bedford Park, Illinois

Patnarin Benyathiar
Department of Food Technology
Mahidol University, Kanchanaburi Campus,
    Thailand

Peter M.M. Bongers (Deceased)
Process Science
Unilever R&D
Vlaardingen, The Netherlands

V. R. (Bob) Carlson (Deceased)
VRC Co., Inc.
Cedar Rapids, Iowa

Tyler Dixon
Mead Johnson Nutrition
Rickett Benckiser Health
Evansville, Indiana

Robert Fox
Synergy Packaging, LLC
Williamsburg, Virginia

Ralph H. Graves (Deceased)
Food & Dairy Consultant
Visalia, California

Jeffrey Merritt
Ecolab
River Heights, Utah

Dharmendra K. Mishra
Department of Food Science
Purdue University
West Lafayette, Indiana

Wilfredo Ocasio
Eurofins Microbiology Laboratories
Fresno, California

Ferhan Ozadali
Mead Johnson Nutrition
Reckitt Benckiser Health,
Evansville, Indiana

Tunc Koray Palazoglu
Department of Food Engineering
Mersin, Turkey

Ben Rucker
CRB Consulting Engineers
Marshfield, Wisconsin

Deepti Salvi
Department of Food, Bioprocessing and
    Nutrition Sciences
North Carolina State University
Raleigh, North Carolina

K.P. Sandeep
Department of Food, Bioprocessing and
    Nutrition Sciences
North Carolina State University
Raleigh, North Carolina

Toni de Senna
Eurofins Microbiology Laboratories
Fresno, California

George N. Stroforos
Department of Food Science & Technology
Agricultural University at Athens
Athens, Greece

Thomas E. Szemplenski
Aseptic Resources, Inc.
San Diego, California

Jason A. Tucker
CRB Consulting Engineers
Marshfield, Wisconsin

Emily Weyl
Center for Food Safety and Applied
   Nutrition
Office of Food Safety
U.S. Food and Drug Administration
College Park, Maryland

Nihat Yavuz
Department of Food Engineering
Canakkale Onsekiz Mart University
Canakkale, Turkey

PART I

Fundamentals and Frontiers, Framework for Regulations, and Marketing

Chapter 1
Aseptic Processing and Packaging: Fundamentals and Frontiers
Jairus R. D. David, Pablo M. Coronel, Josip Simunovic, Ralph H. Graves,
    V. R. (Bob) Carlson, and Thomas E. Szemplenski

Chapter 2
US Federal Regulations for Aseptic Processing and Packaging of Food
Nathan M. Anderson and Emily Weyl

Chapter 3
The US Markets for Aseptically Processed and Packaged Products
Thomas Szemplenski

Chapter 1

Aseptic Processing and Packaging

Fundamentals and Frontiers

Jairus R. D. David, Pablo M. Coronel, Josip Simunovic, Ralph H. Graves,
V. R. (Bob) Carlson, and Thomas E. Szemplenski

CONTENTS

1.1 INTRODUCTION

Aseptic processing and packaging of foods consists of filling sterilized and cooled product into presterilized containers, followed by hermetic sealing with a presterilized closure in a presterilized and continuously decontaminated tunnel or aseptic work zone. The details are illustrated diagrammatically in Figure 1.1.

Aseptic processing and packaging is an attractive and challenging alternative compared to conventional methods of canning foods. Continuous sterilization of heat-sensitive pumpable fluids at ultra-high temperatures, followed by prompt cooling, results in a superior finished product that is shelf-stable for 12–24 months at ambient storage conditions. The finished product can be filled into containers of varying compositions, of different shapes, and with many attractive features. In addition, it is possible to aseptically fill and package small portion packs and bulk containers such as bags, drums, totes, and tankers destined for remanufacturing.

The physical size of flexible and semirigid retail packages ranges from 10 ml to 2 L (0.3 to 64 oz.) and is sold for direct consumption and private households; bag-in-box food service containers size up to 20 L (5 US gallons) are intended for food service and institutional markets; drums and tanks up to 250 L+ (55+ US gallons) are intended for industrial markets for further re-processing and packaging, or direct portion-packaging without further re-processing (Hallstrom, 1979). The recent market intelligence data point to an approximate 80% to 20% volume split between aseptic bulk and aseptic retail packaging, respectively.

Marketing, industrial, and regulatory frameworks based on the principles of aseptic processing and packaging are available and continue to be of interest to the food processors of

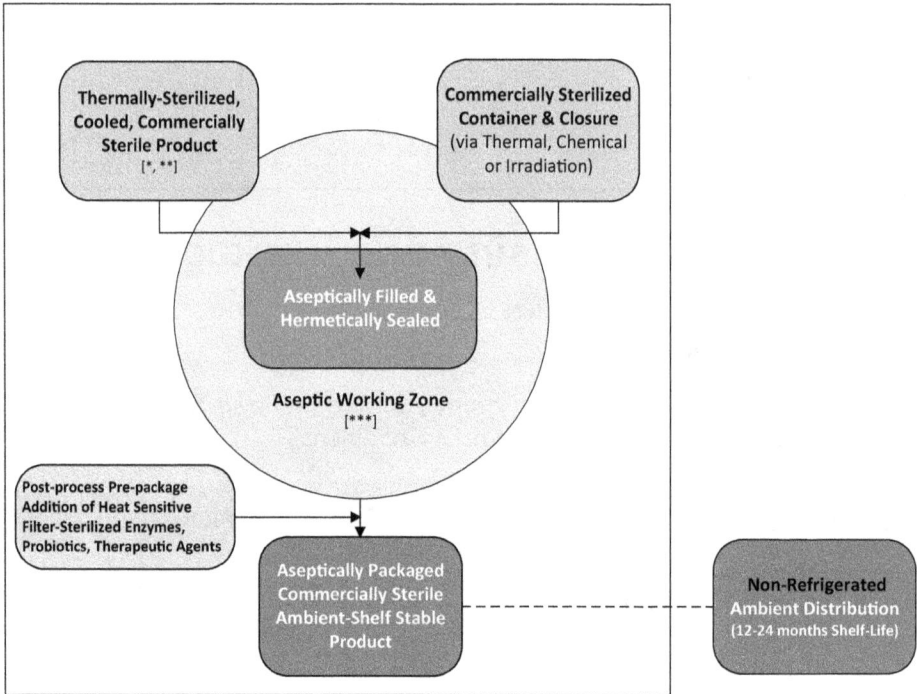

[*] Low Acid, Acidified; Pumpable Fluids, Purees, and Particulates

[**] Target Microorganisms: Mesophilic and Thermophilic Bacterial Spores of Importance – Public Health & Economic Spoilage
    (Certain heat sensitive products may be cold filter-sterilized prior to aseptic packaging)
[***] Most depend on Passive/HEPA sterile air in working zone. Exception: Martin-Dole Aseptic Canner with Active-Sterilizing
Work Zone

Figure 1.1 Factory physics and distribution system for aseptically processed and packaged, non-refrigerated ambient shelf stable products.

ambient, shelf-stable foods in non-refrigerated retail packages, food service containers, and bulk containers.

Aseptic processing represents a key technology in modern thermal processing of food and was ranked the number one innovation in food technology (Reisert et al., 2015; Floros et al., 2010) (Table 1.1).

1.2 FRAMEWORK AND CURRENT STATE

For many years, researchers recognized that the use of high temperatures for short times had potential advantages over conventional thermal processes at lower temperatures for longer times, but there were difficulties in taking advantage of this information. Heat causes reactions in food, some of which are undesirable; the rate of reaction approximately doubles for every increase in temperature of 10°C. (The temperature dependence of the reaction rate is z-value, represented as 18 $F°$ degrees to differentiate this from temperature scale of 18 °F.) In contrast,

TABLE 1.1   COMPARISON OF CONVENTIONAL CANNING AND ASEPTIC PROCESSING AND PACKAGING OF PRODUCTS (DAVID, 2013)

Criteria	Retorting	Aseptic Processing and Packaging of Foods
I. Sterilization		
A. Product		
1. Temperature regime	220°F–250°F	HTST (180°F–220°F) and UHT (260–290°F)
2. Delivery	Unsteady state	Precise—square wave
3. Heat/cool lethality credit	Possible	Isothermal temperature in hold tube only
4. Process calculation		
Fluid	Routine—convection	Routine
Particulate	Routine—conduction or broken heating	Complex
B. Other sterilization (process equipment, container, lid, and aseptic tunnel)	None	Complex; many
C. Energy efficiency	Lower	30% saving or more
II. Quality		
A. Psychophysical or sensory	Mushy; not suitable for heat-sensitive and nutritional products	Superior; suitable for homogeneous heat-sensitive and nutritional products
B. Nutrient loss	High	Minimum
C. Value added	Lower	Higher
Convenience	Shelf-stable	Shelf-stable and other features
Microwaveability	Semirigid containers (bowls and trays)	All non-foil flexible and rigid containers
D. Product quality	Dependent on container size and shape	Independent of container size and shape
III. Production aspects		
A. Container speed	High; 600–1000+ containers per minute	Medium; 500–700 containers per minute is common; higher speeds via multiple lanes
B. Handling/labor	High	Low
C. Downtime	Minimal, typically at seamer and labeler	Resterilization due to sterility loss in sterilizer or filler
D. Versatility/flexibility for the manufacture of a product in different container sizes	Different size containers need different process delivery and/or retorts	Need one or two aseptic fillers to fill different size containers

(Continued)

TABLE 1.1 (CONTINUED)   COMPARISON OF CONVENTIONAL CANNING AND ASEPTIC PROCESSING AND PACKAGING OF PRODUCTS (DAVID, 2013)

Criteria	Retorting	Aseptic Processing and Packaging of Foods
IV. Process deviation		
A. Under processing	Reprocess intact container-in-product after removal of labels if possible	Reprocess product after the destructive opening of containers
B. Survival of heat-resistant enzymes	Rare	Common in certain foods, especially in fluid milk; problem could be overcome, if system is designed properly (refer to Chapter 18)
V. Spoilage analysis		
A. Troubleshooting	Simple; preprocess, inadequate process, and postprocess recontamination	Complex; need to deal with compromises in CIP and sanitation, and aseptic zone and its elements (refer to Chapters 12 and 13)
B. Traceability: container code versus case code	Usually not identical; may differ from 1/2 to 6 hours	Usually, identical
VI. Low-acid particulate processing	In practice	Work in progress by industry consortium and regulatory agencies; data from numerous high-acid particulate systems may be used for design basis of low-acid systems with additional validation and modeling
VII. Postprocess prepackage sterile additives	Not possible	Possible and in practice to add filter-sterilized enzymes (lactase), bioactive compounds, therapeutic agents, and probiotics

typical rates of destruction of bacteria and spores increase tenfold for the same temperature increase. Therefore, processes using higher temperatures for shorter times can achieve commercial sterility with improvements in quality with respect to flavor, color, vitamin retention, and physical properties, as compared to the quality of products from conventional heat processes such as retorting or canning.

Application of this principle was limited by the availability of processes and equipment to apply it in commercial practice. Rapid heat transfer for heating and cooling is readily achieved in liquid foods but not in very viscous or solid foods, which depends on the conduction heat transfer rather than convection. Therefore, early work focused on liquid foods, especially milk and its products. Specialized equipment was developed for applying ultra-high-temperature (UHT) treatments using tubular or other heat exchangers, or steam injection or infusion devices, which were usually coupled with vacuum coolers. The commercial use of such equipment necessitated aseptic packaging after sterilization, which proved to be the most serious limitation.

Early successes with milk and its products increased interest in adapting aseptic processing and packaging to other liquid foods. This book outlines progress with products such as soups, juices, and purees, and current research and development directed toward the challenging problems with foods containing solid particles. The introduction of new aseptic processing and packaging technologies necessitated the evolution of a new body of food laws and regulations, and new or expanded agencies to enforce them. Innovations in the United States have been delayed by the industry's justifiable cautious approach to developing validation procedures for compliance with its own quality and safety standards and those of regulatory agencies (Chapters 2, 15, and 16). Collaboration among industry, industry organizations, public officials, and agencies has been excellent in guiding the development of a rapidly expanding industry based on aseptic processing and packaging.

Aseptic processing is a commercially successful and robust technology. It is approximately 70 years old and began with the invention of the first aseptic line of the heat/hold–fill/hold–cool (HFC) system by Dr. C. Olin Ball (1936). Aseptic technology became a commercial reality in the 1950s, with the installation of Dr. W. M. Martin's Dole (1951) aseptic canner, and the introduction of Tetra Pak cartons by Ruben Rausing in 1952. Martin recognized that the production of nutritionally superior and "home-style" baby foods, heat-sensitive infant formula, and milk products would be possible through the use of short-time, high-temperature sterilization together with aseptic canning (see Appendix 1, US history and evolution, and Appendix 2, William McKinley Martin aseptic canning patents).

In the United States before 1981, the Martin–Dole aseptic canning system was the only aseptic filling and packaging system of commercial importance for milk and milk-based low-acid products in metal cans. This core technology was correctly designated as UHT-sterilized and aseptically packaged process. The pre-1981 market drivers were better quality, low-acid, heat-sensitive products than retorted versions. It was a formidable marketing and research and development challenge to educate consumers to appreciate aseptic technology and differentiate products in Martin (Dole) aseptic metal containers from that of classical canning or retorting. Also in commercial use were 55-gallon metal drums invented in the 1970s by Fran Rica (JBT Company) for bulk packaging of tomato products (Chapter 10).

In 1981, the US Food and Drug Administration (FDA) approval of the food additive petition for use of hydrogen peroxide as a sterilant for food contact surfaces provided the impetus for the introduction of various aseptic filling and packaging systems into the US markets (Chapters 8 and 9). The market was primarily driven by "juice box" technology, which involved high-acid food pasteurization at high-temperature, short-time (HTST) followed by aseptic filling and packaging (Figure 1.2). The post-1981 market drivers were the use of alternate flexible and semirigid packages, consumer convenience, reduced package weight, cost reduction, superior quality and nutrition, and package sterilization by optimal methods.

Aseptic processing and packaging is one of the most dynamic and profitable sectors in the US food and beverage market. The major aseptic packaging segments are cans, plastic bottles, plastic cups, bag-in-box, paperboard, and pouches. In addition, the aseptic processing industry is going through another period of specific growth via plastic, primarily polyethylene and polyethylene terephthalate (PET) containers to the world market. These packages have been in commercial use for several years for packaging high-acid products (such as fruit juices and spaghetti sauces) and refrigerated low-acid products (such as smoothies, and assorted milk beverages), but

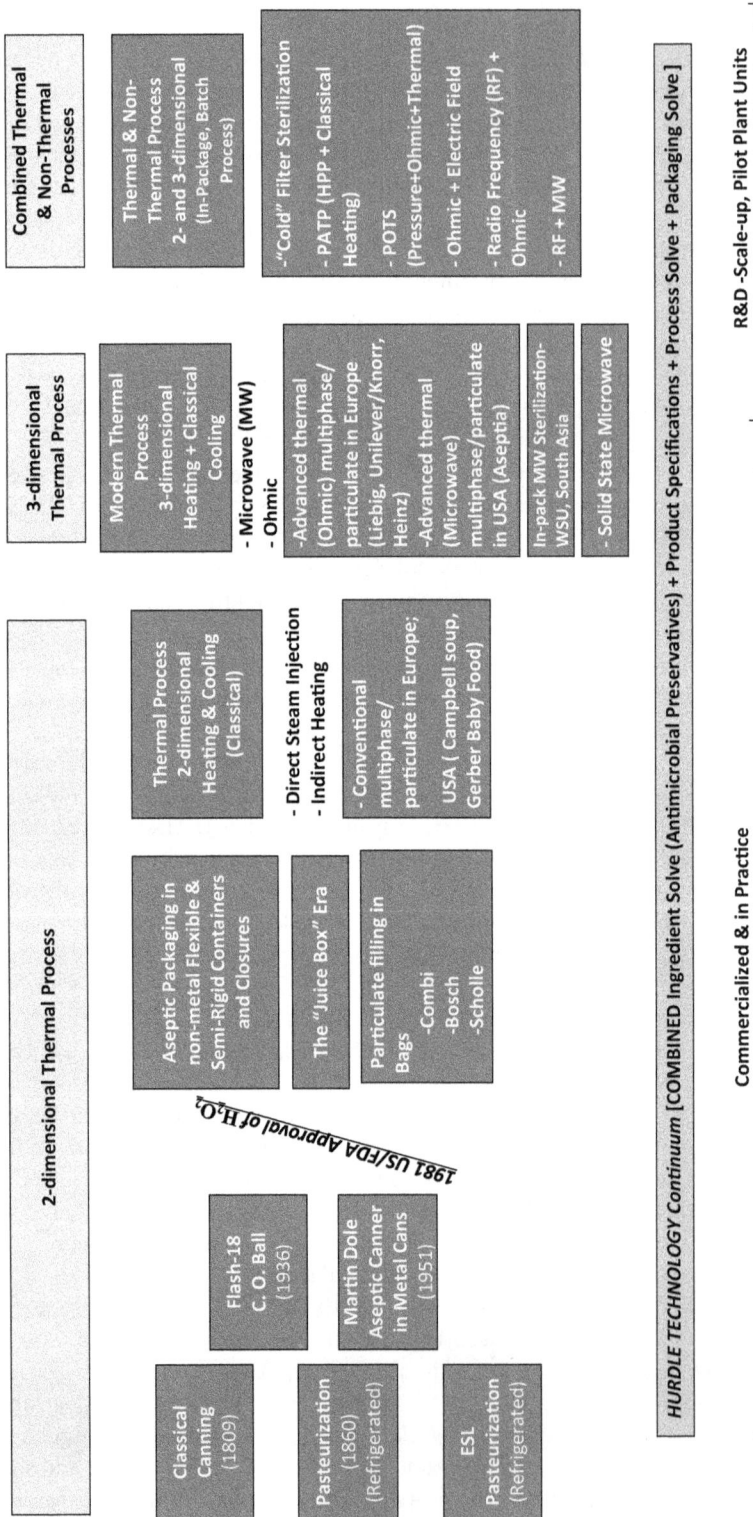

Figure 1.2 Optimization continuum for aseptically processed products, filled and packaged in hermetically sealed containers for non-refrigerated ambient distribution. Shown here are classical two-dimensional and three-dimensional thermal processing technologies, combined technologies, and hurdle technology.

are being used with increasing frequency for low-acid shelf-stable fluid products (such as coffee and protein-fortified nutritional beverages, energy drinks, and sports drinks) to meet consumers' modern lifestyle and demand for ergonomic and environmentally sustainable containers (Chapters 3, 8, 9, and 18).

Today, there are more than 34 manufacturers of aseptic filling equipment worldwide (Appendix 3, Aseptic filler profiles) and more than 600 aseptic systems for the manufacture of retail package and bulk containers in the United States. Contract manufacturers play a crucial role in the introduction of innovative and consumer-convenient new products to the market in a timely and cost-effective manner (Appendix 4, Aseptic contract manufacturers). Contract manufacturers facilitate rapid prototyping, product sensory and specifications development, and for go/no-go business decisions, and speed to market.

1.3 DEPARTURES FROM OPTIMA AND CHALLENGES

It is axiomatic to consider aseptic processing and packaging as the benchmark in optimization for the manufacture of sterile, shelf-stable food. Contrasted with retorting or in-container "terminal" sterilization, in aseptic processing the product and the package are independently sterilized by optimal processes, wherein microbial inactivation and quality factor degradation are also co-optimized. Aseptic systems permit sterilizing the product and the container (flexible, semirigid, rigid) separately and appropriately, without the rate-limiting heat transfer modes, or the attendant's thermal and pressure stress to the container closure and seal integrity, and permitting high-temperature, short-time (HTST), or ultra-high-temperature (UHT) processing of heat-labile products without excessive quality and nutritional degradation, while achieving requisite commercial sterility

Even though the aseptic technology is considered the benchmark in optimization for the manufacture of sterile, shelf-stable foods and beverages, there are several departures from optima or gaps that should be closed in order to leverage the full benefits of quality, nutrition, safety, and convenience. Some of the gaps include potential compromises to sterile work zone, and package seal integrity leading to microbial recontamination, over delivery of designed thermal process, inadequate or slow cooling, and special handling of sensitive ingredients known to contain and protect thermoduric and thermophilic spores during heating.

Aseptic systems that process and package shelf-stable foods are indeed well suited to be hierarchically co-optimized, as there are many competing priorities and numerous critical parameters that must be validated and controlled. But most important, the numerous vectors of recontamination must be recognized for their capacity to cause failure. Although sophisticated control systems help, there is no substitute for training, environmental control, and system validation for defect prevention and food protection (Table 1.2, David, 2013).

The aseptic zone is undoubtedly the heart of the aseptic system and is crucial to the delivery of a sterile product. Although it can be presterilized and validated to be sterile to a sterility assurance level of 10^{-6}, procedures and practices to maintain or monitor its sterility are unavailable. It is assumed to be sterile by virtue of presterilization and the use of sterile, high-efficiency particulate air (HEPA) or laminar flow air. Thus, the work zone is at best, sterile or passive, and not sterilizing or active, and thus not able to overcome recontamination. The vectors of recontamination include filter failures, grow through, and nonsterile air aspiration by the work zone due to loss of laminarity, eddy currents at interfaces, splashing from filling or leaks, and the discharge of finished containers from the zone.

TABLE 1.2   COMPARISON OF DIFFERENT CONTINUOUS PROCESSING METHODS BASED ON OPTIMIZATION HIERARCHY (DAVID, 2013)

Raw Product	Heat and Cool Continuous Processes	Filling and Packaging	Process Technology Designation	Finished Product and Shelf Life	Optimization Hierarchy Index
High-acid food	Continuous pasteurization–fill–hold–cool	Hot fill to sterilize lid and container at ambient conditions	Core hot–fill–hold–cool technology	Shelf-stable juices and acidified foods; commercially sterile	Interim hierarchy
Low-acid food	Continuous UHT sterilization and flash to 255°F	Hot fill to sterilize lid and containers in pressurized room	"Flash 18" process	Shelf-stable, low-acid food in institutional-size containers; commercially sterile	Higher hierarchy
Low-acid food	Continuous HTST pasteurization and cool	Cold fill in clean room; sanitized container	Conventional pasteurization	Refrigerated milk, dairy, and juice products; 2–3 weeks	Lower hierarchy
Low-acid food	Continuous ultra-pasteurization and cool	Cold fill in clean room; sanitized container	Ultra-pasteurization; extended shelf life (ESL) product	Refrigerated milk, creamers, and juices; 6–8 weeks	Interim hierarchy
Low-acid food	Continuous UHT sterilization and cool	Aseptic filling and packaging at ambient	Core aseptic technology; UHT-sterilized and aseptically packaged foods	Shelf-stable milk, sauces, puddings, and other homogeneous food; commercially sterile	Benchmark in optimization for shelf-stable, low-acid, commercially sterile canned foods
High-acid food	Continuous HTST pasteurization and cool	Aseptic filling and packaging at ambient	HTST pasteurized and aseptically packaged foods	Shelf-stable juice box and juices in clear PET bottles; commercially sterile	Benchmark in optimization for shelf-stable, high-acid, and acidified canned foods
Low-acid food	UHT or HTST heat and cool	(a) Aseptic filling and packaging at ambient	UHT or HTST processed and aseptically packaged refrigerated food	Commercially sterile refrigerated puddings for market diversification and repositioning; 6–8 weeks stock rotation cycle	Interim hierarchy
		(b) Non-aseptic; unapproved filler or processes	UHT or HTST processed and non-aseptically packaged or processed refrigerated food	Refrigerated/ESL beverages, puddings; 8–12 weeks	Lower hierarchy

It should be reemphasized that despite subsystem validation to 10^{-6}, overall sterility assurance is governed by the aseptic work zone, wherein maintenance of sterility cannot be assured and monitored, and any recontamination cannot be detected, removed, or inactivated. An exception is the active or sterilizing work zone (using superheated steam), which is an integral part of the first aseptic system—the Martin–Dole aseptic canner (Appendix 1 and 2). However, it is important to note that the Dole aseptic system is somewhat vulnerable when superheated cans are "tempered down" prior to filling to avoid product contact burn-on.

The terminal retort process in canning is a de facto seal integrity tester, and marginal seals usually do not survive the time, temperature, and pressure of the retort scheduled process. In aseptic, not only are the product and package decoupled but also the final seal, made in a sterile work zone, is not trauma tested. In aseptic, the reliability of seal inspections, mandatory incubation holds, sterility tests, and daily review of "Cumulative Daily Pack Record" take on additional significance and attendant due diligence. Thus, seal integrity and maintenance of hermeticity in aseptic processing is a key determinant of final sterility assurance.

Precise control of flow and temperature is essential for the proper design and delivery of an approved thermal process, followed by prompt cooling. Inability to control temperatures within 5 °F (2.5 °C) in the UHT regime can lead to irreversible damage, because this may correspond to a doubling or tripling of process lethality (F_o). In reality, one can notice very significant differences between flavors of milk sterilized for 4 and 4.7 seconds at 280 °F. The small difference of 0.7 seconds can cause a big flavor difference in UHT regimes. Food industry should strive for defining the safety and quality and biofunctionality limits in a scheduled process to prevent departure from process and product quality optima (see Chapter 18, Figure 18.2).

1.4 CURRENT AND FUTURE OPPORTUNITIES FOR OPTIMIZATION

The future of aseptic processing will depend on reducing the departures from process optima, and capturing findings from ongoing research in both basic and applied research leading to the use of both novel thermal sterilization methods (two-dimensional and three-dimensional such as ohmic and microwave) coupled with innovative nonthermal processes, sensors, validation methods, and nanotechnologies. It is important to underscore that no one single technology can replace the shelf-stable capabilities of either classical retorting or aseptic processing. However, many of the innovative thermal and nonthermal processes, sensors, and nanotechnologies can be used either additively or synergistically to build "hurdles" in tandem with an objective to produce superior products with minimal heat-induced damage and at an affordable price (Figure 1.2), (David, 2020; Legan & David, 2021).

Developments on the filling and packaging side of the aseptic technology will continue to be the major driving force in the further expansion of this technology. Emerging package and packaging material sterilization processes like plasma generation and glass lining will expand the number and variety of polymers in use for aseptic packaging as well as available package shapes and sizes (Sandeep et al., 2004).

Larger (institutional and industrial ingredient) sizes of aseptically packaged products currently have the most favorable (low) ratio of packaging material used per unit of product weight and volume, and this advantage will continue to grow, especially for products yet to make a significant impact in the aseptic processing area, such as low-acid particulate products (Appendix 6, Particulate study—single particle, and Appendix 7, Particulate study—multiple-type particles). There is the need to avoid re- or double-processing of previously bulk processed aseptic high value-added commodity products (orange juice, purees of banana, prunes, and other exotic berries; Appendices 5.1–5.4, Examples of typical thermal process calculations) via the development of reliable proper transfer fitment and techniques for ambient repackaging into retail-sized containers.

Uniform quality, reduced need for frozen and refrigerated distribution and storage, as well as progressively more favorable ratios of packaging material used per unit product will positively impact the environmental and economic advantage of this technology over currently dominant conventional technologies. The technology has been diversified in recent years and concepts of aseptic processing and packaging which are somewhat removed from the core of the original technologies are being adopted to liquid and solid foods, both shelf-stable and refrigerated. Several food companies retail their aseptically processed products refrigerated for diversification, volume saturation, market velocity, and repositioning.

Slotting shelf-stable products in the dairy case or in chilled cabinets in supermarkets may be one way of meeting the current consumer demand for "fresh-like" premiumized foods. Furthermore, low-acid and acidified products filled using aseptic fillers and processes that are *not* US/FDA accepted must be refrigerated (Table 1.2). The fact that one can produce finished products via aseptic processing and packaging that are shelf-stable for 12–24 months without the need for use of refrigeration or freezing is a true testament to its environmental sustainability, and this advantage must be fully leveraged to realize savings on energy and carbon footprint. Also, the global refrigerated supply chain or cold chain is rapidly expanding in developing countries. The sustainability impacts of many of these changes are unknown, given the complexity of interacting social, economic, and technical factors (Heard & Miller, 2016).

Developments in conventional tube-in-tube heat exchanger design will also continue. Helical, dimpled, corrugated, and agitated heat exchangers will be introduced into commercial production with increasing frequency (Carlson, 1996). Emerging thermal processing technologies, specifically volumetric heating methods like continuous-flow microwave heating and ohmic/electric resistance heating, will have a major impact on the quality and variety of available aseptic food products in the near future (Coronel et al., 2005). Applications of emerging nonthermal and thermally assisted technologies like ultra-high pressure processing [pressure-assisted thermal sterilization (PATS), Institute of Food Safety & Health (IFSH)], pulsed electric field treatments, irradiation, sonication, thermosonication, and manothermosonication will expand and integrate with other aseptic processing operations in single and multiple concurrent and sequential bactericidal treatment to achieve extended refrigerated shelf life and shelf stability (Sastry, 2014, Figure 1.3; David, 2020, Legan & David, 2021).

Incremental adjustments will be made to all processing stages and equipment to accommodate multiphase products, especially low-acid products containing large particulate components (Appendix 6, Particulate study—single particle, and Appendix 7, Particulate study—multiple-type particles). Particle-compatible pumps, heat exchangers, tanks, back pressure systems, aseptic surge tanks, filler heads, and package transfer fitments will continue to be improved and integrated into aseptic production lines.

Improper removal of heat from bulk sterilized foods can lead to continual degradation of quality as well as nutritional and functional loss. Prompt cooling and an adequate cooling rate are integral to aseptic processing. Emerging cooling methods will be introduced and integrated into aseptic processes over an extended period of time. Evaporative, cryogenic, reverse thermocouple/Peltier cooling and newer volumetric cooling technologies such as magnetic field cooling and ultrasonic cooling will have an impact on the quality and economy of aseptic product processing.

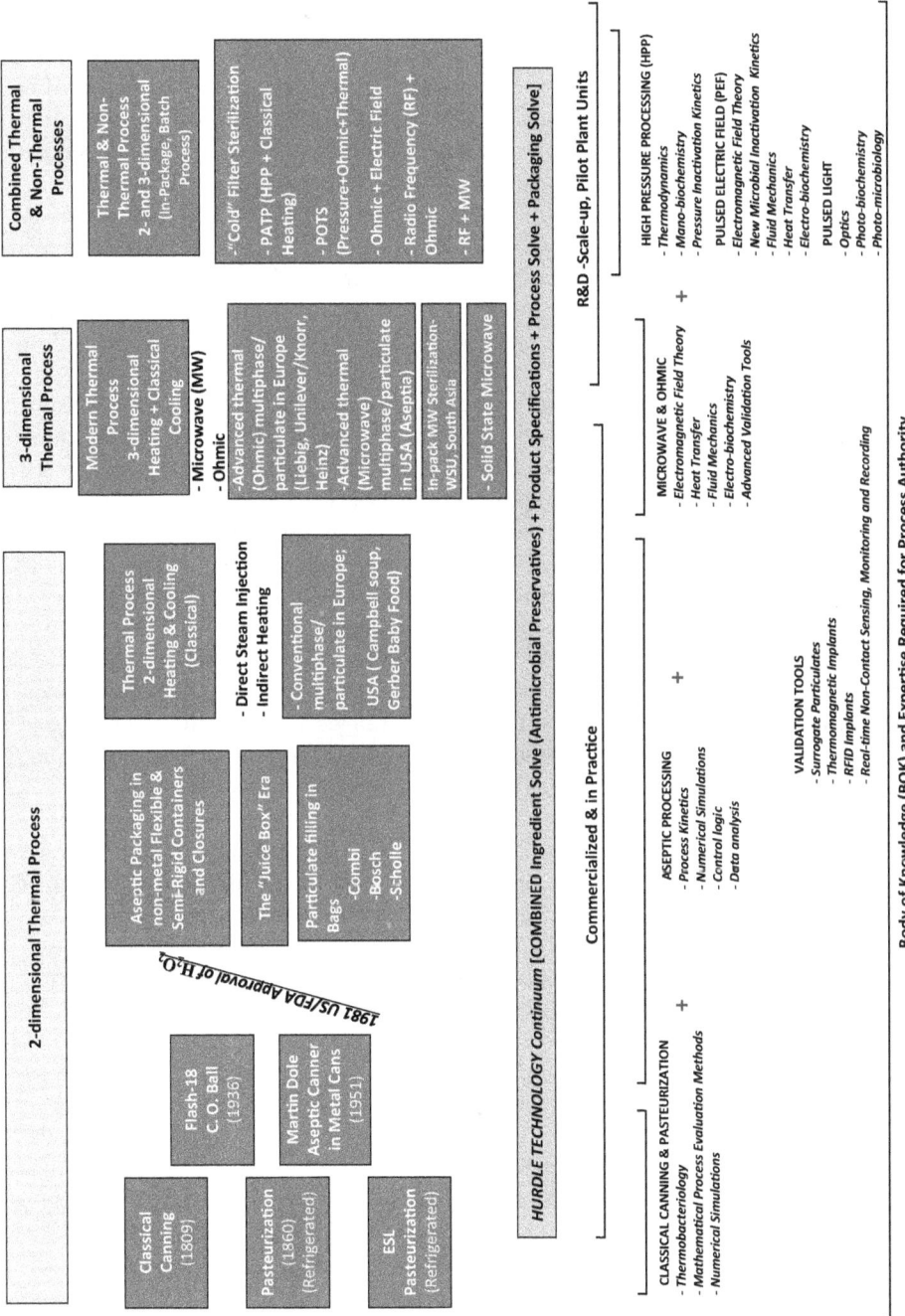

Figure 1.3 Optimization continuum for aseptically processed products, filled and packaged in hermetically sealed containers for non-refrigerated ambient distribution. Also shown is body of knowledge (BOK) and expertise required for process authority for the management of classical two-dimensional and three-dimensional thermal processing technologies, non-thermal processing technologies, combination technologies, and hurdle technology.

Spore-sensitive ingredients of concern are cocoa, tapioca granules, nonfat dry milk (NFDM), carboxymethylcellulose (CMC), starches, sugar, corn, mushrooms, and spices. It is a good practice to monitor, measure, track, and control the mesophilic and thermophilic spore loads of each batch of raw product based on ingredients in a formulation. In addition, these ingredients must be completely hydrated prior to batching to ensure that any spores (if and when present) are fully exposed to a designed and delivered thermal process via a direct or indirect method of heating for the prevention of economic spoilage (Chapters 13 and 16). It is important to note that novel plant-based ingredients that are used to formulate premium clean label natural beverages and food have little history or track record in regard to their interaction with other ingredients. There is the need to develop a good working knowledge on the use of novel plant-based ingredients (both globally sourced and locally sourced) and its impact on viscosity which in turn can determine flow and residence time distribution (RTD) characteristics (Chapter 6) and lethality in hold tube; thermal degradation and fouling behavior; shelf-life stability, and heat resistance and survivability of thermoduric and thermophilic spores via designed thermal process leading to potential microbial spoilage.

The concept of cold, inline sterile formulation wherein one or more HTST/UHT sterile streams are operated in tandem with one or more filter-sterilized streams into an aseptic surge tank, a filler bowl, or a sterile container is sometimes utilized for heat-sensitive ingredients or components. Some heat-labile products are exclusively filter sterilized and aseptically filled and sealed with no heat trauma at all. Sterilization by filtration is called the "cold" method of sterilization since it is the only method that does not rely on either elevated temperature or some other form of energy to destroy microorganisms (Figure 1.3). Sterile filtration does not destroy microbial life, rather it separates microbial life from the rest of the product. There are four primary types of filters used in the parenteral and biopharmaceutical industry: (1) particle filters, (2) microfilters, (3) ultrafilters, and (4) nanofilters. Microfilters are the classic sterilizing filters used in the food, biotechnology, and biopharmaceutical industries. The porosity of microfilters ranges from 0.1 to 10 μm and is used to remove all bacteria, yeast, and colloidal forms. The availability of a wide array of membrane, ceramic, and sintered metal filters is making this process option more commonplace.

New techniques of real-time and postprocess measurement and monitoring will be implemented to accurately and reliably monitor and quantify all lethality delivered to all segments of an aseptic processing system. These emerging techniques and tools will take advantage of the miniaturization of sensing elements and nanotechnology level development currently taking place in other areas of research and development. The absence or inconsistent application of process establishment, monitoring, and validation tools has been one of the most significant hurdles in expanding the range of aseptic processing of more difficult, particularly multiphase food products.

As the locus of research and development and commercialization of aseptic processes moves toward higher loads and larger particulates, sterile formulation, higher throughputs, and extended production runs (72–120 hours) for better overall equipment effectiveness (OEE), the demands for control and monitoring will no doubt increase. Maintenance of sterility and prevention of recontamination are the aseptic equivalents of postprocess contamination (historically the canning industry's Achilles' heel), except that in aseptic they are so intertwined with the system's design and operation that they will require constant, significant, and specialized attention including preventive maintenance, breakdown maintenance, and intermittent cleanup and clean-in-place (CIP) to keep aseptic systems acceptable (Chapter 12).

There is the need to develop sensitive, reliable, and cost-effective nondestructive at-line package integrity tests compatible with current and future line speeds (Chapter 8).

Troubleshooting of microbial spoilage problems often can be resolved using known problem-solving tools (Chapter 13). Further identification and speciation via modern molecular methods is needed for proper root cause(s) analysis (RCA) and definitive corrective action and preventive action (CAPA) (Chapter 14). Tracking sources of microbial contaminants have been a concern, given the uncertainties associated with the integrity and maintenance decontamination of the "sterile working zone" or the aseptic tunnel during normative and extended production runs of 72 to 120 hours (Chapters 13 and 16).

Simplification and streamlining of industrial controls and means of human interaction also need to become a goal to be pursued by upstream suppliers of industrial controls and vendors of highly automated aseptic filler machines. It is important to recognize the widening "technical chasm" between automation with all bells and whistles and human/operators on the floor. People on the floor are central to the successful production of safe and high-quality products, shift-on-shift. However, the technical skills needed to confidently, i.e., without intimidation operate highly complex "Human–Machine Interface" (HMI) may not be readily available in today's work force. Craft/trade schools and local technical community colleges have a crucial role to play in training, continuing education, and preparing the next generation of technical personnel who can comfortably work with automation, Process Logic Controls (PLCs), use of Human Machine Interface (HMI), touchscreens, and types of software. Humans must leverage automation for scale, for maximum efficiency, and for productivity.

The role and responsibilities of process authority (PA) in the context of aseptic processing are paramount in designing and delivery of processes from "front-door to back-door"—ingredients, batching, schedule processes for two-dimensional thermal processing and cooling, barriers, sterile tanks, package integrity, and commercial sterility of finished products. The PA must possess a rigorous body of knowledge (BOK) and background in process kinetics, residence time distribution, numerical simulations, control logic, data analysis, and thermobacteriology (Figure 1.3). The PA managing three-dimensional heating or a combination of two-dimensional and three-dimensional heating must also have a good understanding of electromagnetic field theory, heat transfer, fluid mechanics, electro-biochemistry, and advanced validation tools. In addition, there is the additional complexity, skill set, and BOK required of the PA managing nonthermal technology, or combined thermal and nonthermal technologies, and multifactorial preservation or "Hurdle Technology" (Sastry, 2014; David, 2020; Legan & David, 2021). Finally, there is a need for standardizing BOK and for developing Better Process Control School (BPCS) program and curriculum for training and continuous education for PAs managing three-dimensional thermal technologies, nonthermal technologies, combined thermal and nonthermal technologies, and hurdle technology (Figure 1.3).

1.5 SUMMARY

Aseptic processing and packaging is an attractive and challenging alternative compared to conventional methods of canning of foods. Continuous sterilization of heat-sensitive foods at ultra-high temperatures, followed by prompt cooling, results in a superior finished product, which can be filled into containers of varying compositions, of different shapes, and with many consumer-attractive features. Compared to classical canning, the definitive market advantage of aseptically processed and packaged foods originates from the ability to incorporate several value-added features, such as substantially increased sensory and nutritional qualities, microwaveability, several user-friendly conveniences, and cost saving from the use of semirigid and flexible plastic containers and closures.

Sustainable packaging and longer shelf life are of great importance for the food and beverage industry, along with the cost of environmental benefits in terms of ambient shipping and storage. The cost of refrigeration is higher and is a key driver for companies to invest in aseptically processed and packaged products that are ambient shelf-stable for 12–24 months. This advantage must be fully leveraged to realize savings on energy and carbon footprint. Also, the global refrigerated supply chain or cold chain is rapidly expanding in developing countries. The sustainability impacts of many of these changes are unknown, given the complexity of interacting social, economic, and technical factors (Heard & Miller, 2016).

The future of aseptic processing and packaging of foods and beverages will be driven by customer-facing convenience and taste, use of current and new premium clean label natural ingredients, use of multifactorial preservation or hurdle technology for maximizing quality, and sustainable packaging with claims and messaging (David, 2016, 2020).

REFERENCES

Ball, C.O. 1936. Apparatus for and a method of canning. U.S. Patent 2,029,303, issued February 4, 1936.

Carlson, V.R. 1996. Food processing equipment: Historical and modern designs. In *Aseptic Processing and Packaging of Food: A Food Industry Perspective*, edited by David, J.R. D, Graves, R.H., and Carlson, V.R., Chapter 6, 95–127. Boca Raton, FL: CRC Press, Taylor and Francis Group.

Coronel, P., Truong, V.D., Simunovic, J., Sandeep, K.P., and Cartwright, G.D. 2005. Aseptic processing of sweet potato purees using a continuous flow microwave system. *J. Food Sci.* 70(9):E531–E536.

David, J.R.D. 2013. Thermal processing and optimization. In *Handbook of Aseptic Processing and Packaging*, edited by David, J.R.D, Graves, R.H., and Szemplenski, T., Chapter 11, 167–186. 2nd edition. Boca Raton, FL: CRC Press, Taylor and Francis Group.

David, J.R.D. 2016. What is clean label? A food industry perspective. In Institute of Food Technologists Annual Meeting, Chicago, IL.

David, J.R.D. 2020. Hurdle technology: Multifactorial food preservation for high quality foods. Presented at the 7th Clean Label Conference, 14–16, March 26, 2020, Westin Hotel, Itasca, IL. 2020 Clean Label Conference Proceedings. Global Food Forum, Inc.

Floros, J.D., Newsome, R., and Fisher, W. 2010. Feeding the world today and tomorrow: The importance of food science and technology. An IFT scientific review. *Compr. Rev. Food Sci. Food Saf.* 9(5):572–599. Institute of Food Technologist.

Hallstrom, B. 1979. Aseptic packaging of UHT processed products. In Proceedings of International Conference on UHT Processing and Aseptic Packaging of Milk and Milk Products, 133–138, November 27–29, 1979. Raleigh, NC: Department of Food Science, NCSU.

Heard, B.R., and Miller, S.A. 2016. Critical research needed to examine the environmental impacts of expanded refrigeration on the food system. *Environ. Sci. Technol.* 50(22):12060–12071.

Legan, J.D., and David, J.R.D. 2021. Hurdle technology: Or is it? Multifactorial food preservation for the 21st century. In *Antimicrobials in Foods*, edited by Davidson, P.M., Taylor, M.T., and David, J.R.D., Chapter 21, 695–714. 4th edition. Boca Raton, FL: CRC Press, Taylor and Francis Group.

Martin, W.M. 1951. Apparatus and method for preserving products in sealed containers. U.S. Patent 2,549,216, issued April 17, 1951.

Reisert, S., Geissler, H., Weiler, C., Wagner, P., and Schoning, M.J. 2015. Multiple sensor-type system for monitoring the microbicidal effectiveness of aseptic sterilization processes. *Food Control* 47(2015):615–622.

Sandeep, K.P., Simunovic, J., and Swartzel, K.R. 2004. Developments in aseptic processing. In *Improving the Thermal Processing of Foods*, edited by P. Richardson, Boca Raton, FL: CRC Press.

Sastry, S.K. 2014. Advanced thermal and nonthermal food safety technologies: Academic perspective and future research. Presented at the International Nonthermal Conference, September, 2014. Ohio State University, Ohio.

Chapter 2

US Federal Regulations for Aseptic Processing and Packaging of Food*

Nathan M. Anderson and Emily Weyl

CONTENTS

2.1 INTRODUCTION

Aseptic processing and packaging is defined as "the filling of a commercially sterilized cooled product into presterilized containers, followed by aseptic hermetical sealing, with a presterilized closure, in an atmosphere free of microorganisms" (21 CFR 113). Commercial sterility is the condition that renders the food free of viable microorganisms of public health significance, as well as microorganisms of non-health significance, capable of reproducing under normal non-refrigerated conditions of storage and distribution (U.S. Food and Drug Administration, 2011). Aseptic processing is commonly employed to produce shelf-stable, low-acid (pH > 4.6, water activity > 0.85) foods, but is sometimes used to produce acid and acidified foods having pH ≤ 4.6.

---

* Disclaimer: This document has not been formally reviewed by FDA and should not be construed to represent Agency determinations or policy.

DOI: 10.1201/9781003158653-3

2.2 US FDA REGULATIONS

Aseptically processed and packaged low-acid products, containing little or no meat, are regulated by the FDA Code of Federal Regulations given in Table 2.1. Production of formulated meat or poultry products containing at least 3% raw meat or 2% cooked poultry falls under the regulatory jurisdiction of the US Department of Agriculture (USDA) Food Safety and Inspection Service (FSIS).

2.2.1 Facility Registration and Product Filing

Title 21 CFR part 108—Emergency Permit Control—requires a commercial processor engaged in the manufacture, processing, or packing of thermally processed low-acid foods in hermetically sealed containers to register each food canning establishment and file a scheduled process for each product with the FDA. Guidance from FDA on facility registration and process filing is discussed later in the chapter.

2.2.2 Better Process Control School

All thermal processing operations are to be conducted under the operating supervision of an individual who has satisfactorily completed an FDA-approved course of instruction on the control of thermal processing systems, container closures, and acidification procedures (21 CFR 108.25(f), 108.35(g), 113.10, and 114.10). Several universities and trade associations offer a Better Process Control School (BPCS) each year. A certificate awarded following successful completion to any of these BPCS providing a section on aseptic processing and packaging systems will satisfy the FDA requirement for supervisors of aseptic operations. More in-depth training with "hands-on" laboratory activities specifically tailored to aseptically processed foods is offered by some academic institutions.

TABLE 2.1 US FDA REGULATIONS FOR ASEPTIC PROCESSING AND PACKAGING OF FOODS

Title 21, CFR	Regulatory Aspects
Establishment registration, process filing, and good manufacturing practices	
Part 108	Emergency permit control, facility registration, product filing
Part 113	Thermally processed low-acid foods in hermetically sealed containers
Part 114 .	Acidified foods
Part 117	Preventive controls for human foods
PMO	Grade A dairy products
Aseptic packaging materials as indirect food additives	
Part 171	Petition
Parts 174–179	Materials: resins, coatings, paper, etc.
Part 178	Sterilants, residuals

2.2.3 Process Authority

A process authority is generally described as an individual, or group, an expert in the development, implementation, and evaluation of thermal and/or aseptic processes (IFTPS, 2011b). However, neither FDA nor USDA defines the term process authority, nor do they maintain a list of recognized processing authorities. Processing authorities responsible for aseptic systems need specific knowledge and experience in this area to meet the unique challenges of aseptic processing and packaging operations. Some food processors, consultants, and equipment suppliers can serve as a processing authority or have individuals on staff who serve in this role.

2.2.4 Low-Acid Foods Packaged in Hermetically Sealed Containers

The criteria provided in 21 CFR part 113 are the current good manufacturing practice (CGMP) requirements for thermally processed low-acid foods packaged in hermetically sealed containers. In addition to general provisions, the major emphasis within this part is the requirements and recommendations for the processing and packaging equipment, the role of the process authority, how deviations from a scheduled process are to be handled, and the records that must be maintained. Specific requirements for the aseptic product sterilizer and packaging system are given in 21 CFR 113.40(g). Most of the CGMP requirements apply to the product sterilizer including a temperature recorder and a temperature-indicating device at the outlet of the hold tube, a positive pressure gradient between the sterile and non-sterile product pathways in any product-to-product heat exchanger used in the system, a flow control device, and a means to segregate underprocessed product (21 CFR 113.40(g)(1). In order to accommodate numerous designs and technological advancements, the requirements for the packaging system are more general in nature requiring that the packaging system has a means to operate the system in such a way as to properly establish and maintain commercial sterility, a device that records the critical parameter(s) of the machine and packaging, including the container and closure, sterilization processes, and a timing method (21 CFR 113.40(g)(2)).

Part 113.83 discusses what is required when establishing a scheduled process for low-acid canned foods (LACF). LACF process establishment is rooted in hazard analysis and critical control point (HACCP). The processor conducts a thorough hazard analysis of the aseptic system, identifies the critical parameters, establishes the critical limits (maximum, minimum, or both), develops record-keeping procedures, and establishes a corrective action plan. Microbiological validation of the sterilization of aseptic filling machines and packages, including containers and closures, is needed and several guidelines for conducting these studies are available (IFTPS, 2011a; VDMA, 2006, 2008). When a validation procedure is applied to a system, all of the requirements and assumptions of that procedure must be met. Moruzzi et al. (2000) describes in detail the statistical principles used for biological validation of aseptic systems and offers a statistical spreadsheet tool to determine the likelihood that the data collected from a validation study have satisfied the initial assumptions of the test procedure.

2.2.5 Pasteurized Milk Ordinance

Processors of aseptic, shelf-stable milk or milk-based products must adhere to the requirements imposed by the Pasteurized Milk Ordinance (PMO). However, in a milk plant processing and packaging aseptic Grade "A" milk or milk products, the aseptic processing and packaging system, beginning at the constant level tank and ending at the discharge of the packaging machine, is regulated in accordance with the applicable requirements of 21 CFR 108, 110, and 113. The process authority may provide written documentation that clearly defines additional processes or equipment that are considered critical to the commercial sterility of the product (Item 16p, PMO, 2019).

2.2.6 Acidified Foods

Acidified foods are regulated by Title 21 CFR 114. While some processors manufacture acidified foods on aseptic processing and packaging systems, specific requirements for the processing and packaging systems are not provided in this regulation.

2.2.7 Process Filing Forms

All manufacturers of low-acid and acidified canned foods must file the processing plant location on Form FDA 2541 (FDA, 2020) for a Food Canning Establishment (FCE) number. A five-digit number is assigned to each physical plant location. Each product in each container size and type is filed on the appropriate FDA form, 2541d low-acid retorted foods, 2541e acidified foods, 2541f water activity control or formulation control, or 2541g aseptically processed low-acid foods and assigned a Submission Identifier (SID) number. Detailed instructions for both the paper and electronic versions of these forms can be found on FDA's Acidified & Low-Acid Canned Foods Guidance Documents & Regulatory Information website: https://www.fda.gov/food/guidance -documents-regulatory-information-topic-food-and-dietary-supplements/acidified-low-acid -canned-foods-guidance-documents-regulatory-information.

Aseptically processed low-acid canned foods are filed on Form 2541g. The form requires entries for several sections including product information, container type and size, product sterilizer information, product critical factors, package sterilization system, and the scheduled process.

Aseptic systems consist of two to three separate systems that must be brought to a condition of commercial sterility and maintained sterile during aseptic operation: a product sterilization system, an optional aseptic surge tank, and a package sterilization system. Due to the complexity of these systems, the numerous critical factors necessary to achieve commercial sterility, and the variety of systems, the Supplemental Submission Identifier (SUP SID) provides the filer the ability to document the detailed information for each unique system. While the FDA has provided a basic table format for this submission, many systems are extremely complex and require additional information not outlined in the example table. A discussion regarding critical factors for the aseptic sterilization system can be found in FDA's instructions for the 2541g form in Appendix C.

All acidified foods, *even when aseptically processed and packaged*, must be filed on Form FDA 2541e Food Process Filing for Acidified Method. Since the 2541e form covers several different processing modes, the filer should select the process mode "High Temperature Short Time (HTST)" in Section G, which is the appropriate choice for an acidified product that is aseptically processed and packaged. Section H covers the Container and Container Closure Treatment and the appropriate selection here is "Aseptically Filled." The filer enters the manufacturer name of the aseptic package filler as well as the model number and/or version number. The scheduled process for the product treatment is entered in Section I. Form 2541e differs from Form 2541g in that it does not request the hold tube dimensions or product flow rates.

If the filer of a low-acid food is controlling the outgrowth of spores of public health significance through water activity control or formulation control, the product is filed on Form FDA

2541f *even when aseptically processed and packaged.* Form FDA 2541f requires similar thermal processing information to be entered as Form FDA 2541e.

Form 2541d is for all low-acid retorted foods. This form covers the method of filling a container with low-acid food, hermetically sealing it, and commercially sterilizing the entire container and food contents at the same time in a pressurized vessel. Form 2541d should not be used to file an aseptically processed and packaged food.

2.2.8 Preventive Controls for Human Foods

Longstanding CGMP requirements were modernized and new requirements for hazard analysis and risk-based preventive controls were established in the Preventive Controls for Human Foods (PCHF) rule (21 CFR 117) published in 2015. In general, the rule applies to both domestic facilities and those facilities exporting food for consumption in the United States, but there are exemptions and modified requirements, primarily for the requirements for hazard analysis and risk-based preventive controls. Foods subject to regulation under 21 CFR 113 are exempt with respect to microbiological hazards [Food Safety Modernization Act (FSMA) ref needed]. However, the facility must conduct a hazard analysis for physical and chemical hazards. If physical and chemical hazards requiring preventive control are identified, the facility would need to develop a food safety plan to address those hazards. A similar exemption for acidified foods subject to part 114 was not issued. Thus, acidified foods are subject to the PCHF rule with respect to microbiological, chemical, and physical hazards.

2.2.9 Aseptic Packaging Materials as Indirect Food Additives

With the publication of 21 CFR 178.1005 in 1981, the use of hydrogen peroxide (H_2O_2) sterilization of packaging material was permitted. FDA also set a maximum residual hydrogen peroxide level of 0.5 ppm in the food at the time of packaging on the basis that within 24 hours of packaging the food product, the levels will fall below 0.5 ppm (21 CFR 178.1005; Davis & Dignan, 1983). Since the publication of these amendments to FDA regulations, hydrogen peroxide has become the primary disinfectant used to sterilize packaging materials.

2.3 USDA REGULATIONS

Formulated meat or poultry products containing at least 3% raw meat or 2% cooked poultry falls under the regulatory jurisdiction of the USDA FSIS. All USDA regulated products must be processed following hazard analysis and critical control point (HACCP) systems (9 CFR 417). HACCP plans for thermally processed/commercially sterile products do not have to address the food safety hazards associated with microbiological contamination if the product is produced in accordance with the consolidated canning regulations (9 CFR 431). Detailed requirements for aseptic processing and packaging operations are not included in USDA FSIS canning regulations. However, any system not specifically addressed in the regulations and used for thermal processing must be capable of producing shelf-stable products consistently and uniformly.

Similar to FDA, USDA FSIS requires that a process authority establish the scheduled process and equipment operating procedures and that operators of thermal processing systems and container closure technicians be under the direct supervision of a person who has successfully completed an approved course of instruction

2.4 CONCLUSION

There are numerous FDA and USDA regulations that are applicable to aseptic processing and packaging. These regulations are aimed to ensure the production of safe food. Processors should become familiar with all regulatory requirements applicable to their operations and comply with these regulations.

REFERENCES

9 CFR 417 61. 2018. *Hazard Analysis and Critical Control Point (HACCP) Systems.* Washington, DC: U.S. Government Publishing Office. FR 38868, July 25, 1996, as amended at 62 FR 61009, Nov. 14, 1997, 83 FR 25308, May 31, 2018.

9 CFR 431. 2018. *Thermally Processed, Hermetically Sealed Products.* 83 FR 25308, May 31, 2018. Washington, DC

21 CFR 108. 2016. *Emergency Permit Control.* Washington, DC: U.S. Government Publishing Office. 42 FR 14334, Mar. 15, 1977, 81 FR 49896, July 29, 2016.

21 CFR 113. 2011. *Thermally Processed Low-acid Foods Packaged in Hermetically Sealed Containers.* Washington, DC: U.S. Government Publishing Office. 44 FR 16215, Mar. 16, 1979, 76 FR 81363, Dec. 28, 2011.

21 CFR 114. 1979. *Acidified Foods.* Washington, DC: U.S. Government Publishing Office. 44 FR 16235, Mar. 16, 1979.

21 CFR 117. 2015. *Current Good Manufacturing Practice, Hazard Analysis, and Risk-Based Preventive Controls for Human Food.* Washington, DC: U.S. Government Publishing Office. 80 FR 56145, Sept. 17, 2015.

21 CFR 178.1005. 1981. *Indirect Food Additives: Adjuvants, Production Aids, and Sanitizers; Hydrogen Peroxide.* Washington, DC: U.S. Government Publishing Office. 46 FR 2342, Jan. 9, 1981.

Davis, R.B. and Dignan, D.M. 1983. Use of hydrogen peroxide sterilization in packaging foods. In *Quarterly Bulletin Association of Food and Drug Officials of the United States.* Washington, DC

FDA. 2019. Pasteurized milk ordinance, item 16p. In National Conference of Interstate Milk Shippers, 89. https://www.fda.gov/food/milk-guidance-documents-regulatory-information/national-conference -interstate-milk-shipments-ncims-model-documents. Accessed on June 10, 2021

FDA. 2020a. Form FDA 2541. *Food Canning Establishment Registration.* FDA, 2020. https://www.fda.gov/ food/establishment-registration-process-filing-acidified-and-low-acid-canned-foods-lacf/estab-lishment-registration-process-filing-acidified-and-low-acid-canned-foods-lacf-paper-submissions

FDA. 2020b. Form FDA 2541d. *Food Process Filing for Low-Acid Retorted Method.* FDA, 2020. https://www .fda.gov/food/establishment-registration-process-filing-acidified-and-low-acid-canned-foods -lacf/establishment-registration-process-filing-acidified-and-low-acid-canned-foods-lacf-paper -submissions

FDA. 2020c. Form FDA 2541e. *Food Process Filing for Acidified Method Registration.* FDA, 2020. https:// www.fda.gov/food/establishment-registration-process-filing-acidified-and-low-acid-canned-foods -lacf/establishment-registration-process-filing-acidified-and-low-acid-canned-foods-lacf-paper -submissions

FDA. 2020d. Form FDA 2541f. *Food Process Filing for Water Activity/Formulation Control Method.* FDA, 2020. https://www.fda.gov/food/establishment-registration-process-filing-acidified-and-low-acid -canned-foods-lacf/establishment-registration-process-filing-acidified-and-low-acid-canned-foods -lacf-paper-submissions

FDA. 2020e. Form FDA 2541g. *Food Process Filing for Low-Acid Aseptic Systems.* FDA, 2020. https://www .fda.gov/food/establishment-registration-process-filing-acidified-and-low-acid-canned-foods -lacf/establishment-registration-process-filing-acidified-and-low-acid-canned-foods-lacf-paper -submissions

IFTPS. 2011a. *Guidelines for Microbiological Validation of the Sterilization of Aseptic Filling Machines and Packages, Including Containers and Closures [Document G005.V1].* Guelph, Ontario, Canada: Institute For Thermal Processing Specialists. http://iftps.org/wp-content/uploads/2017/12/aseptic-filler -packaging-validation-G-005-V1.pdf. Accessed on June 9, 2021.

IFTPS. 2011b. *Process Authority Definition [Document WP. 002.V1]*. Guelph, Ontario, Canada: Institute of Thermal Processing Specialists. http://iftps.org/wp-content/uploads/2017/08/Process-Authority -Definition.pdf. Accessed on June 9, 2021.

Moruzzi, G., Garthright, W.E. and Floros, J.D. 2000. Aseptic packaging machine pre-sterilization and package sterilization: statistical aspects of microbiological validation. *Food Control*. 11(1):57–66.

VDMA Fachverband Nahrungsmittelmaschinen und Verpackungsmaschinen. 2006. Aseptic packaging machines for the food industry: Minimum requirements and basic conditions for the intended operation. *VDMA Doc*. 11:1–11.

VDMA Fachverband Nahrungsmittelmaschinen und Verpackungsmaschinen. 2008. Guide to checking the microbiological safety of filling machines of VDMA hygiene classes IV and V. *VDMA Doc*. 12:1–14.

Chapter 3

The US Markets for Aseptically Processed and Packaged Products

Thomas Szemplenski

CONTENTS

3.1 DEVELOPMENT

Even though aseptic processing and packaging was invented decades ago, there was no significant activity in the commercialization of aseptic processing and packaging until the late 1960s and early 1970s when the Dole canning system was used by food processors with foresight. These processors started to aseptically process and package shelf-stable milk, puddings, and soup. At about the same time, Tetra Pak, a Swedish company, introduced its laminated paper–aluminum foil–plastic container to the United States. The system was, at that time, a continuous form–fill–seal system for fluid pasteurized milk and beverages. The container was a tetrahedron. This package was extremely efficient in material use but complicated to pack or stack, and a real challenge to open. The US licensee of this system was the Milliken Company in South Carolina, and in conjunction with Real Fresh of California, the Tetra system was modified to include a chlorine sterilizing bath of the packaging web. This allowed sterilized milk to be filled and sealed aseptically in a hermetically sealed container. These packages were followed by fruit products being aseptically filled using the aseptic bag-in-box system developed by William Scholle in the early 1970s.

In 1981, Tetra Pak returned to the United States with a new and improved container. The basic system remained the same with a web of laminated material being formed, filled, and sealed in a continuous motion. What was different was that after the container was sealed and cut from the web, it was formed and folded into a rectangle or brick. This presented the consumer with a container that looked familiar and suitable, and could be displayed on store shelves.

The real growth in aseptic processing and packaging was experienced starting in the early 1980s. Following Tetra Pak's introduction of the Brik-type package and the Scholle aseptic

DOI: 10.1201/9781003158653-4

bag-in-box filler, other packaging alternatives started to be introduced, as well as other manufacturers of paperboard packaging and bag-in-box aseptic filling equipment. There are now quite a number of aseptic containing alternatives, including plastic cups, coffee creamers, steel drums, form–fill–seal pouches, plastic bottles of various polymers, and even aseptic bulk storage tanks (some of these tanks holding nearly 2 million gallons of aseptically processed acid products). The market growth or driving force of each aseptic packaging alternative is different and will be reviewed in this chapter.

3.2 ASEPTIC METAL CAN MARKET

The Dole canning system was very reliable. It was based on heat for presterilization and maintenance of sterilization of the filler and the metal cans and lids. Filling speeds varied from 30 cans per minute for #10 cans (see Figure 3.1) up to 450 cans per minute for 4-oz. cans.

There was considerable interest and acceptance of aseptic packaging of food into metal cans as Dole eventually supplied more than 60 canning systems, many of which are still in operation. As the learning curve for aseptic packaging using the Dole Canner increased, so did the number of different products that were aseptically filled. Products such as cheese sauces, ketchup, cream-style corn, baby food, eggnog, ice-cream mix, banana puree, tomato paste, dietetic drinks, and sandwich spreads all were aseptically filled into metal cans using the Dole canning system.

The driving force for interest and the growth factor in aseptic packaging into metal cans was the improved organoleptic and nutritional properties of the food being canned. The products no longer had to be overcooked for long periods of time in retorts. Instead of the food products being subjected to a temperature of 250°F from 45 minutes to sometimes 2 hours resulting in overcooking, it now could be homogeneously heated in a matter of seconds from an ambient temperature to around 275°F, held for a short period of time, and then cooled very fast to a filling temperature of between 70°F and 90°F. Aseptic processing and canning resulted in dramatic

Figure 3.1   Aseptically canned pudding (photograph courtesy of Real Fresh, Inc.).

quality and taste improvement in the food being processed and food processors embraced the technology.

The Dole canning system was the only aseptic packaging system that was ever developed for metal cans, other than the aseptic filling system into 55-gallon metal drums invented in the 1970s by Fran Rica for tomato products. Unfortunately for the Dole system, alternative, less expensive aseptic packaging was developed that became the choice of food processors. Aseptic plastic cups, bag-in-box, and pouches are far less expensive than the metal can, so very few other Dole canning systems have been installed in the last 20 years.

3.3 ASEPTIC BAG-IN-BOX

In the early 1970s, William Scholle of the Scholle Corporation, a leading manufacturer of flexible packaging of various polymers, visualized the potential for flexible packaging to replace the expensive, rigid packaging that was being used to transport food products such as tomatoes and other fruit products. Instead of using existing retorting technologies to commercially sterilize the product to be packaged in flexible packaging, he leaned toward applying a relatively new technology that was rapidly developing at the time: aseptic processing and packaging. Scholle engineered and manufactured a prototype of an aseptic filler to fill preformed, flexible bags. He tested and improved upon the initial design at Purdue University's food processing facility in West Lafayette, Indiana.

The first Scholle aseptic filler was developed to fill bags from 1 to 5 gallons. Improvements to the original Scholle were made to the point they now can aseptically fill bags up to 330 gallons. Scholle's prototype aseptic filler was presterilized with steam, superheated water, and chlorine. The flexible bags were and still are sealed and presterilized by gamma radiation. Aseptic bag-in-box packaging was an immediate success (Figure 3.2). Prior to the development of aseptic

Figure 3.2 The Scholle aseptic bag-in-box filler.

bag-in-box packaging, most acid food products such as tomato paste and fruits for pies, yogurt, and so forth were either hot-filled into #10 cans, aseptically filled into metal drums, or frozen in 30-pound plastic pails. Number 10 cans were and still are quite expensive, troublesome to open and dispose of, and yielded loss of products due to residuals, not to mention liabilities due to workers getting cut while handling these containers. Fifty-five-gallon metal drums were very expensive and additionally sacrificed yield at the usage point. Thirty-pound plastic pails were the most common way to transport fruit that was frozen at the growing area and shipped all over the United States for remanufacturing into pies, fruit for yogurt, and so forth. Not only were the pails expensive, but the cost of freezing was additionally pricey.

Economics is the main driving force for the aseptic bag-in-box market. As an example, a brief comparative analysis of aseptic bag-in-box compared to the product packaged into number #10 cans will be presented. Many food products destined for food service are packaged into #10 cans and delivered 6 cans per case. Food service is one of the largest markets for bag-in-box.

Case of #10 cans
- A #10 size can will normally contain approximately 96 oz., therefore a case of #10 cans will usually be about 4.5 gallons
- Although the price of a #10 can will vary depending on the cost of the raw material at the time and the size of the customer based on the number of cans purchased, the average price for a #10 can at the time of this writing is $0.75 each
- $6 \times \$0.75 = \$4.50$ per case (usually about 4.5 gallons)

Five-gallon aseptic bag-in-box
- A 5-gallon, presterilized bag will vary in price depending on packaging materials, barriers, metallization, and so forth, but usually will cost between $0.80 and $1.20. If an average price of $1.00 is used in the calculation, the savings would be $4.50 – $1.00 = $3.50 per case savings
- Adding the price for the corrugated container would increase the price of the flexible bag packaging by as much as $0.20 more. If this is entered into the comparison, the savings would be $3.20 for the aseptic bag-in-box compared to a case of #10 cans, but it should be noted that the bag holds approximately half a gallon more product

An extended economic comparison was generated for an actual food processor. This food processing facility utilizes 9 million #10 cans per year. During the fresh-fruit season, it packages all the products into #10 cans. During the off-season, it opens the #10 cans and reprocesses the product into alternative packaging. To elaborate on the initial comparison:

- 9,000,000 #10 cans divided by 6 cans per case = 1,500,000 cases × 4.5 gallons per case = 6,750,000 gallons of product being processed
- Cost of cans: 9,000,000 × $0.75 = $6,750,000; yearly cost of #10 cans
- Cost of bag-in-box: 6,750,000 gallons divided by 5-gallon bags = 1,350,000 bags @ $1.00 per bag = $1,350,000 yearly cost of bags
- Yearly savings to package product into bag-in-box: $6,750.000 – $1,350,000 = $5,400,000

Additionally, the processor advised that the cost of reopening and disposing of the #10 cans was approximately $1,500,000 annually. Further economic savings to be realized by aseptic bag-in-box compared to rigid metal cans are:

- Reduced space required for bags compared to empty and full metal cans
- Reduced shipping cost

- Reduced liability
- Reduced disposal cost

Based on advantages similar to those described in our comparison, economics is the main driving force for the aseptic bag-in-box market. Other manufacturers of food packaging equipment realized this, and quite a few competitors to Scholle have introduced their versions of aseptic fillers for bag-in-box. There are now at least 10 different aseptic bag-in-box fillers installed in the United States alone with more than 200 aseptic bag-in-box fillers operating. There are also approximately six major suppliers of presterilized aseptic bags of various polymers and oxygen barrier materials. Improvements have been made in aseptic bag-in-box fillers to the extent that several have received US Food and Drug Administration (FDA) validation for aseptically filling low-acid foods, and bag sizes have increased to the point that some of the fillers can fill bags up to 300 gallons.

The market for aseptic bag-in-box is actually two distinct markets: the market for product packaged in smaller bags from 1 to 5 gallons and the market for larger bags packaged in bags from 55 to 300 gallons. The market for smaller bags is generally for products destined for the food service industry. Initially, the market for the smaller bags was for acidified products, such as stabilized fruit for pies and yogurt, but this market has given way to saturation and the larger bags. The market segment for smaller bags that are experiencing growth now is for low-acid food products such as prepared sauces for restaurants, soups, chili, milk, and condiments, like ketchup, salsa, mustard, and single-strength juices. There is hardly a convenience in the United States that does not have a dispenser from Gehl's installed to dispense warm cheddar cheese sauce and chili over nacho chips. These products are aseptically processed and packaged. It is estimated that there are more than 100 aseptic bag-in-box fillers installed filling smaller bags in the United States. This market is not mature. Most assuredly more products will be aseptically packaged into flexible bag-in-box packaging.

There are approximately 140 aseptic fillers installed to fill larger (55–300 gallons) bags. This market is additionally divided into two predominant markets: one for tomato paste and the other for citrus products. The total market for bulk bags in the United States is estimated at 4,500,000 units in 2008. Based on the average price for a bulk bag in 2008, this would amount to an approximate $50 million market for bags.

With approximately 90 aseptic bag-in-box fillers for bulk packaging operation, the larger of the two markets for bulk bags is for tomato products. California grows and supplies most of the tomatoes in the United States and aseptically packages paste and diced tomatoes for shipment to other parts of the country. The product is harvested, condensed, and aseptically processed and packaged in California. It is then shipped to various points around the United States to be reprocessed into ketchup, sauces, soups, and so on. JBT FoodTech is the leading manufacturer of aseptic fillers for bulk bags, although JBT FoodTech does not manufacture the packaging. The demand for aseptic bag-in-box packaging for tomato products continues to grow but at a slower pace than when the aseptic bag-in-box fillers were rapidly replacing nearly all the aseptic drum fillers. Nearly all tomatoes in California that are harvested and packaged to be reprocessed are aseptically filled into the bag-in-box.

The other major market for bulk aseptic bags is in the citrus industry. The citrus industry does not have as many aseptic bag fillers installed as the tomato industry; however, the citrus industry is afforded two crops creating an approximate season of 250 days compared to the 100-day tomato season. In the 1990s, the citrus industry embraced aseptic filling of juice into bulk bags. Almost all the operating aseptic fillers are for the larger 300-gallon bags, and although Scholle and JBT FoodTech have a few aseptic fillers installed for citrus products, the DuPont/Liqui-Box StarAsept filler is the predominant filler being used in the citrus industry.

The market for citrus products being aseptically stored in the bag-in-box has been in steady decline for the past decade. Aseptic bags are rapidly being replaced with aseptic bulk storage tanks capable of aseptically containing millions of gallons of citrus juices. Even at that, industry sources have reported that in 2008, approximately 1.5-million bulk aseptic bags were utilized in the citrus industry (P. Brocher, personal communication, 2008).

3.4 ASEPTIC PAPERBOARD MARKET

In the early 1980s, Tetra Pak returned to the United States with a new and improved package. The basic system remained the same with a web of laminated material being formed, filled, and sealed in a continuous motion. What was different was that after the package was sealed and cut from the web, it was formed and folded into a rectangle or brick. This presented the consumer with a container that looked familiar and suitable to be displayed on the store shelves. The real functional feature was the straw for the smaller containers that were designed to puncture an opening at the specially scored spot. This made the container popular with many consumers who like the portability and ease of consuming whatever the package contained.

Aseptic milk and flavored milks experienced their first real introduction to the mass market in Tetra Pak's Brik-Pak packaging. At that time, the demand for Tetra Pak aseptic filling equipment for acid products such as fruit juices and flavored liquid beverages far exceeded the demand for fillers for milk. It was not until the processing of milk was significantly improved upon with the introduction of steam injection when real growth occurred in the dairy segment.

The acceptance of juices in Tetra Pak Brik packaging was phenomenal. The consuming public embraced the new package and products with a passion as evidenced by the grocery store shelves that were lined with the many different products packaged and offered by the many beverage processors who installed Tetra Pak fillers or had the product copacked at various locations (Figure 3.3).

The market for new Tetra Pak fillers for acid products is somewhat static at the time of this writing due to the many fillers that are already installed and operating. However, as the learning curve in aseptic processing and packaging improved, so has the scope of new products that are being aseptically packaged into paperboard-laminated containers. It appears that new products are being aseptically packaged every day. Products such as soups, broths, nutritional drinks, and many sauces like the ones pictured in Figure 3.4 are all being aseptically packaged into paperboard-type containers.

It appears that the real growth in paperboard-laminated packages is in specialty food and beverages other than juices and fruit-flavored beverages. Tetra Pak now has approximately 200 aseptic fillers installed and operating in the United States. Almost half of these fillers are now for low-acid beverages and other foods. This trend will continue.

The other supplier of aseptic filling equipment for products into paperboard-laminated packages is SIG Combibloc. Unlike the Tetra Pak's form–fill–seal principle of producing sterile packages, the Combibloc filler uses preformed packages that are supplied to the filler, folded flat. The filler then automatically opens, forms, sterilizes, fills, and seals the packages. The packages with the Combibloc filler are sterilized with hydrogen peroxide and hot air. Both the Tetra Pak and Combibloc fillers are FDA validated for aseptically filling low-acid foods, but the Combibloc filler was the first filler used in a commercial aseptic processing system containing a low-acid food

Figure 3.3   Aseptic product in Tetra packaging (photograph courtesy of Tetra Pak).

Figure 3.4   Low-acid aseptic product in Tetra packaging (photographs from sales literature).

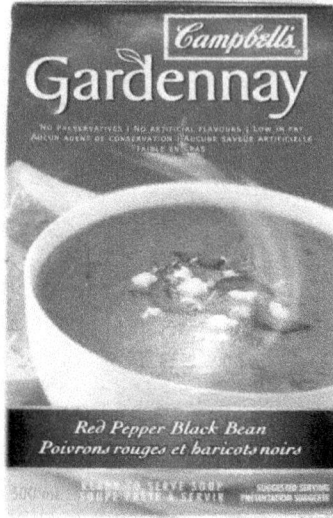

Figure 3.5   Low-acid aseptic product containing low-acid particulates.

with particulates. The installation is located at a Campbell Soup facility in Canada aseptically filling soups with small particulates like the one pictured in Figure 3.5.

Tetra Pak is by far the leading supplier in the world for the supply of aseptic fillers and packaging material. Aseptic packaging at Tetra Pak is dynamic. They now have many different types of fillers designed to aseptically fill into many different packaging configurations, the most recently being a form–fill–seal gable top design. One thing can be counted on with Tetra Pak and that is innovation.

3.5 ASEPTIC PLASTIC CUP MARKET

The market for food products aseptically filled into plastic cups enjoyed tremendous growth starting in the early 1980s. Robert Bosch Company installed the first aseptic filler in the United States for filling food products into plastic cups. This installation was followed by a number of other manufacturers such as Hassia, Metal Box, ERCA, Benco, Gasti, and Hamba. The Gasti and Hamba fillers never received FDA validation for aseptically filling low-acid foods but were used as extended shelf-life (ESL) fillers for refrigerated products.

The introduction of these aseptic fillers for filling into plastic cups all but took away the market for the Dole canning system due to high speed and economics of packaging materials. In all, 26 aseptic fillers and 12 more extended shelf-life fillers were installed starting in the 1980s. Due to excellent engineering, high production speeds, lack of chemical sterilants, and aggressive marketing, Hassia, now OYSTAR Hassia, has become the dominant supplier of aseptic cup fillers in the United States (Figure 3.6). The latest OYSTAR Hassia aseptic cup filler can fill almost 1,700 cups per minute (C. Ravalli, personal communication, 2010).

The first products to be aseptically filled into plastic cups were puddings. The interest in aseptic filling into plastic cups was driven not only by the economics of higher production and less expensive packaging but also by the much improved organoleptic quality of the aseptically

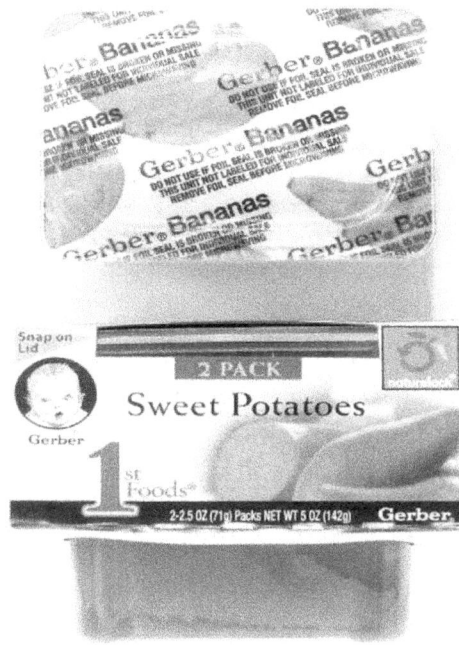

Figure 3.6   Some products aseptically filled using an OYSTAR Hassia filler.

processed pudding. Over the years, the scope of products expanded to include cheese and other sauces, condiments, soup, baby food, and flavored gels. Of late, however, the market for aseptic fillers for cups has softened considerably and not many new aseptic cup fillers have been installed. Manufacturing sources have advised that this is most likely due to the capital-intensive cost of aseptic processing and packaging and saturation.

3.6 ASEPTIC POUCH MARKET

The aseptic pouch market is believed to be in its infancy. This is an underdeveloped market that is expected to explode with activity mainly due to the economic savings in packaging and also due to

- Improved nutritional and organoleptic quality of the food product compared to retorting or hot filling
- Convenience compared to #10 cans
- Less disposal cost
- Less space required
- Less manpower requirement at the end-use point

The economic savings are exceptional and are the main driving force. The predominant market for products to be supplied in pouches is the institutional or food service market for replacing the products in #10 cans. Cans are not only costly but take up considerable space, can be difficult to open and dispose of, and incur liability cases from users cutting themselves while handling them. Aseptic filling equipment and packaging material suppliers in addition to food processors have advised that #10 cans generally cost about $0.75 per can. They additionally have advised that a comparable pouch costs approximately $0.27. That is an overwhelming difference in cost. For each million cans a food processor uses, it would save approximately $480,000 in packaging cost alone.

Another major saving is in floor space. The photograph in Figure 3.7 taken (with permission) at a trade show demonstrates the space required for 832 #10 metal cans compared to the space requirement for 832 pouches in the roll on the bottom left or in one corrugated box on the bottom right of the photograph (B. Pritchard, personal communication, 2010).

Robert Bosch was the first company to develop an aseptic filler for flexible pouches. Bosch supplied a number of aseptic pouch fillers to replace fillers that were using the Dole canning system filling puddings and cheese sauces in #10 cans. The market for cheese sauce exploded with activity, and today there is hardly a convenience store that does not have a dispenser for aseptic cheese sauces for chips.

Inpaco, DuPont/Liqui-Box, Fres-co, OYSTAR Hassia, and Cryovac all have developed aseptic fillers for pouches that operate at varying filling capacities, size of pouches, handling of different particulate sizes and technology, such as fitment attachment offered by Fres-co.

Figure 3.7   Space requirements for pouches compared to cans—each showing 832 packages.

As the technology to aseptically process and receive FDA validation for food products containing particulate matter develops, so will the market for aseptic pouch fillers and packaging material. The food service industry will embrace these food products as a wonderful alternative to not only metal cans but also consistent and organoleptically more palatable foods compared to over- or undercooking foods due to human judgment by restaurant cooking staffs.

3.7 ASEPTIC PLASTIC BOTTLE MARKET

Besides Tetra Pak's paperboard-laminated fillers, the market with the most activity in recent years has been the introduction and installation of aseptic fillers for plastic bottles of polypropylene (PP), high-density polyethylene (HDPE), and polyethylene terephthalate (PET). The beverage industry has embraced the plastic bottle. The first aseptic plastic bottle filler installed in the United States and validated for the filling of low-acid foods was manufactured and installed by Bosch for nutritional beverages. Since then many aseptic bottle fillers have been installed to fill high-acid beverages, whereas others are using their aseptic fillers to fill extended shelf-life of refrigerated dairy products. Many fillers being used in an ESL mode are not FDA validated and therefore cannot be used to fill shelf-stable low-acid beverages.

In all, there are nine manufacturers of aseptic filling equipment with installations in the United States; six manufacturers have received FDA validation. In all, there are approximately 75 installations, aseptically filling high- and low-acid beverages including extended shelf-life products.

These suppliers include:

Bottle Filler Manufacturer	FDA Validation
Robert Bosch	Yes
OYSTAR Hamba	No
KHS	Yes
Krones	No
Procomac	Yes
Serac	No
Shibuya	Yes
Sidel–Rotary	No
Sidel–Linear (Tetra Pak)	Yes
Stork	Yes

Consumer acceptance of the plastic bottle is outstanding and is chipping away at the market for beverages that were previously supplied in paperboard or Brik-type packages. The primary reasons for this acceptance are, but not limited to:

- With plastic bottles, the consumer can see the product
- The bottles fit easier into the cup holders in automobiles
- Bottles are easier to open and easier to reclose
- Bottles can come in many sizes and shapes
- They are easier to recycle

The market for beverages is interesting and very dynamic. It is also quite segmented between high- and low-acid beverages. Initially, beverage processors could justify the capital-intensive aseptic processing and filling system based on the reduced cost of resin compared to a heat-set bottle for hot filling. Initially, this cost difference could be as much as 20% more than a lighter-weight bottle used on the aseptic fillers. Over the years, the blow molders have improved upon the technology and have reduced the cost of the heat-set bottles to the point that it is now economically more attractive to hot-fill, high-acid beverages. In a detailed economic comparative analysis of hot fill versus aseptic including, but not limited to, capital cost for processing and filling equipment, bottle cost, utility costs, and operating costs, it was calculated that the overall difference between the cost to hot fill versus aseptically filling was less than 1 cent. It is no wonder beverage processors are now returning to hot-fill, high-acid beverages.

The market for low-acid beverages is quite different. It is almost impossible to hot-fill, low-acid beverages for shelf stability; therefore, aseptic processing of mostly dairy products and some other high-acid beverages is divided between aseptic shelf-stable beverages and extended shelf-life beverages. Although there are a number of aseptic fillers producing shelf-stable dairy products, this is not a growing market. US and Canadian consumers are accustomed to drinking their dairy products refrigerated and prefer them that way. The market growth for low-acid beverages in plastic bottles is in extended shelf-life products. Extended shelf-life products in most cases are processed the same way as aseptic products with the general exception of two differences: first, with extended shelf-life products, the end product must be distributed refrigerated, therefore the filling temperature is approximately 40°F or less; second, the products are normally not filled using an FDA-validated filler. This does not mean the filler is not clean and sterile. It only means the filler more than likely did not go through the validation process. In both cases, the processing system and filler are both presterilized prior to production. With extended shelf-life dairy products, processors can generally expect a 90- to 110-day refrigerated shelf-life.

Science and Engineering Aspects of Aseptic Processing and Packaging Technologies

Chapter 4

Processing System and Thermal Process Design

Pablo M. Coronel, Josip Simunovic, K. P. Sandeep,
V. R. (Bob) Carlson, and Thomas E. Szemplenski

CONTENTS

DOI: 10.1201/9781003158653-6

4.1 INTRODUCTION

Aseptic processing of pumpable foods (liquids, purees, and liquids with particulates) is a method to produce high-quality shelf-stable products which are then packaged in many different containers. Shelf stability is benchmarked by the shelf life of a product at room temperature without refrigeration, since under those conditions microorganisms can grow relatively fast and at ease. The process must be designed and implemented in such a way that food achieves a state in which it doesn't spoil and doesn't pose a risk to the health of consumers. This condition is called commercial sterility. Commercial sterility is different from complete sterility because the food needs to be palatable to consumers which means that at the same time the food is sterilized, its flavor, color, and nutrient content must be maintained during the entire expected shelf life. Therefore, the balance of food safety and food quality is one of the biggest challenges and objectives of aseptic processing.

In order to ensure food safety, in every step of the process the system and product must achieve and remain in a state of "Commercial Sterility." In the United States, the relevant regulations are contained in the Code of Federal Regulations, chapter 21, and are discussed in Chapter 2. 21CFR113.3(a) defines aseptic processing and packaging as: "Aseptic processing and packaging of foods means the filling of a commercially sterilized cooled product into presterilized containers, followed by aseptic hermetical sealing, with a presterilized closure, in an atmosphere free of microorganisms." Commercial sterility is defined by 21CFR113.3(e) as the condition achieved after processing, thermally or with other means, of food, which renders it free of microorganisms that could reproduce under nonrefrigerated storage and distribution conditions, including spores of pathogenic and spoilage microorganisms. This definition is usually expanded depending on the intended storage conditions of the products to also cover some thermophilic microorganism spores which are not pathogenic. Given the importance of achieving and maintaining the condition of asepticity of product, equipment, and packaging, critical control points (CCPs) become vitally important. A robust HACCP (hazard analysis and critical control points) plan based on the foreseeable risks of contamination needs to be developed, following guidelines by regulations and experienced process authorities. The CCP parameters for an aseptic process must be defined early in the design phase, monitored by proper instrumentation and control system, where alarms and divert systems alert the processor when a parameter is close to the minimal (or maximal) limit of control.

Control systems, in the age of computer-controlled plant, have become increasingly important. Most aseptic plants are now fully automated, with a large number of interconnected programmable logic controllers (PLCs) with human–machine interfaces (HMIs) distributed throughout the plant. Automated control systems require not only qualified personnel for operating the equipment, but also an understanding from management and programmers that all the control systems must be robust to prevent contamination, and at the same time HMIs must be simple for the operators to understand. The operators and programmers are usually on different planes of knowledge and understanding of a system, and in order to prevent the formation of a chasm between both, it is essential to have open discussions about the needs of operators and management. Touchscreens are very helpful and become the face of the system to operators, while searchable databases convey a plethora of information to management. Programmers must make an effort to cater to both sides of the plant, and an effort must be made to simplify the information that operators see while emphasizing the critical parameters for operators.

Thermal processing of foods has two concurrent and competing effects: the first is the heating of foods which prevents spoilage by denaturing enzymes and reducing microbial populations below levels which can harm consumers; and the second which changes the food products by altering flavor, color, viscosity, starch gelation, availability of nutrients, etc. These changes are part of the process and must be taken into account during product development, and consistency of organoleptic characteristics between batches is important for consumer acceptance and also when aseptically packaged products are used as ingredients for other food products. The commercial value of the product and consumer acceptance is a result of these characteristics, known as quality, and a balance must be achieved between safety and quality to ensure that consumers will accept and repeatedly buy the product. This will be discussed in the process-setting section. Aseptic processing is based on continuous thermal processing to achieve commercial sterilization of pumpable products, which must be preceded by the sterilization of the processing system and filler equipment. After these preoperational steps, the products must be stored in aseptic surge or storage tanks and finally packaged in aseptic packaging inside the aseptic area of fillers. With that complexity in mind, equipment and facilities must be built on purpose, applying special standards of design to equipment, as it must not only be reliable and easy to clean but must also achieve a commercial sterility state before any product processing can occur, and during processing the system must be maintained free of microbial contamination. Success in aseptic processing relies on the concurrent optimal operation of each piece of equipment and procedure.

Aseptic processing is usually based on thermal processing of foods, and most processes adhere to all or most of the steps in the following sequence:

- Reception and storage of ingredients
- Handling of solid and liquid ingredients
- Blending tank/formulation of product
- Metering pump for accurately controlling the flow of product through the system
- Continuous heat exchangers to heat the product rapidly to the required sterilization temperature
- A continuous holding tube to hold the product at the set temperature and time to achieve the required thermal treatment
- Continuous heat exchangers to cool the product rapidly to the filling temperature
- Back pressure systems to maintain high temperature and prevent the product from flashing
- Steam-sealed, air-operated valves to direct the product to the desired locations
- Aseptic surge and/or storage tanks to maintain asepticity of the product
- Aseptic mixing or separation equipment
- Aseptic filling equipment

INGREDIENTS

HEATING

MIXING / COOKING

STARCH COOKING

FEED PUMP

HOLDING

TC

COOLING

PV

DIVERT VALVE

STERILE SURGE TANK

FILLING

ASEPTIC ZONE

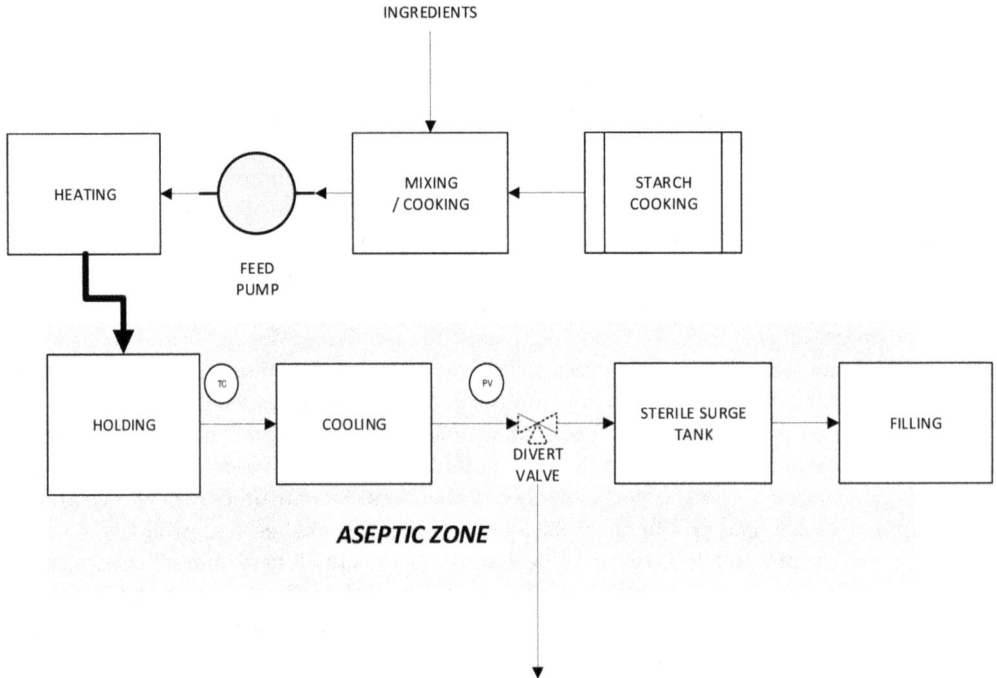

Figure 4.1   Basic aseptic process flow.

- Accurate instrumentation and controls for the system to monitor and document each of the steps that are required for food safety including presterilization of the system, sterilization barriers, and sterilization of the product.

As can be seen from Figure 4.1, the product is first pumped at a set rate by a positive displacement pump, thermally treated in a series of heat exchangers, maintained hot in the hold tube for a set time, and finally cooled to temperatures close to or below ambient. It is clear that after the holding tube, the processed product should be free of microorganisms, and from this point onwards any contamination must be avoided, and thus the product should be protected from contamination from the external environment of the facility. The aseptic state of the product must be preserved which is a requirement for the design and construction of equipment used in any aseptic processing system. The facility must also be built in a way that the environmental microbial load will be low in order to lower the contamination risk, this is called hygienic design and is discussed in Chapter 11.

Equipment for aseptic processing must not only fulfill the hygienic and regulatory requirements used in conventional food processing, such as USDA-FSIS and 3-A requirements, but it must also be capable of sustaining repeated high-temperature sterilization and have measures to ensure maintenance of a sterile state during processing by preventing the ingress of contamination from the environment. While this difference might seem minor, the requirements in design, parts, materials, and maintenance are more stringent as well as the need to monitor and document the conditions of sterility and maintenance of sterility. While the formal regulations for the design of aseptic processing equipment have not been established, the principles that must be followed in the design of aseptic processing equipment have been determined over the course of the evolution of aseptic technology. Associations such as EHEDG (European Hygienic Engineering & Design Group), VDMA (German Association of Packaging Machine Manufacturers), and ISPE (International Society for Pharmaceutical Engineering) have published guidelines for the design of aseptic processing systems.

Aseptic process also relies on good hygienic design of the factory, as discussed in Chapter 11, and trained personnel, who need to be highly motivated to operate such a complicated system. Properly trained and motivated personnel is absolutely critical to the success of an aseptic operation and every effort should be made to recruit, train, and retain the best employees.

Compared to other methods of thermal sterilization of food in pursuit of a shelf-stable product, due to its continuous nature and use of higher temperatures, aseptic processing subjects the product to substantially reduced heating and cooling times, resulting in end products that are most often more nutritious and organoleptically more palatable.

4.2 ASEPTIC PROCESSING ESTABLISHMENT

Aseptic processing is a thermal process, and it relies on temperature for the reduction of microorganisms and spores. Aseptic process is a continuous flow process, in which foods are heated rapidly in a series of connected heat exchangers, the temperature rise is determined by the type of food processed, the target microorganism, and the target ambient storage conditions of the food. Microorganisms of concern are dependent on each product to be processed and must be determined by the safety specialist of the company in conjunction with process authority. Process authority is a role defined as "An individual, or group, expert in the development, implementation and evaluation of thermal and/or aseptic processes" (IFTPS, 2011), and this authority must have knowledge of microbiology, regulations, process calculations and analysis, design and methods of thermal processing studies, and experience to identify and evaluate deviations and spoilage incidents.

The goal of commercial sterility requires a reduction to below detectable levels of pathogens and spore formers that could cause sickness or spoil the food. This is different from terminal sterility in which no microorganism is allowed, in food this terminal sterility leads to products that are not palatable. The processing conditions have been studied in canning (Esty & Mayer, 1922; Stumbo, 1973), where the process of the packaged food has been modeled using first-order kinetics for the destruction of microorganisms by heat as shown in Equation 4.1.

$$\frac{dN}{N} = -k_T \, dt \tag{4.1}$$

Once converted into base 10 logarithms and integrating for a certain time at a fixed temperature, it can be observed that there is a time at which the reduction of population is 90% (1 log cycle). D_T (decimal reduction time) is a direct measure of the resistance to heat of a given microorganism, in a given food, under a set of conditions (temperature, food matrix, pH). Studies of D_T showed that a log–linear relationship existed with temperature within a range of temperature and is called z-value and is also a characteristic of the microorganisms and food.

$$\log_{10}\left(\frac{N}{N_0}\right) = -k_T t$$

$$\log_{10}\frac{1}{10}=-k_{T}D_{T} \tag{4.2}$$

It becomes then necessary to use a reference temperature (T_{ref}) in order to be able to compare results, once that is established it is easy to compare different processes at different temperatures as shown in Equation 4.4.

$$-\int_{ti}^{tf}\left(\log_{10}N\right)=\frac{1}{D_{\mathrm{Tref}}}\int_{ti}^{tf}\frac{dt}{10^{\frac{(\mathrm{Tref}-T(t))}{z}}} \tag{4.3}$$

From this, a sterilization value (F-value) can be derived, which is shown in Equation 4.4:

$$F_{Tref}^{z}=D_{Tref}\left(\log N_{i}-\log N_{f}\right)=\int_{t_{i}}^{t_{f}}\frac{dt}{10^{\frac{Tref-T(t)}{z}}} \tag{4.4}$$

F_{Tref}^{z} is the equivalent number of minutes at a given reference temperature (T_{ref}) to reduce the population of a target microorganism with thermal resistance summarized in the z-value or spores from N_{i} to N_{f} based on a temperature distribution over the production line ($T(t)$). The values of microbial kinetic data (D, and z) can be found in literature such as Holdsworth (1992), Nelson (2010). Most of this data has been compiled and made public into the "Lemgo D- and z-value Database for food" https://www.th-owl.de/fb4/ldzbase/ by the Institute for Food Technology (ILT.NRW) at the OWL University of Applied Sciences and Arts.

In the case of foods with a pH of 4.6 or higher (low acid), these microorganisms of concern include mesophilic organisms such as *Clostridium botulinum* as well as thermophilic organisms such as *Geobacillus stearothermophilus*. *Clostridium botulinum* is a well-known mesophilic spore-forming anaerobic bacteria, which is capable of producing a very potent toxin and has been used as the main target for the sterilization of foods since the works of Esty and Meyer (1922). This process is the reference for any low-acid food process and is known as F_0, where T_{ref}= 121.1C (250°F), z= 10°C (18 F_o), and D_{Tref}=0.24 min. Thus, for a reduction of 10^{12} (12D) spores, a process of F_0>3 is required, also known as Botulinum cook. CCFRA (2008). Other common reference F values are T_{ref}=200°F (93.3C) and z= 16 F_o (9°C) for acid foods (P93 in Europe), T_{ref}=205°F (96°C), and z= 18 F_o (10°C) for acid foods (P96 in Europe), and T_{ref}=212°F (100°C) and z= 16 F_o (9°C), for intermediate or acidified products. It is, however, paramount to agree on the minimal thermal process with a qualified process authority as other factors might impact the thermal requirement.

Following the same logic, quality degradation can be modeled in a similar way, and was named *Cook value* by Mansfield (1962) and shown in Equation 4.5. Depending on the target nutrient and food matrix, the kinetic values may also change and tables of the kinetic parameter can be found in several sources (Toledo, 1991; Holdsworth, 1992) providing D, T_r, and Z_c-values (or rate constant and activation energies) for several quality factors.

$$C(t)=\int_{0}^{t}10^{(T(t')-T_r)/Z_c}dt' \tag{4.5}$$

Compared to in-pack processing (retorting), the heating and cooling of the products happen in a very short time, thus subjecting most food products to high temperature for a short time. This should lead to a higher nutrient and quality retention which can be quantified by observing the effects of temperature. A higher z-value means that the change of DT with an increase of temperature is less pronounced than a reduction with a lower z-value. Looking at the abovementioned

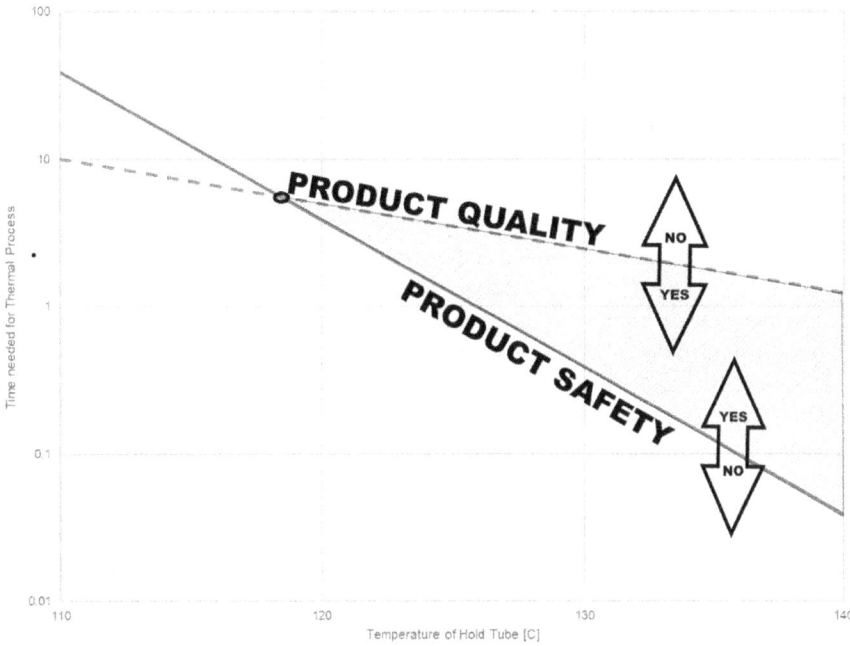

Figure 4.2   Effect of temperature in process time for safety and quality. The shaded zone between the two limits is where safe and high-quality products are produced.

tables, it can be observed that z-values for microbial reduction are between 8 and 14 C while for nutrients they are between 20 and 40 C. This leads us to the HTST paradigm, processing at higher temperatures results in less damage to the quality and higher microbial reduction.

The HTST paradigm is illustrated in Figure 4.2, where the same thermal treatment is applied at different temperatures. A comparison of the two most common safety processes (minimum $F_0 > 3$) for 12D reduction of *C. botulinum* and quality parameters (maximum $C_{100} < 30$) with changing temperature in the hold tube is displayed. Product safety (F_0) minimum line has a slope (z) of 10°C while the maximum quality line has a slope (Z_c) of 33°C, which means that an increase in temperature will reduce the time needed to achieve the same safety level at a much higher rate than for quality. At low temperature, the time needed to achieve a safe product renders a product of unacceptable quality, as temperature increases, there is a point in which both processes are equivalent (115.9°C), and after this temperature is exceeded, the time needed to achieve a safe product is shorter than the time at the same temperature that will render the product quality unpalatable. The area shaded in gray is an area in which quality is acceptable and product is safe, and it becomes wider with increases in temperature, which leads us to the advantage of aseptic processing. Aseptic processes can achieve microbial reduction and at the same time maximize the quality attributes by exposing the product at a high enough temperature for the minimum required time.

It is also noticeable that quality degradation can begin at a much lower temperature than microbial reduction, and it continues to be accrued after the hold tube and in the cooling section. Thus, rapid heating must also be matched with rapid cooling, such that the exposure of the products to a temperature where quality degrades rapidly is minimized.

4.2.1 Considerations for Process Calculations

In order to determine the optimal processing conditions, including time, temperature, and pressure, it is necessary to involve a process authority. Depending on the product characteristics such as pH, viscosity, thermal conductivity, specific heat, solid concentration, size of the solids, etc., and processing requirements and methods, these calculations might be more or less complex. A few examples of process calculations have been compiled in Appendix 5.

All safety calculations are based on the reduction of microorganisms to levels which pose very low risk to product under normal storage conditions. The process must reduce the initial load of microorganisms (bioload) to acceptable levels based on the commercial sterility definition. This is achieved by a combination of minimum time and minimum temperature as shown in Figure 4.2, operational conditions will have to take into account other factors such as stability of utilities, changes in incoming materials, proper hydration of starches, cocoa powder and other dry ingredients, variation in the utilities of the factory, and heat losses to the environment. The control system must be flexible enough and have a fast response to account for all these factors. It is customary to measure temperature at the center of the liquid flow at the exit of the last heat exchanger and at the exit of the hold tube. The temperature at the exit of the hold tube is the one used for safety evaluation and should never be below the minimum.

Bigelow's general method (Equation 4.4) was developed for canning, or in-pack processing, in which a product with a bioload is sealed inside a hermetic container which is then sterilized by heat, cooled, and stored. In this case (Figure 4.3), the initial bioburden comes from the raw food (F), container (C), and lid (L). During production, there might be some growth before the beginning of sterilization which brings us to the initial (N_i) state. Once the food goes into the sterilizer (retort), it must be reduced to a final load (N_f) that is below the maximum allowed for safety (N_{Max}).

Aseptic process must follow a similar procedure for the determination of the heat treatment; however, the process follows a different path, in which each part of the product and container is sterilized separately and assembled in a sterile environment (Figure 4.4). In aseptic processing, the food, container, and closure (lid) are sterilized separately, but also the heating (E) and cooling (Q) equipment as well as the aseptic storage and filler must be sterilized (T), and any gas that comes in contact with the product (A) must also be sterilized. After the product and packaging have been sterilized, every process step must be designed in a way that actively or passively avoids contamination and falls into the aseptic zone of the process and packaging, while food is flowing in a close tube, there are reciprocating valves, rotating agitator shafts, homogenizers, etc., which have the probability of brining contamination. The added complexity must be

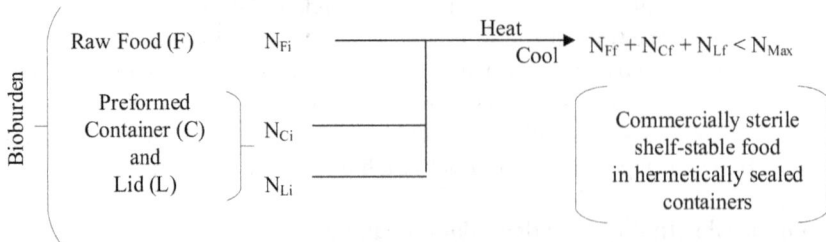

Figure 4.3 Diagram of state for in-pack thermal processing for terminal sterilization or canning (adapted from David, 2013).

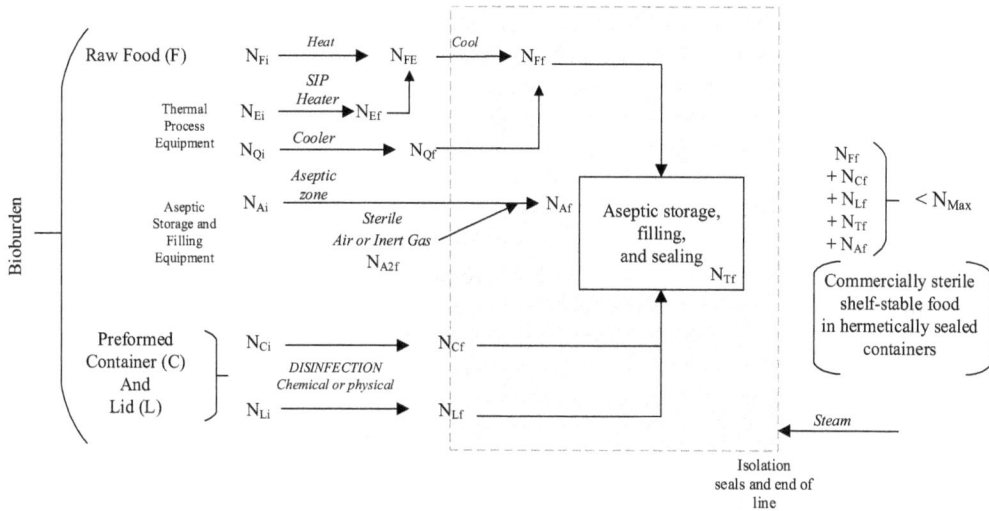

Figure 4.4   Diagram of state for aseptic processing and packaging of food (adapted from David, 2013).

managed and results in the need to use a HACCP (Hazard Analysis and Critical Control Points) plan, which will be explored in Section 4.2.2.

Establishment of the process is based on a combination of temperature and residence time, which must be achieved before reaching the hold tube. It is important to understand the heat transfer that will occur in the heat exchangers, and each type of heat exchanger has advantages and limitations that should be considered. Most heat exchangers work by indirect heating, with no direct contact between the product and heat transfer fluid so that heat is transferred between the product and heat transfer fluid through a wall separating the fluids. Thus, the heat transfer goes from the heat transfer media to the wall, to the fluid part of the product, and then to the surface of any particles and from there to the center by a combination of conduction and convection. This process can be visualized as a series of electrical resistances as shown in Figure 4.5, and the resistances can then be grouped depending on whether they are due to conduction or convection, as shown in Equation 4.6. In order to produce a conservative and safe product, a worst-case scenario is generally considered, taking into account the possibility of a product following the fastest flow path and being the most difficult to heat and thus the slowest to reach the required temperature for sterilization. The thermophysical properties of the fluid and particulates will

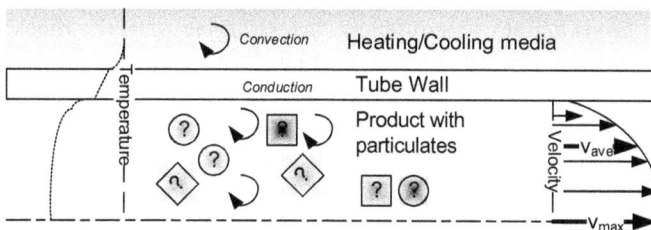

Figure 4.5   Heat transfer in a heat exchanger.

determine the individual resistance of each step and the slowest heating portion of the product should be used as the indicator of the process safety.

$$\frac{1}{R} = \sum_i \frac{1}{R_i}$$

$$\frac{1}{R} = \sum \frac{1}{R_{\text{convection}}} + \sum \frac{1}{R_{\text{condution}}} = \sum \frac{1}{h_m} + \sum \frac{t_s}{\lambda_s} \tag{4.6}$$

Residence time is determined by the flow characteristics of the products along the production system and a ratio of minimum to average residence time is called the flow profile. The flow profile of the product is discussed in Chapter 6 and depends on the rheology of the product at the process temperature. Velocity distribution within the product flow must be taken into account since it determines the rates of convective heat transfer as well as the time different parts of the product spend in the system. This velocity field is very important in the case of products with particulates since it is considered that the fastest moving particle moves with the maximum velocity (v_{max}) of the product. The fastest-moving particle will determine the process time, and the coldest spot in the tube (normally the center) will be the process temperature.

Most food products can be modeled using the Herschel–Bulkley model, where the shear stress (σ) is a function of the flow velocity ($\dot{\gamma}$ the shear rate) such that $\sigma = \sigma_0 + K(\dot{\gamma})^n$. While this may seem complicated, it takes into account that some fluids need force to begin flowing, which is called yield stress (σ_0); and those fluids can react linearly ($n = 1$, Newtonian fluids) or nonlinearly ($n \neq 1$). For any of those combinations, and under laminar flow, the maximum velocity in a straight round tube, as a function of the average flow velocity ($\bar{V}$), can be estimated by Equation 4.7.

$$V_{\text{max}} = \bar{V}\left(\frac{3n+1}{n+1}\right) \tag{4.7}$$

Equation 4.7, when applied to Newtonian fluids ($n = 1$), leads to the well-known ratio $V_{\text{max}} = 2\bar{V}$.

While this equation is valid only on long straight sections of the tube, hold tubes may have bends between sections where some mixing may occur. For conservative purposes, this mixing is dismissed which means that for hold tubes with straight sections and laminar (or transitional flow) the correction factor of 2 must be used. It is important to consider that the hold tube must not be fouled, as fouling will decrease the effective diameter for flow and make the computations invalid. Some fluids have a flow index (n) larger than 1, which means that the product becomes more viscous with shear, in this case it is possible that the region close to the wall has sufficient viscosity to behave like a solid and thus the product will flow through a small portion of the tube. This should be foreseen during product and process design.

Non-Newtonian fluids could use a different correction factor, if the rheological parameters were known at the temperature the product would be in the hold tube, and coiled tubes should consider the changes in flow profile by the formation of secondary flows in the axial direction, known as the Dean effect. These measurements are, however, non-trivial and have not been sufficiently covered in literature; thus, the correction factor of Newtonian laminar flow (0.5) is also used as the most conservative case for laminar flow.

Temperature at the center of the flow can be monitored for the liquid portions, and it is used as the basis for most control systems, as explained the above flow at the center of tubes will be the fastest, and it can be inferred that it will also be the coldest spot. Temperature within particulates can't be measured during the process; and a good understanding of the process and heat transfer needs to be gained for a good modeling and validation exercise as shown in Appendices 6 and 7.

It is important to note that the temperature that defines the safety of products is measured at the end of the hold tube, it is customary to have redundant temperature sensors for such critical measurement, and the control system must take into account that any product that goes below the minimal temperature is unsafe and must be discarded, known as a deviation. Once a system goes into deviation it must be stopped, and it receives a full cleaned in place (CIP) and sterilization in place (SIP). Therefore, day-to-day operations must occur at a higher temperature to avoid such conditions.

Low-acid aseptic process follows the same F_0 recommendations used in canning for public health concerns; however, the calculations are performed mostly for the hold tube only, unless significant lethality could be accrued in the heating section. This presents a very conservative case, since all the lethality that might be accrued in the heating and cooling sections cannot be used. Quality degradation, as explained in the HTST paradigm, begins at a lower temperature than microbial destruction, and the calculations must include heating and cooling. High acid and acidified products are generally processed in a similar manner as those of low acid. However, the lethality integrators are different and dependent on product, pH, etc., resulting in different process temperature requirements.

In order to have products that are safe and of high quality, the thermal process must follow calculations shown previously, but must be produced in a series of equipment units that allows for such production. As shown in Figure 4.4, the process is complex and many parameters need to be controlled in order to maintain the safety of the final product. A methodology to harness this complexity is the use of a HACCP (hazard analysis and critical control points) plan. The hazard analysis (HA) has to be carried out for every product and process and will determine the control points that are critical (CCP) to the system. HACCP plan cannot exist in a vacuum and needs to be supported by a whole facility food safety and defense plan that encompasses cleaning and sanitation, receiving and warehousing, personnel safety and training, distribution, recall, and involves everyone involved in the production of aseptic products.

4.2.2 HACCP and Documentation of Control Points

Food safety is the priority for all food production, especially when shelf-stable products with very long shelf life are involved. A proactive risk-based approach is needed to ensure food safety and prevention of contamination, either intentional or accidental. A food safety and defense plan must be developed from the early stages of design of the whole facility including the production line. A food safety plan must statistically analyze the risks of every step of the process and determine where contamination can take place and where it can be avoided taking into account every aspect of the facility, operation, personnel and handling of materials, and final products. Food defense is an aspect which must analyze when, where, and how an intentional adulteration can happen in the product, as well as possibilities of theft, misuse of chemicals, etc.

The process and packaging of aseptic products are maintained safe by the application of a HACCP-like program, such as the US-FDA regulations for low acid (21CFR113) and acidified (21CFR114) shelf-stable foods. HACCP was developed by NASA for the space program in the 1960s, together with Pillsbury Crop, and US Army Laboratories at Natick. It was adopted by FDA in the 1970s and its use has spread worldwide, being a very important part of the Codex Alimentarius by WHO in 1993. In the United States, an extension of the HACCP is being applied to all the food industry, with the use of risk-based preventive controls (HARPC) to reduce and minimize risks at all levels, which was implemented with the Food Safety Modernization Act (FSMA).

The main advantages of using a proactive system like HACCP are as follows:

- Focuses on identifying and preventing hazards that may render food unsafe
- Is based on sound science
- Allows the producer to have a tailored food safety method, which aligns with the product and process
- Permits more efficient and effective government oversight, primarily because the record-keeping allows investigators to see how well a firm is complying with food safety laws and following practices that reduce the risk of unsafe food over a period rather than how well it is doing on any given day
- Places responsibility for ensuring food safety appropriately on the food manufacturer or distributor
- Helps food companies compete more effectively in the world market
- Reduces barriers to international trade

FDA published guidelines following the report by the National Advisory Committee on Microbiological Criteria for Foods from August 1997 on their website: https://www.fda.gov/food/hazard-analysis-critical-control-point-haccp/haccp-principles-application-guidelines, and they describe HACCP as:

´ "HACCP is a management system in which food safety is addressed through the analysis and control of biological, chemical, and physical hazards from raw material production, procurement and handling, to manufacturing, distribution, and consumption of the finished product. For successful implementation of a HACCP plan, management must be strongly committed to the HACCP concept. A firm commitment to HACCP by top management provides company employees with a sense of the importance of producing safe food."

"HACCP is designed for use in all segments of the food industry from growing, harvesting, processing, manufacturing, distributing, and merchandising to preparing food for consumption. Prerequisite programs such as current Good Manufacturing Practices (cGMPs) are an essential foundation for the development and implementation of successful HACCP plans. Food safety systems based on the HACCP principles have been successfully applied in food processing plants, retail food stores, and food service operations. The seven principles of HACCP have been universally accepted by government agencies, trade associations, and the food industry around the world."

HACCP follows seven basic principles for every product and process, based on the existence of prerequisites such as GMP compliance, cleaning and sanitation programs, allergen programs, and a well-designed facility as described in Chapter 11. These principles are summarized as follows:

1. Conduct a hazard analysis
2. Determine the critical control points (CCPs)
3. Establish critical limits (CL)
4. Establish monitoring procedures
5. Establish corrective actions
6. Establish verification procedures
7. Establish record-keeping and documentation procedures

An unspoken eighth principle is the training of personnel, both in procedures and documentation keeping and in explaining the need and principles of the HACCP and GMP implementation.

Aseptic process and packaging safety is derived directly from HACCP due to the complexity of processing and packaging (Figure 4.4) and the severity of any possible resulting illness, a preventive approach is the only way to ensure that products are safe. The basic HACCP principles can be applied to aseptic processing and packaging. Before the hazard analysis is carried out, it is very important to identify the product characteristics, expected shelf life, expected storage conditions, and to think in the ways that final consumers will use and abuse the product. A basic process flow diagram (Figure 4.6) is needed to analyze all different process steps and possible risks in each of them.

Hazard analysis of the process and packaging needs to be performed, where the risks in each ingredient and process step are identified. These risks must include physical, chemical, microbiological, and allergen risks. The likelihood of each risk and its severity are listed and if needed a process step in which the risk is controlled is identified. Once the risk control points are identified, the most critical of them must be identified (CCP). The CCPs must be documented, including the location (tag on the PLC), physical parameters that will be measured to ensure it is under control (temperature, pressure, speed, etc.), the limits under which the parameter is under control, and the corrective actions describing what happens in case the parameter gets out of the control limits. The CCP definition must also include how often and how the parameters will be documented. An example of a simple CCP definition is found in Table 4.1. Table 4.1 refers to the sterilization of a process line for low-acid aseptic, where an F_0 of 30 or above must be achieved in the farthest point of the system, this point is identified as the inlet to the sterilization cooler (crash cooler) and a temperature sensor (tagged TT312) is recorded by the PLC control system.

FDA regulations took the burden of these analyses and identified many of the CCPs in the aseptic part of the low-acid regulations (21CFR113.40(g)) where the minimal critical steps have been listed as follows:

- Startup: all food contact surfaces must be brought to a state of commercial sterility. This involves the sterilization in place (SIP) of the process line, aseptic surge tank, and filler. SIP is based on a time–temperature combination, and it is customary to use an F_0 value of 30 for the minimal sterilization of low-acid lines, temperature must be measured in the coldest spot of each of the components. Process line cold spot is located at the return of the product after the diversion system (if present) to ensure that all the components downstream from the hold tube are sterilized. Aseptic surge tank requires a validation study to determine the location of the cold spot, and its sterilization also includes the sterilization of the sterile gas filtration system, steam barriers, connecting tubing to filler, and "end of line" (EOL) valve cluster. Fillers have their own sterilization procedure, which is validated and dictated by the maker of such filler. These procedures include sterilization of gas filtration, product contact tubes and nozzles, and the filling zone, which must be clearly defined.

- Filed Process: a combination of hold time and temperature at the end of the hold tube must be prepared with the assistance of a competent process authority and filed into form 2451 with FDA. The critical parameters include flow rate; length and diameter of the hold tube; final temperature at the end of the hold tube; and if direct steam injection is used, initial temperature and volumetric expansion (see Appendix 5 for examples).

- Regeneration: pressure differential between hot product (sterile) and cold product (nonsterile) must be monitored. A regeneration system must have a higher pressure on the hot side, with the intention that if a crack or pinhole is developed in the heat exchanger, the flow will go from sterilized to nonsterilized product and thus prevent contamination.

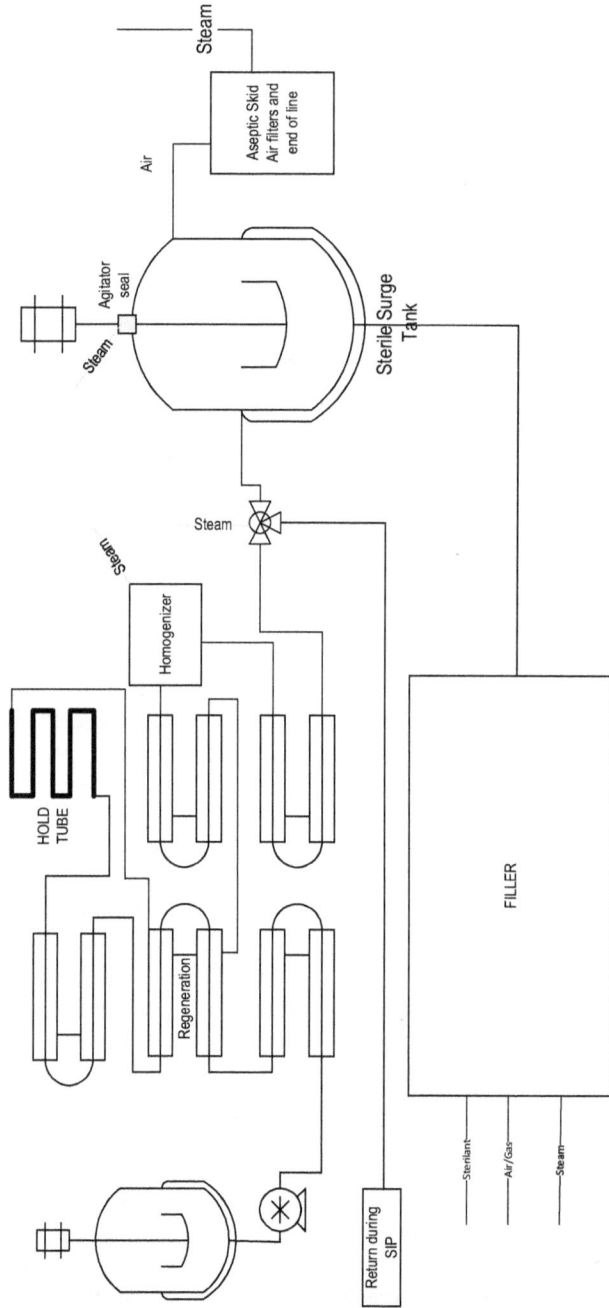

Figure 4.6   Aseptic process flow diagram (using indirect heat exchangers).

TABLE 4.1   EXAMPLE OF CCP DESCRIPTION

Critical Control Point	Hazard	Critical Limit	Monitoring				Corrective Action	Verification	Record-keeping
			What	How	Frequency	Who			
CCP1-pre-sterilization (thermal process line)	Presence of pathogenic bacteria such as *Listeria, E. coli, Salmonella,* or *C. botulinum*	Process line must be sterilized to prevent any microorganism to remain in the cold part of the line	Temperature at the inlet of sterilization cooler (TT312), and time after reaching 121°C	PLC controller screen	During sterilization	Trained operator	Repeat sterilization if temperature is not reached or time above 121°C is not sufficient	Documentation of process establishment; alarm testing yearly	Temperature by PLC safety charts
		Sterility is achieved by keeping the temperature at the farthest point of the line. Inlet of the crash cooler must be above 121°C for 30 minutes					PLC has built in alarm that will restart the countdown if temperature falls below 121°C	Review monitoring, corrective action, and verification records within one business day of preparation	Operator records of the beginning of sterilization, the time when TT312 reaches 121°C, and the time as cool-down begin

- Sterility barriers: equipment downstream from the hold tube with rotating or reciprocating shafts (valves, homogenizers, centrifuges, agitators, etc.) must have barriers such as steam or sterile condensate seals or chemical feed that will prevent the ingress of microorganisms through the moving shafts. The temperature and flow of steam or sterile condensate must be monitored in these seals, and in the case of chemical streams the concentration of sterilant and its flow.
- Diversion system to isolate the aseptic surge tank from the process line must be validated and alarmed, the valves must be isolated with steam seals.
- Aseptic surge tank must have a sterile gas (air or nitrogen) overpressure and protective barriers. Sterile gas must be kept above a preestablished minimal temperature. A gas-sterilizing unit must be kept sterile and monitored, the filters must be replaced according to the preestablished schedule or number of cycles.
- Container sterilization monitored by temperature, concentration, and dwell time of the materials in the sterilant. This is established by the filler manufacturer.
- Maintenance of sterility in the filler by temperature, positive air pressure, and other protective barriers.

Each individual packaging and processing system must be evaluated to ensure that all other possible points of contamination, monitoring system, and alarms are considered. Computer control is now prevalent, and PLCs allow a continuous monitoring of the CCPs, with the ability to automatically document, alarm, and generate corrective actions. Once the CCPs have been defined, it is important to modify the process flow diagram to include each of the CCPs, with reference to the sensors for the critical parameters, as shown in Figure 4.7.

HACCP plans must be validated regularly, to ensure that what is written matched the real operation and documents must also be reviewed to match the operator actions.

Since automated control is prevalent, it is very important to calibrate the sensors to ensure the readings are correct at least once a year. During this time, the line is shut down and should be used for preventive maintenance. This is also a good time to have a verification of alarms and automated controls that maintain sterility, such as steam seal-monitoring alarms, SIP timers, etc.

Documentation of all the CCP and CP parameters is very important, and is part of regulatory compliance, manual records have been the norm; but electronic records need to be pushed forward. Automated control can generate large amounts of data and records that surpass the accuracy of manual control, as long as sensors are calibrated and the whole control system undergoes a validation following guidelines like NFPA 43-L (2002). Complying with electronic record regulation in the United States (21CFR11) requires the implementation of procedures such as electronic signature and data retrieval which are commonplace as the writing of this book, several commercial software LIMS software packages exist and are implemented by the industry.

Record-keeping must follow the regulations in each country, but a rule of thumb is to keep records for any shelf-stable product for three years after the expected shelf life. Electronic storage has the advantage of reducing the footprint and speed of retrieval of such records and is slowly gaining acceptance.

4.3 ASEPTIC PROCESSING EQUIPMENT

Aseptic processing is a thermal process, and as such it needs equipment that will heat the product while it flows (continuous flow). This equipment is generally enclosed, making the whole

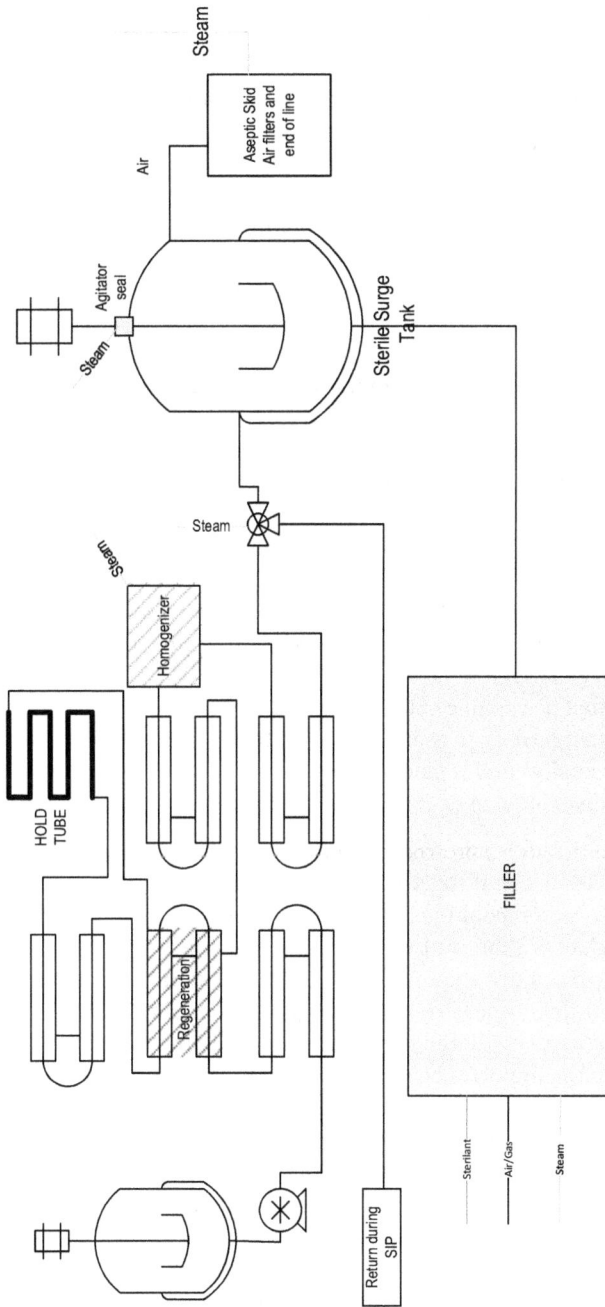

Figure 4.7 Aseptic process flow diagram including CCPs (shaded).

aseptic processing system a closed environment, keeping the product isolated from the environment (Figure 4.1). Since the system is closed, it must be cleaned and sanitized and must achieve a sterile state without disassembly, or if disassembly is needed it must be minimal and should not interfere with the sterilization steps. Like in all other food processing industries, equipment must be of sanitary design, which includes the following:

1. Equipment should be made of materials that will not contaminate food and must be free of any cracks, crevices, or dead legs. The material must be chosen to prevent corrosion and withstand the products to be produced.
2. Equipment must be drainable (no standing water in clean equipment), this includes pumps and instrumentation. Heating and holding tubes must be sloped toward the unprocessed side.
3. Equipment should be capable of being thoroughly cleaned preferably without disassembly or cleaned in place (CIP).
4. Equipment must be capable of being sterilized in place (SIP). Presterilization is usually accomplished with steam or high-temperature water or chemicals so that the equipment achieves a state of commercial sterility.
5. The equipment must be capable of being maintained in a sterile state. Most equipment should be enclosed, and where there is a possibility of contamination (valve stems, agitators, etc.), a barrier of steam, clean condensate, or sterile water must be maintained.
6. Process line should be capable of operating at a greater pressure than the surrounding area or heat exchange media.
7. Process line must be instrumented to monitor all parameters that will maintain safety and quality. Monitoring and recording of such parameters must comply with the requirements of regulations.
8. The equipment must be capable of being maintained in a constant operating mode.
9. The equipment must be easy to repair and maintain, and parts that are subject to wear must be easily accessible and replaceable.
10. The equipment must conform to design, state, and federal regulatory codes if they exist.

Raw materials might go through a pretreatment process (washing, peeling, tempering, etc.), and with other ingredients are brought together in the mixing vessel, special consideration is given to starch, which might need precooking in a special equipment.

The formulated product is then sent to thermal treatment using a feed pump. The product is heated to a temperature above the minimal, calculated as in Section 4.2, in a series of heat exchangers. Once the product is heated to the temperature of process, it is held at that temperature for a minimal time in an insulated tube (hold tube), where the microbial reduction occurs for the most part. Given the importance of this step, both flow rate and temperature are part of a critical control point for the system and any deviation from the minimal conditions requires a shutdown of production. Temperature control (TC) sensors must be located at the end of the holding tube, at least in duplicate and they should feed the control system and the data recorder. Process temperature setpoints and alarm points must be agreed upon in advance with process authority to avoid the possibility of sending the product that is not safe to the filler, both a divert valve to prevent unsterile product to reach the surge tank and filler and a back pressure valve (PV) to prevent boiling at high temperature are present. The product is then cooled and sent to the aseptic surge tank before filling,

Once the product goes through heating and holding, it is free of microorganisms, and extra care must be taken to prevent the introduction of any microorganism or spore back into the system; as it will generate loss of product and could be hazardous to the health of the customers. In Figure 4.1, for illustrative purposes, this high care zone is shaded and called "Aseptic Zone" between the heater and the filler (when packages are closed) this zone of the process must be considered a high care zone, and every piece of equipment should be constructed in a way that it prevents contamination by isolating it from the environment. This isolation is usually achieved by using steam, condensate, sterile water, or sterile air overpressure in valve stems, agitator axles, etc. The aseptic surge tank is usually overpressured with sterile air or nitrogen. With these considerations, each piece of the aseptic processing line needs to be built to comply with these standards and maintained sterility.

4.3.1 Mixing and Cooking Vessel

Delivering a product that is uniform in composition, and distribution of particles to the processing system is important for the consistency of processing and of the final product. Any inconsistency in product density, composition, or phase distribution could potentially affect the overall sterility of the product or result in consumer complaints. If one portion of the batch has a different consistency and percentage of ingredients compared with another, not only will the final product vary but different characteristics of the batch could jeopardize the sterility of the processing system and the final product.

The blending vessel and the balance tank for the supply of product to the aseptic processing system may or may not be the same, and design team must analyze different configurations. These vessels need to be of sanitary design, but don't need to conform to aseptic requirements because they are upstream of the sterile part of the system and the reduction of microbial population will happen afterward.

Tanks have heating and cooling jackets, which need to be designed for the required duty with the help of a competent supplier. It is important to remember to maintain the batches below 10°C (50°F) or above 60°C (140°F) to prevent microbial growth.

Mixing and agitation are very important at this stage depending on the viscosity, presence of thickeners, number of particulates, emulsification, cooking, or other processes needed in the batching stage. The agitator can have several different designs which will impact mixing, hydration, and attrition of particles, and must be designed and constructed properly.

4.3.2 Pumps

The timing or metering pump is one of the most critical pieces of equipment of an aseptic process system. Pumping of the product through the system is extremely critical and must be done at a measurable constant rate in a way that particles are not broken and air is not injected into the product. To ensure that hold time is accurate, the flow through the aseptic system should be as consistent as possible.

Due to the pressures needed to prevent boiling and flashing in aseptic processing, positive displacement pumps are the most used in aseptic processing. Rotary-positive displacement, progressive cavity pumps, and piston pumps are some of the available options for the metering pump.

Rotary-positive displacement pumps (Figure 4.8) can come in several configurations such as lobe, geared, circumferential piston, sine, progressive cavity, and twin screw pumps. The selection of design must consider viscosity, particle concentration, particle size and solid preservation, and final process pressure. When particles are involved, preservation of the particle size,

Figure 4.8   Typical rotary-positive displacement pump (courtesy of Alfa Laval).

attrition, and clearance for the largest possible particle must also be considered. In rotational pumps there might be some slip, and the degree of slip is directly related to:

- Clearances, age, and wear of the moving parts of the pump
- Viscosity of the product being pumped
- Processing pressure drop in the system
- Processing temperature of the product being pumped

Progressive cavity pumps rely on a rotor–stator combination, which pushes the products in the system at a constant rate. These pumps have minimal slip and can pump against high pressures as long as the sizing is done correctly. Progressive cavity pumps can be used with particles, as long as their cavity is properly sized and the attrition rate of particles is considered acceptable in product development.

Piston pumps, on the other hand, have almost no slippage, can pump very viscous products, and can pump very large particles without breaking them, provided that the total pressure against which it is pumping remains within the design specifications. As seen in Figure 4.9, the

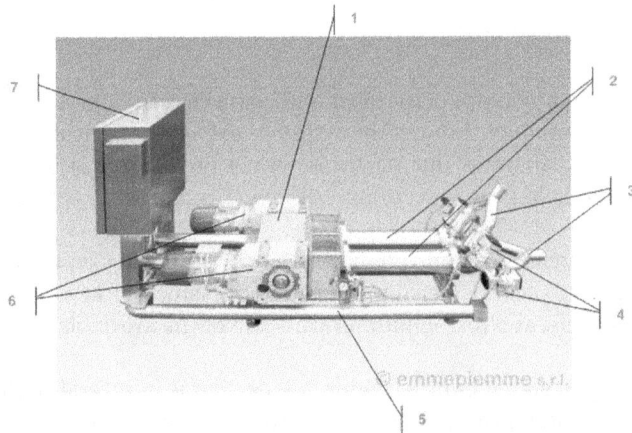

Figure 4.9   Piston pump (courtesy of Emmepiemme SRL).

pump consists of two reciprocating pistons (2) within a stainless steel casing (1). The product is fed through a series of tubes from either local or remote feed tank by a series of tubes (3). The pumping of the product alternates from one piston to the other by using a set of pneumatic valves (4) to obtain a constant flow rate. The pistons are placed on a supporting frame (5) and hydraulic elements or electrical motors (6) are used to move the pistons; a control panel (7) holds all the controls needed by the operator.

When the pistons are pumping, control strategies allowing for the piston pumps will produce a constant flow, as long as the piston pump is fed adequately and has the net positive suction head required to pump the full amount for which it is capable. Piston pumps, when improperly sized or maintained, tend to produce pulses, and means to dampen this pulsation exist and may be necessary for viscous flows.

CIP of positive displacement pumps requires special attention; in many cases the pump needs to be by-passed for CIP in order to have the high flow rates needed. The pump will operate in intervals during CIP to clean the cavity and rotors, but the by-pass allows the CIP to feed the rest of the system at a very high flow rate needed. This by-pass can be permanent (with valves) or temporary around the pump.

4.3.3 Heat Exchangers

Aseptic processing relies of thermal processing at high temperatures to quickly reduce the unwanted microorganisms to levels that will not cause harm without adversely affecting product quality. Thus, it is important to rapidly heat the product to the process temperature, and rapidly cool it back down to temperatures in which the quality will be maintained. This is achieved by the use of heat exchanger units. These units are described in detail in Chapter 5.

The type of heat exchanger used in a process is dictated by the known or foreseeable characteristics of the products that will be produced. In many cases, a combination of different types of heat exchangers is used in the same aseptic processing system to optimize product quality and production efficiency. It would be ideal to have a processing line dedicated to one product family of similar characteristics while in reality production lines must be designed for flexibility, and thus the heating and cooling systems have to be designed to process a range of products of dissimilar characteristics. Meeting this challenge falls under the responsibility of the design engineers who will have to consider the extreme cases of properties in order to maximize efficiency and quality.

There are multiple heat exchanger technologies available for heating and cooling, the more common methods include:

- Direct steam: steam injection or steam infusion
- Plate heat exchangers
- Tubular (single, triple, multitube) heat exchangers
- Scraped surface heat exchangers
- Ohmic heating
- Microwave heating

Each of these technologies can be used in the industry and the selection will depend on the range of products that will be foreseeably made in the process line. An in-depth description of each of these types of heat exchangers is provided in Chapter 5.

4.3.3.1 Steam Injection or Infusion Heaters

Steam injection or infusion is a method of rapid heating of liquids by applying the latent energy of steam to the product. With steam injection, steam is introduced into a continuous flow of product to heat it to the desired sterilization temperature; while in steam infusion the product is made to flow into a chamber filled with high-pressure steam. Both steam injection and steam infusion add water to the product as the steam condenses, giving away its latent heat of vaporization, for process establishment, the amount of water added, and the temperature increase results in thermal expansion of the product, which affects the flow rate in the hold tube, lowering the residence time, and this expansion must be considered as part of the CCP. Steam that condenses into the product becomes part of the product, and this water must be removed or must be considered as part of the formulation, therefore, the quality of the steam added is of critical importance. Steam used in a steam injector or infuser must be of culinary quality, i.e., produced using purified water with no added chemicals that could react with components of the product preferably in a stainless steel boiler and transported in stainless steel tubes. In the United States, culinary steam, per the definitions established by 3-A as the minimum, should be used.

4.3.3.2 Plate Heat Exchangers

The first aseptic processing systems utilized plate heat exchangers for either heating the products, cooling the products, or both as an extension of commercial pasteurizing operations accomplished at lower temperatures (Figure 4.10). Beverages and fairly low viscosity food products such as milk, plant-based milks, thin sauces, and juices still utilize plate heat exchangers in aseptic processing systems. Some of the inherent problems with plate heat exchangers in aseptic processing systems are their pressure limitation and their ability to be cleaned in place (CIP).

Figure 4.10   A typical plate heat exchanger (courtesy of AGC).

Plate heat exchangers were originally developed for use in food applications where they could be disassembled, manually cleaned, inspected, and reassembled into the plate pack. With higher capacity systems, plates are now cleaned in place by washing with various solutions.

Other types of heat exchangers have been designed and perfected which are replacing plate heat exchangers in aseptic processing systems. Not only do the new heat exchanger designs withstand sterilization operations, but they also operate in a reliable manner so sterility is not in question.

4.3.3.3 Tubular Heat Exchangers

Tubular heat exchangers have been used in the chemical industry for a very long time, and they are very efficient for medium-viscosity liquids. The basic idea of a tubular heat exchanger is to have product flowing through a closed tube inside a shell full of heating liquid; and this can take several shapes (Figure 4.11) such as tube in tube, triple tube, tube and shell, coiled tube exchangers, etc. Most aseptic processing systems utilize tubular heat exchangers or a combination of tubular heat exchangers with steam injection or infusion.

Special consideration must be given to helical heat exchangers, in which a continuous tube is coiled around a core which is submerged in a shell filled with thermal exchange fluid (Figure 4.12). This coiled tube is constantly changing direction, and that induces mixing in the product by a phenomenon called the Dean effect; secondary flows are observed in the tube cross-sectional area which increase the mixing of the product and increase the heat transfer rate; on the other hand an increase in pressure drop has been observed. These heat exchangers are used with fluids that have low heat transfer properties and need extra mixing to achieve uniformity of heating.

Tubular heat exchange systems can be very compact and efficient and are commonly sold as a skid to facilitate installation.

4.3.3.4 Scraped Surface Heat Exchangers

In order to increase the heat transfer rates in thick or very thick products, where the core might be cold while the section close to the tube would be hot and burning-on, a special type of heat exchanger was developed in which a blade constantly moves the product. These are known as scraped surface heat exchangers (SSHE) and consist of a shell that is heated or cooled, and inside the shell the product is flowing in a gap where a rotating blade scrapes the outer wall of the exchanger, mixing the product, increasing the heat transfer rate and preventing burn-on. Special products such as shelf-stable puddings and cheese sauces utilized scraped surface heat exchangers (SSHE) for heating and cooling. A diagram of an SSHE is depicted in Figure 4.13, the seals for the rotating scrapes must be sterilized with the system and must be isolated using steam or sterile condensate.

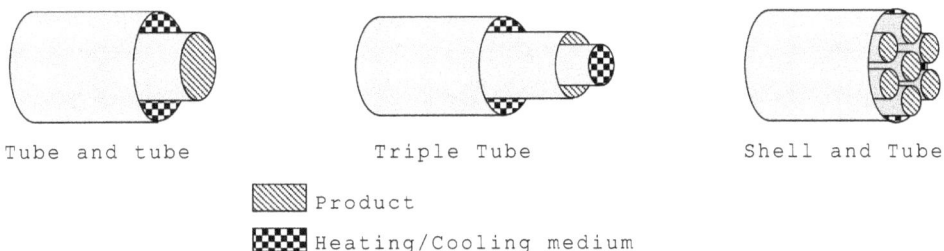

Tube and tube   Triple Tube   Shell and Tube

Product
Heating/Cooling medium

Figure 4.11 Types of tubular heat exchangers.

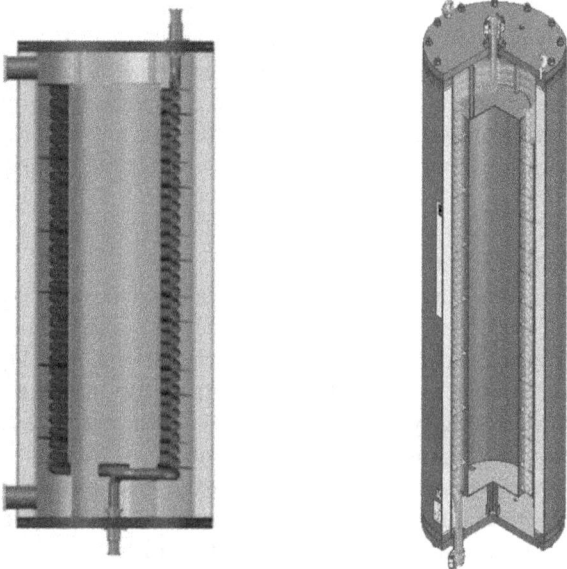

Figure 4.12   A coiled heat exchanger (diagram courtesy of Advanced Process Solutions).

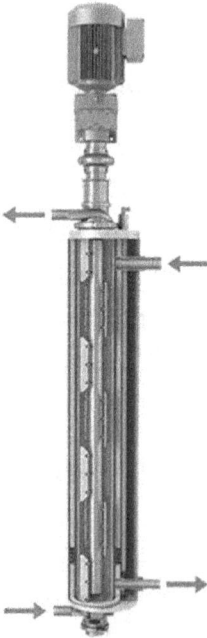

Figure 4.13   Scraped surface heat exchanger (courtesy of Alfa Laval).

4.3.3.5 Ohmic Heating

The ohmic heater systems are a type of volumetric heater in which electrical energy is converted into heat directly in the product. It was originally developed in the United Kingdom by C-Tech Innovation and was further developed by Emmepiemme SRL in Italy, and has made inroads into the thermal process industry. The ohmic heater uses the electrical current that passes directly through the food product being processed to generate rapid, uniform heating (Figure 4.14). Electrodes are located in contact with the food, within tubes made of food-grade materials. The liquid and food particulates are both heated at nearly the same rate and unlike the use of conventional heat exchangers, fouling, or burn-on of the surfaces of the product piping is eliminated.

Several commercial installations utilizing ohmic heating can be found in aseptic processing, including fruits pieces, tomato products (including dices), fruit purees, and especially noteworthy is a line of complex soups with large particulate produced by Unilever which was introduced in 2010 in Europe.

4.3.3.6 Microwave Heating

Continuous flow microwave heating is a method to heat food products volumetrically using microwave energy. Microwave energy is generated from electricity and is applied to a continuous flow of product in a specially designed applicator. The microwave energy is provided by using 2450 and 915 MHz magnetrons. The product is pumped through specially designed tubes and applicators and products are heated uniformly, both liquid and particles are processed aseptically. The lack of a hot surface, combined with fast and uniform heating of the food product, results in high-quality products with no burn-on.

The basic technology currently used in the industry has been developed at North Carolina State University and has been applied to several aseptic products in the market. An illustration showing the pilot plant size microwave heating system is depicted in Figure 4.15. More recent

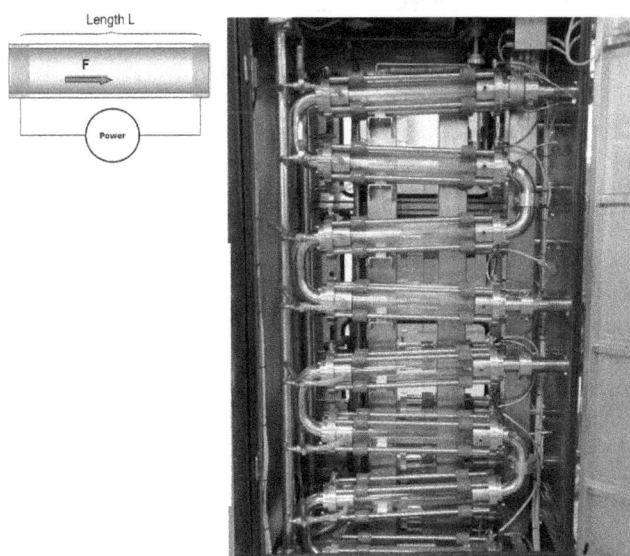

Figure 4.14   Diagram of an ohmic heater (courtesy of Emmepiemme SRL).

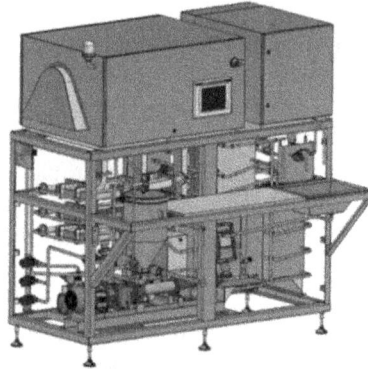

Figure 4.15   A pilot plant size microwave heating system (courtesy of Sinnovatek).

developments of continuous flow microwave technology include solid-state microwave energy generators.

4.3.4 Regeneration

Regeneration is a process technique that consists of exchanging heat between streams of product, so that the cold incoming product cools the hot sterilized product, and in turn, the hot sterilized product heats the cold incoming product, aiming at increasing the energy efficiency of production systems. Regeneration is generally used when plate or tubular heat exchangers are used in the aseptic process. Other types of regeneration include the use of heating or cooling media. The process of regeneration basically results in free or reduced-cost heating and cooling.

When product-to-product regeneration is used, it is a CCP and extreme care must be taken to keep the two continuous flows apart, because the raw incoming product is nonsterile and contains microorganisms that can recontaminate the product that has been sterilized. If direct regeneration is used where a hot product is separated from the cold product that contains high bacteria counts by a relatively thin piece of metal or a gasket, as in plate heat exchangers, even a small leak can cause contamination.

When the heating or cooling media is used for regeneration, the heat exchange is indirect; the hot sterilized product is used to heat water, which in turn heats the raw incoming product, and the water is cooled. Depending on how the system is designed, the hot water next to the cold raw product can be sterile if it is heated to a high enough temperature. In some cases, sterile water can be generated by filtration, designed to remove all, or at least most, of the microorganisms present in water.

Regeneration should be engineered so the pressure on the sterile side of the heat exchanger is always higher than the product on the raw side, and the differential pressure in this stage must be monitored as part of the HACCP plan. If there is ever a leak in the heat exchanger, the leak would be from the sterile product into the raw product or media.

4.3.5 Continuous Holding Tubes

Aseptic processing requires a short hold time of the food at high temperature, generally the sterilization of the food product is only calculated and credited for the holding tube as explained in Section 4.2. As shown in the calculations section, sterilization is calculated at a uniform temperature for a determined hold time, and any credits for microbial reduction in the heat exchangers and connecting piping are generally ignored. The time–temperature combination is a CCP and determines the commercial sterility of the product being processed, this combination of time and temperature is a crucial part in FDA process filings for low-acid aseptic process.

Holding tubes are designed to keep the products at elevated temperature, and are made of a series of continuous stainless steel, insulated, tubes either straight with 180-degree return bends, or coiled. Making coiled hold tubes is an inexpensive method and a long hold can be provided easily without consuming vast amounts of floor space. Several requirements for the design and construction of hold tubes are as follows:

- Hold tubes must be sloped toward the heat exchanger at a minimum slope of 2.1% (1/4″ per foot as per 21CFR113.40 (g)).
- Hold tubes, at least in the United States, must not have heat tracing or any type of heating but can be insulated. Insulation is recommended to prevent heat losses to the atmosphere and minimize the temperature difference between last heat exchanger and the end of hold tube.
- Hold tube must be well instrumented and at least two temperature sensors (one recording and one indicating) should be located at the end of the hold tube and provide feedback to the temperature control system, and the flow diversion system.
- Hold tubes must be maintained at a pressure above the boiling point of the product at the expected hold temperature. This extra pressure prevents "flashing" in which bubbles of steam are formed and collapse creating mechanical damage and disrupting the flow patterns. A pressure sensor is recommended.
- Records of the temperature at the exit of the hold tube are CCP and must be maintained accordingly.

Temperature sensors at the end of the hold tube must be at least redundant, and in the United States, there should be one indicating sensor (showing temperature locally or on the PLC screen) and one recording sensors that make sure that the product is always safe. A pressure sensor is also a good practice to ensure that the pressure is maintained above the flashing temperature of the product, this pressure must be designed to be above the expected maximum temperature in the system and allowances should be made for extreme cases of runaway heating.

In addition to the inactivation of bacteria, many other events may take place in the hold tube, such as the inactivation of enzymes and hydrolyzation of thickeners that increase viscosity. In addition, many adverse reactions can also occur to the product, such as caramelization of sugars, formation of burned odors and flavors, vitamin destruction, color development, and stability changes. Because so many things are happening during the holding, care should be taken to accurately engineer and fabricate the holding tube correctly.

4.3.6 Controls

The basis of aseptic processing is the continuous and rapid heating of food products to a predetermined sterilization temperature, and the holding of the product at this temperature for at least the minimum scheduled time to destroy the unwanted microorganisms followed by the rapid cooling of the product to the filling temperature under sterile conditions. The complexity of aseptic processing and packaging makes all the mentioned areas critical. There is no such thing as almost aseptic, and any small deviation from the designed scheduled process can lead to fatalities. Therefore, all areas of aseptic processing must be accurately controlled and monitored, and records must be kept of all critical parameters. Thus, the control system of the aseptic processing is the system that guarantees a successful operation and a commercially sterile product at all times.

Control systems rely on computer-based control (PLC) and have become very sophisticated. Sensors provide real-time data of different production parameters such as flow, temperature, pressure, brix, viscosity, weight of tanks, steam temperature, opening or closing of valves, etc., and the PLC controls the flow of heating/cooling fluid, opening or closing of valves, agitation, divert valve systems, and monitors for loss of sterility and fouling of the process. A volumetric heating system requires even more advanced controls since the required response times of these technologies are very short.

As previously stated, "the basis for aseptic processing involves the continuous and rapid heating of the product." In order to be continuous, the product must be pumped, therefore, *pumping* is one area within an aseptic processing system that must be accurately controlled. The next step in the definition of aseptic processing is the "rapid heating of the food product to the predetermined sterilization temperature." *Temperature* is then another important part of an aseptic system that must be accurately controlled. Most often when aseptically processing low-acid food products, the temperature required for sterilization is considerably above that of the atmospheric boiling point of water. To reach these temperatures, which may be as high as 300°F (150°C), pressure must be induced in the aseptic processing system that will be above the atmospheric boiling point. Therefore, *pressure control* within the aseptic processing system is vitally important. The product to be processed will generally dictate how the pressure is mechanically induced to the aseptic processing system.

The engineering programmer for the PLC should include interlocks to ensure that should anything happen, that will jeopardize commercial sterility, and since the PLC will react faster than any human, that the system will not allow the product to be sent to the filler, or aseptic tank if present until the system returns to a commercially sterile state. These interlocks should be tested regularly and validated during commissioning and whenever changes to the system are made.

Programmers should also ensure that the HMIs are accessible for the plant personnel and that they are easy to use. It is very important to understand how the program will be used, and have the input of the final users, programmers tend to over-sophisticate the screens and menus, making the control screen too complicated for normal operations.

4.3.7 Sterile Surge Tanks, Barrier Seals, and Automatic Valves

Once the product is sterilized, it must be maintained in a sterile state, as per the HACCP plan, production lines and fillers might have different rates, and fillers sometimes need extra maintenance or short rinse stages and thus a sterile surge tank (aseptic tank) is usually present in the system. This tank should keep products sterile for long periods of time and be able to compensate for the production rate differences between filler and process line. Other barriers are needed, as explained in the HACCP section to ensure that any moving shaft (rotating or reciprocating) such as valves, homogenizer, etc., are maintained free of contaminations. Valves are especially mentioned as several designs are available and are a critical part of the system.

4.3.7.1 Sterile Surge Tanks

Aseptic surge tanks are used to balance the flow from the processing system to the demands of the packaging equipment. Most often aseptic surge tanks are used to process products that will be adversely affected by recirculating and reprocessing. While aseptic surge tanks add no value to the product, they are an additional means of potential contamination, and they are initially very expensive to purchase and operate due to the sophisticated controls necessary to sterilize and maintain sterility in these vessels. Aseptic surge tanks should be avoided if possible.

Aseptic tanks are subject to pressure, steam sterilization, agitation, vacuum, and thus must be designed following the American Society of Mechanical Engineers (ASME) vessel design guidelines for higher than 15 psig (1 bar) pressures. Regardless of the pressure rating, the tank must first be sterilized, which is often done with steam (SIP), and as such the steam pressure and temperature should be considered in the design. After sterilization, the tank is cooled either with air (gas) or with water in the tank jackets. If air is used as the cooling medium it must first be sterilized, usually by filtration. The cooling operation is critical because the steam that is in the tank condenses as the temperature is lowered and the pressure is reduced. Positive pressure must be maintained during cooling to prevent contaminating microorganisms being drawn into the tank from the atmosphere; vacuum conditions must also be avoided to prevent crushing of the tank if it is not designed to withstand vacuum. This pressure is maintained by air or gas that has been sterilized either through a filter or filter–incinerator combination. The pressure is controlled so the maximum pressure desired in the tank is not exceeded.

Potential points of contamination include flanges, connections, and agitator shafts, which must have barrier seals. If the product is not heat sensitive, steam is used as the barrier medium, and temperature and pressure can be monitored continually. If the product is heat sensitive, steam is usually used to initially sterilize the tank and all barriers, followed by a continuous flow of sterile condensate, or a bactericidal solution such as a peracetic acid, hydrogen peroxide, or iodophors. Any contaminants that try to enter the tank should be stopped by such barriers.

Aseptic tanks are generally insulated or are jacketed with a cooling solution to maintain the products cold and prevent thermal damage. Insulation might help to shorten the sterilization time and saves energy while preventing the tank from heating the area.

Pressure in the tank is often used for discharging the product after the tank is cooled, the choice of sterile air or nitrogen depends on the sensitivity of the products to oxidation. With some products, a pump may be used to unload the tank, such pump must be specially designed to have barriers in the axle and prevent any contamination, which is an added expense that is not normally required. A certain amount of air or gas used to maintain or build pressure in the tank is absorbed at the interface of the product and the gas. Consequently, lines feeding the air or gas, filter housings, and pressure control valves should be made of stainless steel of sanitary design, and they should be regularly cleaned.

4.3.7.2 Barrier Seals
Barrier seals are necessary to separate contamination from a product that has been commercially sterilized. They are necessary for any moving parts in the sterile zone (after the hold tube) such as valves, agitators, pumps, homogenizers, divert systems, scraped surface heat exchanger seals, and to keep contaminating bacteria from a vacuum vessel or anywhere the sterilized product is at a lesser pressure than the surrounding area. Any of the microorganisms that might contaminate the food product must pass through the sterilizing media, which prevents such contamination by heat or chemicals.

Barrier seals should have a space sufficient to allow the flow of sterilizing media, such medium is either steam of culinary quality, sterile condensate, or chemical solutions. If the item to be sealed is the plunger of an aseptic homogenizer or a valve stem of an air-operated valve, the sealing distance (barrier) should be greater than the stroke of the pump or of the valve. In this way, no portion of the plunger or valve stems that are in the sterile zone ever move to the atmosphere and become contaminated. Pump plungers and valve stems can move faster than contaminants can be sterilized; therefore, the seal area needs to be of a greater length than the stroke. Any portion of the plunger or valve stem that moves must travel from the sterile product zone into the seal area only. The plunger or valve seal cavity must initially be sterilized with heat that can be conducted to all portions, including the areas between gaskets and sealing flanges, and in cracks and crevices. Heat can travel to these areas and sterilize them. After the heat from steam or water is used to sterilize the area, it can be maintained in a sterile condition by chemicals. This technique is used to prevent damage to heat-sensitive products, such as liquid eggs, and is very useful for scraped surface heat exchangers when used for cooling products. The steam flow, or the concentration of chemicals and flow through the barriers, is a critical control point and must be monitored and documented.

4.3.7.3 Valves

Aseptic valves have been developed as aseptic processing and packaging has evolved. Initially, aseptic processing systems were used with Dole aseptic canning systems and the aseptic valves that were used were furnished with the Dole aseptic canning system. These valves were a combination, primarily of hand-actuated piston-type valves that were maintained sterile by enclosing them in the filling chamber, which was maintained at 265°F (130°C) with superheated steam. Hence, the main requirement of the valves at that point was to be able to withstand such a temperature. Because of this requirement, most of these valves had metal-to-metal seats and the use of elastomers was limited.

The routing valves, which were initially used on the Dole canners in either tee or tee-tee, were piston valves. These valves were used to direct the product to a filler, aseptic tank, or drain supply. The concept used with these valves was having the piston in a barrier material, such as steam, and the barrier was of greater length than the stroke of the valve. Hence, no portion of the valve that was sterile ever moved to the contaminating atmosphere. These valves were available with barrier seals on the connections so if the valve was applied where a vacuum existed, any contaminant that was drawn into the sterile area would first have to move through the barrier. The product was sealed from the barrier with a sanitary seal and the barrier was sealed from the atmosphere with an O-ring seal. Hence, a double seal existed and leaks did not occur.

Piston valves equipped with diaphragms or bellows also exist so that when the piston moves up or down, the diaphragm flexes, and no barrier seal was required. This appeared to be satisfactory; however, the valve would be actuated at high temperatures (up to 300°F) and low temperatures (40°F), and materials that could withstand a number of cycles under these conditions have been developed and are routinely used in pharmaceutical and food industries.

One of the most logical ways to provide a flexible diaphragm is to use a diaphragm-type valve made of stainless steel to sanitary standards where the diaphragm is made of high-grade materials (Figure 4.16). The diaphragms are quite reliable and will withstand a number of cycles and varying temperatures; however, the diaphragms should be replaced as part of the preventive maintenance program. Certain users of these types of valves have gone to the point of using sterilizing solutions on the top of the diaphragm to guard against such failures such as pinholes or cracks. If a failure occurs, then leakage is to the sterile material from the sterilizing liquid, such as Oxonia or iodine. The use of aseptic processing techniques has escalated rapidly in the pharmaceutical industry, and development efforts by valve manufacturers have increased to satisfy this demand. In the future, improved materials and designs are anticipated for larger sizes as needed by the food and dairy industries. Currently, piston-type valves that are normally used in the food industry are available in 1- to 6-in. sizes. Diaphragm valves are available in ½- to 4-in. sizes.

FB6CPM Outlet Valve Series Cut-A-Way

Leak detection port

Air loaded for remote
setpoint adjustments

Guiding above process
area eliminates
particulate generation

OUTLET

Cast Body
(CF3M)
and Dome
(CF8M)

Optimized diaphragm
area for
minimal offset
and extended life

Polished 32 Ra
(0.81 Ra μm) on
all wetted surfaces

INLET

No hold up, fully
drainable in this
orientation

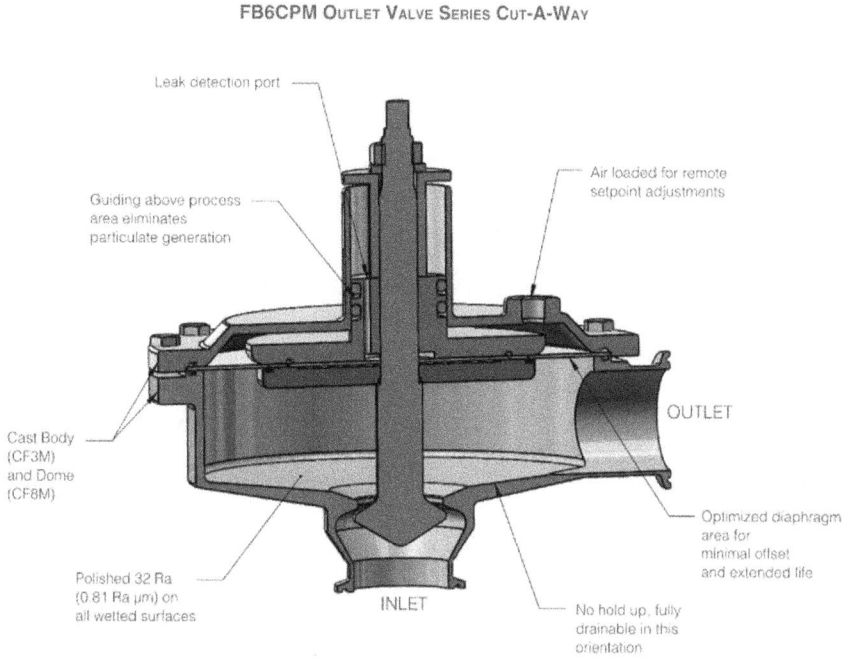

Figure 4.16   Aseptic diaphragm valve (courtesy of Steriflow).

Valves with bellows as means to separate the shaft from product contact are also common, the life of the bellows is finite and the preventive maintenance program should consider the bellows as part of it. Double-seat valves are preferred for isolation and mix-proof of different fluids (Figure 4.17). Bellows are made of food-grade materials, such as stainless steel, titanium, or polymer combinations that provide a trouble-free operation as long as the preventive maintenance program is followed. Actuators are needed for each of the different streams and isolation chambers separate the streams and are flushed with steam of sterilant.

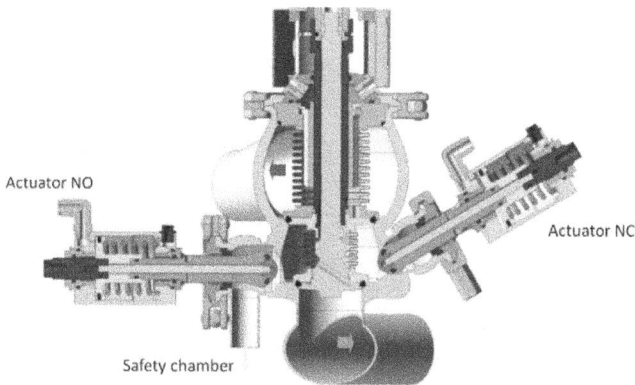

Actuator NO

Actuator NC

Safety chamber

Figure 4.17   Aseptic isolation (mix-proof) valve (courtesy of Evoguard).

4.3.8 Homogenizers

Homogenization is a process needed in several liquid products that could separate, such as dairy, plant-based milks, and other beverages and soups. Homogenization reduces liquid droplet size, especially of fat globules, by subjecting the product to high levels of shear which in turn makes the product color and appearance better. Milk, ice-cream mix, and certain other dairy products are homogenized to reduce the size of the fat globules, whether the products are aseptic or pasteurized. Shear is directly proportional to the pressure, or energy, used to create it. If high pressures are used, homogenizing the product in a pasteurization system that is not part of the aseptic system is desirable. Aseptic homogenizers are expensive to purchase and maintain and are potentially contaminating devices (Figure 4.18).

The general design of aseptic homogenizers is to flush the pistons with steam in a chamber that is of a greater length than the stroke of the piston. In this manner, no portion of the piston that is in the sterile zone ever gets to the nonsterile atmosphere. Other portions of the homogenizer that will be modified include the pressure-adjusting stems used with homogenizing valves. Again, the adjusting valve stem is in a chamber that is of a greater length than the stroke to ensure no portion of the valve stem used for adjusting the pressure that is exposed to the atmosphere. Such steam seals are a CCP and must be monitored and documented.

4.3.9 Filters

Filters are a part of several pieces of equipment for aseptic processing, gases that are in contact with food such as the air used in overpressure of the aseptic surge tank must be filtered to ensure these are sterile, liquids can also be filtered for sterility, this is especially important for thermo-sensitive liquids that are added to thermally processed products, such as vitamins, chemicals, flavors, colors, and enzymes.

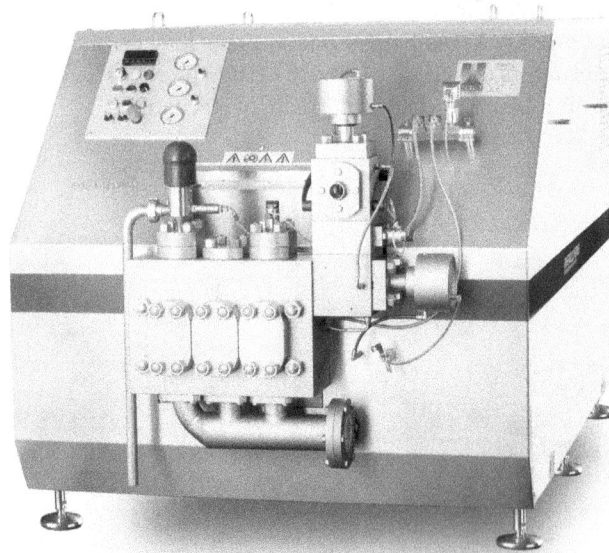

Figure 4.18   Aseptic ready homogenizer (courtesy of Bertolli SRL).

Generally, with gas filtration, the removal of particles is by initial impaction, diffusion, and interception. In liquids, diffusion and initial impaction are unimportant and only interception is involved. There are other factors involved in gas filtration, including the filtered particles acting as filtering media themselves.

Because the technology of filtration is not reviewed in detail, the reader should understand that gas filtration and the filters used are considerably different from liquid filtration that may employ depth filters and that statements made concerning liquid filters are not necessarily applicable to gas filters or vice versa. The statements made by certain manufacturers concerning the ability of their filters to withstand various conditions may not apply. The number of manufacturers having experience in large-volume operations such as those existing in the food or related industries is small but growing.

4.3.9.1 Filter for Gases

Filters used for gases are designed to remove undesirable constituents from the gases, this includes oil, water, undesirable odors and vapors, and particles as well as microorganisms. Filters for gases are made of many different materials that may or may not withstand inline sterilization. Filters that can't stand sterilization must be part of a prefiltration step, followed by sterilizable filter cartridges to ensure that no microbes can make it through. Coalescing filters, charcoal filters, and other prefilters are usually made of heat-sensitive materials.

Sterilizing-grade gas filters normally use borosilicate-type elements that can pass steam through the element without damage. Borosilicate is a hygroscopic material, meaning it will not pass water; therefore, the filter must be arranged in the system so that as steam contacts the element is on the outside of the cartridge rather than on the inside. If steam contacts the inside of the element, condensate will be formed and probably will move through the element. Therefore, after a few cycles, the cartridge will fill with condensate, be exposed to sterilizing temperatures and pressures, and fail prematurely. If the steam is on the outside of the filter, condensate will form and can be trapped off without doing damage. The temperature of the condensate must be adequate for sterilization. The gas can be either on the outside or inside the cartridge as it will pass through the element without damaging it. Sterilization of these filters is part of the SIP, and must be monitored for such a CCP.

Given the exposure to high temperature, pressure, and humidity, filters will fail and leakage will occur; thus filters are rated based on the number of sterilization cycles they can withstand. It is a common practice to have two filters in series to minimize the risk of leakage in case one of them fails, and the preventive maintenance program must include filter replacement at the intervals determined by the filter element manufacturer.

4.3.9.2 Filters for Liquids

Liquids are filtered, primarily, with depth-type filters (although they may be in a cartridge form) that retain the particles that are in the liquid. Usually, they are rated to 0.2, 0.45, and 1.0 μm. This is the maximum size of most organisms that will be in a liquid that must be kept out of the sterile product. Liquid filters will not filter out flavors, colors, viruses, or similar materials that are smaller than 0.2 μm. They are used extensively for purifying water, or other liquids, which may be added aseptically to a product.

Many liquid filters are made of fairly heat-sensitive materials, for example, polypropylene or cellulose acetate and will not withstand temperature above 250°F or 15-psig steam. These filters are often sterilized in a laboratory autoclave where they are not subjected on one side or the other to high pressures. Newer filters are made so that they can be sterilized with steam or a chemical sterilant and are replaced according to a maintenance plan.

The use of liquid filters in an aseptic processing plant is a judgment decision that is based on the quality of the liquid to be used. If the liquid is of questionable quality or is to be used in a critical operation, then it should be filtered.

4.4 UTILITIES

Aseptic processing requires utilities to fulfill each of the steps needed in processing. Water, steam, air, and nitrogen are used in several of the steps of the process. Each of the uses has different requirements which are discussed as follows:

4.4.1 Formulation Water

Water used in the preparation of product is critical in aseptic products, there must be an assessment during development, because plant water may contain undesirable minerals that affect the quality and need to be modified by softening, reverse osmosis, or another preprocessing step. New product formulas are developed in a research lab by using either distilled water or the water from such location. When the same product is made in a different location, the difference in water quality results in a different product.

Potable quality water, which complies with EPA regulations is acceptable for low bacterial load, municipal sources are good; however, in many cases, well water is used and pretreatment must be done on site. First, the dissolved minerals in the water should be analyzed, and depending on the water hardness level, a softening or complete demineralization step might be required. Potable water with chlorine or fluorine might also need a "polishing step" to remove these chemicals and eliminate most extraneous odors and flavors, which is done in a series of filters with activated charcoal. Reverse osmosis, or ultrafiltration, removes chemicals and microorganisms, as well as odors and flavors, making the water of better quality for formulation. This approach is used when companies have plants in different locations as a strategy to maintain organoleptic consistency within the different locations.

Therefore, the water used in the preparation of formulated products should be analyzed to verify it is of the quality necessary to produce the product wanted.

4.4.2 System Sterilization Water

Water used initially to sterilize the system must be of the same quality as water for product formulation. Water is heated through the heating units, and hot water is used to sterilize all the process lines up to the return after the diversion system. The process of initially sterilizing the equipment consists of continually circulating the water and increasing the temperature as it moves through the system. The equipment at the end of the system will be lower in temperature than the water discharging from the heater. If the water or equipment has numerous microorganisms in it or if the microorganisms are very heat resistant, they may not be sterilized. This is of particular concern when acid products are sterilized. The temperatures used during initial sterilization are lower and surviving organisms may grow if the pH is high enough so the condition is nonacid, or organisms in cracks or crevices may survive if they are not contacted with the sterilizing water.

Another concern is that the water may contain various minerals that adhere to the surfaces of equipment causing fouling and scaling. Fouling can be caused by many factors, but it often starts with buildup on the surface of the equipment. After sterilization, these minerals can slough into the product and cause the product to be of a different chemical makeup than that desired. Varying adverse qualities can result, including flavors, viscosities, and color changes. One technique used to reduce fouling is to flush the surface after sterilization with water that contains a food-grade acid. The acid tends to remove the precipitated minerals and make the heat exchange surface clean. By doing this, it will allow for a longer operation before fouling occurs.

4.4.3 Heating/Cooling Water

Water is used as heat transfer media in certain types of heat exchangers, usually in a closed loop system to minimize the waste of water. Among the advantages of using water for heat transfer instead of steam are as follows: temperature can be more accurately controlled and it does not vary in temperature when valves and traps are opening and closing. Chances of burn-on due to hot spots are minimized. Because water is in a closed circuit, undesirable chemicals will be precipitated in the first few minutes and the water will be fairly inactive chemically but it needs to be treated if exposed to the atmosphere (cooling towers).

If the water is used for a single pass and then is used for another function in the plant, such as CIP or rinsing and cleaning raw products, minerals in the water can stick to heat exchanger surfaces and a film will develop, which is called scaling. This often happens in high-temperature coolers if the water is not maintained at the proper pressure to prevent it from flashing and boiling. Besides the heat exchange resistance, the salts may react with the metal of the heat exchanger and eventually cause it to fail.

If a cooling tower is used to maintain the temperature of water used as a coolant, two concerns must be addressed. First, the chemicals used in the tower to keep the water quality must be compatible with the equipment; second, the bacteriological load in cooling water must be managed. Some tower water, particularly in high ambient temperature areas or during summer, can have high bacteriological content and must be reduced to a determined level. There is a possibility of biofilms developing in the coolant side of the heat exchangers, which not only reduces the heat exchange but also can cause contamination. It could get to the product because of a pinhole or stress crack in the heat exchanger or through a gasket. Tower water must then be filtered and chemically treated to ensure safety and performance of the equipment. Any chemicals used must be compatible with the equipment used and prevent corrosion and pitting.

The system used to circulate the water should also be arranged in such a manner that adequate pressures always exist and "flashing" does not occur. The use of tower water coolant is perfectly logical; however, the system for using the tower water must be designed and operated properly. Another consideration with the use of tower water is that it should be removed from the heat exchangers when they are initially sterilized by using an "air blow" or by pumping it back to the tower.

4.4.4 Refrigerated Water

Refrigerated water usually comes from one or two sources. It may be produced in an ice bank, which is a common technique if the process is only operated for part of each day. The other method is to use a continuous heat exchanger, such as a shell-and-tube, which uses a direct-expansion refrigerant, such as ammonia, to cool the water. Often, the equipment used to cool the water is not sanitary and cannot be adequately or properly cleaned. Bacterial buildup can be significant and contamination can result if the aseptic processing coolers were to leak.

It must be considered that cooling water can leak into the product although the pressure of the product may be higher than the cooling water. Venturi effects can literally suck the water into the product although the pressure (product) may be several pounds higher than the water. This problem is increased if a pulsating pump is used to drive the product through the heat exchanger. If the pressure varies, although it may be higher, pulsation on the bacterial cells or spores causes them to be drawn into the product leading to contamination. It must be remembered that bacteria will move and grow against high pressures, and pressures must be exceedingly high (100,000 psi or more) to inactivate them. Even high pressures in the 100,000-psi range will not inactivate certain types of bacteria, enzymes, or spores. To protect against bacterial contamination, refrigerated water should be filtered before being used in the cooling heat exchangers.

Aseptic processing heat exchangers should have thick walls and be arranged so that stresses either during initial sterilization or processing can be withstood and cracks do not develop. Thick walls of 316 or 316L stainless steel are probably the best impediment to the formation of pinholes.

4.4.5 Steam

Steam is used in almost every step of aseptic processing and packaging plants. It is used in the preparation of the product by heating it indirectly through the wall of a preparation vessel or by direct addition to the product being prepared. It is used for heating up products to sterilization temperature, either by direct injection to the product or indirectly by heating through a wall, or by heating the heat exchange fluid. It is used for heating cleaning solutions for CIP and for sterilizing equipment used in processing and packaging operations, for sterilizing barriers, and for maintaining a sterile atmosphere in the barrier area. It is also used for developing heat that causes sterilants to vaporize or breakdown, such as hydrogen peroxide.

When steam is used indirectly to heat the product, i.e., through a heat exchanger wall that is heated by a heating liquid that has been heated by steam in a secondary heat exchanger, it does not have to be of culinary quality. Steam used to directly heat the product-contacting surface, for instance, the wall of a tank, should be of culinary quality using the 3-A definition. This definition requires that no boiler compounds can be used that are not listed or will cause the product to deteriorate or react negatively and generate off colors, have off flavors or odors, or be harmful to the consumer. It is a good practice, however, to consider that any steam used in the plant might get in contact with the product and thus must be made from good quality water and use chemical additives which are approved for food contact. Please note this consideration for CIP, there is a chance that the chemicals used in the steam that heats the CIP solutions may be left on the walls of food contact equipment such as heat exchangers, tanks, aseptic hold tanks, or pipeline walls.

When steam becomes part of the product, such as in the direct injection or infusion during preparation or sterilization process, then steam becomes an ingredient and it should be produced from a reboiler that is sanitary and steam should be made from purified water. Not only are there possible regulatory implications, but the product itself may develop undesirable characteristics from the chemicals that may be in the steam. These chemicals may not be harmful to humans, but they may cause undesirable reactions to occur between the product and the steam.

Steam is generated in boilers, which operate ideally at 150 psi (10 bar) or more. After the steam travels through the various distribution lines and headers, it may lose some pressure at the use point, regulators are used to ensuring a constant pressure of 125 psig (8.5 bar). From the constant distribution point, the steam is controlled to the desired use pressure for the equipment and process required. If such pressure fluctuates, the steam temperature will also fluctuate and might cause the loss of control of the system. One operational consideration when steam is used to heat products indirectly is that as the control valve opens and closes, the temperature of the steam entering the heat exchanger varies. This will cause the temperature of the heated wall to also vary and can cause spots of burn-on when the product touching these spots is overheated.

4.4.6 Air

Air is often overlooked in food processing operations, it surrounds everything and it must be treated as an ingredient when the product is exposed to the environment, and managed properly to prevent cross-contamination throughout the plant. Air is in contact with the product in many instances, from the warehouse to the moment the product is sealed and it can be a source of contamination if not properly managed. Air handling in the process plant is discussed in the hygienic design of the plant (Chapter 11).

Compressed air is a versatile utility, it is used for instrumentation, control, cushioning of equipment or transfer of fluids. Compressed air must also be food grade, which means it must be made in an oil-free compressor, be free of water, desiccated, and filtered to prevent any foreign object contamination. Moisture removal is normally done with a dryer or it can be a filter that removes small amounts of moisture as air passes through it. A coalescing filter is used for this purpose. To remove particles, various filters are used that are generally rated at some efficiency level for a given size of a particle and the volume of gas they will pass at a specified pressure. Because filters for air or other gases are different from filters used with liquids, they should not be interchanged and used for both purposes.

Air used with steam valves, routing valves, and solenoid valves should be clean and dry and should be purified so particles are not present that cause the valves to improperly operate. Air used with instruments or valves should have the moisture removed and should be filtered to remove particles. Air used in direct contact with the product, such as air used to agitate tanks, should be clean and sanitary. This means a specialized filter with small pores should be used, the air in the aseptic zone must be sterilized as well using specialized filters.

In many cases, products contain sterile gases that are part of the product when it is packed, such as a mousse or other aerated product, sterile air is also used in aseptic hold tanks and to maintain positive pressure in aseptic fillers. Sterilization of gases is performed using specialized filters, preferably in conjunction with incineration. If only filters are used at least two sterilizing-grade filters should be used. It is very difficult to tell whether a filter is operating correctly until after a lot is processed, which may be very expensive. The use of filters must be incorporated in the design and particles, not only those that carry bacteria, but other undesirable particles must be removed. Incineration systems are used with filters to provide a record of the temperature to which the gas has been heated for sterilization. This temperature must be chosen so it is adequate for sterilizing the gases, considering the product into which it will be injected.

Sanitary air systems used for delivering air to products should not only be made of stainless steel and sanitary, but should also be cleaned and inspected regularly. Filter housings, pressure-regulating valves, and on/off and modulating valves are available today in stainless steel of sanitary design. These components, along with the distribution system, should be cleaned just as the product lines and control valves in the product distribution system. Usually, the air-handling system should be considered sanitary from the discharge of the check valve of the air-producing system, which is probably non-sanitary (compressor, receiver, dryers, etc.), through the sanitary portion, which is the distribution system, pressure-regulating valves, on/off valves, and lines.

The air system from a sanitary point should be designed in a similar manner to the product-handling system so it can be subjected to CIP.

4.5 ASEPTIC PROCESSING OPERATIONS

Aseptic processing of food and beverages is continuous commercial sterilization of these products preceded by the sterilization of the processing system. Aseptic processing of food is a continuous process requiring strict control. Unlike in the past, today most aseptic processing systems are accurately controlled by programmable logic controllers (PLCs) and use human–machine interfaces (HMIs) with screens. Documentation of the aseptic processes and packaging is also computer generated and is up to the producer to develop protocol and procedures to summarize all the data that is generated by the control system.

As discussed in Section 4.2.2, aseptic operations are complex, and require the use of a well-thoughtout HACCP plan. In addition to the discussion about HACCP presented in that section, here we present some thoughts on operations of aseptic lines gathered from industrial experience and learnings.

4.5.1 Presterilization of the Processing System

The aseptic processing system and aseptic filler are mutually dependent and both must be presterilized prior to processing of food products. Presterilization of the system and filler is usually accomplished by hydrogen peroxide; peracetic acid; superheated, saturated steam; or high-temperature, continuous water flow. This presterilization of the components of the aseptic processing and packaging system is a CCP.

The aseptic processing line is almost always presterilized by steam or high-temperature water flow for a designated period of time. Once presterilized, the system must remain in a commercially sterile state by preventing the ingress of microorganisms with the use of isolation seals and positive pressure inside the sterile zone. The temperature for initial sterilization of the system varies based on the product to be processed, and must be confirmed and agreed to by the process authority and is part of the supplemental information for process filing in the USA.

The sterilization temperature of the equipment in the system is mostly influenced by the pH of the product to be processed. With acid products (<pH 4.6) concern is most generally with the growth of yeast and mold. Yeast and mold will not survive temperatures above 200°F, therefore the sterilization of the system is generally accomplished slightly above this temperature for a designated period of time. Most processors of acid products presterilize the aseptic system with water at somewhere between 235°F and 250°F for 30 minutes. With low-acid products (>pH 4.6) preservation is based on killing the more heat resistant spoilage microorganisms and customarily an F_0 of 30-50 min is used for low-acid products SIP, with some products for infants and immuno-compromised consumer reaching F_0 values in excess of 100 min.

Whether the system is being presterilized for high- or low-acid products, the sterilization water required is always above the atmospheric boiling point, therefore a back pressure valve is required to pressurize the system to maintain the sterilization solution liquid at elevated temperature. All equipment and piping in the system from the end of the final heater to and including the aseptic divert valve at the filler must reach the sterilization temperature for a designated period of time. The back pressure valve, therefore, must be located after the aseptic divert valve at the filler.

Sterilization of the aseptic processing system is generally accomplished either using the main flow control pump for the system or an alternative pump such as a centrifugal pump from the water supply tank that normally is part of the system, through the heat exchangers and holding tube past the aseptic divert valve and back pressure valve back to the water supply tank. Any cooling media in the cooling section of the heat exchangers is drained from the heat exchangers. The water is heated to the sterilization temperature by means of the heating section in the processing system. Once the sterilization temperature is reached at the most remote part of the aseptic system, which in most cases is after the aseptic divert valve at the filler, where a temperature sensor is located to document the CCP, a timer in the control panel starts the necessary time required to deem the system sterile. After the hot sterilization solution passes the control point after the aseptic divert valve it needs to be cooled back down below the boiling point, an auxiliary heat exchanger, also known as crash cooler, is used to cool the sterilization water returning to the supply tank for the system. This recirculation continues until the timer in the control panel indicates the necessary time reach, the commercial sterility state of the system sterile. Once the system reaches commercial sterility, cooling of the system can begin until it reaches the normal fill temperature of the product, and the flow rate of the sterile water has matched the operating flow rate, processing of the product can begin. If, at any time in the sterilization cycle, the temperature drops below the temperature setpoint necessary for sterilization, the PLC ensures that the timer resets and the sterilization cycle must start all over again. This correction action must be documented and the alarm system must be validated and tested periodically.

4.5.2 Loss of Sterility

Once the system has completed the sterilization step, it must be maintained in a state of commercial sterility. Sterility loss happens when the parameters determined in the CCP plan are out of control, e.g., the temperature falls below the safety limit or the flow rate is above the limits set in the HACCP plan. If that happens, the system must change to an unsterile state and the divert system close to prevent contamination of the product in the sterile surge tank or filler. The system will have to undergo resterilization, and depending on the production step a CIP would be needed before SIP.

This deviation from the CCP needs to be documented and a deviation file must be maintained, in this file, the root cause analysis, corrective actions and product disposition should be documented.

4.5.3 Cleaning

Cleaning of an aseptic system is a critical step of preoperation; it is crucial for a successful sterilization of the system and to eliminate the risk of allergen cross-contamination, and is treated in detail in Chapter 12. Generally, a clean-in-place (CIP) is preferred and such a CIP system does not have to be sophisticated and interlocked. However, if cleaning efficiency is determined through manual inspection, or by observing charts, thermometers, and gauges, then an exact procedure must be established, followed by records to ensure cleaning is being accomplished.

Cleaning in place relies on the action of shear on the walls, by fast flow rate, with the addition of chemicals to effectively scrub the walls. The flow rate, temperature, and concentration of the cleaning solution must be tested by a reputable engineer. During commissioning a validation of the cleaning step should be made, where the system is soiled and a CIP cycle is used to verify that the whole system can be cleaned. CIP conditions must be reviewed regularly as long-time operations tend to create harder to clean soil, and normal wear of the equipment creates different conditions than the ones used in the initial validation. CIP conditions must be adjusted when new products are developed, equipment is replaced, or processing conditions change.

The system should be inspected regularly after CIP has been completed. This inspection should be made at preestablished locations where the most difficult to clean components in the process system are identified. It has become common practice to use ATP (adenosine tri-phosphate) detection swabs, to ensure the absence of organic material in such preestablished locations. Allergens swabs are also useful to comply with allergen prevention programs.

During this inspection, the equipment, pipelines, and valves can be examined to ensure they are operating properly, and do not have any cracks or crevices, or gaskets that should be replaced. Gaskets should be taken from joints regularly, inspected, cleaned if necessary, and replaced if bad. It is important to realize that automatic controls with sensors can indicate, control, and record items such as pressure, temperature, and pH; however, they cannot look at a gasket and determine whether it needs to be replaced or cleaned or whether it is hard and brittle. Therefore, it is important that a program be established to inspect critical points at regular times to verify that they are correct.

4.5.4 Preventive Maintenance

The role of preventive maintenance is critical for a successful aseptic operation. Once the system has achieved a validated state; in which all processes and equipment are working and sterility is assured, there is always the need to maintain such state. Replacing consumable parts such as filter elements, gaskets, rubber seals, valve bellows, and even sensors is necessary to maintain the operational status. Preventive maintenance must follow a schedule so that all these parts are operating at peak performance.

Preventive maintenance must be written as a program, which is agreed by and supported by management, allowing enough time for maintenance and money for parts to be ordered. A stock of commonly used or long lead time replacement parts is a good practice to prevent emergencies and to ensure that the machines will be in operational shape with a very short downtime. Maintenance personnel should be well trained not only in mechanical and electrical maintenance, but also in hygienic operation, GMP, and HACCP training. The cost of preventive maintenance is minimal compared to a component failure due to the lack of it. Large batches of food worth millions might need to be destroyed because of a single gasket, worth cents, that was not changed on time.

Maintenance in aseptic processes must be very careful in considering the replacement of parts; and the use of tools, and management must allot sufficient time and expenses for complete maintenance. A maintenance log for each machine, room, and process line needs to be maintained, documented, and reviewed as part of the food safety and quality plan. Replacement parts must be OEM and a like-for-like replacement is preferred, unless a suitable replacement is found which should be approved by the safety team and documented. When improvements are made to the line, or parts that are not OEM are to be installed, it is necessary to have a review of the effect of this modification on the validated state of the equipment. Once the food safety team has approved, the modification in writing it can proceed.

Maintenance personnel operate in all the spaces and rooms of the factory, and they usually have a shop or workspace that is segregated from the production space. Tools are usually shared among the different rooms, and in aseptic processing and packaging this can be problematic. It is recommended to have different sets of tools for clean areas that will serve only a small number of pieces of equipment. These tools must be cleaned and sanitized periodically as any other piece of nonfood contact surface in the facility. Personnel must be very well trained in following GMPs, boot cleaning, hand washing, and the use of clean overall covers must be enforced.

4.6 CONCLUDING REMARKS

The design and operation of aseptic processes require commitment from all involved in the process, with support from the highest level inside a company. Commercial sterility must be maintained at all times, and the use of risk-based analysis and implementation of a food safety plan based on the HACCP principles is a very important step in assuring the aseptic condition of the products made in any facility. A thermal process authority, either internal or external, is needed to establish processes, manage deviations, and assist during new product or process development, as well as validation of the process. Validation is not a one-time activity but a life-cycle management of processes and products.

Personnel is a key part of any operation, and continuous training, positive enforcement of GMPs, and benefits should be implemented to have highly motivated and engaged personnel. Each person in the facility, from the highest to the lowest level, must be part of a culture of food safety and understand how much.

In the near future, the use of electronic records will become widespread, with the implementation of software and hardware that complies with regulatory and security requirements. Industry must be ready for these changes, taking into account the final user of the technology for control and operation as well as the user of the data and regulatory support.

Disciplined, meticulous, and consistent implementation of all safety and control measures are of special significance and importance in the operation of aseptic processing systems and their maintenance in safe operating conditions and delivery of safe, commercially sterile shelf-stable, high-quality products. Maintenance of high discipline of implementation gains in importance over the time of commercial operation of the system due to the fact that the potential issues are more likely to appear after the processing plant has been in operation for prolonged periods of time and varying levels of stress. Additionally, there is also a tendency to gradually relax the more tedious and repetitive control measures after extended periods of problem-free operation. This should also be considered as another potential source of safety hazard and proper measures should be implemented to maintain the high quality and consistency of control measures throughout the operating life of the processing and packaging systems in the facility.

NOMENCLATURE

A	m^2	Surface area
C	minutes	Cook value
C_p	J kg^{-1} °C^{-1}	Specific heat capacity
D	minutes	D-value
E_a	J mol^{-1}	Activation energy
F_0	minutes	Equivalent process time at 121.1°C with a z-value of 10
$F_{Tref/z}$	minutes	Equivalent process time at T_{ref} with z-value a

H	W m⁻² °C⁻¹	Convective heat transfer coefficient
M	kg s⁻¹	Mass flow rate
N		Number of microorganisms
T	s	Time
T	°C	Temperature
T_{ref}	°C	Reference temperature
U	W m⁻² °C⁻¹	Overall heat transfer coefficient
V	m/s	Velocity
W	m s⁻¹	Local axial velocity
Z	m	Streamwise distance
Z	°C	Z-value
Λ	W m⁻¹ °C⁻¹	Thermal conductivity
P	kg m⁻³	Density
M	Pa-s	Viscosity of Newtonian fluids
H	Pa-s	Apparent viscosity of non-Newtonian fluids
Subscripts		
f		Final
o		Initial
max		Maximum
ave		Average
P		Of the product/particle
C		Of the container
A		Of the environment
E		Of the equipment

BIBLIOGRAPHY

21 CFR, Parts 11, 113,114 and 117. 2016. Washington, DC: U.S. Government Printing Office.

3-A Accepted Practices for a Method of Producing Steam of Culinary Quality, 609–00. 1996. *Formulated by International Association of Milk, Food, and Environmental Sanitarians*. United States Public Health Service, and the Dairy Industry Committee.

9 CFR, Parts 308, 318, 320, 327 and 381. 1986. *Canning of Meat and Poultry Products*. U.S. Department of Agriculture.

CCFRA 2008. *History of the minimum Botulinum cook for low acid canned foods*. R&D Report 260. Chipping Campden, UK.

David, J.R.D. 2013. Failure mode and effect analysis, and spoilage. In *Handbook of Aseptic Processing and Packaging*, edited by David, J.R.D, Graves, R.H., and Szemplenski, T., Chapter 13, 203–216. 2nd edition. Boca Raton, FL: CRC Press, Taylor and Francis Group.

Esty, J.R. and Meyer, K.F. 1922. The heat resistance of the spores of B. botulinus and allied anaerobes. XI. *J. Infect. Dis.*:650–664.

Evoguard. 2020. *Component Guide*.

Grocery Manufacturers Association (GMA). 2007. *Canned Food: Principles of Thermal Process Control, Acidification and Container Closure Evaluation.* 7th edition. Washington, DC: GMA Science & Education Foundation.

Holdsworth, S.D. 1992. *Aseptic Processing and Packaging of Food Products.* Elsevier Applied Science.

Homer, C. 1992. More changes. *Dairy Field Magazine*, September 1992, p. 90.

Institute of Food Thermal Process Specialists (IFTPS). 2011. Definition of process authority. Document WP.002. http://iftps.org/wp-content/uploads/2017/08/Process-Authority-Definition.pdf

Janoschek, R., and Du Moulin, G. 1994. Ultraviolet disinfection in biotechnology: Myth vs. practice. *Biopharm Magazine*, January–February 1994, pp. 24–31.

LeBlanc, D.A., Danforth, D.D., and Smith, J.M. 1993. Cleaning technology for pharmaceutical manufacturing. *Pharmaceutical Technology Magazine*, October 1993.

Mansfield T. 1962. High temperature: Short time sterilization. In Food Science and Technology. Proceedings of the 1st International Congress of Food Science and Technology, edited by Leitch, J.M., Vol. IV, 311–316, 18–21 Sept. 1962, London, UK, New York: Gordon and Breach Science Publisher.

National Food Processors Association (NFPA). 2002. Validation guidelines for automated control. *Bulletin* 43-L. Washington DC: NFPA.

Nelson P.E. (editor) 2010. *Principles of Aseptic Processing and Packaging.* 3rd edition. West Lafayette, IN: Purdue University Press.

Stumbo, C.R. 1973. *Thermobacteriology in Food Processing.* 2nd edition, Chapters 10 and 11. New York: Academic Press.

Tichener-Hooker, N.J., Sinclair, P.A., Hoare, M., Vranch, S.P., Cottam, A., and Turner, M.K. 1993 The specifications of static seals for contained operations: An engineering appraisal. *Pharmaceutical Technology Magazine*, October 1993, pp. 60–66.

Toledo, R.T. 1991. *Fundaments of Food Process Engineering.* New York: Chapman & Hall. pp. 315–397.

Zander Filter Brochures. n.d. *1: Ecodry, The Range of Dryers with Performance; 2: Sterile Filters, Aeration Filters and Steam Infusers.*

Chapter 5

Thermal Processing Equipment for Heating and Cooling

George N. Stoforos, Pablo M. Coronel, V. R. (Bob) Carlson, and Josip Simunovic

CONTENTS

DOI: 10.1201/9781003158653-7

5.1 INTRODUCTION

The preceding chapters provided an overview of the basic considerations and equipment of asep-tic processing. As described in the previous chapters, aseptic processing involves the thermal sterilization of the product, packaging material sterilization, and conservation of sterility dur-ing packaging. Aseptic processing is a continuous flow thermal process wherein a commercially sterilized product is filled and hermetically sealed into sterilized containers and closures within a sterile environment. Commercial sterility is defined as that condition achieved by the applica-tion of heat or other sterilizing agents which render the finished food product, packaging, and equipment, free of viable microorganisms (including spores) of public health significance as well as microorganisms of non-health significance, capable of reproducing in the food under normal nonrefrigerated conditions of storage and distribution.

As already has been discussed in previous chapters, an aseptic processing system consists of the following primary components (Kelder et al., 2009; Anderson et al., 2020):

- Bulk feed tank: to mix the ingredients and feed the product to the aseptic processing system.
- Metering pump: to accurately control the flow of product through the system
- Heating system: to heat the product to the required sterilization temperature. There are three basic methods of product heating: direct, indirect, and advanced heating methods:
  - Direct heating systems, which involve the direct contact and mixing of steam and product, can be done in two ways:
    - Steam injection, where steam is injected directly into the product flowing through an injection chamber.
    - Steam infusion, where the product is sprayed into a pressurized steam chamber and sterilized while falling in the form of film or droplets within the chamber.
  - Indirect heating systems, which involve the use of plate, tubular, and scraped sur-face heat exchangers for indirect heating of the product (heat is transferred between the hot surface of the heat exchanger and the product).
  - Advanced heating systems and novel heating technologies have been developed/ employed, which utilize a direct conversion of electromagnetic energy into heat to provide uniform volumetric heating.
    - Microwave heating uses the interactions of molecules in a rapidly changing electromagnetic field. This interaction does not need direct contact with elec-trodes or antennas but requires that the product be transported using sections of tubes that are transparent to microwaves.
    - Ohmic heating converts electrical energy directly into heat by using the Joule effect, utilizing an electrical current passed through the food product, causing product heating. The product must have contact with electrodes in tubes made of insulating material.

- Hold tube: to maintain product sterilization temperature for a specific amount of time that results in commercial sterilization of the final product. The holding tube (length, diameter) is designed to accomplish the required lethality for the product, ensuring the required log reduction of the target microorganisms. Thus, the sterilization temperature of the product must be maintained for the time needed in the holding tube. For that reason, the product flow rate within the hold tube and the product temperature at the exit of the hold tube must be controlled, monitored, and recorded.
- Cooling system: to cool the product to the filling temperature. Depending on the heating method used (direct or indirect) and food characteristics, flash evaporating cooling may be used to evaporate the added water and cool down the product; otherwise, indirect heat exchangers (plate, tubular, or scrape heat exchangers) are the primary cooling equipment in aseptic processing.
- Steam-sealed, air-operated valves to direct the product to the desired locations.
- Aseptic surge tank (optional): to hold the sterilized product before the filling/packaging process

In aseptic processing, the processed food is heated at higher temperatures for a shorter time, compared to other conventional thermal processes, such as canning, resulting in better-finished product quality. Aseptic thermal processing involves three main steps: (i) heating the product to the desired temperature using direct or indirect heating methods, (ii) holding the product at the desired temperature for a specific amount of time to achieve the commercial sterility of the food, and (iii) cooling of the product before filling into presterilized containers, under sterile conditions (Kelder et al., 2009; Anderson et al., 2020).

Although no credit is given to sterilizing the product (microbial inactivation) during heating and cooling cycles in aseptic processing, both remain critical for designing an aseptic processing system. The heating and cooling of the product should be performed as rapidly as possible. Inefficient heating poses a food safety risk that could result from underprocessed food, while slow heating and cooling cycles can lead to overcooked food, resulting in food quality degradation with nutritional and organoleptic losses. For that reason, heating and cooling methods and the associated equipment used are essential in an aseptic processing system, aiming at high heating and cooling rates with a minimal processed finished product. The design and selection of the presterilization heating and subsequent cooling equipment depend on the product characteristics.

This chapter will focus on heating and cooling systems in aseptic processing, describing the current methods/equipment used in an aseptic system, the critical parameters of the system, and finally, provide an overview of the future trends and developments in advanced heating and cooling systems.

5.2 DIRECT HEATING

Direct heating systems are utilized in aseptic processing for rapid volumetric heating of low-viscous foods such as dairy-based beverages. The processed food is heated to the target sterilization temperature by adding and mixing culinary steam into the product. There are two different methods of direct heating: (i) by injecting pressurized steam into the flowing food (steam injection) and (ii) by pumping the food product as a curtain or spray into a vessel filled with pressurized steam (steam infusion) (Lewis & Heppel, 2000; Emond, 2001). The basic principle behind direct heating is the transfer of latent heat of vaporization from the condensed steam to the flowing food. Direct heating provides rapid, uniform volumetric heating resulting in a more efficient heat transfer method than indirect heating systems. However, direct systems' applications are limited to low-viscous homogeneous liquid products (Lewis & Heppel, 2000; Emond, 2001).

In an aseptic processing system with direct heating, the processed liquid food after the balance/mixing tank goes through a preheating system, typically an indirect heat exchanger, such as a plate heat exchanger. The preheater exit product temperature (usually around 165 °F) is critical and should be monitored and recorded. After the preheater, the product is pumped through either a steam injection or a steam infusion system to reach the required sterilization temperature. Due to the condensation of the steam that occurs in both direct heating systems, the product enters the hold tube section with a considerable dilution of the product. The additional condensed steam added to the product in the direct heating step is critical for food quality (dilution of the product) and food safety (Lewis & Heppel, 2000; Emond, 2001). The excess water of the product is removed in the first cooling step, flash cooling, which will be discussed later in this chapter. Furthermore, the added volume of water increases the volumetric flow rate, critical for the residence time (hold time) of the processed foods within the hold tube (sterilization step). Estimation of the corrected product volumetric flow rate in the hold tube depends on the temperature difference between the sterilization temperature and the exit preheater temperature and will be discussed in more detail in the following sections.

5.2.1 Direct Heating–Steam Injection

As mentioned above, in a steam injection system, pressurized steam is introduced into the flowing product. A well-designed steam injection system aims to achieve rapid condensation of the steam and transfer the latent heat of vaporization into the food while preventing the passage of non-condensed steam into the holding tube. In a steam injector system, the steam is introduced into the flowing food in small bubbles or thin sheets to enhance condensation and limit the amount of non-condensed steam entering the holding tube (Lewis & Heppel, 2000; Emond, 2001). A low-pressure difference between the introduced steam and the flowing food is preferable. Having the pressure of the steam higher (~14-15 psig) than the pressure of the product ensures the proper heating (to the sterilization temperature) of the product and enhances the mixing of steam with the product (Lewis & Heppel, 2000). The rapid collapse of the steam bubbles, cavitation, and high steam pressure, hence high temperature, may cause unwanted overheating of the product that is first in contact with the steam. Moreover, the rapid collapse of the steam bubbles and cavitation can cause changes in pressure in the liquid food, making the injector noisy (Lewis & Heppel, 2000). A back pressure valve (high-pressure pump) is usually installed in the system to control the above and prevent the product from boiling or flashing within the unit (Emond, 2001).

Furthermore, a steam injection system should be adequately designed to reduce any potential fouling of the product at the surface of the injector. There are different steam injector systems, having the steam being injected via nozzles or a system of orifices under sharp angles to ensure proper heating and mixing; if the product is introduced close to the injector steam head, the product will start heating up at the surface of the nozzle, which may result in fouling at the surface of the injector, reducing steam flow, and decreasing system's efficiency (Lewis & Heppel, 2000). To overcome the above, the steam injection systems have been designed with the product entering the heating zone under the right angles to the injector, using a venturi product tube; this approach keeps the product away from the injector head and reduces fouling (Figure 5.1).

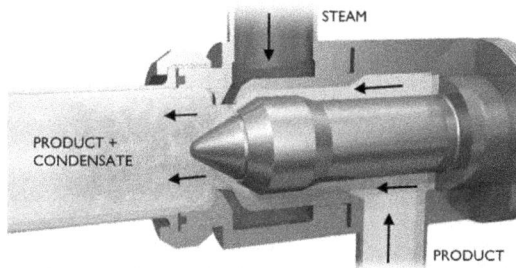

Figure 5.1   Schematic diagram of a steam injector (courtesy of Tetra Pak).

5.2.2 Direct Heating–Steam Infusion

Direct steam infusion systems follow the primary heat transfer principle of steam condensation described for the steam injectors. The only difference is in how product and steam are introduced and mixed. Although both direct heating methods provide an instant heating process, in theory, the steam infusion process can be considered a slightly gentler process (Lewis & Heppel, 2000). That is because the heating process with the steam infuser does not involve the collapse and cavitation of steam bubbles compared to the steam injection case, which may lead to overheating the food, as described above. On the other hand, the mechanical, technical, and control requirements are more complex for the infusion system, as described in the following paragraph.

Similar to steam injectors, steam infusers are capable of processing low-viscous liquid foods. But additionally, with modifications, infusers can be suitable to process products with small particles, such as juice pulps. A typical steam infusion system consists of a pressurized, conical-shaped steam vessel into which the preheated product falls in the form of free-falling strings/droplets, and from there, is pumped to the holding tube (Lewis & Heppel, 2000; Emond, 2001). Different designs of steam infusers are available, mainly varying on the number and size of diffusion holes through which food product enters the steam vessel. A steam infusion chamber may have many diffusion holes through which product as stream or jets is distributed and enters the steam environment without hitting the wall of the vessel until it falls to the conical bottom of the vessel (Emond, 2001). The infuser system may also have a cooling jacket on the bottom cone to minimize product foaming or fouling issues (Figure 5.2).

One of the most critical aspects of an infusion system that should be controlled appropriately is the time the heated product will remain in the pool, the conical shape base of the infuser before it is pumped through the holding tube and the vacuum flash cooling vessel (Figure 5.3) (Lewis & Heppel, 2000; Emond, 2001). This is important to provide precise control over holding times and temperature, significantly impacting final food safety and quality of the product. The commercially available aseptic steam infusion systems have level controllers for that purpose. The level controllers are installed and connected to an infuser discharge product pump, which is mounted directly below the infusion chamber and prior to the holding tube, to control the amount of product entering the infuser and the time the heated product stays within the infuser's pool, ensuring sufficient heating of the product, reducing any potential unwanted overheating of the food, while ensuring adequate pressure in the holding tube (preventing the passage of unwanted bubbles or uncondensed steam). The flow rate and temperature of the discharged product are critical parameters of the process and should be controlled and monitored.

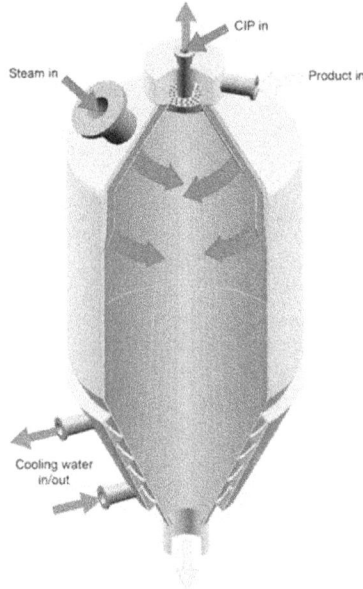

Figure 5.2   Schematic diagram of a steam infusion vessel (courtesy of SPXFLOW).

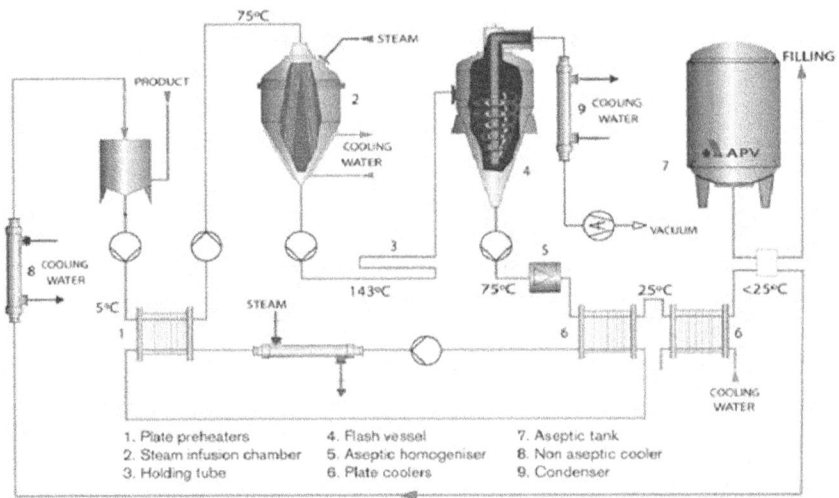

Figure 5.3   Schematic diagram of a complete aseptic direct steam infusion system with all the components: preheating heat exchangers, steam infusion chamber, discharge product pump, holding tube, cooling flash vessel, additional cooling heat exchanger, and aseptic tank (courtesy of SPXFLOW).

5.3 FLASH EVAPORATIVE COOLING

As described above, direct steam heating is a method used in aseptic processing, typically with food products such as dairy products, juice-based drinks, soy products, creams, ice-cream mix, and tomato paste. Direct heating systems utilize culinary steam, either infused or injected into the product to provide rapid and uniform heating. Direct steam heating systems quickly heat the product at high temperatures, minimizing product quality losses (burnt flavors and odors, color, and stability). Using direct steam heating, water is added to the product, and the added steam condenses into the product, which must be removed. The hot product is passed to a holding tube and then to a flash evaporative cooling chamber (or vessel). The flash evaporative cooler has the following purposes: (i) to rapidly reduce the product temperature, (ii) to remove a portion or all the water (steam condensation) that was added to the product during heating, and (iii) to remove undesirable heat-induced volatile components of the product, responsible for the cooked flavor and odors (Kelder et al., 2009; Toledo et al., 2018; Anderson et al., 2020).

This cooling method is based on evaporative cooling principles, having pressure and temperature as critical parameters for the cooling process's effectiveness and the finished product quality. Flash evaporative cooling requires operation under vacuum conditions. Operating at pressures lower than liquid saturation levels results in cooling employing evaporation. By rapidly reducing the pressure of a liquid product with free-standing water, the liquid water quickly evaporates; as the vapor pressure decreases, it results in a depression of its boiling point. Under these conditions, the free-standing water in the product absorbs heat (in the form of the phase change (liquid to vapor) latent heat), resulting in the evaporation of free-standing water while at the same time reducing the temperature of the product. To obtain adequate cooling, the product must be subject to vacuum conditions that quickly reduce the temperature without causing freezing or quality degradation (texture, odor, flavor, etc.).

Typically, flash evaporative cooling occurs in a conical-shaped condenser-equipped vacuum vessel that enables rapid cooling of the product. In some cases, the conical shape and the installed baffles inside the flash cooler help the cooling process by blocking the vaporized water and volatiles to re-enter the processed product. Vacuum conditions within the flash chamber are achieved by removing air from inside the chamber using a vacuum pump. The pump maintains and controls the vacuum conditions to ensure the same amount of water is evaporated as was earlier added by the condensed steam (Figure 5.4) (Toledo et al., 2018). Operating at pressures below atmospheric (vacuum) increases the hazard of contaminants entering the chamber. To prevent contamination from entering the flash cooling system, effective sterile barriers, such as steam seals, should be applied to all potential leakage/contamination sites, such as rotating or reciprocating shafts, stems of aseptic valves. Steam seals at these locations can provide an effective barrier. Still, sterile barriers must be monitored by performing frequent visual inspections or using an indicator monitoring pressure and/or temperature to ensure the seal discharge points' proper performance. If other types of seal barriers are used, there must be a means to permit monitoring and verifying the appropriate functioning of the barrier (Chandarana et al., 2010).

In a flash evaporative cooler, the product temperature is critical to ensure that the finished product is not diluted or concentrated. Products such as milk and dairy products, typically treated with direct steam systems, require accurate control of the product temperature within the flash cooler to correctly flash/remove the appropriate amount of water, equal to the amount of condensation steam added to the product. Lewis and Heppel's (2000) chapter presents a review of the mathematical and experimental studies conducted to accurately estimate the target exit product temperature at the flash evaporative cooler and adequately remove the added water

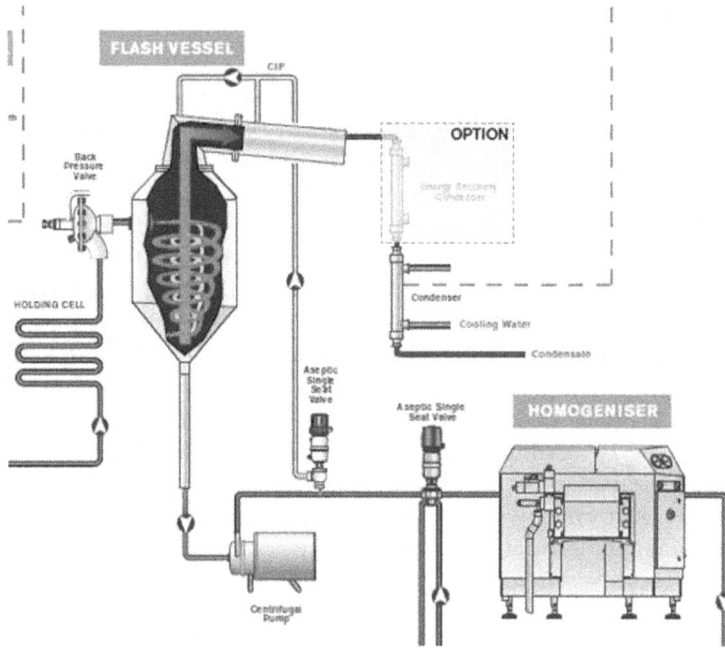

Figure 5.4 Schematic diagram of a vacuum flash cooling vessel; located after the hold tube and prior to the homogenizer (for milk and dairy products) (courtesy of SPXFLOW).

without affecting the quality of the product. In general, flash evaporative cooling temperature, boiling point, should be set 1-3 °C higher compared to the preheater exit temperature to ensure that the correct amount of water is removed, compensating for the amount of added water from condensed steam during heating.

Flash evaporative cooling can be highly efficient, resulting in rapid volumetric cooling, improved finished product quality, and extended shelf life. On the other hand, there are a few disadvantages to direct heating and flash evaporative cooling systems. In general, direct steam–flash cooling systems are very complex in design and operation, making a system like that quite expensive. Finally, vacuum flash cooling systems are limited to low-viscosity products, providing an option for a limited number of different products. The flash evaporative cooling process does not impair flavor and texture. Flash evaporative coolers are used in the initial cooling step of the process; on discharge from the flash chamber, the product goes through a homogenizer (for dairy products) and can be further cooled using additional heat exchangers to deliver the complete cooling process the different types of heat exchangers used in cooling are discussed in the following paragraphs.

5.4 INDIRECT HEAT EXCHANGERS

Indirect heat exchangers are the main heating and cooling methods employed to increase or decrease the temperature of pumpable food products. Depending on viscosity and product characteristics (liquid or multiphase, etc.), different heat exchangers can be used for aseptic processing, namely plate, tubular, and scraped surface heat exchangers (Lewis & Heppel, 2000; Kelder et al., 2009; Toledo et al., 2018; Anderson et al., 2020). The indirect heat exchangers employed for cooling are similar to those used for heating. However, the cooling rate is usually slower compared to heating due to higher product viscosities associated with the cooling temperatures, resulting in a slower flow, an increase in product accumulation at the heat transfer surface, and a lower heat transfer rate (Kelder et al., 2009).

For indirect systems, heat transfer involves the heat exchange between the product and the heating or cooling medium through heat exchanger's surface (usually stainless steel). The heat transfer rate depends on the temperature difference between the product and the heat transfer medium (heating or cooling medium), the flow rate of the product, the flow configuration between the product and the heat transfer medium (co-current or counter-current), the surface area of the heat exchanger, and finally the thermophysical properties of the product. The governing equation of the overall energy transferred Q (W), assuming zero energy transferred by the surrounding environment (Q_L (W)), is given in Equation 5.1:

$$Q = Q_m - Q_L \tag{5.1}$$

where Q is the energy transferred or gained by the food product to the heat transfer medium (Equation 5.2), Q_m (W) is the energy lost or gained by the heating or cooling medium and the food product (and the surroundings/environment) (Equation 5.3), and Q_L is the energy transferred by the surrounding environment ("heat losses"); simplifying by assuming that the heat losses are zero (Equation 5.4).

$$Q = \dot{m}_P \cdot C_{P(P)} \cdot \Delta T_P \tag{5.2}$$

$$Q_m = \dot{m}_m \cdot C_{P(m)} \cdot \Delta T_m \tag{5.3}$$

$$Q_L = 0 \tag{5.4}$$

And hence based on the above, the overall energy transferred to the food product during heating or cooling, Q (W), in an indirect heat exchanger can be calculated by the following equation (Equation 5.5):

$$Q = \dot{m} \cdot C_{P(P)} \cdot \Delta T_P = U \cdot A_{lm} \cdot \Delta T_{lm} \tag{5.5}$$

where $\dot{m}$ is the mass flow rate (kg/s), C_p is the specific heat (J/kg·K), and ΔT (K) is the temperature difference between the inlet and outlet temperature. In the above equations, the subscripts "p" and "m" refer to the food product and heat transfer medium, respectively.

On the right-hand side of Equation 5.5, U (W/m²·K) is the overall heat transfer coefficient, and it depends on material properties, system dimensions, and process parameters (Equation 5.6). A_{lm} (m²) is the total surface area of heat transfer (Equation 5.7), and ΔT_{lm} (K) refers to the logarithmic mean temperature difference between the food product and the heat transfer medium at each end of the examined heat exchanger (ΔT_1; ΔT_2 (K)) (Equation 5.8). The ΔT_{lm} is used to determine the temperature driving force for heat transfer in heat exchangers and arises from analyzing a heat exchanger with a constant flow rate and fluid thermal properties. ΔT_1 and ΔT_2 refer to the temperature difference between the product and the heat transfer medium at the inlet and outlet of the heat exchanger, respectively, and they depend on the flow configuration between the product and the heat transfer medium: (i) co- and (ii) counter-current flow, respectively (Figures 5.5, 5.6). Co-current flow configuration is mostly only when rapid initial

Temperature profile in a co-current flow   Temperature profile in a counter-current flow

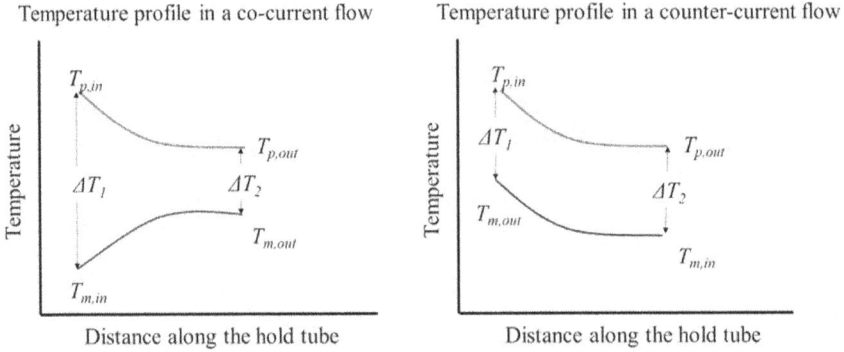

Figure 5.5   Temperature profile across an indirect heat exchanger; on the left co-current heat flow configuration and the right counter-current flow configuration.

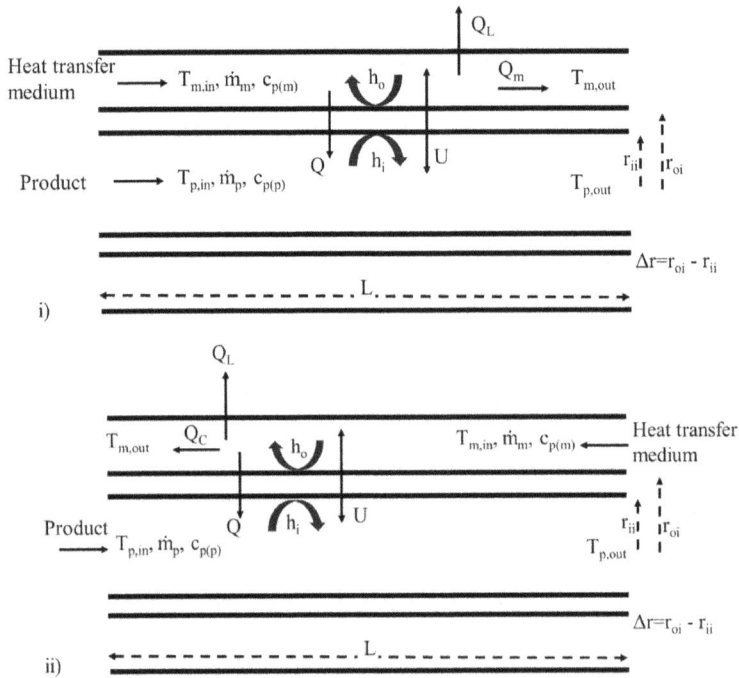

Figure 5.6   Schematic diagram of a double-tube heat cooling exchanger, with the food product (inner tube flowing in (i) co-current, and (ii) counter-current flow with the heat transfer medium (outer tube (annulus)). The subscripts refer to "p" for the product, "m" for the heat transfer medium, "in" for the inlet, "out" for the outlet, "i" for inside, "o" for outside.

heating/cooling is needed. On the other hand, counter-current flow configuration is most used because of the higher heat transfer efficiency.

$$\frac{1}{U \cdot A_{lm}} = \frac{1}{\left(h_i \cdot A_i \right)} + \frac{\Delta r}{\left(k \cdot A_{lm} \right)} + \frac{1}{\left(h_o \cdot A_o \right)} \tag{5.6}$$

$$A_{lm} = \frac{\left(A_o - A_i \right)}{\left[\ln\left(\dfrac{A_o}{A_i} \right) \right]} \tag{5.7}$$

$$\Delta T_{lm} = \frac{\left(\Delta T_1 - \Delta T_2 \right)}{\left(\ln\left(\dfrac{\Delta T_1}{\Delta T_2} \right) \right)} \tag{5.8}$$

where A_i (m²) and A_o (m²) are the surface areas of the outside surface of the inner and outer tube of the examined heat exchanger, respectively (Equation 5.8, 5.9), k (W/m·K) is the thermal conductivity material property of the inner tube of the heat exchanger, and h_i and h_o (W/(m²·K)) are the convective heat transfer coefficients of the convection heat transfer from the food product to the inside surface of the inner tube and from the outside surface of the inner tube to the heat transfer medium, respectively.

$$A_i = 2 \cdot \pi \cdot r_{ii} \cdot L \tag{5.9}$$

$$A_o = 2 \cdot \pi \cdot r_{oi} \cdot L \tag{5.10}$$

where r_{ii} (m) and r_{oi} (m) are the inner and outer radius of the inner tube of the examined heat exchanger, respectively.

5.4.1 Plate Heat Exchangers

Plate heat exchangers have been used for many years in continuous flow thermal processes such as pasteurization and ultra-high temperature (UHT) processing of food products such as milk, ice-cream mix, and fruit juices. Plate surface heat exchangers have been utilized in aseptic processing for heating and cooling or regeneration, which will be discussed in the following paragraphs (Kelder et al., 2009; Toledo et al., 2018; Anderson et al., 2020).

Plate heat exchangers consist of a stack of corrugated metal plates pressed and bonded together on a fixed frame. The plates' construction usually requires plate–plate contact on parts of the corrugations to maintain rigidity of the plates and a uniform plate spacing. In the plate surface heat exchangers, the food product flows over one side of the plates and the heat transfer medium over the other side. The product is introduced through a corner hole into the first plate section and flows vertically through the channel. The heat transfer medium is introduced at the other end of the section and passes, in a similar way, through an alternate plate channel. Gaskets are used to prevent leakage (Figure 5.7) (Lewis & Heppel, 2000; Toledo et al., 2018; Tetra Pak, 2021). The gaskets are typical, made of rubber, seal the plate edges and ports to prevent intermixing between the product and the heat transfer medium. The role of gaskets is also to direct the two flow streams into the respective alternate paths. Plate heat exchangers can be designed in three separate flow directions: co-current, counter-current, and crossflow depending on how effective heat transfer occurs within the system (Lewis & Heppel, 2000; Toledo et al., 2018). As discussed

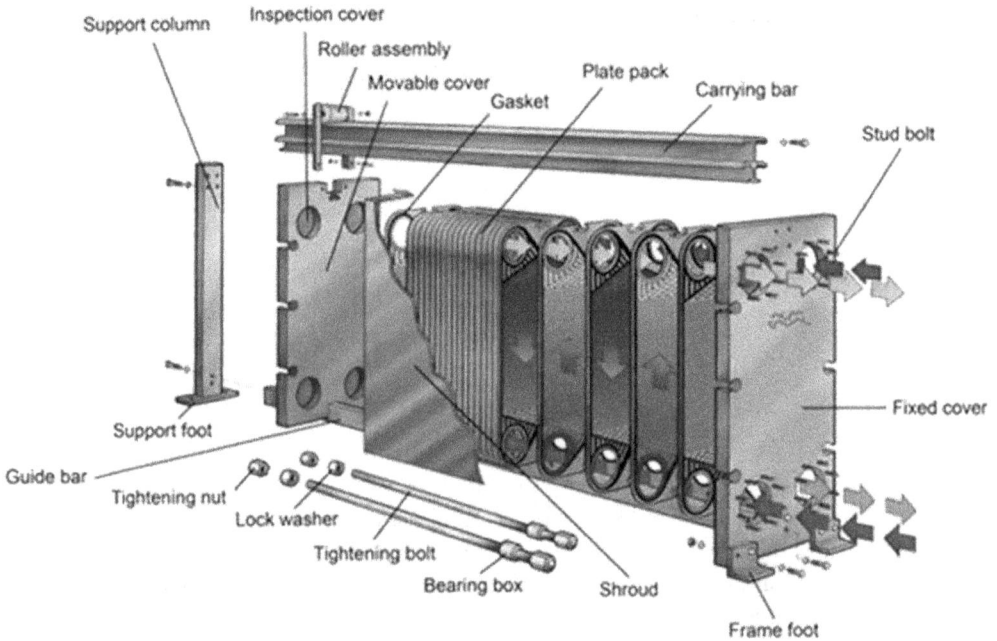

Figure 5.7   Schematic diagram of a plate surface heat exchanger (courtesy of Alfa Laval).

above, plate heat exchangers utilizing co-current flow configuration, both fluids in one singular direction, resulting in more even temperatures, plate surface heat exchangers with countercurrent flow configuration, allow for a higher heat transfer rate. Crossflow plate heat exchangers are in the middle ground between the other two methods. The crossflow method streams fluids at 90-degree angles of one another (Lewis & Heppel, 2000; Toledo et al., 2018; Tetra Pak, 2021).

The advantages of plate heat exchangers for these purposes are well established with recent developments. The gap between the plates can be designed to give high levels of induced turbulence in the flowing streams of product and heat transfer media through suitable profiling of the plates with corrugations, designing narrow flow channel size between the plates, which result in high heat transfer coefficients (Figure 5.8). Plate surface heat exchangers, as an indirect heat transfer method, transfer the thermal energy between the food product and the heat transfer medium through the metal plates, a much larger heat transfer surface area than the other types of indirect heat exchangers (Lewis & Heppel, 2000; Toledo et al., 2018; Tetra Pak, 2021). The heat transfer capacity and efficiency of plate surface heat exchangers can easily be increased by adding more plates. Finally, maintenance of this type of heat exchanger is simple, considering that they can be easily and quickly disassembled for cleaning and inspection.

On the other hand, some inherent problems with plate heat exchangers in aseptic processing systems are their pressure limitation and their cleaning-in-place (CIP) ability. Although with improvements in CIP systems, plate heat exchangers are now not disassembled for routine cleaning. However, it is required to be opened frequently to inspect and monitor plates and parts, checking for pinhole gasket condition, cracks, etc. Furthermore, the high temperatures used in aseptic processing can potentially result in the hot food product burned/fouled onto the plates. CIP operations may not remove all burnt material, adversely affecting the processing parameters such as flow rate and temperature control, resulting in inadequate presterilization of the

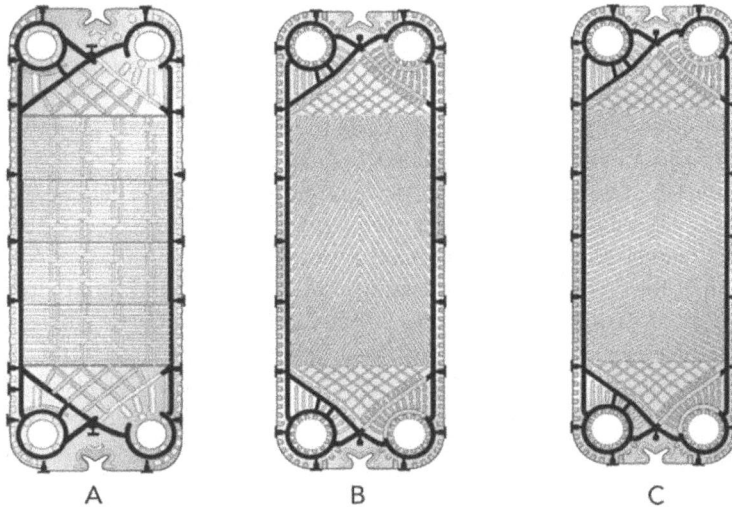

Figure 5.8 Picture of different shapes and corrugations of the plates (A, B, and C) that may be used in a plate heat exchanger depending on the product to be treated and cooling efficiency requirements (courtesy of Tetra Pak).

plates and recontamination of the commercially sterile cold product. That can cause a significant increase in pressure drop. For that reason, plate surface heat exchangers are not used for thermal processing (heating or cooling) of viscous food products or multiphase foods because food particles or fibers of these products will accumulate on these contact points and lead to blockage of the flow channel.

Another issue that can cause mechanical problems using plate surface heat exchangers in aseptic processing is the high temperature/pressure requirements. For example, during the system's presterilization, at temperatures at $\geq 250°F$ for a prolonged duration, the metal plates tend to expand; simultaneously, the frame and the tie bolds holding the system together on the frame do not. That puts enormous stress on the gaskets and thin metal plates, leading to softening and broken gaskets and creating pinholes in the plates. For that reason, frequent maintenance and inspection are highly recommended.

Control and monitoring of differential pressure are essential for plate surface heat exchangers (Anderson et al., 2020). During cooling processing, one side of the plate flows the sterile product while the cooling medium is on the other side. If the side of the sterile product does not have a greater pressure compared to the unsterile side, then in the case of a pinhole in the plate, raw unsterile media can potentially be sucked into the sterile side and hence contaminate the already sterile product. For that reason, proper and precise control of the differential pressure in the plate surface heat exchanger is recommended, so the pressure on the side of the sterile product is always higher compared to the unsterile side. Due to the pressure and temperature limitations, Plate surface heat exchangers are mainly used in aseptic processing of low-viscous, homogeneous liquid food products with no (or very small size) solid particles such as milk and other dairy products and juices. In aseptic processing, newly designed systems usually combine plate surface heat exchangers with other heating and cooling methods/equipment, as discussed in the direct heating systems.

5.4.2 Tubular Heat Exchanger

In aseptic processing, there are many applications where the plate and scraped surface heat exchangers are not suitable, and tubular heat exchangers are preferred. Tubular heat exchangers have a relatively larger flow channel compared to the other indirect options, making them capable of handling high viscous foods and multiphase products with high levels of pulp, fiber, and even solid particulates to a certain size. The maximum particle size depends on the diameter of the tube. On the other hand, tubular heat exchangers have a smaller heat transfer area to the product's volume compared to the other indirect systems, resulting in a slower heat transfer rate (Kelder et al., 2009; Toledo et al., 2018; Anderson et al., 2020; Tetra Pak, 2021). Compared to the other indirect heat exchangers, a higher flow velocity is needed to create efficient heat transfer in a tubular heat exchanger. Although in the past plate heat exchangers and scraped surface heat exchangers were mainly used for processing of low-viscous and high viscous foods, respectively, the improvement in operational conditions such as higher flow rates, additionally with new designs and configurations have significantly increased the use of tubular heat exchangers in aseptic systems.

In tubular heat exchangers, similar to the other indirect systems, the heat transfer between the product and the heat transfer medium occurs through the metal surface (stainless steel 316L) of the tube that separates the two streams (Anderson et al., 2020). Tubular heat exchangers are available in two fundamentally different types: concentric tube and shell tube or multi/mono tube. Concentric tubular heat exchangers may be constructed of either straight or corrugated tubing. They may consist of double, triple, or more concentric tubes with associated entry and exit ports for the product and the heat transfer medium. There are several different designs of such heat exchangers in aseptic processing, namely double-tube heat exchanger, triple-tube, coil-in-shell, and corrugated tube configurations aiming to enhance the heat transfer.

5.4.2.1 Double-Tube Tubular Heat Exchanger

The double-tube or tube-in-tube heat exchanger consists of either two straight smooth or corrugated tubes with different diameters concentrically located (Figure 5.9) (Lewis & Heppel, 2000; Toledo et al., 2018; Tetra Pak, 2021). Typically, the product flows through the inner tube, while the heat transfer medium flows through the annular flow channel. The broad inner tube diameter makes the double-tube heat exchanger suitable to process viscous foods, such as fruit and

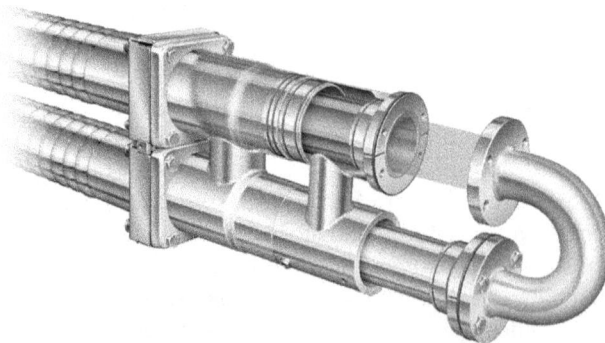

Figure 5.9 Schematic diagram of a concentric tubular heat exchanger (courtesy of Tetra Pak).

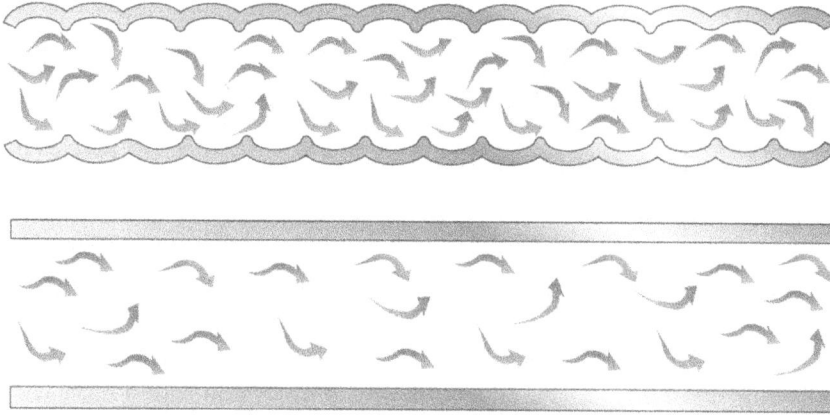

Figure 5.10   Comparison of product flow in a corrugated tube (top) versus smooth tube (bottom). Corrugated tube, turbulence will be much more intense when the surface is corrugated compared to a smooth surface (courtesy of Tetra Pak).

vegetable purees, sauces, and multiphase products with particulates. Double-tube heat exchangers are capable of handling high pressure and temperatures required by aseptic processing (Tetra Pak, 2021). However, limitations come from the potential laminar flow and small surface area limiting the heat transfer capabilities.

The use of corrugated tubes instead of straight-wall pipes has significantly increased in use in food processing. Corrugated surface promotes thermal mixing of the product without compromising product integrity (Tetra Pak, 2021). The corrugations increase the turbulence-inducing flow of the product and hence enhance the heat transfer (Figure 5.10).

5.4.2.2 Triple-Tube Heat Exchanger
Triple-tube heat exchangers consist of three concentric smooth or corrugated tubes of different diameters. In the triple-tube heat exchanger configuration, the product flows through the annular channel between the inner and the intermediate tube. In contrast, the heat transfer medium flows in both the inner and outer tubes (Figure 5.11) (Lewis & Heppel, 2000; Toledo et al., 2018;

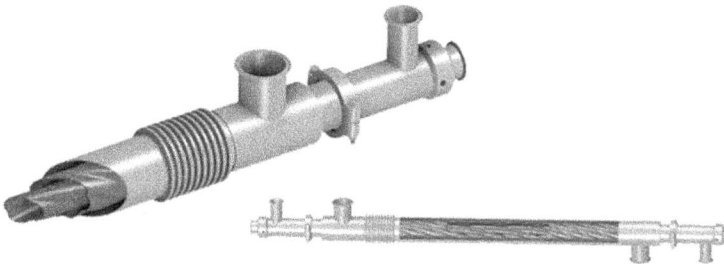

Figure 5.11   Schematic diagram of a triple tube heat exchanger consisted of concentric corrugated tubes of different diameters, with the product flowing through the annular channel between the inner and the intermediate tube (courtesy of Advance Process Solutions).

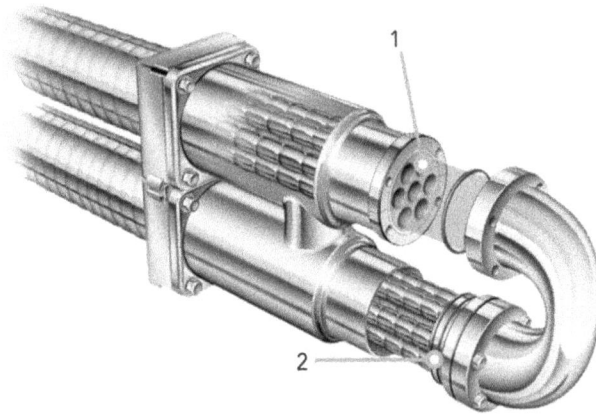

Figure 5.12   Schematic diagram of a multitube tubular heat exchanger. Product tubes (1) surrounded by heat transfer medium. A double O-ring seal at the end connection ensures no leakage (courtesy of Tetra Pak).

Tetra Pak, 2021). Triple tube heat exchangers have a larger surface area than the double tube resulting in a higher heat transfer rate. Triple tube heat exchangers are ideal for viscous products with or without particulates, operating under the high pressure and temperature requirements of aseptic processing (Lewis & Heppel, 2000; Toledo et al., 2018; Tetra Pak, 2021).

5.4.2.3 Multitube Heat Exchanger

The shell-in-tube or multitubes heat exchangers consist of a bundle of thin, smooth or corrugated inner tubes. The food product flows through the inner tubes, while the heat transfer medium flows through the heat exchanger's outer shell (Figure 5.12) (Tetra Pak, 2021). Multi-tubes heat exchangers provide a large heat transfer surface area, providing efficient heating and cooling for homogeneous thin or medium viscous foods, such as juices, milk, and other low-viscous dairy and coffee-based products. Multitube heat exchangers are designed to withstand high pressure and temperature conditions (Tetra Pak, 2021).

5.4.2.4 Coil Tube Heat Exchangers

Coiled tubular heat exchangers consist of a coil-shaped inner tube that carries the product inside a shell tube, the flow channel of the heat transfer medium. The design coil-shaped inner tube at high product velocity creates a secondary flow pattern known as Dean flow (Figure 5.13). In coiled tubes, the secondary Dean vortices improve mixing and increase the heat transfer rate and cooling efficiency. The magnitude of the heat transfer enhancement is based on the Dean number (N_{De}) (Equation 5.11), and it depends on the design of the coil, the flow rate, and the physical properties of the product. A Dean number that exceeds 100 is required to observe any significant improvement (Tetra Pak, 2021). High product velocities usually reach Dean numbers in combination with a tightly coiled tube. Coiled tubular heat exchangers in aseptic processing systems of thin and viscous foods, such as soups, sauces, nutritional drinks, pudding, tomato paste, and fruit or vegetable purees.

$$N_{De} = N_{Re} \cdot \sqrt{\frac{D_t}{D_c}} \qquad (5.11)$$

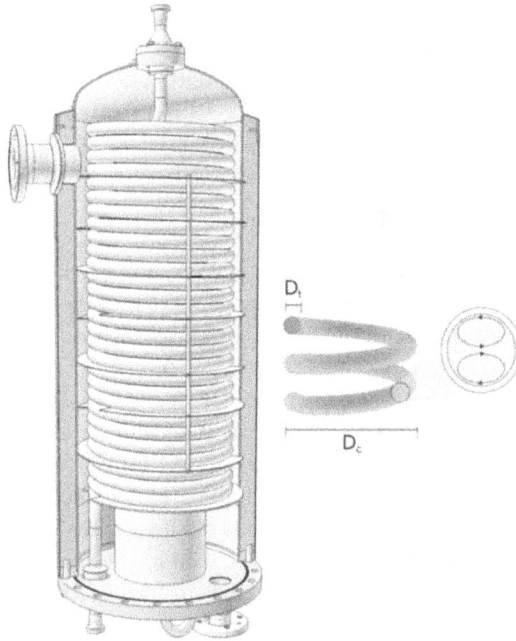

Figure 5.13   Schematic diagram of a coil tube heat exchanger, showing the D_t (m) and D_C (m), the coil-shaped inner tube diameter, and the diameter of curvature of the coil-shaped tube, respectively, used to estimate the Dean flow number (courtesy of Tetra Pak).

where N_{Re} is the Reynolds number, for Newtonian products given by Equation 5.12. D_t (m) and D_C (m) are the coil-shaped inner tube diameter and the diameter of curvature of the coil-shaped tube:

$$N_{Re} = \frac{\rho \cdot \bar{u} \cdot D_H}{\mu} \tag{5.12}$$

where ρ (kg/m³) refers to product density and μ (Pa⊙s) is the dynamic viscosity of the product. D_H (m) is the hydraulic diameter of the tube, for this case is equal to D_t (m), and $\bar{u}$ (m/s) is the average flow velocity of the product.

5.4.3 Regeneration

In aseptic processing, regeneration is a design option generally used for the plate or tubular heat exchangers. Regeneration is the practice where the hot sterile product is used to preheat the cold raw product, and in reverse, the cold raw material serves to cool the hot sterilized food. Regeneration efficiency of up to 95 % can be achieved with modern indirect heat exchanger systems, making the regeneration approach a very efficient energy and cost-saving method (Lewis & Heppel, 2000; Toledo et al., 2018; Tetra Pak, 2021).

When a product-to-product regeneration system is used, it is important to ensure that the raw cold product does not get into the sterile side, resulting in contamination of the hot sterile food. When a regenerator is used, the system must be designed, operated, and controlled, so the pressure on the heat exchanger's sterile side is always higher than the product on the raw side. Greater pressure will ensure that any leakage that may occur in the regenerator, the flow will be only from the sterile side to the raw cold side, eliminating the possibility of contamination of the sterile hot product, something that is required by the PMO, US Food and Drug Administration (FDA) (21 CFR, 1987), and US Department of Agriculture (USDA) (9 CFR, 1986) (Anderson et al., 2020).

5.4.4 Scraped Surface Heat Exchanger

Another conventional type of indirect heat exchanger used in aseptic processing is scraped surface heat exchangers. Scraped surface heat exchangers are mainly used for the thermal process of highly viscous foods such as cheese sauce and puddings or products containing particulates such as certain soups.

The hydraulic drag affects the processing of highly viscous foods due to the fluid's increased viscosity and the film build or fouling created on the tube wall. To enhance the heat transfer and cooling efficiency, scraped surface heat exchangers break the heat resistance build-up film of viscous product. In this type of heat exchanger, the application of mechanical means, continuously scraping the inside surface of the tube, results in rapid heat transfer to a relatively small product volume. The constant blending action accomplished in the scraped surface heat exchanger enhances the heat transfer and product temperature uniformity (Lewis & Heppel, 2000; Toledo et al., 2018; Tetra Pak, 2021).

The scraped surface heat exchangers consist of a concentrically located cylinder located in a driven shaft (mutator) with scraper blades. The cylinder containing the product and the shaft are enclosed in an outside jacket, and the heat transfer medium is supplied to this outside jacket. The food product is pumped through the small open space between the cylinder and the shaft, in contact with the cooled heat exchange wall (Figure 5.14) (Toledo et al., 2018; Tetra Pak, 2021).

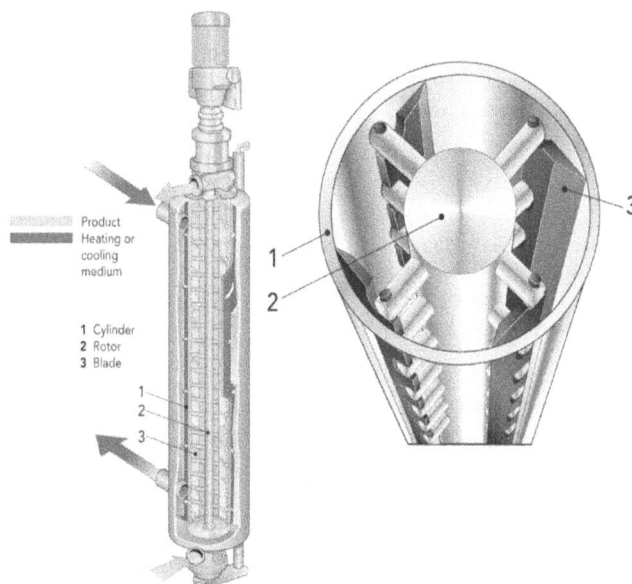

Figure 5.14 Schematic diagram of a vertical type of scraped surface heat exchanger, built from three different parts: (1) cylinder tube, (2) rotor (or mutator), and (3) blades (courtesy of Tetra Pak).

The materials used to build the scraped surface heat exchanger parts must be compatible with the product and with each other. The food-contact surfaces, such as the cylinder's inner tube, should be of a corrosion-resistant material such as stainless steel, chromium-plated nickel, or another suitable alloy. The scraper blades must not cause wear when in contact with the cylinder, and for that reason, stainless steel blades cannot be used, and therefore a softer material such as various types of plastics are used for the scraper blades. Based on the products to be processed, the mutator shaft can be of different sizes. A larger diameter gives a small annular space, typically used to cool viscous liquid foods or multiphase food products with small size particulates.

In contrast, a small diameter shaft can accommodate multiphase foods with large size particulates in the liquid. However, physical damage to the particulates from the blades' rotation is possible (Toledo et al., 2018; Tetra Pak, 2021). This issue can be minimized by slowing the mutator and increasing the number of blades. The mutator's speed should be optimized to get the best compromise between product damage and the cooling heat transfer rate.

Furthermore, there are a few disadvantages to scraped surface heat exchangers used in aseptic systems. For example, the higher capital, maintenance, and operating cost of this type of heat exchanger compared to the other indirect systems. Scraped surface heat exchangers with all the construction requirements and moving parts and seals require additional maintenance and control. For instance, in an aseptic system where scraped surface heat exchangers are used for cooling, seals of the system's moving parts may require an aseptic design, additionally with continuous control and monitoring. Furthermore, the energy requirement of a scraped surface heat exchanger to drive the mutator adds to the system's heat load. It may impact cooling due to viscous dissipation of the product, which may reduce cooling efficiency (Lewis & Heppel, 2000).

Another issue that may be observed using scraped surface heat exchangers comes with the potential mechanical problem where the mutators containing to which pins were welded and then the blades were attached to the pins. When the welds fail, they cause leakage resulting in the shafts being filled with product that accumulates and cannot be cleaned or sterilized. This bacteria buildup can then migrate into the product being processed, jeopardizing the equipment's ability to be sterilized, resulting in contamination of the product. Finally, another issue with using a system consisting of scraped surface heat exchangers is that at the initial high temperatures required for aseptic processing, the plastic blades tend to deteriorate and break, resulting in plastic getting into the product.

Although scraped surface heat exchanger remains a reliable indirect system for highly viscous and multiphase foods, nowadays, many aseptic processing systems are replacing the scraped surface heat exchangers with less expensive tubular heat exchangers. The newly improved designs of the tubular heat exchangers discussed above have enabled the processors to replace the scraped surface heat exchangers with this type of heat exchangers to cool the food products with the same efficiency as the scraped surface heat exchangers, but with lower operational and maintenance cost.

5.5 ADVANCED HEATING AND COOLING TECHNOLOGIES

5.5.1 Advanced Heating Systems

Advanced heating systems are novel heating technologies that have been developed to overcome the limitations of conventional heat exchangers, especially in viscous, low thermal conductivity products, and products with particulates. Advanced heating technologies utilize a direct conversion of electromagnetic energy into heat to provide rapid and uniform volumetric heating. The characteristics of these advanced heating methods are:

- Power can be turned on and off instantly
- Dynamics are very rapid
- It does not rely on contact with hot surfaces or a hot medium
- It is selective, i.e., different materials, or portions of the same food material having different properties will heat at different rates
- It is volumetric, thus more uniform than conventional heating
- It relies on the electrical properties of the materials and not in viscosity or thermal conductivity
- Efficiency of conversion of energy into heat is high

Advanced heating methods benefit from modern process control systems, in which computers with very short response times provide the dynamic control that matches the very fast response times of heating. Feed-forward/feedback control systems need to be used with a large number of sensors to ensure that process stays in control.

Two such methods that have proven to be viable alternatives for industrial applications are ohmic and microwave heating. Ohmic works by heating the product using electrical current that is made to flow inside the product, the resistance of the food to this current flow generates heat. Microwave works by the movement of the molecules of the product in a rapidly changing electromagnetic field that generates heat. Both methods used polymeric or glass tubes for the thermal transfer section and have no hot walls.

Intensive research has been performed on these heating methods, with many scientific papers published that have led to successful industrial applications of ohmic and microwave heating for aseptic processing.

5.5.1.1 Microwave Heating

Microwave heating uses the interactions of molecules with a rapidly changing electromagnetic field. The system operates under continuous flow conditions; this interaction does not need direct contact with electrodes or antennas but requires that the product be transported using sections of transparent tubes to microwaves.

Microwaves are a part of the electromagnetic spectrum that comprises frequencies between 300 and 3000 MHz. Microwaves find their most widespread use in communications, radar, and medical devices, thus in the United States the frequencies in which these microwave ovens operate have been regulated by the Federal Communication Commission. The allocated frequencies for industrial, and household applications are 915 ± 13 MHz and 2450 ± 50 MHz (47CFR18.301, 2004). The 2450 MHz frequency is used mostly in household microwave ovens, while 915 MHz is used in industrial applications.

The dielectric properties of the materials are very important for heating using microwaves, and these properties are product and temperature dependent. The interaction of any material with a rapidly changing electromagnetic field is called permittivity and is expressed as:

$$\varepsilon_c = \varepsilon \left[1 - j\frac{\sigma}{\omega\varepsilon} \right] = \varepsilon' - j\varepsilon'' \tag{5.13}$$

The complex permittivity of a dielectric material consists of two parts, a real and an imaginary part, which are generally expressed as factors of the permittivity of free space ($\varepsilon_0 = 8.85 \times 10^{-12}$ F/m). The real part, called dielectric constant (ε'), relates to the amount of energy that is reflected or transmitted by the material and to the ability of the material to store electromagnetic energy. The imaginary part, called loss factor (ε''), relates to the ability of the material to convert electromagnetic energy into heat (Engelder & Buffler, 1991). Food materials comprise a wide variety of dielectric properties, as compiled by Kent (1987), and Nelson (1991).

These properties are dependent not only on the material and temperature but also on the frequency of the microwaves, making scaling up a difficult step which requires good engineering.

The amount of energy converted into heat (q_{gen}) is a function of the properties, and its application also depends on the time of exposure to the microwave field.

$$q_{gen} = 2\pi f E^2 \varepsilon_0 \varepsilon'' \tag{5.14}$$

Microwave heating has been researched and successfully applied to aseptic processing in several applications, it was initially applied to sweet potato as published by Coronel et al. (2005) which resulted in the first aseptic processing plant using continuous flow microwave which received FDA acceptance in 2007. This plant is located in Snow Hill, NC, and produces vegetable purees as ingredients for the industry. Later a series of aseptic products processed using microwave were launched in the United States, by Aseptia. New validation methods were developed using this technology for processing of products with large particulates, receiving FDA acceptance and are published in Appendix 6 for the first time. Finally, a small R&D and production unit was developed by Sinnovatek. Sinnovatek coupled their unit with an aseptic small pouch filler which uses irradiated packages (Figure 5.15).

Microwave heating has proven its potential and is slowly adopted by the industry. Further research is needed in the interactions of products with changing dielectric properties and heterogeneous products.

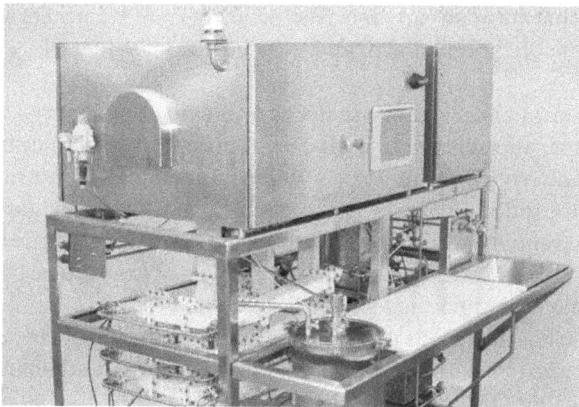

Figure 5.15 Continuous flow microwave processing unit. Nomatic self-contained unit by Sinnovatek, Raleigh, NC.

5.5.1.2 Ohmic Heating

Ohmic heating is a method to heat food materials by converting electrical energy directly into heat by using the Joule effect, utilizing an electrical current passed through a suitable conducting product, causing product heating. The system operates under continuous flow conditions, with the product passing over electrodes in one or more heating tubes.

Ohmic heating involves the passage of an electric current through the material to be heated, this electric current passage involves the motion of charges, which produces molecular agitation, and consequently, heat. Since the heat is generated within the food, heating is not dependent on thermal conductivity, and it is thus much more uniform than with conventional modes of heating.

Ohmic heating was discovered in the nineteenth century and applied initially in the 1930s for electric pasteurization of milk, and later in the 1980s and 1990s for continuous flow sterilization of heterogeneous foods. Currently several aseptically processed and packaged products are commercialized in both bulk and retail.

In practice, ohmic heating requires the food to be in contact with electrodes, across which a potential difference is applied. In order for heating to occur, the material must have some ability to conduct electricity. The rate of energy generation per unit volume (q_{gen}) may be expressed as shown in equation 5.15.

$$q_{gen} = E^2 \sigma \qquad (5.15)$$

where E is the electric field strength, and the σ electric conductivity of the product to be heated. E may be modified either by physical design of the electrode gap, or more easily, by altering the applied voltage or intensity. The electrical conductivity, σ depends on the product and temperature, and may be altered by formulation and pre-process steps to include various ingredients. Electrical conductivity of foods (σ) is critical for ohmic heating, the major factors influencing σ of a liquid food are temperature, frequency, and composition; while for solid particles, the additional effect of microstructure must be considered. Solid–liquid mixtures, need to consider the relative proportions of each component. In general, electrical conductivity increases with temperature; for liquids, the change tends to be linear, while for solids, the trend depends on cell structure. Solids that have already received prior thermal treatment such as blanching or freezing, have no further cell breakdown, and the electrical conductivity changes linearly with temperature.

A detailed treatment of electrical conductivity is provided by Sastry (2005). Ohmic heaters are designed either in parallel plate or in in-line flow configurations. Since many foods are of significant electrical conductivity, the most common commercial design uses annular electrodes as the ends of applicator tubes. In order to prevent electrolysis, high frequency solid state power sources are used with frequencies ranging from 10 to 50 kHz. Since Ohmic heating is performed at frequencies in the kHz range the dependency on frequency needs to be better understood, but observations have shown very little dependency.

Scalability of Ohmic heating needs a good understanding of the underlying phenomena, from lab scale to pilot to industrial scale analysis are possible using mathematical modeling and a solid engineering approach (Figure 5.16)

Ohmic heaters can be found in many industrial applications, especially in tomato products, fruit pieces, purees, and vegetables in brine used as ingredients and packed in bulk. Around 2010, Liebig (Campbell Soup) and Knorr (Unilever) launched a range of aseptically packaged low-acid soups with large particulates processed using Ohmic heaters.

Figure 5.16 Ohmic heating as part of an integrated aseptic processing plant (courtesy of Emmepiemme SRL).

5.5.2 Advances in Cooling Technology

Advanced heating technologies such as continuous flow microwave systems and ohmic heating have enabled food processors to provide uniform rapid volumetric heating, resulting in reduced come-up time, reduction of fouling deposition, and minimization of product quality losses, during continuous thermal processing of viscous and multiphase food products. These include typical dairy products, fruit and vegetable purees, and soups (Coronel et al., 2005; Salengke & Sastry, 2007; Steed et al., 2008; Cullen et al., 2012). However, due to the lack of advanced technologies, cooling of viscous products is still implemented using inefficient conventional cooling methods. During conventional cooling of viscous food products, laminar flow, and low thermal conductivity, characteristic of these materials, lead to a non-uniform, slow cooling process, an increase in operation cost, and degradation of final food quality (Stoforos et al., 2016). Unfortunately, often predicted and anticipated developments of volumetric cooling technologies, such as magnetic field cooling (Sarlah et al., 2006; Kawanami et al., 2011), have been slow and expensive in their progress to large-scale, commercial-industrial applications. Additionally, to enhance the cooling rate of continuous flow thermal processing, food processors utilize cooling media (like ethylene glycol-based coolants) at a very low-temperature range without any significant improvement in cooling viscous foods. Using this cooling approach, an additional problem has appeared, ethylene glycol and cold/chilled water piping can sweat or be covered with ice, resulting in dripping water, with a negative impact on the food production plant (Moerman, 2016).

Although the industrial application of advancements in cooling heat exchangers used for viscous and multiphase foods is rather limited to the recent developments in the corrugated tubular heat exchangers, more research has recently been conducted and focused on the enhancement of continuous flow cooling. One cooling technique that has recently been studied and has shown promising results is thermal mixing. Thermal mixing, temperature equalization within the product, has been proposed as an efficient method to enhance heat transfer and the overall continuous flow cooling process of highly viscous food products (Metcalfe & Lester, 2009; Stoforos et al., 2016).

In a cooling system consisting of a series of tubular coolers, applying thermal mixing, equalization of product temperature at the exit/entrance of each cooling section can significantly reduce the cooling time process. The application of thermal mixing during cooling can potentially result in the enhancement of heat transfer while reducing final product quality degradation (Stoforos et al., 2014). Equalizing the radial bulk product temperature between each cooling section reduces the hot product's temperature flowing at the center of the cooler tube while increasing the temperature of the product flowing close to the heat transfer area at the surface of the heat exchanger tube. The resultant higher temperature difference between the product and the coolant flowing close to the heat transfer surface can potentially increase the cooling rate (Stoforos et al., 2014). While reducing the temperature of the product flowing at the center of the tube minimizes the product quality degradation, considering reduces the time the product stays at the high aseptic process temperatures, a characteristic disadvantage of laminar flow, dominant for viscous food products.

The thermal mixing application as a cooling enhancement method can be easily adapted by the currently installed, operating tubular cooling systems. Thermal mixing can be achieved by installing static mixers or applying any mechanical and rotational means that can efficiently result in product temperature equalization. Since there are applicable in an open duct flow tubular system, application of the above thermal mixing methods can potentially result in less physical damage to food particles than scraped surface heat exchangers. At the same time, no additional CIP requirement would be required. However, the efficiency of thermal mixing and cooling processing of viscous food products depends on rheological food properties. It is affected by the formation and build-up of fouling product layers on the surfaces of processing equipment (heat exchangers, mixing units) (Stoforos et al., 2014, 2016).

Although no significant developments with an industrial application are available for cooling heat exchangers, much new research has focused on developing rapid cooling heat exchangers. Recent studies showed that using a hydrophobic surface-modified tubular heat exchanger could potentially provide a useful tool for rapid cooling (Stoforos et al., 2021). A hydrophobic modified tubular heat exchanger showed that a combination of good conductive material, such as stainless steel, with a food-contact surface with antifouling, hydrophobic surface characteristics can significantly enhance cooling and potentially improve sanitary conditions during the processing of viscous foods. The obtained results are encouraging further research into the development and application of rapid cooling heat exchangers; however, it is in the very preliminary stage with additional research studies that need to be conducted before these modified heat exchangers are in the commercialization stage for industrial applications in aseptic food processing. The lack of any predominant commercial cooling applications allows for future research in the cooling technology area, targeting the enhancement of continuous flow cooling of viscous and multiphase products while minimizing the final food product quality degradation (Stoforos 2014, 2017).

5.6 HOLDING TUBE

Aseptic thermal processing is based on the sterilization of the food product in the holding tube. The holding tube is designed to hold the product at the sterilization temperature long enough to result in commercial sterility. The holding tube is designed to accomplish the required lethality for the product; thus, the product's temperature must be maintained for a specified minimum required time in the holding tube. That minimum residence time must be established so that the hold time/temperature combination assures product's sterility. Because the holding tube is the heart of an aseptic processing system, care should be taken to engineer and fabricate the holding tube accurately. The design and the size, length, and inner tube diameter of the hold tube are critical. The holding tube should be designed to retain the fastest moving food fluid element or particle under a controlled flow rate for a specified time. According to the US FDA, the holding tube must be sloped upwards by ¼ inch per foot, and no additional heat can be applied to the holding tube. However, it can be insulated to minimize heat losses and potential product temperature drop.

Regulatory agencies rely upon process authorities to establish the adequate processing parameters that must be met at the holding tube to assure the commercial sterility of the final product. The critical factors required to establish the proper thermal (time/temperature) process to render the product commercial sterile are:

1) *The thermal resistance of potential pathogen and spoilage microorganisms of concern*: For low-acid foods (pH > 4.6 and water activity > 0.85), the thermal sterilization process should give the required lethality to control *Clostridium botulinum* and other spoilage organisms of non-health significance. While if aseptic processing is used for high-acid and acidified foods (pH < 4.6 and water activity > 0.85), the thermal process for these products is designed to eliminate yeast, molds, vegetative bacteria, and heat resistant enzymes that could spoil the product in addition to controlling of vegetative pathogens (e.g., *Listeria monocytogenes*, *Salmonella* spp, and *Escherichia coli O157:H7*).

2) *Hold tube dimensions*: hold tube dimensions, namely holding tube length and inner tube diameter, are critical parameters in determining the residence time of the product within the hold tube.

3) *Flow rate within the hold tube*: product flow rate is a critical process parameter and should be properly controlled, monitored, and recorded. Typically, in aseptic processing systems, the metering pump, and the flow meter are located prior to the final heating stage and the hold tube. Based on that, process authorities should consider any increase in product flow rate due to thermal and volumetric expansion of the food product that occurs during the heating stage. An increase in the flow rate in aseptic processing systems occurs due to an increase in the volume of the flowing stream. The increase in flow rate then leads to a decrease in the residence time, which poses the danger to result in under-processed food. For all the systems, direct and indirect heat exchangers, an increase in volume may occur due to thermal expansion of the food because of the increase in temperature. In the high operating temperatures (100-160 °C) for aseptic processing, the change in food volume is nearly linear within that temperature range. Using the information from steam tables (specific volume of saturated liquid (m^3/kg)) regarding the inlet and outlet product temperature of the heat exchanger, the increase of food volume can be estimated and hence the % increase in flow rate. Furthermore, in the direct heating systems, in addition to flow rate correction due to thermal expansion of the food, proper adjustment of the flow rate needs to be calculated due to the added volume/mass of steam. Based on mass and energy balances, steam tables (saturated vapor enthalpy (kJ/kg)), and product and steam temperatures, the approximate increase of product flow rate due to volumetric expansion is about 1% for every 10 °F difference between product sterilization temperature and product exit preheater temperature.

4) *Rheological properties of food (flow characteristics)*: Having information on the rheological properties of the food product (viscosity, Newtonian, or non-Newtonian fluid, etc.) is necessary to calculate the Reynolds number and determine if the flow within the hold tube is laminar or turbulent. The decision of laminar versus turbulent flow is critical in determining the maximum velocity of the fastest flowing food element and hence the residence time of the food within the hold tube. If the rheological properties of the food are unknown, it is highly recommended to go with laminar flow assumption as the worst-case scenario. Chapter 6 provides additional information on the subject.

5.7 TEMPERATURE INDICATING DEVICE AND TEMPERATURE RECORDING DEVICE

The product temperature measured at the exit of the hold tube represents the minimum product temperature within the hold tube. The US FDA requires a temperature indicating device (TID) and temperature recording device (TRD) at the outlet of the hold tube. The TID should be a "stand-alone" sensor, having its own independent sensor different from the temperature recording device (TRD) sensor. The TID is the reference instrument for indicating the processing temperature and should be installed in a proper location where it can be easily accessed and accurately be read. A TRD should also be installed to monitor, and continuously record the temperature at the exit of the hold tube, ensuring that the desired product sterilization temperature is achieved and maintained. The TRD sensor must be adjusted to agree as near as possible, but in any case, not higher than the TID. According to the US FDA the TRD chart graduations must not exceed 2°F (1°C) within a range of 10°F (5.5°C) of the processing temperature. A working scale of not more than 55°F per inch (12° C per Cm) is required within 20°F (10° C) of the processing temperature.

5.8 AUTOMATIC FLOW DIVERSION

An automatic flow diversion device, divert valve, may prevent the underprocessed product from entering the sterilized system. Flow diverse valve is usually located at the end of the cooling section to prevent the nonsterile product from entering the aseptic tank and the filler. The flow diversion may be required, for example, when a temperature drop occurs in the hold tube (Anderson et al., 2020).

5.9 BACK PRESSURE VALVE

Back pressure valves regulate a predefined pressure in the aseptic processing system, which includes the heating system, the hold tube, and the cooling system. The desired product pressure is achieved by supplying compressed air to a pneumatic actuator. Back pressure valves typically are installed at the end of the cooling section to assure and maintain the system under positive pressure and prevent product flashing or boiling (i.e., water vapor expanding as steam). Product flashing may cause an increase in product flow rate, reducing the residence time of the product in the hold tube and, hence affecting the delivered process. Furthermore, in direct heating systems, where cooling and water removal from the product occurs within the flash chamber, the backpressure valve's role in the system is vital. A direct heating system should separate the hold tube with a flash chamber with a backpressure valve to prevent flashing from taking place in the holding tube.

5.10 PREPRODUCTION SYSTEM STERILIZATION

Sterilization of the food product is one part of aseptic processing; before introducing food into the system, it must itself be presterilized. Considering that the cooling system is located downstream of the last heater, it must be properly presterilized before production. Prior to the start of the presterilization cycle, any cooling media remaining in the cooling system (heat exchangers, piping, etc.) is either drained or not activated, as in the case of ammonia and Feron. For most systems, presterilization of the aseptic system is accomplished by applying saturated steam or pressurized hot water

(Chandarana et al., 2010). The system is sterilized by maintaining the temperature within all the system parts at or above a specified temperature ($\geq 250°F$) for a sufficient length of time (≥ 30 min) to render it commercially sterile. Achieving and maintaining the time/temperature requirements of the pre-sterilization cycle is achieved with the sterilized medium's recirculation (Chandarana et al., 2010). During the per-sterilization cycle of the aseptic system, the temperature is monitored, at the "colds spot" in the system, typically at the end of the cooling section. The sterilization time should incorporate only when all parts of the system's temperature are above or at the required minimum temperature. Time and temperature during pre-sterilization should be monitored and recorded.

After the pre-sterilization cycle is completed, the system downstream of the hold tube is cooled and prepared to introduce the processed product. The processing system must maintain sterility from the end of the sterilization cycle through the end of production. To achieve that, positive pressure is applied to the system. As described above, to maintain the system's positive pressure, a backpressure device is used, located at the end of the cooling line before the filler. Having the system under positive pressure maintained the product in the system at a pressure usually of 10–15 psi over the maximum product temperature pressure; it prevents the product from flashing and assists in avoiding recontamination of the system (Chandarana et al., 2010; Anderson et al., 2020).

Areas in the aseptic cooling system with any moving parts such as reciprocating shafts (valve stems) should have applied an effective barrier to eliminate any recontamination entry. Typically, steam seals are used for that reason. Steam must continuously be applied in the barriers to have steam form a ring around a shaft or cover the total stroke of a valve to prevent recontamination from these areas. All barriers should be presterilized during the presterilization cycle of the system (Chandarana et al., 2010).

5.11 DIFFERENTIAL PRESSURE

As described in the product-to-product regenerator section, controlling the pressure difference between the sterile and the unsterile side is critical and highly recommended. Pressure sensors should be located at the non-sterile cooling medium inlet of the regenerator (point of highest pressure) and the regenerator's sterilized product outlet (point of lowest pressure). The differential pressure requirement applies to all types of conventional indirect coolers to prevent non-sterile cooling media from recontamination of the sterile food product. The lowest point of the product sterile side's pressure should be greater than 6894.76 Pa (1 psi) compared to the highest pressure point of the unsterile coolant side (Anderson et al., 2020).

The differential pressure devices must comply with the specification listed in the regulation and the nature of the control action taken by the device in the event of improper pressures. Failure to maintain the required pressure differential cause a deviation to the required process and causes the automatic flow diversion to assume the diverted flow position.

5.12 HEAT TRANSFER MEDIA

5.12.1 Steam

Steam is used in many areas of aseptic processing and packaging systems. It is used to prepare the product by heating it indirectly through the heat exchanger walls or by direct addition to the product being prepared. It is used for sterilizing equipment used in processing or packaging operations, sterilizing seals, and maintaining a sterile atmosphere in the seal area. It is also used for developing heat that causes sterilant to vaporize or breakdown, such as hydrogen peroxide.

The steam is used indirectly to heat the product through a heat exchanger wall that will heat water, which in turn is used to heat the product, or used where it will not become a part of the product does not have to be culinary or pure. If the steam is used to heat the product directly or the food-contact surface, for instance, the wall of a tank, it should be culinary. Culinary steam must meet all applicable codes and regulations of the appropriate regulatory agency (e.g., FDA and USDA). Culinary steam requires that no boiler compounds can be used that are not listed or will cause the product to deteriorate or react negatively and generate off colors, have off-flavors or odors, or be harmful to the consumer. If the steam becomes part of the product, such as in the preparation or sterilization process, it should be produced from a sanitary reboiler, and the steam should be made from distilled water. There are possible regulatory implications, but the product itself may develop undesirable characteristics from the chemicals that may be in the steam. These chemicals may not be harmful to humans, but they may cause undesirable reactions between the product and the steam.

Steam generally should be at least ideally ~150 psi at the boiler. After the steam travels through the various distribution lines and headers, it may be drop pressure at the use point. At the use point, it should be regulated at a constant stable pressure. The steam can be controlled to the desired use pressure whether it is in a preparation vat, heat exchanger, aseptic tank, or packaging machine. If the pressure fluctuates, degrees of superheat can vary and cause the heat content of the steam to vary, and the final temperature of the product can vary. Steam should be treated in such a manner that superheat is removed. It is highly recommended that proper standard operation procedure (SOPs) are in place in every aseptic processing plant regarding the quality of the steam used and the process produced/prepared.

5.12.2 Hot Water

Water is used in certain heat exchangers to heat the product for several reasons; the temperature can be more accurately controlled than steam. Hot water compared to steam does not vary in temperature because no valves and traps are opening and closing. When steam is used to heat products, one condition that exists is that as the control valve opens and closes, the steam temperature entering the heat exchanger varies. This will cause the temperature of the heated product to differ, something that is better controlled with the use of hot water. Another advantage of the use of hot water is that it offers a wide range of operating temperatures. However, due to the high operating conditions of aseptic processing, proper pressure control on the hot water side is necessary.

5.12.3 Cooling Media

Chilled water and glycol are the two main coolants used in aseptic processing. Water is preferred as a coolant due to its high heat capacity and low cost. Water is usually used with additives, such as corrosion inhibitors and antifreeze solutions, with the most often used to be ethylene glycol and propylene glycol (ASHRAE, 2010).

Glycol is an organic compound with anti-freezing and anticorrosion properties. Glycol is a water-miscible coolant that is frequently used in heat transfer and cooling applications because mixing it with water can provide various heat transfer characteristics (temperature difference, heat capacity). When it is mixed with water, glycol provides a great cooling medium that operates at a wide range of temperatures. The boiling and freezing points of glycol mixtures are a function of the relative amounts of glycol and water in the mixture (ASHRAE, 2010).

To prepare the cooling medium, heat exchangers, such as shell-and-tube, utilize refrigerants such as ammonia to cool down the cooling medium. During that process, the heat exchanges usually are not sanitary, and they are not thoroughly cleaned and maintained. The above can lead to bacterial buildup that can contaminate the cooling medium and potentially compromise the aseptic processing.

The bacterial buildup in equipment used for the cooling medium's preparation can be significant, which may result in contamination of the cooling medium used. A contaminated cooling medium can result in contamination or adulteration of the food product if the aseptic processing coolers leak through a pinhole or stress crack in the heat exchanger or a gasket. For that reason, frequent chemical and microbial testing of the cooling medium is recommended. Microbial testing of the cooling medium should look for the levels of microorganisms such as coliforms and psychrotrophs. Chemical testing should ensure that cooling water additives and cooling media products are safe for use in aseptic food processing. Finally, it is important to properly monitor and record-keeping all the tests conducted and prove the cooling medium's safety.

5.13 CONCLUDING REMARKS

Heating and cooling systems are a vital part of an aseptic processing system. Heating system is located prior to the hold tube, and it is responsible for heating the product at the required sterilization temperature. Cooling system is downstream from the hold tube, responsible for rapidly cooling down the product prior to filing. An inadequate heating process may pose the risk of under-processed food. While slow heating and cooling may result in an overcooked final product, resulting in degradation of food's nutritional and organoleptic characteristics.

Selection of the proper heating and cooling systems is critical for aseptic processing. The selected heating and cooling systems should result in rapid heat transfer, minimizing any food quality losses (color, texture, etc.) of the finished food product. The heating and cooling systems used in aseptic processing depend on food product characteristics (viscosity, homogenous, or multiphase products). For liquid, low-viscous homogeneous foods, where direct steam heating systems are used, flash evaporative cooling is used for the initial cooling stage. In most cases, flash evaporative cooling is used after the hold tube to provide an initial rapid cooling while also removing the excess condensed water added from the steam. In systems that utilize indirect heat exchangers for heating and cooling, the system will consist of plate, tubular, or scraped surface heat exchangers depending on product viscosity and characteristics. Liquid homogeneous products, low or medium viscosities, can be processed in any of the indirect systems. Highly viscous and multiphase foods are typically processed in scraped surface heat exchangers, while advanced heating technologies are also an excellent option for these products.

To conclude, rapid heating and cooling are essential to ensure food safety and prevent and minimize food quality losses. Although novel advanced heating technologies have been adapted in aseptic processing and have helped the process of more complex and multiphase food products, cooling is still implementing slower conventional methods. The limited number of new industrial rapid cooling systems raises many challenges and opportunities for further research on that field.

NOMENCLATURE

Latin Letters

A: Surface area (m^2)
c_p: Specific heat (J/kg·K)
D_t: Coil-shaped inner tube diameter (m)
D_c: Diameter of curvature of coil-shaped tube
E: Electric field strength (V/m)
f: Frequency (Hz)
L: Total length of heat exchanger (m)
$\dot{m}$:   Mass flow rate (kg/s)
N_{De}:   Dean number
N_{Re}:   Reynolds number
Q: Overall energy transferred during cooling (W)
q_{gen}:   Energy generation per unit volume (J/m^2)
r: Radius (m)
T: Temperature (K)
U: Overall heat transfer coefficient (W/(m^2·K))
$\bar{u}$:   Average flow velocity (m/s)

Greek Letters

ΔT: Temperature difference (K)
ε_0:   Permittivity of free space (8.854x10^{-12} F/m)
ε':   Relative dielectric constant
ε'':   Relative dielectric loss factor
ρ: Density (kg/m^3)
σ: Electrical conductivity (S/m)

Subscripts

in: Refers to the inlet of the heat exchanger or the mixing unit
i: Refers to the inner tube
ii: Refers to the inside surface area of the inner tube
io: Refers to the outside surface area of the inner tube
o: Refers to the outer tube
L: Refers to the energy transferred by the surrounding environment ("heat losses") to the cooling medium
lm: Logarithmic mean difference
m: Refers to the heat transfer medium
p: Refers to the product

REFERENCES

Anderson, N.M., Benyathiar, P., and Mishra, D.K. 2020. Aseptic processing and packaging. In *Food Engineering Series*, edited by A. Demirci, H. Feng, K. Krishnamurthy, 661–692, Chapter 25. Cham, Switzerland: Springer Nature Switzerland AG.

ASHRAE. 2010. *Handbook: Refrigeration (I-P Edition)*. American Society of Heating, Refrigerating and Air-Conditioning Engineers, Inc. (ASHRAE). https://app.knovel.com/hotlink/toc/id:kpASHRAE41/ashrae-handbook-refrigeration/ashrae-handbook-refrigeration

Chandarana D.I, Unverferth J.A., Knap R.P., Deniston M.F., Wiese K.L, and Shafer B. 2010. *Establishing the Aseptic Processing and Packaging Operation, Principles of Aseptic Processing and Packaging*, 135–150, Chapter 7. 3rd edition. West Lafayette, IN: Purdue University Press.

Coronel, P., Truong, V.D., Simunovic, J., Sandeep, K.P., and Cartwright, G.D. 2005. Aseptic processing of sweetpotato purees using a continuous flow microwave system. *J. Food Sci.* 70(9):531–536. https://doi.org/10.1111/j.1365-2621.2005.tb08315.x

Cullen P.J., Tiwary B.K., and Valdeamidis V.P. 2012. *Novel Thermal and Non-Thermal Technologies for Fluid Foods*. London, UK: Food Science and Technology, International Series.

Engelder, D.S., and Buffler, C.R. (1991). Measuring dielectric properties of food products at microwave frequencies. *Microwave World* 12(2):6–15.

Kawanami, T., Hirano, S., Fumoto, K., and Hirasawa, S. 2011. Evaluation of fundamental performance on magnetocaloric cooling with active magnetic regenerator. *Applied Thermal Engineering* 31:1176–1183. https://doi.org/10.1016/j.applthermaleng.2010.12.017

Kelder, J.D.H., Coronel P., and Bongers, P.M.M. 2009. Aseptic processing of liquid foods containing particulates. In *Engineering Aspects of Thermal Food Processing*, edited by R. Simpson, Chapter 3. Boca Raton, FL: CRC Press.

Kent, M. 1987. *Electrical and Dielectric Properties of Food Materials: A Bibliography and Tabulated Data*. London: Science and Technology Publishers.

Lewis M., and Heppell N. 2000. *Continuous Thermal Processing of Foods Pasteurization and UHT Sterilization*. Gaithersburg, MD: Aspen Publishers, Inc.

Metcalfe, G., and Lester, D. 2009. Mixing and heat transfer of highly viscous food products with a continuous chaotic duct flow. *J. Food Eng.* 95(1):21–29. https://doi.org/10.1016/j.jfoodeng.2009.04.032

Moerman, F. 2016. Hygienic design of food processing facilities. *Food Safety Magazine*. http://www.food-safetymagazine.com/ magazine-archive1/octobernovember-2010/hygienic-design-of-foodprocessing-facilities

Mudgett, R.E. 1986. Microwave properties and heating characteristics of foods. *Food Technol.* 40(6):84–93.

Nelson, S.O. 1991. Dielectric properties of agricultural products. *IEEE Trans. Electr. Insul.* 26(5):845–869.

Salengke, S., and Sastry, S.K. 2007. Experimental investigation of ohmic heating of solid–liquid mixtures under worst-case heating scenarios. *J. Food Eng.* 83(3):324–336.

Sarlah A., Kitanovski A., Poredos A., Egolf P.W., Sari O., Gendre F., and Besson Ch. 2006. Static and rotating active magnetic regenerators with porous heat exchangers for magnetic cooling. *Int. J. Refrig.* 29(8):1235–1394.

Sastry, S.K. 2005. Electrical conductivity of foods. In *Engineering Properties of Foods*, edited by M.A. Rao, S.S.H. Rizvi, and A.K. Datta, Chapter 10, 589–639. Marcel Dekker, Inc.

Steed, L.E., Truong, V.D., Simunovic, J., Sandeep, K.P. Kumar, P., Cartwright, G.D., and Swartzel, K.R. 2008. Continuous flow microwave-assisted processing and aseptic packaging of purple-fleshed sweet potato purees. *J. Food Sci.* 73(9):E455–E462. https://doi.org/10.1111/j.1750-3841.2008.00950.x

Stoforos, G.N. 2014. *Acoustic enhancement of continuous flow cooling*. Raleigh, NC: North Carolina State University, Master of Science thesis.

Stoforos, G.N. 2017. *Enhancement of Continuous Flow Cooling of Viscous Foods Using Surface Modified Heat Exchangers*. Raleigh, NC: North Carolina State University.

Stoforos, G.N., Farkas, B.F., and Simunovic, J. 2016. Thermal mixing via acoustic vibration during continuous flow cooling of viscous food products. *Food Bioprod. Process.* 100(Part B):551–559. https://doi.org/10.1016/j.fbp.2016.07.008

Stoforos, G.N., Rezaei, F., Simunovic, J., and Sandeep, K.P. 2021. Enhancement of continuous flow cooling using hydrophobic surface treatment. *J. Food Eng.* 300. https://doi.org/10.1016/j.jfoodeng.2021.110524

Tetra Pak. 2021. Heat-exchangers. https://dairyprocessinghandbook.tetrapak.com/chapter/heat-exchangers

Toledo, R.T., Singh, R.K., and Kong, F. 2018. *Fundamentals of Food Processing Engineering*. 4th edition. Cham, Switzerland: Springer International Publishing AG, part of Springer Nature. 2007, 2018.

Chapter 6

Flow and Residence Time Distribution for Homogeneous and Heterogeneous Fluids

Pablo M. Coronel and Josip Simunovic

CONTENTS

6.1 BASIC CONSIDERATIONS OF RESIDENCE TIME AND FLOW PROFILE

Aseptic processing is carried out in continuous flow, which means flow in closed conduits (circular, rectangular). Flow characteristics of different products are very important as they determine how the product flows as well as the transfer of heat to the products. Process settings and safety must be calculated based on a worst-case scenario, which includes the fastest moving fluid element or solid particle that might be processed for which both the residence time distribution (RTD) of particles and the flow profile of liquids are needed.

Residence time is important as it is the time products are exposed to a certain temperature and has implications for quality and safety. For safety, the fastest moving element or particle needs to be identified and based on that the worst-case scenario determines the minimum residence time for the process (t_{Res}), while for quality the average residence time is used. The basic relations between flow rate, velocity, average, and minimum residence time are shown in Equation 6.1. These calculations are the basis for Food and Drug

DOI: 10.1201/9781003158653-8

Administration (FDA) filings in the United States, with the use of correction factors depending on the flow behavior.

$$A = \frac{\pi\, ID^2}{4} \qquad \text{Area of a circular tube}$$

$$V = AL = \frac{\pi\, ID^2\, L}{4} \qquad \text{Volume of a circular tube}$$

$$\bar{t} = \frac{V}{Q} \; \text{or} \; \bar{t} = \frac{\pi\, ID^2\, L}{4\,Q} \qquad \text{Average residence time}$$

$$U = \frac{Q}{A} = \frac{4\,Q}{\pi\, ID^2}$$

$$U_{Max} = \bar{U} \cdot \xi \qquad \text{Maximum velocity}$$

(6.1)

The average residence time is a function of the volume of the tube (V) and the volumetric flow rate (Q), in the case of a circular tube, the internal diameter (ID) and the length (L) are used as shown in Equation 6.1. For food safety, as discussed above, the worst-case scenario of the fastest moving fluid element or particle should be taken into consideration for the minimum residence time (t_{Res}) of the food in the hold tube. To estimate the minimum residence time of the food in the hold tube, it is necessary to know the following: length (m) and ID (m) of the hold tube and the maximum velocity (U_{max} (m/s)) of the fastest moving fluid element or particle (Equation 6.1). The maximum velocity is a function (ξ) of the average fluid velocity and depends on the flow behavior of the food, as it will be discussed in the following sections.

Ideally, fluids would move through a tube at the same velocity, in a shape called plug flow profile, as shown in Figure 6.1. In reality, as fluids are transported in tubes, the portion in contact with the wall will be subject to shear and move slower than the center, the difference between the velocities in different positions in the tube is called flow profile. The most well-known flow

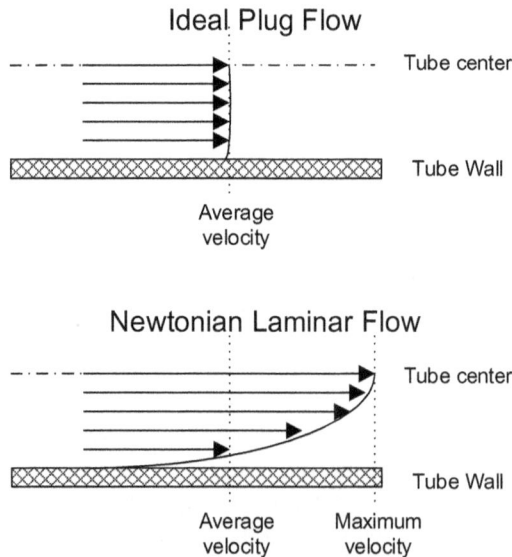

Figure 6.1   Plug and laminar flow comparison.

profile is a laminar flow profile when fluids are moving at a moderate speed, which is parabolic and the portions close to the center move faster than those in contact with the walls as shown in Figure 6.1 (Bird, Stewart Lightfoot, 2001).

The flow of a product in a complex process system is going to be found between these two extremes, with deviations from the ideal depending on the properties of the product, length of the system, types of heat exchangers, bends, expansion and contraction zones, etc.

These flow profiles depend heavily on the rheology of the materials, which is commonly summarized as viscosity. Fluids with high or complex viscosity and fluids containing particulates require more attention than low viscosity fluids which are very well defined in the literature. The viscosity of fluids and pastes needs to be characterized as part of the physical characteristics of the product, preferably in the whole range of temperatures under which they are processed.

Viscosity is defined as the internal resistance of a fluid to flow, and it is defined as the ratio of shear stress (σ) and shear rate ($\dot{\gamma}$) and has units of Pa.s or lb.s/ft^2. However, the cgs unit called centipoise (cp) is widely used, considering that $1 cp = 10^{-3}$ Pa.s. This definition of viscosity can only be applied to fluids that do not change as shear rate changes such as water, milk, or honey, which are called Newtonian fluids. Temperature has a strong effect on the viscosity, and in most fluids, viscosity decreases with temperature making fluids at high temperatures thinner and easier to transport in the tubes and equipment. Starches must have a special mention as their viscosity is greatly affected by temperature, during the cooking state as the solution will thicken when starches are gelatinizing and might break with excessive thermal treatment. Correlations for the temperature dependency of viscosity for a number of products are available. (Steffe, 1996)

Viscosity is measured using an apparatus called rheometer or viscometer. Several types of viscometers are available for liquids: tube type for low viscosity fluids, rotational viscometers which cover a wide range of viscosities, and oscillating viscometers for elastic products. Most of these apparatuses rely on the measurement of torque needed to maintain a certain velocity and small gaps between the cup and spindle. Rotational viscometers, such as the ones sold by Brookfield, Anton Paar, Ika, and other companies are common and provide excellent engineering data. In these viscometers, a spindle of a certain shape is submerged in the liquid and rotates at a given rotational speed (RPM); this speed corresponds to a determined shear rate and the resistance to rotation is measured and returned as viscosity. Other empirical methods exist, such as Bostwick, Ostwald viscometers, and pipe flow viscometers. Products with particulates are a challenge for the measurement and estimation of the rheological properties as the particles collide with the rotating spindle or are too large to fit in the gaps; the methods to determine such viscosity for engineering purposes will be discussed in Section 6.4.

Fluid behavior is categorized between ideal or Newtonian and real fluids (non-Newtonian), the differences between these are very important for process design and control as behaviors of flow, pressure drop, pump, and tube sizing must consider these differences.

6.2 NEWTONIAN FLUIDS

Newtonian fluids such as water, brine, solvents, vegetable oils, milk, and honey have been studied in depth. Viscosity (μ) is constant over the whole range of shear stresses (flow velocity) giving a straight-line stress to the strain equation (Equation 6.2).

$$\sigma = \mu\left(\dot{\gamma}\right) \tag{6.2}$$

The flow of Newtonian fluids through closed conduits (circular, rectangular) has been thoroughly studied in straight and curved pipes. It has been observed that at moderate flow rates, the flow

Newtonian Laminar Flow

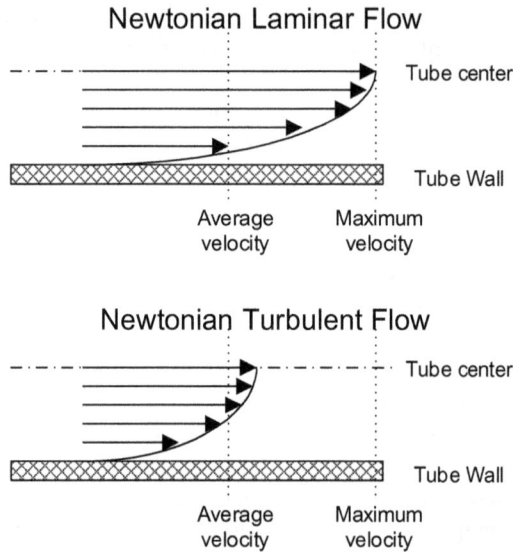

Tube center

Tube Wall

Average velocity

Maximum velocity

Newtonian Turbulent Flow

Tube center

Tube Wall

Average velocity

Maximum velocity

Figure 6.2   Laminar and turbulent flow profiles for Newtonian fluids.

is laminar in which parts of the fluid move in a well-ordered pattern of layers (Figure 6.1), and at high flow rates the flow becomes turbulent in which the fluid moves in a disorganized manner; with a transition between the two states. The flow regimes are determined by the Reynolds number which is the ratio of inertial over friction forces

$$N_{Re} = \frac{\text{Inertial forces}}{\text{Viscous forces}} = \frac{\bar{U}\rho}{\mu/D} \tag{6.3}$$

In straight sections, an N_{Re} less than 2,100 means that the flow profile is considered laminar, and an N_{Re} above 6,000 means that flow is turbulent or chaotic. Between laminar and turbulent flow, there is a transitional type flow, which is not well defined. For certain materials, such as starches and other hydrocolloids, it is recommended to use laminar flow up to $N_{Re} = 10,000$ to be conservative and provide allowances for the temperature-dependent changes in viscosity (Figure 6.2).

Flow in round tubes, known as Poiseuille flow in honor of the French researcher, under such conditions the maximum shear rate at the walls can be written as a function of the volumetric flow rate (Q) and the radius of the pipe (R) as shown in Equation 6.4

$$\dot{\gamma}_{max} = \frac{4Q}{\pi R^3} \tag{6.4}$$

Considering a tube of length L, which is flowing under laminar flow and at a uniform temperature in which a pressure drop ΔP is measured, Equation 6.5 shows the velocity profile in radial coordinates. Integrating this equation and considering the shear at the wall, it can be shown that the radial flow profile ($v(r)$) in the tube is a paraboloid (Equation 6.6)

$$U(r) = \frac{\Delta P}{4\mu L}\left(\frac{D^2}{4} - r^2\right) \tag{6.5}$$

$$\frac{U(r)}{\bar{U}} = 2\left[1 - \frac{r^2}{R^2}\right] \tag{6.6}$$

And thus, the maximum velocity is at the center and it's twice as fast as the average velocity $U_{max} = 2\bar{U}$. This result is important for the residence time of fluid elements, as the central elements will move faster (Steffe, 1996). FDA uses this as a correction factor for laminar flow (0.5), considering that the fastest portion of a fluid will have a residence time ½ of the average portion calculated as in Equation 6.1.

Pressure drop is also well defined based on the friction factor, which is based on the Reynolds number for smooth tubes as $f = 16/N_{Re}$ (round tube)

$$\frac{\Delta P}{\rho} = \frac{2 f L \bar{U}^2}{D} \tag{6.7}$$

A similar analysis can be carried out for the turbulent flow of Newtonian fluids such that

$$\frac{1}{\sqrt{f}} = -3.6 \log\left[\frac{6.9}{N_{Re}} + \left(\frac{\epsilon/D}{3.7}\right)^{10/9}\right] \tag{6.8}$$

And $u_{max} = U^+ U$, where $u^* = \bar{U}\sqrt{\dfrac{f}{2}}$ is known as friction velocity for the viscous sublayer

and $U^+ = 5.5 + (5.756)\log\left(\dfrac{y^+ V^* \rho}{\mu}\right)$ for the turbulent core.

This indicates that the maximum velocity is dependent on the flow velocity as well, as shown in Table 6.1. FDA recommends the use or a correction factor of 0.8 for turbulent flow conditions, which should be proven beyond doubt at sterilization temperatures. As can be observed in Table 1, 0.8 corresponds to $N_{Re} > 5{,}000$.

Flow in curved pipes such as coiled tubes, was studied by Dean (1927), and it was noted that due to the constant changes in direction a secondary flow was developed. Dean observed that the flow profile was flattened when compared to laminar flow and that two counter-rotating vortices were present in the cross-sectional flow as shown in three counter-rotating vortices (Coronel, 2001). This flow behavior resulted in better mixing, increased heat transfer coefficients, and increased pressure drop.

Due to this secondary flow, a modified Reynolds number, which takes into account the diameter of the coil (D) compared to the diameter of the tube (d), known as the Dean number.

$$N_{De} = N_{Re}\sqrt{d/D} \tag{6.9}$$

Dean number can be used for coiled tubes to determine the transition from laminar to turbulent flow, as noted by Ali (2001) and Coronel and Sandeep (2003), the laminar flow regime is stable to

TABLE 6.1 MAXIMUM VELOCITY FOR TURBULENT FLOW AS A FUNCTION OF REYNOLDS NUMBER

N_{Re}	<2100	4,000	10^4	10^5	10^8
$\bar{U}/U_{max}$	0.5	0.790	0.811	0.849	0.907

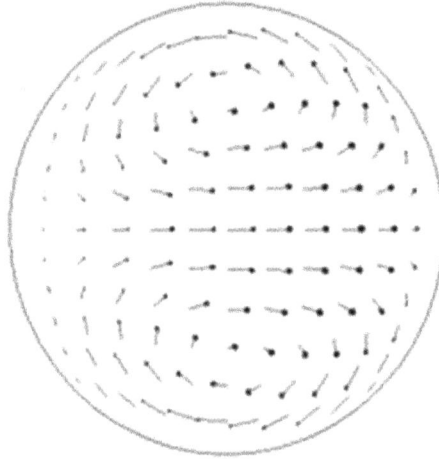

Figure 6.3   Secondary flow in curved tubes.

a larger N_{Re} depending on the curvature. The more pronounced the curvature the higher the critical Reynolds, as shown by Srinivasan (1968)

$$N_{Re(c)} = 2,100 \left[1 + 12 \left(d / D \right)^{1/2} \right] \tag{6.10}$$

For d/D of 0.14, $N_{Re(c)}$ is about 10,000 and if the tubes are coiled tighter I can even be higher.

Pressure drop can be calculated knowing that the friction factor for laminar flow was estimated by Adler (1934) for $200 < N_{Re} < N_{Re(c)}$ as follows:

$$f_c = 6.82 \, N_{Re}^{-0.5} \left(d / D \right)^{0.25} \tag{6.11}$$

The distribution of velocities in the axial direction is a field where more studies are needed. As shown in Figure 6.3, the maximum velocity is not in the center of the tube anymore and the ratio of V_{max} to the average velocity depends on the curvature and N_{Re}. For this type of tube, the correction factor needs to be determined experimentally.

6.3 NON-NEWTONIAN FLUIDS

Non-Newtonian fluids, which represent the majority of food products, have a resistance to flow which changes with shear rate changes, some fluids become thinner (pseudoplastic) while others become thicker (dilatant). Other fluids need some initial force to begin flowing (yield stress) and others change their rheology over time. Some fluids, such as starch suspensions, experience dramatic changes as they undergo processing, changing from one regime to another.

Several rheological models exist for time-independent fluids, and the most common are shown in Table 6.2

The Herschel–Bulkley model or power-law model encompasses several possible behaviors, from Newtonian to quasi-Newtonian to viscoelastic fluids. In this model of fluid, K is called consistency index and the exponent n is the flow behavior index. An important characteristic is the presence of the yield stress (σ_0), which represents a finite stress needed to achieve flow. Below

TABLE 6.2   TYPES OF COMMON FLUIDS

Newtonian	$\sigma = \mu(\dot{\gamma})$	
Bingham plastic	$\sigma = \sigma_0 + \mu'(\dot{\gamma})$	
Herschel–Bulkley	$\sigma = \sigma_0 + K(\dot{\gamma})^n$	
	Shear thinning (pseudoplastic)	$n < 1$ and $\sigma_0 = 0$
	Shear thickening (dilatant)	$n > 1$ and $\sigma_0 = 0$
Casson	$\sqrt{\sigma} = \sqrt{\sigma_0} + \sqrt{K(\dot{\gamma})}$	
Windhab	$\sigma = \sigma_0 + \eta_\infty(\dot{\gamma}) + (\sigma_1 - \sigma_0)\left[1 - \exp\left\{-\dfrac{\dot{\gamma}}{\ddot{\gamma}}\right\}\right]$	

Adapted from Steffe, 1996.

the yield stress, the material exhibits characteristics of a solid and doesn't flow freely, such as ketchup. The yield stress is a very practical value, is not easy to measure, and is well described by Hartnett and Hu (1989). Yield stress must be considered during the design of processes, and products that exhibit yield stress must be maintained in movement to prevent blockages in the process line or filler feeding line where flow can be very slow or in the aseptic hold tank if products are held stationary for a long time.

Fluids that have a yield stress present a critical radius (R_c), which is the radius above which the product will flow as a power-law fluid. Below the critical radius, a very thick layer of fluid can be present if not properly managed which will create an insulating layer and will make the product flow preferentially in the open part of the tubular section.

$$c = \sigma_0 \big/ \sigma_w = R_c \big/ R$$

$$Rc = \frac{\sigma_0 2L}{\delta P}$$

$$(6.12)$$

Viscosity is not as clearly defined as in Newtonian fluids and the apparent viscosity (η) is used. Apparent viscosity is the ratio of shear stress to shear rate and must include the yield stress and the fluid model used.

$$\eta = \frac{\sigma}{\dot{\gamma}}$$

$$(6.13)$$

which for a general Herschel–Bulkley model then becomes:

$$\eta = K(\dot{\gamma})^{n-1} + \frac{\sigma_0}{\dot{\gamma}}$$

$$(6.14)$$

In the case of a Newtonian fluid, apparent viscosity is equal to the viscosity ($n = 1$ and $\sigma_0 = 0$) over the range of shear rates while it will increase for fluids with $n > 1$ (shear thickening) and decrease for fluids with $n < 1$ (shear thinning) as shear rate increases. In non-Newtonian fluids, it is clear that the apparent viscosity is a function of the shear rate, and as such any viscosity measurements should also report the conditions under which they were measured (RPM).

Non-Newtonian fluids present challenges when calculating the N_{Re} due to the variable viscosity of them; however, it can be estimated for the Herschel–Bulkley model as

$$N_{Re} = \frac{U D}{\eta / \rho}$$

(6.15)

$$N_{Re} = \frac{D^n \bar{U}^{2-n}}{8^{n-1} K / \rho} \left(\frac{4n}{1+3n} \right)^n$$

Considering the viscosity variability, the flow profile in the laminar regime of a non-Newtonian fluid can't be assumed to be the same as a Newtonian fluid. For Herschel–Bulkley-type fluids, it can be observed that the value of the exponent n is a determinant of the flow profile, and Steffe (1996) showed that

$$U_{\text{max}} = \bar{U} \left(\frac{3n+1}{n+1} \right)$$

(6.16)

This is important for shear-thickening fluids ($n > 1$) as the factor of V_{max} can be larger than 2 and the fastest particle will have a very short residence time. Figure 6.4 shows a series of theoretical flow profiles for power-law fluids with different flow behavior index (n), the line corresponding to 1 is the Newtonian laminar profile.

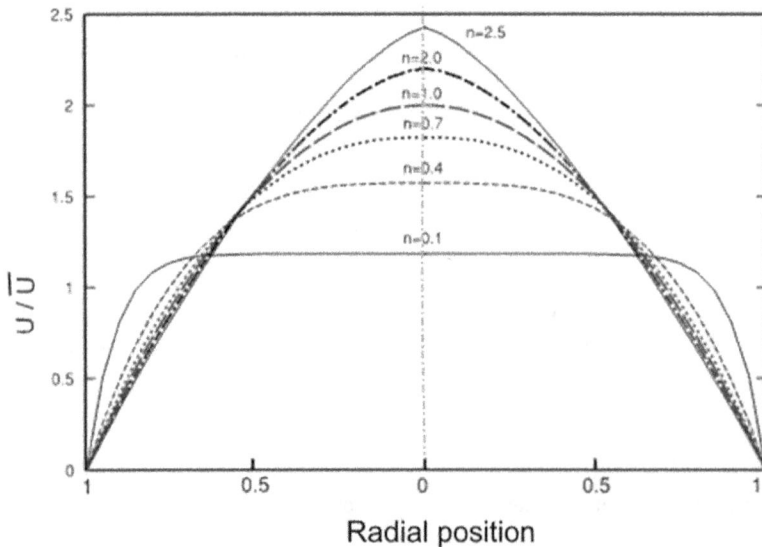

Figure 6.4   Flow profile for power-law fluids based on the flow behavior index (n).

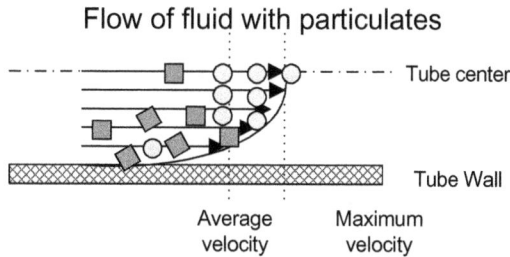

Figure 6.5   Particulate flow showing velocity of carrier fluid and various particle types.

6.4 FLOWS WITH PARTICULATES

The processing of flows with particulates presents many challenges to the designer and processor, among those the need to know how long particles stay in each of the sections of the process line, which together with the heat transfer from liquid to particles determines the thermal process that needs to be applied. Residence time distribution in flow with particulates should be part of the process establishment, as the fastest moving particle will determine the minimum thermal process and the average residence time of the particles will determine the quality loss of the particulates and carrier fluid.

The residence time in the holding tube is of great concern for determining the sterility of the product, and research has been pointed in this direction for both straight and coiled hold tubes under isothermal conditions. In the heating and cooling sections, the pressure drop and heat transfer seem to be more of a concern for researchers. However, both should be taken into account to balance the quality and safety of the final products.

Heterogeneous flows are comprised of two parts, a carrier fluid and a population of particles, of diverse sizes, compositions, and densities (Figure 6.5). Carrier fluid needs to have sufficient density and viscosity to convey the particulate flow and ensure that the suspension is maintained. Particles that stay a very short time in the system are a problem for sterility, but particles that stay a long time will have poor quality. Natural variations in the particles might be a concern in certain cases, but generally, pretreatments are carried out for such products.

The combination of both is part of the hydraulic conveyance; if particles are too dense, they may settle, and if they are too light, they will float at the top of the tubes, and in both cases they will be delayed by friction. Only particles flowing along the center flow line in the tube can be the fastest—their flow velocity is not reduced by friction with the walls—either bottom or top. To have this ability, their effective density will be close to the density of the carrier fluid at the operating temperature—or slightly lower in case the hold tube is inclined up.

Equation 6.17 shows an expression of hydraulic conveying defined as the energy needed to move suspension (carrier fluid **cf**) with solids (**s**) of concentration C moving at a velocity U in a tube of diameter D, where i is the hydraulic gradient measured as the difference in meters of a water column between two points. X and Y are empirical constants that need to be evaluated for each carrier fluid–particle combination.

$$\frac{i - i_{cf}}{C\, i_{cf}} = X \left[\frac{\bar{U}}{\sqrt{gD\left(\dfrac{\rho_s}{\rho_{cf}} - 1\right)}} \right]^{-Y} \tag{6.17}$$

Conveying only gives a hint on whether the particles can be carried by the fluid but doesn't paint the whole picture, and it becomes very complex in products such as soups or stews that have multiple types of particulates of different densities and sizes. A complete review of the process line must be done to ensure that the product will flow freely and that the solids will have little damage, especially at high temperatures; this review must include all the process lines, from pumps to heat exchangers, valves, and instrumentation. Attrition of particles not only clouds the carrier fluid but also leads to cubic particles with round edges which consumers might not appreciate as high quality. Improper design of product or process line can also cause jams of particles, which in turn can be disastrous. From empirical experience, long cylinders such as string beans or long prisms such as shredded carrots must be cut into smaller pieces to avoid blockage and promote flow (Allen, 1981; Kelder, 2009).

From an engineering and modeling point of view, the heterogeneous flows need to be modeled to calculate pressure drops, pumping requirements, and bulk heat transfer. However, rheology of heterogeneous flows is more difficult to measure than homogeneous flows, as the presence of particulates interferes with the conventional rheometers. The rheological properties of heterogeneous flow can be estimated using several methods:

- Pipe flow. Measuring the pressure drop between two points of a straight tube and approximating to one of the known fluid behavior correlations.
- Modified rheometers that can take the interference of particles into account. Inoue et al. (2013) used a modified viscometer using a technique from polymer science. They called it a ball measuring system (BMS).

Inoue et al. (2013) using the modified rheometer found that depending on the concentration of particles and their shape, the flow could be modeled using a power-law fluid. Interactions between particles slow down the ones that could be faster and help move the slower ones when the concentration is high enough, these interactions become dominant and the fluid–particle mix moves in a very similar way that a fluid would at high particle concentration the flow behavior index is small, and the flow is very close to a plug flow. Using a power-law fluid allows for an estimation of the average residence time for quality considerations, design of pressure drop, and heat transfer.

However, due to the variation in shape, consistency, and changes in rheological properties, and the challenge of measuring them at temperatures representative of the aseptic processing, it was recommended to have other methods for validation of the process. Especially, a validation of the worst-case scenario or fastest particle needs to be performed to confirm safety of the product.

Validation of the flow of particles has received attention from researchers for the last decades, studies in straight and coiled hold tubes (Sandeep & Zuritz, 1994, 1997) showed that RTD in helical tubes was narrower than in straight hold tubes, and that the RTD was normally distributed. The same authors ran simulations in Newtonian and non-Newtonian fluids and showed that an increase in specific gravity, tube diameter, or coil diameter resulted in an increase in the residence time of the particles, while an increase in the flow rate decreased the residence time of the particles. An increase in the particle diameter or the flow rate narrowed the residence time distribution (RTD) of the particles, while an increase in specific gravity or the tube diameter increased the RTD of the particles of non-Newtonian fluids (Sandeep et al., 2000).

Other researchers used hard plastic particles (spheres, cubes, cylinders) and high speed video to determine the residence time distribution of particulate flows in clear tubes, both straight (Simunovic, 1998) and coiled tubes (Palazoglu & Sandeep, 2004). These experiments are representative of uniform particle flows at uniform temperatures. The correlations developed can be modified by using an "equivalent diameter" as devised by Morikita et al. (1994). The concept of "equivalent diameter" for nonspherical particles, equates them to a sphere which can then be used to estimate the RTD of flows with a population of oddly shaped particulates, as well as allowing the validation of such complex flow using spherical simulated particles.

Further advances have been made, such as the use of simulated plastic particles with sensors that allow the measurement of residence time in portions of the system, injection and recovery methods and recoverable implants that carry a biological load, which ultimately can be used for a complete process validation as shown in *Appendix 6.*

6.4.1 Residence Time Distribution Measurement of Particulate Flows

Residence time distribution is one of the crucial inputs for the establishment and confirmation of the delivery of the appropriate thermal sterilization process to each and every element of the processed product. Process establishment requires that the "worst case" product segment, or the least treated part of the product, also referred to as the "cold spot," receive at least the minimal process required as established for the system.

For products containing larger particles such as low-acid soups or stews, fruit pieces in syrup, tomato products, etc., determination of the residence time distribution, and consequent determination of the worst-case/least treated (cold-spot) product segment, as well as residence time distribution, can be a very complex and labor-intensive procedure (Larkin, 1997; Shing and Morgan, 2010, Sandeep 1994).

6.4.2 Simulated Particle Design and Application for Conservative (Worst-Case or Cold-Spot Carrier) Validation

For several decades, intensive efforts have been dedicated to the development and implementation of both scientifically and commercially applicable methods and procedures to measure the residence time and residence time distribution of continuously thermally processed particulate products. A solution using simulated particles that carry implants has been developed by researchers at North Carolina State University and applied successfully in several particulate products.

In addition to the residence time measurement techniques and technology, numerous other system elements had to be developed, such as methods and materials used in their fabrication, design principles to achieve the desired properties needed for these particles, and a variety of sensing, real-time in-flow detectable and inert, magnetic, physical, chemical, microbial, enzymatic, or electronic tags and/or implants. Additionally, noninvasive real-time sensing, detection, display, and recording systems had to be developed.

The sensors are based on magnetic particle implants and magnetic field and flux detection, consisting of giant magneto-resistive sensors (GMR) sensor chip assemblies, signal amplifiers, data acquisition hardware, and custom-developed software for sensing, real-time displays, analyses, and digital recording. These particles can also carry microbial-based implants for full biovalidation of the process, as shown in Figure 6.6

When it comes to the real use of these particles in validation, such as the one presented in Appendix 6, there was a need to develop industrially applicable methods and devices for charging the simulated particles into the flow of product, detection systems, and capture at the end of the cooling section. Capturing particles at the end of the cooling section and before the pressure control devices or aseptic hold tank makes the process faster, otherwise the recovery might take hours. Once captured, the simulated particles can be opened, analyzed, and the implanted spore suspension load incubated and evaluated for growth/survival of thermo-resistant spores.

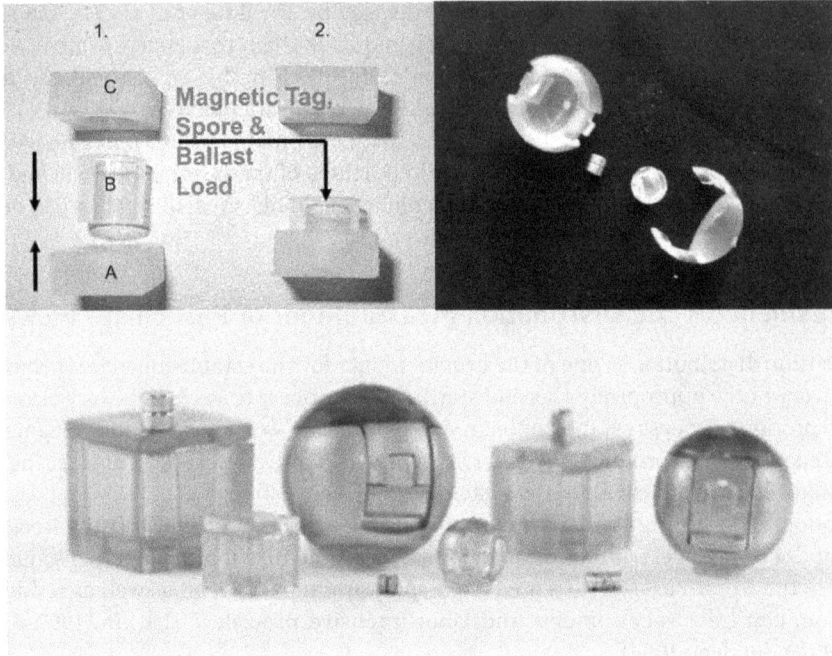

Figure 6.6 Simulated particle picture showing magnetic implants for RTD sensing and recoverable implants carrying a bioload for validation.

The following guideline shows the necessary steps for planning, design, and implementation of particle residence time measurements as developed by researchers. In real applications, several of these steps can take place concurrently, and the sequence can be adapted to one more appropriate for each unique system installation. Future improvements and additional functionality are expected, as well as improved strategies of implementation as this method is adopted by the industry.

Simulated particles are typically fabricated from a combination of thermoplastic polymers with several required and several desired properties. Required properties of the polymers are the ability to be machined, stamped, or injected into a desired particle shape, normally separated into a bottom and top part. Wall thickness can be engineered to match the thermal diffusivity of any product, leaving a hollow cavity in the center. The hollow cavity in the fabricated particle center needs to be large enough to contain and carry real-time detectable tags or post-process recoverable tags and sensing/recording implants. Another required property is the ability to withstand the sterilization temperature without melting or deformation, as well as without leaking carrier fluid into the assembled particles or leaking particle contents/tags/implants. That also defined the required ability of particle segments to be sealed by using a gasket, adhesive or their combination, and subsequently withstand the thermal and mechanical stresses of continuous flow conveyance and thermal sterilization. One of the desired properties is the polymer density which is lower than the density of water or the fluid carrier multiphase product at the temperature within the hold tube.

Based on these requirements, the most frequently used polymers for these applications (based on their application conditions) are polypropylene, polyethylene, and polymethyl-pentene / PMP, also known under its trademark TPX. Among these, PMP has the highest melting point tempera-ture and the lowest relative density compared to water (0.81) but also the highest price.

The main objective of the carrier particle design is to develop a particle with conservative properties in both the residence time and thermal diffusivity. The designer of the simulated/fabricated particles has several parameters to tune to adjust the effective particle density to the critical value. These are the particle shape, size, wall thickness, and the size of the cavity, as well as gasket or adhesive mass and the mass of implanted and sealed tags, implants, and optionally small inert weights (typically small glass beads) to adjust the total mass of the particle to the desired critical level.

As far as residence time, the objective is to design a particle with the highest likelihood of being the fastest when flowing within the real product flow. This is achieved by selecting an appropriate particle shape (typically a sphere since this shape will encounter the least collisions and flow obstructions on its way through the processing system) and adjusting its effective den-sity to a value known as the "critical density." Critical density is defined as the particle density or the particle density range statistically most likely to contain the fastest moving (worst case from the perspective of food safety) individual particle.

Critical density enables the particle to flow faster than the surrounding and competing par-ticles with other density characteristics. If the flow is perfectly horizontal, then a neutral density, i.e., the density of the particle identical to the density of the carrier fluid at the desired (hold tube) temperature will provide the properties necessary for the fastest flow. Since hold tubes are by design and regulations sloped upward, to prevent the flow of non-sterilized material forward in case of system failures, then the realistic critical density value for real processing systems is slightly lower than the density of the carrier fluid at hold tube temperature. This density property can be designed and implemented in particle design as has been done for all process validation filings using the system described.

Figure 6.7 illustrates the definition of the "critical density" property as well as the ability of the particles with critical density characteristics (0.98 to 0.99 relative to the carrier fluid) to achieve up to 5% higher velocity averaged over 27 runs under different flow rates and particle loading levels compared to the neutrally buoyant particles (relative density 1.0 compared to the carrier fluid at hold tube temperature, and up to 10% higher velocity than the heavier/denser particles flowing along the bottom of the tube (Simunovic, 1998).

While the critical density will establish the conservative (fastest moving) particle relative to flow, the conservative properties relative to heat penetration to the contained sensing implants need to also be engineered. That way the particle has both the fastest moving and slowest heat-ing properties which allows it to be used as a worst-case particle or the "cold-spot carrier." This thermally conservative property can be designed by numeric modeling using the conductivity and density characteristics of the polymers and confirmed experimentally. Figure 6.8. Illustrates the basic steps of that procedure.

Thermophysical properties of the ingredients and carrier fluid of the product must be known or estimated to a high degree of accuracy. Preferably, these properties should be measured at a range of temperatures, that includes the highest temperature achieved in the last heater (before the hold tube). When advanced thermal technologies, like continuous flow microwave heating or ohmic heating, are used it is also needed to determine other relevant properties like dielectric properties and electric conductivity. Verification of the conservative design of the simulated particles is carried out by comparing the penetration of heat toward the center of simulated particles when compared to real food particles. A similar approximation as the classical transient heat transfer by conduction developed by Heisler is used to determine the conservative nature of the simulated particle.

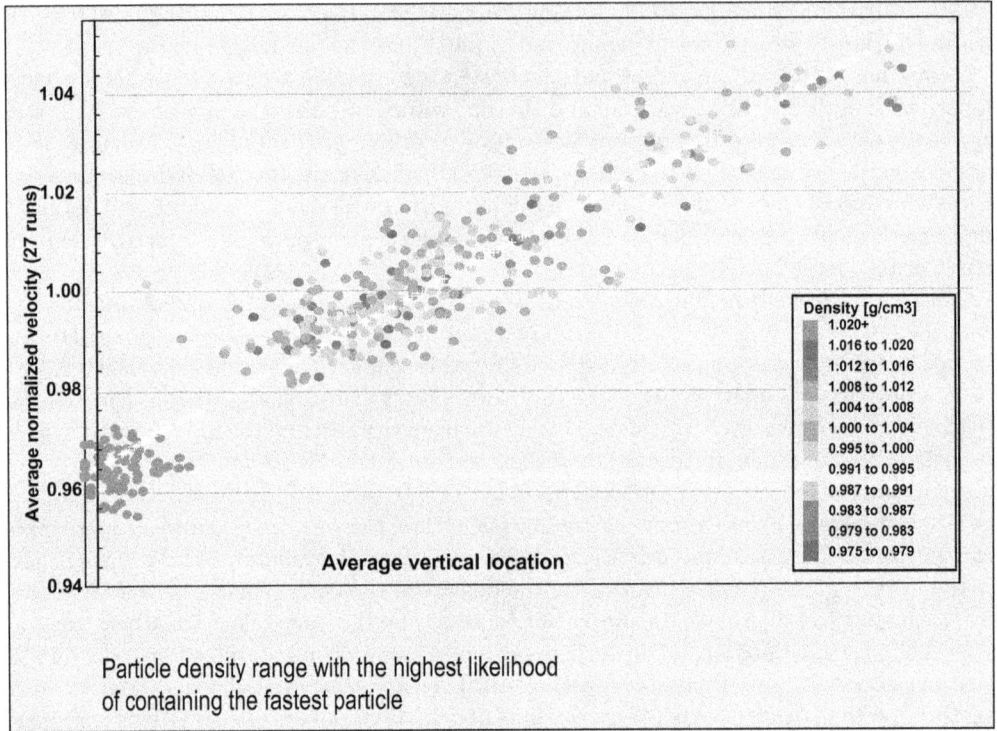

Figure 6.7   Definition of critical density and normalized velocity properties of different density ranges.

Magnetic tags are used to measure the residence time of each particle, and such magnets need to be small and light enough to not negatively affect the conservative flow and thermal characteristics of the assembled particles, but must have a strong enough magnetic field to be detected. All of these requirements are fulfilled by the NdFeB magnets (Neodymium-Iron-Boron) which can withstand temperature levels of up to 150°C without degradation of magnetic field strength and enable detection of implants as small as 0.1 g under realistic processing conditions and robust conservative carrier particle characteristics.

6.4.2.1 Sensors, Data Capture, and Analysis for RTD Calculations

Sensors to determine the residence time of the particles have been developed so that they are noninvasive and can be installed temporarily on the external surfaces of the process tubes, and can be removed after the validation is finished. These sensors are connected to amplifiers which send the signal to a data acquisition system that monitors and analyzes in real time the residence time of each particle. A diagram of this installation is shown in Figure 6.9, including insertion and capture devices.

Insertion of the fabricated particles into a fully operating processing system is a challenge, insertion must not disturb the flow of product, and must overcome the high pressure that is present after the metering pump. Figure 6.10 presents a simple system using three automatic valves controlled with a PLC that allows automated injection of particles in the process line. The valve open and closed states and the implemented flow paths of the product during the insertion and regular operation are illustrated.

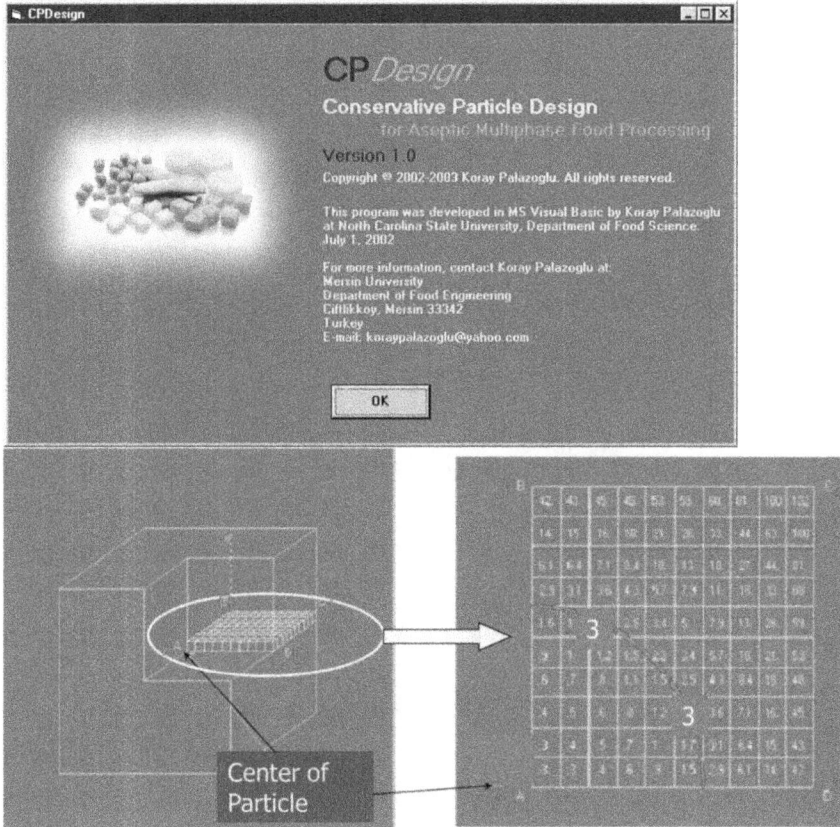

Figure 6.8 Design principle and modeling of conservative thermal properties of fabricated particles.

Each particle to be inserted in the system has magnetic implants, is marked with a unique code that is recorded at the time of insertion, and if particles are inserted at long enough intervals, the residence time of each particle can be monitored by receiving inputs from each of the sensors along the process line. By having the time difference between each sensing station, the residence time per section can be measured for each individual simulated particle. This data, together with the PLC temperature recording of the carrier fluid temperature allows for a detailed reconstruction of the thermal history of each particle. If a statistically significant number of particles are inserted and sensed, both RTD and microbial reduction can be measured and used to establish and later validate the thermal process.

Recovery of the particles is another challenge, particles are fabricated from hard plastics, and can cause problems in the back pressure devices, booster pumps, and filling nozzles. Also, if biological loads are used, all particles need to be accounted for to prevent the possibility of contamination by a bioload that was not flushed out of the system. Previous validation attempts have used alginate cubes (marked by color) that had a bioload but no sensing capabilities, the recovery of such particles from the test product was a labor-intensive and time-consuming activity. The use of magnets is advantageous, as magnetic traps can be used to capture the particles before they reach the aseptic surge tank or the filler, thus simplifying the validation process as shown in Figure 6.11. The use of parallel or in-series magnet traps increases the possibility of a complete recovery of all inserted particles.

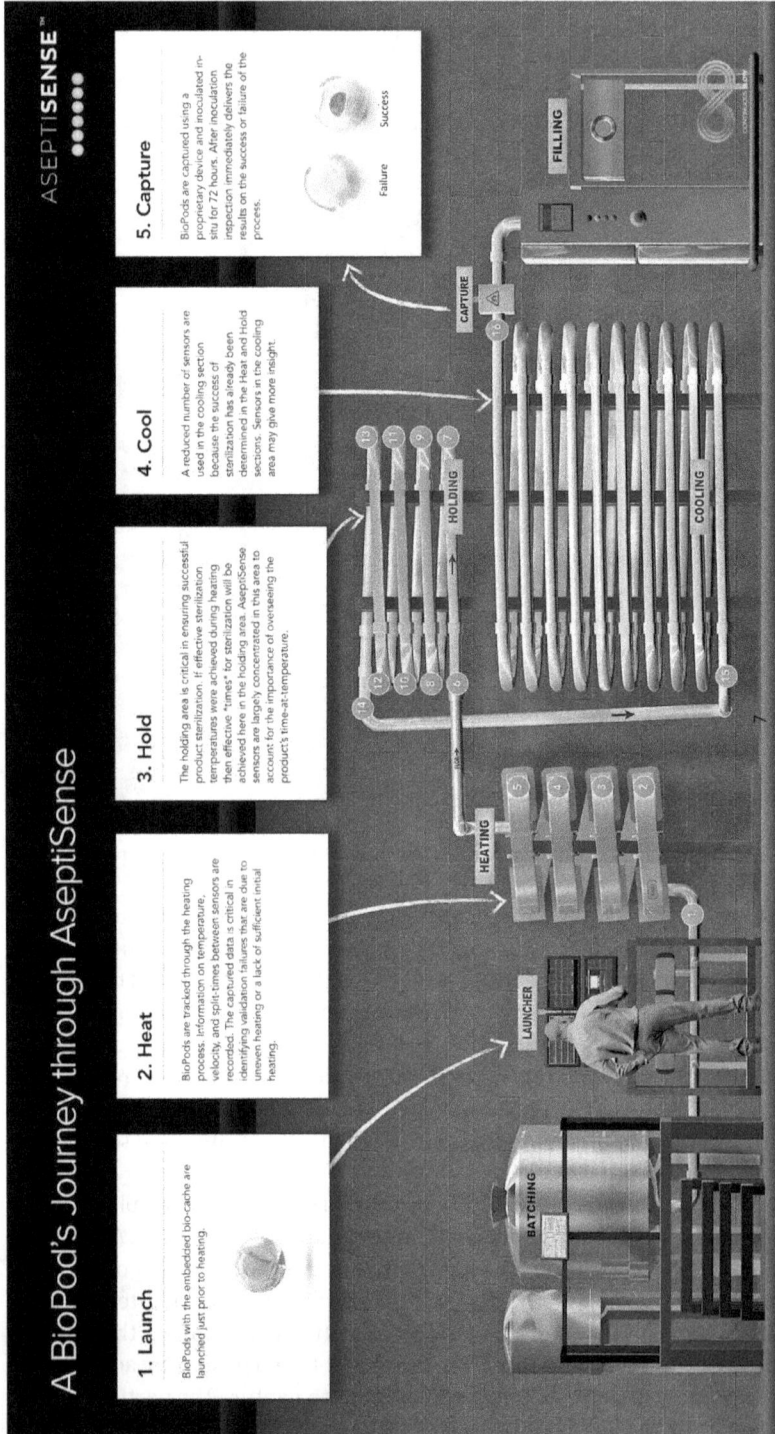

Figure 6.9   Sequence of insertion, sensing and capture, followed by the optional incubation and analysis of fabricated particles.

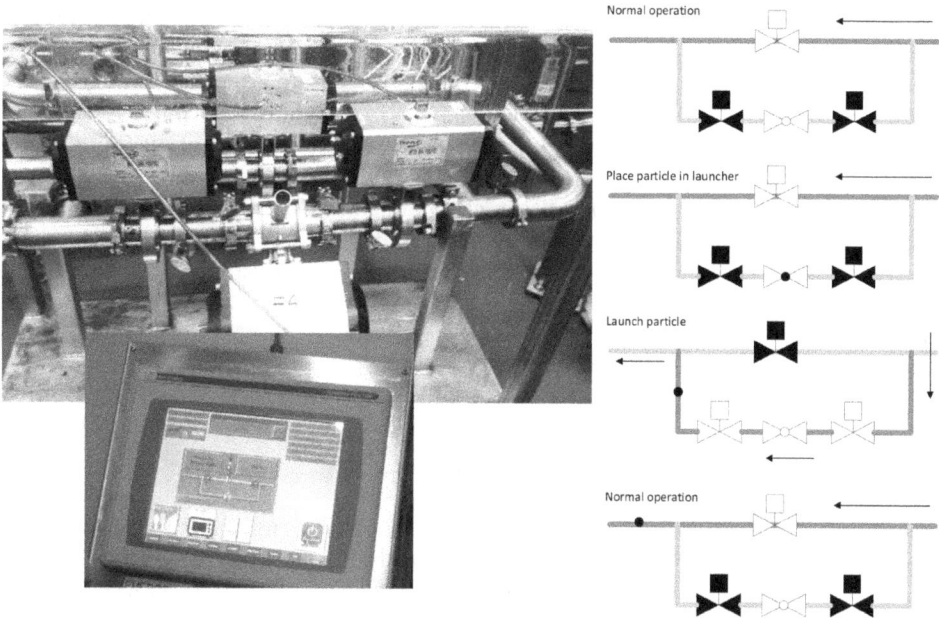

Figure 6.10   Particle insertion device and sequence of operations.

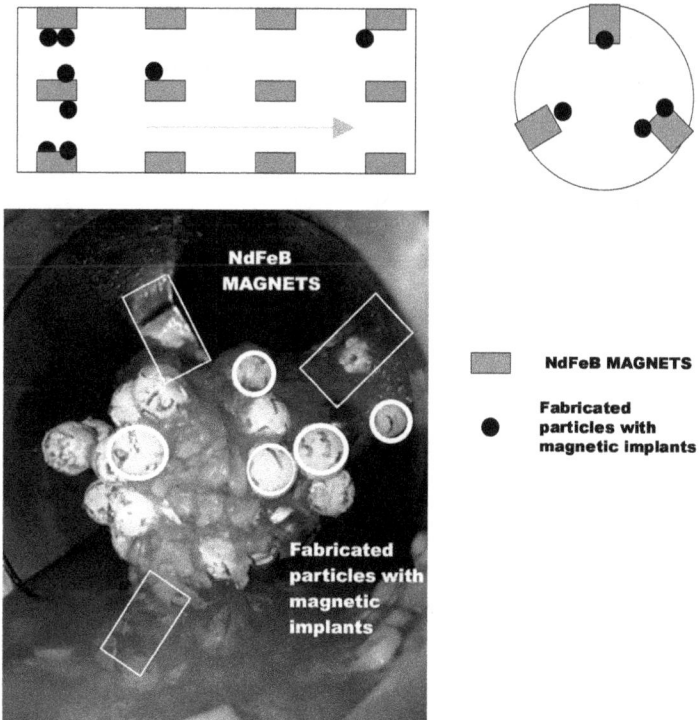

NdFeB MAGNETS

Fabricated
particles with
magnetic implants

Figure 6.11   Magnetic trap used to recover fabricated particles with magnetic implants. Design scheme and trial results.

Data analysis of the residence time data is important, and the ability to produce digital files of the sensor detection events along the process flow path permits a convenient replay of the events and residence time can be analyzed offline, providing insights and information on the process. By doing this, the flow characteristics and temperature of each individual particle can be correlated to reconstruct the thermal history of each particle and fine-tuning of the thermal process can also be achieved.

Implants containing spores or surrogate microorganisms must be recovered and analyzed. It is recommended to use a known initial population of spores or concentration of enzymes (Tucker and Heydon 1998; Tucker et al., 2002), with well-defined thermal resistance. Microbial reduction can then be calculated by incubating and counting the surviving microorganisms or residual enzymatic activity. The use of pH-sensitive spore suspension media can give a visual detection of any surviving spore after incubation.

6.5 CONCLUDING REMARKS

Residence time and flow velocity are very important parameters for the establishment of aseptic processing. Flow of homogeneous fluids needs to be fully understood before establishing the flow, to ensure that the process is conservative and that the fastest portion of the product is processed properly.

FDA recommends using correction factors of 0.5 for laminar flow, and 0.8 for turbulent flow based on Newtonian flow. These factors are a worst case for most real products which are non-Newtonian and flow through bends, expansions, contractions, valves, and other equipment.

RTD of particulate flows is a challenge for the validation of aseptic products with large particulates, methodologies such as the one presented in this chapter are available and are gradually being implemented by the industry.

Several gaps still exist in the understanding of flow profiles and residence time distribution for non-Newtonian flows and particulate flows, which should be explored to make the modeling and validation of aseptic processes more feasible.

NOMENCLATURE

Latin Letters

$\dot{m}$:   Mass flow rate (kg/s)
A: Surface area (m²)
C: Concentration of solids (%)
D or d: Diameter (m)
D_c: Diameter of curvature of coil-shaped tube
D_i: Coil-shaped inner tube diameter (m)
i: Hydraulic gradient (Pa or m of water)
ID: Internal diameter of a tube (m)
K: Consistency index
L: Length (m)
N: flow behavior index
N_{De}:   Dean number

N_{Re}:	Reynolds number
P:	pressure (Pa)
Q:	Volumetric flow rate (m³/s)
r:	Radius (m)
R:	Radius of the tube (m)
T:	Temperature (K)
$\bar{u}$:	average flow velocity (m/s)
U:	Flow velocity (m/s)
U_{max}:	Maximum flow velocity (m/s)

Greek Letters

γ:	Shear strain (s⁻¹)
ΔP:	Pressure difference (Pa)
ΔT:	Temperature difference (°C)
η:	Apparent viscosity (Pa.s)
ρ:	Density (kg/m³)
σ:	Shear stress (Pa)

Subscripts

c:	Refers to a critical parameter
cf:	Refers to carrier fluid
in:	Refers to the inlet of the heat exchanger or the mixing unit
max:	Maximum
s:	Refers to solid particles

REFERENCES

Adler, M., 1934. Strömung in gekrümmten Rohren. *ZAMM-Journal of Applied Mathematics and Mechanics/Zeitschrift für Angewandte Mathematik und Mechanik*, 14(5), 257–275.

Ali, S., 2001. Pressure drop correlations for flow through regular helical coil tubes. Fluid dynamics research, 28(4), 295.

Allen, T., 1981. *Particle Size Measurement*. Chapman & Hall.

Bird, R.B., Stewart, W.E., and Lightfoot, E.N. 2001. *Transport Phenomena*, Vol. 1. 2nd edition. John Wiley & Sons.

Coronel, P., 2001. Pressure drop and heat transfer coefficients in helical heat exchangers (Masters thesis, North Carolina State University).

Coronel, P. and Sandeep, K.P. 2003. PRESSURE DROP and FRICTION FACTOR IN HELICAL HEAT EXCHANGERS UNDER NONISOTHERMAL and TURBULENT FLOW CONDITIONS 1. *Journal of food process engineering*, 26(3), 285–302.

Dean, W.R., 1927. XVI. Note on the motion of fluid in a curved pipe. *The London, Edinburgh, and Dublin Philosophical Magazine and Journal of Science*, 4(20), 208–223.

Hartnett, J.P., and Hu, R.Y. 1989. The yield stress: An engineering reality. *J. Rheol.*33(4):671–679.

Inoue, C., Versluis, P., Coronel, P. and Elberse, J.M.M., 2013. Effect of Particle Phase Volume, Shape and Liquid Phase Concentrations on Rheological Properties of Large Particulate-Liquid Model Food Systems by Using Ball Measuring System. *Journal of Chemistry and Chemical Engineering*, 7(7), 643.

Kelder, JDH, Coronel, P., & Bongers, PMM (2009). Aseptic processing of liquid foods containing solid particles. In R. Simpson (Ed.), *Engineering aspects of thermal food processing* (pp. 49–72).(Contemporary Food Engineering). CRC Press.

Larkin, J.W., 1997. Workshop targets continuous multiphase aseptic processing of foods. Food technology (USA).

Morikita, H., Hishida, K., and Maeda, M. 1994. Simultaneous measurement of velocity and equivalent diameter on non-spherical particles. *Part. Part. Syst. Char.* 11(3):227–234.

Palazoglu, T.K., and Sandeep, K.P. 2004. Effect of tube curvature ratio on the residence time distribution of multiple particles in helical tubes. *LWT-Food Sci. Technol.* 37(4):387–393.

Sandeep, K.P., and Zuritz, C.A. 1994. Residence time distribution of multiple particles in non-Newtonian holding tube flow: Statistical analysis. *J. Food Sci.* 59(6):1314–1317.

Sandeep, K.P., and Zuritz, C.A. 1999. Secondary flow and residence time distribution in food processing holding tubes with bends. *J. Food Sci.* 64(6):941–945.

Sandeep, K.P., Zuritz, C.A., and Puri, V.M. 1997. Residence time distribution of particles during two-phase non-newtonian flow in conventional as compared with helical holding tubes. *J. Food Sci.* 62(4):647–652.

Sandeep, K.P., Zuritz, C.A., and Puri, V.M. 2000. Modelling non-Newtonian two-phase flow in conventional and helical-holding tubes. *Int. J. Food Sci. Technol.* 35(5):511–522.

Shing, R.K., and Morgan, M.T. 2010. Residence time distribution in aseptic processing. In *Principles of Aseptic Processing and Packaging*, Chapter 2. Purdue University Press.

Simunovic, J. 1998 *Particle flow monitoring in multiphase aseptic systems*. NC State University PhD Thesis.

Simunovic, J., Swartzel, K., Farkas, B., and Adams, J. 1995. Particle residence time measurement of single and multiple solid phase particulate products in clear aseptic holding tube segments. In Proceedings of the International Symposium on Advances in Aseptic Processing and Packaging Technologies.

Steffe, J.F. 1996. *Rheological Methods in Food Process Engineering*. Freeman Press.

Tucker, G., and Heydon, C. 1998. Food particle residence time measurement for the design of commercial tubular heat exchangers suitable for processing suspensions of solids in liquids. *Food Bioprod. Process.* 76(4):208–216.

Tucker, G.S., Lambourne, T., Adams, J.B., and Lach, A. 2002. Application of a biochemical time–temperature integrator to estimate pasteurisation values in continuous food processes. *Innovative Food Science Emerg. Technol.* 3(2):165–174.

Chapter 7

Thermal Process and Optimization of Aseptic Processes Containing Solid Particulates

Pablo M. Coronel, Jairus R. D. David, and Peter M.M. Bongers

CONTENTS

DOI: 10.1201/9781003158653-9

7.1 INTRODUCTION

Optimization of aseptic processing is required for a successful and profitable commercial operation (Richardson, 2004). Aseptic products show better organoleptic and nutritional quality than their retorted counterparts due to their shortest exposure to high temperatures. However, aseptic processing plants require a higher capital investment, an operation is more complex and due to the regulations under which it must operate, the required level of skilled workers, quality control, and documentation are generally higher. Taking into account the constraints of safety and regulations, optimization of aseptic processes has both product quality and cost as the two main objectives.

Quality, as perceived by the consumer, should be as high as possible for the price point of the product, quality is especially relevant for the differentiation of products with particulates (Sastry & Cornelius, 2002). Optimization of product quality starts at product development and includes all stages of the supply chain: ingredient selection, transport, storage, handling, process, quality assurance, operator training, etc.

Thermal process is the core of aseptic processing and, as discussed in Chapter 4, the impact of temperature and time on quality is very important. It is necessary to have a discussion on the fundamentals of thermal optimization. Both for homogeneous products and products with particulates. Heat transfer to the liquid will be discussed, starting from a theoretical case of perfect heat transfer to fluid and particles, with recommendations for all the stages of aseptic processes.

The cost of aseptic products, both on the ingredient and conversion, is the second dimension to be optimized. Depending on the price point of the final product, the specifications of ingredients must be written based on the needed quality and their price will depend on market conditions. Ingredient costs might benefit from global sourcing of non-perishable ingredients or the use of frozen seasonal ingredients; frozen ingredients incur high energy costs for storage at low temperatures. It might also be possible to use ingredients that are already aseptically processed and packaged. Bulk aseptic products, such as vegetable purees, tomato pieces, and vegetable pieces, could be processed close to the fields and packaged in large totes for later use, with the advantage of eliminating energy costs of freezing and providing uniform quality.

On the production side, optimization should address efficiency and productivity, with constraints of food safety. Production line and packaging downtime carry penalties in capital utilization as well as product, packaging, and man-hour losses. Thus, extending the run-length, optimizing process control and batch scheduling should be addressed, as they offer improved efficiency without compromising quality. This chapter discusses factors driving both quality and cost optimization and proposes strategies to integrate them into production, keeping in mind, however, that commercial production of aseptic products requires a balance between product quality, safety, and cost which can't be overlooked.

7.2 OPTIMIZATION OF THERMAL PROCESSES

Aseptic processing relies on thermal treatment for the preservation of the products, thermal treatment requires both heating and cooling of products at high rates. Thermal process makes the process energy intensive and requires advanced control strategies for optimization.

Thermal processing uses temperature to kill microorganisms but at the same time affects the quality of final products. This section aims to discuss different approaches on how to optimize thermal processes, with a short review of the definition of quality and safety as affected by the thermal process, and a review of high-temperature, short time (HTST) axioms leading to a theoretical example of optimization for food with particulates.

7.2.1 Thermal Process of Homogeneous Aseptic Products

Homogeneous aseptic products have been available for a long time, and they are mostly produced using thermal treatments. With this process, food is subject to a sufficient combination of temperature and time to destroy pathogenic or spoilage-causing microorganisms, antinutrients such as antitrypsin and lectins, and enzyme systems that cause degradation in the food. A review of the theories and methods to determine a sufficient heat treatment was given by Stumbo (1973). Routinely, low-acid foods are canned or retorted to ensure the destruction of *Clostridium botulinum* spores as the most dangerous pathogen, as well as thermophilic spores that cause spoilage, by heating so that the center of the containerized product receives a heat treatment of 3 to 6 minutes at a temperature of 250°F or an equivalent sterilizing process at another time and temperature combination. The resultant product in hermetically sealed containers is shelf-stable under normal conditions of storage and handling for up to 2 years ("Canned Food" 2007).

When a homogeneous population of viable spores is subjected to a constant lethal temperature, T, the rate of death of spores generally follows a first-order reaction in which the reactant, N (viable spores), decreases exponentially or logarithmically with time, t, and is expressed as

$$\frac{dN}{N} = -k_T \, dt \tag{7.1}$$

This can be manipulated easily for a time interval (Δt) with a known initial population (N_0) and transformed to common logarithms (to base 10) to obtain

$$\ln\left(\frac{N}{No}\right) = -k_T \Delta t$$

$$\log_{10}\left(\frac{N}{No}\right) = \frac{-k_T}{2.303} \Delta t \tag{7.2}$$

When the reduction of population is 1 log cycle (1/10) or 90% of microorganisms, the time needed for such reduction is:

$$\frac{2.303}{k_T} = D_T \tag{7.3}$$

where $D_{\mathrm{T}} = 2.303/k_{\mathrm{T}}$. Thus, a plot of the logarithm of the concentration N of surviving spores versus time is a straight line called a time–survivor curve.

The heat resistance parameter (D_{T}) is a characteristic of the type of spore, pH, and water activity (a_{W}) of the suspending menstruum, bacteriological recovery medium, and laboratory protocol and procedures under consideration, and is found experimentally to vary with temperature according to

$$D_{\mathrm{T}} = D_{\mathrm{Tref}} \cdot 10^{(T_{\mathrm{ref}}-T)/z} \tag{7.4}$$

where D_{T} is the D value at temperature T, T_{ref} is an arbitrary reference temperature, D_{Tref} is the D value at T_{ref}, and z is a temperature dependence characteristic of the microorganism, assumed to be constant over normal processing conditions. The higher the value of z, the less significant the changes in D value with changes in T.

Consider a homogeneous or purified population of viable spores suspended in a hermetically sealed can of food, which is heated and cooled in a retort. The temperature at a given spatial position in the can is a function of time, $T(t)$. The change in spore concentration at that position is found by substituting Equation 7.4 into Equation 7.3. Integrating between terms t_i and t_f when spore concentrations are N_i and N_f, respectively, yields

$$-\int_{ti}^{tf} (\log_{10} N) = \frac{1}{D_{Tref}} \int_{ti}^{tf} \frac{dt}{10^{\frac{(Tref-T(t))}{z}}} \tag{7.5}$$

Evaluating the integral results in the F value or the equivalent thermal treatment in minutes at T_{ref} for the specific microorganism

$$F_{ref}^{z} = D_{Tref}\left(\log N_i - \log N_f\right) = \int_{t_i}^{t_f} \frac{dt}{10^{\frac{Tref-T(t)}{z}}} \tag{7.6}$$

The expression on the left gives the relationship between F and the change in concentration of spores in food, as a result of heating and cooling. N_i is the initial microbial load of pathogenic mesophilic spores of public health significance or other mesophilic and thermophilic spore bioburden in the food established experimentally. N_f is a "safe" probabilistic final concentration of spores established from public health or economic spoilage rate considerations. This forms the basis for the definition of a required minimum F value. Containers and lids also need to be subject to a similar process in which the microbial load is reduced to a safe level.

$$\left(F_{Tref}^{z}\right)_{required} = F_{required} = D_{Tref}\left(\log N_i - \log N_f\right) \tag{7.7}$$

The right-hand side of Equation 7.6 relates F to the time–temperature of the heating and cooling process $T(t)$ experienced by the spores between times t_i and t_f. This forms the basis for the definition of a process F value.

$$\left(F_{Tref}^{z}\right)_{process} = F_{process} = \int_{t_i}^{t_f} \frac{1}{10^{\frac{T_{ref}-T(t)}{z}}} \cdot dt$$

$$\text{Or} \qquad F_{process} = \int_{t_i}^{t_f} 10^{\frac{T(t)-T_{ref}}{z}} \cdot dt \tag{7.8}$$

The subscript T_{ref} on F in Equations 7.7 and 7.8 indicates that the entire integrated time–temperature effect on the spores is equivalent to the time F minutes at the single temperature T_{ref}. The superscript z emphasizes that only one type of spore is considered.

Given the importance of *C. botulinum* for public health and food safety, the subscript F_0 has been given for the processes designed for the destruction of such microorganisms. For this microorganism, $T_{ref} = 250°F$ (121.1°C) and $z = 18 F_o$ (10°C). It is also known that every minute of F_0 results in the inactivation of 10^4 *C. botulinum* spores, which explains the minimum requirement of $F_0 > 3$ to inactivate at least 10^{12} (12D) spores.

Other common reference F values are $T_{ref} = 200°F$ (93.3C) and $z = 16 F_o$ (9°C) for acid foods, $T_{ref} = 212°F$ (100°C), and $z = 16 F_o$ (9°C), for intermediate or acidified products.

Determination of the required F value for a given thermal process is the job of a thermal process authority and must take into account both safety and economic spoilage. This is the most crucial step in the design of any aseptic process. In principle, the most thermoresistant microorganisms should be identified, D_T, T_{ref}, and z can be obtained from literature, a maximum initial load (N_i) and maximal safe final level, or probability of nonsterile units (N_f) for a large production lot estimated. The latter defines the commercial sterility.

The commercial sterility of thermally processed food refers to the absence of disease-causing microorganisms, the absence of toxic substances, and the absence of spoilage-causing microorganisms capable of multiplication under a variety of nonrefrigerated conditions of storage and distribution (Downes & Ito, 2001). Shown in the following equation is the concept of 12D requirement for *C. botulinum* or the "12D Bot cook":

$$F_{required} = D_{Tref}\left(\log 10^{12} - \log 10^0\right) = 12D \tag{7.9}$$

If an existing process has proven to be safe with a certain time–temperature combination, this one can be adapted by knowing the z-value of the most thermoresistant microorganism. The tables of the kinetic factors are available in literature such as Holdsworth (1992) and Nelson (2010). However, a database of kinetic values (D, T_{ref}, z) has been made public by the Institute for Food Technology (ILT.NRW) at the OWL University of Applied Sciences and Arts called the "Lemgo D- and z-value Database for food" https://www.th-owl.de/fb4/ldzbase/.

Some nonpathogenic mesophilic spore-forming bacteria are more resistant to thermal processing than *C. botulinum*, while this is not a food safety risk, it can cause economical losses due to damages to products, spoiled cans, and thus is commonly known as economic spoilage. In this case, it is necessary to use these microorganisms as a basis for the process. These microorganisms are characterized by a D_{250} of about 1 minute and a 12D process would result in a F_o of 12 minutes. A thermal process of F_o of 12 minutes would result in a product of poor quality due to the many heat-induced physical and chemical changes. Many products require a 5D value with a resultant F_o of 5 to 6 minutes (Lund, 2003).

Another factor that must be taken into account is that when shelf-stable foods are stored at high temperatures (100°F–125°F) in tropical or desert areas, where nonpathogenic thermophilic aerobes and anaerobes can cause economic spoilage, the designed thermal process must be extremely severe. The $D_{250°F}$ value for *Bacillus stearothermophilus* and *Clostridium thermosaccharolyticum* is about 4 minutes, and thus a 5D process would result in a F_o of 20 minutes.

When the 5D process is used, it is important to verify that the F_o is at least equivalent to a 12D process based on the most heat-resistant spore-forming pathogen presumed present in the food.

The lethality of a process is the ratio of the F value of the actual process to the F value required for commercial sterility, and the product $F_{required} \times 10^{(Tref-T(t))/z}$ is called the thermal death time, the time required "to destroy all the mesophilic spore-forming microorganisms capable of spoiling the food."

$$\text{Lethality} = L = \frac{\left(F_{T_{ref}}^z\right)_{process}}{\left(F_{T_{ref}}^z\right)_{required}} \tag{7.10}$$

$$= \frac{1}{\left(F_{T_{ref}}^z\right)_{required}} \int_{t_i}^{t_f} \frac{dt}{10^{\frac{T_{ref}-T(t)}{z}}}$$

$$\text{Or} \quad L = \int_{t_i}^{t_f} \frac{dt}{TDT} \tag{7.11}$$

Obviously, the lethality must be at least unity for commercial sterility (Merson & Leonard, 1979). Prevention of postprocess microbial recontamination and maintenance of container hermetic seal are integral to assurance of commercial sterility and shelf life of thermally processed foods.

While most of the initial research was performed in canned foods, in which the product is contained, it can be applied to continuous thermal processes, in which the food is heated and cooled in closed systems using the same set of equations. Calculation of the required F value is the work of a process authority and it involves a deep understanding of the process and the product.

While thermal treatment of the food in a closed system is a process that starts as soon as the food begins heating until it's finished cooling, a conservative estimate must be used and thus the thermal process is accumulated only from the heat exchangers to the hold section until the beginning of cooling. In some cases, thermal process lethality is only claimed in the hold tube, neglecting the effects of temperature in the heating and cooling section. Unless temperature can be observed in several locations in the hold tube, it must be assumed that the temperature at the end of the hold tube must be used and a correction factor based on the flow characteristics of the food in the hold tube applied to the residence time. Laminar flow uses a 0.5 correction factor, while turbulent or non-Newtonian flows use other correction factors.

Credit for the heating section can only be taken if this section is long, or has several steps that are well defined and documented. Changes in incoming product temperature, heating media flow, and fouling can affect the designed thermal process. The accumulated lethality during cooling is generally not claimed in the process, as rapid cooling of different parts of the product might end the thermal process prematurely.

7.2.2 Quality of Aseptic Products

Aseptic process implies the use of thermal processes to reduce the number of microorganisms to achieve commercial sterility and shelf stability, depending on the product and storage conditions the thermal treatment applied can be more or less intense.

The quality of any food product is defined by how closely it satisfies the expectations of consumers during consumption. While factors such as marketing and packaging play important roles in the initial purchase of a product and its positioning, in this chapter quality is limited to intrinsic factors to the product. Quantification of consumer perception and quality is very challenging, as only a few of the parameters can somehow be quantified. Product evaluation involves extensive sensorial analyses where sensorial parameters are scored by trained or untrained panels and data is then broken down and analyzed. However, due to the multi-sensorial experience that is related to the consumption of food, cultural, sensorial, environmental, or even age factors have an influence on consumer acceptance so that high scores on any or some of the sensorial dimensions do not necessarily reflect consumer preference.

For optimization of the thermal process, we can only concentrate on measurable quality parameters that can be linked to the thermal history of the product and be used as indicators, such as:

- Appearance (color, particle size, and shape)
- Texture and consistency (rheology, stickiness, mouthfeel)
- Flavor and odor
- Nutrient content (micro and macro nutrients, with an emphasis on thermolabile compounds such as vitamin C or thiamine)

During thermal processes, many chemical reactions occur, both desirable and undesirable, such as changes in color and degradation of vitamins. While microbial destruction is generally described by using a cumulative lethality of F value, changes in quality parameters are similarly expressed in terms of cooking, or cook value C as developed by Mansfield (1962). Cook value uses a calculation similar to the calculations used in microbial destruction.

$$C(t) = \int_0^t 10^{(T(t')-T_r)/Z_c} dt' \tag{7.12}$$

For any chosen quality parameter, T_r is the reference temperature, and Z_c is the temperature difference required for a tenfold change in denaturation time (similar to z-value of microbial). Specific C values can be used for nutrients, such as chlorophyll, vitamin C, or thiamine; activation and disintegration of hydrocolloids, kinetic parameter tables can be found in several sources (Toledo, 1980, Holdsworth, 1992, David 1985) providing D, T_r, and Z_c-values (or rate constant and activation energies) for several quality factors. However, the industry has defined a few common C values, which are summarized in Table 7.1

It is noticeable that the Z_c value for these integrations is much higher than the one used for microbial reduction, which means that changes in temperature will have a lesser effect on quality than they do on microbial reduction, which opens the door for optimization of processes. However, in the estimation of quality with C value, it is important to give credit to both heating and cooling.

Interpretation of the cook value is less straightforward than microbial reduction, but the cook value can be used in terms of a rough quality indication of the final product. Insufficient C value results in food that can be considered raw without flavor and texture development, while too high C value results in food that is overcooked with detrimental effects on flavor, color, and

TABLE 7.1   COOK VALUES USED IN INDUSTRY

Name	T_r [C]	Z_c [C]	Reference
C_{100}	100	33.3	Lund 1977
C_{121}	121.1	33.3	Holdsworth 1985
C^*	135	31.4	Kessler 1981 for UHT milk processing

texture. The range in which C values go from under to overcooked depends on each food and can be wide or narrow, e.g., vegetable particulates range between 5 and 20 minutes of C_{100}, after such range all texture is lost and vegetables become soft and even break into puree.

Optimization of homogeneous flows by balancing microbial destruction and temperature has been discussed by Holdsworth (1992), Banga (2003), and others, but optimization of quality of aseptic processes with particles has received less attention.

7.2.3 The HTST Paradigm and Homogeneous Flow Optimization

Summarizing the above thermal treatment effects on microbial and quality, we can see that all these effects can be modeled using first-order kinetics. Similar parameters are used for describing both, mainly D_T, T_{ref}, and z as described in Equation 7.8.

Applying Equation 7.8 for the inactivation of a microorganism over a range of temperatures, it becomes obvious that as temperature increases, the time needed to achieve the required lethality diminishes. For quality parameters, the same behavior is expected when Equation 7.12 is observed in the same way. Using the two most common safety ($F_0 > 6$) and quality parameters ($C_{100} < 30$) assuming constant temperature in the hold tube, it can be observed in Figure 4.2. These figures explain the HTST paradigm that allows high-quality aseptic processing: HTST (high-temperature, short time) and UHT (ultra high temperature) processes are achieved by exposing the product at a high enough temperature to minimize the time a product is exposed to such temperatures in order to maximize the quality. Product safety (F_0) has $z=10°C$ while quality parameters have higher Z-values, such as thiamine loss ($Z = 22.5$ min), discoloration ($Z = 24$ min), and texture loss ($Z = 33°C$). At low temperature, the time needed to achieve a safe product is larger than the maximum time required to maintain quality attributes, as temperature increases, there is a point in which both processes are equivalent and after this temperature the time needed to achieve a safe product is shorter than the maximum time at the same temperature that will render the product quality unacceptable. In Figure 4.2, the series of time–temperature combinations that result in a safe and high-quality product is shown as a shaded area.

In real-life operations, the HTST paradigm must be reevaluated to maintain the balance between safety (microbial) and the maximum allowable reduction of nutrients (quality), by considering operational limitations, such as the maximum temperature equipment can withstand, or the minimum and maximum practical flow rates in the equipment. These considerations can be added as constraints that further limit the combinations of time and temperature that can be used. The addition of operational considerations, such as the maximum pressure that the individual equipment can safely support, the maximum temperature allowed by SHE-Q, and the minimum and maximum rate of filling which determines the flow rate and hold time completes the study. Taking all these constraints into consideration, a similar graphic to the one shown in Figure 7.1 can be developed for each individual process line. The time–temperature combinations which fall inside the shaded gray area in Figure 7.1 are much smaller than in Figure 4.2 and help narrow the design of the process.

UHT processing has been concentrated in the dairy industry. However, other products such as low-acid viscous liquids and high-acid particulate foods are also similarly processed. The benefits of aseptic processing stem from UHT sterilization of foods. These benefits depend on a high temperature being maintained for only a few seconds, as time–temperature response for microbial inactivation and for chemical reactions are different. In other words, heat treatment of products at much higher temperatures for only a few seconds can achieve sterilization with greatly reduced product sensory and nutritional damage. The benefits of UHT sterilization cannot be

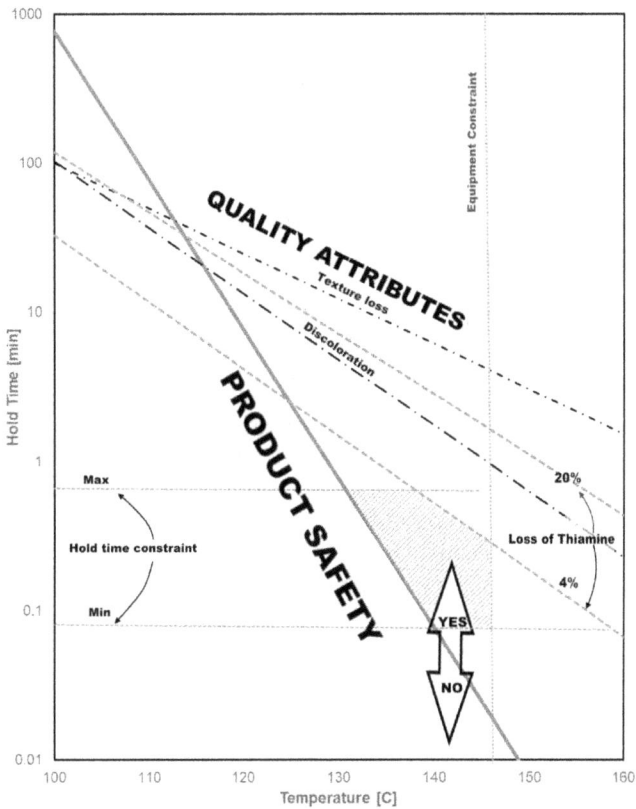

Figure 7.1 Determination of working conditions during aseptic processing, taking into account quality, equipment, and microbiological safety constraints.

realized in conventional canning or terminal sterilization due to long process times needed to achieve high temperatures. Come-up times (CUT) associated with transient or unsteady-state heat transfer can account for significant thermal degradation yielding overcooked or mushy product. The other overriding mechanical constraint is the inability of containers to withstand high internal pressure in retorts corresponding to temperatures in the UHT regimes. In aseptic processing, this is overcome by partitioning UHT sterilization of raw food product from that of container and lid. Compared to raw product, container and lid may experience different and milder methods of sterilization. The containers and lids are sterilized using steam, hydrogen peroxide, heat of coextrusion, or irradiation where approved. Also, other sterilants have been approved and are used (National Food Processors Association, 2004). The resistance to heat transfer in UHT applications is not determined by container shape or dimension as it is in canning. This results in more uniform product quality over hours of production run, independent of container shape and dimension (David, 2013).

Finally, the cooled sterile product is filled into sterile containers and hermetically sealed in a presterilized and continuously decontaminated tunnel or aseptic zone. Similar to canning, the products are shelf-stable with a shelf life of about one to two years at ambient temperatures. The shelf life is a function of the barrier characteristics of the container in terms of loss of moisture and oxygen transmission, which may possibly cause physicochemical changes, and not necessarily due to contaminating microorganisms. Storage at high temperatures may dramatically reduce the shelf life of heat-sensitive products or promote economic spoilage due to the presence of thermophilic anaerobes.

7.3 COMPARISON OF CONVENTIONAL CANNING AND ASEPTIC PROCESSING AND PACKAGING OF FOODS

7.3.1 Comparison of Conventional Canning and Aseptic Processing and Packaging of Foods

During the past three decades, there has been a significant market growth and associated technical development in the area of aseptic processing and packaging of foods. Compared to classical canning, aseptic processing is best suited for heat-sensitive and nutritional foods and beverages in order to obtain a finished product with better sensory qualities and higher nutrient retention. In addition, it is possible to fill finished products into different container types and sizes, with features attractive to consumers and manufacturers. It is possible to aseptically fill and package portion packs, retail size cartons and bottles, and bulk transport containers and industrial drums, totes, and so on, destined for remanufacturing into retail and institutional and industrial containers.

Salient features distinguishing conventional canning and aseptic processing and packaging technology are summarized in Table 7.2.

7.3.2 Some Advantages of Aseptic Processing and Packaging of Foods

Compared to canning, aseptically processed product easily lends itself to value-added features, such as increased quality, more attractive packaging, sustainable storage and distribution, microwaveability, user-friendly convenience, and cost saving from the use of plastics and paper.

7.3.2.1 Nutritional Quality

The literature is replete with kinetic data that point to better nutritional quality. Examples are shown in Tables 7.3 and 7.4 (Wilhelmi, 1988).

Also, research done at the University of California at Davis (Luh, 1970) compared basic differences in chemical and quality changes of four vegetable purees—strained carrots, peas, corn, and green beans—processed by aseptic and retort methods. Overall, the aseptic canning process appeared to be superior to the retort method. The aseptic products were of better organoleptic quality and higher in vitamin retention. Similar efforts have been carried out by Centers for Advanced Processing and Packaging Studiers (CAPPS), US Army, and others with similar results.

7.3.2.2 Sensory Quality

Aseptically processed foods have typically increased qualities such as color, flavor, and taste as well and consistent quality in the whole package. Typical retorted tomato soup in metal cans sells for about 1.6¢/oz, compared to 3.6¢/oz for typical aseptically processed tomato soup in aseptic cartons. Aseptically processed foods command a higher price due to substantially increased discernible qualities, such as color, flavor, and taste (T. E. Szemplenski, of Aseptic Resources, personal communication, 2010).

TABLE 7.2   COMPARISON OF CONVENTIONAL CANNING AND ASEPTIC PROCESSING AND PACKAGING OF FOODS (DAVID 2013)

Criteria	Retorting	Aseptic Processing and Packaging of Foods
I. Sterilization		
A. Product		
1. Temperature regime	220°F–250°F	HTST (180°F–220°F) and UHT (260–290°F)
2. Delivery	Unsteady state	Precise—square wave
3. Heat/cool lethality credit	Possible	Isothermal temperature in hold tube only
4. Process calculation		
Fluid	Routine–convection	Routine
Particulate	Routine–conduction or broken heating	Complex-requires modeling
B. Other sterilization (process equipment, container, lid, and aseptic tunnel)	None	Complex; many
C. Energy efficiency	Lower	30% saving or more
II. Quality		
A. Psychophysical or sensory	Mushy; not suitable for heat-sensitive and nutritional products	Superior; suitable for homogeneous heat-sensitive and nutritional products
B. Nutrient loss	High	Minimum
C. Value added	Lower	Higher
Convenience	Shelf-stable	Shelf-stable and other features
Microwaveability	Semirigid containers (bowls and trays)	All non-foil flexible and rigid containers
D. Product quality	Dependent on container size and shape	Independent of container size and shape
III. Production aspects		
A. Container speed	High; 600–1000+ containers per minute	Medium; 500–700 containers per minute is common; higher speeds via multiple lanes
B. Handling/labor	High	Low
C. Downtime	Minimal, typically at seamer and labeler	Resterilization due to sterility loss in sterilizer or filler
D. Versatility/flexibility for manufacture of a product in different container sizes	Different size containers need different process delivery and/or retorts	Need one or two aseptic fillers to fill different size containers

(Continued)

TABLE 7.2 (CONTINUED)   COMPARISON OF CONVENTIONAL CANNING AND ASEPTIC PROCESSING AND PACKAGING OF FOODS (DAVID 2013)

Criteria	Retorting	Aseptic Processing and Packaging of Foods
IV. **Process eviation**		
A. Under processing	Reprocess intact container-in-product after removal of labels if possible	Reprocess product after destructive opening of containers
B. Survival of heat-resistant enzymes	Rare	Common in certain foods, especially in fluid milk; problem could be overcome, if system is designed properly (refer to Chapter 15)
V. **Spoilage analysis**		
A. Troubleshooting	Simple; preprocess, inadequate process, and postprocess recontamination	Complex; need to deal with compromises in CIP and sanitation, and aseptic zone and its elements (refer to Chapters 12 and 13)
B. Traceability : container code versus case code	Usually not identical; may differ by 1/2 to 6 hours	Usually, identical
VI. **Low-acid particulate processing**	In practice	Possible and commercial both in retail and bulk packaging
VII. **Postprocess prepackage sterile additives**	Not possible	Possible and in practice to add filter sterilized enzymes (lactase), bioactive compounds, therapeutic agents, and probiotics

TABLE 7.3   DIFFERENCE IN VITAMIN C RETENTION BETWEEN RETORTED AND ASEPTICALLY PROCESSED TOMATO SOUP (DAVID 2013)

Sample	Vitamin C (mg/100 g)	Vitamin C Retained Compared to Raw Product
Raw product	27.8	—
Aseptically processed	25.3	91%
Retort processed	16.3	59%

Source: Wilhelmi, F., 1988, Soups and sauces: the aseptically packed feast ready for the table, *Dragoco Report*, 3, pp. 63–77.

TABLE 7.4   DIFFERENCE IN THIAMIN RETENTION BETWEEN RETORTED AND ASEPTICALLY PROCESSED CHICKEN SOUP (DAVID 2013)

Sample	Thiamin (mg/100 g)	Thiamin Retained Compared to Raw Product
Raw product	0.11	-
Aseptically processed	0.09	82%
Retort processed	0.03	27%

Source: Wilhelmi, F., 1988. Soups and sauces: the aseptically packed feast ready for the table, *Dragoco Report*, 3, pp. 63–77.

7.3.2.3 Sustainability in Storage and Distribution

Aseptically processed and packaged foods are shelf-stable, which means that storage and transport doesn't need to be refrigerated.

Refrigerated storage requires 400–650 kWh/m^2 per year which generates 150–250 kg of CO_2 per year, by storing the products in ambient warehouses thus footprint is minimized. Transportation uses even more energy, as trucks and refrigerated containers are less insulated and subject to solar radiation. The savings in energy and the reduced carbon footprint need to be quantified as part of the sustainability plan of each company, but this savings is substantial.

7.3.2.4 Package Convenience and Microwaveability

Aseptically packaged foods use plastic or carton containers in an infinity of shapes. Both for retail and institutional use, the convenience of a package that is light, easy to open, and easy to dispose of is a big advantage. Most retail aseptic product uses carton boxes, which have several advantages, such as filling the shelves without empty spaces, the ability to have easy open caps, and easy disposal.

Microwaveability is another value-added feature that can be easily incorporated into flexible and semirigid non-foil containers commonly used in aseptic filling operations. This is a consumer-friendly feature that can command two to three times the price of that for similar products in metal cans. As of 1988–1989, the market niche for shelf-stable microwaveable products was estimated to be $2 billion. The projection for 1994 was $4.4 billion (T. E. Szemplenski, Aseptic Resources, personal communication, 2010). One of the pending improvements to aseptic packaging is recyclability; the number of layers makes recycling a very complex task.

7.3.3 Comparison of Processing Methods

With the advent of ultra-pasteurization for refrigerated dairy products, there has been confusion in usage of the term UHT sterilization as applied to dairy products among producers and consumers, as evidenced by publications and advertisements in a number of trade and scientific journals. The name UHT should not be used for ultra-pasteurized products. In the following sections, the different methods of heat treatments—pasteurization, ultra-pasteurization, and canning—are compared with respect to process time–temperature, target microorganisms, shelf life, packaging and refrigeration requirements, and regulatory ordinances.

7.3.3.1 Pasteurization

Pasteurization is a mild heat treatment process for milk and fluid foods to specifically inactivate certain pathogenic vegetative microorganisms with low heat resistance. The usual minimum time–temperature combinations are low-temperature, long-time (LTLT) 145°F for 30 minutes or high-temperature, short-time (HTST) 162°F for 15 seconds. The process can be delivered by either batch or continuous heating. It is important to note that the heat treatment is not intended or sufficient to inactivate all spoilage-causing vegetative cells or any heat-resistant spores, if present. This fact determines a short keeping-quality period up to about two to three weeks under refrigerated conditions (less than 45°F). In other words, the finished product (low acid) is not commercially sterile. However, pasteurized high-acid fluid packaged via the hot–fill–hold method in a hermetically sealed container may yield a commercially sterile shelf-stable product.

Pasteurization is warranted for high-acid foods such as juices and beverages, and low-acid refrigerated foods such as milk and dairy products that have a rapid turnover in commerce. Pasteurized milk and dairy products are regulated primarily by Grade "A" Pasteurized Milk Ordinance (2009).

7.3.3.2 Ultra-Pasteurization

Ultra-pasteurization refers to pasteurization at very high temperatures of 280°F or above for 2 seconds or longer. The objective is similar to a pasteurization process and further extends the shelf life of the product. However, this high-temperature process is sufficient to destroy a greater proportion of spoilage microorganisms leading to extended shelf life (ESL) of about six to eight weeks under refrigeration, compared to two to three weeks for traditionally pasteurized products. This process is also known as ESL technology. Ultra-pasteurization is not UHT sterilization as the processed product is not commercially sterile and shelf-stable. This process has been effectively used for half-and-half, chocolate and flavored milks, other dairy products, and non-dairy creamers in portion pack cups and tabletop containers. Ultra-pasteurized dairy products are defined and regulated by the Grade A PMO. Ultra-pasteurization gives producers great latitude to regulate the stock rotation, product velocity, or turnover rate, and reduction of spoilage returns due to postdated conventionally pasteurized products.

7.3.3.3 Conventional Canning

Compared to pasteurization, canning is a severe form of heat treatment to specifically target and inactivate mesophilic spores of *C. botulinum* and prevent economic loss in low-acid foods. The heat resistance of *C. botulinum* is characterized by a $D_{250°F}$ of 0.21 minutes and a z of 18 F_o. The food must experience at least a 12D cycle reduction at the "cold point" to be made commercially sterile. This is also referred to as F_o of 3 minutes. In practice, a thermal process of equal to or greater than F_o of 3 minutes is delivered in closed pressure cookers or retorts above atmospheric pressure. Canning is sufficient to inactivate heat-resistant enzymes, microorganisms, and mesophilic spores that cause spoilage under normal room temperature storage conditions.

The food to be canned is filled in metal cans, glass jars, retortable semirigid plastic containers, or pouches, followed by double seaming or heat sealing with a cap or lid. Such containers are heated and cooled in a pressurized batch or continuous retort.

This type of conventional canning is also referred to as in-container terminal sterilization because the thermal process is delivered after filling and sealing operations. Both container and product experience an identical method of sterilization. Heterogeneous high- and low-acid particulate foods such as chunky fruits and soups with meats and vegetables can be processed by this method (9 CFR, 1986). The technology of containers and seaming and capping are very well established. Canned products are regulated by the Code of Federal Regulations (21 CFR, 2016).

7.3.3.4 Refrigerated Aseptic Products

Several food companies retail their aseptically processed products such as puddings, refrigerated, or chilled for diversification, volume saturation, market velocity, and repositioning. Placing shelf-stable products in the dairy case or in chilled cabinets in supermarkets may be one way of meeting the current consumer demand for "fresh-like" premium foods. Interestingly, refrigeration circumvents the regulatory requirements that govern thermally processed low-acid, shelf-stable foods. Low-acid vegetable and acidified products filled using aseptic fillers and processes that are not approved by the US Food and Drug Administration (FDA) or US Department of Agriculture (USDA) must be refrigerated throughout the supply chain.

7.3.3.5 Comparison of Continuous Processing Methods Based on Optimization Hierarchy

It is axiomatic to consider aseptic processing and packaging as the benchmark in optimization, for the manufacture of sterile, shelf-stable canned food. Compared with retorting or in-container sterilization, aseptic product and package are independently sterilized by optimal processes; microbial inactivation and quality factor degradation are also co-optimized given first-order kinetics. Aseptic systems permit sterilizing the product and container separately and appropriately, without the rate-limiting heat transfer modes, or the attendant thermal and pressure stress to the container, and permitting HTST or UHT processing of very heat-labile products, without excessive quality factor degradation, while achieving requisite commercial sterility.

The technology has been diversified in recent years, and concepts of aseptic processing and packaging that are somewhat removed from the core of the original technologies are slowly being adapted to liquid and solid foods, shelf-stable, or refrigerated. Table 7.5 shows a comparison of different processing methods reflecting some key differences between the core of UHT processing and aseptic packaging technology with that of other adaptations and commercial concepts. In the United States before 1981, the Dole aseptic canning system was the only aseptic filling and packaging system of commercial importance for milk and milk-based low-acid products in metal cans. This core technology was correctly designated as a UHT-sterilized and aseptically packaged process. The pre-1981 market drivers were better quality, low-acid, heat-sensitive products than retorted versions. It was a formidable marketing and research and development challenge to educate consumers to appreciate aseptic technology and differentiate products in Dole aseptic metal containers from that of classical canning or retorting.

In 1981, the FDA approval of the food additive petition for use of hydrogen peroxide as a sterilant for food contact surfaces provided the impetus for the introduction of various aseptic filling and packaging systems into the US market. The market was primarily driven by "juice box" technology, which involved high-acid food pasteurization at HTST followed by aseptic filling and packaging. This process was designated as an aseptically processed and aseptically packaged process or an aseptically processed and packaged product. The post-1981 and current market drivers are the use of alternate packages, consumer convenience, light-weighting, cost reduction, superior quality, and package sterilization by optimal methods. Today, any food product heated continuously either in UHT, HTST sterilization, or pasteurization temperature regimes, followed by aseptic filling and packaging is generally designated as an aseptically processed and packaged product.

TABLE 7.5   COMPARISON OF DIFFERENT CONTINUOUS PROCESSING METHODS BASED ON OPTIMIZATION HIERARCHY (DAVID 2013)

Raw Product	Heat and Cool Continuous Processes	Filling and Packaging	Process Technology Designation	Finished Product and Shelf Life	Optimization Hierarchy Index
High-acid food	Continuous pasteurization–fill–hold–cool	Hot fill to sterilize lid and container at ambient conditions	Core hot–fill–hold–cool technology	Shelf-stable juices and acidified foods; commercially sterile	Interim hierarchy
Low-acid food	Continuous UHT sterilization and flash to 255°F	Hot fill to sterilize lid and containers in pressurized room	"Flash 18" process	Shelf-stable, low-acid food in institutional-size containers; commercially sterile	Higher hierarchy
Low-acid food	Continuous HTST pasteurization and cool	Cold fill in clean room; sanitized container	Conventional pasteurization	Refrigerated milk, dairy, and juice products; 2–3 weeks	Lower hierarchy
Low-acid food	Continuous ultra-pasteurization and cool	Cold fill in clean room; sanitized container	Ultra-pasteurization; extended shelf life (ESL) product	Refrigerated milk, creamers, and juices; 6–8 weeks	Interim hierarchy
Low-acid food	Continuous UHT sterilization and cool	Aseptic filling and packaging at ambient	Core aseptic technology; UHT-sterilized and aseptically packaged foods	Shelf-stable milk, sauces, puddings, and other homogeneous food; commercially sterile	Benchmark in optimization for shelf-stable, low-acid, commercially sterile canned foods
High-acid food	Continuous HTST pasteurization and cool	Aseptic filling and packaging at ambient	HTST pasteurized and aseptically packaged foods	Shelf-stable juice box and juices in clear PET bottles; commercially sterile	Benchmark in optimization for shelf-stable, high-acid, and acidified canned foods
Low-acid food	UHT or HTST heat and cool	(a) Aseptic filling and packaging at ambient	UHT or HTST processed and aseptically packaged refrigerated food	Commercially sterile refrigerated puddings for market diversification and repositioning; 6–8 weeks stock rotation cycle	Interim hierarchy
		(b) Non-aseptic; unapproved filler or processes	UHT or HTST processed and non-aseptically packaged or processed refrigerated food	Refrigerated/ESL puddings; 6–8 weeks	Lower hierarchy

In the last column of Table 7.5 is an optimization hierarchy index based on co-optimization of competing priorities, optimal method of sterilization of product package and aseptic zone, aseptic zone integrity, finished product shelf stability and/or commercial sterility, and requirement of additional hurdles, such as refrigeration required for pasteurized and ultra-pasteurized foods (K. S. Purohit, Process Tek, personal communication, 2008).

7.4 ASEPTIC PROCESS CALCULATIONS FOR PARTICULATE FLOWS

Particulate flows are made of a liquid carrier and a population of particulates of different sizes, shapes, and composition. Each particulate has different thermophysical properties, dielectric properties, and microbial loads; the carrier fluid must be dense and viscous enough to carry all the particulates and maintain buoyancy during the process.

7.4.1 Calculating Cumulative Thermal Process in Fluids and Particulates

As discussed in Chapter 4, the calculation of cumulative lethality or F value for a single fluid element or solid particle that has uniform temperature is relatively easy and, in many processes, it is done based on the worst-case scenario of using the temperature at the end of the hold tube. However, in the real process assuming a uniform liquid velocity and temperature would be incorrect. Temperature profiles exist in the cross-section of heat exchangers, from the heated walls toward the center as well as changes that might happen over time due to fouling. The extreme cases manifest a fast-moving cold core (in heaters and holding tubes) or a hot core (in the cooling section). Volumetric heating, either ohmic or microwave, tends to eliminate the heating problem; however, cooling still uses conventional heat transfer; and the search for rapid and uniform aseptic cooling is the source of research.

For products that contain particles, the calculation of F_0 becomes more complex, as the temperature inside the particles is difficult to measure, and the heat transfer to the center of particles is also a complex chain of heat transfer as sketched in Figure 7.2. A heat transfer medium is on the outside of the wall of the heat exchanger and transfers its energy by convection to the wall, then transfer of heat is by conduction through the wall of the heat exchanger, and through the

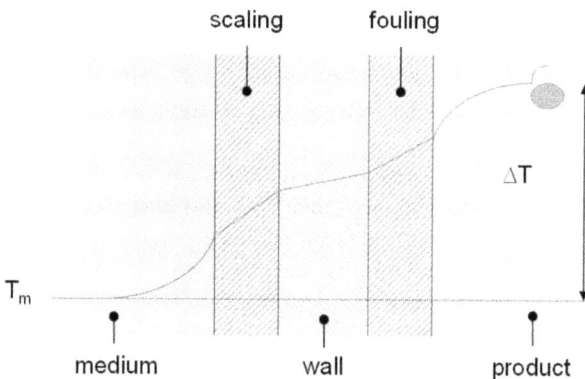

Figure 7.2   Heat transfer cascade during indirect heating of products.

fouling layers on both sides, the product liquid phase gets its energy by convection, and reaches the surface of the particles, where it is transferred by conduction.

In particulates, it becomes necessary to calculate F_0 based on a worst-case particle that has both the largest thermal resistance and the shortest residence time, and which has experienced the lowest external liquid temperature. The concept of this is that if the worst-case particle is properly processed every other particle will, but this conservative approach results in over-processing in the liquid and most particles. Design of the aseptic system based on the lethality accrued only in the holding tube neglects the often-considerable lethality accumulated during heating and cooling. Due to the complex heat transfer cascade, the liquid part of a product can receive a considerable thermal treatment in the heating section (Awuah et al., 2004, Kelder et al., 2002, Jung & Fryer, 1999), while the center of particles continues building lethality in the first part of the cooling section (Sandeep, 1999). Emerging thermal methods, such as microwave or ohmic heating, can help by maximizing the heat transfer toward the center of the particles and minimizing the time the fluid part is exposed to high temperatures.

Observation in real flows showed that a particle will not always be the fastest, as it may radially migrate to streamlines of lower velocity and higher temperature (e.g., move closer to the wall of the heat exchanger) or might be slowed down by friction with other particles or with the walls (Simunovic 1998, Palazoglu 2004). Further mixing effects might be related to mixing actions induced by the geometry (e.g., Dean effect in bends), or by particle–particle, fluid–particle, or particle–wall interactions.

Lethality in particulate flows, while difficult to measure in real time, can be predicted by mathematical modeling. This modeling is more complex than the modeling of homogeneous flows, as both liquid and several types of particles must be tracked simultaneously. The theoretical best way to obtain accurate predictions of the thermal process would be to model the aseptic system in a full CFD three-dimensional analysis. For heterogeneous flows, there would be a need to account for the different types and loading of particles and track the spatial and thermal trajectories over the processing system, as well as the two-way coupling between liquid and particles. In principle, this type of modeling could be predictive and would require little experimental/ analytical work, as the heat transfer coefficients would follow from the simulations. The models should also account for temperature-dependent properties of the particles (thermophysical) and the liquid (rheology), including changes in phase and effects of temperature on hydrocolloids. The 3D modeling will simulate the impact of bends, free convection, and shear applied, as well as volumetric heating for both liquid and particle as long as the temperature-dependent dielectrical properties are also known. The simulation would generate a 3D matrix of temperature on space and time in such a transient flow; this temperature data together with microbial and quality destruction kinetics can provide predictive microbial and quality estimations throughout the system. In a very short time, several worst-case scenarios could be simulated which can be used for process design and optimization.

Unfortunately, the computational capacity needed for such modeling is significant and needs specialized software and highly trained personnel, both now and in the foreseeable future and is discussed in Chapter 4.2. Thus, simpler models have been developed and commercialized to have an accurate calculation of lethality and quality reduction, which can include the contribution of heating and cooling as well and provide initial optimization cases. Several commercial software packages are available from JBT, Tetra Pak, etc.

All future improvements in model resolution will come at the expense of the experimental effort required to obtain data on residence time distribution, heat transfer correlations or rheological and thermal properties of liquids and particles. While modeling and supporting experimental effort may yield hugely improved predictions, these may never fully replace validation of the actual formulation and thermal process.

7.4.2 Thermal Treatments for Heterogeneous Products

As mentioned earlier, high-temperature processes are known to result in the inactivation of microorganisms and their spores at higher rates than the destruction of quality factors. High temperatures applied for short times produce the same aseptic effect as longer process times at lower temperatures, whereas the destruction of quality factors is dramatically reduced. This approach has become known as the high-temperature short-time (HTST) paradigm (Holdsworth, 1992).

In well-mixed liquids of a uniform temperature, the HTST paradigm is constrained by several practical limitations. First, at very high wall and product temperatures, nonlinear fouling and browning reactions may result in products of unacceptable quality and in short run times between cleaning, affecting the economy of the operation (Simpson, 1974). Second, the maximum achievable medium temperature is limited by the equipment and factory utilities (e.g., steam pressure available). Third, boiling and flashing in the system is generally avoided, which requires maintenance of high pressure in the system proportional to high product temperatures. Fourth, high temperatures require extremely short holding times and accurate control of these may no longer be guaranteed. Finally, the validity of microbial destruction kinetics beyond the temperature range for which these were originally obtained is unreliable.

The HTST approach should also be used with caution, besides the cases mentioned above, where residence time distributions and temperature distributions might not be uniform in either homogeneous (Jung & Fryer, 1999, Kelder et al., 2002) or heterogeneous flows (Sandeep, 2000). In such cases, the final product may exhibit an unacceptable distribution in quality factors.

Mathematical modeling can be used to determine a set of temperature–time combinations for optimal product quality (i.e., a minimized and uniform cook value). A simple model that reconstructs the temperature evolution of both liquid and particles of a heterogeneous food along a tubular aseptic system can be used. The underlying equations follow the heat transfer cascade as shown in Figure 7.2 and allow for an axial development of the liquid temperature (Equation 7.13), two-way energy transfer between the liquid and the fraction of representative particles (Equation 7.14), and a one-dimensional temperature distribution in cylindrical particles of infinite length (Equation 7.15).

$$\rho_f Cp_f A_f \frac{DT_f}{Dt} = U_{mf} l_f \left(T_m - T_f \right) + n_p h_{fp} A_p \left(T_p^s - T_f \right) \tag{7.13}$$

$$-\lambda_p \frac{\partial T_p^s}{\partial n} = h_{fp} \left(T_f - T_p^s \right) \tag{7.14}$$

$$\rho_p Cp_p \frac{\partial T_p(r)}{\partial t} = \lambda_p \nabla^2 T_p(r) \tag{7.15}$$

In Equation 7.13–Equation 7.15, ρ, C_p, and λ are the constants and have the usual meaning of density, heat capacity, and thermal conductivity; t indicates the time, and T indicates the temperature. The subscripts m, f, and p indicate the heat transfer medium, and the products liquid and particles, respectively, and the superscript s the surface of the particle. In Equation 7.13, the substantial derivative (Bird et al., 2006) is used to calculate the energy flow to or from the liquid, where A_f is the tube cross section, l_f the tube perimeter, n_p the specific number density of the particles, and A_p the surface area of the representative particle. The overall heat transfer coefficients between the heating medium and the fluid (resistance through the heat exchanger wall) and between the fluid and the particles are given by U_{mf} and h_{fp}, respectively.

With this model, tubular heater and cooler were modeled to establish the boundaries of the cook values at different flow and heat transfer scenarios. The case of "instantaneous heating" is presented here to illustrate the key factors that impact the thermal processing of heterogeneous products.

The scenario of instantaneous heating might apply to particulate flows subject to steam injection where the liquid is heated rapidly by condensing steam. This case has a very rapid heat transfer into and through the liquid, followed by a much slower process of heating the particles via conduction. For this scenario, both heat transfer coefficients U_{mf} and h_{fp} are set to very high values, making conduction throughout the particle the limiting factor.

To evaluate industrially relevant processing conditions, 10% mass fraction cylinders of four different diameters were chosen, at a product mass flow of 5 tons per hour in 2″–1.83″ ID (50 mm–ID 46.5 mm) tubes. In the heater, a constant wall temperature of 140°C was maintained, and in the cooler a wall temperature of 10°C was assumed. Heating and cooling sections were dimensioned such that the core F_0-value reached at least 2.5 minutes. No hold tube is used, even though in a real system this will increase the thermal treatment.

As an example, Figure 7.3 shows the ideal average and center temperatures for a particle of 15 mm diameter under the instant heating and cooling conditions lined up above. As expected, the liquid temperature follows the medium temperature almost immediately in both heating

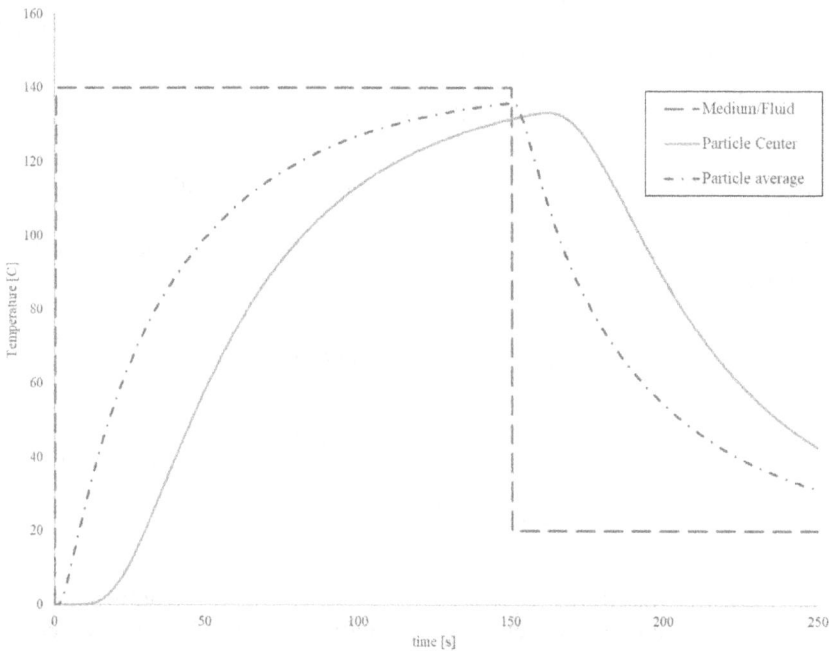

Figure 7.3   Ideal thermal process profile for a product with large particulate (15 mm) with instantaneous heating and cooling of the carrier fluid.

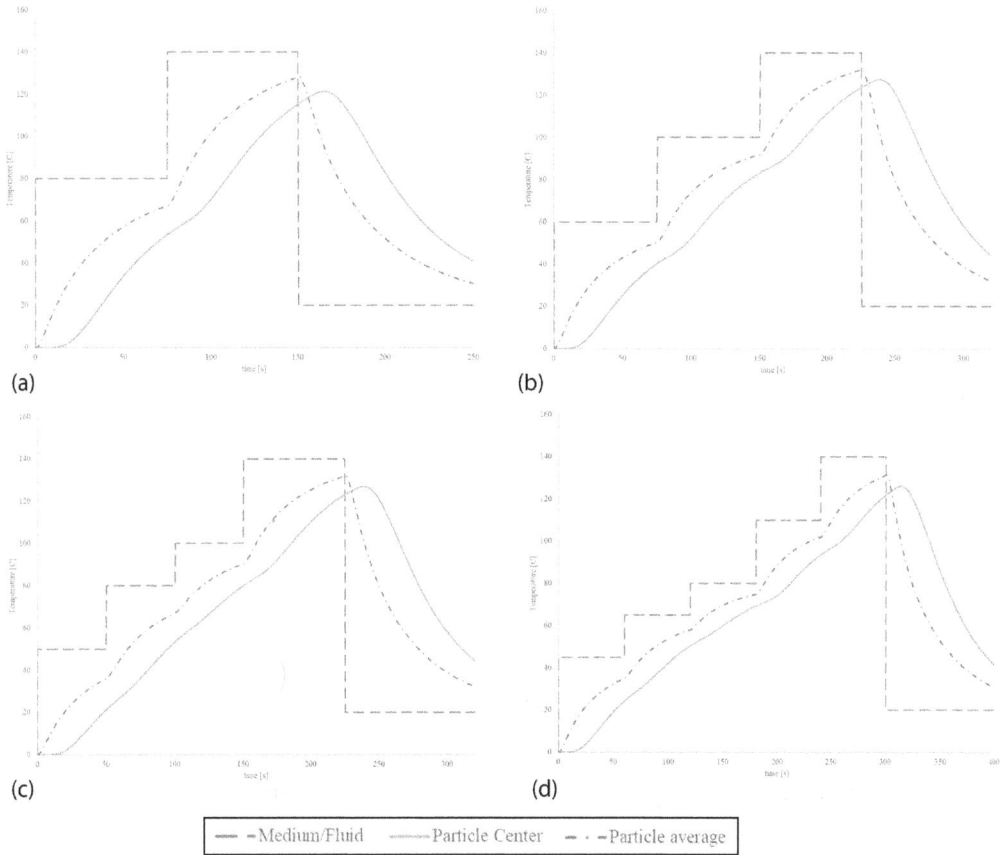

Figure 7.4 Ideal thermal process profile for a product with large particulate (15 mm) with instantaneous heating and cooling of the carrier fluid considering different equilibration steps 1(a), 2(b), 3(c), and 4(d).

and cooling. Heat transfer between fluid and particle surface is also rapid and the surface temperature of the particle equals that of the fluid, both in the heater and in the cooler.

Even though the surface heats instantaneously, the particle core needs to heat through conduction and large differences in temperature are observed between the center and the surface.

This process can be improved by giving the particles time to equilibrate at intermediate temperature levels by adding lengths of tubing that are not heated. Figure 7.4 a–d shows the different temperatures for 1-, 2-, 3-, and 4-step equilibration process.

While equilibration increases the residence time of particles, the temperature non-uniformity inside particles decreases. As a result, the distribution of the thermally affected quality factor becomes narrower. This simulation was repeated for other particle sizes (6, 12, and 20 mm). Figure 7.5 shows the temperature evolution throughout the particles for the four particle sizes and the five processing scenarios (0–4 step equilibration) in three cook values.

In Figure 7.5, the five upper traces represent the cook value of the surface of the particle, and the lower five the cook value at the center. The numbers indicate the equilibration steps from 0 to 4. The solid reference line shows a C value of 20 min, beyond which most vegetables have lost

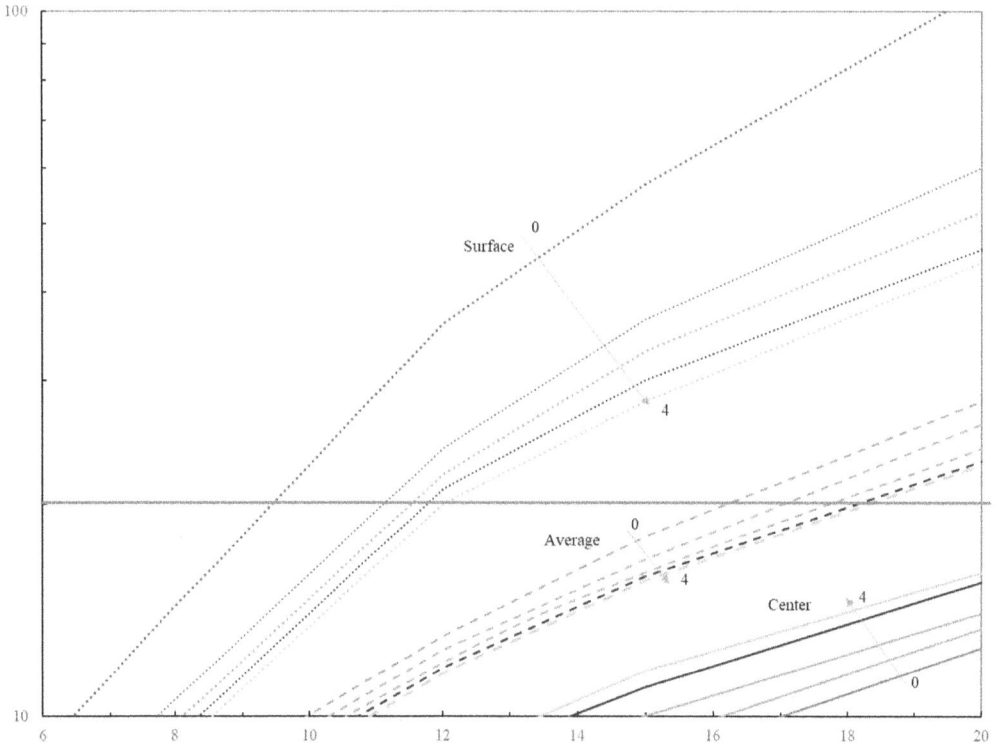

Figure 7.5   Comparison of cook values for particles of 6, 12, and 20 mm with 0 to 4 steps of equilibration.

textural integrity. Surface cook of particles in a process without equilibration surpasses this value for particles larger than about 6 mm in diameter. If equilibration is employed properly (e.g., in the four-step process), particles could be processed up to 12 mm in diameter. This value is consistent with practically observed maximum particle sizes in conventional tubular aseptic processing.

Despite the simplified nature of the "Instantaneous heating" case, it illustrates the complexity of the process and provides several hints as to procedures and methods to improve product quality in aseptic processing of heterogeneous products:

1   —The use of fresh ingredients at chilled or ambient temperature reduces the thermal degradation of quality when compared to frozen ingredients. It also improves the temperature uniformity and spread in quality factors in particulates.
2   —Large temperature differentials between the medium and the product should be avoided throughout the heating stage. It is preferable to have several heating stages operated at increasingly higher temperatures to allow for the particles to equilibrate thermally. The maximum sterilization temperature to be used is a function of the optimal driving force between the medium and the particulates.
3   —Residence time distribution of the particles plays a significant role, the more uniform the flow the less intense heating process.
4   —Cooling rates must be maximized as far as practically and economically feasible.

5 —The limiting factor for thermal process in conventional systems is the thermal path of the largest particle.

In Section 7.5, these guidelines will be used for practical improvements on each step of the thermal process. In addition, the role of ingredients and auxiliary equipment (e.g., pumps) is discussed from the point of product quality.

7.5 ASEPTIC QUALITY OPTIMIZATION

As a case study, a sample process will be followed from ingredient selection to final handling, following the steps shown in Figure 7.6.

7.5.1 Ingredients

Aseptic processing uses high temperature which inactivated microorganisms and spores very quickly but also produces changes in quality attributes of the products such as texture, color, and nutritional content. Aseptic systems are continuous flow processes, which can also cause further textural degradation and particle breakage due to shear, contact with the wall, or collisions with other particles. This breaking of the particles not only affects the appearance, causing rounded corners but also leaches particulate material to the carrier liquid, blending flavors and clouding the broth.

To make high-quality products, ingredients must also be sufficiently robust to survive the harsh thermal treatment posed the aseptic process. The cook profile of meats must match that of the aseptic process and may require pre-cooking or frying. Hydrocolloids and starches should provide the right consistency throughout the process and in the finished product, and this often means the use of or modified high-temperature starches or blends of starches which are activated in different temperature ranges.

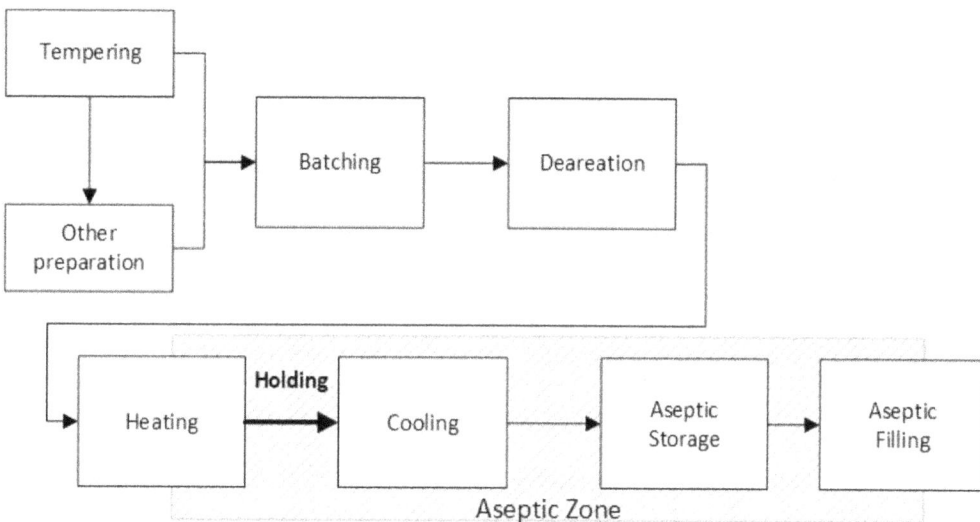

Figure 7.6 Steps needed for aseptic processing and packaging.

Vegetable particles are the most delicate ingredients to include in aseptic heterogeneous products. Vegetables should be harvested at a predetermined ripeness and processed immediately to maintain nutrients and optimal texture. Several pretreatments to increase the firmness of vegetables are available, such as immersion in Ca^{+2} salts (Floros et al., 1992). Seasonality of fresh vegetables can be a limiting factor, which is overcome by the use of frozen vegetables. Vegetables are either frozen conventionally or using the Individually Quick Frozen (IQF) process, which results in a higher quality product.

While frozen vegetables allow for a year-round operation, in many cases the quality of frozen vegetables is not the same as that of the fresh ones. Prior to freezing, vegetables are blanched to inactivate enzymes and prevent color and nutrient degradation, but blanching causes softening, which is aggravated by the water expansion during freezing. During thawing, the cellular damage sustained may cause significant loss of "firmness," and vegetables lose water (drip-losses) making them more vulnerable to the detrimental impact of the heat treatment (Fellows, 1988).

Finally, the particle size distribution of the vegetables will affect the quality and careful selection of the particles help in maintaining a more uniform thermal path. If the thermal path is kept small but the particulates are long in the other dimensions (parallelepipeds or cylinders), the perceived size by the consumers can still remain quite large.

Frozen vegetables may benefit from a tempering step (steam, microwave, or radiofrequency) to improve product quality and batching. Tempering and thawing may also be required to meet safety and quality standards and must be performed in a controlled fashion.

During batching, clumping of frozen particles and flotation are a well-known phenomenon. Even when the ingredients are dumped slowly, large particle loadings (e.g., for "rich" formulations) will likely overstretch the heating duty of the batching vessels. This is especially relevant in the case of cold-batching (please refer to Section 5.2). Tempering brings the frozen product to a temperature in which some of those are avoided.

7.5.2 Batching

In principle, the less the product is exposed to temperature, the less the quality it loses; thus, the operating temperature of the batching vessels should be as low as possible. However, due to safety concerns and to allow activation of thickeners such as starch, the operating temperature should be above 70°C. Control of this temperature must be tight in order to limit heat damage. Among the possible strategies to optimize this step are using multiple small cook vessels and preventing stoppages. Batching vessels might also have several jacket sections that are controlled independently to avoid burning of material around the surface of the vessel during loading and draining.

7.5.3 Heating

In order to minimize the exposure to high temperature, the heating rate of the product should be maximized while product hold-up volume in the system minimized. Also, the thermal path for the whole product must be kept at a practical minimum. The heating rate is proportional to the inverse of the square of the thermal path for viscous liquid products. In tubular heaters, this can be achieved by decreasing the tube diameter to the smallest possible size. However, the tube diameter is in practice dictated by the particle size, particle damage due to shear and blocking of the tubes must be avoided. Conscious use of bends or continuously coiled heaters may also serve to reduce the thermal path. A second method to improve heat transfer rates is to use static mixers.

Considering all of the above, it is clear that rapid heat transfer solutions such as steam injection or scraped surface heat exchangers (SSHEs) are not optimal for the processing of heterogeneous flows with particles exceeding certain dimensions. Of course, such equipment may still be used to successfully heat (viscous) liquids, but in the case of large particles different technologies are required. This can be solved by using volumetric (ohmic or microwave) heating, in which the whole product, fluid, and particles are heated uniformly.

Volumetric heating has been successfully applied for the production of soups with large particulates that are available in the market both in Europe and in the United States.

7.5.4 Holding Tube

As discussed in previous sections, the maximum product temperature in the holding section is limited by several practical limitations. Additionally, the optimization of the quality should occur along the whole temperature treatment, while ensuring a sufficient F_0 in the holding tube (or in the high-temperature section of the process if this can be justified). In practice, optimal liquid and particle core temperatures range between 125°C and 138°C, and these products show significant quality improvements over conventional retorting.

Residence time distribution, especially the particles in the heater and the hold tube, that is narrow enables the use of a system that is less conservative (that is, smaller). A previously proposed solution to give particles a sufficient and well-defined residence time was the Rota-Hold concept developed by Stork Food & Dairy systems (Holdsworth, 1992), but using coiled holding tubes (Sandeep, 2000) or holding tubes consisting of many bends to promote chaotic cross-sectional mixing may be employed to a similar effect.

7.5.5 Cooling

Final cooling is critical in optimizing the quality of particulate products, especially since the product liquid structure has fully formed and particulates have softened. In the cooling section, contrary to the heater, both heat transfer coefficients and temperature difference between product and medium should be maximized, by using chilled water or glycol facilities. However, some caution should be exercised to avoid freezing or fat crystallization at the walls, or very nonuniform axial velocity distributions due to large differences in viscosities.

7.5.6 Aseptic Storage

Aseptic storage tanks are usually insulated or have a small cooling capability. They also have agitation which can prove important for quality of heterogeneous products. Settling and separation of multicomponent heterogeneous products must be avoided and the agitator inside the aseptic storage tank can be used for this purpose. However, this agitator must be gentle to prevent shearing and damaging the particles.

Prevention of oxidation can also be achieved in the aseptic storage tanks, by overpressuring them with nitrogen instead of filtered air.

7.5.7 Minimizing Shear Damage

Process in continuous flow generates shear in all the steps of the process, and this shear can affect particulates through attrition and breakage, leading to a reduction in particle size, unsharp corner, and "clouding" of otherwise clear broth. The rate of shear damage is at its maximum when the product is hot (at the end of heating, and holding) and in the first stage of the cooling section when particulates have softened due to the thermal treatment. Components of the process line need to be designed systematically to minimize the shear exerted on the particulates and the liquid structure. In practice, this means using short tubes of large diameters, avoiding sharp transitions in flow cross sections and providing gentle agitation when required (e.g., in the aseptic hold tank).

Pumps and back pressure devices require special attention in heterogeneous flows. Pressures in heterogeneous products may run as high as 30 bar (450 psi) or more and positive displacement pumps are required to deliver such heads. Rotary, reciprocal, screw, and lobe pumps can be used with different suppliers offering pumps that can be used for delicate particulates. Pump selection should also account for particle thermal history and the viscosity of the continuous phase.

Back pressure devices are used to prevent boiling at high temperatures. For homogeneous products, spring valves and diaphragms are used which essentially generate a pressure drop over a narrow slit. These devices are unsuitable for particulate products due to the high amount of shear and the probability of blockage of the narrow slit. In these cases, back pressure can be generated having a second positive displacement pump at a lower speed. Other options are especially designed back pressure devices (Cartwright, 2004), or using multiple aseptic tanks.

Aseptic hold tanks can be used as a back pressure device to avoid particle damage. In such operation, at least two aseptic tanks are operated in the alternating mode under a sterile gas blanket (e.g., nitrogen). The first tank is filled up to the target level, upon which the flow is diverted to the next (pressurized) aseptic tank. Next, the pressure is released from the first tank and its content fed into the filler. This mode of operation eliminates the need for a back pressure device which is incompatible with heterogeneous products; however, it is more complex in terms of operation, control, and hygiene.

7.6 ASEPTIC COST OPTIMIZATION

Cost of aseptic products is a big concern for producers, while in Section 14 we aimed to optimize from a quality perspective, this section is concerned with obtaining cost benefits by optimizing ingredients and production. Starting at a global scale we can optimize the sourcing of some ingredients, then we'll look at factory scale production scheduling, and finally, proper control to ensure safe processing, while at the same time creating conditions for uninterrupted production. Such run-length extension unlocks the full potential of the economies of scale provided by aseptic processing.

7.6.1 Ingredient Sourcing

Ingredients can be obtained from various sources to obtain cost benefits. Ambient stable ingredients like oils, flavors, powders, thickeners, and vegetable pastes are sourced on the global market. The same is true for frozen ingredients such as meats and vegetables, although sourcing may be more local due to the costs incurred by the frozen supply chain.

Fresh ingredients such as vegetables, dairy components, and fresh meat have a very limited shelf life even when kept refrigerated, and their use is generally constrained by geographic proximity. Depending on the season, fresh produce might be considerably cheaper than frozen, but the seasonality of crops would limit production to a few weeks per year leading to very low capital utilization and high product storage cost. Clearly there has to be a balance between capital expenditure, work–force utilization, cost of ingredients, and warehousing, so that production may (partially) be steered by the availability of seasonal ingredients as is the case for tomatoes and fruits. However, for most aseptic factories, the main vegetable ingredients are used in the frozen state.

7.6.2 Factorywide Optimization

Aseptic processes have the advantage of delivering a wide range of products in a variety of product formats. Products of very different colors, viscosity, and pH might be produced in the same factory, with special consideration to food safety and the presence of allergens. Allergens pose a unique challenge, as these are classified in groups that cannot be processed simultaneously with other products and require especial cleaning and verification cycles. Down time of the production lines might be the result of changes from one product family to another depending on the difference in the specific characteristics (e.g., color or flavor), products considered organic, or the presence of allergens. Especially after products containing allergens, a complete CIP and SIP cycle (including additional validation) is required. In addition to changes in product, changes in product format can be time consuming as the aseptic filler and packing machine might require adjustments and possibly re-assembly.

Generally, factories share resources and utilities for multiple manufacturing lines. Typical examples of shared resources are the CIP/SIP system, waste treatment, compressed air, and heating/cooling utilities. To reduce investment in such systems, sizing is not based on peak demand: for example, the CIP system is not designed to provide CIP to all lines at the same time.

Optimizing the sequence of product and format on each line, and sharing auxiliary resources between lines and can be achieved in two ways. The first option is to create sufficient idle time on the production lines, at the cost of low capital utilization. The second option is to apply multistage scheduling to maximize the overall factory capacity. Bongers and Bakker (2006) applied the latter methodology to ice-cream manufacturing lines to the effect of significantly higher utilization and less unexpected shut-downs and product waste. This approach is expected to have a positive impact on industrial food manufacturing in general and will benefit most aspects of the product costs.

7.6.3 Optimal Control Strategies

Digital control systems are now widespread in factories as they provide a reliable and efficient control system. Aseptic control systems have the purpose to maintain the specified time/temperature profiles despite disturbances and ensure safe products (Hasting, 1992). The speed of reaction of the control system is dictated by the fastest critical step, in the case of volumetric heating response times in milliseconds are needed.

A general controlled process setup is depicted in Figure 7.7 in a simplified manner. A certain control point (set point) is desired at the end of a process step, and the controller compares the value of a process variable (PV) to the set point, applying a correction (Δe) to the process equipment in the form of a real manipulated variable (MV), which can be a certain voltage or the order to open or close a valve. The system reacts to disturbances acting upon the different stages of the thermal process such as:

- Flow variations
- Variations in product properties (including dielectric properties)

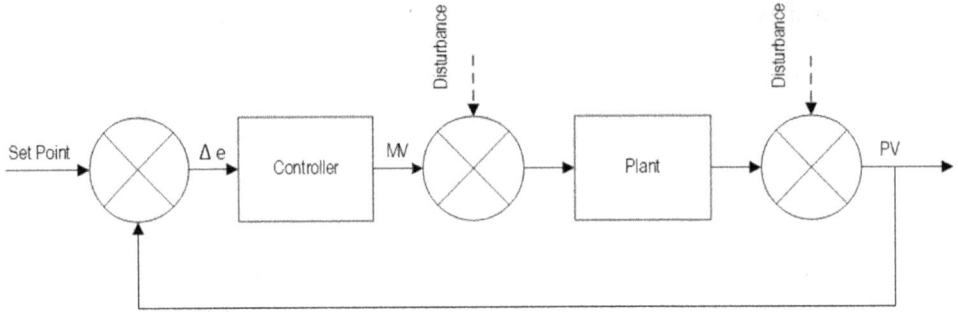

Figure 7.7   Simplified process control system.

- Fluctuations of the heating/cooling utilities
- Fluctuation in incoming temperature
- Wear of the control actuators and control valves (Ruel, 2000)
- Poor controller tuning (Buckbee, 2002)
- Fouling of the heat exchanger (both for product and medium side)

Apart from disturbances, bias in the temperature or flow sensors may result into under-processed products which are unsafe, calibration of the control instruments and system is required as part of the food safety plan and preventive maintenance. A measured temperature history in an untuned system is shown in Figure 7.8. Due to disturbances discussed before, considerable fluctuations are apparent; and the process is out of control. In factory operation, the action limit

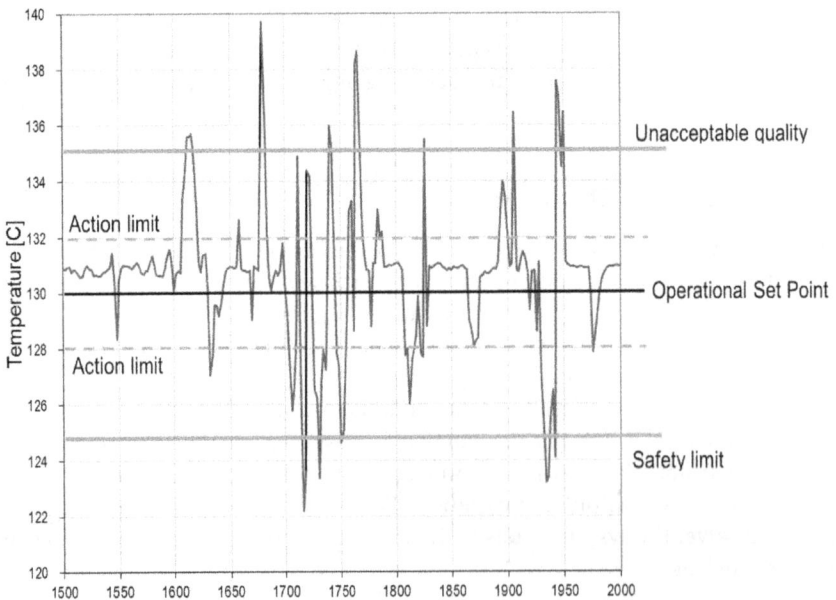

Figure 7.8   Uncontrolled system.

of temperature will prevent product from entering the aseptic holding tank leading to waster; and if the temperature went below the safety limit, the whole system is now "unsterile" and should go into a CIP/SIP cycle.

Before tuning the control, it is obvious that temperatures can fluctuate outside of the action limits, and even the safety limit. Depending on these fluctuations, the temperature set point must be well above the safety limit in order to prevent loss of time for a CIP/SIP cycle. As a consequence, the maximum temperatures occurring in the exchanger are also higher than originally design, leading to product with lower quality attributes and more rapid fouling.

To reduce temperature fluctuations, tuning of the control system control is required before starting processing. A Proportional-Integral-Derivative (PID) controller is the most popular feedback controller used in the process industries (Stephanopoulos, 1984). PID-controller parameters are:

- P = Proportional gain: feedback is proportional to the error measured
- I = Integral time: eliminate offset between set point and process variable
- D = Derivative time: reduce the time to reach the set point

Tuning a control loop is the adjustment of its control parameters to values for the desired control response. This can be achieved most effectively using a process model (in some shape or form), and setting P, I, and D based on the model parameters. Manual tuning has proven inefficient, inaccurate, and often dangerous, and most modern industrial facilities use PID tuning and loop optimization software. These software packages will gather the data, apply process models, and suggest optimal tuning. Tuning the PID parameters for the process as shown in Figure 7.8 resulted in stable operation (Figure 7.9).

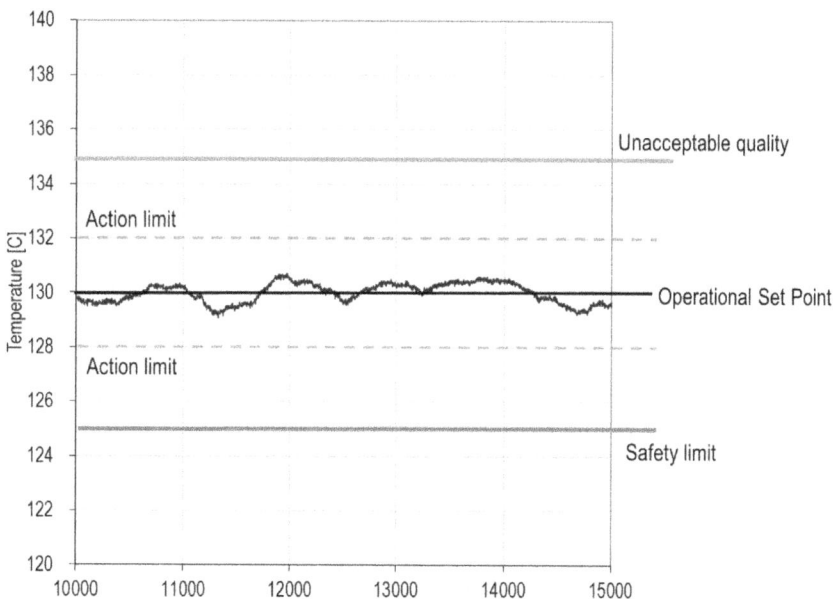

Figure 7.9   System under control.

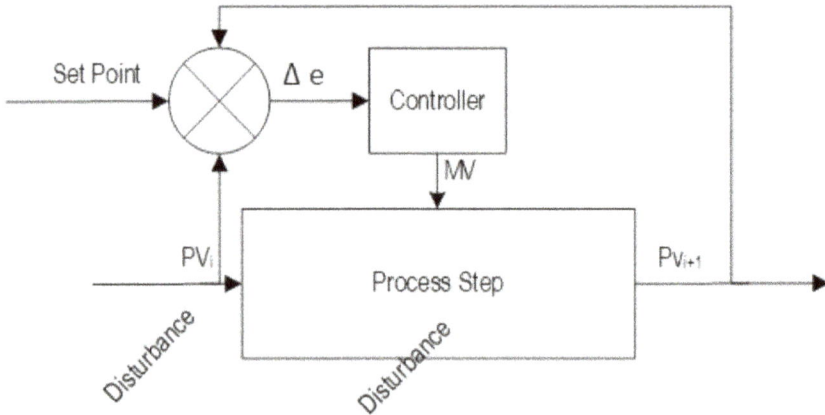

Figure 7.10   Simplified feed-forward control.

One of the drawbacks of feedback control is that the deviations need to be substantial before the controller can respond. When a disturbance can be measured in the process variable (PV_i) before the process step, feed-forward control (Figure 7.10) can compensate more rapidly and can thus be more effective in maintaining the final value pf this variable (PV_{i+1}) closer to the set point. These controllers need to be able to handle several inputs of the same variable from different points in the process (PV_{i-1}, PV_i, PV_{i+1}, ...) and convert them into an MV that is able to control the output rapidly and precisely. Feed-forward controllers are especially useful for rapid control needed in volumetric heating.

Building on both concepts of feedback and feed-forward control, more advanced control strategies may be devised. Temporary failures downstream from the continuous part of the aseptic system (filler, packaging machines, etc.) cannot always be buffered by the aseptic tank(s) and the desired flow through the heat exchangers may vary to prevent stoppage and product wastage. In modern aseptic systems, flow variations are possible within a limited range. However, when the temperature set-points are not modified accordingly, under- or overprocessing results.

To overcome this, Negiz et al. (1996) and Schlesser et al. (1997) introduced the concept of multi-variable control in HTST pasteurization systems to simultaneously control product flow (and hence residence time in the holding tube) and product temperature. This concept can be extended to maintain a constant lethality rate and optimize product quality attributes, and accommodate for production requirements (e.g., intermediate cleaning, filler availability).

7.6.4 Run-Length Extension

When products are run in a continuous line, burning and thermal degradation in the wall is unavoidable, but it should be minimized wherever possible. Such deposits on the walls of the heat exchangers either affect the flow performance of the equipment by increasing the required pressure head, or they diminish the efficiency of heat transfer. Fouling may also present a micro-biological hazard in the accumulation of materials in non-hygienic parts of the line, or create quality problems by having off-colors and flavors or when large deposits are removed from the wall and are carried with the product.

Fouling is the result of a combination of several physicochemical processes which are accelerated at high temperatures such as crystallization, corrosion, caramelization, polymerization, Maillard reactions, and protein denaturation. In many cases products processed have a large variety and complexity of the food materials, fouling can be a complex residue composed of carbohydrates, proteins, fats and minerals, and the composition and properties of the deposits can be very different from those of the individual food ingredients.

The final result of fouling can be a deposit which can be observed by a pressure drop on the product side, or temperature (or pressure) of steam as a result of the reduced heat transfer. However, it would be better to detect fouling in earlier stages of its development. The development of fouling can be illustrated by the following stages: induction, transport, attachment, build-up, and aging. During induction and attachment very few changes in pressure and heat can be observed, while during build-up and aging these changes can be observed easily (Figure 7.11). Detecting fouling in these early stages is not an easy task and research is being carried out in the subject by several groups (Belmar & Fryer, 1993, Kim & Lund, 1998).

Cleaning of aseptic lines is required to remove the fouling that has accumulated and return the system to the initial state. Normally limits on pressure drop and steam temperature trigger the shutdown of production and the start of a cleaning cycle. Modern plant equipment can be cleaned in place (CIP) where equipment is cleaned either by flow of the cleaning solutions or spraying such chemicals onto walls (e.g., cook vessel). CIP requires a parallel system to convey cleaning solutions and rinsing water at the speed and temperature needed for a proper cleaning. However, some equipment requires opening and (manual) cleaning.

Cleaning solutions include caustic, acid, and detergent solutions applied in a different order depending on the fouling deposits. Milk protein fouling has been studied extensively and the cleaning of such deposits is performed by water rinse, caustic cleaner, acid rinse, and water rinse. Cleaning is performed in that order since the proteins are more susceptible to an alkali environment, whereas the salt deposits dissolve in acids. Other products may require different cleaning regimes since their deposits are composed of different elements (Gillham, 1999, Bansal & Chen, 2006).

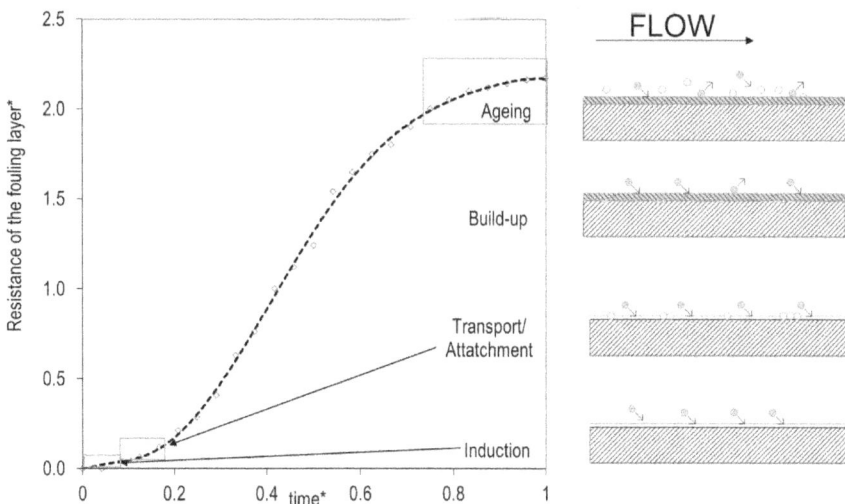

Figure 7.11   Fouling development stages.

Cleaning cycles are optimized in terms of utilization of cleaning solutions, energy usage, and water consumption, but the interruption of production remains costly both in terms of lost production time, waste, and energy. Though cleaning chemicals may be reused to save cost and protect the environment, they represent a potentially hazardous waste stream. Cleaning in principle solves the problem of fouling, but preventing it should have a high priority.

Fouling can be reduced by the adequate setting of the operating parameters. This includes a moderate and approximately constant temperature differential between the product and the medium side. For this reason, counter-current pressurized water is mostly used as the heating medium where direct application of steam would lead to burn-on of the product. It is also vitally important that the process is well controlled, as large temperature fluctuations imply high maximum temperatures, and can cause severe burn-on and fouling. In some cases, direct (steam) heating may be applicable, as the heat is transferred directly into the bulk of the product by condensation. However, steam-injection points should be carefully designed and well-insulated to avoid direct contact of the product with very hot surfaces associated with the steam supply.

Flow velocity is also an important factor, as an increase in velocity generally decreases the rate of fouling by increasing shear, turbulence and preventing that the fouling deposits "stick" to the wall. Second, minimizing the heat transfer resistance in the liquid as compared to the tube wall decreases the temperature at the metal–product interface, which decreases fouling rates. When fouling is severe, and other solutions do not apply, use of an agitated heat exchanger (SSHE) may be considered. Such heat exchangers continuously remove deposits from the walls, and can keep the heat transfer rate sufficiently high for longer times, provided the deposits do not impair the final product quality.

Careful formulation, selection, and pretreatments of the ingredients may also mitigate fouling. Dairy ingredients may be replaced by oils and proteins having a vegetal origin, or by denatured versions of those ingredients such as pretreated (e.g., pasteurized) cream to inactivate the main fouling component β-lactoglobulin (Bansal & Chen, 2006). Protein-rich foods can be pretreated to prevent fouling, and starches and gums can be prehydrated and precooked to the desired values of viscosity (Lelieveld et al., 2005).

An aseptic process line can also be designed to minimize the severity or impact of fouling. An example of the first approach is to include intermediate holding tubes where fouling components (such as certain dairy proteins) are given a residence time to inactivate or denature (De Jong, 1997). The latter approach can be embodied in the design of a heat exchanger having a sufficiently large surface area. Deposits will now be spread over a larger surface area, and the run length can be extended before cleaning is required.

Finally, networks of heat exchangers are used in the sugar industry, where fouling is so severe that in intermediate, in-production cleaning is scheduled to certain portions of the heat exchanger network (Smaili et al., 1999).

7.7 FUTURE TRENDS

Aseptic processing of heterogeneous aseptic food products can also be grouped according to quality and cost optimization. To improve product quality, novel technologies such as volumetric heating can bring significant benefits. Operational changes in the process allow for a successful application of such technologies, which should be driven by the required properties of the final product as perceived by the consumer. One of such operational changes is the use of "split-stream" processes, where the heat treatment could be optimized for the different product components, which could be aseptically assembled in the final pack.

Rapid cooling is one of the fields in which more research is needed, and where optimization can be fruitful.

From a cost optimization perspective, it is expected that production will be increasingly global and may shift between production locations to follow the seasonal supply of raw materials. Alternatively, production facilities may acquire mobility to maximize capital utilization. Finally, though the economy of scale is one of the selling points of aseptic processing over retorting, current trends indicate a demand for smaller economic production run-lengths to cater to shifting consumer preferences. Ideal aseptic systems should therefore combine the product quality of aseptic processing with the flexibility of the traditional retorting.

NOMENCLATURE

A	m^2	Surface area
C	minutes	Cook value
Cp	$J\,kg^{-1}\,°C^{-1}$	Specific heat capacity
D	minutes	D value
Ea	$J\,mol^{-1}$	Activation energy
F_0	minutes	Equivalent process time at 121.1°C
h	$W\,m^{-2}\,°C^{-1}$	Heat transfer coefficient
l_f	M	Tube perimeter
m	$kg\,s^{-1}$	Mass flow rate
n_p	-	Number density of particles
R_r	-	Relative reaction rate
t	s	Time
T	°C	Temperature
T_r	°C	Reference temperature
T_s	°C	Starting temperature of phase transition
T_e	°C	End temperature of phase transition
U	$W\,m^{-2}\,°C^{-1}$	Overall heat transfer coefficient
w	$m\,s^{-1}$	Local axial velocity
z	m	Streamwise distance
Z	°C	Z-value
ΔH	$J\,kg^{-1}$	Enthalpy of melting
λ	$W\,m^{-1}\,°C^{-1}$	Thermal conductivity
ρ	$kg\,m^{-3}$	Density
Subscripts		
f		Of the fluid
m		Of the medium
p		Of the product/particle
w		Of the wall

REFERENCES

9 CFR, Parts 308,318,320,327 and 381. 1986. *Canning of Meat and Poultry Products*. U.S. Department of Agriculture.

21 CFR, Parts 11, 113,114 and 117. 2016. Washington, DC: U.S. Government Printing Office.

Awuah, G.B., Economides, A., Shafer, B.D., and Weng, J. 2004. Lethality contribution from the tubular heat exchanger during high-temperature short-time processing of a model liquid food. *J. Food. Proc. Eng.* 27:246–266.

Bansal, B., and Chen, X.D. 2006. Effect of temperature and power frequency on milk fouling in an ohmic heater. *Food Bioprod. Process.* 84(C4):286–291.

Banga, J.R., Balsa-Canto, E., Moles, C.G., and Alonso, A.A. 2003. Improving food processing using modern optimization methods. *Trends Food Sci. Technol.*, 14(4):131–144.

Barigou, M., Mankad, S., and Fryer, P.J. 1998. Heat transfer in two-phase solid-liquid food flows: A review. *Trans. IChemE pt. C* 76:3–29.

Belmar-Beiny, M.T. and Fryer, P.J., 1993. Preliminary stages of fouling from whey protein solutions. *Journal of Dairy Research*, 60(4):467–483.

Bhamidipati, S., and Singh, R.K. 1994. Thermal time distributions in tubular heat exchangers during aseptic processing of fluid foods. *Biotechnol. Prog.* 10:230–236.

Bird, R.B., Stewart, W.E, and Lightfoot, E.N. 2006. *Transport Phenomena*. 2nd edition. Hoboken, NJ: John Wiley & Sons.

Bongers, P.M.M., and Bakker, B.H. 2006. Application of multi-stage scheduling. In ESCAPE 16 Proceedings.

Braud, L.M., Castell-Perez, M.E., and Matlock, M.D. 2000. Risk-based design of aseptic processing of heterogeneous food products. *Risk Anal.* 20(4):405–412.

Buckbee, G. 2002. Poor controller tuning drives up valve costs. *IEEE Contr. Syst. Mag.* 15:47–51.

Cacace, D., Palmieri, L., Pirone, G., Dipollina, G., Masi, P., and Cavella, S. 1994. Biological validation of mathematical modeling of the thermal-processing of particulate foods: The influence of heat-transfer coefficient determination. *J. Food Sci.* 23(1):51–68.

Cartwright, G.D. 2004. Apparatus and method for controlling flow of process materials. Patent US2004/001335.

Canned Food: Principles of Thermal Process Control, Acidification and Container Closure Evaluation. 7th edition. 2007. Washington, DC: GMA Science & Education Foundation.

David, J.R.D. 1985. *Kinetics of inactivation of bacterial spores at high temperature in a computer-controlled reactor*. Ph.D. dissertation, University of California at Davis.

David, J.R.D. 2013. Thermal processing and optimization. In *Handbook of Aseptic Processing and Packaging*, edited by David, J.R.D, Graves, R.H., and Szemplenski, T., Chapter 11, 1–8. 2nd edition. Boca Raton, FL: CRC Press, Taylor and Francis Group.

De Jong, P. 1997. Impact and control of fouling in milk processing. *Trends Food Sci. Tech.*, 8(12):401–405.

Downes, F.P., and Ito, K., eds. 2001. *Compendium of Methods for the Microbiological Examination of Foods*. 4th edition. Washington, DC: American Public Health Association.

Fellows, P.J. 1988. *Food Processing Technology, Principles and Practice*. Woodhead Publishing Ltd.

Floros, J.D., Ekanayake, A., Abide, G.P., and Nelson, P.E. 1992. Optimization of a diced tomato calcification process. *J. Food Sci.* 57(5):1144–1148.

Grade "A" Pasteurized Milk Ordinance. 2009. *Revision*. U.S. Department of Health and Human Services, Public Health Service, Food and Drug Administration (FDA).

Gillham, C.R., Fryer, P.J., Hasting, A.P.M., Wilson, D.I. 1999. Cleaning in place of whey protein deposits: Mechanisms controlling cleaning. *Trans. IChemE pt. C*, 77:127–136.

Hasting, A.P.M. 1992. Practical considerations in the design, operation and control of food pasteurization processes. *Food Control* 3:7–32.

Holdsworth, S.D. 1992. *Aseptic Processing and Packaging of Food Products*. Elsevier Applied Science.

Jung, A., and Fryer, P.J. 1999. Optimizing the quality of safe foods: Computational modelling of a continuous sterilization process. *Chem. Eng. Sci.* 54:717–730.

Kelder, J.D.H., Ptasinski, K.J., and Kerkhof, P.J.A.M. 2002 Power-law foods in continuous coiled sterilizers. *Chem. Eng. Sci.* 57:4605–4615.

Kim, J.C. and Lund, D.B., 1998. Milk protein/stainless steel interaction relevant to the initial stage of fouling in thermal processing. *Journal of Food Process Engineering*, 21(5):369–386.

Lelieveld, H.L.M., Mostert, M.A., and Holah, J. editors. 2005. *Handbook of Hygiene Control in the Food Industry*. Woodhead publishing Ltd.

Liao, H-J., Rao, M.A., and Datta, A.K. 2000. Role of thermo-rheological behavior in simulation of continuous sterilization of starch dispersion. *Trans. IChemE pt. C* 78:48–56.

Luh, B.S. 1970, May. *Physicochemical Differences of Pureed Vegetables Packed by the Aseptic and Retort Processes*. Davis, CA: University of California at Davis, for United States Steel.

Lund, D.B. 1977. Design of thermal processes for maximizing nutrient retention. *Food Technol.* 31(2):71–72, 74, 76–78.

Lund, D.B. 2003. Heat processing. In *Physical Principles of Food Preservation*, edited by M. Karel, O.R. Fennema, and D.B. Lund, Chapter 3. 2nd edition. New York: Marcel Dekker.

Mankad, S., Branch, C.A., and Fryer, P.J. 1995. The effect of particle slip on the sterilization of solid-liquid food mixtures. *Chem. Eng. Sci.* 50(8):1323–1336.

Mankad, S., and Fryer, P.J. 1997. A heterogeneous flow model for the effect of slip and flow velocities on food sterilizer design. *Chem. Eng. Sci.* 52(12):1835–1843.

Mateu, A., Chinesta, F., Ocio, M.J., Garcia, M., and Martinez A. 1997. Development and validation of a mathematical model for HTST processing of foods containing large particles. *J. Food Protect.* 60(10):1224–1229.

Merson, R.L., and Leonard, S.J. 1979. *Principles of Thermal Processing: FST 150 Class Notes*. Davis, CA: Department of Food Science and Technology, University of California at Davis.

Morgan, M.T., Lund, D.B., and Singh, R.K. 2010. Design of the aseptic processing system. In *Principles of Aseptic Processing and Packaging*, edited by P.E. Nelson, Chapter 2. 3rd edition. Purdue University Press.

National Food Processors Association (NFPA). 2004. Technical overview of alternate sterilants for use in aseptic processing. In National Food Processors Association Workshop, Arlington, VA, June.

Negiz, A., Cinar, A., Schlesser, J.E., Ramanauskas, P., Armstrong, D.J., and Stroup, W. 1996. Automated control of high temperature short time pasteurization. *Food Control* 7(6):309–315.

Nelson, P.E. (editor) 2010. *Principles of aseptic processing and packaging*, 3rd edition, edited by P.E. Nelson, Chapter 3. West Lafayette, IN: Purdue University Press.

Richardson, P. 2004. *Improving the Thermal Processing of Foods*. Woodhead Publishing Ltd.

Ruel, M. 2000. How valve performance affects the control loop. *Chem. Eng. Mag.* 107(10):13–18.

Sandeep, K.P., Zuritz, C.A., and Puri, V.M. 1999. Determination of lethality during aseptic processing of particulate foods. *Trans. IChemE pt. C.* 77:11–17.

Sandeep, K.P., Zuritz, C.A., and Puri, V.M. 2000. Modelling non-Newtonian two-phase flow in conventional and helical-holding tubes. *Int. J. Food Sci. Technol.*, 35:511–522.

Sastry, S.K., and Cornelius, B.D. 2002. *Aseptic processing of foods containing solid particulates*. Hoboken, NJ: John Wiley & Sons Ltd.

Schlesser, J.E., Armstrong, D.J., Cinar, A., Ramanauskas, P., and Negiz, A. 1997. Automated control and monitoring of thermal processing using high temperature, short time pasteurisation. *J. Diary Sci.* 80:2291–2296.

Simpson, S.G., and Williams, M.C. 1974. An analysis of high temperature/short time sterilisation during laminar flow. *J. Food. Sci.*, 39:1047–1054.

Singh, R.K., and Morgan, M.T. 2010. Residence time distribution in aseptic processing. In *Principles of Aseptic Processing and Packaging*, edited by P.E. Nelson, Chapter 3. 3rd edition. Purdue University Press.

Skjöldebrand, C., and Ohlsson, T. 1993. A computer simulation program for evaluation of the continuous heat treatment of particulate food products. Part 1: Design, *J. Food Eng.* 20(2):149–165.

Smaili, F., Angadi, D.K., Hatch, C.M., Herbert, O., Vassiliadis, V.S., and Wilson, D.I. 1999. Optimization of scheduling of cleaning in heat exchanger networks subject to fouling: Sugar industry case study. *Food Bioprod. Process.* 77(C2):159–164.

Stephanopoulos, G. 1984. *Chemical Process Control, An Introduction to Theory and Practice*. Prentice Hall.

Stumbo, C.R. 1973. *Thermobacteriology in Food Processing*, Chapters 10 and 11. 2nd edition. New York: Academic Press.

Toledo R. (editor) 1980. *Fundamentals of food process engineering*. Westport, CT: AVI Publishing.

Weng, Z. 1999. AseptiCal software: A mathematical modeling package for multiphase foods in aseptic processing systems. In *Advanced Aseptic Processing and Packaging*. Davis, CA: Department of Food Science and technology, UC Davis.

Wilhelmi, F. 1988. Soups and sauces: The aseptically packed feast ready for the table. *Dragoco Rep.* 3:63–77.

Chapter 8

Aseptic Filling and Packaging for Retail Products and Food Service

Patnarin Benyathiar, Dharmendra K. Mishra,
Thomas E. Szemplenski, and Jairus R.D. David

CONTENTS

DOI: 10.1201/9781003158653-10

8.1 INTRODUCTION

Aseptic processing and packaging are a methodology of food preservation where product is commercially sterilized by heating in a heat exchanger and holding at specific time and temperature before cooling. The sterilized cold product is then filled in a sterilized container and sealed hermetically to produce a shelf-stable product which does not need refrigeration for storage and distribution. Low-acid food products such as milk, high-protein beverage, soup, and purees are aseptically produced under the regulation FDA 21 CFR Part 113. On the other hand, acidified products such as juices can also be commercially produced using aseptic technology under FDA 21 CFR Part 114.

The development of aseptic technology grew out of a desire to preserve the beverage quality of milk. The ability to preserve milk before had required the complete alteration of product that came from the cow, goat, or sheep either by drying, condensing, or coagulating. All these alterations maintained the nutritional qualities of milk as a food, but it no longer was the refreshing beverage it was originally.

Aseptic processing and packaging were initially invented with the first patent (No. 2,029,303) for filling food into the steel metal cans on February 4, 1936, by C. Olin Ball. The first commercial operation for metal can filling was established with Dole aseptic canning system. Even though the metal cans are not widely used in aseptic processing today, they formed the basis of aseptically packaged food products with the Dole canning system. The Dole fillers were very reliable and worked on the heat as a method of sterilization of the cans, lids, and the filler presterilization and maintenance of sterility. The cans used in aseptic packaging are of the size #10 that can run up to a speed of 30 cans per minute and 4 oz cans that can run 450 cans per minute.

However, it was not until the 1980s that the aseptic manufacturing picked up significant pace mainly due to the acceptance of peroxide as a sterilizing agent by the USFDA. In early 1980, Tetra Pak introduced the first Brik-type paperboard packaging into the United States. Since then, there have been several advancements in the aseptic packaging materials and size of the final package. Currently, the most widely used retail aseptic packaging materials are high-density polyethylene (HDPE), polyethylene terephthalate (PET), paperboard, and high-impact polystyrene (HIPS). Commercial fillers can run at high speed from 300 to 1,200 packages per minute depending on the type of food and packaging material type and size. Recent advancements in material sterilization include the use of E-beam sterilization.

Sustainability has been at the forefront of the aseptic industry and innovative solutions are needed to resolve the issue of packaging material sustainability. Major food manufacturers have made bold moves in this area and have made changes toward sustainable packaging. This includes the use of renewable, compostable, and recyclable material, which helps reduce the carbon footprint in the entire production and distribution chain.

8.2 ASEPTIC PACKAGING

8.2.1 Packaging Material

To select a suitable packaging material for aseptic food, it is important to understand the fundamentals of aseptic processing and material properties. Unlike other food processing, the aseptic package is separately sterilized prior to being filled with sterile food in the sterile zone. The packaging for aseptic products, thus, must withstand the sterilization agents and methods. Several types of packaging materials, including metal can, paper, and plastic, have been used in aseptic

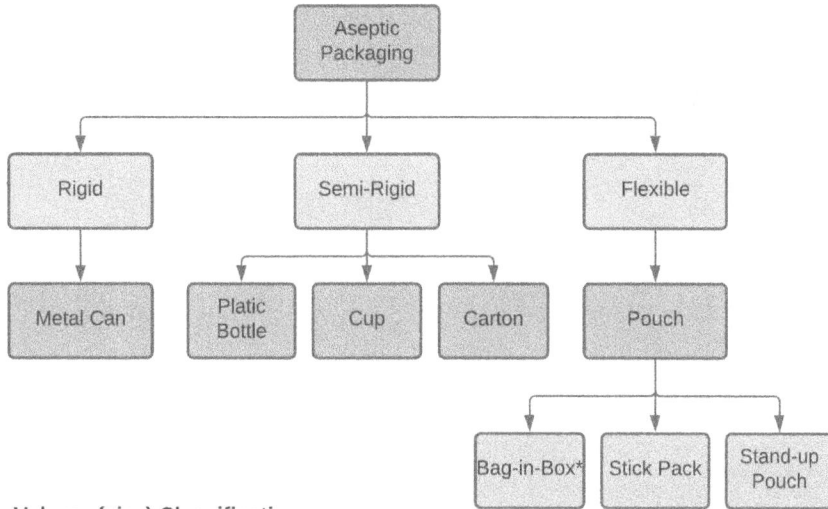

Volume (size) Classification
Retail: 4.5 - 96 oz
Food Services: 1 - 5 gal
Bulk: 10 gal - 1 million gal
*Bag-in-Box inclused the flexible bag inside and semirigid box on the outside

Figure 8.1   Classification of aseptic packages based on rigidity, shape, and volume (size).

food processing, as shown in Figure 8.1. Plastic and carton board, however, are the typical packaging material due to their flexibility in forming containers and lightweight.

Steel metal cans have been initially used as aseptic packaging prior to other materials. The first patent for aseptically packaged can milk was given to Nielsen in Denmark in 1921. Later, the first commercial aseptic canning system was started by the Dole Company in the 1960s. The metal cans with the size range from 4.5 oz (202×214) to 105 oz (603×700, the #10 can) and the canners end with a high-temperature plastisol liner, manufactured from can manufacturers, are used. The aseptic canning system has mainly produced low-acid foods, including pudding, tomato soups, tomato paste, cheese sauces, banana puree, apple sauce, eggnog, sandwich spread, dairy products, and nutritional beverages. Currently, the Dole aseptic canning system is still operating.

Paperboard beverage carton is an unbleached and/or bleached paperboard, which is laminated with layers of plastics (such as polyethylene) and aluminum foil to be impermeable to liquids, gases, and light. The combination of multilayer materials provides good moisture and gas barrier properties, strength, stability, and safety for the product. For convenience purposes, the straw hole or closure is designed to provide ease of use and consumption. These packages are usually used for dairy-based beverages, juices, stock, and sauces.

A multilayered web (Figure 8.2)

Outside layer-polyethylene (for protection of fiberboard from moisture in the environment)

- Structural layer-paper (typically one side preprinted)
- Adhesive layer (or tie layer helps to bond foil and paper, minimizes fiber penetration)
- Barrier layer-aluminum foil (barrier for light, gas, and moisture)

**Layers of aseptic
carton package**

Inside
package

Outside
package

Polyethylene

Polyethylene

Aluminium foil

Polyethylene

Paper

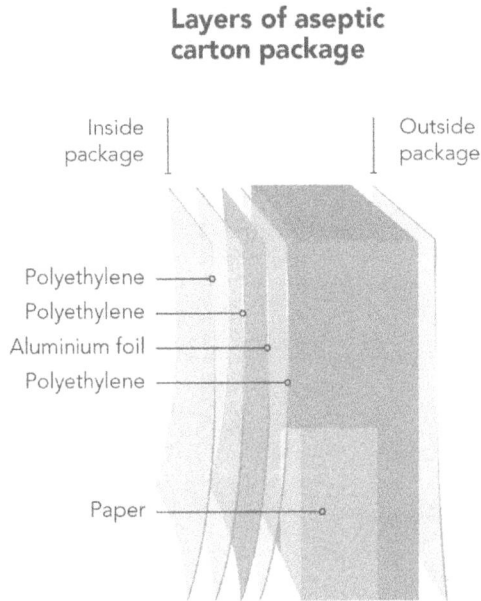

Figure 8.2 A typical material structure in preformed, paper-based barrier cartons. (Courtesy of Tetra Pak International S.A.)

- Adhesive layer
- Food contact layer-polyethylene

Thermoplastic materials used in aseptic packaging can be in the form of bottles, pouches, and cups. They can be a prefabricated container or roll stock (web) to form the container using form/fill/seal (FFS) packaging machine. To maintain the food quality and extend its shelf life, barrier materials of thermoformed packages have been chosen for the benefits that they provide. For plastic bottles, high-density polyethylene (HDPE), polypropylene (PP), and polyethylene terephthalate (PET or PETE) are general materials for aseptic beverages. HDPE is commonly used as a preformed plastic bottle, while PET can be used in the form of either a premade bottle or PET preform, which needs another step to blow to be a bottle (Figure 8.3).

Aseptic cups can be a premade or roll stock material which depends on the design of aseptic filling machines. Barrier thermoforming sheets such as high-impact polystyrene (HIPS), PP and PET are used to form plastic cups.

For the pouch style, multilayer films with barrier materials are prefabricated pouches such as the Ecolean pouch, Scholle CleanPouch, and the bag-in-box (BIB) pouch. Generally, bag-in-box pouches and Scholle CleanPouch have a spout or fitment attached to the pouch to provide convenience for food consumption and serving purposes. The web material can also be formed into pouches such as stick-pack or sachet style in an aseptic form-fill-seal machine, which can be either a vertical or a horizontal style.

The multilayer structure of the Ecolean pouch provides protection from light, gas, moisture, and other environmental conditions. It consists of polypropylene to provide excellent printing properties, mineral-filled polypropylene for rigidity, ethylene vinyl alcohol (EVOH) for oxygen barrier, carbon black to protect light, and polyethylene for sealing layer.

Figure 8.3 PET preform and PET bottles. (Courtesy of Amcor.)

Prefabricated Ecolean pouches are provided as roll stock from the Ecolean supplier, as shown in Figure 8.4.

As the name implies, a bag-in-box package is a sealed plastic bag stored inside a corrugated box. The unique part of this packaging bag style is the attachment of a spout to the bag. Thus, the bag-in-box package includes three main parts: (1) bag, (2) spout and fitment, and (3) box, as illustrated in Figure 8.5. The bag of bag-in-box package generally consists of several layers, including metallized polymeric films. Flexible film is formed as a pillow pouch style with usually four-side seals. On the surface of the bag, the spout, which is a plastic-framed opening part, is directly sealed into the bag. It provides an entry point for filling the food product into the bag. Bag-in-box spout is generally covered by a fitment. An outer package to hold the flexible bag inside is a box, which is a corrugated cardboard box. Bag-in-box pouch is usually a multilayer material that can be constructed based on the product protection requirement. It can include (1) polyester or polyethylene, (2) nylon/linear low-density polyethylene (nylon/LLDPE) films laminate or co-extruded, (3) barrier layer such as metallized lamination or EVOH, and (4) foil laminate layer.

8.2.2 Retail Packaging

The type, style, and size of aseptic packaging can vary considerably from small sachet up to bulk container depending on consumer preference, market demand, and logistic distribution.

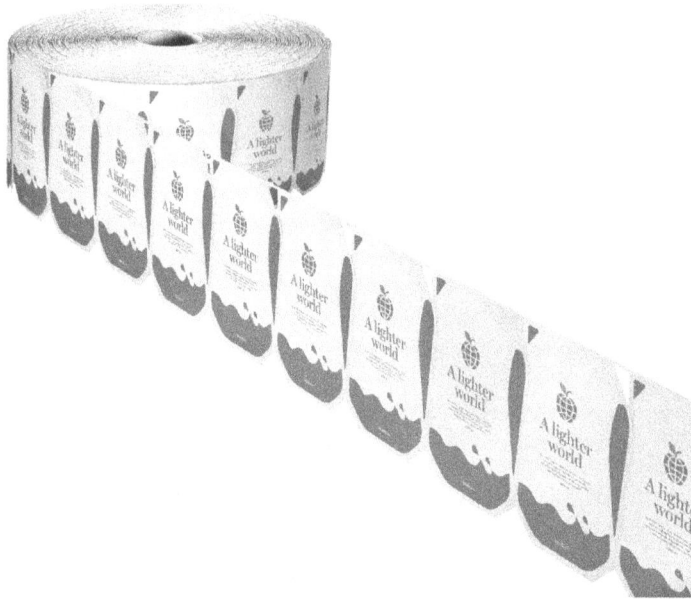

Figure 8.4   Prefabricated web-stock of Ecolean pouch. (Courtesy of Ecolean.)

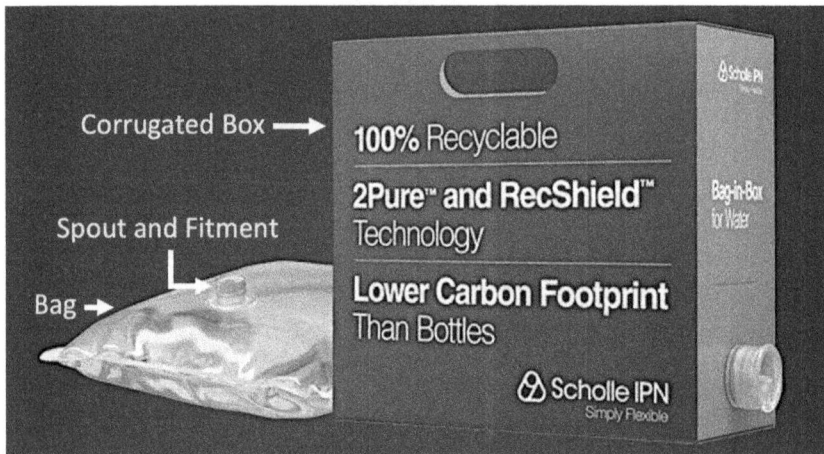

Figure 8.5   Aseptic bag-in-box (BIB) packaging. (Courtesy of Scholle IPN.)

For retail market size, aseptic packaging, including metal can, carton, cup, bottle, and pouch, can range from single-serving size to family-sized package in order to provide convenience to customers. As a single-serve portion (6.76–17 oz or 200–500 ml), an aseptic carton usually has a prepunched straw hole or pull tab sealed with aluminum foil and a drinking straw or screw caps for reclosing application are normally provided for serving on-the-go consumer (Figure 8.6). One liter (32 fluid ounces or 946 ml) is also available for family size for products such as milk, plant-based milk, and coffee.

Figure 8.6 Portion-sized beverage cartons with (*from left to right*) straw hole, pull tab, and screw cap. (Courtesy of Tetra Pak International S.A.)

Figure 8.7 Ecolean transparent (*left*) and opaque (*right*) pouches with SnapQuick.

Ecolean stand-up pouch (Figure 8.7) is designed like a pitcher with an air-filled handle, which helps to hold and pour the food product. With the unique feature of SnapQuick™, closure provides convenience for reclosing packages by folding and snapping after use. The capacity of Ecolean air aseptic pouch for the retail market is available from 125 to 1,000 ml. Moreover, pouches are offered as transparent (consumers can see the product inside) and opaque graphic pouches. They also are microwavable pouches.

Scholle introduced CleanPouch aseptic (Figure 8.8) for on-the-go consumption with the size from 0.35 to 70.39 fluid ounces (10–2,000 ml). With a special design spout, it can be processed as packaged aseptic food for both high-acid and low-acid foods such as coffee, sour cream, fruit and vegetable juice, pudding, coconut oil, sauce, milk, soy milk, and other dairy products.

For the retail-size bag-in-box pouch, the pouch capacity is available in sizes of 1–20 liters. The design of the fitment or spout depends on the usage, pouch size, and the design of the aseptic filling machine. With the unique design of an aseptic filler, the pouches are normally purchased from the same filler manufacturer. The dispensing requirement is also a criterion when selecting

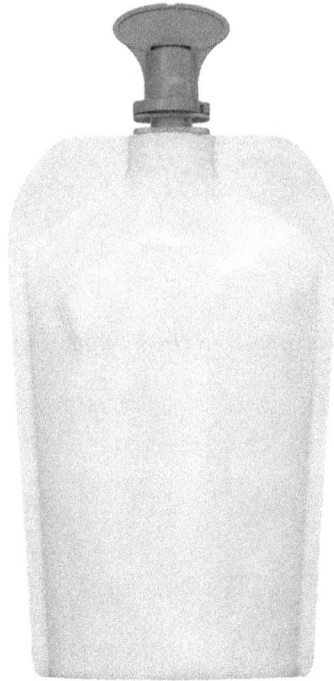

Figure 8.8   Scholle CleanPouch aseptic.

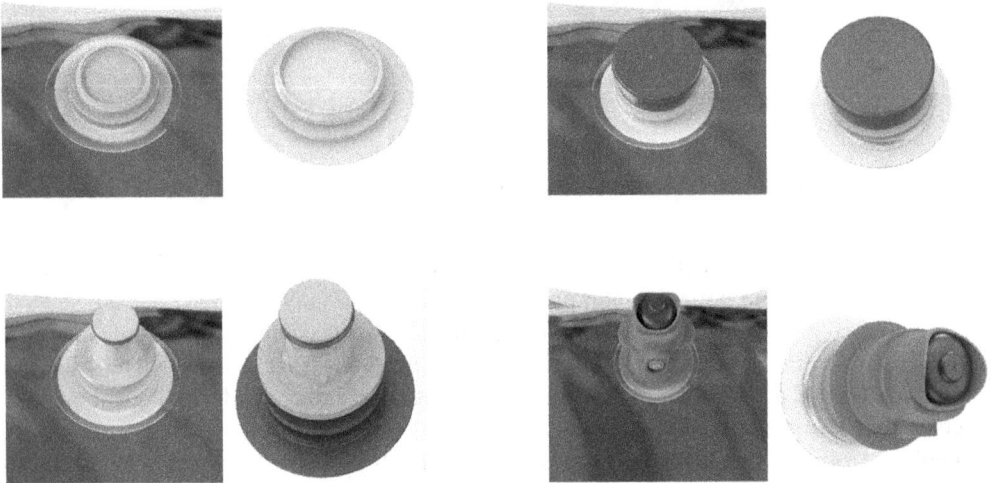

Figure 8.9   Spout and fitment styles for aseptic bag-in-box package. (Courtesy of Scholle IPN.)

a specific bag-in-box pouch. The aseptic filling manufacturers for the bag-in-box machine are Rapak, Scholle, LiquiBox, Alfa Laval, and Omve pilot filler.

Several styles of spouts and fitments for the aseptic bag-in-box package have been developed in order to securely maintain aseptic during the filling operation, storage, and distribution. The design selection depends on the customer requirement and the end-use. Bag spouts can be covered by a dispensing mechanism. They could be a simple removable cap, plastic lidding film, dispensing tap, and connector and hoses that connect to the spout with the outside dispensing equipment (Figure 8.9).

Aseptically packaged food in a thermoformed cup in the retail market also varies depending on the commercial design of the mold to create the cup shape and size, and convenience for customer consumption and serving. The small serving portion can range from a liquid creamer (0.38 fluid ounces or 11.24 ml) to a cold brew coffee cup (1.35 fluid ounces). Baby food and pudding cup size is the most common in this category, for example, Gerber baby food has a cup size from 2 to 5 oz (57–113 g). The variety of aseptic food in the thermoformed cups is shown in Figure 8.10. The aseptic cup is normally sealed with lidding film and sometimes it is covered with a plastic lid on the top of the lidding film to provide extra protection during storage and distribution. For example, Gerber baby food has a plastic lid as a resealable closure after consumption.

Bottles are the most common choice for beverages such as juice, high-protein drinks, coffee, tea, and milk-based products. A variety of aseptic beverage bottles are available on the shelves of grocery retail with the size ranging from 8 oz (227 g) for single-serve bottles and 32 oz (907 g) for multi-serve bottles. The choices of bottle closure are based on the filling manufacturer and food producer to concern about product security. A plastic screw cap with a security band is the

Figure 8.10 Aseptic food in the thermoformed cups for retail market. (Courtesy of Heartland Food Products Group and Gerber Products Company.)

Figure 8.11   #10 metal can.

primary choice for bottle closure. A lidding film can also be used as another sealing protection that is normally combined with Snap-on closure.

8.2.3 Packaging for Retail Food Service

Aseptically packaged foods are also very convenient for serving customers at a retail food service such as fast food restaurants and hotels. Metal can size 603×700 (a diameter of 157 mm and length of 178 mm), which is known as the #10 metal can (Figure 8.11), with a capacity of 105 oz (2976.7 g) is a primary packaging material used for aseptic food service. Currently, aseptic food in the metal cans is still produced by the Dole canning system, which is the first to commercialize metal cans in an aseptic format. Many aseptic foods in a metal can in the food service market are tomato soup, tomato sauce, cheese sauce, and pudding.

Bag-in-box package is suitable packaging for retail food service due to its lightweight, maximum space, and capacity for distribution. Also, the unique attachment of fitment on the bag can provide ease of opening and ease of serving. Some fitment styles such as connector and hose style help in dispensing the food or beverage for serving directly. An example is concentrated coffee products in a bag-in-box package that is attached to a coffee machine at a hotel where the concentrated coffee is mixed with hot water in the right proportion for dispensing coffee to the customers. These bags are available in various sizes and vary from 5 to 20 liters. Bags are a much preferred option as compared to the #10 cans due to the price differences and space requirement for cans. They can also provide a reduction in disposal and shipping cost.

In the early 1970s, William Scholle, an entrepreneur with tremendous foresight, visualized the potential for flexible packaging to replace the expensive, rigid packaging that was being used to transport food products such as tomatoes and other fruit products. Many years prior to this vision, Scholle developed flexible pouches of various propylene derivatives for packaging battery acid. Even today, virtually all battery acid is packaged, transported, and sold in flexible packaging invented by Scholle. Scholle felt that he could use the same packaging technology that was being used to package battery acid to package shelf-stable food products. Instead of using existing retorting technologies to commercially sterilize the product to be packaged in flexible packaging, he leaned toward applying a relatively new technology that was rapidly developing at the time: aseptic processing and packaging.

8.2.4 Tamper Proof

Tamper-evident features provide a barrier to the packaged product and identification against counterfeiting and tampering to alert the consumers of possible safety concerns before purchasing or consuming products. Different food products require different evidence of tampering. In aseptic food and beverage, tamper proof seal can be used in the form of closure with a tamper-evident band or security ring, and aluminum sealing film, which is welded to the packaging.

Seal liners are commonly used to create a hermetic seal on the packaging container or between the packaging body and the cap, which is normally a screw cap style or hinged closure. A sealing film lid can be a plastic film or plastic laminated with aluminum foil. For the bottle container, the foil liner is heat-sealed and pressed on the mouth of the plastic bottle by conduction sealing. Seal liner is commonly used for aseptic thermoformed packages. An additional resealable lid can be provided for extra protection, for example, a 3.5-oz plastic cup of aseptic Gerber baby food.

Plastic screw caps with a breakable band or security ring as a tamper-evident band are commonly used for beverage cartons, plastic beverage bottles, and spouted flexible pouches. The closure design for the carton recently consists of a flange as a base to attach with the carton body and a screw cap with an integrated cutting ring. When the cap is twisted or opened for the first time, the bridges are broken, and the security ring is separated from the screw cap. In the case of the bottle packaging, the security ring remains in place on the bottle neck.

For an aseptic carton, a foil seal liner is usually used as a barrier seal on the carton packaging and is affixed with a resealable closure. Tetra Pak has designed "HeliCap" (Figure 8.12), which has a distinctive tamper-evident closure with a tamper-evident ring and an internal cutting mechanism for the foil seal. In this case, consumers don't need to remove the foil seal or the pull-tab

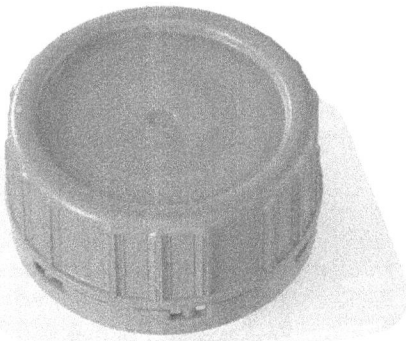

Figure 8.12 Tamper-evident plastic screw cap from HeliCap Tetra Pak (*left*) and CombiMaxx from SIG Combibloc (*right*). (Courtesy of Tetra Pak International S.A. and Combibloc.)

from the paperboard carton. A similar mechanism "CombiSwift" and "CombiMaxx" screw caps (Figure 8.12) are used in the Combibloc carton packages.

8.3 STERILIZING MECHANISMS

To sterilize packaging material, it can be done in many ways which depends on the design of aseptic filler, type of packaging materials, cost consideration, and regulatory requirement. There are four main sterilizing agents commercially used: heat, chemicals, radiation, and irradiation. They can be used independently or in combination.

8.3.1 Heat

Sterilization by heat is another common method for aseptic packaging. Heat sterilization is a simple method; however, it may not be suitable for some thermoplastic materials whose heat distortion temperatures are below the high-pressure steam temperatures.

8.3.1.1 Wet Heat

Steam is an effective sterilant demonstrated by the fact that it has been in commercial use longer than any other sterilant. The high temperature of steam is needed for the sterilization in a short time. Since the temperature used is above normal boiling conditions, it requires the use of pressure and hence a pressure chamber.

In 1950s, the Martin-Dole aseptic canning process used superheated steam at 220–226°C under pressure for 35–45 seconds as sterilant for tinplate and aluminum cans and lids (Davies, 1975; Larousse & Brown, 1996). Superheated steam is mainly used for the sterilization of aseptic metal can in Dole aseptic canning system.

To sterilize thermoforming sheets such as high-impact polystyrene (HIPS) in the OYSTAR Hassia thermoform-fill-seal aseptic filler, incoming rolls of plastic sheets are heated on the top (food contact) surface with culinary steam (160°C or 320°F) sufficient to sterilize the contact surface. The sheet is then indexed to the forming area where it is heated to its forming temperature and then indexed to the forming station where it is formed into the desired container shape. The next index moves it into the filling area followed by the sealing station and indexed out of the sterile zone. Container sterilization starts with the steam sterilization of the sheet's surface and continues through the forming, filling, and sealing stations after which containers exit the sterile zone. Lidding material is similarly sterilized prior to entering the sterile zone and sealed to the container.

In older versions of the bag-in-box aseptic filler, steam has also been used with chlorine to sterilize the spout of the aseptic bag-in-box prior to filling the sterilized food product inside the bag. Recently, this sterilization method has been replaced with the vapor peroxide sterilization of the spout.

Hot water (sometimes acidified) has been used commercially for high-acid products. The CrossCheck system, developed and patented by Mead Packaging (US patent 4,152,464), was commercialized by a joint venture of Mead Packaging and Rampart Packaging. CrossCheck was used successfully by Seneca Foods for single-serve applesauce for more than 10 years. Preformed containers were carried through a hot water bath at 82°C (180°F), exiting the bath in an inverted position (Figure 8.13). The heat seal flanges and inside of the containers were dried by the flow of warm sterile nitrogen as they rotated to an upright position prior to filling. Sterile nitrogen also

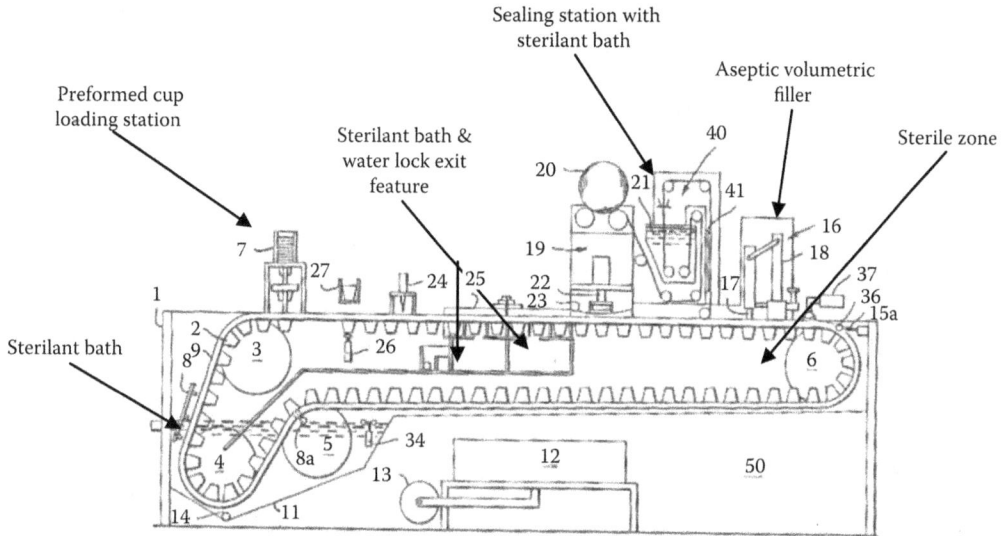

Figure 8.13   Cross-check aseptic deposit, fill, and seal system.

maintained a slight positive pressure on the inside of the sterile zone. The sterile zone was pre-sterilized by using culinary steam. Cups were filled volumetrically, sealed, and exited through a sterile water lock. Mead and Rampart stopped supporting the technology in 1990.

8.3.1.2 Dry Heat
Hot air is a dry heat method, which has been used to sterilize laminated paperboard cartons. The high temperature of 315°C (599°F) is used to sterilize packaging containers in order to achieve the surface temperature of packaging material of 145°C (293°F) for 3 minutes (Reuter, 1993; Toledo, 1988). This method is suitable for high-acidic product with a pH <4.6 such as fruit juice and tomato soup.

8.3.1.3 Heat Extrusion Process
In the process of forming plastic into container shape, heat extrusion can be used as a sterilizing agent to sterilize packaging surfaces. OYSTAR Erca-Formseal aseptic filler uses the heat of the forming packaging cup and lidding web coextrusion to sterilize the two webs. The bottom web is typically made up of polystyrene/adhesive/ethylene vinyl alcohol/adhesive/low-density poly-ethylene/polypropylene (PS/adhesive/EVOH/adhesive/LDPE/PP). The top web is typically made of PP/LDPE/adhesive/aluminum foil. In both structures, the edges of the webs are trimmed as they enter the sterile zone. The polypropylene layer of the bottom web is stripped away and exits the sterile zone. The remaining web has the newly exposed sterile surface and is indexed into the heating, forming, and filling stations. The top web, similarly having its polypropylene layer trimmed and removed as it entered the positive pressure sterile zone, is indexed to the sealing and trimming stations. Sterile bottles, which are extruded, blown with sterile air, and then her-metically sealed on the top of the bottle under sterile condition, are also used as preformed asep-tic bottles. Before the bottles enter into the aseptic zone of an aseptic filling machine, the outside of the bottles is sterilized, and the seal area is cut to fill sterile food product into the bottle.

The sterilization method from the heat of the extrusion process is suitable for high-acid food (pH<4.6) since the heat distribution of extrusion is not uniform. It requires a combination method with chemical sterilization such as hydrogen peroxide or peracetic acid to sterilize aseptic packaging for low-acid food (pH >4.6).

8.3.2 Chemical

The use of chemicals to sterilize aseptic packaging can be done alone or in combination with other sterilization methods such as heat and radiation. Chlorine was the primary sterilizing agent for aseptic packaging in the 1960s. After the US Food and Drug Administration (FDA) approved hydrogen peroxide (H_2O_2) to be used as sanitizer on food-contact surfaces of packages in 1981, it has become a widespread sterilizing agent for aseptic packaging. In 2002, peracetic acid (PAA) also received approval from FDA to be used as the sterilant. Chemical sterilants can be applied in different forms such as liquid, vapor, and spray. The type and form of chemical sterilants to be used depend on the design of aseptic filling machine.

Hydrogen peroxide (H_2O_2) is a clear, slightly viscous oxidizing agent that has been the chemical sterilant of choice for most packaging equipment manufacturers for paperboard cartons, preformed plastic bottles, and pouches. It is identified in the Code of Federal Regulations under 21 CFR 178.1005 Part (e) (1) may be safely used to sterilize polymeric food-contact surfaces. It also meets the specifications of the Food Chemicals Codex (3rd ed., 1981, page 146–147). Hydrogen peroxide is an effective sterilant when used in concentrations of 30–35% and an activation temperature above 100°C or 212°F (for vapor generation), and then followed by hot air drying at 60–125°C (140–257°F). According to FDA regulation for the sterilization of food packaging material by hydrogen peroxide, the concentration of hydrogen peroxide residual present in distilled water packaged under production condition must be no greater than 0.5 part per million of hydrogen peroxide (0.5 ppm). The hot air substantially increases sporicidal activity in addition to dissipating the residual hydrogen peroxide.

Peracetic acid (CH_3CO_3H) is a colorless, liquid organic compound that is highly corrosive. It is accepted by the FDA for sanitizing and disinfecting (21 CFR 178.1005–1010). It is always sold in solution with acetic acid and hydrogen peroxide to maintain its stability. The concentration of the acid as the active ingredient can vary depending on its application. Peracetic acid has an advantage over hydrogen peroxide in that it can be used at lower temperatures (40°C or 104°F). This is a benefit for aseptic packaging applications using materials like PET and LDPE, which have low heat distortion temperatures. Sterile water, thus, can be simply used to rinse the packaging container rather than the use of hot air to ensure a final residual of peroxide level less than 0.5 ppm as per FDA regulation since hydrogen peroxide is a principal component of PAA.

8.3.3 Radiation

8.3.3.1 Ultraviolet Radiation

Ultraviolet (UV) spectrum with wavelengths of 248–280 nm, known as UV-C, is the most effective for disinfection (optimum efficiency at 254 nm) by damaging the DNA of microorganisms (Falguera et al., 2011; Gómez-López et al., 2012). In aseptic packaging, UV-C is commercially used in combination with hydrogen peroxide. To achieve most effectiveness, packaging material should have a smooth surface, simple structural design (less corners or angles), and no dust to avoid the shading effect of surfaces.

8.3.3.2 Infrared Radiation

Infrared (IR) rays in the wavelength (λ) 0.8–15×10^6 can be used for surface sterilization of packaging materials and laminated aluminum lidding films (Ansari & Datta, 2003). After contacting with the surface of packaging materials, the radiation energy is converted into heat, which causes the increase in temperature on the packaging surface. The maximum temperature, thus, should be lower than 140°C (284°F). In a similar manner to the UV sterilization, infrared radiation is suitable only for the smooth surface without shadow effect from packaging geometries and any dust particles.

8.3.3.3 Pulsed Light

Pulsed light (PL) is a non-thermal technology, which has gained more attention in food processing and packaging sterilization. In 1970, the short pulsed UV light and inert-gas flash lamps to eliminate microorganisms was discovered in Japan. Subsequently, Pulsed light technology was developed for surface decontamination of packaging materials in 1990 (Dunn et al., 1991). This technology was first adopted for food processing and handling by FDA (Code of Federal Regulation, CFR: 21CFR179.41) in 1996 (FDA, 1996). Pulsed light has a wide wavelength range of 200–1,100 nm, which includes spectrum of UV (200–400 nm), visible (VIS) light (400–700 nm), and near-infrared (NIR) (700–1100 nm) (Elmnasser et al., 2007; Palgan et al., 2011), and a duration flash range of 1 μs to 0.1 second (Dunn, 1996). It provides an energy density in the range of 0.01–50 J/cm² (Condon et al., 2014). As compared to other sterilization methods such as heat and chemical treatments, pulsed light sterilization is a faster process and leaves no residues on food surface material.

Like other light sterilization methods, the efficiency of pulsed light can be limited by (1) opacity of packaging material, (2) insufficient light exposure, and (3) shadow effect from packaging geometries, material surfaces, and dust particles. The efficacy of pulsed light to inhibit microorganisms is limited by the visual transparency of objects. The germicidal effect of pulsed light is based on the light wavelength of lambda (<400 nm) and UV (<280 nm) (Levy et al., 2012; Woodling & Moraru, 2007; Chen et al., 2015). Opaque packaging material can block the penetration of pulsed light; however, the light is still able to penetrate through some packaging materials such as HDPE, LDPE, LLDPE, PP, and nylon with their natural color. The penetration of pulsed light is also limited to the surface roughness of the laminated layers of polymer, paperboard, and aluminum in beverage carton, which is greater than a micrometer (μm), compared to surface roughness of LDPE and HDPE (Ringus & Moraru, 2013; Chen et al., 2015).

8.3.3.4 Cold Plasma

Non-thermal plasma (NTP), known as cold plasma, is an emerging non-thermal decontamination technology, which has been widely used in food packaging as an alternative source for surface sterilization and disinfection methods without leaving any residual. Plasma has been defined by Irving Langmuir in 1928 as "the fourth state of matter (beyond the three distinct phases: gas, liquid, solid), which is an ionized state of gas." It is an electrically charged gas, mixed between negative electrical charge and positive charged ion, which can be generated by exposing a gas to an electric field either constant (direct-current field) or alternating amplitude (high-frequency field) (Misra et al., 2011; Ratner et al., 1990; Thirumdas et al., 2015). Many electrical devices, including dielectric barrier discharge (DBD), cascaded dielectric barrier discharge (CDBD), resistive barrier discharge (RBD), corona discharge (CD), glow discharge (GD), radio frequency discharge (RFD), and atmospheric pressure plasma jet (APPJ), can generate cold plasma.

Cold plasma can inactivate a wide range of microorganisms, including vegetative and spore-forming bacteria, yeast, mold, and virus. The use of cold plasma at 200 W can reduce more than 3.5 log of *Bacillus subtilis* within 5 minutes (Hury et al., 1998). The CDBD plasma treatment on PET foil can inactivate *Aspergillus niger, Bacillus atrophaeus, Bacillus pumilus, Clostridium botulinum* type A, *Clostridium sporogenes, Deinococcus radiodurans, Escherichia coli (E.coli), Staphylococcus aureus,* and *Salmonella mons* (Muranyi et al., 2007). Low-pressure microwave plasma can reduce 10^5 CFU of *B. atrophaeus* and 10^4 CFU of *A. niger* on PET bottles in less than 5 seconds (Deilmann et al., 2008).

Due to the low temperature and fast method, cold plasma can be used as a surface sterilization for various packaging material such as plastic bottles, cups, pouches, lidding films, and, especially, heat-sensitive materials such as polyethylene (Vesel & Mozetic, 2012). Moreover, cold plasma can be used to remove the residual of H_2O_2 from the packaging container after sterilizing with H_2O_2 (Thirumdas et al., 2015; Jacobs & Lin, 1987).

8.3.4 Irradiation

Three irradiation sources that are proved by the FDA to use for the sterilization process are gamma, electron beam, and X-ray. In aseptic packaging, gamma and electron beam are common treatments to sterilize packaging containers, especially those that cannot withstand high temperature of heat sterilization or those that have a special shape/design that is hard to sterilize by other methods. Irradiation dose of 25–30 kilo gray (kGy) is the most effective and commercially used for sterility assurance. However, it depends on the type of polymers. Even though it can effectively inactivate microorganisms and prevent foodborne illness, irradiation causes the property changes to the polymeric packaging materials. The cost and regulatory consideration are also another concern. Guidelines for radiation sterilization can be found in 21 CFR Part 178.1005.

Gamma rays are electromagnetic energy, which is emitted from radioactive forms of the element cobalt (Cobalt 60, ^{60}Co) or of the element cesium (Cesium 137, ^{137}Cs). With the high penetration, it is commonly used to sterilize aseptic packaging, especially one with complicated shape and/or special design such as bag-in-box and drum liner from Rapak and Scholle.

Electron beam (E-beam) is also commonly used to sterile preformed pouches with hermetic seals such as the Ecolean aseptic pouch. Low-energy E-beam (80–150 keV) can be used for the surface sterilization of preformed bottles and closures (Haji-Saeid et al., 2007). Recently, the aseptic filling machine with the use of E-beam technology as an in-line sterilization method to sterilize packaging surfaces has been commercialized. Tetra Pak has applied E-beam sterilization to sterilize the paperboard carton as it runs through the filling machine. This form-fill-seal paperboard carton, Tetra Pak E3, is developed to replace the use of hydrogen peroxide bath sterilization process. Moreover, Shibuya aseptic filler has also developed the first commercial E-beam sterilization system as a dry decontamination for the preformed bottles. This technology helps to replace the use of the traditional chemical sterilization process for aseptic packaging.

8.4 FILLER TYPES

8.4.1 Retail Fillers

8.4.1.1 Aseptic Paperboard

Carton packaging has been used for aseptic liquid food and beverages such as soup, broth, juice, milk, spreadable cheese, and other dairy products. There are two main types of aseptic carton-filling machine: (1) form-fill-seal (FFS) system and (2) fill-seal system (preform filler). Many shapes of carton packages are provided for product differentiation and consumer choice, including brik base, brik slim, brik with gable top, cuneiform, square, diamond, pyramid (triangle), and pillow pouch.

In an aseptic FFS system, a roll stock of paperboard sheet with printed graphic design is fed into the aseptic machine for sterilization method with a bath system of 30–35% H_2O_2 or a wetting system of 15–35% H_2O_2. Squeezer rollers and hot air are applied to remove liquid peroxide. The sterile paperboard sheet is then folded to form into a tube shape and then sealed longitudinally by heat sealer, called longitudinal seal strip (LS strip) (Figure 8.14). Sterile food is filled inside the formed tube and the bottom tube is transversely sealed. Nitrogen (N_2) or other inert gas is normally flushed to the container headspace. After sealing and exiting from an aseptic zone, the top and bottom parts are folded down to shape as a carton form. The cap applicator then

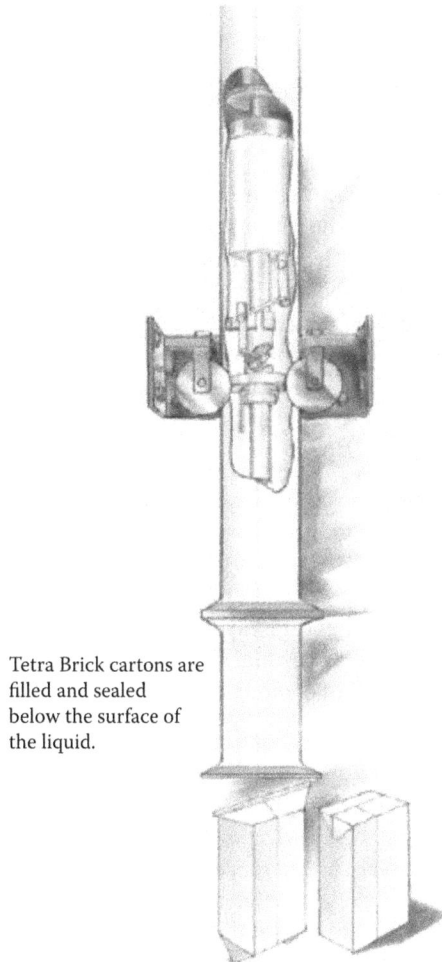

Tetra Brick cartons are filled and sealed below the surface of the liquid.

Figure 8.14   A typical roll-fed process schematic.

Figure 8.15   Tetra aseptic form-fill-seal carton-filling machine. (Courtesy of Tetra Pak International S.A.)

glues a closure on the top of the carton at the precise location. The leading producers of roll-fed, paper-based packaging systems and materials are Tetra Pak (Figure 8.15), Division of Tetra-Laval, Elopak, subsidiary of the Ferd Group of Norway, and GA Pack, a new Chinese supplier of paper-based packaging for roll-fed aseptic packaging systems.

Efforts to minimize the effects of energy-related cost increases during the 30-plus years that the brik pack has been in commercial use produced a 30% reduction in aluminum foil thickness and a 20% increase in board stiffness while reducing package weight by 15%. Although costs have been positively affected by these changes, other aspects of the carton have been adversely affected. These would include:

- Thinner aluminum foil increases the potential for pinholes to develop.
- Stiffer board increases the potential for pinholes to develop because of stiffer fiber content.
- Increased thickness of the adhesive material between the foil, the board, and the poly-ethylene (PE) layer, and the foil and PE layers on the inside of the carton, to minimize/eliminate fiber-induced pinholes through the foil and PE layers increasing the risk of organoleptic impact as a result of chemical migration.

For preform carton-filling machine, paperboard sheet is already die cut, creased, and formed as a flat cardboard sleeve from packaging manufacturer so that the bodies are preassembled and distributed in the flat. Materials used to produce the containers are similar in construction to those used to produce the roll-fed, brik-style cartons. Differences are usually found in the thickness of the fiberboard to compensate for increased carton stiffness required for larger container capacities. Pinhole issues and solutions to them are similar to those of the roll-fed laminate.

After loading into the filling machine, the flat paperboard blank is opened, and the bottom of the container body is sealed. The tabs that result from the fin seal are folded over the bottom seam or up along the sides of the container body. The gable top of the package is prefolded as a preparatory step to ensure for folding and sealing. Ultrasonic heating is used to apply the closure

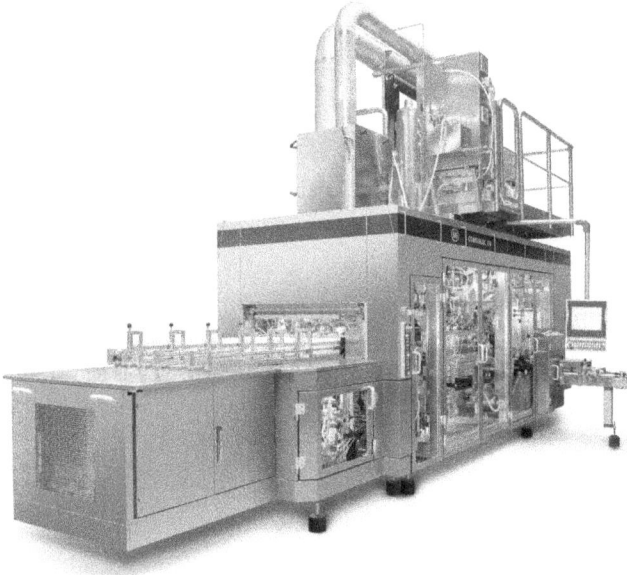

Figure 8.16   SIG Combibloc aseptic carton-filling machine. (Courtesy of SIG Combibloc.)

on the paperboard container. The inner carton is then sprayed with hot H_2O_2 vapor and then exploded with UV-light. Lastly, hot air is applied to remove the H_2O_2 residues prior to filling the sterile food product into the package. Commonly, N_2, which is an inert gas, is injected to eliminate oxygen (O_2) from headspace. The gable top is prefolded for the final top folding and then heated with the hot air to prepare for sealing with the heat sealer.

SIG Combibloc (hereafter Combibloc) is another manufacturer of aseptic filling equipment and supplier of packaging for beverages into composite cardboard, polyethylene, and aluminum packaging (Figure 8.16). Combibloc's home office is in Germany with sales and services in the United States and other countries. Combibloc has many variations of aseptic filling equipment that can fill package sizes of 125–2,000 mL at varying flow rates of up to 24,000 pph. Unlike Tetra Pak's form-fill-seal technique, Combibloc's packaging is preformed into individual packages.

Both the roll-fed and preformed aseptic fiberboard packages are typically sterilized by hydrogen peroxide when packages are to be used with low-acid foods. Packages to be used with high-acid foods can include sterilants such as peracetic acid–based sterilants in addition to hydrogen peroxide. Other methods that have been used, to a far lesser degree, include UV-C, gamma, and electron beam radiation.

8.4.1.2 Aseptic Plastic Cup

Thermoformed plastic cups are typically used for an individual serving of products such as puddings, baby foods, dairy creamers, cold brew coffee, yogurts, jelly, and long-life entrées. These aseptic products can be produced either by plastic cup thermoform-fill-seal machine or preformed cup-filling machine. Laminated lidding film can also be used as a custom-printed roll stock or pre-die-cut lids. In 2015, the IMA Group acquired majority stakeholder in ERCA, Hassia, Hamba, and Gasti and the holding company was called IMA Dairy & Food Holding GmbH.

Figure 8.17 Bosch aseptic thermoform fill-seal cup filler. (Courtesy of Robert Bosch Packaging Technology GmbH.)

Like the aseptic form-fill-seal carton-style container, the plastic sheet/roll-fed plastic is commercially sterilized using the acceptable methods of hydrogen peroxide, peracetic acid, or steam. The sterilization method is based on the design of the filler manufacturer. After sterilizing, a thermoplastic sheet is heated to soften the polymeric material and then formed to the container shape using the thermoforming process. The cups are filled with sterile food product and then sealed with sterile lidding film, which is sterilized by either H_2O_2, PAA, or steam in the same manner as the roll sheet.

In the 1970s, two new aseptic linear fillers for form-fill-seal cups were introduced to the US market: the German-manufactured Bosch and the French-manufactured Erca/Conoffast (now OYSTAR Erca). Bosch filler used various polymers, including PP, to form the cups, and the Conoffast used a combination of PS, PP, and PET with a barrier of EVOH. The technology to sterilize the cups was vastly different. The Bosch filler sterilizes the roll-fed packaging material with hydrogen peroxide, whereas the OYSTAR Erca uses either hydrogen peroxide or heat from coextrusion process. Both fillers gained market acceptance as a less expensive alternative to the metal can. The food products such as puddings, cheese sauces, tomato sauces, and flavored gels, which have been aseptically canned with the Dole system, are currently filled into plastic cups. These aseptic fillers enjoyed market acceptance and success, and both are still being used to aseptically fill food products. Figure 8.17 illustrates the Bosch aseptic thermoform fill-seal cup-filling machine.

Hassia, a German manufacturer of form-fill-seal fillers for plastic cups, was introduced to the United States in 1980 as a high-speed aseptic cup filler. Currently, the P500 Hassia filler (Figure 8.18) can fill at a speed of 1,800 cups per minute for hygienic mini portions of jam, honey, sauces, and spreads. The sterilization of the plastic roll stock and lidding film is accomplished by steam.

Aseptic preformed plastic cup usage appears to be widely spread in Europe; however, this aseptic system is not widely used in the United States. OYSTAR Gasti's Dogaseptic (now IMA

Figure 8.18 Hassia thermoform fill-seal cup machine. (Courtesy of Hassia Verpackungsmaschinen GmbH.)

Dairy & Food) aseptic preformed cup system was introduced for producing aseptic food in Europe. This system utilizes stem to sterilize the inside of the cup. The high intensity of UV-C light is additionally used to sterilize the container flange and interior surfaces. Other aseptic fillers for preformed cups available on current market include Ampack Ammann (German), Benco (Italian) and Metal Box (the United Kingdom).

Ampack Ammann is a German manufacturer of aseptic filling machines for preformed plastic cups available in two systems: rotary filler and linear filler. The production speed can reach up to 60,000 cups per hour with anti-spilling functionality for the linear filler system. Many foods are aseptically produced in premade cups such as yogurts, dairy products, puddings, and other desserts. Ampack has been manufacturing aseptic filling machines since 1978 and has more than 130 aseptic filler installations, which are mostly in international regions. In the United States, Ampack Ammann is represented by Evergreen Packaging, which is located in Cedar Rapids, Iowa.

Benco is an Italian manufacturer of aseptic cup filler. Hydrogen peroxide is used as sterilization agent to sterilize the packaging material while the combination of steam and hydrogen peroxide is used to presterilize the filling machine. Benco filler can produce aseptic low-acid food such as pudding, cheese sauce, and fruit-based gel. The production rate of the Benco filler is slow compared to other aseptic filling system in the same category.

Metal Box is a manufacturer located in the United Kingdom. In the 1980s, it introduced an aseptic preformed cup-filling system. Metal Box was purchased by FMC FoodTech, which has changed the name to JBT FoodTech. Despite the relatively slow production rate of 9,600 cups per hour, Metal Box has been chosen to commercially install in several food manufacturers to process cheese sauce, oatmeal, pudding, and other desserts. At present, only one Metal Box aseptic cup filler at a commercial installation in the United States is located at Minnesota to mainly produce cheese sauce.

8.4.1.3 Bottle Packaging

The use of plastic bottles has increased tremendously in the past decade. This is due to the approval of hydrogen peroxide and PAA sterilization of the packaging materials. Hydrogen peroxide and PAA have a short sterilization time depending on several factors such as concentration, temperature, and flow rate. It allowed the manufacturers to install bottle fillers that could run more than 1,000 bottles per minute. Thermoplastic materials such as PET and HDPE are commonly used in aseptic beverages such as juice, nutritional beverages, high-protein drinks, plant-based milk, and dairy-based beverages that are shelf stable. There are many filler installations that are not FDA-compliant and are running in the extended shelf-life (ESL) mode, and the products need refrigerated distribution and storage. There are two primary systems of aseptic bottle-filling machine: (1) fill-seal preformed bottle system and (2) blow mold-fill-seal bottle system.

Like other fabricated packaging materials, bottles are usually premade at the packaging manufacturer site with either extrusion blow molding such as HDPE and PP bottles or injection stretch blow molding (ISBM) for PET bottles. After feeding into the aseptic filling machine, preformed bottles are sterilized with either vapor hydrogen peroxide (VHP) or peracetic acid (PAA) depending on the design of filling machine system by the manufacturer. When VHP is selected as a sterilant, hot air is applied to remove peroxide residual from the bottles as required by the regulation, but if PAA is applied to sterilize the bottles, which is normally for PET bottle, sterile water is required to rinse the bottles prior to filling the food product.

Robert Bosch Company was the first manufacturer to introduce an aseptic filler for beverages in the bottle package. In the late 1970s or early 1980s, two bottle fillers from Bosch were installed to produce nutritious aseptic beverages (Figure 8.19). Bosch received FDA validation at this installation for aseptically filling low-acid beverages into plastic bottles. The production rate of these fillers had a relatively slow filling rate compared to the fillers being supplied by the latest

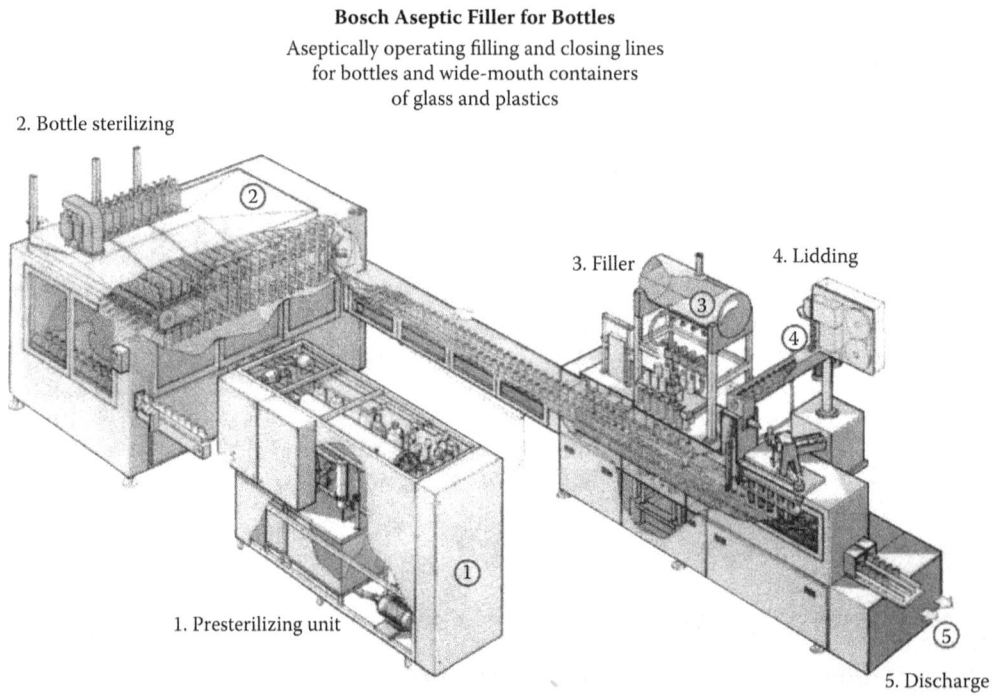

Figure 8.19   Bosch aseptic bottle filler. (Diagram from Bosch published literature.)

Figure 8.20   Stork linear aseptic bottle filler. (Courtesy of JBT Corporation.)

aseptic bottle filler manufacturers today. Bosch, thus, has not supplied any other aseptic bottle filler in the United States after the initial installation.

JBT (Stork) aseptic filler (Figure 8.20) is a linear aseptic filler and can aseptically fill different polymeric materials such as HPDE, PP, and PET. The filler speed ranges from 12,000 to 30,000 bottles per hour. Vapor hydrogen peroxide is used for sterilization of internal preformed bottle and the external neck area of the bottle. Sterile hot air is applied to remove peroxide residual. The lidding material is also sterilized in the same manner as the body bottle. The sterilized bottle is then filled with the sterile product and sealed with the sterile lid material or screw cap.

Shibuya Hoppmann installed its first aseptic bottle filler in 1994 and then over 160 filling machines have been installed to produce aseptic beverages such as juices, teas, coffee, milk, and other dairy products. The Shibuya aseptic bottle filler (Figure 8.21) can be used with both PET and HDPE bottles, and it has received FDA validation for aseptically filling low-acid beverages. The Shibuya bottle filler has a high production rate with filling speed up to 54,000 bottles per hour. Shibuya has also developed rotary aseptic bottle fillers that use the dry decontamination method of vapor peroxide or most recently the E-beam. The use of vapor peroxide is similar to other aseptic systems. However, when E-beam is used as a sterilizing medium, the bottles are rotated 180 degrees to make sure that the entire surface of the bottle is sterilized prior to filling and sealing. E-beam can achieve 6-log reduction and has fewer critical points to monitor and control as compared to the peroxide sterilization. Moreover, the use of E-beam also helps reduce energy and water usage to sterilize and remove the chemical residuals from packaging containers. It also substantially reduces required floor space.

Figure 8.21   Shibuya aseptic E-beam filler. (Courtesy of Shibuya Hoppmann Corporation.)

Aseptic bottle-filling machine from Ampack Ammann, a German manufacturer, is a linear filler. Ampack also has a rotary bottle filler, but it can only process for extended shelf-life (ESL) products. Most Ampack fillers are installed internationally, and they have yet to be commonly used in the United States.

Hamba, which is one of the well-known suppliers of aseptic cup-filling machines with linear fillers, has also developed a linear aseptic bottle filler. Hamba BK 20010A is an aseptic (UHT) milk-filling machine with production speed of 12,000 bottles per hour for 1 liter bottle. The Hamba manufacturing facility has recently been relocated to the OYSTAR Hassia facility in Ranstadt, Germany. It is marketed in the United States through the OYSTAR facility in New Jersey where management, sales, service, and spare parts are located.

GEA Procomac S.p.A., an Italian manufacturer, was the first developer of aseptic bottle fillers in 1994. GEA is now one of the world's leading suppliers of aseptic bottle-filling machines and is widely used in the United States. With the design of a rotary aseptic filling machine for beverages, GEA aseptic bottle filler can be used with both HDPE and PET bottles for producing both high-acid and low-acid beverages. Furthermore, it can also process beverages with small particulates. GEA fillers can fill bottles sizes of 250 ml to 2 liters. Depending upon the bottle size and configuration, the fillers can fill at a production rate of up to 800 bottles per minute. GEA has a testing facility for potential users to test their products prior to purchasing in Parma, Italy. In the United States, GEA facility for sale, service, and spare parts is located in Hudson, Wisconsin. There are several types of aseptic filler. For example, GEA aseptic filling Whitebloc (Figure 8.22) uses the hydrogen peroxide–based sterilization technology, while GEA aseptic filling Modulbloc and GEA ECOSpin (Figure 8.23) use PAA-based sterilization technology.

Serac is another world's leading supplier of aseptic bottle-filling machines and extended shelf-life (ESL) filling machines. Based on the bottle sizes and machine configuration, Serac filler can fill up to, for example, 48,000 bottles per hour for 1 liter bottle. Several plastic materials including HDPE and PET bottles can be used with the Serac filling machine (Figure 8.24). Most

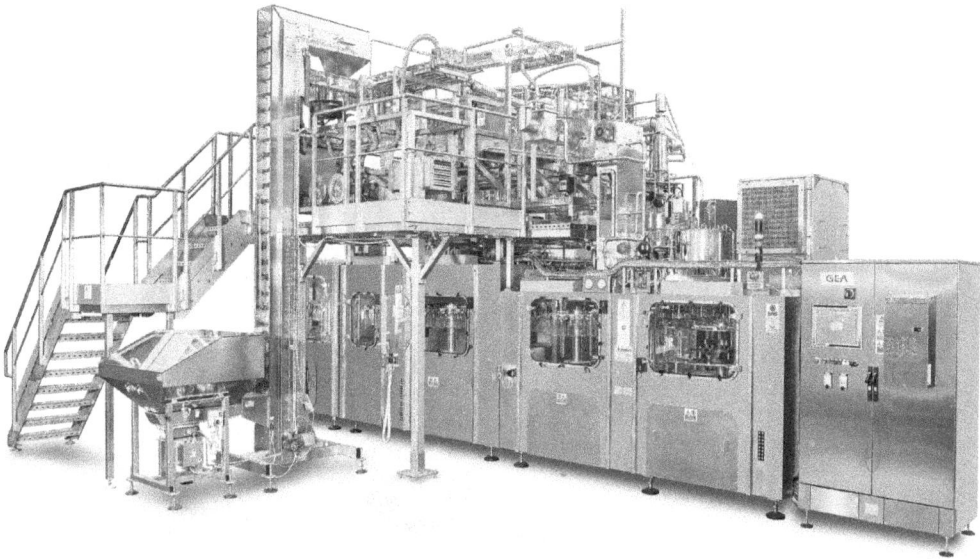

Figure 8.22   GEA aseptic filling Whitebloc machine for plastic bottles. (Courtesy of GEA.)

Figure 8.23   GEA aseptic filling ECOSpin machine for plastic bottles. (Courtesy of GEA.)

Serac aseptic fillers are manufactured at the home office in France. In the United States, Serac sales and service facilities are located in Carol Stream, Illinois, for both aseptic and ESL fillers.

A recent advancement in aseptic bottle-filling technology is blow mold-fill-seal PET bottle system for aseptic beverage which uses dry preform decontamination method. Two companies, Sidel and Krones, have advanced this technology in the United States for milk and other sensitive beverages.

Krones, a German manufacturer, has a major sales and service organization located in Wisconsin, the United States. Krones aseptic filling machines can process aseptic high-acid and

Figure 8.24   Packaging bottles for aseptic beverage being filled on Serac fillers. (Courtesy of Serac USA.)

low-acid beverages and extended shelf-life (ESL) foods for low-acid beverages. Krones can fill bottles of sizes up to 2 liters and have production rates of up to 55,200 bottles per hour.

Sidel was founded by a French manufacturer to develop packaging technology for the beverage industry. Their first processed beverages, including milk, wine, and water, were filled in polyvinyl chloride (PVC) bottles and later in HDPE and PET bottles. Sidel joined Tatra Laval to offer two styles of aseptic filling machines for plastic bottles: (1) rotary system, and (2) linear system to produce a variety of aseptic beverages such as juices, teas, milk, other dairy products, and flavored water. The aseptic rotary filler can process aseptic high-acid and ESL products, while a linear filler, Tetra Pak LFA-20 (Figure 8.25), received FDA validation for processing aseptic low-acid beverages. The linear fillers, however, have a much lower production rate than the rotary fillers, which can fill up to 60,000 bottles per hour. The sterilization method of Sidel's rotary filler utilizes the wet decontamination with hydrogen peroxide, while Sidel's (Tetra Pak LFA-20) linear filler uses hydrogen peroxide to sterilize the plastic bottles. Several plastic materials, including HDPE, PET and PP, can be used as aseptic packaging materials.

Sidel introduced aseptic Combi Predis™ blow-fill-seal filling machine as a rotary system (Figure 8.26). This filler has a unique sterilization method for PET bottles that helps to reduce the cost and the amount of chemical residuals. PET preform from packaging supplier is used for this aseptic filling machine to blow to be a bottle for aseptic beverage. The machine utilizes dry decontamination sterilization with peroxide at the temperature of 120–140°C (248–284°F) for sterilizing the PET preform before transferring to the blowing station. Heat from the oven helps to remove the residual peroxide. At the blowing station, PET preform is blown using 0.01 μm filtered air in the mold to take the shape of the final bottle. After the blowing station, the bottles are transferred by the neck into the sterile filler to be filled and then capped with a capping machine. These systems are more complicated as compared to the other bottle fillers due to the addition of preform and blowing function in the same machine.

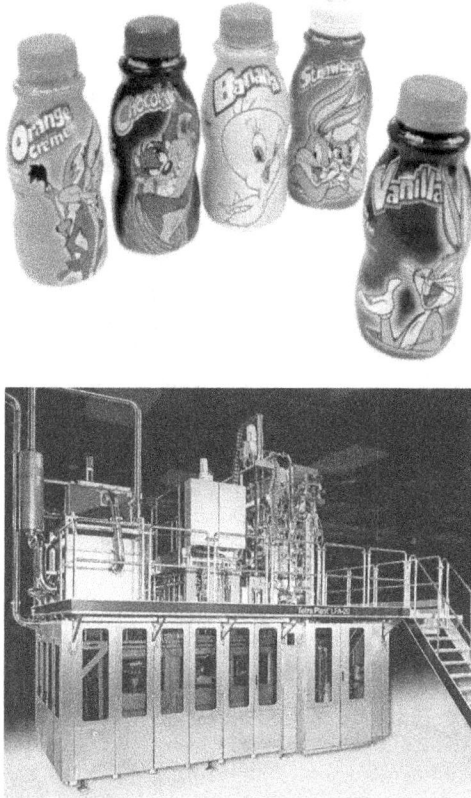

Figure 8.25   Sidel/Tetra Pak LFA-20 linear aseptic bottle filler (*top*) and sample of some aseptic product (*bottom*) filled on the LFA-20 filler. (Courtesy of Tetra Pak International S.A.)

Figure 8.26   Sidel aseptic blow-fill-seal filling machine. (Courtesy of Sidel.)

8.4.1.4 Pouch Packaging

Flexible aseptic packaging can be in the form of pillow pouch, stand-up pouch, stand-up pouch with spout, stick packs, and small sachet. Flexible pouches are normally multilayered materials, which are constructed of different barrier materials such as PET, PP, LLDPE, EVOH, and aluminum foil. Pouches generally range in size from 200 ml up to 10 liters. Two types of aseptic pouch-filling machines are used: (1) form-fill-seal machine (vertical and horizontal machines) and (2) fill-seal preformed pouch machine. The common method to sterilize polymeric film is the use of warm 35% hydrogen peroxide for disinfection and hot air drying for elimination of peroxide residual. Due to the lightweight material, ease of opening, and ease of disposal, aseptic pouch-filling machine has gained more attention in the United States for many types of food products such as puddings, cheese sauces, chili, dairy products, juices, ice cream mix, and tomato products such as ketchup and paste.

In a similar way to process the form-fill-seal aseptic filling carton and cup machine, the roll stock film is fed through inside the machine. The film material is sterilized with a heated H_2O_2 bath and then blown with sterile hot air to remove peroxide residual as required by regulation (less than 0.5 ppm). The sterile film material is promptly formed as a pouch/sachet to be ready to fill in the sterile food products, and then hermetically sealed by heat sealer before exiting the aseptic zone. There are many form-fill-seal aseptic pouch-filling machines available on current markets such as Fresco, Cryovac, and Hassia.

In the 1970s, the first aseptic pouch form-fill-seal machine was manufactured by Robert Bosch GmbH, a German manufacturer of food packaging equipment, to fill puddings and cheese sauces, which were previously filled into #10 cans. After Bosch's introduction, Inpaco, a manufacturer of pouch fillers located in Nazareth, Pennsylvania, also introduced a pouch filler in an aseptic food market. Inpaco was later purchased by Liquibox in Worthington, Ohio.

Fresco is another aseptic form-fill-seal pouch-filling machine which was introduced in the aseptic food industry by Fresco System, a Telford, Pennsylvania, corporation (Figure 8.27). It can

Figure 8.27   Fresco aseptic form-fill-seal pouch filler. (Courtesy of Fres-co Inc.)

Figure 8.28 Hassia S800A for stick pack. (Courtesy of Hassia Verpackungsmaschinen GmbH.)

fill either lay flat or stand-up pouch styles, and the pouch size ranges from 1-gallon single-lane pouches up to 30 pouches per minute and 1/2-oz multilane pouches up to 500 pouches per minute. Filling speeds depend on pouch size, product characteristics, and types of film materials. Fresco filler can be used to aseptically fill high-acid, low-acid, and particulate food.

OYSTAR Hassia pouch filler has designed to fill aseptic food in stick pack pouches up to 6 inches wide and 10 inches long at a filling rate of 10 pouches per minute. Hassia S800A (Figure 8.28) is an aseptic form-fill-seal machine for stick-pack configuration with the filling speed of 24,000–50,400 stick packs per hour depending on the fill volumes and products such as skim milk, sour cream, and pudding. Hydrogen peroxide is used as sterilant for this filler system.

Cryovac Food Packaging, a division of Sealed Air Corporation, manufactured aseptic form-fill-seal pouch filler that can aseptically fill high-acid, low-acid, and particulate food with the particulate size of more than 1 inch. The first Cryovac aseptic pouch filler was installed in Europe and also introduced to the US market.

In aseptic filler for preformed pouches, flexible pouches are formed and hermetically sealed under clean environment at packaging manufacturers. Premade pouches are normally presterilized by irradiation and then delivered to the food manufacturing as a roll stock. Prior to entering to aseptic zone in the aseptic filler, the external surface of pouches needs to be sterilized by normally hydrogen peroxide.

The use of pouches has several advantages over the traditional rigid packages: (1) less use of space, (2) less disposal cost, and (3) improved sustainability. Filler manufacturer Ecolean (Figure 8.29) produces a single-serve stand-up pouch which includes a built-in air handle for easy handling for consumers. These pouches are sterilized with E-beam and are rolled up in a stock. Since the internal food-contact surface of the pouch has already been sterilized, the individual

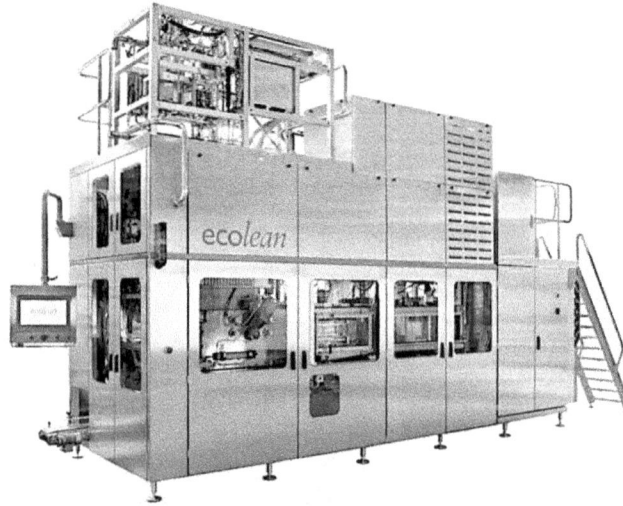

Figure 8.29   Ecolean aseptic filling machine. (Courtesy of Ecolean.)

pouch from the roll stock is again treated externally with vapor peroxide in the aseptic filler. The sterile pouch is then cut open to fill with sterile product and sealed to form a hermetic seal.

Another type of pouch filler that was recently launched in the market is SureFill from Scholle IPN (Figure 8.30). These pouches have a spout at the top and are hermetically capped. They are manufactured in a clean environment to avoid any contamination. Flexible pouches are then stacked on a rail system that are sent for gamma irradiation sterilization before being shipped

Figure 8.30   Scholle Surefill aseptic pouch filler. (Courtesy of Scholle IPN.)

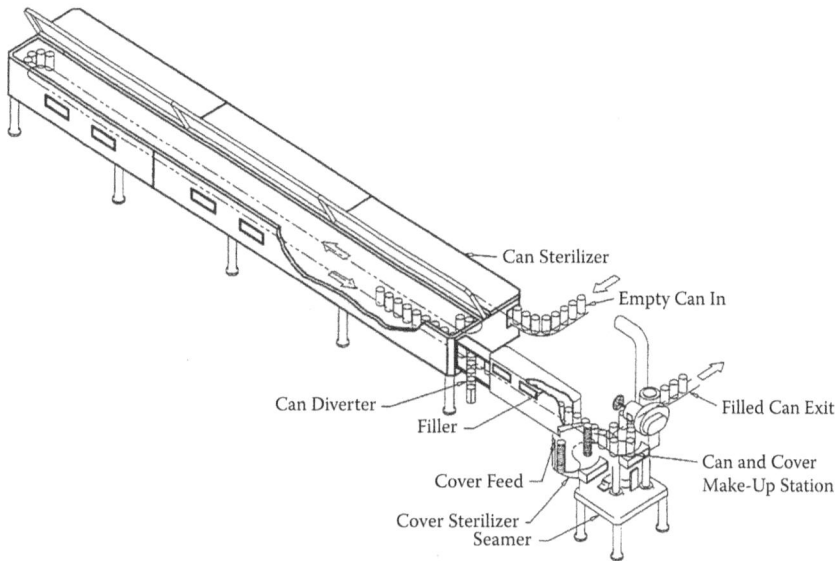

Figure 8.31   A Dole aseptic canner.

to the food manufacturers. At the food manufacturer site, these rails are loaded in the machine and the cap area is sterilized with vapor peroxide before opening the cap and filling with sterile product. The pouch is then hermetically sealed with the cap and released from the filler as a finished product.

8.4.2 Aseptic Filler for Food Service

For food service, aseptic food is normally produced and packaged into #10 metal can (603 × 700) or bag-in-box of 5–20 liters. Due to the advantage of no refrigeration requirement, large-volume aseptic products can be safely stored in ambient storage. This helps to reduce the refrigerated storage spaces and energy requirements. Only aseptic can filler and bag-in-box filler are described in this section.

The Dole aseptic canning system (Figure 8.31) has been in commercial applications since the 1960s. Even though semirigid packaging materials such as plastic and paperboard cartons have occupied the market share of rigid packaging material for aseptic food, approximately 40 Dole canners are still in operation. Dole aseptic canning is the only system available to fill aseptically processed foods into metal cans. The Dole system is currently owned and operated by the Graham Corporation. The system consists of four main components: (1) can sterilization, (2) filling section, (3) can lid sterilization, and (4) sealing section. Unlike traditional canning processing, all four components are connected with an integrated network of instruments. After loading into the aseptic canning machine, metal cans are transferred into a stainless steel, double-insulated tunnel for sterilization. Cans are subjected to superheated culinary steam at 220–226°C (428–439°F) for sterilization. The steam is superheated by electric heaters supplied with the system. The temperature requirement depends on the material used in a two-piece or three-piece metal can. The cans are moved through the tunnels and the speed of the conveyors is varied to establish the overall capacity of the system. The treatment is approximately 45 seconds until the surface of the metal can reaches 224°C (435°F), which is the temperature required for the destruction of heat-resistant bacteria. The superheated steam is more lethal than dry hot air and requires less time to destroy the bacteria.

Sterile cans are continuously transferred to the filling station. Cold sterile water is used to cool the can exterior in order to reduce the temperature of hot cans. At this point, the sterile food is filled into the can container by one of several specially designed filling mechanisms. The basic model is the slit filler, which consists of a thick-walled, stainless steel pipe that is slit along the lower side. This pipe has a second inner pipe that has multiple parts and provides consistent flow to the filling system. The annulus between the pipes is approximately 1/2 inch. The sterile product is fed into the inner pipe and flows out to the slit in the outside pipe. The slit is approximately 1/4 inch wide and 6 inches long. The sizes of slit vary depending upon the type of product and speed of filling. With the slit filler, the cans pass underneath a slit opening in a tube-type filler.

While metal can is sterilized and filled with sterile food, the metal can lid is also sterilized by transferring into the sterilization unit through one of several mechanical devices. They are completely enveloped in superheated steam for approximately 80 seconds. At the discharge, the lids are sterile and are conveyed into the closing machine. Both regular-style metal lid and easy-open lids can be used in the unit.

After finishing from the filling station, the cans are moved to the sealing station, which is the last component of the Dole system. To accommodate the sterile container filled with cold sterile product and the seaming of the filled containers, the double seam sealer is enclosed and heated with superheated steam. The enclosure permits initial sterilization of the equipment and maintains a completely sterile atmosphere in the area where the sterile container, lid, and product come together. The continuous flow of superheated steam creates a positive pressure inside the can seamer, preventing external, bacteria-laden air from entering.

Bag-in-box package is a fabricated flexible multilayer metalized packaging material with an attachment of specific designed fitment. Due to the complexity in sterilization of the internal surface of the bags, the premade bags are presterilized by irradiation, which is normally gamma irradiation prior to using in the aseptic filler. Irradiated bag can be shipped to the food manufacturing facility in a lay-flat condition in the shipping boxes as an individual bag or web of the bags depending on the feeding style of the filler. Before filling the bags with sterile product, the fitment of irradiated bag is again sterilized either with steam or vapor hydrogen peroxide.

In the late 1950s, William Scholle developed and patented a more efficient means of containing battery electrolyte using flexible packaging constructed of polyethylene (PE). In later years, Scholle engineered and manufactured a prototype of an aseptic filler for preformed bags into which the products such as tomatoes and fruit could be filled. The prototype filler was installed and improved upon at Purdue University in West Lafayette, Indiana, with the assistance of Dr. Phil Nelson, Food Science faculty, and graduate students. At the time, Purdue University already was improving upon aseptic processing techniques in the food processing facility on campus. With an aseptic processing system already in place, Purdue was a logical place to prove Scholle's aseptic filler for flexible packaging.

Originally, Scholle aseptic bag-in-box filling machine was presterilized with steam, hot water, and chlorine. Currently, the presterilization of the aseptic zone is performed with hydrogen peroxide. The sterile zone and product contact parts of the aseptic fillers are subjected to high temperatures (approximately 121°C or 250°F) for 30 minutes to presterilize the filler prior to production. The preformed bag, which is manufactured of multilayer of polymers with a fitment and closure, is presterilized with gamma irradiation. The bags are manually or automatically inserted into the sterile filling chamber and then automatically resterilized with vapor hydrogen

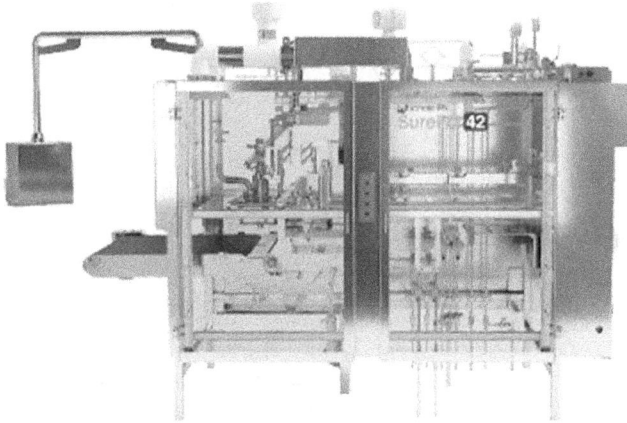

Figure 8.32 The Scholle SureFill 42 aseptic bag-in-box filler. (Courtesy of Scholle IPN.)

peroxide. The closure is automatically removed, and the filling valve is then inserted into the bag while the vacuum inside the bag is pulled. The sterile product is volumetrically filled into the bags by a flowmeter. After filling, there is an optional nitrogen flush before the cap is applied to hermetically seal the bag. The bag is then ejected from the sterile filling zone. Scholle IPN aseptic bag-in-box filling machine is shown in Figure 8.32. It can be used to fill the particulate food products such as diced and sliced fruits, as depicted in Figure 8.33.

Liquibox is another aseptic bag-in-box filler manufacturer. The filling machine can fill the food products from 1.5 to 20 liters (0.5–5 gallons) for food service application. The fitment style of Liquibox bag-in-box is hermetically sealed with the top membrane, which is made of the film material. High-temperature steam sterilization is used to sterilize the filling interface and outer surface of the top membrane prior to the filling process. The filling plunger pierces the top membrane to fill the food products into the sterile bag by passing through the partially sealed internal membrane on the bottom of the gland. To hermetically seal, the internal membrane is then heat-sealed to the gland from below. The steam is applied to flush the gland prior to releasing the bag from the aseptic zone.

8.5 PACKAGING INTEGRITY TEST

Packaging integrity is one of the most important aspects for food manufacturing and distribution to the markets for consumers. Any critical defect in packaging or seal failures can lead to spoilage or a food safety risk for the consumers. Food manufacturers take extensive precaution to avoid such a situation. Packaging integrity is a measurement of packaging quality and ability in order to ensure security of product and the protective capacity of package. Generally, package integrity must be validated before the packages are used in the commercial production. In aseptic processing, many packaging tests can be conducted depending on the type of packaging materials and protection requirement for the food products. Testing can be accomplished by either destructive or non-destructive methods. The destructive method is a test to tear down the material to determine its properties, while the non-destructive method is a non-invasive inspection.

Figure 8.33   Aseptic product in aseptic bags. The outer packaging layer has been removed to show the particulate identity.

Leak test is one of the most important measures for all packaging types. It can be achieved by many choices of tests such as bubble bath test, dye penetration, and electrolytic conductivity test, which are destructive tests. There are some non-destructive tests for leak testing, for example, vacuum decay (in-line or off-line inspection) and high-voltage leak detection (HVLD). Burst test can be used for package seal integrity testing. It is also used for setting the seal parameters of the heat-sealing method, including temperature, dwell time, and pressure. To ensure the seal quality and ease of opening of the caps/closures, a torque test can be conducted using a torque tester, which is available as a manual device or automated torque tester. For the in-line inspection system, vision sensors can help to detect the missing or improperly applied lids or caps on the packaging containers.

Methods for measuring oxygen concentration in the packaging headspace are also important to maintain the quality and shelf life of food products. The test can be conducted using conventional gas headspace analyzer, which is a destructive test. Oxygen headspace can also be measured using a non-invasive device, which uses the optical fluorescence oxygen analyzer to read the sensor that is attached inside the wall of the package headspace.

There are opportunities to further enhance the in-line non-destructive testing to provide fast response for the operation efficiency. This will help to minimize the defective packages to the market and reduce the chances of spoiled products in the hands of consumers.

8.6 REGULATIONS FOR PACKAGING

From a food safety perspective, there are concerns of food additives that can be inadvertently added to the product from the packaging material. In an FDA guidance document to industry, it has provided recommendations for a food-contact surface (FCS), which is primary packaging for food in aseptic processing and packaging technology. The concentration of FCS can be measured in the food or in a food simulant. Although the FCS testing in real food is acceptable, it is sometimes hard for some analytes to measure in real food due to the complexity of the food. Food simulants can be used based on the nature of food product for testing the concentration of FCS from packaging materials to food under certain test conditions. A worst-case approach should be used for testing the FCS into food, such as maximum temperature and time that the product could undergo during its shelf life.

For the recycled materials, the requirement from FDA is that the recycled food-contact surface should behave similar to the virgin material of suitable purity and meet existing specification for the virgin material for food-contact surface, 21 CFR Parts 174 through 179. Maximum level of chemical contaminant must be less than 0.5 parts per billion (ppb) dietary concentration, which is considered to be negligible risk for a contaminating compound from the recycled material.

FDA provides the guidance for using irradiation as sterilization mechanism in food and packaging materials, listed in regulation 21 CFR 179.45. The common irradiation dose for the prepackaged foods ranges from 10 to 60 kilo gray (kGy) depending on the types of packaging materials. Besides, the typical irradiation dose used to sterilize the medical devices is 25 kGy. As a safety aspect, the food and packaging interaction of irradiated packaging materials should be studied since the irradiation can affect the material properties depending on the material types.

8.7 SMART PACKAGING FOR ASEPTIC PRODUCTS

There are two forms of smart packaging: active and intelligent packaging. Active packaging aims to maintain or enhance the food quality and shelf life of the product by slowing down the microbial growth or chemical reactions, while intelligent packaging is the technology that communicates with consumers throughout distribution and retail stores. Active and intelligent packaging can also be combined in the same product package. Even though the use of smart packaging currently is not a common practice in the aseptic industry, there are future opportunities to incorporate this technology in aseptic food products. Some of the areas for using smart packaging are (1) quality and shelf life, (2) communication, (3) enhanced product functionality, and (4) anti-counterfeiting.

Active packaging helps with reduction of the food deterioration by incorporating active compounds on the packaging component. More specifically, in aseptic products, antioxidants can be added to the packaging material which reacts with oxygen headspace to reduce the oxidative reactions over the product shelf life. For example, special design of the cap with antioxidants or incorporation of antioxidants in the packaging layer as an oxygen barrier.

One of the recent innovations in food packaging is the use of indicators for food quality, which is a part of intelligent packaging. There are several aspects of food quality that can be communicated using indicators to monitor the quality of the packaged food such as freshness and spoilage. Currently, there are labels available with incorporated color indicators that will change based on the food quality change. Another development is in the area of thermal indicator. Thermochromic ink has been developed as an indicator to monitor the color change when product is exposed to a certain selected temperature during storage. The thermochromic technology has been applied for the wine and beer products. It has the potential to be used in aseptic beverages.

Smart packaging to enhance product functionality is based on packaging design innovation. It can be applied to the food products that require special functionality such as sensitive vitamins, minerals, and nutritional ingredients at the consumer end. For example, special cap design was developed by Emmi, a Swiss company, to store the sensitive vitamins and minerals as a dry tablet separate from the milk serum until it would be mixed in product at the time of consumption (O'Sullivan et al., 2008).

The use of intelligent packaging technology helps manufacturers to communicate with the consumers. Consumers can receive the pertinent information about the product such as the source of raw ingredients, manufacturing location, company profile, and anti-counterfeiting information. It can also help retail stores to manage inventory tracking and product quality control. Radio frequency identification device (RFID), near-field communication (NFC), and quick response (QR) code are the tools used to accomplish the desired functionality.

8.8 SUSTAINABILITY IN ASEPTIC PACKAGING

Packaging from commercial food manufacturing has received global attention recently due to the enormous packaging waste and consequently pollution to the environment. Food and beverage industry is in need of innovative solutions to reduce the carbon footprint and packaging waste problem. Sustainable packaging has become the main drivers of food manufacturers and many manufacturers have set a goal to reduce the environmental impact.

Sustainable packaging should be beneficial, safe, and healthy for consumers, and it should meet market criteria for performance and cost. Clean production technology is important for manufacturing of the packaging materials. While looking for sustainable packaging, it is critical to balance the food protection with the environmental impact.

Reduce, reuse, and recycle are fundamental aspects of packaging sustainability. Source reduction and lightweight help to address the sustainability concern by minimizing uses of the packaging material. It further helps to reduce the burden on the distribution chain and reduce the greenhouse gas emission.

Recycling is another systemic solution to achieve sustainable waste reduction of petroleum-based plastic. Less than 20% of global plastic waste is currently being recycled. Single-use plastic has become one of the challenging environmental issues. PepsiCo. and Coca Cola, the largest user of PET bottles for beverage business, have focused their effort on sustainable packaging by improving the carbon footprint of their packaging material. The solution includes the reduction of the post-consumer waste and recycling, thereby minimizing the environmental impacts. The vision is to reuse disposable plastic bottles and to reduce them from municipal solid waste (MSW) and environment. Recycled PET (rPET) bottles, which are made from recycled PET flakes or pellets, have become an eco-friendly packaging solution. rPET bottles have seen commercial applications for water, soda, and juice. This trend may continue for aseptic beverages. The use of post-consumer recycled (PCR) materials, therefore, can be an alternative packaging material for the aseptic industry.

To consistently drive toward sustainable packaging, bio-based packaging materials, derived from renewable resources, can be a promising approach to reduce plastic pollution. Tetra Pak has developed a paperboard carton package by replacing synthetic plastic layers with plant-based

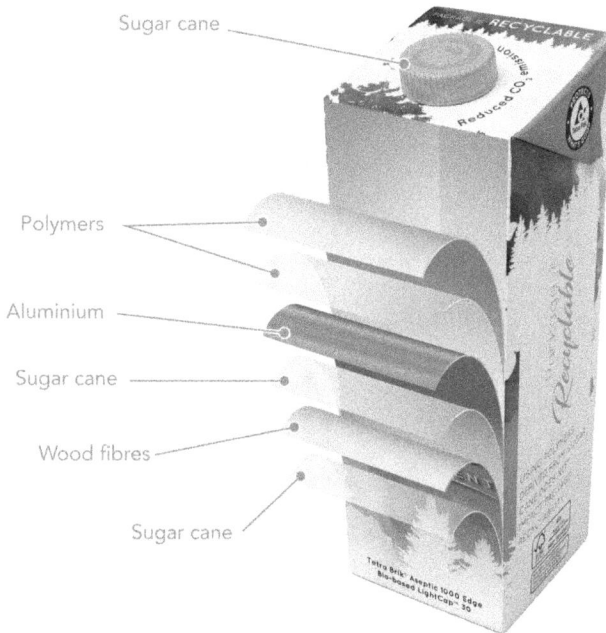

Sugar cane

Polymers

Aluminium

Sugar cane

Wood fibres

Sugar cane

Figure 8.34   Tetra Pak plant-based beverage carton. (Courtesy of Tetra Pak International S.A.)

plastic material from sugarcane (Figure 8.34), which has low environmental impact by not using finite fossil sources. Moreover, the closure material is also plant-based plastic.

With the continuous effort to use the renewable packaging materials for sustainability, Tetra Pak has introduced Tetra Brik Aseptic 1000 Edge, which contains bio-based LightCap™ 30 for shelf-stable products. Using natural color variation on the package provides an authentic-looking container.

The innovation of paper bottles for the industry is also potential for future aseptic beverages. Tetra Pak has developed a carton bottle, Tetra Pak Evero, to be used with an aseptic filling machine. It has been used for milk, plant-based milk, and brewed coffee. The new generation of the carton bottle also consists of a plant-based plastic layer. Coca-Cola has also developed a paper bottle, which was commercialized in Hungary.

SIG also introduced the sustainable carton packages: SIGNATURE 100, SIGNATURE FULL BARRIER, Combibloc ECOPLUS, SIGNATURE CIRCULAR, and SIGNIA (Figure 8.35). SIGNATURE 100 carton is derived from 100% forest-based renewable material. It is also an aluminum-free carton. SIGNATURE FULL BARRIER is made from 95% forest-based material and contains an ultrathin layer of aluminum foil to protect the sensitive food products such as orange juice from oxygen and light.

For the Combibloc ECOPLUS carton, it is a non-foil-based barrier carton. The foil layer has been replaced with a nylon-6-based barrier polymer. This innovation provides significant energy-saving and carbon footprint reduction, which reduces carbon dioxide (CO_2) emission by 28%, compared to conventional Combibloc cartons in the same format. SIGNATURE CIRCULAR is a carton beverage package, which contains recycled plastics, while SIGNIA is made from unbleached paperboard cartons. Besides, the new design of CombiMaxx closure with a visible,

Figure 8.35 Eco-friendly carton packages from SIG Combibloc (*from left to right*): SIGNATURE 100, SIGNATURE FULL BARRIER, Combibloc ECOPLUS, SINATURE CIRCULAR, and SIGNIA cartons. (Courtesy of SIG Combibloc.)

tamper-evidence ring also supports the sustainable packaging solutions by reducing the plastic use by 4.5% compared to the previous closure, CombiSwift. These innovative beverage cartons help to reduce the carbon footprint in the environment.

Furthermore, SIG Combibloc and Tetra Pak have developed and introduced paper straws with straight, U-shaped, and telescopic styles in 4–6-mm diameter to replace the use of plastic straws. The drinking paper straw is wrapped and attached to the package of the small-size carton pack for a single-serve portion. Since it is derived from renewable resources and can be easily recyclable, this paper straw can be disposed of with carton after consumption without creating the packaging waste in the environment. The production and use of paper straws also delivers a lower carbon footprint and less environmental issues in air, soil, and ocean than that of plastic straws. This represents a solution to sustainable packaging.

The packaging innovation has led to the development of a plant-based plastic bottle as a "green" bottle. The material is derived from the renewable resources, including switch grass, pine bark, corn husks, and additional agricultural by-products such as orange and potato peels and oat hulls. The bottle is chemically identical to standard PET. The plant-based bottle is currently in use by PepsiCo and Coca-Cola for water, soda, and ketchup.

The sustainability in the aseptic market using pouches is also a focus area. Ecolean, one of the single-serve aseptic pouch manufacturers, also targets packaging waste reduction. Due to the absence of an aluminum layer in the pouch structure, flexible pouch material can be easily recycled. Moreover, carbon footprint is minimized by the lightweight of film material and the use of less raw materials and energy in production and transportation.

Many biodegradable and compostable packaging materials are developed and used in some commercial food products such as polylactic acid (PLA) and poly(3-hydroxybutyrate-*co*-3-hydroxyvalerate)—PHBV. These materials, however, have not yet been used in aseptic packaging. Extensive research is needed for material improvement, sterilization, and application for shelf-stable packages (Benyathiar et al., 2021; Benyathiar et al., 2016).

REFERENCES

Ansari, I.A., and Datta, A.K. 2003. An overview of sterilization methods for packaging materials used in aseptic packaging systems. *Food Bioprod. Process.* 81(1):57–65.

Benyathiar, P., Selke, S., and Auras, R. 2016. The effect of gamma and electron beam irradiation on the biodegradability of PLA films. *J. Polym. Environ.* 24(3):230–240.

Benyathiar, P., Selke, S.E., Harte, B.R., and Mishra, D.K. 2021. The effect of irradiation sterilization on poly (lactic) acid films. *J. Polym. Environ.* 29(2):460–471.

Chen, B.-Y., Lung, H.-M., Yang, B.B., and Wang, C.-Y. 2015. Pulsed light sterilization of packaging materials. *Food Packag. Shelf Life* 5:1–9.

Condon, S., Alvarez, I., and Gayan, E. 2014. *Non-Thermal Processing: Pulsed UV Light.* In Carl A. Batt, Mary Lou Tortorello, Encyclopedia of Food Microbiology (Second Edition), (pp. 974–981). Academic Press.

Davies, E.S. 1975. Aseptic canning method. Google Patents.

Deilmann, M., Halfmann, H., Bibinov, N., Wunderlich, J., and Awakowicz, P. 2008. Low-pressure microwave plasma sterilization of polyethylene terephthalate bottles. *J. Food Prot.* 71(10):2119–2123.

Dunn, J. 1996. Pulsed light and pulsed electric field for foods and eggs. *Poul. Sci.* 75(9):1133–1136.

Dunn, J.E., Wayne Clark, R., Asmus, J.F., Pearlman, J.S., Boyer, K., Painchaud, F., and Hofmann, G.A. 1991. Methods for preservation of foodstuffs. Google Patents.

Elmnasser, N., Guillou, S., Leroi, F., Orange, N., Bakhrouf, A., and Federighi, M. 2007. Pulsed-light system as a novel food decontamination technology: A review. *Can. J. Microbiol.* 53(7):813–821.

Falguera, V., Pagán, J., Garza, S., Garvín, A., and Ibarz, A. 2011. Ultraviolet processing of liquid food: A review: Part 2: Effects on microorganisms and on food components and properties. *Food Res. Int.* 44(6):1580–1588.

FDA. 1996. Code of Federal Regulations 21CFR179. 41. *Food Drugs* 21:179.

Gómez-López, V.M., Koutchma, T., and Linden, K. 2012. Ultraviolet and pulsed light processing of fluid foods. In *Novel Thermal and Non-thermal Technologies for Fluid Foods*, 185–223. Amsterdam: Elsevier.

Haji-Saeid, M., Sampa, M.H.O., and Chmielewski, A.G. 2007. Radiation treatment for sterilization of packaging materials. *Radiat. Phys. Chem.* 76(8–9):1535–1541.

Hury, S., Vidal, D.R., Desor, F., Pelletier, J., and Lagarde, T. 1998. A parametric study of the destruction efficiency of Bacillus spores in low pressure oxygen-based plasmas. *Lett. Appl. Microbiol.* 26(6):417–421.

Jacobs, P.T., and Lin, S.M. 1987. Hydrogen. US Patent 4,643,786.

Larousse, J., and Brown, B.E. 1996. *Food Canning Technology.* Weinheim, Germany: Wiley-VCH.

Levy, C., Aubert, X., Lacour, B., and Carlin, F. 2012. Relevant factors affecting microbial surface decontamination by pulsed light. *Int. J. Food Microbiol.* 152(3):168–174.

Misra, N.N., Tiwari, B.K., Raghavarao, K., and Cullen, P.J. 2011. Nonthermal plasma inactivation of foodborne pathogens. *Food Eng. Rev.* 3(3–4):159–170.

Muranyi, P., Wunderlich, J., and Heise, M. 2007. Sterilization efficiency of a cascaded dielectric barrier discharge. *J. Appl. Microbiol.* 103(5):1535–1544.

O'Sullivan, M.G., and Kerry, J.P. 2008. Smart packaging technologies for beverage products. In *Smart Packaging Technologies for Fast Moving Consumer Goods*, (211–232). Wiley: Hoboken, New Jersey

Palgan, I., Caminiti, I.M., Muñoz, A., Noci, F., Whyte, P., Morgan, D.J., Cronin, D.A., and Lyng, J.G. 2011. Effectiveness of high intensity light pulses (HILP) treatments for the control of Escherichia coli and Listeria innocua in apple juice, orange juice and milk. *Food Microbiol.* 28(1):14–20.

Ratner, B.D., Chilkoti, A., and Lopez, G.P. 1990. Plasma deposition and treatment for biomaterial applications. In *Plasma Deposition, Treatment and Etching of Polymers*, 463–516. Cambridge, MA: Academic Press.

Reuter, H. 1993. Processes for packaging materials sterilization and system requirements. In *Aseptic Processing of Foods*, edited by H. Reuter, 155–165. Lancaster, PA: Technomic Publishing Company, Inc.

Ringus, D.L., and Moraru, C.I. 2013. Pulsed Light inactivation of Listeria innocua on food packaging materials of different surface roughness and reflectivity. *J. Food Eng.* 114(3):331–337.

Thirumdas, R., Sarangapani, C., and Annapure, U.S. 2015. Cold plasma: A novel non-thermal technology for food processing. *Food Biophys.* 10(1):1–11.

Toledo, R.T. 1988. Overview of sterilization methods for aseptic packaging materials. In *Food and Packaging Interactions*, edited by J. Hotchkiss, 94–104. Washington, DC: American Chemical Society.

Vesel, A., and Mozetic, M. 2012. Surface modification and ageing of PMMA polymer by oxygen plasma treatment. *Vacuum* 86(6):634–637.

Woodling, S.E., and Moraru, C.I. 2007. Effect of spectral range in surface inactivation of Listeria innocua using broad-spectrum pulsed light. *J. Food Prot.* 70(4):909–916.

Chapter 9

Aseptic Packaging Materials and Sterilants

Robert Fox

CONTENTS

9.1 PRODUCT REQUIREMENTS

All products have specific package requirements necessary to keep the product in a safe and acceptable condition for its expected shelf life. In order to maintain the integrity of the package, these requirements may include functional barriers to light, oxygen, and moisture as well as mechanical properties such as the container's distortion temperature, stiffness, and impact properties.

Figure 9.1 provides a broad view of the oxygen tolerance of various products found in shelf-stable and refrigerated packages. Low-acid particulate products that are aseptically packed present a US Food and Drug Administration (FDA) challenge in the United States, but can be found in the European Union and other countries outside the United States. For that reason, they are included in this chart.

Some products have a need for a high degree of protection from the effects of ultraviolet (UV) light. A UV inhibitor can be added to minimize product degradation or the container can have an opaque layer. Most products have a need for a high degree of protection from moisture gain or loss. The next section discusses materials that have been identified as capable of providing suitable protection from either the ingress or the egress of oxygen or water vapor.

DOI: 10.1201/9781003158653-11

Oxygen Sensitivity and Fill/Process Temperatures for Foods and Beverages

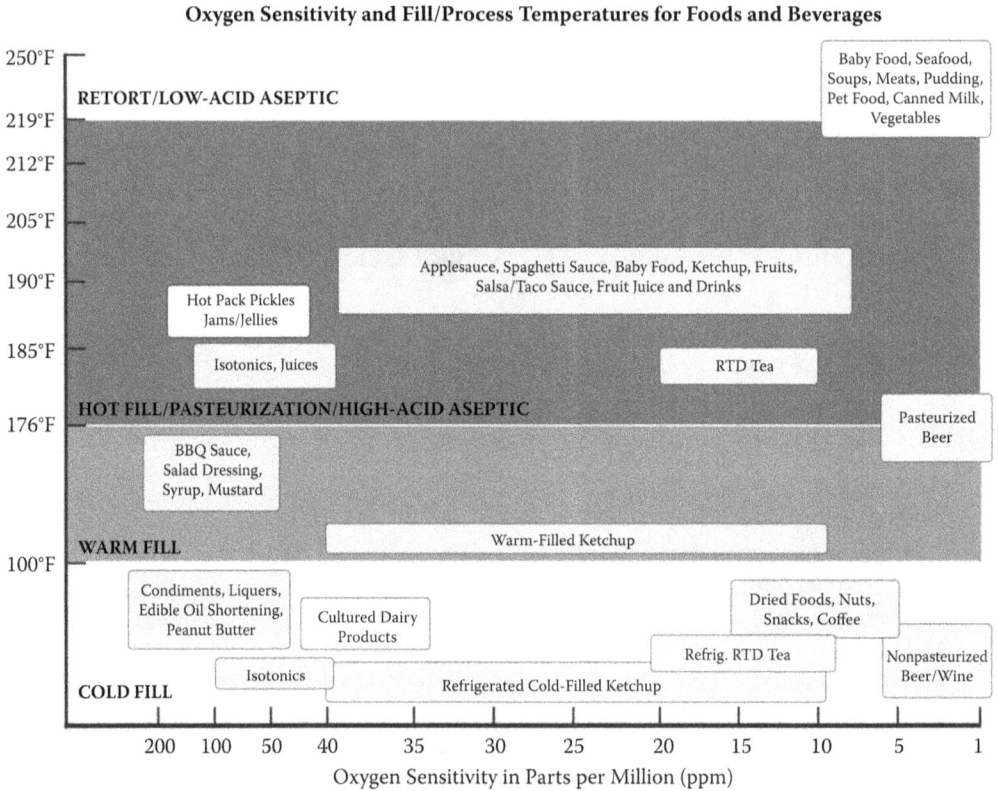

Figure 9.1   Oxygen tolerance of food groups.

9.2 MATERIALS

Thermoplastic materials used in sheet production for both aseptically packed non-barrier and barrier thermoformed packages have been chosen for properties that they provide.

9.2.1 Non-Barrier Sheeting

Non-barrier sheeting can be made of a single material but usually has two or more layers of different materials to reduce cost. Materials are chosen so that the outer layer(s) provide notch sensitivity resistance, cold temperature impact strength, and improved heat seal capability. The inner core material provides stiffness.

9.2.2 Barrier Sheeting

No single material can supply all of the properties required to produce a barrier container at an acceptable cost. Different materials are brought together in a layered structure (typically five to seven layers) to provide an adequate level of protection to the product at an acceptable cost.

Oxygen barriers of all materials can be significantly improved by the use of nanoclays, active scavengers, and vacuum deposition of metal or glass. Heat distortion can also be improved by the addition of inorganic fillers. In actuality, within a given material, changing any one property by the introduction of an additive influences all the other properties.

The following chart lists some of the common materials used in packaging and their respective barrier and physical properties:

Material	Density	Oxygen Barrier[a]	MVTR[b]	Heat Distortion[c] (°F)	Flex Modulus[d]
LLDPE	0.91–0.94420	1.0–1.5	<100	40–105	
HDPE	0.95–0.96	160	0.4	149–176	145–225
PP	0.90–0.91	150–240	0.7	125–250	170–250
HIPS	1.06	226	10.0	165–200	160–390
EVOH	1.1–1.2	.02	3.8–6.5	>250	–
PET	1.29–1.40	90.0	1.0–1.3	167	350–450
NYLON 6	1.12–1.14	2.6	16.0–22.0	311–365	405
PLA	1.23–1.25	>150	>10.0	104–150	350–450

[a] cc/mil/100 in^2/24 h at 65% RH/20°C.
[b] grams/mil/100 in^2/24 h at 90% RH/100°F; MVTR: moisture vapor transmission rate.
[c] ASTM D-648 at 66 psi.
[d] ASTM D-790 at 103/73°F.

9.3 STERILIZING AGENTS

Heat, chemicals, radiation, and ultraviolet light have been used independently or in combination with sterilants for aseptic packaging. Regulatory requirements and cost considerations have put a practical limit on the number of sterilants in commercial use.

9.3.1 Heat

Steam is an effective sterilant demonstrated by the fact that it has been in commercial use longer than any other sterilant. Although steam is not suitable for some thermoplastic materials whose heat distortion temperatures are below the high-pressure steam temperatures required, it has had a recent revival in the OYSTAR Hassia aseptic form–fill–seal (FFS) packaging system (Figure 9.2). Incoming rolls of plastic sheeting are heated on the top (food contact) surface with culinary steam (320°F) sufficient to sterilize the contact surface. The sheet is then indexed to the forming area where it is heated to its forming temperature and then indexed to the forming station where it is formed into the desired container shape. The next index moves it into the filling area followed by the sealing station and indexed out of the sterile zone. Container sterilization starts with the steam sterilization of the sheet's surface and continues through the forming, filling, and sealing stations after which containers exit the sterile zone. Lidding material is similarly sterilized prior to entering the sterile zone and sealed to the container.

9.3.2 Hot Water

Hot water (sometimes acidified) has been used commercially for high-acid products. The CrossCheck system, developed and patented by Mead Packaging (US patent 4,152,464), was

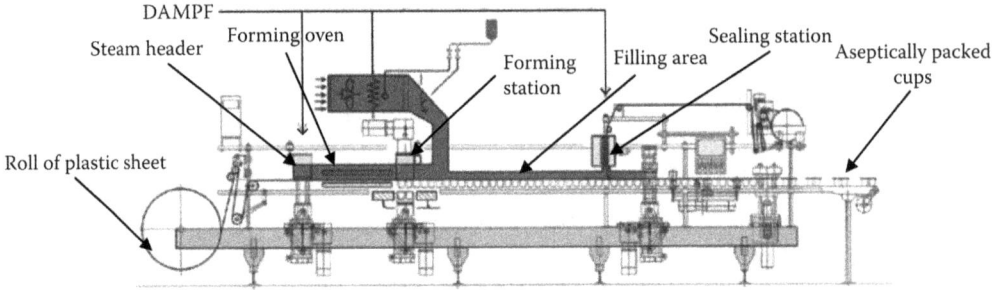

Figure 9.2   OYSTAR Hassia aseptic form–fill–seal system schematic.

commercialized by a joint venture of Mead Packaging and Rampart Packaging. CrossCheck was used successfully by Seneca Foods for single-serve applesauce for more than 10 years. Preformed containers were carried through a hot water (180°F) bath, exiting the bath in an inverted position (Figure 9.3). The heat seal flanges and inside of the containers were dried by the flow of warm sterile nitrogen as they rotated to an upright position prior to filling. Sterile nitrogen also maintained a slight positive pressure on the inside of the sterile zone. The sterile zone was presterilized by the use of culinary steam. Cups were filled volumetrically, sealed, and exited through a sterile water lock. Mead and Rampart stopped supporting the technology in 1990.

9.3.3 Neutral Aseptic System (NAS)

Manufactured by OYSTAR Erca-Formseal, the system uses the heat of the forming and lidding web coextrusion to sterilize the two webs. The bottom web is typically made up of polystyren

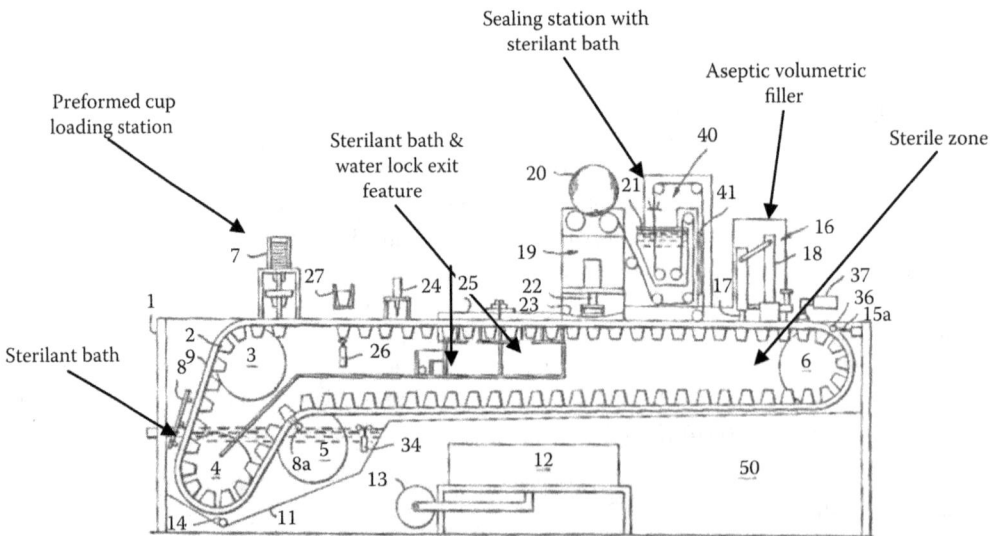

Figure 9.3   Cross-check aseptic deposit, fill, and seal system.

e–adhesive–EVOH–adhesive–LDPE–PP. The top web is typically made of PP–LDPE–adhesive–aluminum foil. In both structures, the edges of the webs are trimmed as they enter the sterile zone. The polypropylene layer of the bottom web is stripped away and exits the sterile zone. The remaining web has the newly exposed sterile surface and is indexed into the heating, forming, and filling stations. The top web, similarly having had its polypropylene layer trimmed and removed as it entered the positive pressure sterile zone, is indexed to the sealing and trimming stations.

9.3.4 Chemical Sterilants

Hydrogen peroxide (H_2O_2) is a clear, slightly viscous oxidizing agent that has been the chemical sterilant of choice for most packaging equipment manufacturers. It is an effective sterilant when used in concentrations of 30–35% followed by hot air at 60–125°C. The hot air substantially increases sporicidal activity in addition to dissipating the residual hydrogen peroxide.

Hydrogen peroxide identified in the Code of Federal Regulations under 21 CFR 178.1005 may be safely used to sterilize polymeric food-contact surfaces identified in paragraph (e)(1) of the regulation. Hydrogen peroxide also meets the specifications of the *Food Chemicals Codex* (3rd ed., 1981, pp. 146–147).

Peracetic acid (CH_3CO_3H) is a colorless, liquid organic compound that is highly corrosive. It is always sold in solution with acetic acid and hydrogen peroxide to maintain its stability. The concentration of the acid as the active ingredient can vary depending on its application. It has an advantage over hydrogen peroxide in that it can be used at lower temperatures (40°C). This is a benefit for aseptic packaging applications using materials like polyethylene terepthalate (PET) and LDPE, which have low heat distortion temperatures. It can also be used as an aseptic bottle rinse, spray, or mist without the need for a secondary sterile water rinse. Peracetic acid is accepted by the FDA for sanitizing and disinfecting (21 CFR 178.1005-1010).

9.3.5 Radiation

Irradiation has been evaluated for several forms of packaging and found to be an effective solution for large-capacity bags such as Scholle's bag-in-box aseptic packaging system. The dominant method of sterilizing hermetically sealed bags, pouches, and drum liners is by electron beam irradiation. Gamma irradiation has been evaluated and is not used for these types of applications due to its initial installation costs and regulatory considerations and potential damage to polymeric packaging materials. Scholle, the leader in bag-in-box aseptic packaging systems, has developed closure features and filling equipment that maintain asepsis during the filling operation. Guidelines for radiation sterilization can be found in 21 CFR Part 178.1005.

9.4 PACKAGING SYSTEMS

9.4.1 Dole Aseptic Canning

The Dole aseptic canning method differs from other aseptic packaging methods in that the packaging materials are subjected to culinary steam as the primary method of container sterilization. Containers are subjected to surface temperatures of 215.6–218.3°C (420–425°F) and cover temperatures of approximately 210–212.8°C (410–415°F). Today, the high temperatures required in their container sterilization restrict the choice of materials to metal two- or three-piece cans. Work is ongoing to develop a CPET container that would withstand the temperature requirements of the process and provide a cost-effective replacement with the benefit of microwaveability.

9.4.2 Preformed Thermoformed Containers

Aseptic preformed plastic container usage appears to be on the rise in Europe with recent installations of OYSTAR Gasti's Dogaseptic aseptic packaging system. The system utilizes hydrogen peroxide as the sterilizing agent in the aseptic system.

OYSTAR Gasti also has a Dogatherm preform aseptic cup filling system that uses steam to sterilize the inside of the cups and a long-life version that utilizes high-intensity UV-C light for sterilizing the container flange and interior surfaces.

9.4.3 Form–Fill–Seal

Form–fill–seal includes four main methods of container manufacture:

1. Paper-based packages such as fiberboard cartons and "brick packs."
2. Thermoformed plastic packages, commonly referred to as cups and trays.
3. Injection and extrusion blow-molded bottles.
4. Bag-in-box and large-volume pouches.

The best known of the aseptic FFS packages is the paper-brick-style package that utilizes a multilayered web (Figure 9.4) consisting of an

- Outside layer—polyethylene (for protection of fiberboard)
- Structural layer—paper (typically one side preprinted)
- Adhesive layer (bonds foil and paper, minimizes fiber penetration)
- Barrier layer—aluminum foil
- Adhesive layer
- Food contact layer—polyethylene

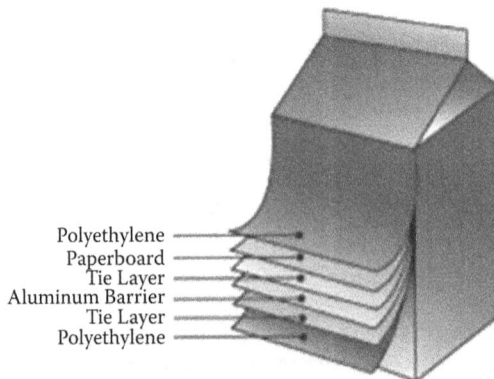

Polyethylene
Paperboard
Tie Layer
Aluminum Barrier
Tie Layer
Polyethylene

Figure 9.4   A typical material structure in preformed, paper-based barrier cartons.

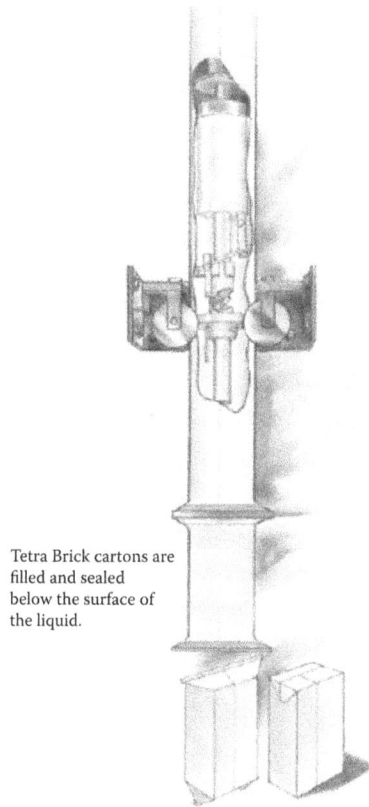

Tetra Brick cartons are filled and sealed below the surface of the liquid.

Figure 9.5   A typical roll-fed process schematic.

Paper-brick and carton-style packages are produced by two different methods. These packages are usually used for dairy-based beverages, juices, stock, and sauces.

1. *Roll-fed, paper-based systems* (Figure 9.5)—The leading producers of roll-fed, paper-based packaging systems and materials are Tetra Pak, Division of Tetra-Laval, Elopak, subsidiary of the Ferd Group of Norway, and GA Pack, a new Chinese supplier of paper-based packaging for roll-fed aseptic packaging systems.

   Efforts to minimize the effects of energy-related cost increases during the past 30-plus years include the commercial use of brick pack that has resulted in a 30% reduction in aluminum foil thickness and a 20% increase in board stiffness while reducing package weight by 15%. Although costs have been positively affected by these changes, other aspects of the carton have been adversely affected:
   - Thinner aluminum foil increases the potential for pinholes to develop.
   - Stiffer board increases the potential for pinholes to develop as a result of stiffer fiber content.
   - Increased thickness of the adhesive material between the foil, the board, and the PE layer, and also between the foil and PE layers on the inside of the carton, to minimize/eliminate fiber-induced pinholes through the foil and PE layers increases the risk of organoleptic impact as a result of chemical migration.

2. *Preformed cartons*—Preformed brick pack and gable top, fiberboard cartons are used in either high- or low-acid aseptic filling systems. The cartons are manufactured so that the bodies are preassembled and distributed in the flat. They are opened just prior to the filling process with the container body being fin sealed at the bottom. The tabs that result from the fin seal are folded over the bottom seam or up along the sides of the container body. Materials used to produce the containers are similar in construction to those used to produce the roll-fed, brick-style cartons. Differences are usually found in the thickness of the fiberboard to compensate for increased carton stiffness required for larger container capacities. Pinhole issues and solutions to them are similar to those of the roll-fed laminate.

Recently, SIG Combibloc, a unit of the Rand Group, commercialized a non-foil-based barrier carton. The EcoPlus carton replaced the foil barrier layer with a nylon-6-based barrier polymer. It was introduced in the European Union and is expected to replace its foil-based barrier board materials over the next few years with positive environmental advantages.

Combibloc contracted the Institute for Energy and Environmental Research, Heidelberg, Germany, to investigate some of the environmental impacts of the new EcoPlus carton and compare its findings to those of traditional foil-based cartons of similar sizes. Figure 9.6 summarizes the institute's findings. The comparison of EcoPlus carton and cCap closure was made against Combibloc's similar standard "cb3" carton with a cSwift closure.

The net result of the institute's study confirmed the polymeric barrier carton has significant energy-saving and carbon footprint advantages over the foil-based traditional carton. It appears that foil's advantage of low tensile strength, useful in opening features that require puncturing the foil and inner polymer contact layer of the carton by straw insertion or leveraged spout opening, may have been met with the polymeric structure.

The Net Results of *"cb3 EcoPlus"* Are	More	Loss
	Favorable than those of *"cb3"* (=100%)	
Acidification	by 22%	
Climate change	by 20%	
Aquatic eutrophication		by 12%
Terrestrial eutrophication	by 7%	
Summer smog (POCP)	by 19%	
Human Toxicity—PM10	by 20%	
Fossil resource consumption	by 22%	
Use of nature—forestry		by 8%
Total primary energy demand	by 15%	
Nonrenewable primary energy demand	by 20%	
Transport intensity—lorry	by 3%	

Left column:
"cb3 EcoPlus w/cCap" has lower indicator values (i.e., a more favorable performance) than *"cb3 w/cSwift"*
Right column:
"cb3 EcoPlus w/cCap" has higher indicator values (i.e., a less favorable performance) than *"cb3 w/cSwift"*
Note: Percentages shaded in gray are smaller than 10% and thus considered insignificant.

Figure 9.6   An environmental comparison of paper/polymeric versus paper/foil barrier materials used in food packaging.

Both the roll-fed and preformed aseptic fiberboard packages are typically sterilized by hydrogen peroxide when packages are to be used with low-acid foods. Packages to be used with high-acid foods can include sterilants such as peracetic acid–based sterilants in addition to hydrogen peroxide. Other methods that have been used, to a far lesser degree, include UV-C, gamma and electron beam radiation.

The environmental impact of the paper-based packages has not lived up to its expectation as there is no good reuse of this scrap material (sustainability) other than as a source of energy through incineration or as a component in the production of extruded lumber replacement. In purposeful incineration, toxic by-products are reduced compared to traditional fossil fuel resources found in other packaging materials. Additionally, the bulk of the package is comprised of a renewable resource wood pulp that helps to keep its carbon footprint small.

Thermoformed plastic packages are commonly referred to as cups and trays. These packages are typically used for an individual serving of products such as puddings, baby foods, dairy creamers, yogurts, and long-life entrées.

1. *Sheet/roll-fed systems:* These systems are widely used today as all plastic containers are being commercially sterilized using the acceptable methods of hydrogen peroxide, peracetic acid, and steam (product/package dependent).
2. *Preformed container aseptic systems:* Not widely used in the United States, preformed aseptic container systems are commercial in other parts of the world. Sterilization methods include chemical, steam, and ultraviolet light.
   - Injection and extrusion blow-molded bottles: These packages are typically used for dairy-based beverages and juices.
   - Preformed injection blow-molded bottles typically utilize preforms in the manufacturing process. While many systems are in commercial operation today, the following are the two primary methods of sterilizing the container prior to filling:
     – *Dry preform decontamination method:* Sterilants are deposited onto the internal wall of each preform. Preforms are then heated in the forming oven until they reach their forming temperature (greater than 100°C), which activates the H_2O_2, sterilizing the interior of the container. Containers are then blown into their final shape and transported to the filling and sealing stations in a sterile overpressure environment.
     – *Wet-sterilization process:* This process uses hydrogen peroxide or peracetic acid-based sterilants. Bottles are formed and the sterilant is dispensed into the formed container. An activation temperature of 100°C is required for hydrogen peroxide and 40°C for peracetic acid. The low activation temperature of peracetic acid-based sterilants is well suited to PET as it is well below the thermal distortion temperature of the polymer.
   - Extrusion blow-molded bottles are typically provided to the aseptic filling system in bottle form. If the bottle is in a closed-top format, its interior is sterile. This is not the case if the bottle is open topped.
   - Closed-top aseptic filling process: The closed-top process blow molds bottles using sterile air and maintains sterility prior to filling by pinching closed the bottle above the container finish at the time the bottle is produced. The combination of the polymers' temperature during the extrusion and blow-molding process, in combination with the sterile air used to blow the bottle into its desired shape, guarantees a sterile container. The top above the finish is removed in a sterile, controlled environment prior to being aseptically filled.

- Open-top aseptic process: Open-top blow-molded bottles have the top section above the finish removed during the blow-molding process. Sterilization of these types of containers is identical to the wet-sterilization process identified for injection blow–molded containers.

Bag-in-box and pouch containers are typically used for bulk packages of foods to reduce storage or shipping costs. Products shipped or stored in this manner are usually transferred aseptically to smaller packaging formats for retail distribution. Large-volume (typically ½ to 1 gallon) pouches are also used for retail and institutional sizes and are commonly used to package wine, cheese sauce, and ketchup and tomato sauces, as well as dairy-based products.

Large volume bag-in-box and pouches are fabricated of coated, metalized, laminated, or co-extruded films. Due to their size and lack of being able to maintain shape independently, they are sterilized using radiation. Radiation sterilization allows bags and large-volume pouches to be sterilized in a closed, hermetically sealed condition. Access fitments such as molded openings or valves are attached with their upper and lower openings sealed with a plastic or foil membrane. These fitments are welded to the interior or exterior of the bag or pouch prior to final sterilization.

Presterilized bags are stored and shipped to filling sites in a lay-flat condition. At the filling site, the bags are placed in a box or container and the filling fitment is suspended at the top of the

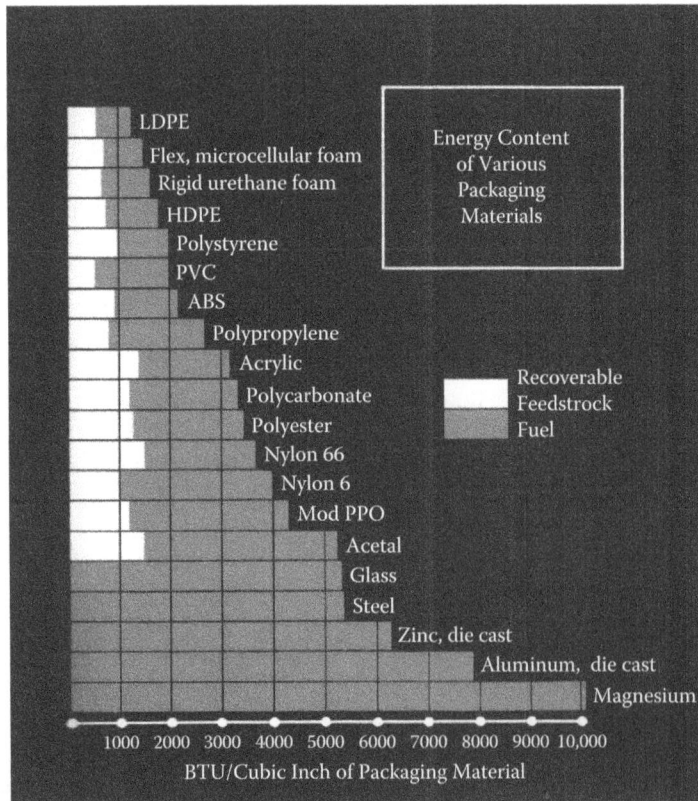

Figure 9.7 Recoverable energy components of various packaging materials.

box or container to minimize any air entrapment during filling. The container-opening feature is connected to a sterile filling head/valve and the exterior of the opening feature is sterilized prior to the aseptic filling head puncturing the membrane seal across the top of the fitment. Once filled, the opening feature is resealed and a protective overcap applied.

Smaller bags and pouches are fabricated and aseptically filled in the same manner as rolled-fed, paper-brick-pack-type packages.

9.5 ENVIRONMENTAL CONSIDERATIONS

Sustainability is a main consideration when looking at packaging options today. Energy conservation is also a main concern as it relates to sustainability and the environment. Figure 9.7 shows the recoverable and non-recoverable energy components of various materials used in packaging. It becomes easy to see which materials will have the largest carbon footprint and which materials will be the most environmentally friendly. Assuming that all packaging materials will end up in the municipal waste stream at some point in time, it is important to look at the long-term effects of that reality.

Typical Heat Content of Materials in Municipal Solid Waste (MSW) (Million Btu per Ton)	
Materials	**Million Btu per Ton**
Plastics	
Polyethylene terephthalate[c,e] (PET)	20.5
High-density polyethylene[e] (HDPE)	19
Polyvinyl chloride[c] (PVC)	16.5
Low-density polyethylene/ Linear low-density polyethylene[e] (LDPE/LLDPE)	24.1
Polypropylene[c] (PP)	38
Polystyrene[c] (PS)	35.6
Other[e]	20.5
Newspaper[c]	16
Corrugated cardboard[c,d]	16.5
Mixed paper[e]	6.7
Glass	0
Steel	0
Aluminum	0
b: Energy Information Administration, Renewable Energy Annual 2004, "Average Heat Content of Selected Biomass Fuels" (Washington, DC, 2005).	
c: Penn State Agricultural College Agricultural and Biological Engineering and Council for Solid Waste Solutions, Garth, J. and Kowal, P. Resource Recovery, Turning Waste into Energy, University Park, PA, 1993.	
d: Bahillo, A. et al. *Journal of Energy Resources Technology*, "NOx and N_2O Emissions during Fluidized Bed Combustion of Leather Wastes," Volume 128, Issue 2, June 2006, pp. 99–103.	
e: Utah State University Recycling Center Frequently Asked Questions.	

Figure 9.8   Recoverable energy content of packaging materials found in municipal waste streams.

	Glass	PE	PET	Alu	Steel
Container Type					
Mass [g]	325	38	25	20	15
Mass/Volume [g/liter]	433	38	62	45	102
Energy/Mass [MJ/kg]	14	80	84	200	23
Energy/Volume [MJ/liter]	8.2	3.2	5.4	9.0	2.4

"Embodied Energy of Drink Containers" from the ImpEE resource on "Recycling of Plastics." A study from the Cambridge–MIT Institute.

Figure 9.9 Energy content per volume (1 liter) of common rigid packages.

Aseptic packaging deals primarily with only a few of the materials shown: LDPE, high-density polyethylene (HDPE), polystyrene, polypropylene, polyester, steel, and aluminum.

Figure 9.8 shows the benefit of thermoplastic containers in terms of recoverable Btus available to aid in the incineration process with the additional benefit of providing heat to power steam-driven electrical generation systems.

Those materials that do not have a recoverable energy component require energy from other sources to move, distribute, or recondition or reshape them for future use. Figure 9.9 compares the energy cost per volume (liter). With energy costs continually rising, the energy cost advantages have slipped away from all of the traditional packaging materials except steel and that is expected to change to the benefit of plastics in the near future.

Chapter 10

Aseptic Bulk Packaging

Thomas Szemplenski

CONTENTS

10.1 ASEPTIC BAG-IN-BOX

In the early 1970s, William Scholle, an entrepreneur with tremendous foresight, visualized the potential for flexible packaging to replace the expensive, rigid packaging that was being used to transport food products such as tomatoes and other fruit products. Many years prior to this vision, Scholle developed flexible pouches of various propylene derivatives for packaging battery acid. Even today, virtually all battery acid is packaged, transported, and sold in flexible packaging invented by Scholle. Scholle felt that he could use the same packaging technology that was being used to package battery acid to package shelf-stable food products. Instead of using existing retorting technologies to commercially sterilize the product to be packaged in flexible packaging, he leaned toward applying a relatively new technology that was rapidly developing at the time: aseptic processing and packaging.

By the early 1970s, the process for aseptically processing food products was proven and commercially operating at approximately a dozen food processing facilities, sterilizing such food products as tomato and fruit pastes, puddings, and dairy-based products. At each of these existing facilities, the aseptically processed product was being aseptically filled into rigid containers using either the Dole aseptic canner or the FranRica aseptic metal drum filler. The can sizes being filled on Dole equipment ranged in sizes from 4 ounces up to the #10 cans that held approximately 96 ounces. The aseptic drum filler aseptically filled product into 55-gallon steel drums.

Scholle's vision was to utilize aseptic processing methods to fill larger quantities for industrial and commercial uses into less expensive flexible packages. At the time, tomato paste and fruit purees and concentrates were hot filled into #10 cans, aseptically packaged into 55-gallon metal drums, or, in the case of fruit products, alternatively filled into plastic pails and frozen for shipment to the end user for reprocessing. All of these methods of packaging were expensive, as was the cost of refrigeration when plastic pails were used.

Scholle engineered and manufactured a prototype of an aseptic filler to fill preformed bags. The prototype was then installed at the food processing laboratory at Purdue University in West

DOI: 10.1201/9781003158653-12

Lafayette, Indiana, and a mutually dependent aseptic processing system was simultaneously installed so that a sterile product could be delivered to the filler. Scholle, together with Dr. Phil Nelson, who was then a professor at Purdue, was able to make the necessary modifications to the prototype filler to successfully package high-acid (<pH 4.6) food products into presterilized, preformed bags. The bags were then, and continue to be, presterilized by gamma radiation.

The vision Scholle had about the flexible market was extremely accurate, for soon after the Purdue modifications and numerous customer trials, the market exploded with interest in this new technology and subsequent commercial installations. The first installations were for tomato paste that was being processed in California and shipped to the East Coast to be remanufactured into sauces and ketchup. Prior to the Scholle aseptic filler, almost all this product was being aseptically filled into very expensive 55-gallon metal drums. Today, almost every installation of aseptic drum fillers has been replaced by aseptic flexible packaging. Most tomato paste today is being aseptically filled into bags that range in size from 55 gallons up to 330 gallons (Figure 10.1).

Many improvements to the original Scholle filler have been made since the early days at Purdue. The Scholle filler can now fill preformed bags up to 330 gallons and it is no longer limited to high-acid foods. The US Food and Drug Administration (FDA) has validated the Scholle filler to now fill low-acid foods (>pH 4.6), and in fact, the most recent low-acid filler is a continuous web filler that is capable of filling up to 15 bags per minute. Over the years, other companies have manufactured aseptic bag-in-box fillers; however, the Scholle filler remains the dominant market leader for aseptic fillers and supply of preformed flexible bag packaging. Since the formative days for the Scholle aseptic filler in the 1970s, hundreds of aseptic fillers for preformed bags have been installed all over the world and the list of products being aseptically filled into the bags ' continues to increase.

Scholle Aseptic Packaging System

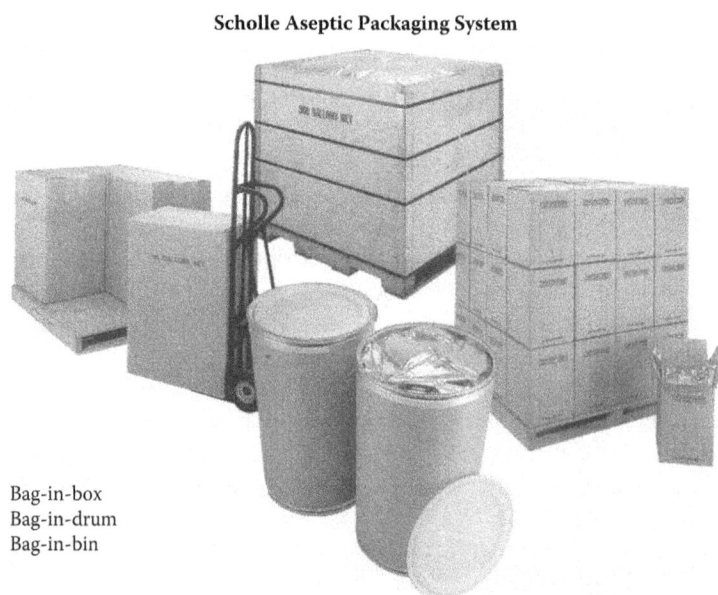

Bag-in-box
Bag-in-drum
Bag-in-bin

Figure 10.1 Various types of aseptic preformed bag packaging. (Photograph from a Scholle brochure.)

The tomato paste market for aseptic bag fillers was followed by the market for fruit for yogurt and other fruit-based products and citrus products. Some of these products were filled into 3- and 5-gallon bags and others into 55-gallon bags.

10.2 ASEPTIC BULK CONTAINER

Following aseptic bag-in-box packaging was the development of reusable aseptic stainless steel containers for high-acid products. These containers are usually manufactured in sizes of 200–300 gallons and are constructed of either 304 or 316 stainless steel (Figure 10.2). The following are some of the advantages of this type of aseptic storage:

- One-time charge for packaging
- Stainless totes are easily sterilized with steam
- Ability to remove partial product with remaining product staying sterile
- No oxygen or light penetration through flexible packaging
- Ability to nitrogen flush the head space
- Ability to be stacked high (Figure 10.3)
- Able to be cleaned in place
- Easily transported by forklift and pallet jacks

Products such as diced tomatoes, citrus juices and concentrates, fruit purees, and stabilized fruit for yogurt are some of the products being aseptic packaged and delivered in stainless steel

Figure 10.2   Stainless steel aseptic tote (800–100 liters). (Photograph courtesy of CCR Containers.)

Figure 10.3   Aseptic totes stacked. (Photograph courtesy of CCR Containers.)

totes. There are a number of suppliers of stainless steel totes. These manufacturers have advised that they will either sell or lease the totes to processors.

10.3 ASEPTIC BULK STORAGE

While working with Scholle to commercialize aseptic bag-in-box technology, Nelson partnered with several other synergistic companies to design and develop equipment to store massive quantities of fruit and vegetable products for long periods at room temperature in bulk-holding tanks. The more paramount companies that Nelson partnered with included Bishopric Products (now called Enerfab) and FranRica (now called JBT FoodTech). The aseptic processing system is generally the same that is used to process the products for bag-in-box; however, Nelson and his new partners developed a means of epoxy-lining carbon steel tanks and sterilizing them so that they could be filled and stored with aseptically processed acid products (<pH 4.5) for long periods of time.

This innovative group proved the concept with two 100-gallon stainless steel and three carbon steel epoxy-lined tanks at Purdue University's food processing laboratory. The first products that they were able to store aseptically were tomato products followed closely by fruit juices, such as apple and grape. Eventually, they found that food-grade, epoxy-lined carbon steel tanks were easier to sterilize and considerably more economical than stainless steel tanks.

Commercialization followed with the installation of a 40,000 bulk aseptic storage tank for tomatoes and several 250,000–600,000-gallon storage tanks for apple and grape juices. Later, the technology was introduced for citrus products. Further research dictated that although the citrus products were commercially sterile, refrigeration after filling into the tanks enhanced the shelf life. Today, the citrus industry has embraced the aseptic bulk storing of refrigerated (35°F)

Figure 10.4 Aseptic bulk storage tanks in construction, 1 million gallons each. (Photograph courtesy of Enerfab.)

citrus juices as there are now more than 300 million gallons of citrus juice stored in approximately 330 bulk tanks utilizing this technology in the state of Florida alone (Figures 10.4 and 10.5). The largest aseptic storage tank holds 1.8 million gallons; however, tanks of 2.1 million gallons of volume are currently under construction. This technology developed by Nelson and his associates, in addition to the facilities installed within the United States, now have international

Figure 10.5 Refrigerated room being built around field fabricated tanks.

installations in Brazil, Belgium, and Spain. Others are sure to follow (R. Brocker, personal communication, 2010).

There are many benefits of aseptic bulk storage, including, but not limited to:

- Reduced storage cost
- Flexibility, the ability to aseptically blend different juices
- Improved product quality
- Inactivation of microbes and enzymes for a stable product
- Improved vitamin C retention and other light-sensitive products

Unlike other smaller (500–20,000 gallons) aseptic storage tanks that are presterilized by steam, the larger tanks cannot be sterilized by steam due to their massive size. Presterilization of the larger tanks is done by initially filling the tanks with the chemical sterilant iodophor, supplied by Klenzade and other chemical companies servicing the food industry. Once the tanks are sterilized, they can be maintained in a sterile state for long periods of time. In fact, at an installation in Florida, one tank has been continuously used for 6 years without it having to be resterilized.

During sterilization, the large tanks are completely filled with the sterilizing iodophor solution as depicted in Figure 10.6, supplied by JBT FoodTech. Once the tank is deemed sterile, the iodophor is removed and replaced by filtered sterile nitrogen and held under pressure during the filling, storage, and removal of product, as shown in Figure 10.7.

There are now more than 565 aseptic bulk storage tanks installed throughout the world storing more than 473 million gallons of product. Most of these tanks are 1-million-gallon storage tanks aseptically holding citrus juices and concentrates, grape juices, and tomato products. What started out as an idea that Dr. Nelson and his associates had more than 25 years ago has turned into a technology that the food industry has embraced and will continue to embrace as more and different food products will be aseptically stored in larger tanks.

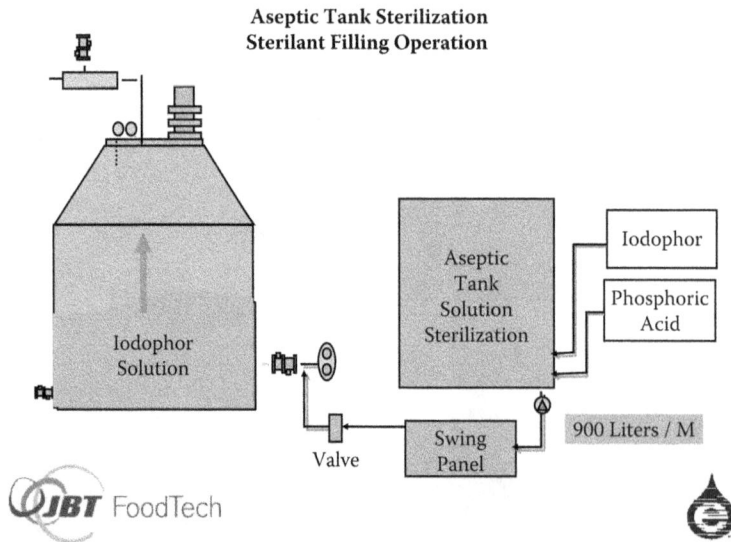

Figure 10.6 Presterilization of aseptic bulk storage tanks (Sterilant Filling Operation). (Diagram courtesy of JBT FoodTech.)

Aseptic Tank Sterilization Sterilant Emptying Operation

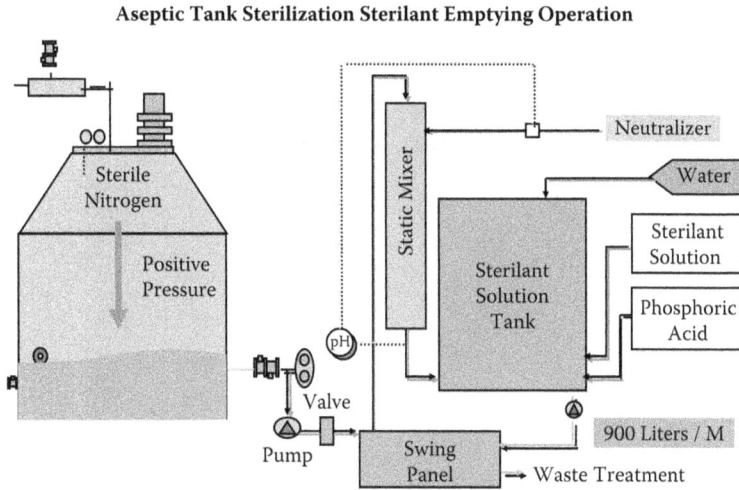

Figure 10.7   Presterilization of aseptic bulk storage tanks (Sterilant Emptying Operation). (Diagram courtesy of JBT FoodTech.)

10.4 ASEPTIC OCEAN LINER TRANSPORTATION AND STORAGE

A number of years after aseptic bulk storage was commercialized, Nelson was contracted to develop a means of aseptically transporting large quantities of citrus by means of an ocean liner. This successful technology would facilitate the transportation of citrus products from Florida and South America to Europe. Nelson visualized using the already developed and proven aseptic bulk storage tanks installed inside the hull of a large ocean liner.

In 1993, the first ship dedicated to transporting aseptic processed and stored citrus was manufactured and put into use. The *Ouro do Brazil* was put into service to ship citrus juices for Citrosuco Paulista of Brazil to Europe, Japan, and the United States. Built in Norway, the first

Figure 10.8   An aseptic bulk transfer ship. (Photograph courtesy of Citrus Coolstores, Inc.)

ship was 564-feet long and enclosed 16 vertical tanks that each held 200,000 gallons of product. In all, 3,200,000 gallons of product can be shipped (P. Nelson, personal communication, 2009).

Since the *Ouro do Brazil,* several other larger ships have been built using this technology, carrying up to 8 million gallons of product in each vessel. No doubt more ships will be manufactured for the purpose of shipping aseptically stored product from the source to points all over the world (Figure 10.8).

Chapter 11

Design of Facility, Infrastructure, and Utilities

Pablo M. Coronel, Ben Rucker, and Jason A. Tucker

CONTENTS

11.1 BASIC CONSIDERATIONS OF HYGIENIC DESIGN

Food processing plant design is a key component for the safety and quality of the products made. Design of a plant includes not only the building materials, equipment layout, and utilities but it also defines the flow of product, personnel, and environmental conditions. Hygienic design of factories and equipment has improved the conditions under which foods are made, allowing for minimal preservation by reducing the microbial burden in the plant, extending the life

DOI: 10.1201/9781003158653-13

of equipment by providing environmental conditions, and giving employees excellent working conditions which help in motivation and retention. If a product has lower microbial load at the start of the process, it is possible to reduce the heat treatment needed to make the product safe. Further care must be taken to avoid recontamination or cross-contamination of products with microorganisms or allergens, which must be considered in the facility design.

Plant design affects the safety of the products for a very long time, usually buildings outlive equipment and are in constant change, production lines are modified or added, new equipment is introduced and requirements from the regulatory and consumer side evolve, pushing buildings beyond the initial design, and thus it is necessary to have flexibility built into the plant, equipment, and utilities design. Buildings need to be designed and constructed keeping safety as a goal, i.e., the ability to produce high-quality, high-safety products for a long time. Hygienic design of buildings is a philosophy which needs to be well understood by engineers, designers, and management during the development and construction of plants and processing lines. Hygienic design must examine every detail of the construction to determine the points where sanitation could fail and cause contamination of products. This contamination can be microbial, with microorganisms which taint the food and can cause health or spoilage issues, allergens which can cause adverse reactions, chemical such as the cleaning agents which can render the food unsafe, physical such as glass, chips of building, or dirt, and lately, due to the increased consumer pressure, cross-contamination can also happen by mixing certified products (organic, non-GMO, religious diet, etc.) with non-certified products. Any of these types of contamination which can result in at best economical damage to the company and at worst in a public health issue or an outbreak should be avoided. Hygienic design is very important when expansions, remodeling, or new process lines are incorporated in existing building, to prevent the situations where the considerations of the initial design could be compromised. When new walls are built or old ones are removed, allowances should be made to ensure that the plant will perform as well as previously and that no failures in sanitation will happen.

At a minimum, hygienic design must comply with all rules and regulations of the regions where the plant is built such as FDA or EU, together with the special requirements for the products being made such as dairy, meat, or organic products. Aseptic processing, being one of the most advanced and complex processes, requires a high level of design in every part of the process, incorporating the use of high-quality materials, air handling, and specialized sanitation conditions. It is important to pay attention to all these at the beginning of the design, where the involvement of quality control personnel and specialists in HACCP, sanitation, and quality must be involved. Decisions taken at the initial part of the design of a facility determine the ability to produce high-quality foods and have a large impact on the final cost of the project. When improper decisions are taken, the cost of fixing the improper design is much larger than any savings that could be obtained by not following the hygienic design principles, resulting in a pay me now or pay me later scenario which must be avoided (Lelieveld, 2005).

Hygienic design goes hand in hand with cleaning and sanitizing, as both are prerequisites to produce safe food products. Hygienic design looks at the cleanability of different parts of the plant, harboring of microorganisms, and flow of products, air, and personnel to prevent the movement of "dirty" particles into cleaner environments which could contaminate the finished products. Good sanitary design incorporates several objectives that are beneficial in the long term and result in savings over the lifetime of the facility such as:

- Efficiency of cleaning programs
- Faster cleaning

- Prevention of contamination and cross-contamination
- Personnel hygienic practices
- Zoning and segregation of plant spaces
- Flow of goods and people
- Flow of air and pressurization
- Construction that is durable
- Long-term value

All of these efforts begin with a well designed and constructed building, where every detail is considered for the final purpose, followed by the acquisition of well-designed equipment and installation.

11.2 PLANT DESIGN AND SITE SELECTION

Safety of aseptic processing and packaging begins by the selection of the site and construction of the facility, well before any products are made. The selection of a site is a complex process in which economical and human factors are generally considered but factors such as odors and airborne contaminants from land and factories surrounding the site, distance to wildlife areas, prevailing winds, etc., should also be considered as part of the hygienic design of the factory. Sites close to the agricultural fields or landfills where dust and odors can blow into the factory need to be avoided or managed. Loading docks should be placed such that wind can't blow trash and contaminants toward the open doors.

11.2.1 Exterior Considerations

11.2.1.1 Landscaping

Any plant must be built on a site with easy landscaping to avoid pests, and if wildlife or swamps are close-by, fences are needed to keep the animals away. Sanitary landscaping should deprive rodents, birds, and reptiles from food and hiding places; ponds must be avoided when possible or located at a distance that will avoid pests and odors. Around the building, a paved or grass free strip must be maintained, at least 1 m (40") wide and 100 mm (4") deep. When paving is not possible, a thick landscape liner and small round gravel are recommended. This strip allows the QA and Safety personnel to inspect the area for rodent activity and is a good area to place traps and baiting stations.

Grass should be kept short and inspected for pest activity regularly. Shrubs should be placed at least 3 m (10 ft) away from the walls and they should be placed in a way that a person can walk around and between them to inspect for pests. Trees should be located in a way that limbs are far enough from the facility so that pests can't jump toward the facility; trees that are not friendly for bird nesting are preferred.

Parking lots, driveways, and dumpster areas should be paved and sloped so that water drains away from the facility, even during the strongest foreseeable storms. There should not be any standing water around the facility to reduce areas attractive to birds, rodents, or insects. The dumpster area can be problematic, it must be located in a paved area, sloped toward a drain, and with easy access to the collection vehicles. It is good practice to have a hose station or at least water access to clean this area regularly.

Perimeter should be fenced, to avoid the access of people and animals to the compound. Perimeter must be accessed regularly and should be considered the first line of defense against pests, with bait boxes and repellent devices placed at the perimeter. Tall grass or bushes must be avoided at the fence, and a grass-free strip at the fence is also recommended. Access to the facility and site should be considered, with cameras for surveillance around the facility and in areas where little traffic is expected (storage, bulk silos, etc.), ingress into the facility should be protected with electronic locks on doors to ensure that only authorized personnel is admitted into the facility and to certain restricted areas. Regulatory compliance in this area requires attention to maintain increased security on site in order to avoid intentional adulteration or bioterrorism.

Exterior light should be sufficient for surveillance, especially parking lots and any yard and outside areas where pests and unwanted visitors could hide. It is preferred to have yellow lights (sodium-vapor or yellow LED) which don't emit UV and thus don't attract insects at least at every entry point to the facility, including personnel and visitor doors, loading docks, and maintenance doors.

11.2.1.2 Exterior Walls

Exterior walls must provide protection against the ingress of pests, including rodents, insects, and birds, and must be built in a way that prevents nesting. Exterior walls can be made of several materials, such as concrete (pores or tilt-up), cement blocks, metal, insulated metal panels (IMP), or a combination of some of those materials.

The choice of material depends on the weather, cost, time, and expected durability, and must take into account the sealing and caulking of the joints of materials to prevent the ingress of insects. Painting or other finishes will help maintain the wall weather-proof and any insulation should be sealed with insect-proof and rodent-proof materials. Rodent proofing is very important and it should be an integral part of construction. Rodents can climb, jump, and like to burrow under the foundation and penetrate the building through openings in the floor. During construction, the wall footers should flange the whole perimeter around the building at least 30 cm (12") at right angles with the foundation and with a depth of at least 60 cm (24"). Special care should be taken in expansion joints and floor drains which are the most vulnerable areas for rodent ingress. As an extra rodent deterrent, the whole building perimeter should also have a grass-free strip at least 1 m (40") wide and 100 mm (4") deep. This strip should be lined with thick landscape sheets to prevent weeds and filled with small round gravel, round gravel (pea gravel) doesn't bridge and prevents burrowing (Graham, 1991).

11.2.1.3 Roof

Roof of the plant is an area where several critical activities are carried out, such as the air inlet to the process and gas venting, as well as supporting air-handling equipment, and should not be overlooked. Roof must be constructed in a way that prevents standing water and harboring of pests and contamination. Smooth materials such as membrane roofs or cement tiles are preferred, and should project at least 30 cm (12") outside of the walls of the facility.

Flat roofs need to be constructed in a way that prevents standing water, with a curb that extends above the room, and equipped with downspouts for rain that project toward the outside of the facility. These spouts must be able to handle the worst foreseen storms without overflowing. Insulation materials should not be placed on the inside of the curb, as maintenance is very difficult. Roofs must be maintained and cleaned regularly to prevent the appearance of cracks, standing water, and the growth of vegetation. Sloped roofs should be inspected for cracked tiles and must project enough to avoid rainwater to splash back into any opening of the facility.

Roofs must be able to withstand penetrations, usually for air inlet, and gas outlet, and some utilities and roof penetrations must be sealed and inspected regularly to prevent water leaks.

Air-handling equipment is, in many cases, placed on the roof. Air-handling equipment is heavy and thus the structural and architectonic design of the roof must allow for such weight, and have allowances for future expansions and modifications of a facility.

Air inside a food processing facility must be considered as an ingredient, and as such it must be free from contaminants and foul odors. Air inlet penetrations in the roof must be made in a way that avoids such contamination, and air inlets must be placed away from gas vents, chimneys, or flumes and on the side so that wind could not blow contaminants toward it. The same considerations should also be given to the air-handling equipment.

11.2.1.4 Loading Docks

Loading and unloading is one of the most common activities in food facilities. Loading docks should be built in a way to allow this transfer of materials to be performed in an efficient and safe way. Loading docks must be built in a way that prevents the ingress of pests and contamination while facilitating the task of loading and unloading. Driveway to the docks must be sloped away from the factory. Some old factories still have sunken loading docks, which can flood and become a contamination point if the drains are not maintained in top shape.

Walls of the loading dock should be made in a way that prevents rodents from climbing. Rodents can climb concrete walls with amazing ease, and as such a strip of smooth material is recommended. It is usual to see rubber bumpers and retaining hooks in docks, and those should be inspected regularly in order to avoid pest harborage. Self-leveling docks can be challenging, as the pit used for the dock leveling can become a contamination and pest-harboring point. Seals on the loading docks must provide a tight fit to the truck. Mice only require 5 mm (7/32") of space to enter. Inflatable seals are available and seal brushes must be avoided.

Loading doors are usually rapid vertical-lift or roll-out doors with durable outside surface to effectively isolate the facility when closed. They must be cleanable, either using wet or dry cleaning, and any housing of the mechanism should be inspected to prevent the harboring of insects.

11.2.1.5 Entry Points

Personnel entry doors must be considered as part of the food safety and food security design. Only the authorized personnel should enter into the facility, and thus unwelcome guests and pests should not be allowed. Exterior doors must not open directly into any production space, but into a foyer, hallway, or locker room. The preferred material is metal, with insulation and polymer seals to prevent entry to any insects or rodents. Personnel doors must have entrance control methods, such as electronic locks or turnstiles to ensure that only authorized personnel are coming into the facility.

Visitor doors must lead to a foyer where visitors are made to sign in, and are made aware of the rules they must follow while visiting the plant. A checklist of GMP and procedures can be pre-printed or made into a poster. Maintenance doors should be locked and used only when needed.

Personnel doors must have lights above them, preferably yellow so as to not attract insects, as well as air knives or other additional procedures to keep the insects out.

11.2.2 Interior

The inside of the facility must be built in a way that not only looks good, but also facilitates cleaning and helps maintain the microbial loads at very low level. Every detail of the design must be examined, including floors, ceilings, walls, curbing, hallways, personnel sanitation, air-handling, and zoning. Interior design must have the input from a qualified industrial architect, food safety practitioner, and operations expert, to consider the flow of personnel and materials to ensure that contamination and post-contamination are prevented.

As it can be observed in the regulatory section, the inside of the plant must comply with several requirements; floors, walls, and ceilings should be constructed in a way that may be adequately cleaned, kept clean, and kept in good repair (21 CFR117.20-b (4)).

11.2.2.1 Floors

Floors are the most abused surface in a food processing facility, and they must fulfill a large number of requirements. Floors must be built to withstand the weight of equipment (calculated with full load), cleaning cycles, hot and cold liquids, steam, spills, etc., must be resistant to shatter due to temperature shock or impact, must be waterproof, must be slip-resistant, and must be sloped to convey any water toward the drains.

Even though the floors are not food-contact surfaces, they have a very large impact on the microbial load in the environment, personnel safety, and overall appearance of the factory. Floors are the destination of most dirt and contamination due to the effects of gravity, and as such they must receive attention both in design and during operation.

Floors are made of two basic parts, the base and the flooring. The base is generally a concrete slab, either freshly poured or preexisting in the building. The quality of the base has a long-term effect on the final quality of the floor. Concrete slabs must be made to withstand the mechanical stress and weight of the equipment, the expansion and contraction of the soil, and must be built to isolate the building from groundwater. Newly poured concrete slabs must be cured for at least 28 days.

Flooring materials used in food processing can be of two types, ceramic tiles or polymer-based floors (resin coated). Each of them has advantages and disadvantages and those must be discussed beforehand.

Ceramic tiles are used in the dairy industry, and they are highly resistant to heat and mechanical stress. The main disadvantage of ceramic tiles is the need for grouting, which if it is not applied properly can result in cracks and harboring of contaminants. Resin-based joints (epoxy) can be used, and that solves the issue (Lelieveld, 2005).

Polymer-based floors have become more popular, as they constitute a continuous surface with no joints, and have high resistance to stress (Figure 11.1). Different polymer combinations are possible, and a complete evaluation needs to be made before the application. These floors consist of a binder and a resin, the binder is either sand or some type of cement, and the resin is the polymer. The aggregate is used to level and create sloping toward drains, as well as to shape the corners with walls. Additionally, a topcoat might be added for durability or for slip resistance. These floors need to have a gas-removal step during installation.

11.2.2.2 Walls

Walls are the second most abused surface of a food processing building; besides having to support equipment, they are washed down with hot and cold water and are subject to chemical exposure. Walls are penetrated and patched, and can be a source of contamination if not properly taken care of.

Interior walls should be strong, impermeable, insulated, flat and smooth, easy to clean, and resistant to impact, wear, and corrosion. In several spaces, walls are wet cleaned, in which case they must also be resistant to direct water spraying, scrubbing, and cleaning chemicals.

The materials of construction of internal walls should be carefully considered. Insulated metal panels in which insulation is sandwiched between metal panels are a good material, as

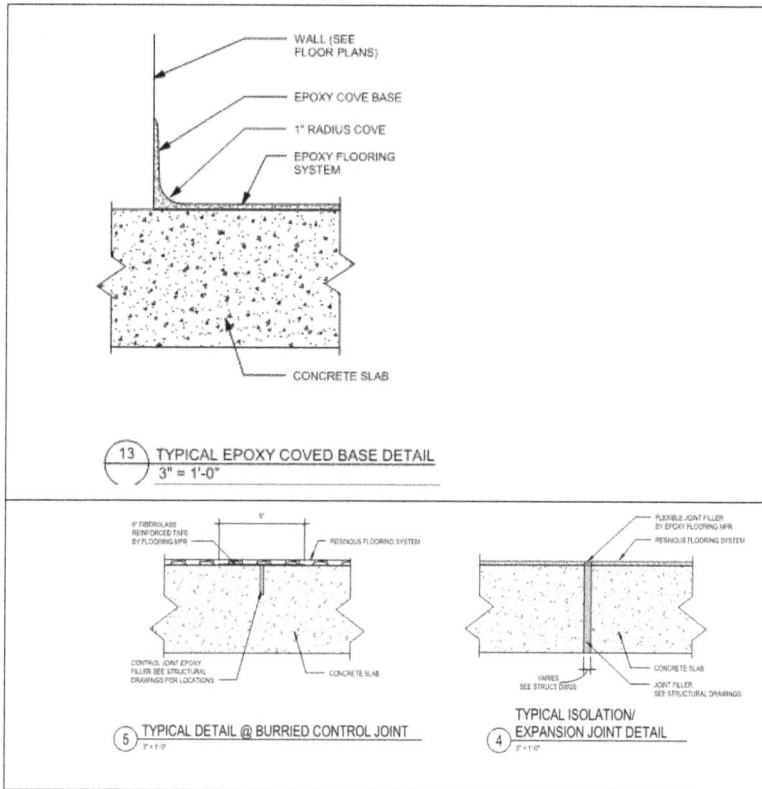

Figure 11.1 Typical details for epoxy floors used in food facilities.

well as tiled walls. In existing facilities, where the walls are uneven, rusting, or peeling, a spray-on insulation covered with polymers (resins) is also a possible solution.

Curbs are required when walls are made of insulated metal panels. This separates the panel from the floor and protects the panels against water, splashing, and impact from forklifts and hand trucks. Curbs are a short piece of concrete wall with height dependent on the design, but at least 50 cm (20"), which are coved, and where a metal channel holds the panels in place. The channel and joints are filled with caulk and covered with the same resin as the floor, thus creating a sealed area where no moisture or microorganisms can penetrate into the wall base. (Figure 11.2)

Designed penetrations in walls are made for several reasons, tubes and transport belts pass from room to room, anchoring points, flumes and gas exhaust, etc. These penetrations can be pathways where contamination in the form of moisture, dirt, pests, or microorganisms can pass from one side to the other; and must be sealed. Wall penetrations should be made watertight to avoid contaminating the interior of the panels where Listeria and Molds can grow. Scutcheons and wall boots are solutions to this problem.

Scutcheons (Figure 11.3.a) are stainless steel plates which are attached to the wall and caulked through, with tubing welded on both sides. These scutcheons can be for a single penetration or a stainless plate can be used for several tubes. When penetrations are made through CMU or IMP walls (Figure 11.3 b and c), some extra considerations are needed. Wall boots such as PipeTite

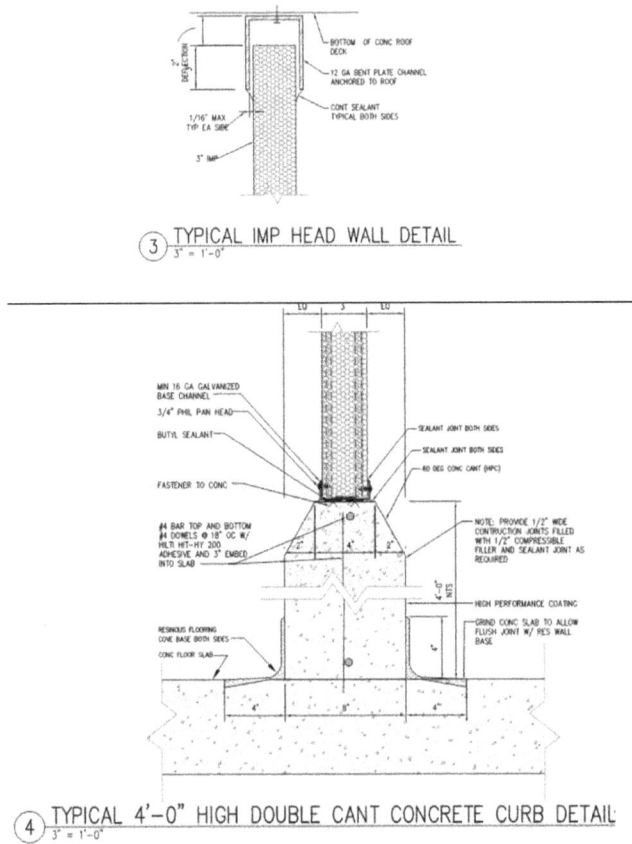

Figure 11.2   Typical IMP partition wall detail.

form a flexible seal around the tube that is water-tight and prevents contamination from one side to the other. These can be used both in walls and in ceilings.

11.2.2.3 Ceiling

Ceilings separate the roof from the production spaces and are as important as walls and floors to maintain food safe. In few cases, the interior of the roof is the ceiling, especially in concrete buildings, but in most cases the roof and ceiling are different, and the ceiling isolates the production space from steel rafters and acts as a second barrier to any external contamination. Having a ceiling area allows for utilities to be run outside of the process area but also requires a ceiling strong enough for maintenance activities and personnel to walk on. This area needs to be well-ventilated to avoid excessive moisture and condensation and should be inspected regularly.

Requirements for food process ceilings include cleanability. Ceilings must be waterproof and be able to hold lights, sprinklers, and other utilities. Lighting is usually part of the ceiling, and the ceiling must be a good reflector of light.

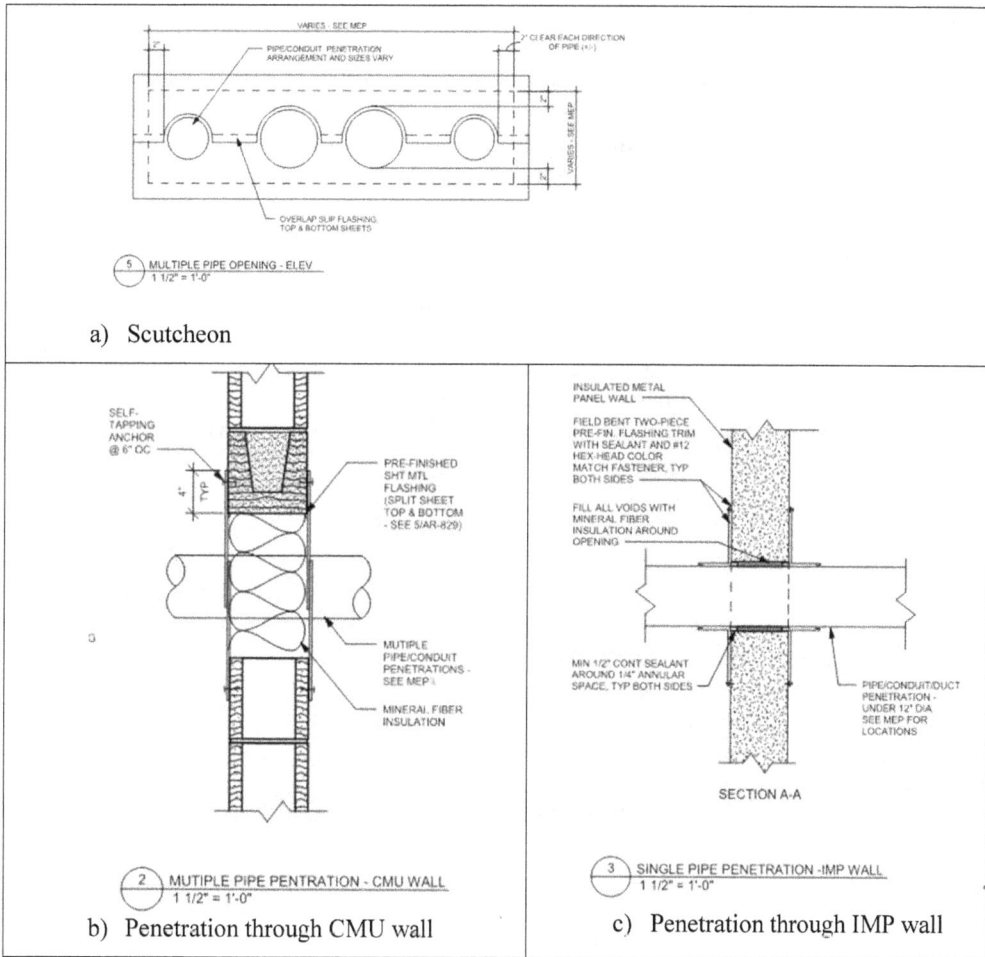

Figure 11.3 Typical wall penetration details.

Ceilings can be false or walk-on. False ceilings are made of tiles which must be water-resistant and have a gasket around them to make the whole installation waterproof. If the area is going to be wet-washed and sanitized, the ceiling must be built to withstand the pressure of a hose. Once panels are removed for maintenance, care must be taken to replace the gaskets so that no water can leak during washing.

Walk-on ceilings are made of IMP panels, and joints must be caulked. The advantage of walk-on ceilings is that most utilities can be placed on the outside of the food processing area and maintenance is performed without the need to replace tiles. This creates a second barrier to the external contamination. Penetrations in walk-on ceilings must be treated like wall penetrations, with wall booths being the preferred method.

11.2.2.3.1 Lighting

Interior lighting must be sufficient for plant to maintain a certain minimum level of illumination. Food contact surfaces, where product is open and manual work occurs, require higher levels of illumination than warehouse or shipping areas and local codes must be observed. Low UV emission lights are preferred to avoid attracting insects.

Light fixtures are generally part of the ceiling, either affixed to the ceiling or embedded into it. Light fixtures should not have ledges where dust or contaminants can accumulate and must be able to withstand cleaning. Fixtures and bulbs must be shatterproof or preferably enclosed in a shatterproof cover as part of the glass exclusion policy of any facility. Light fixtures must be inspected for cracks and covers replaced as plastics tend to become brittle after several years.

Light from metal halide lamps and fluorescent bulbs have been common in the industry, but LED lights are quickly replacing them due to their lower energy consumption, higher light output and uniformity, and longer useful life.

Adequate lighting surrounding equipment, modules, skids, platforms, and mezzanines must be reviewed during the design phase. The ability to visually inspect and maintain all areas of the process is foundation to food and personal safety. The equipment packages can be overlooked for adequate lighting and proper fixtures.

11.2.2.4 Drains

Floor drainage system design and installation should be approached on a systemic basis. Too often consideration is paid only to the visible parts of the system, while concealed components are often overlooked. Connecting a sanitary drain to an improperly selected drainage piping system offers little to no protection against potential biological contamination, and a physically or chemically compromised piping system.

Drains and waste piping are critical parts of any building, can make the difference between success and failure, and are present in every plant where wet processing is used, and other areas where product is stored or wet sanitization methods are employed. Floor drains must rapidly convey any water or product that falls to the floor, while preventing contamination with microorganisms that might grow in them. It is important to minimize any open distance from equipment effluent points to the floor drain. If personnel are walking through soiled effluent on the facility floor, this has the potential to be tracked throughout the facility. Best practice is to keep drains away from major walkways and design drain placement in accordance with the equipment effluent points or tie the drain points from the equipment to a common drain system. If this is done, the drain system will need to have an inspection ability. One of the most common harborage points for *Listeria* are the floor drainage systems.

Selecting the correct Waste System materials is equally as important. Too often these materials are subject to the Value Engineering process and can easily be overlooked. Waste System materials must meet the correct pressure, temperature, and chemical resistance while providing the best possible protection from the formation of biofilms, which can be detrimental and foster the growth and propagation of opportunistic pathogens. There are a wide variety of Waste System materials available and there is no single material that will fit every application. Each system should be thoroughly evaluated to select the most suitable drainage materials.

Besides the local plumbing codes, floor drains in food facilities must help prevent any standing water on the floor. The most common designs found in the food processing facilities are trench drains and area drains. Trench drains must be sloped at 1–2% and have rounded or coved bottom, the end of the trench drain usually has a sieve or grate to collect solid material. Trench drains are covered with grates. Grates must be resistant to traffic and forklifts and be easy to clean. It is good practice to keep any processing or packaging equipment away from the top of the drain to avoid aerosols contamination.

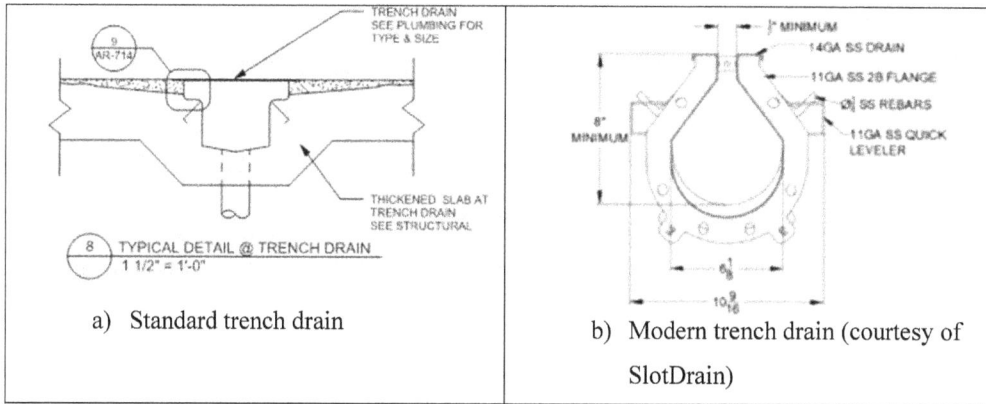

a) Standard trench drain

b) Modern trench drain (courtesy of SlotDrain)

Figure 11.4   Typical trench drain design and modern trench drain design that minimizes aerosols.

Trench drains require special attention to avoid aerosols coming from them during cleaning, sanitation, or even operation. Grates are hard to clean underneath and must be inspected and scrubbed clean regularly. Trench designs must incorporate well-considered and thoughtful design characteristics that will help to mitigate the risk of contamination due to harborage points and translocation of contaminants. These designs need to provide desirable flow characteristics, correct load bearing, effective solids handling, reduce or eliminate damage to flooring components due to thermal expansion and contraction, and need to be easily maintained and sanitized (Figure 11.4).

Area drains are preferred in aseptic processing, a single round or square drain, which are located at set spacing in the process area. These drains and the connected piping system must be sized for the worst possible case (e.g., a whole tank dump).

Installation of area drains is more involved, as the floor must be sloped toward the area drain at 2% (1/4" per foot) creating patterns in the room to cover the whole area (Figure 11.5). Area drains must have a basket or filter to intercept solids and prevent clogging the drain lines. Reinforced slabs are needed where drains will be placed to keep the slope in the room after many years.

Area drains are easier to clean, and grates and baskets must be inspected and scrubbed clean regularly.

Careful attention must be paid to the connection methodology between the Drains and Piping System. There are a multitude of drain outlet connection types and waste piping materials. The connection must be made utilizing a third-party fitting that is listed and approved by Code for each material or incorporates proprietary connection methods, such as push-fit, to ensure that there is no risk of leakage or disconnection which can lead to contamination or costly damage to the facility.

Selecting the correct grate materials and types for each application is essential. There are a multitude of grate types available that address specifics such as load-bearing characteristics, hygienic requirements, and even gas-tight drains where physical barriers to prevent the ingress of sewer gasses and containments or to provide spill containment are desirable.

11.2.2.5 Equipment Anchoring

Anchoring of equipment to the floor can make a difference in the cleanability and sanitary state of a factory. Microbes tend to harbor in cracks and small spaces between the feet of equipment

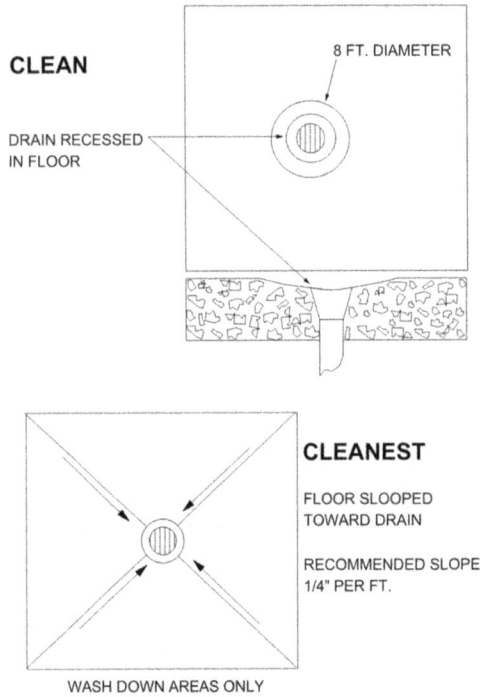

Figure 11.5   Typical area drain arrangements. Courtesy of Blucher drains.

and floor. Equipment that is bolted cannot have visible screw-threads, as these become a harboring place for bacteria.

Several solutions are available, as shown in Figure 11.6. A coved base (Figure 11.6a) is preferred, in which equipment is set on a raised area, which is sloped after the equipment is installed, covering the screw threads and creating an area where water can't stand and dirt will not accumulate. If the bolts are to be passed through the leg, a piece of steel must be welded and grouted to make the leg watertight, as one continuous piece of metal (Figure 11.6b).

11.2.2.6 Personnel Facilities: Locker Rooms and Bathrooms

Personnel is a key part of any aseptic operation, and motivation and retention are tied together. Building a facility in which personnel will not only be comfortable, but feel appreciated and well cared for can increase motivation. Besides continuous training, personnel hygiene is an integral part of any hygienic design and food safety program. Facilities include locker rooms, bathrooms, breaking rooms, changing rooms, handwashing, foot disinfection, and gowning. In many cases, these facilities are also used as a personnel entrance, and as such they must be very clean to remind employees that this is a hygienic operation. Cleaning programs in restrooms and locker rooms should be stringent to ensure that these spaces are always in the best possible shape.

Restrooms must be built so that they don't open directly to a production area, but toward a locker room or at least a hallway. Floors must be covered in tiles, or sealed concrete with coved wall joints for easy cleaning. Toilets, urinals, sinks, and individual partitions should not touch the floor but be wall- or ceiling-mounted to permit thorough cleaning and hose bibs for cleaning

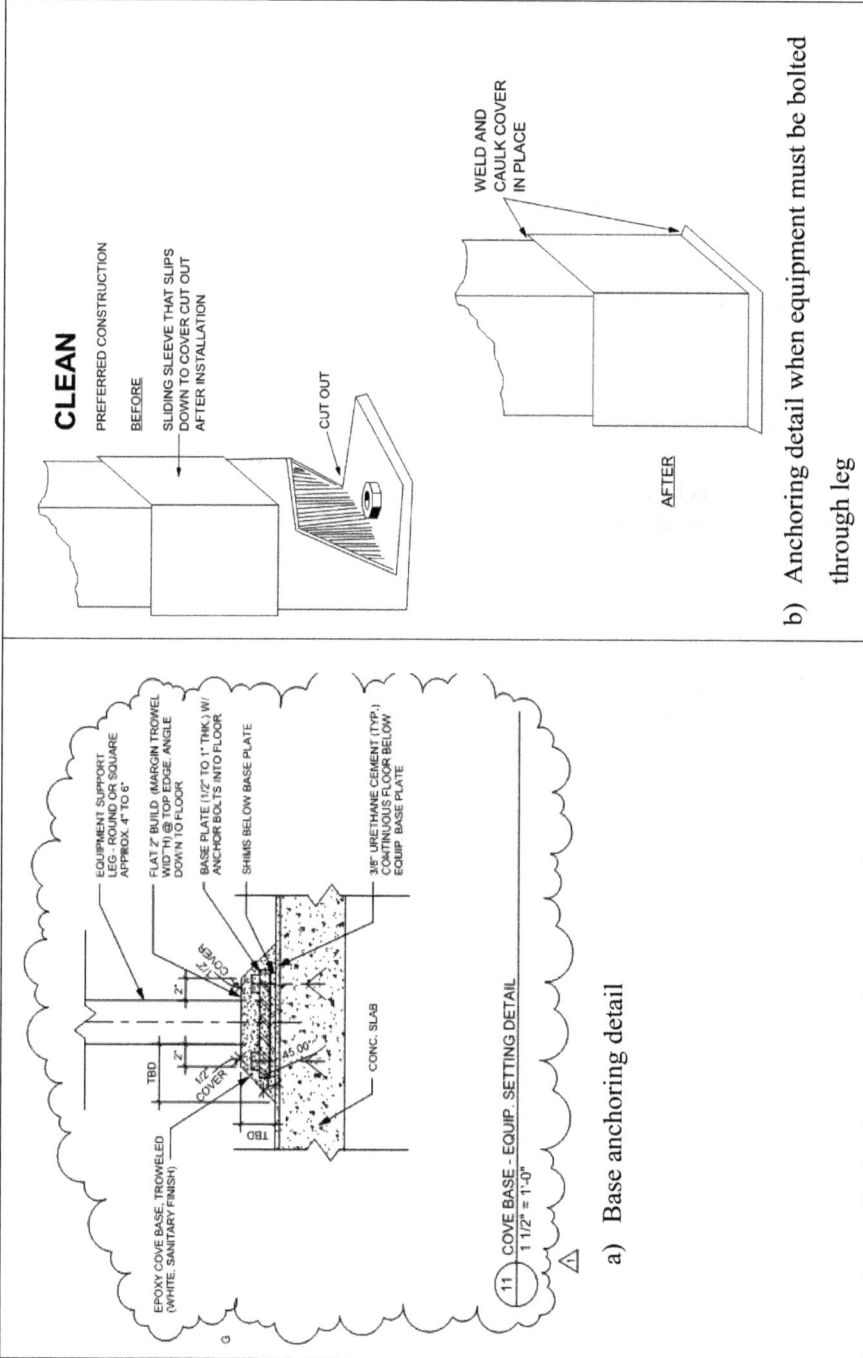

CLEAN

PREFERRED CONSTRUCTION

BEFORE

SLIDING SLEEVE THAT SLIPS
DOWN TO COVER CUT OUT
AFTER INSTALLATION

CUT OUT

WELD AND
CAULK COVER
IN PLACE

AFTER

b) Anchoring detail when equipment must be bolted
through leg

EQUIPMENT SUPPORT
LEG - ROUND OR SQUARE
APPROX. 4" TO 6"

FLAT 2" BUILD (MARGIN TROWEL
WIDTH) @ TOP EDGE. ANGLE
DOWN TO FLOOR

BASE PLATE (1/2" TO 1" THK.) W/
ANCHOR BOLTS INTO FLOOR

SHIMS BELOW BASE PLATE

3/8" URETHANE CEMENT (TYP.)
CONTINUOUS FLOOR BELOW
EQUIP BASE PLATE

CONC. SLAB

45.00°

COVER

2"

TBD

1/2"
COVER

2"

TBD

EPOXY COVE BASE, TROWELED
(WHITE, SANITARY FINISH)

11 COVE BASE - EQUIP. SETTING DETAIL
1 1/2" = 1'-0"

a) Base anchoring detail

Figure 11.6 Typical equipment anchoring details. (a) Base anchoring detail and (b) anchoring detail when equipment must be bolted through leg.

are recommended. Restrooms must have a fan that operates continuously when the plant is in operation and vents to the outside of the factory. Sewage lines from restrooms should be separated from the plant sewage to minimize the risk of backflow. Handwashing facilities with warm water and soap must be present and signs reinforcing the need for washing hands must be attached to the walls. Knee or electronically activated faucets, with paper towels for hand drying and a large covered trash receptacle are preferred. Air dryers are not recommended, as they create aerosols and recirculate the dirty moist air.

Handwashing is one of the most important activities in food plants, since it keeps microorganisms and allergens from contaminating the foods and must be practiced when entering a production area, after handling food, or touching non-product contact surfaces which are not sanitized. Handwashing stations must be placed at the entrance of every production room, and employees, management, and visitors must use them before entry. Rails or metal chains directing the flow of people and signage reminding people to wash their hands are a good practice. Handwashing sink must have warm water and soap, with knee or electronically activated faucets. Paper towels with touchless dispensers and a large covered trash receptacle are preferred. Further disinfection of the hands with alcoholic gel can also be used.

The use of gloves does not preclude the employees from washing their hands. Gloves are only effective if they are applied on clean hands.

Locker rooms must be cleaned and well illuminated; and lockers must be mounted on legs that keep them at least 15 cm (6") away from the floor for cleaning underneath. Their tops must be slanted at least 30° to prevent dirt from accumulating and the back must also be separated from the wall to enable cleaning. Benches if provided must be made of stainless steel or hard plastic that is easy to clean and sanitize. Locker rooms must be well ventilated and exhaust fans should vent to the outside of the facility.

Break rooms are used for eating and resting, and must have solid surface tables, chairs, countertops, and cabinets, which are cleanable. Vending machines and refrigerators should be on rollers for cleaning behind them periodically.

Cleaning and sanitation schedules of all personnel facilities must reflect the shift schedule, so every person who comes into the factory finds a very clean environment to work, change, and eat. This has a very positive effect in the personnel motivation toward hygiene and improves the compliance with all hygienic programs.

11.2.3 Equipment Design

All equipment used in an aseptic process must comply with the requirements of design for aseptics. Aseptic processing can be seen as an extreme case of food processing, in which foods are free of pathogens and spoilage microorganisms, so the products are shelf stable for a very long time.

Like all other food processing, equipment must be of sanitary design, with several additional requirements regarding cleaning, sanitation, sterilization, and prevention of contamination:

1. Equipment should be made of materials that will not contaminate food and must be free of any cracks, crevices, or dead legs. Material must be chosen to prevent corrosion and withstand the products to be produced.
2. Equipment must be drainable (no standing water in clean equipment); this includes vessels, pumps, and instrumentation. Heating and holding tubes must be sloped toward the unprocessed side.

3. Equipment should be capable of being thoroughly cleaned preferably without disassembly or cleaned in place (CIP).
4. Process lines should be capable of operating at greater pressure than the surrounding area or heat-exchange media.
5. Process lines must be instrumented to monitor all parameters that will maintain safety and quality. Monitoring and recording of such parameters must comply with the requirements of regulations.
6. The equipment must be capable of being maintained in a constant operating mode.
7. The equipment must be easy to repair and maintain, and parts that are subject to wear must be easily accessible and replaceable.
8. The equipment must conform to design, state, and federal regulatory codes if they exist.
9. Equipment must be capable of being sterilized in place (SIP). Sterilization adds complexity, as the equipment must not only be clean, it must be guaranteed that every surface of the equipment is exposed to a sufficient treatment.
10. The equipment must be capable of being maintained in a sterile state. Most equipment should be enclosed, and where there is a possibility of contamination (valve stems, agitators, etc.), a barrier of steam, clean condensate, or sterile water must be maintained.

All of the equipment must be in a commercially sterile state before starting production, and this presterilization is usually accomplished by high temperatures with steam or high-temperature water, or by using chemicals. Presterilization is normally a CCP of any HACCP plan, and as such every effort should be designed and constructed to guarantee that every surface reaches a minimum temperature (minimum amount of chemicals) and is kept at such temperature level for a required time. When thermal sterilization is used, be it with water or steam, the coldest spots in the line need to be determined and validated. First, all the coolant must be drained from the cooling heat exchangers, and temperature must be raised in all the process line. Dead legs, air pockets, steam condensation, and coolant pooling must be avoided. When chemical sterilization is used, monitoring of the concentration of chemicals in the process line and validation of the chemical coverage of every surface is a must.

All equipment after sterilization must be able to prevent the product from being contaminated again, and separate the product from the environment. This means ensuring that no microbes can creep into the product line by proper sealing of the equipment. This is especially complex when moving equipment is involved, such as valves, pumps, homogenizers, agitators, etc.

Static seals are used in non-moving parts, such as sight glasses, tank covers, instrumentation, and in tube-to-tube fittings. Most of these are made of elastomeric gaskets, which must be made of materials suitable for food contact, and able to maintain resiliency after being exposed to temperature extremes and changes and harsh cleaning chemicals. Gaskets are a very important part of the food safety and are not always given the attention they need; exposure to hot and cold temperatures, oils, and cleaning chemicals reduces their resiliency, and they must be replaced periodically as part of the preventive maintenance program. This program must track and document the replacement of gaskets before the end of their useful life, which must be determined as part of the validation of the process line and needs to be reevaluated when changes to product, process, or equipment are made. Gaskets that are not resilient any more, or that have been overtightened and touch the product can become microbial harboring places and become the culprit of catastrophic contamination of aseptic products, especially low acid.

Dynamic seals are needed for moving equipment parts, such as pump shafts, homogenizers, and agitators. These seals must provide protection against the ingress of microorganisms which can be 0.1 μm or smaller in size. Double seals with an internal gap that can be flushed with an antimicrobial fluid is needed; this antimicrobial can be steam, hot water, sterile condensate, or water with an antimicrobial agent. Another solution for moving parts is to have a magnetic drive for rotating parts or a membrane (bellows) for reciprocating movement. Membranes and bellows need to be inspected for pinholes and also require regular replacement.

The materials of construction for equipment and seals are very important, as they not only act like the prevention barrier to microbial growth but must also be inert to the products and chemicals used, and be able to withstand the extremes during cleaning, sanitation, and processing.

11.2.3.1 Materials of Construction

Food contact surfaces must prevent the food contamination, by not harboring microorganisms and not holding allergens and other contaminants. This includes metals and non-metals which are used in different stages of processing and packaging. Metals such as stainless steel, or aluminum and plastics such as PTFE, rubber, and other elastomers can be used.

11.2.3.1.1 Stainless Steel

The main material used in the food processing equipment is stainless steel, with many different grades and applications, from tabletop surfaces to tubes, vessels, and pumps. The method to make steel resistant to corrosion (stainless) was discovered in the late 19th century and consists of the addition of at least 12% chromium during the alloying process. The presence of these "impurities" allows the formation of a very thin layer, 4–10 nm, of chromium oxide on the surface of the metal, which is non-porous, impervious to many chemicals, and is self-healing in the presence of oxygen. Other additional alloying elements such as nickel, molybdenum, manganese, titanium, and silicon are also added to change the properties of the steel and make it more resistant to chemicals, heat or cold, or even to withstand marine environments.

Types of stainless steel receive different denominations depending on the country. In the United States, ASTM (American Society for Testing and Materials) recognizes both the AISI (American Iron and Steel Institute) and UNS (Unified Number System), while in Europe Euronorm numbers are replacing DIN (Deutches Institut fur Normung) or BS (British Steel) codes (see Table 11.1).

The most common types of stainless steel used in the industry are called Austenitic Stainless. These steels are non-magnetic and have a good corrosion resistance for most food applications. From this group, AISI 316L and AISI 304L are the most familiar materials, and should have at least 18% Cr and 10% Ni. 304L is used in low-humidity or low-temperature conditions, such as beer brewing, milk tankers, or for tables, shelves, and kitchen sinks. 316L covers three specifications: low (1.4404), medium (1.4432), and high quality (1.4435), with different Cr and Ni contents as seen in Table 11.1. The highest-quality steel should be used for process equipment. 316L is useful under most food applications; however, high salt foods or hot brines corrode it rapidly, the chloride ions remove the chromium oxide layer and make the steel vulnerable. For these applications, different types of steel are needed.

Super austenitic (904L) or Duplex (2205) steels are recommended when aggressive food products are processed, such as hot brine, slow-moving products with high salt content, or corrosive foods such as vinegar.

Stainless steel must be polished to create a smooth surface, which doesn't have any crevices that could harbor microorganisms. Surface must be polished to average surface roughness (R_A) of 0.8 μm (32 micro-inch), which has been proven to be cleanable. Studies with immobilized bacteria showed that wiping the surface with detergent was sufficient to remove them when R_A was less or equal to 0.8 μm (Verran et al., 2001; Hilber et al., 2003). Stainless steel also shows wear over time, and needs to be maintained and checked for pits,

TABLE 11.1   COMMON STAINLESS STEEL TYPES

Euronorm No.	ASTM AISI	ASTM UNS		DIN EN		BS
1.4301	304	S30400	X5	CrNi	18-10	304S31
1.4306	304L	S30403	X2	CrNi	19-11	304S12
1.4401	316	S31600	X5	CrNiMo	17-12-2	316S31
1.4436	316		X3	CrNiMo	17-13-3	316S33
1.4432	316L		X2	CrNiMo	17-12-3	316S13
1.4435	316L		X2	CrNiMo	18-14-3	316S13
1.4406	316LN	S31653	X2	CrNiMoN	17-12-2	316S61
1.4462	2205	S31803	X2	CrNiMoN	22-5-3	318S13
	2205 UR45N	S32205				
1.4410	2507	S32750	X2	CrNiMoN	25-7-4	–
1.4539	904L	N08904	X1	NiCrMoCu	25-20-5	904S13
1.4547	254SMO/F44	S31254	X1	CrNiMoCuN	20-18-7-4	–
1.4571	316Ti	S31635	X6	CrNiMoTi	17-12-2	320S31

crevices, thermal shock, and signs of corrosion. This is especially important in welded sections, where if welds are not properly done, thermal abuse, roughness, and mechanical stress can weaken the steel.

Welding stainless steel is a very important step as the heat and improper welding techniques can cause intergranular corrosion. Proper weld joint preparation and fit up of the weld joint is imperative for a successful weld. Areas that have gaps or dirt will provide irregular weld temperature, improper penetration, and an inconsistent heat-effected zone on the weld. Welding to hot or cold also has an effect on the heat-effected zone of the weld. The heat-effected zone is where most weld failures start. The weld failure is typical on the leading edge/or toe of the weld. It is important to reference the AWS for acceptable joint preparation and visual inspection aids. When choosing a proper fabrication and welding partner for your project, it is important to review the partners quality plan, welding certifications, in-house inspection abilities, and inspection certifications. It is also recommended to request weld coupons for all weld types, sizes, and alloys; per welder performing the actions, and prior to allowing the welder to perform any work on the project, in the plant or on site. Protective inert gas must be used for welding of stainless steel and the welds must be uniform and clean. There are different mixes and levels of quality for inert back-up gases. These different options are typically selected via the welding type and operation. Welding guidelines such as ASME, CRN, PED, EHEDG Document 35, USDA, or AWS D18 must be followed.

11.2.3.1.2 Other Metals

Aluminum is widely used in food preparation and cooking due to its lightweight, thermal conductivity, and relatively high corrosion resistance. In the aseptic processing industry, aluminum is found sometimes in frames, trays, and equipment used before thermal processing. However, for processing, it is not recommended as acids, chemicals, and wet cleaning can corrode it.

11.2.3.1.3 Plastics

Plastics is the name given to materials made of large-chain organic compounds (polymers), which are used extensively in food-contact surfaces. Plastics have a large variety of physical and chemical characteristics and are used in hoses and plastic tubes, gaskets, seal guides, and scrapes, and even advanced plastics can be used in thermal processing when advanced heating methods are used.

Plastic hoses are used extensively in processing, due to their flexibility and several material combinations are available. Materials must be chosen depending on the operating conditions (temperature, pressure, corrosivity, fat content, etc.). It is important to keep the hoses dry, and avoid cross-contamination by preventing the use of hoses in different areas of the process. Hoses must be inspected for cracks or pinholes and replaced periodically.

The most frequent use of plastic materials in aseptic processes is to prevent metal-to-metal contact (gaskets and seals). These materials must be elastic to fill the voids, thus called elastomers, and should do so under both hot and cold temperatures. Elastomers must then be temperature-resistant, and also chemically inert to resist cleaning chemicals, fats, acids, etc.

The following are the most used materials for elastomer gaskets:

- Silicone: Based on polydimethyl vinyl silicone, these materials have very good temperature resistance and excellent compression resistance but poor wear and short-term resilience. Compatible with most food materials and a wide range of temperatures. Lacks resistance to acids, alkalis, and steam.
- Nitrile (Buna-N): Improved version of natural rubber, with better heat-aging characteristics. Good temperature, wear, compression and short-term resilience. Compatible with most food materials and temperatures below 120°C (250°F), but in practical use, temperature below 50°C (120°F) is recommended for long-term use.
- EPDM: Ethylene-propylene rubbers. Very good wear and compression resistance, moderate short-term resilience. Compatible with most food materials except oils and able to resist a wide range of temperatures.
- PTFE: Polytetrafluoroethylene gaskets are very resistant to chemicals, steam, and extreme temperatures. They can be used for sealing tanks but not for pipes, as the material is not resilient.
- Fluorocarbon polymers are made to resist extreme environments, such as high oil, acid, or salt liquids and high temperatures. Good wear and excellent compression resistance, moderate short-term resilience.

Specialty plastics are also used in moving parts inside fillers and other equipment, such as PEEK or MEEK.

11.2.3.2 Hygienic Design Standards

Hygienic equipment design for food processing equipment has multiple avenues of specialization. Government regulations offer broad requirements on factors such as segregation and cleanability. A series of private companies have developed detailed standard outlines of how equipment should be made to achieve hygienic design.

In principle, each standard is outlining a physical means to remove foreign material and curb microorganism growth. Smooth surfaces and open seal designs are employed along with minimizing ledges, crevices, and pinch points to reduce harborage points. This reduces the accumulation of food and water and makes it easier for cleaning solutions to remove it and dry.

Although the standards have common goals of cleanability, one should be cautious mixing equipment design to different standards. With each standard being developed from different sources with different methods of review, the overall system design philosophies can be different leading to direct contradictions.

- 3-A SSI (3-A Sanitary Standards, Inc.): It is a non-profit organization dedicated to providing education and maintaining standards and practices for the design and fabrication of hygienic equipment and systems related to food industries. Originally developed for dairy, 3-A Standards are cited in the FDA PMO, USDA Dairy Guidelines, and each state's dairy regulations as a means of reaching compliance. Its use has grown beyond exclusively dairy as a general guide for hygienic equipment design, but its high specificity of each piece of equipment limits its applications in other industries.
- EHEDG (European Hygienic Engineering & Design Group): It is a non-profit organization that operates as a global network from all sections of the food industry. EHEDG has standards, offers training and outline testing methods, all with the goal of promoting hygienic food production. Based in Europe and connected to members in over 60 countries. 3-A SSI and EHEDG have announced cooperation between their organizations to reach common standards, but currently with their different histories have specific design features that do not overlap.
- NAMI (North American Meat Institute): It is a trade association of companies from the meat processing industry. It functions as a network to aid companies in improved operations and awareness and interaction of legislation change.
- NSF is a non-profit organization dedicated to the development of public health standards, product testing, and in other ways aid in the practices of public health. NSF standards have a history of use in consumer contact areas along with being applied to manufacturing industries such as beverage and meat processing. Their use continues to grow.

11.3 ZONING AND FLOW OF MATERIALS AND PERSONNEL

Food factories have been compartmentalized for decades; factories are separated into specific areas for a number of reasons: protection from the exterior; separation of dry and wet areas; temperature needs; segregation of raw and finished products; and separation of areas with mechanical services or health and safety issues (boiler, compressor, water treatment).

Hygienic zoning means separating products according to the risk of contamination, products that will receive further processing will have a lower risk than exposed final product, which will not receive any more processing. Zones are created so that those zones where hygiene is more critical are isolated from other areas. Plants are usually divided into basic-, medium-, and high-hygiene zones. It is a good practice to separate finished products from raw materials and non-production areas. This isolation is a combination of physical barriers (walls), traffic control (movement of personnel and materials), Good Manufacturing Practices (GMPs), and air handling.

As discussed in Section 11.2, the exterior and interior design of the factory must be made to protect the food from general contamination. The outside of the factory protects the inside from pests, outside odors, unwanted people, and external contamination. Concurrently, the inside of the plant is designed to segregate personnel, offices, and different production zones. The production areas of the factory need to be segregated, and this can be done following several methodologies such as wet and dry areas. Areas where water or steam are present all the time can lead to the generation of aerosols and must be further segregated, such as tray cleaning and any other cleaning areas.

Hygienic zoning requires an approach based on risk assessment of the ingredients, processes, and products in each area. Ideally, this assessment is performed in the design stage and is used to design the production spaces, and the way materials, personnel, and air will flow. The basic idea is that materials will flow as their risk increases while the personnel and air will follow the opposite flow, i.e., personnel of a "dirty" area should not be able to go to a clean area without going through a prescribed cleaning protocol and the air should blow from the clean areas outward, as shown in Figure 11.7.

Storage is the first zone, in which materials are stored based on the requirements for each material. That means refrigeration or frozen storage if products need to be kept cold, and separate storage for vegetables and fruits, meats, or fish. Products that contain allergens must be stored in a separate area to prevent cross-contamination, and especially powders must be stored in the lowest rack level to prevent them from going airborne. Packaging materials must be stored separately.

Low-risk areas are found where the risk of contamination is low or where the ingredients will be subject to a subsequent kill step; examples of this are dry ingredients weighing, produce washing, fruit peeling, and blending etc. Dry areas must be kept dry, moisture control of the air is very important to prevent condensation and quality degradation.

Medium-risk areas are those in which products that will receive further treatment or a kill step are found, and where the risks of chemical or physical contamination are manageable. These areas should be separated from the outside of the facility by never having doors that open

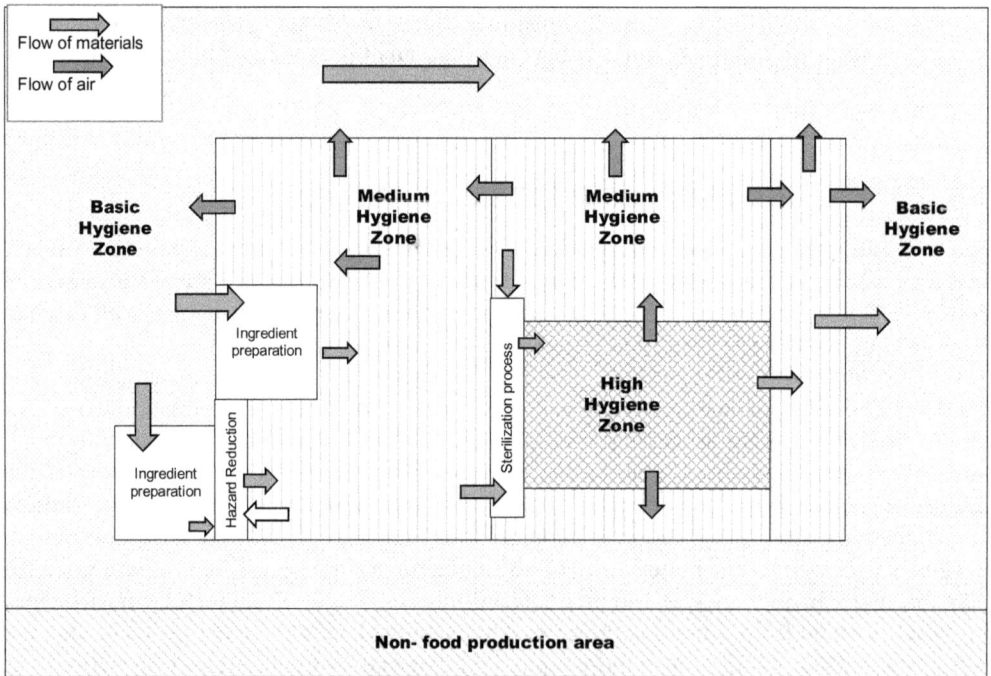

Figure 11.7 Hygienic zoning example. Material flow denoted by filled arrows and airflow.

directly to the outside. Air used in these areas should be filtered to avoid risk of physical contamination and must be kept at a pressure higher than the low-care areas. Walls should be washable and drains should be fabricated in a way that waste is carried away from the products, and aerosols are not easily generated during cleaning.

High-risk areas are where the final products are exposed and will not have any more treatment. These areas in an aseptic process are sparse and are usually enclosed inside the filler aseptic zone. In these areas, commercial sterility should be reached and maintained. These areas should be able to be steam sterilized and withstand sterilization chemicals such as H_2O_2. Air must be sterile, either by incineration or passed by a set of filters, and pressure should be high enough to prevent recontamination toward the aseptic filler nozzles.

Transition of materials from the low- to medium-risk zones needs to be controlled. The use of vestibules in which the materials are transferred from one area to the other is a good option. The concept of a vestibule combined with an airlock is used in ready-to-eat foods and is applicable for aseptic plants as well, as a method to minimize the risk of contamination and avoid disruption of the flow of air in the medium- and high-risk areas. Pallet jacks and carts for the medium-care areas must be separated from the general purpose carts and lift trucks to prevent the wheels from tracking dirt. If these pallet jacks or carts are electrical, a charging station should be placed in the vestibule to prevent the truck from leaving.

Personnel plays a very special part in zoning, as the movement of people across different zones has to be controlled to avoid tracking dirt from one side to the other by Good Manufacturing Practices (GMP). Continuous training in GMP is necessary to maintain the personnel focused and motivated on the hygienic requirements. Among the basic GMO needs are the use of personal protection equipment, and disciplined repeated handwashing.

Before entering the plant, personnel must comply with several requirements, which are set by management, and these usually include the following:

- Use clean clothes, preferably provided by the company.
- Use clean closed-toe shoes, captive shoes are preferred.
- Remove all jewelry, including watches, earrings, rings, pendants, etc.
- Cell phones are not allowed in the production areas and must be left in the locker room.
- Cover hair with a hair net.
- Cover facial hair with a beard net.
- Use eye or ear protection if needed.
- No food or drink is allowed on production floor. Drinking water is usually provided.
- Wash hands after gowning and before entering the production floor.
- If gloves are needed, they must be kept cleaned and disinfected.

Handwashing is one of the most important practices in food processing; dirt under the fingernails can contaminate large batches of product. Proper handwashing must include the use of warm water, antibacterial soap, and paper towel drying. Hands must be washed before entering into a food processing area, or when going from one area to the other; after using the toilet, after touching nose or face, and after eating or smoking. Hands should also be washed after handling contaminated materials, sanitizers, etc. The use of gloves does not preclude the requirement for handwashing. Handwashing stations should be placed strategically around the factory, and it is recommended to have motion-activated or knee-activated faucets, with warm water and soap. Hand drying is better with paper towels, with motion-activated dispensers. Trash cans located after the drying stations help maintain the area clean. It is important to make a habit of washing hands when entering any production space, and the example demonstrated by management is crucial in developing this culture.

Hygienic zoning also helps to maintain the quality of products with special label declarations such as "non-GMO," "allergen-free," or with a religious certification. Changes in clothes or gowns of personnel, as well as washing might be needed if personnel move into this area.

Zoning is also an integral part of the food defense, to prevent the intentional adulteration of foods by preventing the ingress of unwanted people. Unwanted people inside the production zone include media, competitors, and people with ill intentions who might intentionally contaminate the product. In the 21st century, the risk of intentional contamination and bioterrorism is a reality, and all care must be taken in zoning, training of personnel, and isolation of vulnerable areas to prevent it.

11.4 AIR-HANDLING

Air is everywhere in a food production facility, it touches all materials, personnel, and equipment, and as such must be considered as an ingredient. Air can be the carrier of small particles for very long distances, and especially small liquid and solid particles called aerosols, which can carry microorganisms and allergens. Care must be taken for air to not carry dirt, allergens, aerosols, odors, and chemicals, which can contaminate the food. Also, excessive moisture in air helps microbial growth and must be avoided; and temperature of the air makes a big difference for the comfort of production personnel and the ability of microorganisms to grow.

Air-handling for the factory is important for all of the above reasons. First, the humidity and temperature of the factory should be kept relatively low. Humidity must be reduced to avoid condensation on surfaces such as cold tubes, ceilings, and shelves, and personnel comfort is best kept at moderate temperatures. Air in the factory must be filtered to a level that will provide sufficient filtration for the product to be made, this must be agreed to in the design phase and validated after the factory has operated for a few months. Filters have to be monitored according to the required standards, and filter replacement must be a part of the preventive maintenance program. It is important to place the filters in an accessible place, especially if the facility is located in a location where cold weather is expected, so that filter maintenance can be performed even in the coldest months. Air-handling systems must also be cleaned regularly, and depending on the expected severity of soiling, CIP nozzles might be installed inside the ducts and low-point traps to prevent condensation are therefore a good practice. Air suction into the factory is a factor that is often overlooked, it must be away from flues, exhausts from other systems, and facing in a way that nearby odors will not be blown directly into it.

Control of aerosols is a big part of hygienic air-handling, which encompasses microbial and allergen control. Hygienic zoning will determine the required air-handling for each room, and a map of pressures and flow should be drawn at the design phase. High-risk zones must have a positive air pressure when compared to medium risk, and then to low-risk zones, with the goal of moving any contaminant away from the highest-risk zones. This map is only valid within certain limits, and when changes are made to any facility where the air circulation might be affected, therefore, allowances should be made for the foreseeable future.

Several recommendations have been made for a successful air management:

- Ensure that adequate airflow is blown into any room, to maintain the positive air pressure. Ten to twelve changes of air per hour is the minimum flow recommended.
- Ensure that air extraction does not affect the positive air pressure.
- Ensure that room will be maintained under positive air pressure by simulating the worst leakage possible during design (i.e., open all doors).

- Prepare for expansion by having enough power for the air-handling unit in case additional vent hoods or extractors are added to the area.
- Do not position air extracts or returns in higher risk areas close to the doorways of lower risk areas
- Design the air-handling system to be easy to maintain and accessible for cleaning and maintenance

It is important to provide an inclusive design review of the HVAC systems with all other system operations. Equipment placement, equipment duties, and overall operations can affect the airflow, or specific operations of the equipment or supporting equipment. For example, a cold air return duct in line with an elastomer-type labeler can create many blind issues, or point of use dust collection working against the air mass balance required for the area.

To these considerations, and due to the allergen management needs, it must be added that air-handling might need independent units of flow per hygienic zone, to effectively isolate the flow of allergens. Allergens require also increased filtration, in order to be able to prevent the dispersal of smaller particles.

Air-handling units are usually placed on the roof, and due to their weight, they must be designed with allowances for future expansion. Adding or retrofitting air-handling in an existing facility can be expensive, if additional work is needed for structural integrity due to the location and weight of these units.

11.5 CONCLUDING REMARKS

Aseptic processing and packaging is one of the most complex types of food processing, and when properly managed, it produces shelf-stable products of the highest possible quality. Design of the production facility plays a major role in ensuring that products are sterilized by improving cleanliness and lowering the microbial load of the environment.

Facility design needs to pay attention to many small details, which should be handled by a competent engineering group. This initial design effort should not be underestimated and sufficient time and expenses should be allowed; any decisions taken at the design stage will affect the performance of the processing and packaging equipment, potentially for decades.

No design is effective without the help of a well-trained and motivated personnel. Design includes considerations for employee comfort and can make the work environment more attractive. Personnel training and proper hygienic practices will ensure that the environment and processing is maintained as initially designed.

Further recommended reading includes 3-A, EHEDG, FDA, and USDA guidelines and local codes.

REFERENCES

3-A Accepted Practices for a Method of Producing Steam of Culinary Quality, 609–00. 1996. Formulated by International Association of Milk, Food, and Environmental Sanitarians, United States Public Health Service, and the Dairy Industry Committee.

American Welding Society Guideline AWS D18.1/D18.1M. 2020. *Specification for Welding of Austenitic Stainless Steel Tube and Pipe Systems in Sanitary (Hygienic) Applications.*

American Welding Society Guideline AWS D18.2. 2020. *Guide to Weld Discoloration Levels on Inside of Austenitic Stainless Steel Tube.*

American Welding Society Guideline AWS D18.3/D18.3M. 2015. *Specification For Welding Of Tanks, Vessels & Other Equipment in Sanitary (Hygienic) Applications.*

European Hygienic Engineering and Design Group EHEDG. 2006. Document 35. *Hygienic Welding of Stainless Steel Tubing in the Food Processing Industry.*

European Hygienic Engineering and Design Group EHEDG. 2014. Document 44. *Hygienic Design Principles for Food Factories.*

European Hygienic Engineering and Design Group EHEDG. 2020. Document 54. *Hygienic Testing of Hygienic Weld Joints.*

Graham, D.J. 1991. Sanitary Design:A Mind Set. *Dairy, Food and Environmental Sanitation. 11, 338, 454, 533, 600.*

Lelieveld, H.L.M., Mostert, M.A., and Holah, J. 2005. *Handbook of Hygiene Control in the Food Industry.* Boca Raton, FL: CRC Press.

US DEPARTMENT OF AGRICULTURE, FOOD SAFETY AND INSPECTION SERVICE, 2021, 9CFR, Parts 307,318,320,327 and 381. Washington, DC: U.S. Government Printing Office.

U.S. Department of Agriculture. 1986. *Canning of Meat and Poultry Products.*

US Food and Drug Administration. Food and Drugs Title 2021, 21CFR, Parts 11 (Electronic records; electronic signatures), 113 (Thermally processed low-acid foods packaged in hermetically sealed containers),114(Acidified foods) and 117(Current good manufacturing practice, hazard analysis, and risk-based preventive controls for human food). Washington, DC: U.S. Government Printing Office.

Chapter 12

Cleaning and Sanitization for Aseptic Processing Operations

Jeffrey Merritt

CONTENTS

DOI: 10.1201/9781003158653-14

12.1 INTRODUCTION

Cleaning and sanitization procedures have to be considered as an *integral* part of food production. No matter how well the production plant may be designed, if cleaning and sanitation are not possible or not done properly, it is impossible to produce high-quality and safe products. Food production is intertwined with cleaning and sanitization and are mutually inclusive.

Cleaning and sanitization are one of the most important aspects of food manufacturing. This has been made amply clear by announcements of food recalls stemming from contamination with bacteria and foreign matter (CDC, 2021).

Cleaning and sanitization of aseptic processing and packaging unit operations is complex and should get attention the same or more as in the food and beverage industry using canning or "terminal sterilization" technologies.

This chapter provides principles and procedures for cleaning and sanitization for manufacturing operations focused on aseptically processed and packaged fluids, and ESL-refrigerated products. Also, discussed are concepts of CIP, COP, SIP, AIC, validation and verification, passivation, maintenance, change control management, and environmental cleaning.

This chapter is intended to be an overview. Further reading of other references and publications cited in the Reference section is recommended to understand the scope, design, delivery of cleaning, and sanitization for consistent production of high-quality safe products day to day. Collaboration and leveraging of the institutional knowledge of vendors of cleaning and sanitization chemicals is paramount, and is necessary, as this is a team sport (IAMFES, 1996, Katsuyama, 1993, Loncin and Merson, 1979, Troller, 1983, ASTM, 1994, Ecolab, 2022).

12.2 EQUIPMENT PREPARATION AND SET UP (EPSU)

As we start discussing the Cleaning and Sanitation of Aseptic and ESL equipment, it is important to understand that cleaning is the first step in producing safe, quality food. It is not the last step.

As we talk with manufactures of these products, we hear a common theme that is "CIP is down time and cost us production time." This is incorrect. There are several factors that will increase uptime:

- Clean equipment
- Proper and consistent preventive maintenance programs (PM)
- Employee hygiene
- Employees that are trained and receive continuous training
- Clear concise and continuous communication between management and the employees

Following the above guidelines can help increase the number of salable products.

There is an old saying: "Pay me now or Pay me later." Taking the time to do it right the first time, every time, is less expensive than waiting for the failure to trigger corrective action and resolution of the issue. If we do it correctly the first time and follow the pattern of doing it right, we have less spoilage; in addition, the equipment runs correctly, which means you have more consistent and sustained runs.

We should not accept that we are going to have spoilage and build this loss into your budget. Spoilage should not be accepted; it needs to be resolved and fixed. If we don't do it right the first time, you end up with more cost, due to spoilage, unplanned downtime because preventive maintenance was skipped and, in some cases, it could result in a recall of product.

In this chapter, we will discuss these topics and how they influence your operation. A clean, properly maintained system is a reliable system. As we go forward, it is important to remember that the cleaning of the UHT, sterile tanks, and fillers are generally the same for aseptic as for ESL. Some ESL fillers will require more hand-cleaning of pumps, valves, screens, or filters. When we get into the second and third steps, there may be more differences on the sterile tanks and the fillers, we will discuss more as we proceed.

There are five steps in getting the equipment ready for production of safe and high-quality product, as shown in Figure 12.1.

Step 0: Preventive maintenance

> Step 0 is the preventive maintenance and prerequisite step. This step is to make sure you are ready to start the process up. This step requires that you have a PM (preventive maintenance) program set up and you are following it. It also requires that you have prerequisites in place. These prerequisites should be used to help train and guide your employees. Some of these prerequisites would include the following:
> - Proper design of equipment and plant
> - Continuous education, learning, and training for all employees
> - GMP polices
> - SSOP documents
> - SOP documents

Step 1: Cleaning

> Step 1 is the cleaning of your equipment, remember this is your first step not your last step. With a properly cleaned system, you start the production with the best chance of success. Cleaning is the task of removing all soils from the equipment. This includes both internal and external cleaning of the equipment. We need to remember this is not the step where the button is pushed, and you walk away from it. This is the time that you need to monitor what is going on. Are there any leaks? Are the chemicals being added,

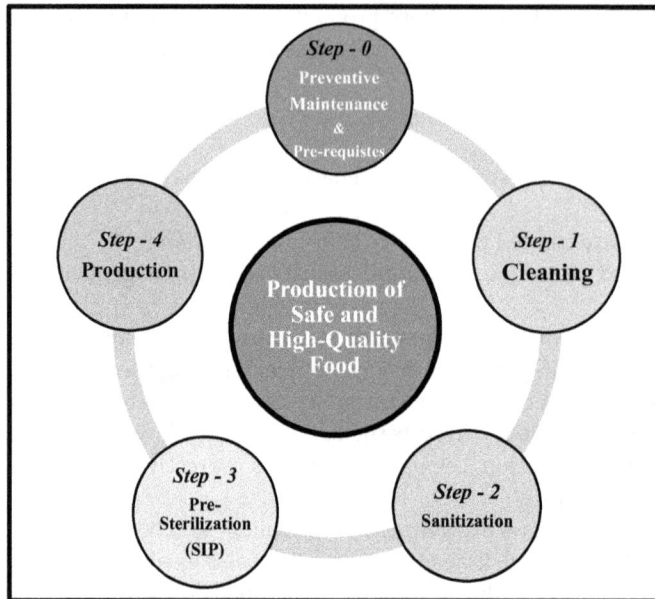

Figure 12.1   Chart to illustrate the five steps in equipment preparation and setup (EPSU) required for production of safe and high-quality safe product. Step 0: Preventive maintenance and prerequisites. Step 1: Cleaning ⎨ CIP & COP. Step 2: Sanitization. Step 3: Presterilization (SIP). Step 4: Production.

and do you have the correct concentration? This is one of the best times to determine if the equipment is functioning correctly.

Step 2: Sanitization

In step 2, you reduce or eliminate microorganisms that may be left on the clean surface. Don't confuse this with sterilization as that comes later. In this step, you are getting the equipment ready so that when you do sterilize, it will have the highest probability of being successful.

Step 3: Presterilization (SIP)

Step 3 is presterilization (sometimes called SIP—sterilization-in-place—some will call it sterilization in progress). In this step, you bring the equipment that is going to run the product to the level of sterility needed for the process you are going to be running. Aseptic SIP is a regulated step (21CFR113.40(g)) and the goal is to bring all equipment and food-contact surfaces to a state of commercially sterility, customarily an F_0 of 30–50 minutes for running low-acid aseptic products. For ESL, the level of commercial sterility is not defined by the regulation and is left to each processor to design and deliver. All equipment must be brought to this state, including the hold tube, homogenizer, cooling section, the sterile or storage tank, the filler, and all the connecting tubing.

Step 4: Food production

In Step 4, you are bringing the product you are running to the level of sterility that you have determined you want for this product. If you are producing an aseptic product you are going to reduce or eliminate the microorganisms to the level that you do not need to refrigerate the product. If you are doing an ESL product, you are using refrigeration to help with your shelf life. Your equipment producing this product will reduce the microorganisms to a level that you will reach the shelf life that you have set for this product.

12.3 PRINCIPLES OF SANITATION

We perform sanitation procedures to produce safe quality food and beverage. Sanitation is the process by which the equipment and environment is cleaned of dirt, pathogens, and bacteria from product-contact surfaces, as well as incidental product-contact areas, the air, the exterior of equipment, and the area the equipment is in (walls, floors, ceilings). Having a proper sanitation program is a major factor in producing hygienic and safe food. This is done using automated and manual cleaning techniques that will use chemicals to clean and sanitize all areas. They may use steam, hot water, or chemicals to sanitize and sterilize the equipment. Sanitation also includes the proper hygiene of the personal.

12.3.1 Principles of Sanitation

The principles of sanitation start with personal hygiene and equipment that is designed to be able to be cleaned and sanitized. Sanitation starts before the equipment is designed and built. The engineering and design will ensure that it follows 3A/EHEDG standards. These two groups look at practical and theoretical design to make sure the equipment is cleanable to ensure the production of safe, quality food. As the equipment is designed, it looks to ensure that there are no dead legs (area that cannot get sufficient flow or turbulence to clean). The surfaces need to be smooth and free from sharp angles. It needs to ensure that any vessel, pipe, or tube in the system can be cleaned and fully drainable (preventing pooling or low-flow areas that will affect cleaning).

The issue comes after design and is during the installation. At times pipes are moved or changed due to space constraints. These changes are needed due to product specifications, or lack of utilities. As a result, sometimes no thought is put into how this is going to affect cleaning. This includes but is not limited to changing pipe diameters, extra elbows, and pipe being cut off and capped due to relocation or change of flow; older equipment being repurposed for new products or processes; the proper materials being used. Is the material correct for the process being used? For example, using 304 stainless steel in a hold tube to produce a product with a higher salt content at temperatures over 250°F will cause pitting in the pipe which then creates harbor areas that are more difficult to clean.

As the equipment is built and installed, we must look if it is cleanable? Also, answer the following questions: (1) Can the necessary flow rates be achieved? (2) Are the spray balls located correctly, and can the correct flow and pressure to each spray ball be achieved and maintained? (3) Do the valves pulse and clean both sides of the seats? These and other questions will be answered as we go through this chapter.

12.3.2 Personal Hygiene

Before starting the cleaning and sanitizing of the machine, we need to make sure that those who are involved in the cleaning and sanitizing are following the correct GMP. This starts at the management level; they must be the example and follow the GMP policies in all aspects, including handwashing. Hands must be cleaned and washed frequently. If you change the task you are doing, you need to wash your hands again. Hand cleaning facilities must be easily accessed and should be in every room. Maintenance personnel need to be sure they have removed all the grease and other contaminants from their hands and clothes prior to working on equipment in the aseptic or ESL process (UHT processors, sterile tanks, fillers). Tools need to be clean and sanitized as well prior to use in these areas.

If using disposable gloves, they need to change them frequently. Just like your hands, if you pick something up from the floor, those gloves are no longer clean. Gloves if used should be put on for that task and then disposed of before going to the next task. Don't wear gloves for long periods of time as they are then just as dirty as not washing your hands. Improperly using gloves can lead to many environmental contamination issues. Treat them the same as your hand.

Clothing: If your uniform becomes overly soiled, you should change it (especially if dealing with allergens or around sterile environments); this will lead to contamination issues. If hats are worn, they need to be clean and free of contaminants. Cloth ball caps are hard to keep clean and should be avoided. If they are allowed, then you need to establish how you are going to keep them clean. Bump caps and hard hats can be cleaned and sanitized daily or if they get dirty.

Footwear: Your feet travel throughout the facility and whatever you walk through is tracked to your next location. Clean and/or sanitize your footwear whenever you enter any area of the plant. If you don't have a captive shoe program, they need to be cleaned when you enter the plant. Each area you go into needs some sort of protection to sanitize and or clean the bottom of the footwear as you enter this area. But regardless of what you use, they are only as good as they are maintained and controlled.

Foaming system that sprays sanitizer or cleaner on the floor is used to keep a layer of foam on the floor. This foam must be walked through to enter the area. It is good, but only in wet areas as it produces wet floors. Dry sanitizers that you walk through are good for dry areas, but it requires that the floor around them be swept up frequently. Foot baths are used in many places but if they are used, they must be cleaned and recharged at least every shift and usually more often for them to be effective.

- Captive boot policy works well in areas where the employee is restricted in their movements. And footwear is not allowed out of these areas. You still need to clean your footwear daily.

Hair and beard nets: These need to be replaced daily or if they get dirty. Need to be properly worn.

12.4 CIP (CLEAN-IN-PLACE)

It is important to remember that it does not matter whether it is ESL or Aseptic: the systems will clean the same. The principles of CIP consist of four factors:

- *Flow*
    This is the mechanical action that is created to produce the scrubbing effect usually in a pipe that is turbulence.
- Concentration
    It is the % of chemical to water ratio based on the chemical used and the quality of the water, product being cleaned, and the equipment being cleaned.
- Time
    It is the time a chemical is in contact with the equipment being cleaned.

- Temperature

    It implies the correct temperature for the chemical being used and the equipment being cleaned.

These four factors if used correctly will clean the equipment, each factor is adjusted for the product and the equipment. If you cannot get one of them, you must make it up with the other three. But there is a limit to each factor and other factors that may contribute to poor cleaning. We want everyone to remember that the chemical is only a small part of the fluid going through the line. Water is over 95% of the fluid and it is important to understand that the makeup of the water can affect the cleaning, for both good and bad. If you are using untreated water, it could be high in minerals. This affects the cleaning because now the chemical must overcome the water before it can clean the organics of the products. This reduces the effectiveness of the cleaning or will require the use of additives to condition the makeup of the water and could also result in the overuse of chemicals. During the cleaning process, we will discuss each one of the factors to look at what happens. At the end of the chapter are a few tables covering the most common factors: time, temperatures, flow, and concentrations for the different types of equipment.

12.4.1 Flow (Scrubbing or Mechanical Force to Remove the Soil) for UHT Systems

In UHT processing, one of the items that cause them to be more difficult to clean and control is a condition known in the industry as fouling. Fouling occurs on the product to metal surfaces due to types of ingredients, the flow of the product, and the heat which causes the product to burn or to caramelize the sugars. The longer you run, the thicker and thicker the fouling layers get. In addition, as one switches from product to water and back during the runtime, layers of minerals might occur and that can turn a simple soil into a complex soil. To reduce this fouling, the frequency that an AIC is performed is important. Running at the designed flow rate (do not slow down the system to reduce product to water transitions) and running at the correct temperatures for the product are important to help control the fouling of the equipment.

One factor that is sometimes hard to achieve is getting enough flow to clean an aseptic or ESL processor. Without the proper flow, it becomes very difficult to remove the layers of fouling. Many times, they are set up for the pipe size to get the flow and this generally will not be enough to clean the equipment. I have followed and found that if I can get 1.5× the flow of maximum product flow, the system will generally clean at a minimum and would result in 30% overproduction flow rates. On systems that have Flash Vessels or Steam Infusion Chambers, they will have spray balls that need to have the proper flow rate to them. This usually involves a valve and an orifice to create the proper flow and pressure as this will be pulled off the mainstream.

Without the proper flow, it may not be possible even with chemical additives to get the proper clean. When designing, testing, and validation of the system, it is important to ensure you're getting the proper flow. The reason behind needing the higher flow can be related back to the temperatures the product is being processed at along with the type of product. In addition, especially on direct systems, there is more equipment that is involved than disrupt the flow due to spray balls or flow patterns that require a portion of the flow to split off.

Hold tubes usually have more fouling due to the high temperatures, this results in needing some way to penetrate this complex soil to get it to release from the stainless piping. This can be done by flow or the addition of additives to help penetrate the soil. Even with the additives, you

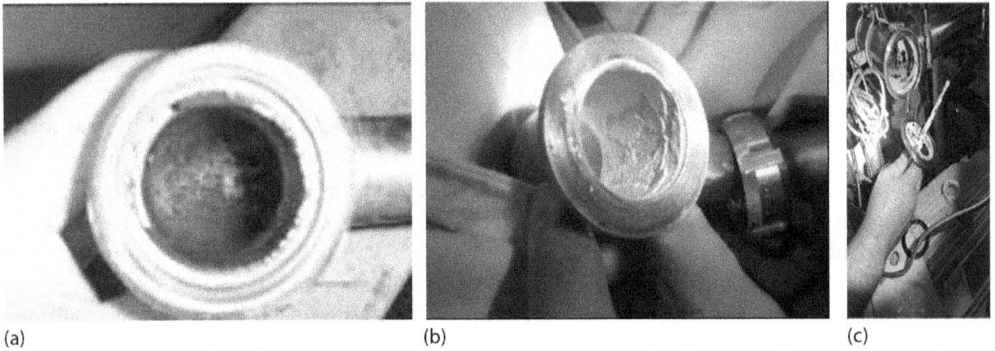

Figure 12.2 Pictures of ineffectively cleaned hold tubes in a UHT sterilizer: (a) Middle of the hold tube; (b) End of the hold tube; (c) "Legal" transducer at the exit of the hold tube.

need enough flow and chemicals to carry it out of the system. These pictures show hold tubes and a transmitter cleaned with the improper flow and chemicals (Figure 12.2).

- The equipment was able to be cleaned once we were able to get the flow up and the proper concentration of chemicals. We did not need additives on this one.
- On an indirect system, you have more heating area to get to the elevated temperatures and as result more pressure drop through the system that must be compensated for. A system that is not properly balanced between the media side and the product side (delta T) will result in higher spots of burn that will take more flow or additives to remove. Figure 12.3 shows an example of a plate that had a hot spot and began to plug off. These become very difficult to clean and at times they may have to be opened and hand cleaned.

Figure 12.3 Final plate in an indirect heat exchanger showing product burn-on due to high ΔT (temperature difference) between heating medium and product.

12.4.2 Chemical Concentration

12.4.2.1 UHT Processor

There are many factors that go into determining the necessary concentration to clean the processor. Concentration will start at 2 3% active caustic and will need to be adjusted up or down based on your product mix. Also, this will change depending on your chemical; if it is commodity caustic, it may go up, and if it is a built caustic cleaner with additives, it may go down. If you are running a thin juice-type beverage, it will not need as much caustic as a system running a plant-based product or a protein drink. Concentrations will need to be adjusted accordingly. More chemicals can be added, but remember you will get to a level that this will reduce the effectiveness of the clean.

Acid will start at 0.7–1.5% and will need to adjust up or down based on your product and water makeup.

CIP is performed mainly using temperature to control. CIP is performed by controlling the temperature of the wash either at the preheater or the hold tube. If on a direct heating system, they also run the flash vessel to remove that extra water. Most systems start the step (phase) with a temperature at the end of the system to ensure that they have the proper temperature throughout the full system. In addition, manual titration of the chemical is a necessary, part to ensure that it is at the proper concentration.

On most of them that have been built in the last ten years, they have a conductivity meter but on some it is only for reading and not controlling. On others they use that conductivity to add chemicals and to hold the timer if the concentration is low on start up. Not all of them use them to add more chemicals if the concentration drops. The author recommends a conductivity meter if the OEM can do it.

Manual cleaning of the outside of the equipment is generally not done during the CIP but prior to it and should be done on a regular frequency. During the CIP, this is the best time to walk the system and check for leaks so that they can be repaired prior to SIP.

12.4.2.2 Sterile Tank

The concentration will start at 1.5–2.5% active caustic and will need to be adjusted up or down based on your product mix. Also, this will change depending on your chemical. If it is commodity caustic, it may go up, and if it is a built caustic cleaner with additives, it may go down.

The concentration in the sterile tank is lower than in a UHT system due to absence of heating in the tank. Acid will start at 0.7–1.5% and will need to adjust up or down based on your product and water makeup.

It is important to do an acid step on a sterile tank. Most sterile tanks are sterilized using steam and if you are not using acid over a period, the tank will start turning slightly brown. If the culinary steam filters fail or are not maintained, this can increase this issue.

12.4.2.3 Fillers

On fillers, the concentration will start at 1–2% active caustic and will need to be adjusted up or down based on your product mix. Also, this will change depending on your chemical. If it is commodity caustic, it may go up, and if it is a built caustic cleaner with additives, it may go down.

The acid will start at 0.7–1.5% and will need to adjust up or down based on your product and water makeup.

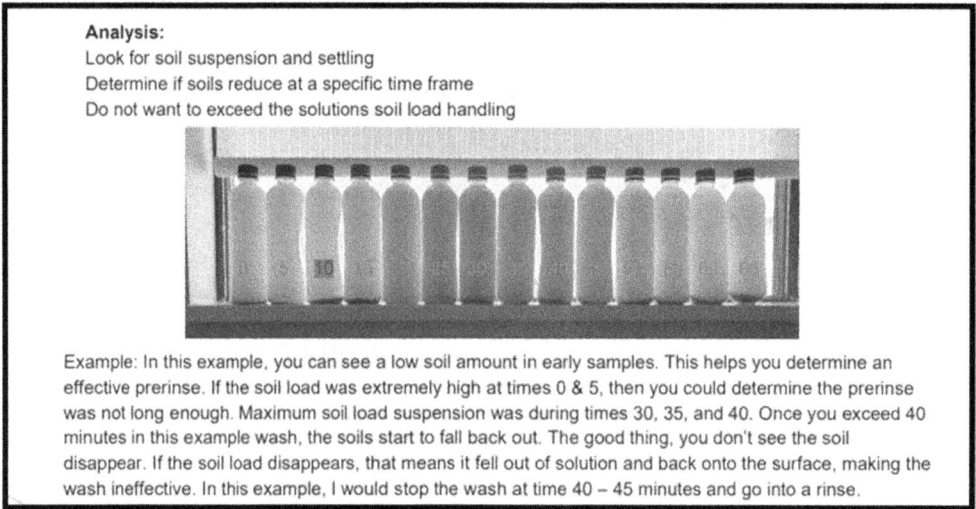

Analysis:
Look for soil suspension and settling
Determine if soils reduce at a specific time frame
Do not want to exceed the solutions soil load handling

Example: In this example, you can see a low soil amount in early samples. This helps you determine an effective prerinse. If the soil load was extremely high at times 0 & 5, then you could determine the prerinse was not long enough. Maximum soil load suspension was during times 30, 35, and 40. Once you exceed 40 minutes in this example wash, the soils start to fall back out. The good thing, you don't see the soil disappear. If the soil load disappears, that means it fell out of solution and back onto the surface, making the wash ineffective. In this example, I would stop the wash at time 40 – 45 minutes and go into a rinse.

Figure 12.4 Quick analysis to determine when to stop alkali wash in the cleaning cycle. In this illustration, the sweet spot is 40 minutes, to stop the cleaning cycle. Any cycle time exceeding 40–45 minutes is wasteful and counterproductive.

12.4.3 Time

As you determine your times, it is important to remember that there is a limit to the time and the effectiveness of cleaning.

You want the right amount of time needed to ensure that the chemical can do its job. Normally you would like to see two to three passes of the chemical through the system. Too long of a time may cause you to redeposit the soils back onto the equipment. With the test that was performed below, samples were taken every 5 minutes to determine when it stops removing solids from the equipment (Figure 12.4).

You can see at around 35–45 minutes the level of suspended solids did not increase, so adding more time will not take any more solids out, so it is time to rinse.

- Actual times need to be established during the validation of the CIP on each piece of equipment.

12.4.4 *Temperature for Cleaning UHT Systems*

This factor can be based on the capability of the chemical and the type of product you are running. *When cleaning aseptic or ESL equipment, it is strongly recommended that you do not use chlorine or a chlorinated product.* This equipment will usually clean at temperatures that will depend on the OEM and may vary from 185°F to 295°F.

There are different schools of thought on what is the best temperature to clean UHT processing equipment. Depending on OEM, the caustic temperature will vary from 185°F to 295°F. When using a high temperature, it is generally the first caustic that is at a higher temperature and is used to remove the outer layers of fouling. The second caustic is generally around 185°F, and the acid should be between 150°F and 160°F (Tables 12.1 and 12.2).

TABLE 12.1   EXAMPLE OF A SINGLE CAUSTIC WASH USED FOR CLEANING A UHT STERILIZER PRODUCING THIN VISCOSITY AND LOWER SOLIDS PRODUCT (MILK)

		UHT Cleaning Single Caustic			
Phase	Chemical	Concentration	Temperature	Time	Flow
Pre-rinse	Water	Not checked	70°–100°F	*7 minutes	****1.5 times the flow of maximum production
Wash	**Caustic NaOH or KOH	2–3% Active caustic	185°F	*45 minutes	****1.5 times the flow of maximum production
Rinse	Water	6.1–7.8 pH	120° F	*15 minutes	****1.5 times the flow of maximum production
Wash	Phosphoric/nitric blend	***0.7–1.5%	150°–160°F	*25 minutes	****1.5 times the flow of maximum production
Rinse	Water	6.1–7.8 pH	Ambient	*15 minutes	****1.5 times the flow of maximum production

*Time may need to be adjusted depending on validation results.
**Chemical restrictions may apply based on local requirements and adjusted based on validation.
***Acid may need to adjust based on water conditions.
*****Flow rate needs to be as fast as possible and as a minimum 30% above maximum production flow rate.

- Temperatures for cleaning sterile tank
  - Most sterile tanks will clean at these temperatures
    - Caustic (non-chlorinated) 175–185°F
    - Acid 140–155°F
- Temperatures for cleaning fillers
  - Caustic (non-chlorinated) 160–175°F
  - Acid 140–155°F

12.4.5 For Sterile Tanks

The flow for the sterile tank during CIP is going to be controlled by the flow needed for the spray balls. This flow is very important to know if it is too high or too low for the spray balls as it will not clean the tank walls or agitators.

Most (not all) sterile tanks use air pressure to return the solution back to the CIP room. If this air pressure is too high, you will blow a lot of air into the lines causing conductivity issues as well as damage to lines due to the surges of water and air. If it is too low, the level will build in the tank. The OEM will need to spend time to get this correct on start up. There are some that are putting in a CIP return pump at the tank, by means of a flow panel, so it can be out of the system during sterilization and production. This makes balancing the system easier as well as more consistent during the CIP (Table 12.3).

TABLE 12.2   UHT CLEANING DOUBLE CAUSTIC FOR PRODUCING PRODUCTS WITH THICKER VISCOSITY AND HIGHER SOLIDS (SUCH AS HIGH-PROTEIN ENERGY DRINK)

		UHT Cleaning Double Caustic			
Phase	Chemical	Concentration	Temperature	Time	Flow
Pre-rinse	Water	Not checked	70°–100°F	*7 minutes	Production flow rates maybe 10 gpm higher
Wash	**Caustic NaOH or KOH	2–3% Active caustic	285°–290°F or 185°F (depending on OEM)	*25 minutes	Production flow rates maybe 10 gpm higher (if lower temperature go with higher flow)
Rinse	Water	7.1–8.5 pH	120°F	*7 minutes	****1.5 times the flow of maximum production
Wash	**Caustic NaOH or KOH	2–3% Active caustic	185°F	*45 minutes	****1.5 times the flow of maximum production
Rinse	Water	6.1–7.8 pH	120°F	*15 minutes	****1.5 times the flow of maximum production
Wash	Phosphoric/ Nitric blend	***0.7–1.5%	150°–160°F	*25 minutes	****1.5 times the flow of maximum production
Rinse	Water	6.1–7.8 pH	Ambient	*15 minutes	****1.5 times the flow of maximum production

*Times may need to be adjusted depending on validation results.
**Chemical restrictions may apply based on local requirements and adjusted based on validation.
***Acid may need to adjust based on water conditions.
*****Flow rate needs to be as fast as possible and as a minimum 30% above maximum production flow rate.

12.4.6 For Fillers

The flow necessary to do the internal and external CIP of a filler is going to be set by the OEM. It is important that this flow can be met and sustained through the CIP. Many fillers come with the CIP system as part of the filler and others will require CIP be sent from a CIP skid.

12.4.6.1 For ESL Fillers

The internal cleaning for all the fillers ESL or aseptic is going to be similar. On some of these fillers, they will have to disassemble equipment and hand-clean and/or clean in a COP tank. Some of the items that will have to be hand cleaned are listed below:

TABLE 12.3   TYPICAL WASH USED FOR STERILE SURGE TANK

Sterile Tank Full Clean

Phase	Chemical	Concentration	Temperature	Time	Flow	Air Pressure
Rinse	Water	Not checked	80°–100° F	*7 minutes	Based on spray balls	23–38 PSI
Wash	**Caustic NaOH or KOH	1.5–2.5% Active caustic	175°F	*30 minutes	Based on spray balls	23–38 PSI
Rinse	Water	6.1–7.8 pH	120°F	*15 minutes	Based on spray balls	23–38 PSI
Wash	Phosphoric/ nitric blend	***0.7–1.5%	150°–160°F	*20 minutes	Based on spray balls	23–38 PSI
Rinse	Water	6.1–7.8 pH	Ambient	*15 minutes	Based on spray balls	23–38 PSI

*Time may need to adjust during validation and may not include the time to pulse valves in steam barriers.

** Chemical restrictions may apply based on local requirements and adjusted based on validation.

*** Acid may need to adjust based on water conditions.

****Air pressure to be set at commission based on distance and no CIP return pump.

- You will have to disassemble the fill nozzles and screen or filter packs.
- You will then send them to an area so that they can autoclave them and some companies will have installed a second set (to rotate the first one with) which has already been cleaned and autoclaved. These are installed after CIP of the filler prior to being sterilized for production.
- The rest of the carton path through the machine must be hand-scrubbed and sanitized.
- There may be other items that need to be disassembled and hand-cleaned. These instructions will come from the OEM (Table 12.4).

As you might have observed, on the equipment we listed above, we do not have a sanitizer step. This equipment is going to be sterilized so that we do not need to do a sanitizer.

- On some of the ESL equipment, it will get a sanitizer used on it as a step to help with the sanitizing of the equipment later getting it ready for production.

TABLE 12.4   TYPICAL INTERNAL WASH FOR DOWNSTREAM FILLERS—ASEPTIC AND ESL

Filler Internal CIP					
Phase	Chemical	Concentration	Temperature	Time	Flow
Pre-rinse	Water	Not checked	70°–100°F	*7 minutes	Set by OEM
Wash	**Caustic NaOH or KOH	1–2% Active caustic	165°F	*35 minutes	Set by OEM
Rinse	Water	6.1–7.8 pH	120°F	*15 minutes	Set by OEM
Wash	Phosphoric/ nitric blend	***0.5–1.5%	150°–160°F	*15 minutes	Set by OEM
Rinse	Water	6.1–7.8 pH	Ambient	*15 minutes	* Set by OEM

*Time may need to be adjusted depending on validation results.
**Chemical restrictions may apply based on local requirements and adjusted based on validation.
***Acid may need to adjust based on water conditions.

12.4.7 Pulsing of Valves

Most of these systems will have steam barriers, valve clusters, and individual valves that require pulsing. Valve pulsing allows both sides of the seat and the stem to be properly cleaned. Many OEM will pulse the valves throughout the program and that is okay. Just watch the CIP system to determine if this is causing extra time due to the pulsing (the times may need to be adjusted).

If they pulse steam barrier valves during the cycle, this can lose concentration of chemical and increase CIP as the chemical must be replaced. It may cause delays due to freshwater being added and this will cause a temperature fluctuation that could cause a hold on the timer.

To avoid delays in the CIP, one practice that is being used is that at the end of each phase or cycle, they will pause the timer and do all the valve pulsing. This allows for more consistent CIP times and the valves can be pulsed as the need may be. The cleaning of these valves during the pulsing requires that they pulse long enough, and enough times, at the correct time to clean the burnt product off them.

This sequence for on/off time must be validated at the start up or commission of the machine and is rechecked on a frequency to ensure it is working. I would recommend it is checked at least every quarter. Changing temperatures, products, or mechanical changes will require it to be revalidated.

We have discussed the four factors that are involved in cleaning a system, and if that was all that was needed, it would be very easy to clean. But to clean a processor and get it cleaned there are other factors that will come into play. We will lay out some of the issues, but it is important to understand that every UHT processor at every plant will be different due to some of the reasons discussed below:

Design: A system that is not properly designed and tested may not clean and every time a change is made in the system, it may impact how it cleans. The ability to inspect areas to ensure proper cleaning is important.

Water: The water used in every plant is different due to the water makeup and local conditions. If you are using untreated water (well or city), the minerals that are in this water will affect the effectiveness of the chemicals and may require the addition of additives to condition the water so that the chemicals can do their job. Hard water if not properly handled will leave a buildup on the stainless steel, it will make it look dull and may cause staining. Having a regular water analysis done will help to understand what is happening. Where possible, using treated water (soft water, RO, or Nano) will help reduce or eliminate these problems. High levels of chlorine in your water system can also contribute to the staining and possible pitting of equipment.

The type of product: Milk, juice, plant-based, high protein are examples and each one may require a different concentration of chemicals to clean. If your system only has one CIP program, then it needs to be set for your hardest product. If you add a new product, then you should do a CIP validation after that product is run to test if the equipment is becoming clean.

12.4.8 Ingredients

The ingredients of a product can change how a system needs to be cleaned. Stabilizer overuse, improperly blended (creating fisheyes, pockets of improperly hydrated product), or the type of stabilizer may cause more fouling and "burn on" in the system. They may also lead to micro-hits in your finished product due to the inability to sterilize the product that is not blended correctly.

Titanium dioxide when used is extremely hard to remove as it is heavy and unless you have enough flow to carry it out, it will leave a white residue on the stainless steel. This is used as a processing aid that will help make the product brighter (white). You may not even know it is in your product as sometimes it is used to bind the stabilizer and your stabilizer company may not let you know this. If you cannot get the proper flow to help carry it out, then more aggressive cleaning with chemicals designed to remove titanium dioxide will need to be used. (Your chemical supplier will need to help you with this and handle it on a case-to-case basis.)

Proteins: As products come out with higher protein, it means it needs to be validated that the chemicals and cleaning program that is being used will remove them. These types of products will cause fouling in the hold tube and in the preheater. As this fouling happens, it coats the probes controlling the temperature, and you will see an increase in steam usage to compensate for the fouling in the hold tube. This results in the product fouling even more onto the hold tube. This is generally observed as the difference in the temperature from the beginning of the hold tube and the end of the hold tube drifts apart. When you get over a 4°F difference between the beginning and the end of the hold tube, this indicates that you need to act with an AIC to help remove this layering.

Plant-based products create different cleaning issues and each one will need to be evaluated and validated to ensure proper cleaning.

Temperature: The temperature on most products is determined by a Process Authority. The high temperatures at which the product runs for aseptic and ESL create a different challenge compared to an HTST. What the temperature does to different ingredients will challenge how we CIP. One of the problems is to maintain longer runs when the system is slowed down, but the temperature is not or cannot be reduced. This causes more "burn on" in the hold tube which then creates another very difficult layer of soil that must be removed. If this is not completely removed during CIP, it creates a layer that cannot be sterilized. As you run, the old product may sluff off and reveal an unsterilized area which will then contaminate the product you are running.

During CIP, it is important to watch the different temperature probes to ensure you are getting the temperature and time in all areas that you need for CIP. Work with your OEM and chemical supplier on what the temperatures in these different areas should be.

12.5 CLEAN-OUT-OF-PLACE (COP)

One piece of equipment that is underused to help in the overall cleaning and sanitization process is the clean-out-of-place (COP) tank. The COP tank is used to clean small parts and equipment that you are not able to CIP. It may include but is not limited to the following equipment:

- Fill tubes for batching tanks
- Buckets used to weigh ingredients
- Tools used to measure ingredients
- Parts of a filler that need to be hand-dissembled and cleaned
- Gaskets, or other rubber parts
- Clamps
- Short pipes and hoses (pipes and hoses may need to be hand-cleaned on the inside with a brush prior to COP. Pipes and hoses over 3 feet should be evaluated to determine if they are being cleaned completely).

Just like a regular CIP, it is important to keep allergen concerns and controls in place to prevent any cross-contamination. In loading a COP tank, it is important to not overload as the mechanical force is limited and you want to be sure all parts are receiving the most flow and mechanical action that they can. COP can be accomplished by following the 11 steps given below.

1st step: Each part or piece of equipment needs to be rinsed to remove the easy-to-remove product residue.

2nd step: Load all the parts and items to be cleaned into the tank, items that float or may be pulled into a pump should be loaded into a perforated basket that has a perforated lid on it and placed into the COP tank. If pipes are put into the COP tank, they need to be shorter pieces. Pipe over 3-feet long may need to be brush-cleaned, on the inside, prior to being put into the tank as they will not get enough flow through them to effectively clean them.

3rd step: Fill the COP tank to a level to cover all the items that you have placed into the COP tanks and above any spray bar lines in the tank.

4th step: Start the pump on the system to start it circulating. Check that you can see the flow in the tank that is creating some turbulent action.

5th step: Add the chemical you are going to use at the level that your chemical representative suggests.

6th step: Heat the solutions to the required cleaning temperature. This temperature raise can be controlled on new COP tanks, compared to older ones. It is important to monitor this temperature to avoid overheating, and charts may be used to estimate the heating time. If using an alkali, the temperature of the COP required would be between 150°F and 160°F. *For safety, you do not want this temperature above 180°F.*

7th step: Let the above step run for 30 minutes.

8th step: Shut the heat off and turn the water on. During this step, you want to overfill the COP tank to flush all the solids you have pulled off and that are in suspension to flow over the top for the first 5 minutes of the rinse. This will help reduce the chance of the solids you have removed from being redeposited onto your clean equipment.

9th step: After you have overfilled the COP tank for 5 minutes, shut the water off and drain all the water out of the tank.

10th step: You now need to rinse off the chemical, so turn the rinse water on and let it fill the COP tank up over the parts in the tank. Then leave the water on to keep the level above the parts start draining and rinsing until the chemicals have been removed. This will generally take 7–10 minutes.

11th step: Shut the water off and let the parts drain. If you are not doing an acid step, then you are finished; however, if you are going to do the acid step, then repeat from step 3 on to only replace the alkali with acid and reduce the temperature to 140°F.

After the COP tank is empty and rinsed, then all parts need to be removed and either put back into service or stored on a clean parts rack until needed. As parts are put back into service, each part will need to be sanitized by hand prior to use. COP if properly used can save time and labor on cleaning parts. It is not a storage place for parts after they are cleaned.

12.6 SIP (STERILIZATION-IN-PLACE)

SIP is step 3 in the process (see Figure 12.1), following preventive maintenance, cleaning, and sanitization in this order. *SIP should only be done on clean equipment; one cannot sterilize a dirty piece of equipment.*

SIP consists of three steps:

1. Heat the system up to the necessary temperature needed to consider it sterile, which is >250°F.
2. Hold it at that temperature for the required time, which is 30 minutes. If the temperature drops below 250°F, then the timer must be restarted at "0," after it is >250°F.
3. Cool down the areas of the equipment that do not need to remain above 250°F.

SIP must be performed on any piece of equipment that is going to come in contact with product after the product has been sterilized. The method that is used can be steam, superheated water, or chemical.

12.7 ASEPTIC INTERMEDIATE CLEAN (AIC)

Aseptic Intermediate Clean (AIC) is an intermediate flush or clean of the system under sterile conditions. They are not as effective as a full CIP, but what they do is remove some of the outer heavy layers of fouling so that you can continue production without needing to do full blown CIP and SIP of the entire system. If they are done at proper intervals for the product you are running, it will help reduce the issues that occur at CIP. AIC generally takes 45 minutes to 1 hour, compared to about 6 hours for a full CIP and SIP. They can be set up to do at regular intervals, but it is important to remember that some products you may run will foul the system quicker than others, so you need to identify them to help determine when an AIC is needed. Some places that can help you determine this are the control valve for your steam or hot water, an increase in the output is a sign of fouling. There is difference in temperatures from the beginning to the end of the hold tube; the farther they are apart, the more fouling there is. AIC is normally done with just caustic but in some circumstances, you may need to do an additional acid AIC, especially if you have very hard water. *The key is to add the chemical very slowly to not disrupt the temperature or the flow* (Table 12.5).

Another issue comes when a system loses sterility and to save time, they do not do a CIP before they sterilize the equipment. This increases the burn that is already in the hold tube and

TABLE 12.5   TYPICAL ALKALI ASEPTIC INTERMEDIATE CLEAN (AIC) WASH FOR ALL UHT STERILIZERS ON EXTENDED RUNS

UHT AIC Steps					
Phase	Chemical	Concentration	Temperature	Time	Flow
Rinse	Water	Not checked	At production temperatures	5 minutes	**At production flow rates
Wash	*Caustic NaOH or KOH	2–3% Active caustic	At production temperatures	25 minutes	**At production flow rates
Rinse	Water	6.1–7.8 pH	At production temperatures	15 minutes	**At production flow rates

All temperatures, flows, and pressures must remain under sterile parameters.
* Chemical restrictions may apply based on local requirements and adjusted based on validation.
**Flow rates may be reduced by up to 5 gpm to prevent loss of sterility due to flow spike.

causes areas that normally do not see the hot product (like your cooling sections) to have product residue in them that will burn on. This is an ineffective sterilization and could lead to major micro-issues over time. Doing an AIC after sterilization does not remove the issues that have been created, as an AIC does not have the effectiveness of a full CIP. It is strongly recommended that if a product has been run or the system has been set for a long period of time (over 24 hours), a full CIP should be performed prior to sterilization.

12.7.1 Concentration of Chemical

The concentration of the chemical should remain consistent throughout the step. What can happen is that due to water added, valves pulsing, or other issues, the concentration is significantly reduced during the step. Depending on the issue, this can happen quickly. As a result, instead of a 30-minute wash at concentration, you only get 10–15 minutes at concentration. On a direct system, it can happen from the addition of steam to the product, at the flash vessel due to not controlling level and or pulsing of valves especially at steam barriers. It is important to titrate your system not just at the beginning or middle of your cycle, but at the end occasionally to ensure you are not losing concentration. Most UHT processors do not have a probe for conductivity and as result a CIP can be run with little or no chemical. Sterile tanks and fillers generally have a conductivity control. Just remember, conductivity is a verification of titration.

Monitoring of the cleaning both internally and externally: During the CIP, it needs to be monitored, it is not the time to push a button and forget. Knowing what the system is doing and verifying will help ensure that it is properly cleaned.

12.7.2 Sterile Tanks

The sterile or storage tanks are generally cleaned from a CIP skid and this skid uses the return temperature and the conductivity sensor to start the time in each step and to hold if they fall out of range. If the tank does not use a CIP return pump but uses air pressure, the system can start up if it has to pressurize the tank. Manual titration of the chemical is a necessary part to ensure that it is at the proper concentration.

The system will generally incorporate throughout the wash special steps to pulse all the valves that are being cleaned with this system. I have found that by putting this at the end of each step, it saves time as the system does not continually add water to make up for what is lost. You do the same number of pulses for the same time but less holds for low temperature or conductivity.

Manually cleaning the outside of the equipment is generally not done during the CIP but rather prior to it and should be done on a regular frequency. During the CIP, this is the best time to walk the system and check for leaks so that they can be repaired prior to SIP.

12.7.3 Fillers

The program and the control for the CIP of the fillers is set up by the OEM. They generally are holding and controlling by temperature and conductivity. Manual titration of the chemical is a necessary part to ensure that it is at the proper concentration.

External cleaning. Most of the fillers have what is called a car wash for the external cleaning (this is inside the enclosure but not the internal product contact areas). This will be used to help clean the star wheels, outside of the filler bowl, and other areas. It will generally use a soft metal foaming cleaner and will rinse it off as well. This does not mean you do not need to clean inside manually as well as clean and check filters and screens. The manual scrubbing may be time-consuming. But if it is not done, the level of contamination builds up and you can get issues, especially around the cap or sealing area. This needs to be done every CIP.

12.8 VALIDATION AND VERIFICATION

12.8.1 Validation

Validation is the process we use to determine if a CIP or cleaning program is effective. Validation is not a one time and done process. There are changes that will drive that a new validation be performed.

- Validation should be done on any new equipment.
- Validation should be done on any modified equipment, and any replacement equipment that is not like for like.
- Validation should be done on any program change of the CIP on the system.
- Validation should be done on every product or product group.
- Validation should be done when ingredients are changed or modified on any product formulation.

After revalidation, document all changes made, and record all results.

Five steps in revalidation

- (1) The first step is to run through the program on water, checking to see if pumps are turning the correct way, valves are pulsing when they are programed to, flow is going where it is designed to. Instrumentation is working and there are no leaks or bad welds.
- (2) The second step is to run the program with the chemicals and checking the same as in the first step.

- (3) The third step requires getting the equipment dirty (equipment should be run for a normal cycle time before cleaning and validation.) Then run the CIP program and check all items from the 1st step.
- (4) The fourth step starts at the conclusion of the CIP. You will need to open the equipment to see effectiveness of the clean and check the following:
  - Visual inspection
  - ATP (adenosine triphosphate) will test for organic residue
  - Doing micro-testing (standard plate count [SPC] is one method)
  - Doing allergen testing if needed
- (5) The fifth step repeats the process starting at step 3 at least three times to check that the CIP performs the same each time. If any failures, you will need to fix the reason for the failure and then repeat the steps.

12.8.2 Verification

Verifications need to be done so that every CIP and CIP program is checked at least every year (if no changes have happened). This is the process which is used to check that the cleaning is being performed as and when it was validated. This process should happen a minimum yearly. If cleaning fails a verification, then you need to determine what may have caused the failure.

Verification uses visual, ATP, and CIP records to check and verify that the system is working, this would include titration checks as well.

12.9 PASSIVATION

Modern food and pharmaceutical processing use equipment that are made up of stainless steel. Stainless steel is preferred to maintain a high level of performance and to undergo minimum corrosion. Passivation is an important surface treatment that contributes to the corrosion resistance of stainless steel used for food-contact surfaces.

Stainless steel gets its corrosion resistance from a thin, durable layer of chromium oxide that forms at the metal surface and gives stainless steel its characteristic "stainless quality." The passive film on a stainless surface consists of a mixture of oxides of iron, chromium, and, if present, molybdenum. The chromium oxide film will form on the metal (stainless steel air dries for 72 hours) if the stainless steel is clean and dry. Further exposure to air does not yield additional corrosion protection.

Complete passivation cannot be achieved if product-contact surfaces are not clean or if they contain surface defects including rust. The different oxides and their relation to passivation is not yet fully understood.

The complete passivation process consists of mechanical cleaning, degreasing, inspection, the actual passivation by immersion or spraying, and rinsing. Passivation is done by immersion or spraying, depending on the size of the piece, with a solution selected from ASTM A 380. In addition to the standard nitric acid solution, there are a number of different solutions appropriate for all grades of stainless steel, including 200, 300, and 400 series. An oxidizing acid such as nitric acid, used for passivation, dissolves any high carbon tramp steel and assures a uniform clean surface that results in the consistent formation of the chromium oxide film.

12.9.1 Passivation Process

12.9.1.1 Some Preliminary Considerations

Before we get into the passivation, we want to remind everyone that safety is the priority to always keep in mind. So, any entry into the tank must be done by someone trained and certified to do confined space entry. The lock out tag out program must be followed, and all PPE (personal protective equipment) be used. Before passivation can start, it is important to know your water makeup to ensure that chlorides are not above 75 ppm as chlorides can be very damaging to stainless steel and it could affect the passivation process.

Passivation is a widely discussed topic and not always agreed upon how to do, what chemical to use, and how often it needs to be done. I am laying out the passivation that I have found to be most successful and if it is performed as it was designed, at the correct temperature and the correct concentration, it will not damage the equipment. Passivation is not a once and done type of process, there are conditions that cause a passivation to have to be redone:

- Time. Over time and continuous use, the chromium oxide wears down
- Incorrect cleaning. Cleaning with the wrong chemical or the wrong temperature for a chemical can damage the tank and will cause staining and rusting.
- Ingredients that are high in chlorides will attack the stainless steel and will have to be treated.
- High-chloride products may require a more resilient stainless steel such as AL6N; there are others as well and will need to be researched.
- Passivation is the process by which a stainless steel piece of equipment is treated with an acid blend to help reduce staining, corrosion, and improve the life of the metal by creating a thin lay of chromium oxide on the surface that helps protect the stainless steel. This is accomplished by the circulation of a passivation acid then draining the equipment and allowing it to air dry. It is during this drying time that the level of chromium oxide is developed. It has been determined that 72 hours of dry time gives the most successful layer.

12.9.1.2 Passivation Process

Let's talk about how this process is done and what is required. I will be using information that I have learned over the years. I will also be referencing some information from Ecolab®. For the best success, please work with a competent chemical supplier to help guide you.

Passivation is usually done on Silos, Tanks, and other vessels. Before passivation can happen, the equipment must be clean. Some of the problems you may face or described below:

- *Rust.* If rust is present, it must be removed either with a de-staining process (this may take several tries to remove). If the rust is too deep in the metal, it may require buffing to help to get it out.
- Stainless may have tramp iron in it from the manufacturing process.
- It could have oil that have been applied after process and prior to shipping to help prevent the tramp metals from rusting during transportation.
- There may be polishing dust that is electrostatically charged to the metal.
- There may be adhesives from the protective film used by manufacturers.

Each of these items needs to be addressed and removed. The tramp metals should be dissolved during the passivation process if the right passivation acid is used. The other items must be removed prior to the passivation. To remove most of them, a caustic cleaner designed to clean out these types of contaminants needs to be used. I have had good success with a product from Ecolab® called Accomplish™ as it was designed for this purpose.

Removing the black polishing dust is a little more difficult as it is electrostatically charged to the metal. To determine if you have this, take a white towel or white paper towel and wipe an area inside the vessel. Then examine the towel under normal light outside of the vessel and you will see the black dust if it is present. To remove it, you will need to clean the vessel with a quaternary ammonium cleaner that will help break the charge so that it can be removed.

12.9.1.3 Cleaning and Passivation

The following is the order needed to be followed for the cleaning and then the passivation. *Remember it should be done under the direction of your chemical supplier and in conjunction with your OEM recommendations.*

- First determine if you have black polishing dust (this is explained in the paragraph before this section. If you have black polishing dust, then the quaternary ammonium circulation will need to be done first. Use under the direction of your chemical supplier and in conjunction with your OEM recommendations. The cleaning to remove the black polishing dust is generally done at temperatures below 100°F; concentrations of 3–4% by volume; circulate 10–15 minutes. *Also be alert as to how this will impact your wastewater. You will need to know what methods will be used to address the disposal of the chemical.*
- After the quaternary ammonium circulation, rinse and drain.
- Then use a caustic cleaner like the Ecolab® Accomplish™ and circulate. Use under the direction of your chemical supplier and in conjunction with your OEM recommendations. Also be alert as to how this will impact your wastewater and what methods will need to be addressed for the disposal of the chemical. The temperature of the caustic is 150°F; Concentration 10% by volume; Circulate 1 hour
- After the final rinse, investigate the tank to see if it is clean. One sign that it is clean is that there will be no water beads. If water beads are present, then the cleaning needs to be redone.
- After the Accomplish™ has been rinsed and drained, use towel method to check that the polishing dust has been removed. If it has and the vessel looks clean, you may continue to the passivation.
- If not, repeat the process starting with the quaternary ammonium until it is clean.

There is another method using citric acid. You need to realize that the citric acid may not dissolve the tramp metal that may be in the tank. In addition, the citric acid will need to be used with EDTA which will help suspend the particles but may not dissolve them and they could redeposit. In addition, it will require longer circulation.

Another commonly used passivation process consists of using passivation acid from Ecolab®, and this is a nitric/phosphoric blend. The nitric will dissolve the tramp metal during the process. Which chemical you use will be up to you, your chemical supplier with input from your OEM.

Recommendations from Ecolab® if using their chemical:

- Temperature 140°F; concentration 50% by volume of passivation acid; circulate once at temperature and concentration for 1 hour; rinse; drain; let dry for 72 hours.
- This process of passivation on a clean vessel will give you the maximum chromium oxide formation on your stainless steel.

The ASTM International has procedures for the testing of chromium oxide. They are not generally used in the food industry due to the cost and the chemicals that they use.

The testing for iron left in the tank is a test that you can do and is an indication that the passivation was successful, remember this is qualitative not quantitative.

12.10 MAINTENANCE

The maintenance of the equipment is one of the key factors to ensure that the cleaning, sanitizing, and the production of products at the facility is consistent batch to batch.

Best-in-class maintenance program can be implemented by a two-stair step process: (1) preventive maintenance and (2) breakdown maintenance. A well-planned and executed maintenance program is one of the easiest and best ways to increase uptime by reducing unplanned downtimes. Following a preventive maintenance (PM) program can also help reduce contamination, because gaskets, O-rings, and pump seals are changed before they fail and as a result do not become a harborage or ingress into the system.

A maintenance program includes a proper inventory and inventory control to ensure parts are on hand when needed. The OEM will generally give a list of suggested parts to keep on hand as well. Remember that if you do not have the correct part on hand, this could cost an extended downtime while sourcing the correct part either locally/domestically or internationally.

12.10.1 Preventive Maintenance

Preventive maintenance is a plan that creates a plan for inspection, by which equipment is rebuilt or replaced as needed. A plan is set up for each piece of equipment to be checked on a frequency that is established to be repaired before it fails. The plan should be mapped out and cover every gasket, O-rings, valve, pump, and instruments in the system. It should also include a calibration program for all instruments. To start with, the frequency is set for a short interval and then the equipment is evaluated. You can extend the time as you do an evaluation and if no evidence of failure is seen. You would shorten the time if the equipment failed before the next inspection time.

Each piece of equipment needs to be on a PM program (these are some items that may be included on a PM program):

- Gaskets for each area will be different based on location.
  - This includes plate gaskets.
  - It includes the O-rings on the fillers and the proper frequency to replace them.
- Pressure gauges checked for cracks are missing stainless cover.
- Instruments, are they working as designed?
- Valves
- Back pressure valves
- Pumps
- Spray balls
- Steam traps
- Preventive maintenance includes calibration of all instruments to ensure they are working and reading correctly.

12.10.2 Breakdown Maintenance

Breakdown maintenance is where equipment is allowed to fail before fixing. This type of mainte-nance is generally more expensive and time-consuming than preventive maintenance.

This leads to what is sometimes called firefighting, which is fix the problem after it fails instead of preventing the problem before it occurs.

Several limitations of relying on breakdown maintenance are as follows:

- Generally, it happens during production resulting in lost product and production time.
- Increased risk of contamination due to leaking areas in the product area where it can affect the product, for example, gaskets, pump seals, pressure gauges, and pressure sensors.
- On pressure gauges/sensors may get a crack in the diaphragm or the bottom foil may come off. This can lead to product contamination and may not be discovered soon enough, so it can be fixed, or replaced. As a result, large quantities of product can be affected and put on hold for testing and disposition.
- There is a risk you do not have the correct part on hand requiring extended down time or using non-standard replacement part that may or may not fix the problem or void the warranty with the OEM.

12.11 CHANGE CONTROL MANAGEMENT PROGRAM (CCM-P)

Every company should have a Change Control Management Program Council that can review all changes and their risk versus benefit, including unintended collateral consequences. This pro-gram documents and tracks all changes during and after implementation. It also sets up a clear path that all changes must go through the CCMP Council and all departments involved and/or impacted by this change. A change control procedure lays out a planned response to any change or update; below are some examples of what a change control should have done on it.

12.11.1 Program Change

Any time any program is changed, it needs to be validated as to how the change impacted the operation, especially the cleaning of the equipment. This includes but not limited to:

- Increasing or decreasing timed steps.
- Changing of valve pulsing.
- Sequence of the steps.
- Removing or adding of a step.
- When any of these changes are made, then you need to do a new validation that it is still cleaning or running to the necessary standard and expectation.
- Part replacement if the part you changed is not like for like. If you change a part and you have to change it with a different part that is not exactly the same, you need to evaluate that change:
  - If you change a pump, it must be of the same horsepower, the same RPM, the same impellor size, the same inlet and outlet. If any of these changes, then you need to revalidate the system.
  - If you change a valve and the stroke length is the same, does it function exactly like the one replaced, if not you need to validate.

12.11.2 Ingredient Change

Sometimes it is felt that an ingredient can be substituted or replaced with another without validation. This may have profound impact on product itself, process, package, and stability of product during its shelf life and life cycle.

- Ingredients are changed for cost-saving. The new ingredient should be analyzed for all risks—microbiological, allergenicity, foreign matter, and chemical contaminants.
- R&D Product Development should make prototypes with "old and new" ingredients to understand the performance while batching, processing, and finished product, and shelf life stability. Any shifts in viscosity, pH, TSS should be reviewed carefully.

12.11.3 Procedure Change

All new or modified procedures when implemented should be tracked and proven out:

- If an SSOP is changed, you need to validate that you are getting the same level or higher effectiveness than with the old one. Changing an SSOP without a validation process for saving time or money can be a recipe for disaster.
- If you change any SSOP or a SOP before it becomes the new standard, it must be validated.

12.11.4 *Chemical Change*

The control and use of chemicals are critical to the cleaning of the equipment and the safety of the employees. Where possible, the number and types of chemical need to be limited to what is necessary. There must be a solid policy in the filling of bulk systems to prevent wrong chemicals in the tank. Validation of every chemical line with water before chemical to ensure that is going where it was designed to go and not leaking.

If any chemical is changed. it is extremely important to have a change control so that you can ensure the old chemical is removed, all lines completed rinsed and tested prior to the new chemical being introduced. All lines and containers clearly marked and if chemical changed all labels, change to match the chemical.

12.12 ENVIRONMENTAL CLEANING

Environmental cleaning should be part of each plants Master Sanitation Program that is tracked.

12.12.1 Environmental Monitoring

Every plant needs an environmental monitoring program. This program will not only cover the inspection of an area but dictate what should be happening in that area. If the plant is dealing with allergens, then this plan for environmental monitoring and cleaning becomes extremely crucial as it can allow cross-contamination if it is not followed.

Environmental is not just the air quality, it is the condition and use of every area of the plant. This includes a routine plan where different areas are swabbed and checked at a predetermined time interval to ensure that each area is getting the attention it should. The check should include swabbing for micro, visual, and smell.

- This includes the walls and floors.
  - The frequency of the cleaning of walls and floors.
  - The cleaning and checking of anchorage points on all equipment to prevent any harborage areas.
  - This would include controlling the foot traffic through sensitive areas.
  - Cleaning of the drains to prevent buildup of bacteria.
- *The ceiling area, lights, and overhead pipes*

Some of the most overlooked and least cleaned areas in the plant or ceiling area, lights, and overhead pipes. These areas get a lot of moisture, and dust buildup and as a result become high areas of growth. What is forgotten is that if these areas are left unattended, this layer of dust and moisture build up and then it will flake or drip off and fall into product, footpaths, and become a source of contamination. These need to be part of a Master Sanitation List that is followed, and each area cleaned based on the conditions in that area.

12.12.2 Air Quality

The air quality is not only the air that is brought into the facility but is a factor especially on fillers and filler rooms. The plant must have a plan where the air quality is tested in every room of the plant on at least a quarterly basis and in some it may be daily due to the sensitive nature of the area.

A lot of the fillers in ESL use HEPA filters for the air that is going into the machine. This air must meet the standard the OEM set up for their filler. This air must be checked on a frequent basis to ensure you are getting the proper filtration of the air used for preventing contamination of the product prior to the package being sealed. Several factors will go into the life of these filters and will dictate how often to change them.

- The construction of the room
- The condition of the room
- The humidity in the room
- The foot traffic in the room
- The personal hygiene of the personals in the room
- Recommendation by OEM for replacement interval or cycles of presterilization–for example, 25 cycles of use

12.12.3 Zoning and Segregation

It is encouraged that areas of the plant be zoned or segregated based on the sensitivity of the room and the area. When you zone or segregate areas, you must also control access into these areas, limit foot traffic into each area to only those who need to be in them.

- Raw receiving should be separated from pasteurized. Employees working in raw should not go into any other area.
- Blending separated from processing.
- Fillers should be separated from other areas.
- Garbage and waste areas isolated and controlled.
- Allergens need to be controlled all the way from receiving until they leave the plant.

12.12.4 Other Items

One of the concerns we want to address is that the cleaning of the plants and equipment require a large amount of water and chemicals. The environment is all our responsibilities, and each facility needs to look at every area of the plant to be sure they are doing what they can. It should include audits by their chemical vendors/partners and or outside parties to help identify and help the companies to reach these sustainable goals.

The control of the chemicals being disposed of needs to be addressed. Only using the amount of chemical that is needed and not overusing chemicals is very important. Many companies have their own wastewater systems that will ensure that the water being rejected from the plant is within the specifications of the location/municipality they are in. This may require a tank to buffer pH prior to it being sent into the waste stream.

The conservation of water should always be looked at. This requires looking at rinse times to ensure that they are correct and not running excess water to the waste system. It requires that equipment be run according to the OEM to ensure that things like pump seals are being controlled and shut off when not in use.

12.13 SUMMARY

Cleaning and sanitization procedures must be considered as an *integral* part of food production. No matter how well the production plant may be designed, if cleaning and sanitization are not possible or not done properly, it is impossible to produce high-quality and safe products consistently. Cleaning and sanitization are intertwined with production, and they are inseparable.

Cleaning and sanitization are one of the most important aspects of food manufacturing. This has been made amply clear by announcements of food recalls stemming from contamination with bacteria and foreign matter.

Cleaning and sanitization of aseptic processing and packaging unit operations is complex, and should be handled with care, and with commitment with top-down mandate. Case count is important. But "quality case count" is much more important, as it is salable product impacting company's bottom line.

There is reawakening of the importance of sanitation personnel and their impact on quality of food produced. Sanitation personnel can make or break a sanitation program. A plant can have the best systems, controls, and products, but if the sanitation job is done poorly, there could be problems. It is important that the sanitation personnel in the plant have the same skills, knowledge, and performance objectives as the middle and top management to get the job done. Continual education, training, and seminar at the plant level is imperative for production of high-quality safe products.

12.14 DEFINITIONS

Breakdown maintenance: Breakdown maintenance, sometimes called firefighting, means that you run equipment until it fails and then fix or replace when that happens. This usually causes unplanned down time and is generally more expensive than preventive maintenance.

CIP (clean-in-place): Circulating a cleaning solution through pipelines and large equipment using a system of pumps and sprays to automatically clean these systems. Some hand work is required.

Cleaning: Cleaning is using, flow (mechanical force), chemical concentration, time, and temperature to remove soil from the surface of the equipment. It can be done manually or using an automated cleaning system. It removes all traces of fats, solid materials, and product or other residues from equipment and other surfaces.

COP (clean-out-of-place): Usually done in a long rectangular sink, 1.5–2 feet deep and 1.5 feet wide, filled with hot soapy water circulated in the tank or sink by a pump mounted to the tank's or sink's base or by rapidly bubbling air through the water. Hand-cleaning is done first for parts in a COP tank or sink.

Fisheyes: Fisheyes in a product represents product that is not mixed correctly; it has a wet outer surface and on the inside a pocket of dry powders that heat will not be able to penetrate.

Fouling: Fouling is the process that happens on the metal surfaces, especially the processing equipment, caused by product sticking or being burnt to the metal. May be called "Burn On."

Intermittent clean-up: Intermittent clean-up is sometimes called an AIC (aseptic intermediate clean) or short cleans for the equipment under sterile conditions to remove the heavy soils for increased production time.

Potable water: Water that is clean and safe to drink.

Preventive maintenance: Preventive maintenance is a program that services, repairs, and/or replaces equipment on a routine basis to help prevent unplanned breakdowns. Will include but not limited to valves, pump, gaskets, and instruments.

Pre-sterilization: Pre-sterilization is the method used to sanitize or sterilize the equipment to a point that it can produce products that will not spoil and be free of pathogens for the shelf life of the product.

- ESL product (product stored under refrigeration) that will not spoil and be free of pathogens for the shelf life of the product.
- Aseptic product (product that is stored without refrigeration) that will not spoil and be free of pathogens for the shelf life of the product

Product contact surface: Any surface of a processing, filling, or packaging machine, valves and piping, tank walls, conveyors, etc. that actually touches the product.

Sanitation: Sanitation is the process of cleaning and sanitizing the equipment to reduce or eliminate the microbiological risks from the equipment.

Sanitizing: Chemical or heat treatment to kill germs. Includes rinsing, soaking, spraying, or wiping with a sanitizing solution. All items to be sanitized must be thoroughly cleaned. Sanitation may also be done by heat.

Water hardness – Concentration of sanitizers available for cleaning can be lowered due to reaction with calcium and magnesium salts present in water. For rinsing, only soft water (0–3.5 grains per gallon [gpg] or 0–60 ppm) of acceptable microbiological standards may be used. Water hardness will increase the sanitizer costs.

12.15 ACRONYMS

1. AIC – Aseptic intermediate clean
2. ASTM – Formally known as the American Society for Testing Materials
3. ATP—adenosine triphosphate
4. CDC – Center for disease control

5. CFR – Code of Federal Regulations
6. CIP – Cleaning-in-place
7. COP – Cleaning out place
8. EDTA – Ethylenediaminetetraacetic acid
9. ESL – Extended shelf life
10. GMP – Good Manufacture Procedure
11. HEPA – High-efficiency particulate air (filter)
12. HTST – high temperature short time
13. OEM – Original equipment manufacturer
14. PPE – personal protective equipment
15. SIP – Sterilization-in-place, some will call it sterilization-in-progress
16. SOP – Standard Operation Procedure
17. SPC – standard plate count
18. SSOP – Sanitation Standard Operation Procedure
19. UHP – Ultra-high pressure
20. UHT – Ultra-high Temperature

REFERENCES

Anonymous. 1994. Cleaning and descaling stainless steel parts, equipment and systems. ASTM A 380.94. American Society for Testing and Materials (ASTM), PA. ASTM. www.astm.org

Centers for Disease Control and Prevention (CDC). 2021. *List of Selected Multistate Foodborne Outbreak Investigations. Burden of Foodborne Illness: Findings | Estimates of Foodborne Illness.* Atlanta, Georgia, USA: Centers for Disease Control and Prevention CDC.

Ecolab® (For more knowledge or excess to these documents contact Ecolab®).
      a. Advisor Documents for Passivation of Stainless Steel
      b. Advisor Documents for the Removal of Black Polishing Dust

IAMFES Pocket Guide to Dairy Sanitation. 1996 Des Moines, IA: International Association of Milk, Food, Environmental Sanitarians. Des Moines, Iowa.

Katsuyama, A.M. 1993. *Principles of Food Processing Sanitation.* 2nd edition. Washington, DC: The Food Processors Insititute (FPI), National Food Processors Association.

Loncin, M., and Merson, R.L. 1979. Cleaning, disinfection, and rinsing, Chapter 10. In *Food Engineering: Principles and Selected Applications*, 310–329. New York: Academic Press, Inc.

Troller, J.A. 1983. *Sanitation in Food Processing.* New York: Academic Press, Inc.

Risk-Based Analyses for Attaining "Validated State" for Production of Commercially Sterile Shelf-Stable Products and Guidance for Quality Assurance, Microbiological Food Safety, and Regulatory Compliance

Chapter 13

Microbiology of Aseptically Processed and Packaged Products

Toni de Senna, Shirin J. Abd, Wilfredo Ocasio, and Jairus R.D. David

CONTENTS

DOI: 10.1201/9781003158653-16

13.1 INTRODUCTION

While the demand for aseptically packaged foods in North America continues to grow (Mordor Intelligence, n.d.), recalls and outbreaks involving aseptically packaged foods and beverages in the United States are extremely rare. It is noticeable that a list of selected multistate food-borne outbreaks posted by the Centers for Disease Control and Prevention (CDC), spanning from 2006 to 2021, failed to identify a single outbreak caused by an aseptically packaged food (Centers for Disease Control and Prevention (CDC), 2021). Aseptic processing and packaging provides for one of the safest and most stable food processing technologies available today. A listing of over 1,100 recalls and voluntary product withdrawals of FDA-regulated products spanning the period of 2017–2021 did not include a single aseptically processed product due to microbial contamination. Recalls of these products for reasons related to microbial contamination are extremely rare but have been reported in prior years for both low-acid shelf-stable products and low-acid extended shelf-life products, in some cases involving pathogenic organisms and millions of units impacted (Food Safety News, 2016; Hartman, 2009).

While recalls and outbreaks due to microbiological contamination are rare in these categories, tens of thousands of units are discarded every year due to spoilage detected prior to product release to distribution channels. This chapter aims to provide an overview of the microbiological risks, the relevant characteristics of common microorganisms, and the potential causes of spoilage in aseptically processed products. The contents of this chapter focus on low-acid, shelf-stable, aseptically packaged foods, but it also touches on other products using similar technologies, such as refrigerated, extended shelf-life products and high-acid, shelf-stable products. For the purpose of this chapter, aseptic process technology includes all processes that rely on continuous heating followed by delivering a predetermined process lethality in a hold tube, continuous cooling and filling the processed products in an aseptic or hygienic filler.

13.2 MICROBIOLOGICAL RISKS ASSOCIATED WITH ASEPTICALLY PROCESSED FOOD

The types of microorganisms of concern to aseptically processed foods include bacteria, yeast, and mold. The organisms may be divided into two broad categories: food safety risks and spoilage risks. Food safety risks typically include bacterial pathogens capable of causing illness, while spoilage risks may include non-pathogenic bacteria, yeast, and mold.

Identifying the microorganisms that may be of concern to low-acid canned foods (LACF) requires the consideration of many factors:

- The intrinsic characteristics of the food
- Organisms that are historically linked to certain raw ingredients or food groups
- The severity of the food safety/spoilage issue caused by the organism

13.2.1 Food Safety Risks

The microbiological food safety risks associated with aseptically processed foods can be divided into two categories based on their ability to form spores. Spore-forming bacterial pathogens are bacteria that may exist in one of two forms—as a vegetative cell or as a spore. As a vegetative cell, the pathogen is actively growing and is generally not resistant to environmental stresses. The cell

TABLE 13.1   ESTIMATED NUMBER OF CASES OF FOODBORNE ILLNESS AND ESTIMATED COST IN THE UNITED STATES; BASED ON DATA FROM 2018 (U.S. DEPARTMENT OF AGRICULTURE (USDA) ECONOMIC RESEARCH SERVICE (ERS), 2021).

Pathogen	Estimated Number of Cases	Estimated Costs ($ million)			
		Direct Medical Costs	Productivity Loss	Premature Death	Total Cost
Clostridium perfringens	966,000	$62.7	$252	$69.3	$384
Campylobacter	845,000	$542	$1,580	$63.8	$2,180
Nontyphoidal *Salmonella*	1,030,000	$387	$3,670	$87.7	$4,140
Shiga toxin-producing *Escherichia coli* O157	63,200	$41.7	$6.08	$263	$311
Non-O157 Shiga toxin-producing *Escherichia coli*	113,000	$16.9	$7.85	$6.92	$31.7
Listeria monocytogenes	1,600	$170	$52.2	$2,970	$3,190
Total	**3,019**	**$1,220**	**$5,570**	**$3,460**	**$10,240**

may enter a spore state when the environment is unfavorable for growth, during which time it is dormant and encased by a protective coat that renders the spore resistant to heat, extreme pH, chemicals, and radiation. When introduced to favorable conditions, the spores germinate into vegetative cells and continue to grow. Non-spore-forming bacterial pathogens do not possess the ability to form spores, and therefore exist only as vegetative cells.

It is estimated that there were approximately 9 million cases of foodborne illness in 2018 in the United States, with an economic impact of approximately $17.6 billion. The four bacterial pathogens most commonly identified as the cause of foodborne illness are *Clostridium perfringens*, *Campylobacter* spp., nontyphoidal *Salmonella*, and Shiga toxin-producing *Escherichia coli* (STEC) (USDA ERS, 2021) (Table 13.1).

13.2.1.1 Spore-Forming Bacterial Pathogens

Spore-forming bacteria in general are ubiquitous and can be isolated from soil, water, and dust. As such, spores may enter a product from raw ingredients (especially dry or powdered ingredients), the packaging materials, or the production environment (Baumgardner, 2012; Behling et al., 2010; Postollec et al., 2012). While in their dormant spore state, spore-forming bacterial pathogens are typically not harmful to the general population (it should be noted that the spores may pose a serious health hazard to very young infants and persons with a weakened immune system). However, the spores germinate under favorable conditions and, as vegetative cells, they grow and produce toxins which are then consumed and cause illness. Therefore, growth of the spore-forming pathogens in the food before ingestion is usually required to cause illness. Spore-formers most often linked to foodborne illnesses include *Clostridium botulinum*, *C. perfringens*, and *Bacillus cereus*.

13.2.1.1.1 C. botulinum

C. botulinum is a Gram-positive, anaerobic, rod-shaped, spore-forming bacteria that produces a potent neurotoxin (lethal dose of 10^{-9} mg/kg of body weight (Rosow & Strober, 2015)) during growth. The neurotoxin causes the disease botulism which causes flaccid paralysis of the muscles. If left untreated, death occurs when asphyxia is caused by paralysis of the respiratory muscles (U.S. Food and Drug Administration, 2012).

Botulism related to food consumption can be caused in two ways. Foodborne botulism is caused by the ingestion of preformed botulinum toxin, which occurs when *C. botulinum* is allowed to grow in a food before it is consumed. Symptoms usually appear within 18–36 hours of ingesting the toxin (U.S. Food and Drug Administration, 2012). Infant botulism results when *C. botulinum* spores are ingested and are able to colonize the large intestine and produce toxins within the gut. The incubation period of infant botulism is estimated to be between 3 and 30 days. Infants less than 12 months are particularly susceptible to infant botulism because the microflora of their gut has not yet been fully established. Adults with an altered gut microflora may also develop a disease similar to infant botulism (i.e., adult intestinal toxemia) (Rosow & Strober, 2015; U. S. Food and Drug Administration, 2012). It is estimated that *C. botulinum* causes less than 100 foodborne illnesses a year. However, botulism is extremely severe, with an 82% hospitalization rate and a death rate can be as high as 18% (Acheson, 2009; Scallan et al., 2011; U. S. Food and Drug Administration, 2012).

Growth of *C. botulinum* and toxin production in the food before ingestion is usually required to cause illness for the general population. Bloated packaging (from the production of hydrogen gas), a putrid or butyric (vomit) odor, a decrease in pH (from the production of organic acids), and changes to the texture of the food may be signs of *C. botulinum* growth in LACF.

The control/destruction of *C. botulinum* is a foundational part of regulations and practices surrounding LACF because this pathogen can form heat- and chemical-resistant spores, can grow in an anaerobic environment at ambient temperature, and causes an extremely severe disease.

13.2.1.1.2 C. perfringens

C. perfringens is a Gram-positive, anaerobic (but aerotolerant), rod-shaped, spore-forming bacteria that produces an enterotoxin during sporulation. The disease caused by *C. perfringens* typically takes a mild gastroenteritis form, which includes watery diarrhea and abdominal cramps; the illness is self-limiting (recovery within 24 hours) and rarely fatal for the general population. Very rarely, *C. perfringens* causes a more severe disease called enteritis necroticans (i.e., pig-bel disease) with symptoms such as abdominal pain and distention, diarrhea (sometimes bloody), vomiting, and patchy necrosis of the small intestine (Lindström et al., 2011; U. S. Food and Drug Administration, 2012).

Foodborne illness by *C. perfringens* is often linked to high protein foods (e.g., meat and poultry) that have been stored at 15–60°C for more than 2 hours. *C. perfringens* is capable of rapid growth under these conditions and can quickly reach high population levels under optimal conditions (Acheson, 2009). Illness occurs when a large number of vegetative cells and/or spores ($>10^6$ cells or spores/g of food) are ingested, and which then sporulate in the gut, producing enterotoxin. Symptoms are often observed within 16 hours of infection (Lindström et al., 2011; U. S. Food and Drug Administration, 2012). *C. perfringens* is one of the most common causes of foodborne illness in the United States, with an estimated 966,000 cases a year. However, because of the mildness of the symptoms, it is assumed that the incidence of *C. perfringens* illness is underreported. The hospitalization rate for *C. perfringens* infection is low (0.6%) and the death rate is estimated to be less than 0.1% (Lindström et al., 2011; Scallan et al., 2011).

As the infectious dose is high, substantial growth of *C. perfringens* in the food before ingestion is required. Bloated packaging (from the production of hydrogen gas), a butyric (vomit) odor, and a decrease in pH (due to the production of organic acids) are typically signs of *C. perfringens* growth in LACF.

13.2.1.1.3 B. cereus

B. cereus is a Gram-positive, facultative anaerobic, rod-shaped, spore-forming bacteria that produces two types of enterotoxins during growth, each causing illness that is generally mild. Emetic illness is caused by the consumption of a *B. cereus* toxin present in the food prior to consumption (preformed toxin); starchy foods like rice and pasta are commonly implicated. Symptoms of vomiting and nausea with occasional diarrhea usually appear 1–6 hours after toxin ingestion and typically last for up to 24 hours. The other toxin causes diarrheal illness and results from ingested cells/spores colonizing the small intestine and producing toxins. Abdominal cramps, watery diarrhea, and occasional nausea occur 8–16 hours after the cells/spores are consumed and may continue for 24 hours or more. Both types of illness require large infectious doses (emetic illness: 10^5–10^8 cells/g of food; diarrheal illness: 10^5–10^7 cells or spores ingested) (Acheson, 2009; Griffiths & Schraft, 2017; U. S. Food and Drug Administration, 2012). *B. cereus* is the causative agent for an estimated 63,000 cases of foodborne illness in the United States each year; however, underreporting is expected due to the mildness of the illnesses. In addition, the hospitalization rate for *B. cereus*–related illness is 0.4% and the death rate is 0% (Scallan et al., 2011).

Growth and toxin production in the food before ingestion are typically required for *B. cereus* to cause illness. A decrease in pH may accompany *B. cereus* growth in LACF (Table 13.2).

13.2.1.2 Non-Spore-Forming Bacterial Pathogens

Non-spore-forming bacterial pathogens include those that are not capable of forming spores and thus exist only in a vegetative state. In the United States, *Campylobacter* species (estimated 845,000 cases per year) and non-typhoidal *Salmonella* species (estimated 1,000,000 cases per year) are responsible for the greatest number of foodborne illness cases. Shiga toxin-producing *Escherichia coli* (STEC; including O157 and non-O157 strains) are also common causative agents of foodborne illness (estimated 175,000 cases per year). In addition, *Listeria monocytogenes* (estimated 1,500 cases per year) is also considered a major foodborne pathogen because of its ability to persist in a production environment (Behling et al., 2010) and the severity of the disease it may cause (Scallan et al., 2011). In general, a relatively low number of cells of these non-spore-forming pathogens is required to cause illness. Therefore, growth in the food may not be required to cause illness; the presence of these pathogenic organisms, even in low numbers, may be sufficient to pose a food safety risk (Figure 13.1).

13.2.1.2.1 Campylobacter *spp.*

Campylobacter spp. are Gram-negative, microaerophilic, spiral bacteria that are part of the gut microflora of chickens and other avian species. In the context of food, *Campylobacter* spp. are strongly associated with raw chicken and poultry.

Of the species of *Campylobacter*, *Campylobacter jejuni* is responsible for up to 90% of the foodborne illness cases associated with this genus. *C. jejuni* causes illness (called campylobacteriosis) by invading the intestinal epithelial cells. Symptoms of campylobacteriosis may appear 2–5 days after infection, and typically include fever, inflammation, diarrhea, abdominal cramps, and vomiting. In the majority of cases, the disease is self-limiting and typically lasts 1–2 weeks. The infectious dose of *C. jejuni* may be as low as a few hundred cells, but is generally thought to be

TABLE 13.2   SUMMARY OF THE LIMITING CONDITIONS FOR GROWTH OF SPORE-FORMING AND NON-SPORE-FORMING PATHOGENS (U.S. DEPARTMENT OF HEALTH AND HUMAN SERVICES; FOOD AND DRUG ADMINISTRATION; CENTER FOR FOOD SAFETY AND APPLIED NUTRITION, 2018).

	Organism	Minimum Water Activity	Minimum pH	Maximum pH	Minimum Temperature (°C)	Maximum Temperature (°C)	Oxygen Requirement
Spore-forming pathogens	Clostridium botulinum, proteolytic	0.935	4.6	9	10	48	Anaerobe[1]
	Clostridium perfringens	0.93	5	9	10	52	Anaerobe[1]
	Bacillus cereus	0.92	4.3	9.3	4	55	Facultative anaerobe[2]
Non-spore-forming pathogens	Campylobacter jejuni	0.987	4.9	9.5	30	45	Micro-aerophile[3]
	Salmonella spp.	0.94	3.7	9.5	5.2	46.2	Facultative anaerobe[2]
	Pathogenic Escherichia coli	0.95	4	10	6.5	49.4	Facultative anaerobe[2]
	Listeria monocytogenes	0.92	4.4	9.4	-0.4	45	Facultative anaerobe[2]

[1] Requires the absence of oxygen.
[2] Grows with or without oxygen.
[3] Requires limited levels of oxygen.

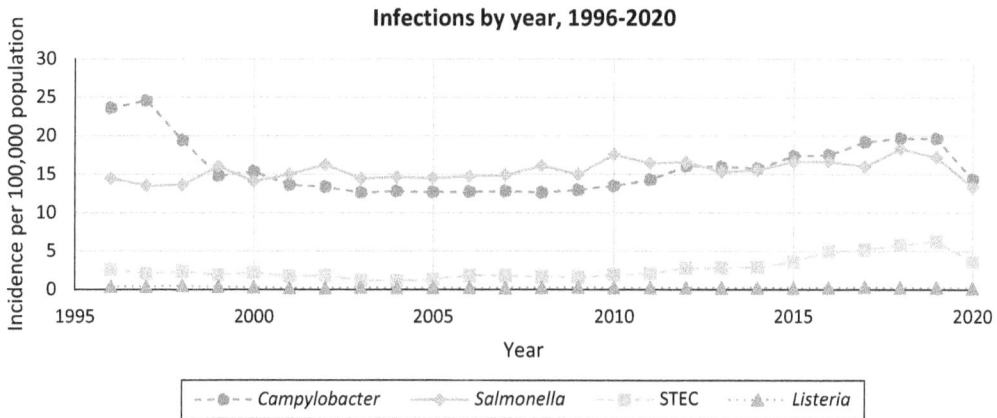

Figure 13.1 Number of infections by year by *Campylobacter*, *Salmonella*, Shiga toxin-producing *Escherichia coli* (STEC), and *Listeria*, 1996–2020, as reported by FoodNet (Centers for Disease Control and Prevention, 2019).

closer to 10,000 cells (Acheson, 2009; U. S. Food and Drug Administration, 2012; Young et al., 2007). Illness caused by *Campylobacter* spp. has an estimated 17% hospitalization rate and a death rate of 0.1% (Scallan et al., 2011).

13.2.1.2.2 Nontyphoidal Salmonella spp.

Nontyphoidal *Salmonella* spp. are Gram-negative, facultative anaerobic, rod-shaped bacteria that are commonly found in the guts of a wide variety of animals including poultry. *Salmonella enterica* is the species of nontyphoidal *Salmonella* that is most commonly associated with food-borne illness, and this species is further divided into >2,500 serotypes (Acheson, 2009; U. S. Food and Drug Administration, 2012). The three serotypes most commonly linked to *Salmonella* food-borne illness outbreaks are Enteritidis, Typhimurium, and Newport (Brown et al., 2017; Jackson et al., 2013). From 1998 to 2008 in the United States, eggs and chicken were the most commonly implicated commodities, followed by pork, beef, fruit, and turkey (Jackson et al., 2013). In addition to these commodities, *Salmonella* is of particular concern in low-water activity products and ingredients (e.g., powders, spices, nuts, flours, confectionary, etc.) because of its resistance to desiccation, allowing it to persist in low-water activity environments (Beuchat et al., 2013; Podolak et al., 2010; U. S. Food and Drug Administration, 2017b).

Ingested cells of nontyphoidal *Salmonella* spp. cause disease by invading the lining of the intestine and proliferating. The infectious dose is dependent on the health of the host and the *Salmonella* strain, but may be as low as one cell (1–3 *Salmonella* CFU/g in chocolate outbreak; 0.04–0.45 CFU/g in potato chips outbreak) (Podolak & Black, 2017). The symptoms of a nontyphoidal *Salmonella* infection may appear 6–72 hours after exposure and include nausea, vomiting, abdominal cramps, diarrhea, fever, and headache typically lasting 4–7 days (U.S. Food and Drug Administration, 2012). The estimated hospitalization rate is 27%, with a death rate of 0.5% (Scallan et al., 2011).

13.2.1.2.3 Shiga Toxin-Producing Escherichia coli (STEC)

Shiga toxin-producing *E. coli* are Gram-negative, facultative anaerobic, rod-shaped bacteria capable of producing Shiga toxin. STEC associated with human illness may also be referred to as enterohemorrhagic *E. coli* (EHEC) or verocytotoxic *E. coli* (VTEC). *E. coli* O157:H7 is the serotype responsible for a majority of foodborne STEC illness. Six serogroups (i.e., O111, O26, O121, O103, O145, and O45) account for most of the non-O157 illnesses in the United States (U. S. Food and Drug Administration, 2012). STEC have been historically linked to the intestinal tracts of ruminants (cattle, sheep, and goats). Food products and ingredients associated with STEC contamination include meat products, fresh produce (particularly sprouts and leafy greens), and water (Acheson, 2009; Kaper et al., 2004). Recently, outbreaks of non-O157 STEC in flour have caused the food industry to consider the possibility of STEC contamination of low-water activity ingredients (Centers for Disease Control and Prevention, 2019; U.S. Food and Drug Administration, 2017a).

The infectious dose of STEC varies by strain, but is generally considered to be low (<100 cells). STEC cause illness by attaching to the intestinal epithelium (while remaining extracellular), causing localized damage and producing Shiga toxin. Symptoms of an STEC infection may range from mild diarrhea to hemorrhagic colitis, which is characterized by severe cramping and bloody diarrhea. Hemorrhagic colitis symptoms usually develop within 3–4 days (may range from 1 to 9 days) and last for 2–9 days. However, hemorrhagic colitis may lead to the more severe condition, hemolytic uremic syndrome (HUS), in 3–7% of cases. HUS is thought to be caused by kidney damage resulting from Shiga toxin and can eventually result in renal failure (Acheson, 2009; Kaper et al., 2004; U.S. Food and Drug Administration, 2012). The hospitalization rates of *E. coli* O157:H7 and non-O157 *E. coli* are estimated to be 46% and 13%, respectively, with death rates of 0.5% and 0.3% (Scallan et al., 2011).

13.2.1.2.4 L. monocytogenes

L. monocytogenes is a Gram-positive, facultative anaerobic, rod-shaped bacteria that is a major concern for the food industry because it is salt-tolerant, able to grow at refrigeration temperatures, is known to persist in the production environment, and is one of the leading causes of death due to foodborne illness (Behling et al., 2010; Ferreira et al., 2014; U.S. Food and Drug Administration, 2012). *L. monocytogenes* has been historically linked to refrigerated ready-to-eat (RTE) products (e.g., deli meats and other fully cooked meats, prepared salads and meals, soft cheeses and dairy products, etc.) and frozen foods (both RTE and not RTE), but has also been the cause of outbreaks and recalls in fresh produce and nuts (Acheson, 2009; Buchanan et al., 2017).

The infectious dose of *L. monocytogenes* may be as high as 10^9 cells for the general population or as low as $<10^3$ cells for vulnerable individuals (e.g., the elderly, pregnant women, immunocompromised individuals, and infants) (Acheson, 2009; U.S. Food and Drug Administration, 2012). Once ingested, *L. monocytogenes* invades intestinal cells and multiplies. It is also able to spread directly from cell to cell, allowing *L. monocytogenes* to infect tissues while being protected from the host's defenses. *L. monocytogenes* may cross the intestinal barrier and spread to the blood, liver, and spleen and then may go on to cross the blood–brain barrier. Notably, *L. monocytogenes* is capable of infecting the placenta of pregnant women and hence the fetus (Cossart & Toledo-Arana, 2008; Disson & Lecuit, 2013). The severity of the disease caused by *L. monocytogenes* infection is dependent on the susceptibility of the host. Non-invasive gastrointestinal illness usually occurs in healthy individuals within three days of exposure; symptoms are generally mild and can include fever, muscle aches, nausea, and sometimes diarrhea. The invasive form is much more severe and can result in septicemia, meningitis, and in pregnant women can cause abortion or stillbirth (Acheson, 2009; U.S. Food and Drug Administration, 2012). Symptoms of invasive *L. monocytogenes* infection may take several weeks to appear (Buchanan et al., 2017).

As previously mentioned, foodborne illness caused by *L. monocytogenes* is relatively infrequent (estimated 1,500 cases per year) in comparison to other foodborne pathogens. However, this organism is considered a major foodborne pathogen because of its ability to persist in a production environment and the severity of the disease it may cause. The hospitalization rate is an estimated 94% with a death rate of 15–30% overall and up to 50–80% for a more severe invasive infection (Scallan et al., 2011; U.S. Food and Drug Administration, 2012).

13.2.2 Spoilage Risks

As mentioned earlier in this chapter, outbreaks and product recalls are a very rare occurrence in aseptically packaged foods. However, microbial contamination resulting in economic spoilage occurs more frequently. Fortunately, in this category, most spoilage problems are caught prior to product release as quality holds at ambient or above temperatures, and analytical testing as a condition of product release allows for capturing spoilage problems prior to product release. Nevertheless, economic spoilage in this category is a serious problem due to large production runs and the complexity of diagnosing spoilage problems. In our experience, the majority of the spoilage problems found in this category are caused by organisms of the genus *Bacillus* as described below. However, other non-spore-forming organisms such as lactic acid-producing bacteria, acetic acid-producing bacteria, yeasts, and molds can also be encountered, particularly when the contamination occurs on the "cold" side of the product lines or filling operation.

13.2.2.1 Spore-Forming Spoilage Bacteria

Due to the severity of the sterilization processes used, spoilage of aseptically processed products is relatively rare. However, when spoilage does occur, spore-forming bacteria are often the cause. Species of spore-forming bacteria are capable of transitioning from an actively growing vegetative cell to a dormant spore state that may persist for extremely long periods of time and are typically resistant to heat and chemical treatments. The optimum growth temperature of the bacteria can generally inform the relative resistance of the spore. Spores of mesophilic bacteria (optimum growth temperature near 25–40°C) are typically less resistant than the spores of thermophilic bacteria (optimum growth temperatures >40°C). Usually, the sterilization methods used in aseptic processing are designed to target mesophilic spore-formers and thus may not be sufficient to destroy the spores of thermophiles.

The recovery of heat-labile (vegetative) organisms from a spoiled aseptically processed food may be an indication of a failure of the sterilization of the product/packaging materials or an inability to maintain sterility. As the sterilization processes would not destroy thermophiles, the presence of thermophilic spores in an aseptically processed product may be expected. However, the growth of thermophilic spore-formers is prevented by maintaining the product temperature below 40°C. Thermophilic spoilage may occur if the product experiences temperature abuse.

Mesophilic and thermophilic species of the genera *Bacillus* and *Clostridium* are the spore-forming bacteria most commonly associated with the spoilage of aseptically processed products due to their ubiquity and high resistance to sterilization methods. Growth of *Bacillus* spp. typically occurs without gas production and spoilage may be detected by a drop in pH (due to the production of organic acids) and other organoleptic changes to the product. In contrast, *Clostridium* spp. growth is usually accompanied by the production of gas (e.g., hydrogen gas and carbon dioxide), which may swell the product packaging, in addition to compounds that may affect the organoleptic characteristics of the product (e.g., organic acids, enzymes, and compounds with a putrid, butyric, or fecal odor).

Spore-formers are commonly isolated from soil, dust, and water and may be introduced to a product through any number of vectors, including raw ingredients, the production environment, and personnel. Dry ingredients such as spices, herbs, egg powder, and cocoa powder contain a variety of spore-forming bacteria; species commonly isolated include *Bacillus licheniformis*, *Bacillus subtilis*, *Bacillus pumilus*, *Bacillus thuringiensis*, non-pathogenic *B. cereus*, and *Bacillus coagulans* (Lima et al., 2011; Mathot et al., 2021; Postollec et al., 2012). These bacilli are also often found in raw milk and dairy-based ingredients, in addition to *Clostridium* spp. such as *Clostridium tyrobutyricum*, *Clostridium halophilum*, and *Clostridium sporogenes* (Buehner et al., 2015; Coorevits et al., 2008; Julien et al., 2008; McHugh et al., 2017; Postollec et al., 2012; Scheldeman et al., 2005).

Due to the severity of the sterilization processes used, spoilage of aseptically processed products is relatively rare. However, some highly resistant spore-formers may be able to survive the sterilization processes and go on to spoil the product. *B. coagulans*, *B. licheniformis*, *Clostridium pasteurianum*, and *Clostridium butyricum* have been associated with the spoilage of aseptically processed tomato paste (Palop et al., 1996b; Tribst et al., 2009) and are also capable of growth in a low-acid environment. Spoilage of UHT milk by *Geobacillus stearothermophilus* (formally *Bacillus*; obligate thermophile), *B. licheniformis*, *B. pumilus*, *B. subtilis*, and *Bacillus sporothermodurans* have been reported. *B. sporothermodurans* in particular is an interesting spoilage concern for UHT milk because it may grow to a maximum level of ~5 log CFU/ml in milk without affecting pH or other organoleptic characteristics. In addition, this organism grows poorly on the recovery media typically used for dairy products and is a poor competitor in relation to other organisms that may be part of the background microflora. Therefore, *B. sporothermodurans* spoilage can be difficult to detect (André et al., 2017; Scheldeman et al., 2006; Westhoff & Dougherty, 1981).

13.2.2.2 Non-Spore-Forming Spoilage Bacteria

Non-spore-forming spoilage bacteria include a wide variety of organisms. Three categories of spoilage organisms pertinent to aseptically processed foods are lactic acid bacteria (LAB), acetic acid bacteria (AAB), and coliforms. Non-spore-forming bacteria are extremely susceptible to the sterilization methods used in aseptic processing. Therefore, recovery of non-spore-forming bacteria from a spoiled product is a strong indication of postprocess contamination, either from a breach in the aseptic zone or a hermetic seal failure.

LAB is a broad group that may be generally described as Gram-positive rods or cocci that produce lactic acid as the main product of fermentation. Other metabolic by-products include other organic acids, carbon dioxide, proteolytic and lipolytic enzymes, and ethanol. LAB are mesophilic, but some strains may be able to grow at refrigeration temperatures. In addition, growth may occur at pH as low as 2.9–3.5. This group includes many genera of bacteria, the core of which are *Lactobacillus*, *Streptococcus*, *Pediococcus*, and *Leuconostoc*. Additional genera considered to be part of the LAB group continue to evolve; *Aerococcus*, *Alliococcus*, *Carnobacterium*, *Dolosigranulum*, *Enterococcus*, *Globicatella*, *Lactococcus*, *Lactosphaera*, *Oenococcus*, *Tetragenococcus*, *Vagococcus*, and *Weissella* have also been included (Salfinger & Tortorello, 2015; Salvetti et al., 2022; Sperber & Doyle, 2009). Members of the LAB group are used extensively by the food industry, particularly for the production of fermented vegetables (sauerkraut, kimchee, pickles), fermented beverages (wine, kombucha), dairy products such as cheese and yogurt, and sourdough bread. In addition, some species of LAB are used as probiotics. However, unintentional or uncontrolled growth of LAB can lead to spoilage. Growth of LAB can be detected by a drop in the pH of the product and may be accompanied by gas production, a "fermented" odor or other off odors, and textural changes.

AAB are a group of organisms belonging to the family Acetobacteraceae. AAB are strictly aerobic, Gram-negative or Gram-variable rods that can oxidize sugars and alcohols to produce acetic acid and other metabolites (e.g., other organic acids and carbon dioxide). AAB prefer mesophilic temperatures, but growth may also occur at refrigeration temperatures. In addition, the pH lower limit for growth is 3.0–4.0. The AAB group includes the genera *Acetobacter*, *Acidomonas*, *Ameyamaea*, *Asaia*, *Bombella*, *Commensalibacter*, *Endobacter*, *Gluconacetobacter*, *Gluconobacter*, *Granulibacter*, *Komagataeibacter*, *Kozakia*, *Neoasaia*, *Neokomagataea*, *Nguyenibacter*, *Saccharibacter*, *Swaminathania*, *Swingsia*, and *Tanticharoenia*. Similar to LAB, AAB are also utilized by the food industry, and are best known as part of the production of vinegar and fermented beverages (e.g., wine, kombucha). However, undesirable growth of AAB may be detected by a drop in product pH, the development of "sour" or "vinegar" flavors/odors, and gas production (Gomes et al., 2018; Sengun & Karabiyikli, 2011; Sperber & Doyle, 2009).

Coliforms are commonly used by the food industry as an indicator organism to assess the sanitary conditions of food manufacturing. Bacteria considered to be coliforms are Gram-negative, facultative anaerobic rods that ferment lactose to produce gas and acid within 48 hours at 35°C. Based on this definition, the genera *Citrobacter*, *Enterobacter*, *Escherichia*, and *Klebsiella* may be considered coliforms. Fecal coliforms (also called thermotolerant coliforms) are a subset of coliforms that are also used by the food industry as an indication of fecal contamination; analyses for fecal coliforms are done at 44.5°C instead of at 35°C. *E. coli* is the most commonly isolated fecal coliform; however, species of *Klebsiella* may also be recovered. The optimum temperatures for coliforms are near the mesophilic range; however, growth may occur between -2°C and 50°C depending on the strain. The pH range for growth is 4.4–9.0. Spoilage by coliforms may be detected by a decrease in product pH (from the production of organic acids), gas production, off-flavors/odors, and changes in texture (Baylis, 2006; Feng et al., 1998).

13.2.2.3 Spoilage Fungi

Fungi are a group of eukaryotic organisms that include yeast and mold. The food industry utilizes yeast and mold for the production of a variety of foods, such as fermented foods and beverages (e.g., pickles, kimchi, beer, wine, kombucha, yogurt, cheese, etc.), thus making them an essential part of food manufacturing. However, the unintentional introduction and/or uncontrolled growth of yeast and mold can lead to food spoilage. Yeast and mold are capable of growing in a wide range of pH and water activity, and therefore may be problematic for many types of products. Spoilage yeast and mold can be found throughout the environment and may be found in the air, water, plant materials, and soil. Molds in particular are able to form spores that become airborne for widespread dispersal. Contaminants may be introduced to food products by the raw ingredients, dust, production facility, air-handling system, and production equipment (Hernández et al., 2018; Moss, 2006; Pitt & Hocking, 2009). Spoilage by yeast and mold in aseptically processed foods may be indicative of a hermetic seal issue in the packaging materials.

Yeasts are single-celled fungi that reproduce via budding. Common and notable genera of spoilage yeast are listed in Table 13.3. Most yeasts are able to ferment carbohydrates, making it possible to grow with and without oxygen. Yeasts typically favor temperatures of 20–30°C, with very few species capable of growth above 40°C. The optimum pH for most yeast is 4.5–7.0, but some species may grow at pH as low as 2.0–2.5. The water activity range for yeast is typically 0.90–0.95, with some osmo- and xerotolerant species (e.g., species of *Debaryomyces*, *Zygosaccharomyces*, and *Saccharomyces*) capable of growth in water activity as low as 0.65–0.85 depending on the solute. In addition, the preservative-resistance of species of *Zygosaccharomyces* (*Zygosaccharomyces bailii* and, to a lesser extent, *Zygosaccharomyces lentus* and *Zygosaccharomyces rouxii*) make

TABLE 13.3   SUMMARY OF SPOILAGE ORGANISMS' CHARACTERISTICS.

Spoilage Organism		Optimum Temperature Conditions	Oxygen Requirements	Characteristics of Growth	Common Species	Additional Notes
Spore-forming bacteria	*Bacillus* spp.	Mesophilic temperatures	Strict aerobe or facultative anaerobe	Growth without gas production, acid production, additional organoleptic changes	*Bacillus licheniformis, Bacillus subtilis, Bacillus sporothermodurans* (in UHT milk)	Also known as "flat sour"
	Clostridia spp.	Mesophilic temperatures	Anaerobe	Growth with gas production, putrid or butyric odor, additional organoleptic changes	*Clostridium sporogenes, Clostridium butyricum, Clostridium pasteurianum*	Gas production may cause packaging to swell
	Bacillus spp.	Thermophilic temperatures	Strict aerobe or facultative anaerobe	Growth without gas production, acid production, additional organoleptic changes	*Geobacillus stearothermophilus, Bacillus coagulans*	Also known as "flat sour"
Non-spore-forming bacteria	Lactic acid bacteria	Mesophilic temperatures	Strict aerobe or facultative anaerobe	Growth with gas production, acid production (e.g., lactic acid), fermented odor, additional organoleptic changes	*Lactobacillus* spp., *Streptococcus* spp., *Pediococcus* spp., *Leuconostoc* spp.	Used throughout the food industry in the production of fermented foods
	Acetic acid bacteria	Mesophilic temperatures	Strict aerobe	Growth with gas production, acid production (e.g., acetic acid), sour or vinegar flavors/odors, other organoleptic changes	*Acetobacter* spp., *Gluconacetobacter* spp., *Gluconobacter* spp.	Used in the food industry for the production of vinegar and fermented beverages
	Coliforms	35–45°C	Facultative anaerobe	Fermentation of lactose to produce gas and acid, other organoleptic changes	*Citrobacter, Enterobacter, Escherichia,* and *Klebsiella*	Used as an indicator organism when assessing the sanitary condition of the production environment

(Continued)

TABLE 13.3 (CONTINUED)   SUMMARY OF SPOILAGE ORGANISMS' CHARACTERISTICS.

Spoilage Organism		Optimum Temperature Conditions	Oxygen Requirements	Characteristics of Growth	Common Species	Additional Notes
Spoilage fungi	Yeast	Mesophilic temperatures	Facultative anaerobe	Growth with gas production, acid production, ethanol production, fermented or yeasty odor, additional organoleptic changes	*Brettanomyces* spp., *Candida* spp., *Pichia* spp., *Rhodotorula* spp., *Saccharomyces* spp., *Zygosaccharomyces* spp.	Can grow at pH as low as 2.0–2.5, some species may grow at water activity as low as 0.65
	Mold	Mesophilic temperatures	Aerobe	Growth may be visible as a fuzzy or powdery mat, musty or stale odor, acid production, ethanol production, other organoleptic changes	*Aspergillus* spp., *Penicillium* spp., *Rhizopus* spp., *Mucor* spp., *Geotrichum* spp., *Fusarium* spp., *Alternaria* spp., *Cladosporium* spp., *Eurotium* spp., *Byssochlamys* spp.	Can grow at pH 3.0–8.0, some species may grow at water activity as low as 0.6

this genus particularly problematic in food (Fleet, 2011; Martorell et al., 2007; Pitt & Hocking, 2009). Yeast spoilage is typically characterized by the production of carbon dioxide (resulting in bloated containers), organic acids, ethanol, and off-flavors and odors (e.g., fermented/ethanolic, vinegary, buttery, yeasty, etc.) (Fleet, 2011; Pitt & Hocking, 2009; Stratford, 2006).

In contrast to yeast, mold may be multicellular and form filamentous, branching structures. The mold genera commonly associated with the spoilage of food products include *Aspergillus*, *Penicillium*, *Rhizopus*, *Mucor*, *Geotrichum*, *Fusarium*, *Alternaria*, *Cladosporium*, *Eurotium*, and *Byssochlamys* (Sperber & Doyle, 2009). Molds are generally thought of as aerobic organisms; however, some species (e.g., species of *Byssochlamys*, *Mucor*, *Rhizopus*, *Fusarium*, and *Geotrichum*) may be capable of growth in the presence of little to no oxygen. Generally, molds favor ambient temperature, although growth of some species (e.g., species of *Fusarium*, *Cladosporium*, and *Penicillium*) may occur at refrigeration temperatures or lower. The optimum pH for mold growth is approximately 5.0, but occurs over a wide range (3.0–8.0) and may be low as pH 2.0 for some species. Most molds grow at water activity above 0.85; notable exceptions include moderately xerophilic species of *Aspergillus*, *Penicillium*, and *Eurotium* (growth at water activity as low as 0.70) and the extreme xerophile *Xeromyces bisporus* (growth at water activity range of 0.61–0.96) (Dagnas & Membré, 2013; Moss, 2006; Pitt & Hocking, 2009; Sperber & Doyle, 2009). Mold spoilage may be obviously visible in the form of a mycelial mat (may appear fuzzy or powdery) on the surface of the product or adhering to the packaging interior, seams, or closures. Mold growth may also be characterized by a "musty" or "stale" odor, the production of organic acids and ethanol, and changes to product texture due to enzymatic metabolites (Gutarowska, 2010; Pitt & Hocking, 2009; Sperber & Doyle, 2009). Although mold growth is generally not considered a food safety risk, some are capable of producing toxins. Some species of *Aspergillus*, particularly *Aspergillus flavus*, and *Penicillium* are recognized as safety concerns because of their ability to produce mycotoxins (Hocking, 2006; Moss, 2006; Pitt & Hocking, 2009).

A subset of mold that requires consideration when discussing spoilage fungi are heat-resistant molds. Heat-resistant molds are associated with the spoilage of many types of foods, including pasteurized products, because of their ability to form heat-resistant ascospores that survive the pasteurization process. Therefore, these types of heat-resistant organisms are typically excluded from finished products through proper management of sanitation programs, ingredient specification programs, etc. rather than destruction by a severe thermal treatment. Heat-resistant molds of the genera *Byssochlamys*, particularly *Byssochlamys fulva* and *Byssochlamys nivea*, are most commonly implicated in the spoilage of food products. Other heat-resistant molds that have been associated with food spoilage include *Neosartorya fischeri*, *Talaromyces flavus*, and *Eupenicillum* Ludwig (Moss, 2006; Pitt & Hocking, 2009; Sperber & Doyle, 2009; Tournas, 1994).

13.3 MITIGATION OF THE MICROBIOLOGICAL RISKS ASSOCIATED WITH ASEPTICALLY PROCESSED FOODS

The processes utilized in aseptic processing are designed to eliminate the microbiological food safety and spoilage risks associated with LACF in order to render the food and its packaging commercially sterile. As defined in the Code of Federal Regulations Title 21, Part 113—Thermally Processed Low-Acid Foods Packaged in Hermetically Sealed Containers (21CFR113, 2011):

- "Commercial sterility" of thermally processed food means the condition achieved—
  - By the application of heat which renders the food free of—

- Microorganisms capable of reproducing in the food under normal non-refrigerated conditions of storage and distribution; and
- Viable microorganisms (including spores) of public health significance; or
- By the control of water activity and the application of heat, which renders the food free of microorganisms capable of reproducing in the food under normal non-refrigerated conditions of storage and distribution
- "Commercial sterility" of equipment and containers used for aseptic processing and packaging of food means the condition achieved by application of heat, chemical sterilant(s), or other appropriate treatment that renders the equipment and containers free of viable microorganisms having public health significance, as well as microorganisms of non-health significance, capable of reproducing in the food under normal non-refrigerated conditions of storage and distribution.

The sterilization methods commonly used in aseptic processing include thermal treatment for product sterilization and chemical sterilants for equipment and packaging sterilization. In addition to heat, microfiltration may also be used to sterilize product. Other methods that may be used in aseptic processing on the equipment and packaging include steam, dry heat, ultraviolet light (UV), and electron beam/gamma radiation. The remainder of this section will focus on the most common sterilization methods: thermal treatment for the product and chemical sterilants for the equipment/packaging (Table 13.4).

13.3.1 Thermal Treatment

In aseptic processing, thermal treatment is most commonly used to achieve commercial sterility of the food before it enters the packaging and to sterilize the aseptic processing equipment

TABLE 13.4   COMMON STERILANTS USED IN ASEPTIC PROCESSING AND PACKAGING APPLICATIONS.

Equipment/Package	Sterilization Method
UHT equipment	Hot water (121°C)
Aseptic surge tanks, homogenizers, product-contact surfaces	Steam (121°C)
Aseptic filling area	H_2O_2, PAA-based sterilant, steam
Sterile gas (air, N_2) provision	Microfiltration, incineration (dry heat)
Sterile water provision	Microfiltration
Bags (bag-in-box, BIB fillers)	Steam, H_2O_2, ethylene oxide, chlorine solution + heat, gamma irradiation
Preformed plastic cups, pouches, cartons, PET and HDPE bottles	H_2O_2, H_2O_2 + heat, ultraviolet rays, H_2O_2 + steam, citric acid + heat, electron beam radiation, ethylene oxide
Lid and/or body roll stock (form-fill-seal, FFS fillers)	H_2O_2, H_2O_2 + heat, PAA-based sterilants, electron beam radiation
Steel/aluminum cans and lids	Superheated steam
Pouches	H_2O_2, H_2O_2 + heat

H_2O_2: hydrogen peroxide; PAA: peracetic acid.

before use. The application of heat to reduce populations of microorganisms in foods is widely used in the food industry, and various mathematical models have been developed to describe the relationship between the death rate (i.e., heat resistance) of a specific organism in a defined matrix and time. These models can be used to predict the level of microbial kill of a target organism achieved at a specific time and temperature. However, caution must be taken when using mathematical models because the heat resistance of a microorganism in a given food product may be influenced by a number of different variables, including the organism strain under evaluation and composition of the product of interest (e.g., formulation and ingredients, fat content, pH, and water activity).

13.3.1.1 Kinetics of Microbial Destruction

Decimal reduction time (D-value) is a numerical representation of heat resistance and is defined as the time required at a certain temperature to inactivate a microbial population in a specific medium by 90% (1 log cycle). The microbial inactivation rate as a function of time using first-order kinetics can be expressed as Equation 13.1:

$$\mathrm{Log}\left[N/N_i\right]=t/D \tag{13.1}$$

where:

N = survivors population following the thermal treatment
N_i = initial microbial population
t = the inactivation time
D = decimal reduction time (D-value)

Figure 13.2 graphically illustrates the concept of D-value for an organism at a constant temperature of 121°C. In this example, the population levels of the selected organism were reduced by 1 log (from 6 logs to 5 logs) within 1 min (from 2 to 3 min) of thermal treatment. Therefore, the

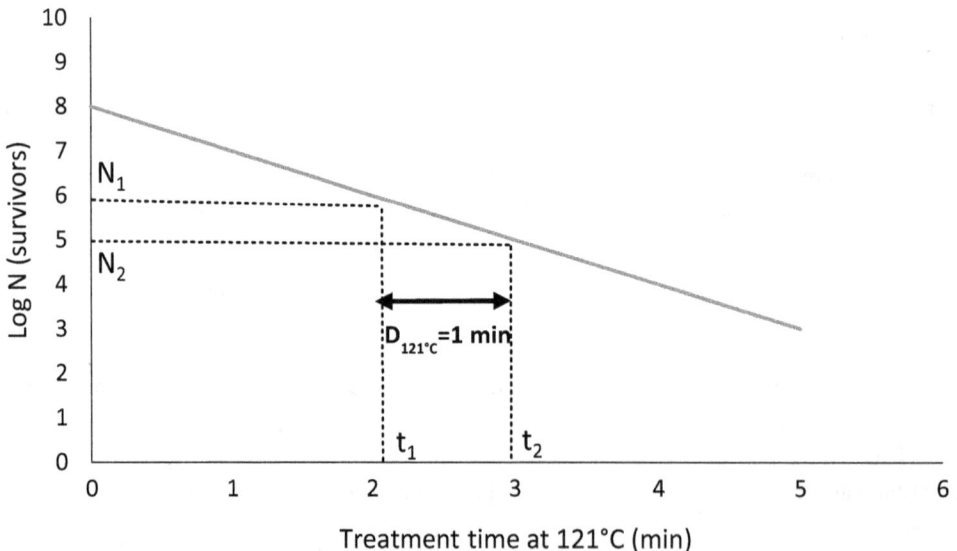

Figure 13.2   Description of the D-value. N_1 and N_2 are survivors at t_1 and t_2, respectively.

$D_{121°C}$ of the selected organism is 1 min. The larger the D-value, the higher the thermal resistance of the organism. For example, if the $D_{121°C}$ value of organism A and organism B are 10 and 1 min, respectively, organism A is ten times more heat resistant than organism B at 121°C.

The z-value serves to express the effect of temperature on the rate of microbial inactivation. The z-value is defined as the change of temperature required to change the D-value tenfold (1-log). Mathematically, the z-value can be expressed as Equation 13.2:

$$z = T_1 - T_2 / \log D_1 - \log D_2 \qquad (13.2)$$

where:

D_1 = D-value at temperature T_1
D_2 = D-value at temperature T_2

Figure 13.3 graphically illustrates the concept of z-value for a selected organism. The D-value of the selected organism is 10 and 1 min at 110°C and 120°C, respectively. An increase of the treatment temperature by 10°C changes the D-value by 1-log; therefore, in this example the z-value is 10°C. To convert the z-value from Centigrade to Fahrenheit, the z-value at °C is multiplied by 1.8.

13.3.1.2 Industry Standard Thermal Treatments (*F*-Value Concept)

The F-value term is commonly used in the industry to express the time required to achieve a *desired log reduction* of an organism at a defined temperature. For example, the canning industry has adopted a 12-log reduction of proteolytic *C. botulinum* spores to adequately address the risk associated with this pathogen in LACF products, including foods expected to have a high microbial load (e.g., beans, garlic, etc.). This 12-log reduction process against *C. botulinum* is generally known as "minimum health" process. A minimum health process of 3 min at 121°C (or equivalent) is based on classical D- and z-values for this organism ($D_{121°C}$ = 0.20–0.25 min, z = 10°C). This process is also known in the industry as F_0 = 3 min, where F-value (3 min) is the time required to achieve the target log reduction at the specified temperature (121°C). A more robust thermal process of 5 min at 121°C (or equivalent) is used to achieve commercial sterility and is based on the classical D- and z-values for *C. sporogenes* PA 3679 (a thermal surrogate for *C. botulinum*;

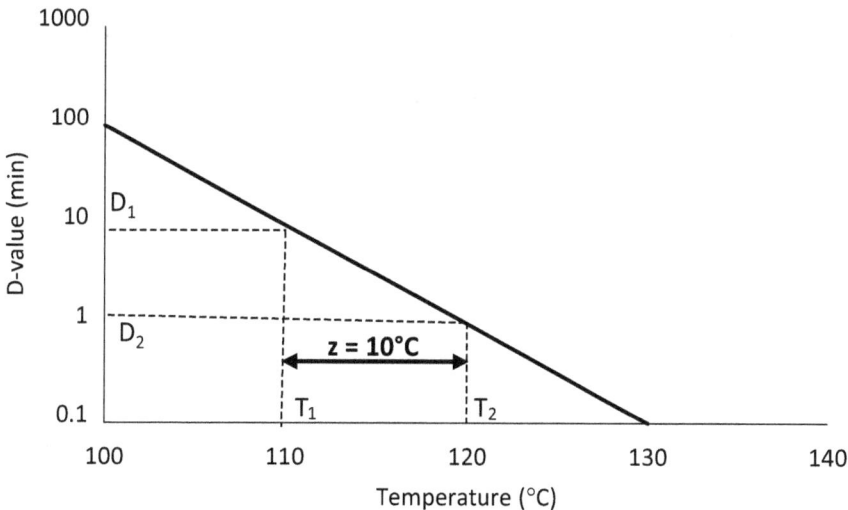

Figure 13.3 Description of the z-value. D_1 and D_2 are D-values at T_1 and T_2, respectively.

$D_{121°C} = 1.0$ min, $z = 10°C$). Similar to the minimum health process, this "commercial sterility" process is known in industry as $F_0 = 5$ min, where the F-value (5 min) is the time required to achieve the target log reduction at the specified temperature (121°C).

The D- and z-values discussed above have been historically used by industry for a wide range of LACF. Diao et al. (2014) completed a meta-analysis of hundreds of D- and z-values for *C. botulinum* and *C. sporogenes* PA 3679 available in the scientific literature. The values included in the meta-analysis were generated in matrices such as liquid laboratory media, milk and dairy products, meat and meat products, fish and seafood products, oil and cream, starch-based products (e.g., rice, pasta), and fresh, canned, or frozen vegetables; all matrices had a pH of 4.5 or greater. The meta-analysis found that the mean D-value for *C. botulinum* was 0.19 ± 0.11 min with a mean z-value of $11.3 \pm 0.3°C$. For *C. sporogenes*, the mean D- and z-values were 1.28 ± 0.69 min and $11.1 \pm 0.2°C$, respectively. The overall results of this meta-analysis show that the widely accepted industry standards align well with the information in the scientific literature.

Similar to the food processes described above, widely accepted industry practices also exist for the sterilization of equipment surfaces sterilized with saturated steam or pressurized water. For example, the UHT system (excluding the surge tank and fillers) is sterilized-in-place (SIP) with pressurized water at 121°C. The pressurized water is circulated through the system for 30 min. The surge tank and surge tank air filters are sterilized with saturated steam at 121°C for 30 min, as are the product-contact surfaces of the fillers up to the filler nozzles. It should be noted that these equipment treatments appear more severe than those applied to the food itself in spite of the fact that these surfaces are cleaned extensively prior to sterilization. The reason for this severe treatment is the difficulty in assessing the actual temperature of all the massive product-contact surfaces throughout the aseptic line.

13.3.1.3 Designing a Thermal Treatment

History and experience have shown that the industry standard processes, as detailed above, are effective in achieving commercial sterility of a wide variety of LACF. However, factors such as an unusually high spore load in raw ingredients, improper hydration, or miscalculations on hold tube residence times may render the industry standard processes insufficient to achieve commercial sterility. For example, historically products containing cocoa have been processed with a higher lethality target (121°C for 8–9 min or equivalent, $z = 10°C$). Deviation from the industry standard processes (e.g., to optimize the thermal treatment) requires a robust evaluation of the thermal resistance of the pertinent microorganisms in the specific matrix to ensure the production of a safe product. In addition, characteristics of the product—e.g., naturally occurring or added bioactive substances or water activity that may preclude the growth of some organisms—may mitigate the risk associated with specific safety or spoilage hazards. Therefore, a competent Processing Authority should be consulted when designing a thermal process to ensure that the following are appropriate:

- The target organism
- Thermal process performance criterion
- Thermal process parameters

The parameters of the thermal process should be accompanied by a robust rationale to support their selection based on scientific evidence such as peer-reviewed scientific literature, industry standard practices, recommendations from regulatory agencies, and laboratory studies executed using the product (e.g., challenge studies, thermal resistance studies).

13.3.1.3.1 Heat Resistance of Pertinent Microorganisms

Selection of the appropriate target organism for the process is critical when designing a thermal process. The process should target the most heat-resistant pathogenic organism pertinent to the product and be robust enough to deliver adequate reduction to achieve commercial sterility (i.e., destroy pertinent spoilage organisms).

13.3.1.3.1.1 Spore-Forming Bacteria

Thermal processes for aseptic LACF products target the most resistant pathogenic and pertinent spoilage organisms. Spore-forming bacteria are the targeted organisms of these thermal processes because of the high heat resistance of the spore forms they are capable of producing. Hence, it is extremely important to gain a thorough understanding of the thermal resistance of these organisms in the context of the specific food matrix studied. Table 13.5 lists microbial destruction values of pertinent pathogenic spore-forming microorganisms in various buffer and food matrices. The thermal resistance values of representative spoilage organisms capable of producing spores are presented in Table 13.6. The thermophilic spoilage organism *Geobacillus stearothermophilus* presents the highest thermal resistance values reported (Mikolajcik & Rajkowski, 1980). However, thermal destruction of these thermophiles is not required by pertinent regulations as they are excluded from the definition of commercial sterility. Hence, thermal processes focus on the destruction of mesophilic spore-formers, such as *C. sporogenes*. While this organism is not pathogenic, it is capable of growing in LACF products under normal conditions. Thermal processes aimed to destroy this organism comfortably exceed the process necessary to destroy 12 logs of *C. botulinum* spores.

13.3.1.3.1.2 Non-Spore-Forming Bacteria

In the case of aseptically packaged, low-acid, shelf-stable products, the thermal treatments necessary to render the product commercially sterile are many orders of magnitude greater than needed to inactivate non-spore-forming bacteria. However, LACF products that are filed under the FDA's "water activity/formulation controlled" provisions (FDA Form 2541f), refrigerated extended shelf-life or high-acid products that are aseptically or hygienically packaged often receive processes targeting non-spore-forming organisms. This is possible because the presence of preservatives, intrinsic parameters of the food, or storage conditions prevent the proliferation of spore-forming organisms.

When compared to the spore-forming bacteria, non-spore-forming bacteria are much less heat resistant. Sörqvist (2003) completed a review of the scientific literature regarding the heat resistance of *Salmonella* spp., *E. coli*, and *Listeria* spp. in high pH liquids (pH: 6–8; e.g., defined laboratory media, liquid milk products, liquid egg products). In general, *L. monocytogenes* was the most resistant of the three pathogens with a $D_{60°C}$ of 1.45 min. The heat resistances of *Salmonella* spp. and *E. coli* were similar at $D_{60°C}$ of 0.40 and 0.65 min, respectively. Gill and Harris (1982) determined $D_{60°C}$ for *C. jejuni* in peptone yeast extract broth to be 0.26–0.95 min. Similar results were reported by Lahou et al. (2015); $D_{60°C}$ of 0.40 min was reported in Bolton broth.

Relative heat resistance within strains of STEC is dependent on the matrix and strain. Monu et al. (2015) investigated the heat resistance of five strains each of *E. coli* O157:H7 and non-O157 *E. coli* individually in phosphate-buffered saline and as multi-strain cocktails in spinach homogenate. In saline, heat resistance varied by strain; $D_{58°C}$ for *E. coli* O157 ranged from 0.53 to 0.80 min and ranged from 0.60 to 1.19 min for the non-O157 strains. Heat resistance of the *E. coli* O157:H7 and non-O157 *E. coli* cocktails in spinach homogenate was higher than in saline; $D_{58°C}$ was 1.95 and 2.01 min, respectively. Vasan et al. (2013) determined the heat resistance of several strains each of O157 and non-O157 *E. coli*

TABLE 13.5   HEAT RESISTANCE OF SPORE-FORMING PATHOGENIC BACTERIA.

Organism	Matrix	Temperature (°C)	*D*-Value (Min)	*z*-Value (°C)	Reference
Clostridium botulinum spores (non-proteolytic type F)	Crabmeat	77 79 82 85	9.50 3.55 1.16 0.53	NR[1]	(Lynt et al., 1982)
	Buffer pH 7.0	77 79 82	1.66–6.64 1.03–2.12 0.25–0.84	NR[1]	
Clostridium botulinum spores (proteolytic type F)	Crabmeat	102 104 107 110 121	5.07 4.02 2.14 1.35 0.18	NR[1]	
	Buffer pH 7.0	99 102 104 107 110 121	12.19–23.22 5.35–12.02 3.55–6.33 2.09–3.33 1.45–1.82 0.14–0.23	NR[1]	
Clostridium botulinum spores (type B)	Mushroom puree	110 116	0.49–0.99 0.12–0.39	NR[1]	(Odlaug et al., 1978)
	Buffer pH 7.0	110 116	1.02–1.38 0.11–0.35	NR[1]	
Clostridium botulinum (type A) strain A16037	Tomato juice pH 4.2	104 110 116	5.9–6.1 1.5–1.6 0.4	9.4	(Odlaug & Pflug, 1977)
	Buffer pH 7.0	104 110 116	16.2–17.6 4.3–4.5 1.3–1.4	9.9	
Clostridium botulinum strains 62A and 213B	Buffer pH 7.1	113	1.7–1.8	NR[1]	(Alderton et al., 1976)
Bacillus cereus spores (linear model)	Rice	91 95 99	62.4 39.7 12.8	12.4	(Juneja et al., 2020)
Bacillus cereus spores strain ATCC 7004	Buffer pH 7.0	100	0.45	7.5	(Mazas et al., 1998)
	Milk pH 6.6	100	0.27	7.9	
	Tomato pH 4.1	100	0.075	8.8	

(*Continued*)

TABLE 13.5 (CONTINUED)   HEAT RESISTANCE OF SPORE-FORMING PATHOGENIC BACTERIA.

Organism	Matrix	Temperature (°C)	D-Value (Min)	z-Value (°C)	Reference
Bacillus cereus spores strain ATCC 4342	Buffer pH 7.0	100	1.74	8.1	(Mazas et al., 1998)
	Rice pH 6.8	100	0.44	8.7	
	Tomato pH 4.1	100	0.21	9.9	
Bacillus cereus spores strain ATCC 9818	Buffer pH 7.0	103	4.76	9.9	(Mazas et al., 1998)
	Beans pH 6.1	103	2.25	10.2	
Bacillus cereus spores cocktail	Pork luncheon roll	85 90 95	29.5 10.1 2.0	8.6	(Byrne et al., 2006)
Bacillus cereus spores strain AV TZ415 (linear model)	Distilled water	85 90 95 100	16 3.9 0.94 0.22	8.1	(Fernández et al., 1999)
Bacillus cereus spores strain AV TZ421 (Linear model)	Distilled water	90 95 100 105	40 11 2.5 0.60	8.0	
Psychrotrophic *Bacillus cereus* spores strain NZRM 984	Beef slurry pH 6.5	70 80 90 100	2.3 1.4 1.0 0.42	42.4	(Evelyn & Silva, 2015)
	Skim milk pH 6.5	70 80 90 100	8.6 4.3 3.2 1.5	40.7	
Clostridium perfringens cocktail	Turkey slurry pH 6.0	99	17.7–23.2	NR[1]	(Juneja & Marmer, 1996)
Clostridium perfringens cocktail	Pork luncheon roll	90 95 100	30.6 9.7 1.9	8.3	(Byrne et al., 2006)

[1] NR: not reported.

TABLE 13.6   HEAT RESISTANCE OF SPORE-FORMING SPOILAGE BACTERIA.

Organism	Matrix	Temperature (°C)	D-Value (Min)	z-Value (°C)	Reference
Clostridium sporogenes	Buffer pH 7	110	12.8	10.3	(Ocio et al., 1994)
		115	5.2		
		118	3.3		
		121	0.95		
	Mushroom extract pH 6.7	110	8.5	10.2	
		115	1.7		
		118	1.2		
		121	0.67		
	Asparagus pH 6	110	20.4	9.3	(Silla Santos et al., 1992)
		115	5.7		
		118	2.1		
		121	1.5		
	Acidified asparagus pH 4.5	110	7.9	13.1	
		115	3.1		
		118	3.1		
		121	1.3		
	Buffered pea puree pH 7	110	25.3	12.2	(Cameron et al., 1980)
		113	15.0		
		116	8.9		
		118	5.2		
		121	3.1		
	Buffered pea puree pH 6	110	24.5	10.6	
		113	1.07		
		116	6.5		
		118	3.8		
		121	2.0		
	Buffered pea puree pH 5	110	16.7	9.2	
		113	7.3		
		116	3.7		
		118	2.0		
		121	1.0		
Bacillus coagulans	McIlvaine buffer pH 4	105	1.7	10.5	(Palop et al., 1999)
		108	0.88		
		111	0.49		
		117	0.13		
		123	0.029		
	McIlvaine buffer pH 5	105	2.3	10.6	
		111	0.50		
		125	0.029		

(Continued)

TABLE 13.6 (CONTINUED)   HEAT RESISTANCE OF SPORE-FORMING SPOILAGE BACTERIA.

Organism	Matrix	Temperature (°C)	D-Value (Min)	z-Value (°C)	Reference
	McIlvaine buffer pH 6	108	1.8	10.0	(Palop et al., 1999)
		111	0.81		
		117	0.22		
		126	0.028		
	McIlvaine buffer pH 7	107	4.2	8.9	
		111	1.7		
		112	1.2		
		117	0.30		
		120	0.16		
		123	0.064		
		126	0.031		
	Tomato pH 7	105	5.8	9.4	
		111	1.5		
		117	0.32		
		123	0.073		
		129	0.019		
	Asparagus pH 7	108	4.4	8.8	
		111	1.7		
		117	0.38		
		120	0.12		
		126	0.033		
Bacillus licheniformis	Enzyme matrix pH 8.0–8.6	90	2.5–9.9	9.3–17.4	(Luo et al., 2016)
	McIlvaine buffer pH 4	93	0.48	10.8	(Palop et al., 1996b)
		96	0.26		
		99	0.23		
		102	0.081		
		108	0.031		
	McIlvaine buffer pH 5	93	2.3	8.5	
		99	0.52		
		105	0.093		
		111	0.019		
	McIlvaine buffer pH 6	99	1.5	7.5	
		105	0.17		
		111	0.038		
	McIlvaine buffer pH 7	99	4.2	6.9	
		102	2.2		
		111	0.11		
	Tomato pH 7	99	3.6	7.0	
		105	0.60		
		111	0.12		

(*Continued*)

TABLE 13.6 (CONTINUED)   HEAT RESISTANCE OF SPORE-FORMING SPOILAGE BACTERIA.

Organism	Matrix	Temperature (°C)	D-Value (Min)	z-Value (°C)	Reference
	Asparagus pH 7	99 102 108 111	3.3 1.2 0.20 0.10	7.1	(Palop et al., 1996b)
Sporolactobacillus nakayamae	Buffer	70 75 80	10.5– 25.2 3.4–9.3 1.5–3.5	11.6– 11.9	(Bozkurt et al., 2016)
Bacillus stearothermophilus[1] strain TH24 (NCDO 1096)	Water	121	8.7	9.3–16.9	(David & Merson, 1990)
Bacillus stearothermophilus[1] (spore type NCA 1518)	Soy milk (soybeans extract)	126 128 129 131	0.36 0.24 0.16 0.088	8.3	(Shih et al., 1982)
Bacillus stearothermophilus[1]	Milk protein base formula	115 121 125	18.5 3.6 1.1	7.7	(Mikolajcik & Rajkowski, 1980)
	Soy protein base formula	115 121 125	26.1 3.6 1.3	7.6	
	Bi-distilled water pH 7	115 118 121 125	23.3 9.0 3.1 0.93	NR[2]	(Fernandez et al., 1994)
	Bi-distilled water pH 6.7	115 118 121 125	13.5 5.6 1.8 0.58	NR[2]	
	Mushroom extract neutral pH	121 130 137	1.59 0.16 0.016	9.3	(Rodrigo et al., 1997)
	Mushroom extract pH 6.2	121 130 137 140	1.13 0.14 0.020 0.0097	9.3	
	Mushroom extract pH 5.3	121 130 137 140	0.66 0.11 0.021 0.0076	10.0	

(Continued)

TABLE 13.6 (CONTINUED)   HEAT RESISTANCE OF SPORE-FORMING SPOILAGE BACTERIA.

Organism	Matrix	Temperature (°C)	D-Value (Min)	z-Value (°C)	Reference
	CaCl$_2$ solution (2%)	121	5.5	8.0	(Rodrigo et
		130	0.31		al., 1997)
		140	0.019		

[1] *Bacillus stearothermophilus* was reclassified as *Geobacillus stearothermophilus* after the reference article was published.
[2] NR: not reported.

(O157, O26, O45, O103, O111, O121, and O145) in brain heart infusion broth. Very few statistical differences were found between strains. However, $D_{58°C}$ of the *E. coli* O157:H7 (ranged from 0.95 to 1.42 min) were typically higher than those of the non-O157 strains (most were ~0.65–0.90 min) and the strains were generally ranked by relative heat resistance as follows:

O157 > O103 > O111 > O26 > O121 > O45 > O145

Therefore, though influenced by the matrix and strain, the heat resistances of *E. coli* O157:H7 and non-O157 *E. coli* appear to be similar (Table 13.7).

Literature suggests that, in a low-acid environment, the Gram-positive pathogen *L. monocytogenes* is more heat resistant than the Gram-negative ones. Therefore, the non-spore-forming pathogens may be ranked by relative heat resistance as follows:

L. monocytogenes > (STEC, *Salmonella* spp., and *C. jejuni*)

Similar to the pathogenic non-spore-forming bacteria, Gram-positive non-spore-forming spoilage bacteria are generally more heat resistant than the Gram-negative ones. Therefore, *D*- and *z*-values for Gram-positive LAB only are presented in Table 13.8 as this group would be expected to be more heat resistant than the Gram-negative AAB and coliforms.

13.3.1.3.1.3 Spoilage Fungi
Of the yeasts commonly associated with the spoilage of food products, *S. cerevisiae* typically is the most heat resistant (Put et al., 1976; Shearer et al., 2002). Similar to yeast, most species of spoilage mold (with the exception of ascospore-producing heat-resistant species) demonstrate very little heat resistance and are comparatively less heat resistant than *S. cerevisiae* (Shearer et al., 2002). The destruction of heat-resistant molds, however, requires a comparatively severe thermal process (Table 13.9).

13.3.1.3.2 Impact of Specific Product on the Thermal Treatment
The intrinsic characteristics of the product (e.g., pH, water activity, viscosity, homogeneity, presence of preservatives, or other inhibitory compounds) may influence the ability of the thermal process to produce a commercially sterile product. Greater heat resistance may be observed in matrices with a higher fat content, whereas naturally occurring bioactive compounds in the product (e.g., antioxidants in fruits and vegetables, lysozyme in eggs and milk, nitrates in spinach, beets, radishes, and celery) may decrease heat resistance. The interactions of the intrinsic parameters of the food matrix are complex and can be unpredictable. Therefore, evaluation of the heat resistance of pertinent microorganisms in the specific product may be required when designing a thermal process.

TABLE 13.7   HEAT RESISTANCE OF NON-SPORE-FORMING PATHOGENS IN LOW-ACID MATRICES.

Organism	Matrix	Temperature (°C)	D-Value (Min)	z-Value (°C)	Reference
Listeria monocytogenes	Chicken gravy	50	119–195	5.2–6.1	(Huang et al., 1992)
		55	39.4–79.1		
		60	3.13–7.07		
		65	0.19–0.48		
	Liquid egg yolk pH 6.3	60	1.34	6.1	(Schuman & Sheldon, 1997)
		61	0.89		
		62	0.58		
	Raw skim milk	52	43.6	5.8	(Bradshaw et al., 1987)
		63	0.47		
		69	0.06		
	Raw whole milk	52	28.1	6.3	
		63	0.33		
		69	0.05		
	Raw cream	52	28.5	6.8	
		63	0.51		
		69	0.10		
	Non-homogenized, whole bovine milk	57	15.2	6.7	(Engstrom et al., 2021)
		60	2.28		
		63	0.64		
		66	0.33		
	Brain heart infusion broth	55	9.2–30.2	4.4–5.7	(Aryani et al., 2015)
		60	0.58–4.1		
		65	0.075–0.57		
Shiga toxin-producing *Escherichia coli* (STEC)	Non-homogenized, whole bovine milk	60	1.04	6.0	(Engstrom et al., 2021)
		63	0.27		
		67	0.12		
E. coli O121	Muffin batter	60	42.0	5.0	(Michael et al., 2020)
		65	7.5		
		70	0.4		
Salmonella	Liquid whole egg	54	5.70	4.1	(Jin et al., 2008)
		56	0.82		
		58	0.27		
		60	0.17		
	Liquid egg white	54	1.51	4.0	
		56	0.42		
		58	0.19		
	Liquid egg yolk	60	0.28	4.3	(Schuman & Sheldon, 1997)
		61.1	0.16		
		62.2	0.087		

(Continued)

TABLE 13.7 (CONTINUED) HEAT RESISTANCE OF NON-SPORE-FORMING PATHOGENS IN LOW-ACID MATRICES.

Organism	Matrix	Temperature (°C)	D-Value (Min)	z-Value (°C)	Reference
Salmonella	Muffin batter	60	38.4	5.2	(Michael et al.,
		65	7.2		2020)
		70	0.5		
Campylobacter jejuni	Chicken	56	0.51–0.67	8.7–10	(Al-Sakkaf, 2021)
	Bolton broth	60	0.30–0.54	NR[1]	(Lahou et al., 2015)
	Peptone yeast extract broth	60	0.26–0.95	NR[1]	(Gill & Harris, 1982)
Campylobacter jejuni (composite of five strains)	Ground chicken	51	9.27	NR[1]	(Blankenship & Craven, 1982)
		53	4.89		
		55	2.25		
		57	0.98		

[1] NR: not reported.

The composition of the matrix may influence heat resistance of microorganisms in unpredictable ways. Moussa-Boudjemmaa et al. (2006) found that the heat resistance ($D_{96°C}$) of B. cereus in phosphate buffer and carrot extract, both adjusted to a pH of 5.2 with citric acid, were significantly different at 3.7 and 6.0 min, respectively. Humphrey et al. (1990) found substantial differences in the heat resistance of S. enterica ser. Enteritidis in three types of raw egg products; $D_{55°C}$ in homogenized egg, egg yolk only, and albumen only were 6.4, 21.0, and 1.5 min, respectively. Thus, the complex nature of food matrices must be carefully considered when designing a thermal process (Table 13.10).

The heat resistance of microorganisms generally decreases as pH decreases. Palop et al. (1996b) and (1999) investigated the influence of pH on the heat resistance of B. licheniformis and B. coagulans spores, respectively, in McIlvaine buffer. At a pH of 7, $D_{99°C}$ for B. licheniformis was 4.2 min and, at a pH of 5, $D_{99°C}$ was approximately eight times lower at 0.52 min. Similarly for B. coagulans, $D_{111°C}$ at a pH of 7 and 5 were 1.7 and 0.5 min, respectively. The heat resistance of S. enterica, E. coli O157:H7, and L. monocytogenes in Tryptic Soy Broth (TSB) at various pH levels was determined by Mazzotta (2001) and again heat resistance decreased with decreasing pH. The magnitude of the impact of pH was greatest for E. coli O157:H7 and L. monocytogenes; for both pathogens, $D_{60°C}$ at a pH of 7 was ~2.25 and ~0.75 min at a pH of 4. The difference in heat resistance was less dramatic for S. enterica; $D_{60°C}$ at pH levels of 7 and 4 were ~1.5 and ~0.5 min, respectively.

In addition to pH, the type and mixture of organic acids present may also have an impact on heat resistance. Palop et al. (1996a) evaluated the heat resistance of B. coagulans in homogenized tomato and homogenized asparagus adjusted to a pH of 4 using various organic acids. When adjusted with acetic and citric acid, $D_{111°C}$ in the homogenized tomato was 0.12 and 0.27 min, respectively, and was 0.084 and 0.45 min in the homogenized asparagus. In contrast, again considering the work of Moussa-Boudjemaa et al. (2006), acidulant type did not have a significant effect on the heat resistance of B. cereus in carrot extract adjusted to a pH of 4.5; $D_{96°C}$ for acetic and citric acid were 3.82 and 3.70 min, respectively.

TABLE 13.8   HEAT RESISTANCE OF NON-SPORE-FORMING SPOILAGE ORGANISMS IN VARIOUS MATRICES.

Organism	Matrix	Temperature	D-Value (Min)	z-Value (°C)	Reference
Lactobacillus sake	Diluent pH 6.2	57	0.88	NR[1]	(Franz & and von Holy, 1996)
		60	0.66		
		63	0.54		
Leuconostoc mesenteroides	Diluent pH 6.2	57	0.58	NR	
		60	0.52		
		63	0.34		
Lactobacillus curvatus	Diluent pH 6.2	57	0.38	NR	
		60	0.26		
		63	0.24		
Lactobacillus plantarum	Reconstituted skim milk	54	3.14	NR	(Jordan & Cogan, 1999)
		56	1.35		
	Apple juice pH 3.4	60	0.36	15.9	(Tajchakavit et al., 1998)
		70	0.14		
		80	0.02		
	Liquid food model pH 7	52	1.27	8.9	(Augusto et al., 2011)
		55	0.61		
		58	0.27		
		61	0.12		
Lactobacillus paracasei	Reconstituted skim milk	60	22.5	NR	(Jordan & Cogan, 1999)
		65	3.71		
		68	0.32		

[1] NR: not reported.

In general, the heat resistance of microorganisms increases as the water activity of the matrix decreases. However, the impact of water activity is dependent on the organism and the solute affecting water activity. Although LACF products have a relatively high water activity by definition (i.e., >0.85), heat resistance at the lower end of the spectrum may be greater than when closer to 1.00. The impact of the water activity of the matrix on the heat resistance of *B. cereus* spores was investigated by Coroller et al. (2001). In pure water (pH 7), $D_{95°C}$ for *B. cereus* spores was ~4 min. When water activity was adjusted to 0.9 with glycerol, glucose, and sucrose, $D_{95°C}$ was approximately 11.2, 10.5, and ~27 min, respectively. Sumner et al. (1991) investigated the influence of water activity on the heat resistance of *S. enterica* and *L. monocytogenes* in a sucrose solution. When the water activity of the solution decreased from 0.98 to ~0.90, $D_{65.6°C}$ for *S. enterica* and *L. monocytogenes* increased from 0.29 to 4.8 min and from 0.36 to 3.8 min, respectively. Zhang et al. (2018) defined the heat resistance of *Aspergillus flavus* in peanut kernels adjusted to various water activity levels. Similar to the trend observed for bacteria, the heat resistance of *A. flavus* also increased as water activity decreased. $D_{62°C}$ was 3.34 and 6.18 min at water activities of 0.921 and 0.846, respectively.

13.3.2 Chemical Sterilants

Chemical sterilants are used widely in aseptic filling applications for the purpose of sterilizing packaging materials and the filler aseptic zone. Hydrogen peroxide and peracetic acid

TABLE 13.9   HEAT RESISTANCE OF SPOILAGE FUNGI, INCLUDING HEAT-RESISTANT MOLDS.

Organism	Matrix	Temperature	D-Value (Min)	z-Value (°C)	Reference
Saccharomyces cerevisiae	Buffer pH 7	58 60 62 64	1.5–3.0 0.56–1.3 0.03–0.24 0.01–0.08	3.3–3.9	(Garza et al., 1994)
	Peach puree pH 3.9	60 62	0.18–0.53 0.03–0.13	3.3–4.0	
	Citrate buffer pH 3–4	60	1.3–2.8	3.5–4.0	(Shearer et al., 2002)
	Apple juice pH 3.4	50 60 70	0.96 0.17 0.032	13.4	(Tajchakavit et al., 1998)
Penicillium citrinum	Citrate buffer pH 3–4	60	0.009–0.016	3.8–4.6	(Shearer et al., 2002)
Zygosaccharomyces rouxii	Citrate buffer pH 3.5–4.0	60	0.008–0.039	2.1–3.3	
Penicillium roquefortii	Citrate buffer pH 3–4	60	0.20–0.29	3.6–4.0	
Aspergillus niger	Citrate buffer pH 3–4	60	0.38–0.45	3.3–3.7	
Byssochlamys nivea[1]	Malt extract broth pH 6	85 90 95	15.7 3.35 1.14	7.1	(Samapundo et al., 2018)
	Cream 10% fat	84 88 92	0.60–0.75 0.13–0.15 0.026–0.032	6–7	(Engel & Teuber, 1991)
Byssochlamys fulva[1]	Grape juice	88	11.3	NR[2]	(Kotzekidou, 2014)
Byssochlamys spectabilis[1]	Buffer pH 6.8	85	47–75	NR[2]	(Kotzekidou, 2014)
Aspergillus hiratsukae[1]	Glucose solution	87 90 95	3.7–7.7 1.5–1.8 0.40–0.30	5.7–8.3	(Berni et al., 2017)
Aspergillus neoglaber[1]	Glucose solution	87 90 95	13.5 3.5 0.30	4.8	
Aspergillus thermomutatus[1]	Glucose solution	87 90 95	4.9 1.9 0.30	6.6	

(Continued)

TABLE 13.9 (CONTINUED)   HEAT RESISTANCE OF SPOILAGE FUNGI, INCLUDING HEAT-RESISTANT MOLDS.

Organism	Matrix	Temperature	D-Value (Min)	z-Value (°C)	Reference
Talaromyces trachyspermus[1]	Blueberry and grape juice	75 78 80 82	90.9 20.8 12.4 2.6	4.7	(Tranquillini et al., 2017)
	Buffered glucose solution	75 78 80 82	50.0 13.2 5.1 1.6	4.7	
Talaromyces bacillisporus[1]	Blueberry and grape juice	82 85 88 91	44.4 11.9 2.7 1.2	5.6	
	Buffered glucose solution	82 85 88 91	60.9 15.5 4.1 1.2	5.2	
Eupenicillium javanicum[1]	Pineapple juice	80 85 90	19.8–32.0 5.0–21.3 1.5–4.8	8.6–12.1	(Evelyn et al., 2020)
Talaromyces flavus[1]	Strawberry filling pH 3.5	85 88 91	47.1–52.0 3.5–14.3 3.9–11.7	5.2–5.3	(Beuchat, 1986)
Neosartorya fischeri[1]	Strawberry filling pH 3.5	85 88 91	19.4–45.0 4.4–11.2 <2.0	3.2–5.0	

[1] Heat-resistant mold.
[2] NR: not reported.

(PAA)–based sterilants are widely used for this purpose. In general, spores of *C. botulinum* are considered to be the pathogen most resistant to the methods used sterilize packaging materials. A notable exception is the resistance of *B. cereus* to PAA-based sanitizers (Blakistone et al., 1999). Therefore, as new sterilization methods become available, the selection of an appropriate target organism must be carefully assessed.

As with the product, the packaging materials of an aseptically processed food must also be commercially sterile (21CFR113.3, 2011). Many factors should be considered when selecting the parameters for the package and equipment sterilization process:

- The relative resistance of pertinent pathogens and spoilage organisms
- Format of the chemical sterilant (e.g., vapor, fog, liquid)
- Characteristics of the packaging/surface to be sterilized
- The ability of the packaging or equipment material to tolerate the sterilization process

TABLE 13.10 HEAT RESISTANCE OF PROTEOLYTIC *C. BOTULINUM* AND *C. SPOROGENES* AT A TEMPERATURE OF 121°C (VALUES FROM BROWN ET AL. (2012)).

Organism	Matrix	*D*-Value (min)	*z*-Value (°C)
Clostridium botulinum strain 62A	Phosphate buffer pH 7.0	0.080–0.31	8.1–11.6
Clostridium botulinum strain 213B	Phosphate buffer pH 7.0	0.055–1.43	8.3–11.0
Clostridium botulinum Various strains	Phosphate buffer pH 7.0	0.07–0.364	9.0–14.1
Clostridium botulinum strains 62A and 213B	Whole milk pH 6.3	0.07	7.9–8.9
Clostridium botulinum Various [B]	Mushroom puree pH 6.4	0.05	NR[1]
Clostridium botulinum strain 62A	Peas puree	0.09	8.3
	Rice puree pH 7.0	0.12	8.6
	Spaghetti puree pH 7.0	0.11	8.3
	Distilled water	0.051	8.5
Clostridium sporogenes PA 3679	Phosphate buffer pH 7.0	0.19–3.50	9.5–14.0
	Artichoke puree pH 5.2	0.36	8.3
	Asparagus puree pH 5.9–6.7	0.70–1.80	8.9–9.3
	Green beans puree pH 5.2 6.1	0.54–2.00	10.0–14.8
	Beans, snap brine pH 5.2	0.79	12.8
	Beets puree pH 6.2	0.69	10.1
	White and yellow corn puree pH 6.8–7.1	0.91–1.72	9.4–11.7
	Cream pH 6.3	1.09	12.8
	Custard pudding	0.33	9.6
	Milk skim/whole pH 6.2–6.5	0.59–1.20	10.8–14.1
	Mushroom extract pH 6.7	1.50	9.6
	Peas puree pH 5.0–7.0	1.0–3.1	9.0–12.2

(Continued)

TABLE 13.10 (CONTINUED)   HEAT RESISTANCE OF PROTEOLYTIC *C. BOTULINUM* AND *C. SPOROGENES* AT A TEMPERATURE OF 121°C (VALUES FROM BROWN ET AL. (2012)).

Organism	Matrix	*D*-Value (min)	*z*-Value (°C)
Clostridium sporogenes PA 3679	Mash potato pH 6.3	0.61	10.0
	Pumpkin	0.40–1.50	9.4–11.1
	Spinach puree pH 6.2	2.3	12.7
	Sweet potato puree pH 5.6	0.7	8.9
	Distilled water	0.80–1.20	9.8–10.6
	White sauce	1.23	9.2

[1] NR: not reported.

- The definition of an acceptable performance criterion (i.e., selection of target organism, surrogate organism, and log reduction target)
- The suitability of the package–sterilant interaction from a toxicological and chemical stability standpoint

Toledo et al. (1973) studied the antimicrobial efficacy of various concentrations of H_2O_2 at different temperatures and determined the *D*-values of selected spore-forming bacteria and *Staphylococcus aureus* at 24°C and an H_2O_2 concentration of 26%. The results of the study indicated that the relative resistance of the tested organisms, ranked from most resistant to least resistant, is as follows:

B. subtilis SA 22 > *B. subtilis* var. *globigii* > *B. coagulans* > *Bacillus stearothermophilus* > *C. sporogenes* PA 3679 > *S. aureus*

Blakistone et al. (1999) evaluated the efficacy of a PAA-based sanitizer (Oxonia Active at 2% concentration) in an aqueous phase at 40°C on the inactivation of pathogenic and spoilage spore-forming organisms. The results of the study indicated that resistance varied between the tested organisms and that relative resistance to the PAA-based sanitizer (ranked from greatest to least) is as follows:

B. cereus > *B. subtilis* A > *B. stearothermophilus* > *B. subtilis* var. *globigii* > *B. coagulans* > *C. sporogenes* PA 3679 > *C. butyricum* > *C. botulinum*

The results of these two studies demonstrate that the relative resistance of organisms to chemical sterilants may be influenced by a number of factors, including sanitizer type and temperature. For example, *B. stearothermophilus* is more resistant to PAA-based sanitizer than *B. coagulans* while it is less resistant to H_2O_2 than *B. coagulans*. Therefore, the use of chemical sterilants must be evaluated for each application in order to ensure the selection of appropriate parameters.

The inactivation kinetics of sterilants such as a PAA-based sanitizer or H_2O_2 may vary when applied as a fog or a solution. Hayrapetyan et al. (2020) studied the effectiveness of 0.06% PAA and 12% H_2O_2 applied as a fog or as a solution on the inactivation of *G. stearothermophilus* on

TABLE 13.11   SPORICIDAL EFFECT OF PERACETIC ACID (PAA) AND H_2O_2 ON *GEOBACILLUS STEAROTHERMOPHILUS* DSM5934 (ATCC 7953) ON STEEL SURFACE APPLIED AS A FOG OR AS A LIQUID (HAYRAPETYAN ET AL., 2020).

Disinfectant	Contact Time (Min) for 4-Log Reduction
0.06% Solution PAA fog	8.7–8.8
0.06% Solution PAA solution	3.3–3.4
12% H_2O_2 fog	67.0–74.1
12% H_2O_2 solution	46.8–47.7

stainless steel surfaces. The results of the study indicated that the time required to achieve a 4-log reduction of *G. stearothermophilus* was longer when the sterilants were applied as a fog than as a solution because a longer shoulder in the inactivation curve was observed for the fog applications (Table 13.11).

13.4 CAUSES OF MICROBIOLOGICAL FAILURE IN ASEPTIC PROCESSING AND PACKAGING

Aseptic processing and packaging is an effective and robust technology where failures due to microbiological contamination are rare. Nevertheless, when failures do occur, the high complexity of aseptic lines makes it extremely difficult to diagnose a root cause. The authors estimate that in aseptic processing, more than half of the spoilage incident investigations fail to identify the source of failure with a high degree of certainty. In most cases, spoilage is sporadic in nature and the investigation discovers multiple minor deficiencies that are promptly corrected. Following the investigation and correction of all identified deficiencies, the spoilage problem often disappears, at least temporarily. This situation leaves several questions unanswered: Was one of the deficiencies the main cause of spoilage? Was the combined effect of the various deficiencies the cause of the problem? Is the root cause of the problem extremely sporadic and return to normal is independent of the corrective actions taken? In a real-life example, a sporadic spoilage problem affecting a form-fill-seal aseptic line at a dairy processing plant was initially tracked down to a malfunctioning valve passed the filler at the end of the line just prior to the drain outlet. The malfunctioning valve rendered the clean-in-place (CIP) process ineffective at this point and product debris was found on the aseptic side of the valve. Microorganisms matching the spoilage incident were found on the non-sterile side of the valve. The situation was corrected and yet the problem reappeared again a few weeks later. Eventually it was discovered that a stainless steel floater on one of the aseptic fillers had been exposed to manual cleaning followed by sanitizing using an inappropriate sanitizing solution at high concentration. This action caused pitting corrosion creating microchannels that allowed small amounts of product to enter the floater and creating a niche for contaminants.

In rare and most difficult cases, the problem repeats itself at occasional intervals creating a chronic condition that has resulted in complete decommission of aseptic lines and even plant closures with the unfortunate accompanying financial and employment losses. It is for these reasons that strong emphasis should be placed in prevention of these unfortunate events rather than on the difficult and often fruitless effort to diagnose a specific cause. This section will touch on some of the most common causes of spoilage identified during more than seven decades of combined experience by these authors.

13.4.1 Hermetic seal failures

Perhaps the most common cause of occasional spoiled containers in aseptic production is the failure of the hermetic seal. For shelf-stable, low-acid products, FDA regulations require detailed closure inspections at "intervals of sufficient frequency to ensure proper closing machine performance and consistently reliable hermetic seal production" (21CFR113.60). In addition to the frequency at which these measurements must be conducted, the processor must select a test method adequate for the given package closing technology. Most "form-fill-seal" aseptic filling machines rely on induction sealing and the efficacy of the seal is confirmed using well-established and detailed teardown and dye penetration methods as recommended by the manufacturer. In other cases, such as pressure seals (used in "bag-in-box, BIB" fillers and screw-capped plastic bottles) methods such as vacuum immersion, application torque, removal torque, pull-up measurements, etc. are commonly used. Regardless of the method of the post-filling hermetic seal checks, the most effective preventive measure for this type of contamination is proper and timely preventive maintenance of the closing machine.

Microbial contamination because of hermetic seal failure is often characterized by the presence of a heterogenic variety of microorganisms in the food product. This includes both microorganisms capable of generating heat-resistant spores (most commonly of the genus *Bacillus*) as well as non-spore-forming organisms such as yeast, lactic, or acetic acid-producing bacteria, and/or those commonly categorized as coliforms. Additionally, molds are often detected in these situations, as hermetic seal failure allows for the oxygen-rich atmosphere necessary for prolific growth by these organisms.

13.4.2 Failure Caused by Poor Valve Design or Malfunction

Only valves designed for the specific purpose should be used in aseptic lines. Such valves must be of a design that allows for proper cleaning (CIP) and sterilization (SIP). Various recognized organizations provide standards for the design of such valves, i.e., 3-A Sanitary Standards (3-A Sanitary Standards Inc., n.d.) and European Hygienic Equipment Design Group—EHEDG (European Hygienic Engineering & Design Group (EHEDG), n.d.). Aseptic valves must provide protection against penetration of outside contaminants into the sterile product or aseptic environment. In addition to primary barriers such as effective valve seat seals, aseptic valves should provide secondary or redundant protection. Secondary protection may consist of steam barriers, sterile condensate barriers, chemical barriers, bellows, etc. Furthermore, aseptic valves must provide mechanisms for detection of seal integrity failures that result in product leaks or steam barrier temperature drop. These detection systems may consist of temperature sensors capable of detecting steam barrier failures or "weep-holes" to allow visual detection of product leaks. In addition to design or secondary protection issues, improper valve logic may also cause problems resulting in microbial contamination. Valves must be in their proper position during the various cleaning, sterilizing, and production modes. During cleaning and sterilization, valves must cycle properly to allow exposure of all pertinent surfaces to the cleaning or sterilizing media. In a case investigated by these authors, spoilage was caused by a malfunctioning valve between the pasteurizer and buffer tank on the filler. The valve was not moving appropriately due to a broken pneumatic line, and as a result the CIP solution and sterilant could not reach the product lines or the affected tank.

From a microbiological standpoint, failures in valve design or function can result in unpredictable spoilage microflora. Failures related to low-steam barrier temperature or the cleaning or sterilization of aseptic valves may present spoilage by a homogeneous population of a spore-forming organism. This is because faulty steam barriers or inadequate cleaning or sterilization may still provide sufficient "selective pressure" whereby heat-labile cells are destroyed but heat-resistant spores are selectively allowed to survive and contaminate the product. On the other hand, a pinhole on the protective bellow of a reciprocating shaft may result in contamination from a variety of heat-labile microorganisms. As is the case with most causes of microbial contamination described in this section, the most effective measure in avoiding valve malfunction issues is an effective preventive maintenance program.

13.4.3 Failure Caused by Poor Design or Inappropriate SIP/CIP Practices for Product Lines

Sterile product lines must be designed, constructed, and installed in a manner that allows for proper cleaning and sterilization. Dead ends, poor welding, and improper joints or couplings are problems that will interfere with proper cleaning and sterilization. Inappropriate cleaning agents or higher than necessary concentrations of CIP or sanitizing solutions may result in pitting and subsequent production of microchannels or pinholes in the cooling plates (or tubes) of heat exchangers or the walls of jacketed aseptic surge tanks. The same is true regarding the use of chlorine or other oxidizing agents in the cooling media of heat exchangers. Once the integrity of the stainless steel surface is compromised, asepsis may not be maintained over long production runs even when the system is designed to maintain the sterile product at a pressure higher than the cooling media as discovered by these authors in multiple spoilage investigations. In one case, it was discovered that pinholes were present in the plates of a heat exchanger that had not been checked for integrity for 7 years! In this case, in spite of records that demonstrated that the product versus cooling media differential pressure was maintained appropriately, sporadic spoilage occurred periodically until the plates were replaced. In another case, pinholes in a tubular cooler were caused by heavy chlorination of tower water with similar spoilage results. Similarly, deterioration of gasket materials in plate heat exchangers may result in exposure of sterile product to non-sterile cooling media.

Depending on the temperature of the product or the concentration of sanitizing agents in cooling media, these types of failures may result in the presence of a pure culture of spore-forming, heat-resistant organisms or a mixed microflora of both heat-resistant and heat-labile microorganisms. Once again, proper design of lines and cleaning procedures, including composition of chemical compounds and preventive maintenance, play a critical role in the prevention of these problems. Periodical (annual or biannual) check for pinholes on heat exchangers via pressure testing is highly recommended as part of a sound preventive maintenance program.

13.4.4 Failure of Sterile Air Overpressure or Unidirectional Sterile Airflow to Protect Filler

These authors believe that a common problem resulting in sporadic spoilage in aseptic fillers is the failure to protect the filling area from penetration of outside contaminants. This problem may be tracked to various root causes as follows: The efficacy of the critical factors used to protect the integrity of the filling area during production (e.g., aseptic zone overpressure, airflow, machine speed, etc.) was not properly validated upon installation; The filler environment presents varied atmospheric conditions such as room pressure drops or strong air drafts (sometimes due to opening and closing of room doors) that result in turbulent air currents that overpower the filler's protection; Improper installation, defective gasket material, or warping of windows on aseptic fillers may create unexpected breaches to the integrity of the aseptic zone. In one filler evaluated by these authors, a plastic window in the aseptic zone of a "form-fill-seal" filler was located in near proximity to the package sealing heads. The heat generated by the sealing heads caused the window's plastic material to warp in a manner that broke the window seal and allowed penetration of outside contaminants. The problem was solved by replacing the existing windows with windows made with a more heat-tolerant plastic polymer. Spoilage caused by this type of failure often includes the presence of mold spores as they are a common airborne contaminant. A good tool in preventing and diagnosing spoilage problems caused by this type of failure is conducting periodic particles exclusion (smoke) tests. These tests consist of appropriately inserting a line connecting to a particle counter at various locations in the filling area and using a particle generator to create an atmosphere of defined size particles around the filler. An effective functioning aseptic barrier inside the filler will prevent the ingress of these particles. An ineffective barrier will allow particle penetration in large numbers that will then be detected by the particle counter.

13.4.5 Failure Due to Plugged Sterilization Nozzles in Aseptic Filler

The aseptic zone in form-fill-seal, linear bottle, and rotary bottle fillers is most often sterilized by spraying a chemical sterilant (i.e., hydrogen peroxide) onto the aseptic zone of the filler. During the preproduction sterilization cycle (presterilization), the sterilant is applied through atomizing nozzles strategically placed throughout the aseptic zone. These nozzles are susceptible to partial or complete plugging, which can interfere severely with the efficacy of the sterilization cycles. Plugging may be caused by inappropriate passivation of the lines or contamination with foreign substances after proper passivation. Commonly, this type of failure is caused by the use of the wrong grade of sterilant for the atomizing function. Hydrogen peroxide preparations intended for use in the hot hydrogen peroxide bath sterilization of packaging materials are unsuitable for use in spraying nozzles. These hydrogen peroxide preparations often contain stabilizers that can precipitate out of solution if used in spraying or atomizing applications, creating solid deposits on the lines and interfering with proper sterilant flow. Therefore, careful attention must be applied to the selection of sterilant preparations that are appropriate for the intended use. In addition, proper function of sterilizing nozzles must be assessed at frequent intervals during the sterilization cycles.

13.4.6 Failures Caused by the Presence of Highly Heat-Resistant Microorganisms

The high temperatures used in UHT processing of low-acid, shelf-stable products packaged aseptically (e.g., $\geq 138°C$ for 6–8 seconds) make survival of common mesophilic spoilage organisms an extremely rare occurrence. However, the spores of thermophilic organisms such as *G. stearothermophilus*, if present in sufficient numbers, are capable of overwhelming standard UHT processes. *D*- and *z*-values for these organisms are presented in Table 13.6 in this chapter. In the event that thermophilic spores are present in the finished product, germination will only occur if the product is exposed to abusively high temperatures (>40°C). Spoilage by these organisms is normally characterized by the production of acid (drop in pH) without swelling of the container (no gas production).

As stated above, most of the failures associated with highly heat-resistant microorganisms in low-acid aseptic products are caused by thermophilic spore-forming microorganisms. A notable but rare exception is a recently characterized mesophilic spore-former. *Bacillus sporothermodurans* is a mesophilic organism capable of producing highly heat-resistant spores that can survive UHT processes (Pettersson et al., 1996). This organism grows only to relatively low numbers in milk (10^5/ml after 5 days at 30°C) and it causes no noticeable signs of spoilage. Furthermore, it grows poorly or not at all in some non-selective microbiological media (i.e., nutrient agar); hence, it is likely missed in routine microbiological testing consisting of plating in non-selective media. On the other hand, in recent years, it has become more common for aseptic processors to conduct novel testing procedures such as flow-cytometry testing. As result, previously missed contaminants that grow poorly in the product and in recovery media are now more frequently detected.

High-acid products that are aseptically packaged, such as tomato products, are processed at much lower temperatures and faced a wider variety of potential issues as a result of heat-resistant organisms, acid-tolerant organisms such as heat-resistant molds (HRM), *Alicyclobacillus* spp., *B. coagulans*, *C. pasteurianum*, and *C. thermosaccharolyticum*.

Exceptionally highly heat-resistant organisms, such as those described above for both high-acid and low-acid shelf-stable products, are relatively rare, but it is important to note that the thermal processes used in the respective applications are insufficient to destroy them if present in the raw ingredients in large numbers. In some cases, the product intended use requires that the thermal processes be increased to provide destruction of highly heat-resistant thermophilic organisms. For example, beverages intended for hot-vending machines in Japan commonly receive processes at >145°C for 6–8 seconds.

In most other cases, prevention of this type of failure depends on selection of high-quality ingredients (that are periodically evaluated for the presence of highly heat-resistant organisms) and proper handling of the finished product during storage, distribution, and retail. It is also worth mentioning that in recent years, there has been a dramatic increase in the use of new and exotic ingredients and formulations. These ingredients and formulations may change the composition and numbers of the microbial burden introduced into the UHT processes with yet unknown impact.

13.4.7 Failure Due to Poor Hydration of Critical Ingredients

In the opinion of these authors, this is likely one of the most common causes of spoilage in aseptic products. Some ingredients widely used in the manufacturing of low-acid, shelf-stable products are very difficult to properly hydrate. The best-known example of this type of ingredient is cocoa. Chocolate-flavored products have been implicated in spoilage issues at a disproportional rate as compared to other flavors produced on the same lines. It is widely estimated that a large proportion of these incidents are related to the difficulty of hydrating this ingredient. Improperly hydrated ingredients result in the presence of dried portions of product commonly referred to in the industry as "fish-eyes." Heat-resistant spores trapped in these dried particles may survive common UHT processes and cause spoilage. The dried heat resistance of these organisms is dramatically higher than that under wet conditions. For example, the D-value for *C. sporogenes* at 121°C in neutral pH phosphate buffer (wet condition) is approximately 1.0 min, while under dry conditions (dried spores on a metal surface) the D-value increased to approximately 7 min (Collier & Townsend, 1956).

For years, a common industry standard has been to hydrate a slurry consisting of the cocoa ingredient, water, and sugar prior to combining other ingredients. While the specific hydration procedures used in the industry vary to a degree (and it is often considered proprietary), traditionally most processors meet the minimum conditions recommended by the National Food Processors Association (NFPA) decades ago. These hydration conditions consist of heating the cocoa slurry to a minimum temperature of 88°C for 30–45 min with high shear circulation. Unfortunately, in some instances, processors have reduced these conditions to an extreme level with disastrous consequences. Furthermore, empirical evidence captured by these authors appears to indicate that the addition of some exotic plant-based ingredients acts synergistically with cocoa in protecting spores from the lethal effects of the UHT process. Spoilage caused by this type of failure is characterized by the presence of spore-forming organisms, often a single species. An event under investigation at the time of this writing appears to be attributable to hydration issues and the spoiled product displays a pure culture of a single *Bacillus* spp. In this case, an exotic plant-based ingredient was identified as occasionally carrying high spore levels (as high as 10^5 spores/g aerobic mesophilic spores). This ingredient appeared to act synergistically with the chocolate ingredient to increase the probability of spoilage. Prevention of this type of failure should focus on providing the necessary time, temperature, and agitation to the hydration procedure. This is particularly important when using novel ingredients or ingredients that are particularly difficult to hydrate or that may carry a high spore load.

13.4.8 Other Potential Causes of Microbiological Failures and General Observations

In recent years, these authors have seen an increase in the number of spoilage incidents related to aseptically packaged products that our team has investigated. We believe that this increment is due to the synergistic effects of various relatively new practices that collectively result in the types of spoilage previously discussed in this section. Among these practices, we can identify the following:

Increasingly complex installations:

Ideally, an aseptic processing and packaging operation should be a simple linear installation with one UHT processing system feeding a single aseptic surge tank and aseptic filler. However, in an effort to increase productivity, more processors have installed highly complex systems that allow increased flexibility in terms of equipment utilization. For example, a processor may have multiple UHT processing systems, each processor may be able to feed any one of multiple aseptic surge tanks and each of these tanks may be capable of feeding multiple aseptic fillers. While this type of configuration is very attractive in terms of the potential utilization of the lines, the highly complex valve clusters and extensive line lengths provide increased opportunities for potential failures. This complexity increases the possibility of cross-contamination across various lines and makes spoilage investigations much more difficult to complete successfully.

Extended runs:

Increasing productivity is a reasonable and worthwhile goal of every food processing operation. In aseptic processing, the requirements for cleaning, sanitation, and sterilization of numerous complex types of machinery require extensive downtime (often more than 8 hours) following a production run. Hence, extending production runs is a highly efficient and attractive tactic for increasing productivity in this market segment. In recent years, we have seen a jump from the initial customary 8–10-hour production runs (for low-acid, aseptic lines) to 72 hours or longer. Manufacturers invest considerable time and effort in assessing and validating increments in the length of production runs. Unfortunately, regardless of the careful approach used prior to initiating an increase in the production run, such increments always come at the price of a higher risk of failure due to various unavoidable vulnerabilities such as product burn-on on heaters, valve operation failures, opportunity for biofilm creation, stagnant product in lines, etc.

Use of novel and exotic ingredients and product formulations:

Perhaps the fastest growing segment in the aseptic market space is the production of plant-based beverages. Plant-based foods are poised for explosive growth according to a recent market report (Henze & Boyd, 2021). This explosive growth includes plant-based beverages (dairy substitutes) such as soy, rice, almonds, oats, peas, etc. In many cases, these new products include exotic functional ingredients resulting from proprietary extraction methods or new ingredients from abroad. The physical behavior of these ingredients in the thermal process and their interaction with other ingredients, such as cocoa, is not yet fully understood. However, it has been reported (private communications) that some ingredients display higher than normally expected spore loads and rates of spoilage, particularly when used in cocoa-containing products. In fact, a large proportion of the spoilage cases we have investigated in recent years involve plant-based beverages, particularly chocolate flavors, and normally present spoilage caused by pure cultures of *Bacillus* spp., including in some cases *B. cereus* (a pathogen). In another case investigated by the authors, the contaminating organism caused only minor or unnoticeable changes in a high-acid aseptically packaged beverage. The contamination was only discovered following extensive microbiological testing done on the product as a result of an unrelated foreign object investigation. The spoilage organism was identified as *Asaia bogorensis*—a Gram-negative, rod-shaped organism. This organism was only capable of growth to approximately 10^4–10^5 CFU/ml in the product and did not affect the sensory characteristics of the product. The organism has been reported as a natural flora of flowers in tropical climates of Thailand, Indonesia, and Senegal (Bassene et al., 2020) where it has been associated in symbiotic relationship with mosquitoes of *Anopheles*. In addition, it has been reported as a spoilage organism in fruit-flavored water (Kregiel et al., 2012).

Market growth and technology maturity:

As mentioned above, the aseptic market, particularly fueled by the popularity of new plant-based foods, is expanding rapidly. At the same time, aseptic technology, now more than four decades since the initial filings of chemically sterilized aseptic fillers with the FDA, is maturing into a more conventional practice. Because of these developments, aseptic lines are becoming more affordable with more manufacturers and first-time end users entering the low-acid, shelf-stable aseptic market. The availability of personnel with extensive knowledge and experience in aseptic processing and packaging has not matched the rate of this explosive growth. It is therefore imperative that new entries into aseptic processing reach out to experts in the field and constantly provide for appropriate training of those designing, installing, and operating their aseptic lines.

More intense and sensitive microbiological testing of finished product:

The regulations governing the production of low-acid, shelf-stable aseptic products are based on the same principles as HACCP programs. They focus on prevention by identifying hazards, assigning critical control points (critical factors) and associated critical limits, requiring monitoring, verification, and record-keeping functions. Finished product testing is an effective verification tool, but its limitations are widely understood. For many years, aseptic processors followed the same approach as conventional canning where microbiological testing as a product release requirement is extremely rare. However, in recent years, microbiological testing as a condition of product release has become the norm in aseptic processing. Microbiological finished product testing for aseptic shelf-stable products is particularly onerous. This is because the product must meet the threshold of commercial sterility. Conventional microbiological methods based on plating are highly limited in their sensitivity (hence the potential for false negatives) and yet have a high propensity for false positive results due to laboratory contamination during the opening of the package and plating and incubation procedures. Currently, alternative methods such as flow-cytometry or ATP-based methods have allowed for more extensive testing with a smaller chance of false positives due to laboratory cross-contamination. On the other hand, these methods are susceptible to food matrix interference and must be calibrated and validated for the specific product application. In previous years using conventional methods, processors would have missed low-level contamination events where the contaminating microorganism had little impact on the product's intrinsic parameters or sensorial attributes, particularly if the contaminant was a strict anaerobic organism (as most conventional plating at the processing plant level is done in non-selective media under aerobic conditions only). Hence, it is reasonable to estimate that more contamination events are detected as new methods that are more effective have been introduced. For example, contamination by an organism such as *B. sporothermodurans* would have likely gone unnoticed using conventional plating as this organism grows to low numbers and does not cause noticeable changes to the product.

13.5 SUMMARY

Aseptic processing and packaging is an effective and robust technology where failures due to microbiological contamination are rare. Nevertheless, when failures do occur, the high complexity of aseptic lines makes it extremely difficult to diagnose a root cause. Tracking sources of microbial contaminants has been a concern, given the uncertainties associated with the integrity and maintenance decontamination of the "sterile working zone" or the aseptic tunnel during normative and extended production runs up to several days. However, recent advances in the development of molecular subtyping methods have provided tools that allow more rapid and highly accurate determinations of these sources, for proper root cause(s) analysis (RCA) and definitive corrective action and preventive action (CAPA).

Microbiological test data should be interpreted with caution by a cross-functional team consisting of a microbiologist with experience in plant ecology, ingredient bio-burden, microbial physiology; environmental engineer, thermal processing specialist, Equipment Preparation and Setup (EPS), maintenance, sanitation and cleaning, and research and development staff.

REFERENCES

21 CFR 113. 2011. *Thermally Processed Low-acid Foods Packaged in Hermetically Sealed Containers.* Washington, DC: U.S. Government Publishing Office. 44 FR 16215, Mar. 16, 1979, 76 FR 81363, Dec. 28, 2011, 21 CFR 113.83. https://www.accessdata.fda.gov/scripts/cdrh/cfdocs/cfcfr/CFRSearch.cfm?fr=113.83. Accessed December 6, 2021.

3-A Sanitary Standards Inc. n.d. 3-A sanitary standards, Inc. https://www.3-a.org/. Accessed December 1, 2021.

Acheson, D.W.K. 2009. Food and waterborne illness. *Encycl. Microbiol.*:365–381.

Al-Sakkaf, A. 2021. Thermal inactivation of New Zealand *Campylobacter jejuni* strains in chicken under dynamic conditions. *J. Food Eng.* 301:110540.

Alderton, G., Ito, K.A., and Chen, J.K. 1976. Chemical manipulation of the heat resistance of Clostridium botulinum spores. *Appl. Environ. Microbiol.* 31:492–498.

André, S., Vallaeys, T., and Planchon, S. 2017. Spore-forming bacteria responsible for food spoilage. *Res. Microbiol.* 168:379–387.

Aryani, D.C., den Besten, H.M.W., Hazeleger, W.C., and Zwietering, M.H. 2015. Quantifying variability on thermal resistance of Listeria monocytogenes. *Int. J. Food Microbiol.* 193:130–138.

Augusto, P.E.D., Tribst, A.A.L., and Cristianini, M. 2011. Thermal inactivation of *Lactobacillus plantarum* in a model liquid food. *J. Food Process Eng.* 34:1013–1027.

Bassene, H., Niang, E.H.A., Fenollar, F., Doucoure, S., Faye, O., Raoult, D., Sokhna, C., and Mediannikov, O. 2020. Role of plants in the transmission of Asaia sp., which potentially inhibit the Plasmodium sporogenic cycle in Anopheles mosquitoes. *Sci. Rep.* 10:7144.

Baumgardner, D.J. 2012. Soil-related bacterial and fungal infections. *J. Am. Board Fam. Med.* 25:734–744.

Baylis, C.L. 2006. Enterobacteriaceae. In *Food Spoilage Microorganisms*, edited by Blackburn, C.deW., 624–667. Cambridge, UK: Woodhead Publishing Limited.

Behling, R.G., Eifert, J., Erickson, M.C., Gurtler, J.B., Kornacki, J.L., Line, E. Radcliff, R., Ryser, E.T., Stawick, B., and Yan, Z. 2010. Selected pathogens of concern to industrial food processors: Infectious, toxigenic, toxico-infectious, selected emerging pathogenic bacteria. In *Principles of Microbiological Troubleshooting in the Industrial Food Processing Environment*, edited by Kornacki, J.L., 5–62. New York: Springer Science & Business Media.

Berni, E., Tranquillini, R., Scaramuzza, N., Brutti, A., and Bernini, V. 2017. Aspergilli with Neosartorya-type ascospores: Heat resistance and effect of sugar concentration on growth and spoilage incidence in berry products. *Int. J. Food Microbiol.* Elsevier 258:81–88.

Beuchat, L.R. 1986. Extraordinary heat resistance of *Talaromyces flavus* and *Neosartorya fischeri* ascospores in fruit products. *J. Food Sci.* 51:1506–1510.

Beuchat, L.R., Komitopoulou, E., Beckers, H., Betts, R.P., Bourdichon, F., Fanning, S., Joosten, H.M., and Ter Kuile, B.H. 2013. Low-water activity foods: Increased concern as vehicles of foodborne pathogens. *J. Food Prot.* 76:150–172.

Blakistone, B., Chuyate, R., Kautter, D., Charbonneau, J., and Suit, K. 1999. Efficacy of Oxonia active against selected spore formers. *J. Food Prot.* 62:262–267.

Blankenship, L.E., and Craven, S.E. 1982. *Campylobacter jejuni* survival in chicken meat as a function of temperature. *Appl. Environ. Microbiol.* 44:88–92.

Bozkurt, H., David, J.R.D., Talley, R.J., Lineback, D.S., and Davidson, P.M. 2016. Thermal inactivation kinetics of *Sporolactobacillus nakayamae* spores, a spoilage bacterium isolated from a model mashed potato-scallion mixture. *J. Food Prot.* 79:1482–1489.

Bradshaw, J.G., Peeler, J.T., Corwin, J.M., Hunt, J.M., and Twedt, R.M. 1987. Thermal resistance of Listeria monocytogenes in dairy products. *J. Food Prot.* 50:543–544.

Brown, A.C., Grass, J.E., Richardson, L.C., Nisler, A.L., Bicknese, A.S., and Gould, L.H. 2017. Antimicrobial resistance in Salmonella that caused foodborne disease outbreaks: United States, 2003–2012. *Epidemiol. Infect.* 145:766–774.

Brown, J.L., Tran-Dinh, N., and Chapman, B. 2012. Clostridium sporogenes PA 3679 and its uses in the derivation of thermal processing schedules for low-acid shelf-stable foods and as a research model for proteolytic Clostridium botulinum. *J. Food Prot.* 75:779–92.

Buchanan, R.L., Gorris, L.G.M., Hayman, M.M., Jackson, T.C., and Whiting, R.C. 2017. A review of Listeria monocytogenes: An update on outbreaks, virulence, dose-response, ecology, and risk assessments. *Food Control* 75:1–13.

Buehner, K.P., Anand, S., and Djira, G.D. 2015. Prevalence of thermoduric bacteria and spores in nonfat dry milk powders of Midwest origin. *J. Dairy Sci.* 98:2861–2866.

Byrne, B., Dunne, G., and Bolton, D.J. 2006. Thermal inactivation of Bacillus cereus and Clostridium perfringens vegetative cells and spores in pork luncheon roll. *Food Microbiol.* 23:803–808.

Cameron, M.S., Leonard, S.J., and Barrett, E.L. 1980. Effect of moderately acidic pH on heat resistance of Clostridium sporogenes spores in phosphate buffer and in buffered pea puree. *Appl. Environ. Microbiol.* 39:943–949.

Centers for Disease Control and Prevention (CDC). 2019. Outbreak of E. coli infections linked to flour. https://www.cdc.gov/ecoli/2019/flour-05-19/index.html. Accessed on February 12, 2019.

Centers for Disease Control and Prevention (CDC). 2021. List of selected multistate foodborne outbreak investigation. https://www.cdc.gov/foodsafety/outbreaks/multistate-outbreaks/outbreaks-list.html. Accessed on November 4, 2021.

Collier, C.P., and Townsend, C.T. 1956. The resistance of bacterial spores to superheated steam. *Food Technol.* 5:447–481.

Coorevits, A., De Jonghe, V., Vandroemme, J., Reekmans, R., Heyrman, J., Messens, W., De Vos, P., and Heyndrickx, M. 2008. Comparative analysis of the diversity of aerobic spore-forming bacteria in raw milk from organic and conventional dairy farms. *Syst. Appl. Microbiol.* 31:126–140.

Coroller, L., Leguérinel, I., and Mafart, P. 2001. Effect of water activities of heating and recovery media on apparent heat resistance of Bacillus cereus spores. *Appl. Environ. Microbiol.* 67:317–322.

Cossart, P., and Toledo-Arana, A. 2008. Listeria monocytogenes, a unique model in infection biology: an overview. *Microbes Infect.* 10:1041–1050.

Dagnas, S.P., and Membré, J.M. 2013. Predicting and preventing mold spoilage of food products. *J. Food Prot.* 76:538–551.

David, J.R.D., and Merson, R.L. 1990. Kinetic parameters for inactivation of *Bacillus stearothermophilus* at high temperatures. *J. Food Sci.* 55:488–493.

Diao, M.M., André, S., and Membré, J.M. 2014. Meta-analysis of D-values of proteolytic Clostridium botulinum and its surrogate strain Clostridium sporogenes PA 3679. *Int. J. Food Microbiol.* 174:23–30.

Disson, O., and Lecuit, M. 2013. In vitro and in vivo models to study human listeriosis: Mind the gap. *Microbes Infect.* 15:971–980.

Engel, G., and Teuber, M. 1991. Heat resistance of ascospores of *Byssochlamys nivea* in milk and cream. *Int. J. Food Microbiol.* 12:225–233.

Engstrom, S.K., Mays, M.F., and Glass, K.A. 2021. Determination and validation of D-values for Listeria monocytogenes and Shiga toxin-producing Escherichia coli in cheese milk. *J. Dairy Sci. American Dairy Science Association.* 104:12332–12341.

European Hygienic Engineering & Design Group (EHEDG). n.d. EHEDG. https://www.ehedg.org/. Accessed December 1, 2021.

Evelyn, C., Muria, S.R., Adella, L., and Ramadhani, R. 2020. Thermal inactivation of Eupenicillium Javanicum ascospores in pineapple juice: Effect of temperature, soluble solids and spore age. *J. Phys. Conf. Ser.* 1655:012020.

Evelyn, E., and Silva, F.V.M. 2015. Thermosonication versus thermal processing of skim milk and beef slurry: Modeling the inactivation kinetics of psychrotrophic Bacillus cereus spores. *Food Res. Int.* 67:67–74.

Feng, P., Weagant, S.D., Grant, M.A., and Burkhardt, W. 1998. Bacteriological Analytical Manual, 8th edition, Revision A. Chapter 4. Enumeration of Escherichia coli and the coliform bacteria. https://www.fda.gov/food/laboratory-methods-food/bam-chapter-4-enumeration-escherichia-coli-and-coliform-bacteria. Accessed on November 19, 2021.

Fernández, A., Ocio, M.J., Fernández, P.S., Rodrigo, M., and Martinez, A. 1999. Application of nonlinear regression analysis to the estimation of kinetic parameters for two enterotoxigenic strains of Bacillus cereus spores. *Food Microbiol.* 16:607–613.

Fernandez, P.S., M.J. Ocio, T. Sanchez, and Martinez, A. 1994. Thermal resistance of *Bacillus stearothermophilus* spores heated in acidified mushroom extract. *J. Food Prot.* 57:37–41.

Ferreira, V., Wiedmann, M., Teixeira, P., and Stasiewicz, M.J. 2014. Listeria monocytogenes persistence in food-associated environments: Epidemiology, strain characteristics, and implications for public health. *J. Food Prot.* 77:150–170.

Fleet, G.H. 2011. Yeast spoilage of foods and beverages. In *The Yeasts, a Taxonomic Study*, edited by Kurtzman, C.P., Fell, J.W., and Boekhout, T., 53–63. 5th edition. Amsterdam, Netherlands: Elsevier Inc.

Food Safety News. 2016. Bolthouse Farms recalls millions of drink because of illness. https://www.food-safetynews.com/2016/06/127981/. Accessed on November 28, 2021.

Franz, C.M.A.P., and von Holy, A. 1996. Thermotolerance of meat spoilage lactic acid bacteria and their inactivation in vacuum-packed vienna sausages. *Int. J. Food Microbiol.* 29:59–73.

Garza, S., Antonio Teixidó, J., Sanchis, V., Viñas, I., and Condón, S. 1994. Heat resistance of Saccharomyces cerevisiae strains isolated from spoiled peach puree. *Int. J. Food Microbiol.* 23:209–213.

Gill, C.O., and Harris, L.M. 1982. Survival and growth of *Campylobacter fetus* subsp. *jejuni* on meat and in cooked foods. *Appl. Environ. Microbiol.* 44:259–263.

Gomes, R.J., Borges, M.deF., Rosa, M.deF., Castro-Gómez, R.J.H., and Spinosa, W.A. 2018. Acetic acid bacteria in the food industry: Systematics, characteristics and applications. *Food Technol. Biotechnol.* 56:139–151.

Griffiths, M.W., and Schraft, H. 2017. Bacillus cereus Food Poisoning. In *Foodborne Diseases*, edited by Dodd, C.E.R., Aldsworth, T., Stein, R.A., Cliver, D.O., and Riemann, H., 395–405. 3rd edition. Oxford, UK: Academic Press.

Gutarowska, B. 2010. Metabolic activity of moulds as a factor of building materials biodegradation. *Polish J. Microbiol.* 59:119–124.

Hartman, B. 2009. "Slim Fast" Recalls All Shakes, Diet Drinks. https://abcnews.go.com/Politics/slim-fast-recalls-millions-shakes-diet-drinks/story?id=9251524. Accessed on November 28, 2021.

Hayrapetyan, H., Nederhoff, L. Vollebregt, M., Mastwijk, H., and Nierop Groot, M. 2020. Inactivation kinetics of *Geobacillus stearothermophilus* spores by a peracetic acid or hydrogen peroxide fog in comparison to the liquid form. *Int. J. Food Microbiol.* 316:108418.

Henze, V., and Boyd, S. 2021. Plant-based foods market to hit $162 billion in next decade, projects Bloomberg Intelligence. Available at https://www.bloomberg.com/company/press/plant-based-foods-market-to-hit-162-billion-in-next-decade-projects-bloomberg-intelligence/. Accessed November 28, 2021

Hernández, A., Pérez-Nevado, F., Ruiz-Moyano, S., Serradilla, M.J., Villalobos, M.C. Martín, A., and Córdoba, M.G. 2018. Spoilage yeasts: What are the sources of contamination of foods and beverages? *Int. J. Food Microbiol.* 286:98–110.

Hocking, A.D. 2006. Aspergillus and related teleomorphs. In *Food Spoilage Microorganisms*, edited by Blackburn, C.deW., 451–487. Cambridge, UK: Woodhead Publishing Limited.

Huang, I.P.D., Yousef, A.E., Marth, E.H., and Matthews, M.E. 1992. Thermal inactivation of Listeria monocytogenes in chicken gravy. *J. Food Prot.* 55:492–496.

Humphrey, T.J., Chapman, P.A., Rowe, B., and Gilbert, R.J. 1990. A comparative study of the heat resistance of salmonellas in homogenized whole egg, egg yolk or albumen. *Epidemiol. Infect.* 104:237–241.

Jackson, B.R., Griffin, P.M., Cole, D., Walsh, K.A., and Chai, S.J. 2013. Outbreak-associated *Salmonella enterica* serotypes and food commodities, United States, 1998–2008. *Emerg. Infect. Dis.* 19:1239–1244.

Jin, T., Zhang, H., Boyd, G., and Tang, J. 2008. Thermal resistance of Salmonella enteritidis and Escherichia coli K12 in liquid egg determined by thermal-death-time disks. *J. Food Eng.* 84:608–614.

Jordan, K.N., and Cogan, T.M. 1999. Heat resistance of Lactobacillus spp. isolated from Cheddar cheese. *Lett. Appl. Microbiol.* 29:136–40.

Julien, M.C., Dion, P., Lafrenière, C., Antoun, H., and Drouin, P. 2008. Sources of Clostridia in raw milk on farms. *Appl. Environ. Microbiol.* 74:6348–6357.

Juneja, V.K., and Marmer, B.S. 1996. Growth of Clostridium perfringens from spore inocula in sous-vide turkey products. *Int. J. Food Microbiol.* 32:115–123.

Juneja, V.K., Osoria, M., Hwang, C.A., Mishra, A., and Taylor, T.M. 2020. Thermal inactivation of Bacillus cereus spores during cooking of rice to ensure later safety of boudin. *LWT: Food Sci. Technol.* 122:108955.

Kaper, J.B., Nataro, J.P., and Mobley, H.L.T. 2004. Pathogenic Escherichia coli. *Nat. Rev. Microbiol.* 2:123–140.

Kotzekidou, P. 2014. Byssochlamys. In *Encyclopedia of Food Microbiology*, edited by Batt, C.A., and Tortorello, M.L., Vol. 1, 344–350. 2nd edition. Amsterdam, Netherlands: Elsevier Ltd.

Kregiel, D., Rygala, A., Libudzisz, Z., Walczak, P., and Oltuszak-Walczak, E. 2012. *Asaia lannensis*-the spoilage acetic acid bacteria isolated from strawberry-flavored bottled water in Poland. *Food Control* 26:147–150.

Lahou, E., Wang, X., De Boeck, E., Verguldt, E. Geeraerd, A., Devlieghere, F., and Uyttendaele, M. 2015. Effectiveness of inactivation of foodborne pathogens during simulated home pan frying of steak, hamburger or meat strips. *Int. J. Food Microbiol.* 206:118–129.

Lima, L.J.R., Kamphuis, H.J., Nout, M.J.R., and Zwietering, M.H. 2011. Microbiota of cocoa powder with particular reference to aerobic thermoresistant spore-formers. *Food Microbiol.* 28:573–582.

Lindström, M., Heikinheimo, A., Lahti, P., and Korkeala, H. 2011. Novel insights into the epidemiology of Clostridium perfringens type A food poisoning. *Food Microbiol.* 28:192–198.

Luo, Z., Tucker, G., and Brown, H. 2016. Empirical manipulation of the thermoinactivation kinetics of *Bacillus amyloliquefaciens* and *Bacillus licheniformis* α-amylases for thermal process evaluations. *Innov. Food Sci. Emerg. Technol.* 38:272–280.

Lynt, R.K., Kautter, D.A., and Solomon, H.M. 1982. Differences and similarities among proteolytic and nonproteolytic strains of *Clostridium botulinum* Types A, B, E and F: A review. *J. Food Prot.* 45:466–474.

Martorell, P., Stratford, M., Steels, H., Fernández-Espinar, M.T., and Querol, A. 2007. Physiological characterization of spoilage strains of *Zygosaccharomyces bailii* and *Zygosaccharomyces rouxii* isolated from high sugar environments. *Int. J. Food Microbiol.* 114:234–242.

Mathot, A.G., Postollec, F., and Leguerinel, I. 2021. Bacterial spores in spices and dried herbs: The risks for processed food. *Compr. Rev. Food Sci. Food Saf.* 20:840–862.

Mazas, M., Lopez, M., Gonzalez, I., Gonzalez, J., Bernardo, A., and Martin, R. 1998. Effects of the heating medium pH on heat resistance of Bacillus cereus spores. *J. Food Saf.* 18:25–36.

Mazzotta, A.S. 2001. Thermal inactivation of stationary-phase and acid-adapted Escherichia coli O157:H7, Salmonella, and Listeria monocytogenes in fruit juices. *J. Food Prot.* 64:315–320.

McHugh, A.J., Feehily, C., Hill, C., and Cotter, P.D. 2017. Detection and enumeration of spore-forming bacteria in powdered dairy products. *Front. Microbiol.* 8:1–15.

Michael, M., Acuff, J., Lopez, K., Vega, D., Phebus, R., Thippareddi, H., and Channaiah, L.H. 2020. Comparison of survival and heat resistance of Escherichia coli O121 and Salmonella in muffins. *Int. J. Food Microbiol.* 317:108422.

Mikolajcik, E.M., and Rajkowski, K.T. 1980. Simple technique to determine heat resistance of *Bacillus stearothermophilus* spores in fluid systems. *J. Food Prot.* 43:799–804.

Monu, E.A., Valladares, M., D'Souza, D.H., and Davidson, P.M. 2015. Determination of the thermal inactivation kinetics of *Listeria monocytogenes*, *Salmonella enterica*, and *Escherichia coli* O157:H7 and non-O157 in buffer and a spinach homogenate. *J. Food Prot.* 78:1467–1471.

Mordor Intelligence. n.d. North America aseptic packaging market: Growth, trends, COVID-19 impact, and forecasts (2021–2026). Accessed November 28, 2021. Available at https://www.mordorintelligence.com/industry-reports/north-america-aseptic-packaging-market-industry.

Moss, M.O. 2006. General characteristics of moulds. In *Food Spoilage Microorganisms*, edited by Blackburn, C.deW., 401–414. Cambridge, UK: Woodhead Publishing Limited.

Moussa-Boudjemaa, B., Gonzalez, J., and Lopez, M. 2006. Heat resistance of Bacillus cereus spores in carrot extract acidified with different acidulants. *Food Control* 17:819–824.

Ocio, M.J., Sánchez, T., Fernandez, P.S., Rodrigo, M., and Martínez, A. 1994. Thermal resistance characteristics of PA 3679 in the temperature range of 110–121°C as affected by pH, type of acidulant and substrate. *Int. J. Food Microbiol.* 22:239–247.

Odlaug, T.E., and Pflug, I.J. 1977. Thermal destruction of Clostridium botulinum spores suspended in tomato juice in aluminum thermal death time tubes. *Appl. Environ. Microbiol.* 34:23–29.

Odlaug, T.E., Pflug, I.J., and Kautter, D.A. 1978. Heat resistance of Clostridium botulinum Type B spores grown from isolates from commercially canned mushrooms. *J. Food Prot.* 41:351–353.

Palop, A., Raso, J. Condon, S., and Sala, F.J. 1996a. Heat resistance of Bacillus subtilis and Bacillus coagulans: Effect of sporulation temperature in foods with various acidulants. *J. Food Prot.* 59:487–492.

Palop, A., Raso, J., Pagan, R., Condon, S., and Sala, F.J. 1996b. Influence of pH on heat resistance of *Bacillus licheniformis* in buffer and homogenised foods. *Int. J. Food Microbiol.* 29:1–10.

Palop, A., Raso, J., Pagan, R., Condon, S., and Sala, F.J. 1999. Influence of pH on heat resistance of spores of *Bacillus coagulans* in buffer and homogenized foods. *Int. J. Food Microbiol.* 46:243–249.

Pettersson, B., Lembke, F., Hammer, P., Stackebrandt, E., and Priest, F.G. 1996. *Bacillus sporothemodurans*, a new species producing highly heat-resistant endospores. *Int. J. Syst. Bacteriol.* 46:759–764.

Pitt, J.I., and Hocking, A.D. 2009. *Fungi and Food Spoilage*. 3rd edition. New York: Springer Science+Business Media, LLC.

Podolak, R., and Black, D.G. 2017. *Control of Salmonella and Other Bacterial Pathogens in Low-Moisture Foods*. West Sussex, UK: John Wiley & Sons Ltd.

Podolak, R., Enache, E. Stone, W., Black, D.G., and Elliott, P.H. 2010. Sources and risk factors for contamination, survival, persistence, and heat resistance of Salmonella in low-moisture foods. *J. Food Prot.* 73:1919–1936.

Postollec, F., Mathot, A.-G., Bernard, M., Divanac'h, M.-L., Pavan, S., and Sohier, D. 2012. Tracking spore-forming bacteria in food: From natural biodiversity to selection by processes. *Int. J. Food Microbiol.* 158:1–8.

Put, H.M., De Jong, J., Sand, F.E., and Van Grinsven, A.M. 1976. Heat resistance studies on yeast spp. causing spoilage in soft drinks. *J. Appl. B* 40:135–152.

Rodrigo, F., Fernández, P.S., Rodrigo, M., Ocio, M.J., and Martínez, A. 1997. Thermal resistance of *Bacillus stearothermophilus* heated at high temperatures in different substrates. *J. Food Prot.* 60:144–147.

Rosow, L.K., and Strober, J.B. 2015. Infant botulism: Review and clinical update. *Pediatr. Neurol.* 52:487–492.

Salfinger, Y., and Tortorello, M.L. 2015. *Compendium of Methods for the Microbiological Examination of Foods*. 5th edition. Washington, DC: American Public Health Association.

Salvetti, E., Torriani, S., Zheng, J., Lebeer, S., Gänzle, M.G., and Felis, G.E. 2022. *Lactic Acid Bacteria: Taxonomy and Biodiversity Encyclopedia of Dairy Sciences*. 3rd edition. Amsterdam, Netherlands: Elsevier Inc.

Samapundo, S., Vroman, A., Eeckhout, M., and Devlieghere, F. 2018. Effect of heat treatment intensity on the survival, activation and subsequent outgrowth of *Byssochlamys nivea* ascospores. *Lwt.* Elsevier 93:599–605.

Scallan, E., Hoekstra, R.M., Angulo, F.J., Tauxe, R.V., Widdowson, M.A., Roy, S.L., Jones, J.L., and Griffin, P.M. 2011. Foodborne illness acquired in the United States: Major pathogens. *Emerg. Infect. Dis.* 17:7–15.

Scheldeman, P., Herman, L., Foster, S., and Heyndrickx, M. 2006. *Bacillus sporothermodurans* and other highly heat-resistant spore formers in milk. *J. Appl. Microbiol.* 101:542–555.

Scheldeman, P., Pil, A., Herman, L., De Vos, P., and Heyndrickx, M. 2005. Incidence and diversity of potentially highly heat-resistant spores isolated at dairy farms. *Appl. Environ. Microbiol.* 71:1480–1494.

Schuman, J.D., and Sheldon, B.W. 1997. Thermal resistance of Salmonella spp. and Listeria monocytogenes in liquid egg yolk and egg white. *J. Food Prot.* 60:634–638.

Sengun, I.Y., and Karabiyikli, S. 2011. Importance of acetic acid bacteria in food industry. *Food Control* 22:647–656.

Shearer, A.E.H., Mazzotta, A.S., Chuyate, R., and Gombas, D.E. 2002. Heat resistance of juice spoilage microorganisms. *J. Food Prot.* 65:1271–1275.

Shih, S.-C., Cuevas, R., Porter, V.L., and Cheryan, M. 1982. Inactivation of *Bacillus stearothermophilus* spores in soybean water extracts at Ultra-High Temperatures in a scraped-surface heat exchanger. *J. Food Prot.* 45:145–149.

Silla Santos, M.H., Nuñez Kalasic, H., Casado Goti, A., and Rodrigo Enguidanos, M. 1992. The effect of pH on the thermal resistance of Clostridium sporogenes (PA 3679) in asparagus purée acidified with citric acid and glucono-δ-lactone. *Int. J. Food Microbiol.* 16:275–281.

Sörqvist, S. 2003. Heat resistance in liquids of *Enterococcus* spp., *Listeria* spp., *Escherichia coli*, *Yersinia enterocolitica*, *Salmonella* spp. and *Campylobacter* spp. *Acta Vet. Scand.* 44:1–19.

Sperber, W.H., and Doyle, M.P. 2009. *Compendium of the Microbiological Spoilage of Foods and Beverages*. New York: Springer.

Stratford, M. 2006. Food and beverage spoilage yeasts. In *Yeasts in Foods and Beverages*, edited by Querol, A., and Fleet, G., 335–379. Berlin: Springer.

Sumner, S., Sandros, T., Harmon, M., Scott, V., and Bernard, D. 1991. Heat resistance of Salmonella typhimurium and Listeria monocytogenes in sucrose solutions of various water activities. *J. Food Sci.* 56:1741–1743.

Tajchakavit, S., Ramaswamy, H.S., and Fustier, P. 1998. Enhanced destruction of spoilage microorganisms in apple juice during continuous flow microwave heating. *Food Res. Int.* 31:713–722.

Toledo, R.T., Escher, F.E., and Ayres, J.C. 1973. Sporicidal properties of hydrogen peroxide against food spoilage organisms. *Appl. Microbiol.* 26:592–597.

Tournas, V. 1994. Heat-resistant fungi of importance to the food and beverage industry. *Crit. Rev. Microbiol.* 20:243–63.

Tranquillini, R., Scaramuzza, N., and Berni, E. 2017. Occurrence and ecological distribution of Heat Resistant Moulds Spores (HRMS) in raw materials used by food industry and thermal characterization of two Talaromyces isolates. *Int. J. Food Microbiol.* 242:116–123.

Tribst, A.A.L., Sant'Ana, A.D.S., and de Massaguer, P.R. 2009. Review: Microbiological quality and safety of fruit juices: Past, present and future perspectives. *Crit. Rev. Microbiol.* 35:310–339.

U.S. Department of Agriculture (USDA) Economic Research Service (ERS). 2021. Cost estimates of foodborne illness (2020). https://www.ers.usda.gov/data-products/cost-estimates-of-foodborne-illnesses/. Accessed on March 11, 2021.

U.S. Department of Health and Human Services; Food and Drug Administration; Center for Food Safety and Applied Nutrition. 2018. Hazard analysis and risk-based preventive controls for human food: Guidance for industry, draft guidance. https://www.fda.gov/downloads/Food/GuidanceRegulation/GuidanceDocumentsRegulatoryInformation/UCM517610.pdf. Accessed on May 16, 2019.

U.S. Food and Drug Administration (FDA). 2012. Bad bug book: Handbook of foodborne pathogenic microorganisms and natural toxins. http://www.fda.gov/downloads/Food/FoodborneIllnessContaminants/UCM297627.pdf. Accessed on January 1, 2016.

U.S. Food and Drug Administration (FDA). 2017a. Final summary: FDA investigation of multistate outbreak of shiga toxin-producing E. coli infections linked to flour. https://www.fda.gov/Food/Recalls OutbreaksEmergencies/Outbreaks/ucm587435.htm. Accessed on January 19, 2018.

U. S. Food and Drug Administration (FDA). 2017b. Risk profile: Pathogens and filth in spices. https://www.fda.gov/downloads/Food/FoodScienceResearch/RiskSafetyAssessment/UCM581362.pdf. Accessed on February 26, 2018.

Vasan, A., Leong, W.M., Ingham, S.C., and Ingham, B.H. 2013. Thermal tolerance characteristics of non-O157 shiga toxigenic strains of Escherichia coli (STEC) in a beef broth model system are similar to those of O157:H7 STEC. *J. Food Prot.* 76:1120–1128.

Westhoff, D.C., and Dougherty, S.L. 1981. Characterization of Bacillus species isolated from spoiled Ultrahigh Temperature processed milk. *J. Dairy Sci.* 64:572–580.

Young, K.T., Davis, L.M., and DiRita, V.J. 2007. *Campylobacter jejuni*: Molecular biology and pathogenesis. *Nat. Rev. Microbiol.* 5:665–679.

Zhang, S., Zhang, L., Lan, R., Zhou, X., Kou, X., and Wang, S. 2018. Thermal inactivation of *Aspergillus flavus* in peanut kernels as influenced by temperature, water activity and heating rate. *Food Microbiol.* 76:237–244.

Chapter 14

Risk-Based Analyses and Methodologies

Ferhan Ozadali

CONTENTS

DOI: 10.1201/9781003158653-17

14.1 INTRODUCTION AND BACKGROUND

In this part of the book, the reader had the opportunity to learn about microbiology as one of the foundational topics to understand validation and food safety concepts. Starting from this chapter and following with the next two chapters, the information flow for the reader will be from the identification of the risks and mitigation plans to development of validation protocols for the "validated State" of the aseptic systems, and finally to maintaining the validated state with food safety and quality management systems. With this information flow structure, this chapter will guide the reader to understand the "risk" concept and its prerequisite characteristics for the validation and quality/food safety management systems. Also, most of the commonly used risk analysis methodologies and the mitigation plans that may help determine the potential risks and their prioritization for the company's actions to develop risk management strategies will be discussed in this chapter. Identifying the risks correctly well in advance in the product/process life cycle would help in designing realistic validation execution plans which are based on worst-case scenarios for the aseptic food production systems. Since aseptic processing and packaging systems are technologically complex compared to conventional processing systems and the existing expertise is very limited, the risk analysis of the aseptic systems may be complicated if not enough proactive attention is devoted to this process.

When we consider and study the product life cycle, it would be clear that early detection of any potential risks would be a vital part of a sound business continuity strategy. If one of the wings of an airplane comes off in the air, what is the likelihood that this plane would crash? Although there are some examples of airplanes landing with one wing, the probability of a crash in this situation is extremely high. Therefore, running proactive "what if" failure scenarios would allow us to identify risks and develop mitigation plans to prevent potential undesirable occurrences.

As we know, production efficiencies are extremely important, and any line downtimes would create organizational stress due to capacity constraints, additional operational cost, and potentially cutting customers' order fulfillment. Especially for the aseptic systems, any major food safety concerns due to spoilage will require extensive root-cause analyses and lengthy line downtimes. Can those incidences be prevented? Absolutely risk-free systems, especially for technologically complex setups such as aseptic lines, do not exist. The author's intent in this chapter is to explain the meaning of the "risks" for the aseptic systems prior to discussing the risk-based validation (Chapter 15) and sharing widely used effective tools and methodologies by identifying such risks well in advance during the design phase to prevent expensive and hazardous consequences.

14.2 RISK-BASED APPROACH

We need to start discussing the problem statement and define the risk-based approach with its social, economic, and technical components. Food processing systems are becoming more complex for overcoming the needs in the market to stay more competitive with the development of more challenging products. These challenges may be coming from innovative and sometimes very controversial ingredients or due to cutting-edge novel technologies to address specific quality attributes or marketing strategies (e.g., aseptic products with discrete particles). Some of the companies demonstrate double-digit growth in some of the emerging markets, which drives urgent decisions for large capital investments or working with third-party contract manufacturers. As one can imagine, large capital investments to address the market demand in a noticeably brief time will bring along major risks. Although working with Third-Party Manufacturers (TPM) or Contract Manufacturing Organizations (CMO) has some advantages in terms of commercialization timelines, this approach may have many risks since the organizational culture for aseptic processing varies from one company to another.

In addition, regulatory requirements, and related risks for the execution of the project may change from country to country. Larger organizations have more harmonized internal standards that are aligned with regional and global regulatory requirements. Some of the regulatory agencies for different countries have a Memorandum of Understanding (MOU) between them to harmonize their standards toward specific product categories. Those memorandums are extremely useful guidance documents to address some of the regulatory risks for a specific country or region well in advance to prevent them from appearing during commercialization.

If we need to single out one major risk that would be the most crucial factor for the probability of success to commercialize aseptic products in a particular market, that would be the "people." They are considered as the most important assets of the aseptic processing and packaging systems. Trained and competent personnel to support the aseptic technologies and complexities is the most crucial factor which may be the biggest risk in some regions due to lack of expertise and competent resource pools. This risk is commonly ignored until the time of product launch, and it is generally too late to address this risk without profound consequences.

14.2.1 Why Do We Need Standardization?

A disciplined approach to determine the risks well in advance in the product life cycle is critical for the sustainability of the validated aseptic systems. If the approach is not predetermined and companies start picking and choosing the requirements and acceptance criteria randomly or based on certain economic or cultural pressure for the validation of aseptic systems, the consequences would be very confusing and extremely dangerous in terms of brand protection. Prior standardization of the requirements and risk-based approach to determine the acceptance criteria for the "*validation state*" of the manufacturing businesses are critical for sustainable business continuity and brand protection.

Even for large organizations, this is not an easy task. Regulatory and compliance requirements are extremely strict and very well-established in countries such as the United States and Canada. Large companies align their internal standards and requirements with most of these regulatory authorities and their published requirements. In other regions and/or countries in the world, there may be less strict regulatory requirements, and this may cause some resistance to additional perceived burdens without proper explanations and justifications. More importantly, those additional requirements may require additional capital investment to comply with the requirements prior to any validation activities. Even larger global equipment suppliers such as Tetra Pak and John Bean Technologies (JBT) may have different equipment and process designs for different regions of the World.

14.2.1.1 Why Do We Need This While It Is Not a Regulatory Requirement in Some Counties/Regions?

This is the most common question that Process Authorities, Quality, and Food Safety representatives come across during the execution of the aseptic commercialization projects. In the absence of very prescriptive authoritarian regulatory requirements (e.g., US FDA and 21 Code of Federal Regulations (CFR) as outlined on www.fda.gov), it would be critical to communicate the scientific evidence and examples of incidences to regional enablers. This approach would help avoid the need for a top-down and force-fitted execution. For the establishment of a sustainable operation, it is critical to communicate risks clearly, educate functions, and influence the culture strategically to support aseptic manufacturing.

Aligning internal standards and regulatory requirements with global standards would guide us to be more proactive with proper risk analyses and risk-based validation. Early detection of potential issues with "what if" scenarios will enhance the probability of success for the aseptic system commercialization.

Food Safety and Quality requirements are universal, and should be based on science and common sense. Consistently executed global standards are critical to highlight a company's and industry's stand on Food Safety and Quality which is translated to consumer trust and higher profitability.

Predicting the risks of failure is the most important task to address food safety, quality, and operational concerns. Implementing the mitigation plans proactively for those risks is critical but not always easy since every company has a different culture. It is always easier to implement changes in case of reacting to a major crisis. However, it is much harder to justify those changes based on predictions, especially if they require additional resources and/or capital investment. Companies that care about brand protection should not operate without a system in place to constantly look for potential risks well in advance before they become a problem. There are two sayings in the industry that explain this cultural gap: *"Pay me now or pay me later"* and *"In this case, failure is not if, it is when."*

There are several tools that can be used to predict potential failures and prepare in-advance mitigation plans to address these risks early in the development and validation stages. Therefore, this chapter will go into detail about the risk definitions and risk analyses tools prior to covering Aseptic Validation and Quality/Food Safety Management chapters of this book section. Risk analysis tools should be used throughout the life cycle of the product and process development.

14.2.2 What Is The Risk?

The risk can be defined as the combination of the probability of occurrence of harm and the severity of that harm to all stakeholders in the food development and consumption chain.

The risk can be mitigated through the level of scientific understanding of how the formulation and manufacturing process factors affect food safety, efficacy, and product quality; and the capability of process control strategies to prevent or mitigate the risk of producing an unsafe/poor-quality product.

Classic risk-management tools along with well-engineered process/product development studies help focus on and affect the product's process and quality predictability.

14.2.3 Why Do We Need Risk Analyses of the Aseptic Systems?

The fact is that a system without a risk does not exist! Risks may be higher or not readily obvious with new hurdle technology implementations (e.g., blow-fill-seal (BFS) aseptic bottle filling systems). During the last decade, the industries (e.g., Food) have slowly shifted to a risk-based, predictive, proactive, and preventive approach to validate their product and process development activities more realistically and effectively. This did not happen easily or smoothly! Industries learned this the hard way and from time to time, with catastrophic failures. As one can imagine, unfortunately, sometimes catastrophes like crashing airplanes and collapsing bridges are good motivators to force us to think about the risks well in advance. This relatively new risk-based approach puts additional emphasis on the true meaning of validation (Chapter 15) as a Good Manufacturing and Engineering Practices beyond just complicated documentation programs. Validation is not seen as a snapshot of processes four weeks before the launch-day anymore! At least, this is the hope and message the author of this chapter is trying to convey to our readers.

The shift in mindset to a more scientifically driven and risk-based framework has transformed expectations for system validations. A risk-based validation identifies key parameters, which drive process variation, and controls well in advance in the development process. Classic risk-management tools along with well-engineered process/product development studies help focus on predictability of the system failures and related risks.

With the incorporation of complex novel technologies and other innovative approaches, concerns over microbiological food safety and potential public health risks have been elevated for the need for an alternative approach to prediction of potential risks. This is beyond a regulatory requirement! This is a necessary approach for "brand protection"! Global food regulations vary from one country to another and are often vague in terms of prescriptive requirements. For instance, the current US regulations governing food safety do not target a process endpoint with specific criteria. The US FDA's low-acid canned foods regulations (US FDA, 21 CFR part 113) require "commercial sterility" which is, in definition, a combination of several requirements. Similarly, the US FDA Juice Hazard Analysis Critical Control Point (HACCP) regulation (US FDA, 21 CFR 120) stipulates a 5-log pathogen reduction without prescribing the details of how the company can achieve this requirement with so much variability in the industry. The brand owner's name is on the label and owns the safety of their products, not the regulatory agencies! Therefore, satisfying non-specific and generic minimum requirements of the regulatory agencies may not satisfy fulfilling the appropriate common-sense risk mitigations.

14.2.4 Risk-Averse Culture Development for Food Systems Should be the Focus of Company-Wide Mission and Vision!

Risk-averse does not mean avoiding innovation! For the growth-based company strategies, innovative product and technology development cannot be degraded for the sake of avoiding the potential risks. It is vital for the company's health to have both innovation with inherent risks and a system to handle those risks in a very systematic way, well in advance in the development process. Commercial pressure for the *"speed-to-market"* goals is inevitable to launch products before the competition. This may create an enormous amount of pressure on different functions within the organization and execution teams such as engineering, quality, and process authority. As is expected, short-cutting some of the necessary steps under such external pressure and stress may increase the probability of occurrence of the related risks. Therefore, policy and procedures based on internal requirements that are fully aligned with the region/county's regulatory agencies can be used for effective and timely communications with related functional groups to minimize the mentioned stress levels by advance planning.

During the establishment of the risk-averse culture, there are certain watchouts based on experiences and learnings from past failures. It should be clearly understood that governmental regulatory agencies do not "approve" the system filings. They only review and provide the applicants (Original Equipment Manufacturers—OEMs) with a *"Letter of No Objection (LONO)"* with a clear disclaimer that they can always come back and question the validity of the system. Therefore, regulatory LONO is not a guarantee for risk-free processing. In the United States, all the OEMs who are providing the food industry (e.g., aseptic processing and packaging of low- and acidified food manufacturing) with equipment should file their design with the FDA for a LONO prior to selling their units/systems. Brand owners or producers are responsible for the filing of their products with the regulatory agencies based on the safety information obtained from the OEMs and additional validation documentation.

Relying upon said experts or dilettantes who do not have the necessary competencies to analyze new systems and technologies for potential risks may teach the company a good lesson the hard way to be more careful with those assignments. For instance, the knowledge and experience level of internal and external Process Authorities (PA) vary from one individual to another or from one company to another company. As they are a critical part of the decision-making with a tremendous amount of trust and influence power, selection of an appropriate PA is extremely critical for proper risk management. Although they are all respected professionals and/or companies, any potential conflicts and pressures should be eliminated from their work to allow them to advise freely and with confidence. In case external PAs are hired by the equipment suppliers to support the projects, a clear reporting structure regardless of the payment terms must be outlined in the agreements. Even if an OEM is hiring and paying the PA, the output of their work must go to the system owner directly without being filtered through the OEM.

Another watchout is regarding the equipment suppliers that are an integral part of the supply chain and should be considered as true partners. There may be a tremendous amount of trust as they are experts in designing/building their equipment and may have established relationships with key individuals within the company. The suppliers' know-how and experiences that have accumulated throughout the years about their equipment and processes are certainly critical assets for the successful execution and maintenance of the food manufacturing systems. However, equipment suppliers may not be able to consider all the operational variables and potential risks. Therefore, their preferred cookie-cutter approach may not identify all the potential risks. Their proud feelings, previous experiences, and regulatory LONOs (e.g., FDA Master Filing for equipment design) may blindside companies that do not have structured risk-averse cultures and heavily rely on the suppliers for technical competencies. Any overlooked risks may be costly for the companies depending on the risk level of the product category. It is important to mention that the financial obligations of the suppliers are limited to the purchase agreement signed at the beginning of the project and may not cover all the issues. Therefore, the author strongly recommends a detailed risk analysis with subject matter experts at the beginning of the project to identify potential risks that may cause project and launch delays. A shared ownership of these risks in case they occur would keep all the parties accountable and responsible to find remedies and actions toward the mitigation plan.

If we need to highlight one of the top risks for the aseptic industry, the lack of fully implemented "change management" should be the one. Changes are inevitable! Every change is a potential risk if they are not evaluated through a structured process by subject matter experts. Change management is the keystone for the "validated state" of the aseptic systems at every stage of the product life cycle. During the design stage, the scope, specifications, or certain parts may have to be changed as part of the product life cycle. Each change may lead to catastrophic consequences and any mitigations further down the road would be more costly. Any changes made after the system qualifications without an implemented change management program would also put a tremendous level of risk on the brand. The criticality of change management and its close connection to the validated state of the aseptic systems are explained in detail in Chapter 16.

14.3 WHAT SHOULD BE THE APPROACH TO MITIGATE RISKS?

So far in this chapter, the focus has been highlighting the vitality, type, and magnitude of the risks and their consequences for the aseptic processing and packaging systems. Once the risks are identified, creating mitigation plans would be very straightforward. Mitigation plans do not eliminate the risks completely. However, it will help organizations to make them visible, prioritize, and take necessary precautions.

Risk analysis is not a one-time event and should be embedded in the product development and commercialization stage-gate process to identify them and minimize the potential unwanted consequences of the identified risks throughout the process. It is critical to create open and collaborative discussions among the subject matter experts, suppliers, and even with the regulatory agencies to minimize potential risks. When it comes to food safety, we are in this all together! Discussions with the regulatory agencies during the preliminary stages of the project would minimize any unexpected surprises and risks surfacing at the last minute. When it comes to the qualification and validation protocols, one size does not fit all! It is highly recommended to challenge the validation protocols for the new systems based on the worst-case scenarios which may be different for each company, site, and production line.

One of the best ways to mitigate the potential risks is creating a learning organization with the development of a proper level of training programs and competencies within the ranks of the company. Responsibility and accountability toward the quality and food safety of the company's brands belong to the brand owner. Proactively determined potential risks and related mitigation plans can only be managed if the company develops internal programs to facilitate learning and developing competencies toward a risk-averse culture.

The best way to go over the potential risks and come up with educated predictions about the production line's performance is to develop an *"operational readiness"* analysis well in advance in the process/product life cycle. One of the recommended approaches would be developing a *Fast Forward First Day Readiness* (F³DR) playscript which would help us fast-forward and take us to the future, on the first day of production, and review all the possible production scenarios, including the first runs of the production. This analysis is extremely important to force every function to think about all the potential risks with "what if" scenarios and corresponding mitigation plans. This concept is further discussed below in this chapter.

14.3.1 Determination of Food Safety Objectives (FSO) as Part of the Risk Management

A simplified way to evaluate the risks related to food safety is proposed to determine the achievable objectives and communicate them scientifically. The following equations define the components of the FSO as defined by Anderson et al. (2011):

$$H_0 - \sum R + \sum I \leq \text{FSO} \tag{14.1}$$

Here,

FSO is the maximum frequency and/or concentration of a hazard in a food at the time of consumption that provides or contributes to *the appropriate level of protection* (ALOP). The ALOP is used to describe the risk level that is deemed as an appropriate or a tolerable public health goal (PHG) and must conform with regulatory policies and public opinion.

H_0 is the initial bioburden of the raw materials of a product

$\sum R$ is the cumulative reduction in the level of the hazard

$\sum I$ is the cumulative increase in the level of the hazard

With this equation (Equation 14.1), we can assign a FSO to any process with the assumption that every product will come with some inherited risks (H_0), we apply some processes to reduce the risks under our control ($\sum R$), and unfortunately, incremental hazards/risks may be added ($\sum I$) during the processing. When we consider all these factors, the result should be equal to or less than our FSO. This method of analysis can easily be applied to aseptic processing and packaging systems to quantify the acceptance criteria for successful validation. For instance, sterilization performance criteria (PC) can be determined based on FSO principles. If one of the risks is related to sterility, we need to determine the PC that is based on an appropriate FSO, which in turn is based on an acceptable *Probability of Producing a Non-sterile Unit* (PNSU). PC is expressed as a *log reduction* (LR) for the most resistant microorganism (target or index microorganism). Therefore, wherever the "log reduction" is mentioned, it is also important to know the product characteristics and the resistance of the target organism of interest for the proposed process.

It would be better if this concept is explained with an example that is related to aseptic process validation. If we want to determine the FSO for the sterilization of aseptic packaging by using vapor hydrogen peroxide (VHP), we need to go through several assumptions and calculations to determine the required log reduction for specific target organisms of concern that pose public health risks. For this exercise, we will assume we have a new target microorganism called *Bacillus ferhanus* (not an actual organism, it was a made-up name for this exercise) with public health concerns and risks.

If we want to achieve PNSU = 1×10^6 for *B. ferhanus* that provides a PNSU > 1×10^9 for *Clostridium botulinum* and vegetative pathogens, we need to identify assumptions and justifications for our analysis. Our first assumption is that PNSU of 10^6 is widely used as an FSO in pharmaceutical and food applications. Also, severity presented by *B. ferhanus* is much lower than *C. botulinum* in terms of public health concerns. We assume that the target strain displays high resistance as compared to other tested strains. It is critical to highlight that all the validation acceptance criteria would be determined based on this analysis and risk levels are set according to the highlighted assumptions and hazards. The next step is to determine the PC for validation based on the risks (personal conversations and exchanges with Dr. Wilfredo Ocasio).

14.3.1.1 What Is The PC and How Do We Calculate the Required Log Reduction?

If we are aiming for FSO that is equal to PNSU of 10^6 (1 positive per 10^6 units) and our PC equal to or more than *4-log reduction of B. ferhanus with the assumption that* 4D provides a PNSU of 10^6 when the initial contamination level is expected to be equal to *or less than 1 spore per 100 containers* (communications with Dr. Wilfredo Ocasio).

14.3.1.2 Log Reduction Calculation

To calculate the target log reduction, we need to make more assumptions. Let's assume that a bio-burden survey of empty bottles showed that incidence of spores is less than 1 in 500. Based on this survey, we can conservatively assume one *B. ferhanus* spore/100 bottles or 1×10^2 / bottle.

Thus,

$$LR = \log N_o - \log N_F \qquad (14.2)$$

where N_o is the initial load of the target organism and N_F is the number of organisms after the treatment.

$$LR = \log(1 \times 10^{-2}) - \log(1 \times 10^{-6})$$

$$LR = -2 - (-6) = \underline{4}$$

Therefore, the LR needed to achieve a 10^6 PNSU is equal to 4, which becomes our PC. Please note that this risk analysis is just an example and may be different for different situations and organisms.

When we are planning for "Microbiological Validation" of an aseptic or any other food/pharmaceutical manufacturing system, we need to utilize surrogate organisms and the calibration of these organisms is an extremely critical step to determine the appropriate performance and acceptance criteria. The correlation between the resistance of surrogate and the target organism must be established for each sterilant (e.g., heat, hydrogen peroxide, e-beam, peracetic acid—PAA). For example,

Bacillus atrophaeus spore crops which are commonly used for hydrogen peroxide sterilization (both hot liquid and vapor forms) are approximately 1.5× more resistant than *C. botulinum*. Therefore, in validation process, a 4-log reduction of *B. atrophaeus* represents a 6-log reduction of *C. botulinum*. It is critical to document the calibration data prior to any validation tests and must be performed for each crop prepared by your supplier and provided to you with a certificate.

14.4 SYSTEMATIC APPROACH IS THE KEY

When we are creating validation strategies and plans, it is critical to determine clear methodologies and tools for risks and related mitigation action plans. A systematic approach to developing strategies for identifying the risks and methods to either mitigate or keep them under control is vital to achieving and sustaining the *"validated state"* of the aseptic manufacturing systems. There are many ways to identify risks. Among many tools, developing a Model HACCP to be applied throughout the organization is a great starting point to understanding overall hazards to an aseptic system. Once established the general understanding of the risks/hazards and related mitigation plans, a more detailed risk analysis for the specific processes and critical equipment must be executed at distinct stages of the product life cycle (PLC). A fitting example of common tools is Failure Mode Error Analysis (FMEA). This method can be applied at various stages of the process through the V-model (Figure 14.1) as explained in the next chapter. In addition to implementation of the risk analysis and validation based on these risks, it is equally important to put a system in place to validate the effectiveness of the plan, review it regularly, and verify it.

V-Model was originally developed and used by the German government with the "Das V Modell" title as an official project management tool (Childs, 2019). This model is widely used in several industries, including software development, nuclear, and pharmaceutical. It has been adapted to graphically demonstrate the validation of food manufacturing (e.g., aseptic processing and packaging) systems. The shape of the diagram has a specific reason. The left side of the "V" identifies the requirements, specifications, and design criteria along with the initial risk analysis steps. The right side of the diagram identifies steps to qualify and validate the requirements listed on the left side. Since Chapter 15 deals with the validation of the aseptic systems, this approach is used as the main guidance to organize the validation activities within the product/process life cycle.

14.4.1 Operational Readiness and Risk Management

Before diving into risk determination and management tools, it is important to discuss the operational readiness of the aseptic processing and packaging systems. When the time comes to execute validation and transition to salable product manufacturing, any flaws and issues would be extremely costly since the commercial commitments would have been made and manufacturing would be under a tremendous amount of pressure of delivering within the launch time frame. The level of stress may create an elevated burden on individuals and/or on the overall organization. Therefore, determination of the risks well in advance and mitigating them prior to the launch date is extremely important for operational readiness.

The *Fast Forward First Day Readiness* (F³DR) analysis is important to set the start-up and ramp-up goals and create a pathway toward a smooth operation on the first day and following 99 days (about three and a half months) to minimize consequences and failures.

A Transition Team with members from the project, business, supply chain, and production facility need to work on a *"Transition Plan"* to minimize the identified risks prior to start-up and production

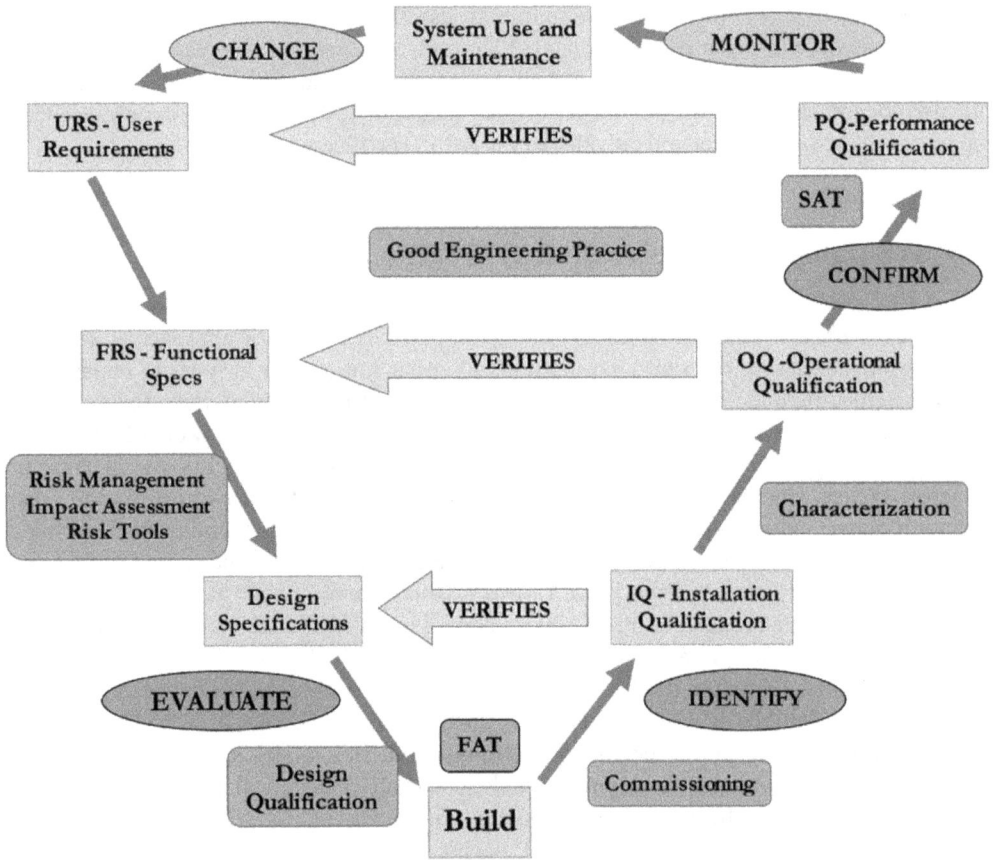

Figure 14.1 The V-diagram outlines the steps to determine the system specifications and design criteria along with the risk analyses on the left side of the diagram and the corresponding qualification steps on the right side of the diagram. At the top of the diagram, the aseptic system achieves the "Validated State" within the total life cycle for the equipment, production line, or the system.

ramp-up. Small operational hiccups are quite common, and they can be overcome relatively easily and in a brief time. However, any food safety and regulatory issues arise right before the launch date or during the operational ramp-up period which may turn into an economically catastrophic event. For lines utilizing innovative technologies, despite the suppliers' commitments and promises, the risks of consistently delivering successful sterility tests are at elevated levels. Any issues arising without any unassignable root causes may take an awfully long time to resolve. Therefore, through the F3DR analysis, several "what if" scenarios must be considered well in advance to eliminate such burdens on the organization and achieve predetermined operational readiness goals.

Still "risks" in mind, commercial launch planning with clear Key Performance Indicators (KPIs) is certainly a good assurance toward successful implementation. The risks for consumers and the business must be considered separately with corresponding risk-based sampling plans. The level of confidence that determines the probability of producing a non-sterile (or a defect for any other manufacturing system) unit (PNSU) needs to be documented with a well-designed production plan, including acceptance batches. Since the consumer risk levels and the nature of the aseptic products determine confidence level in food safety, proper sampling and incubation plans for zero-spoilage without any unassignable root cause must be part of the validation plan. Acceptance batches as they may be considered as "salable" products should be run on a fully validated line according to previously agreed protocols. Expectations and policies of a conservative brand-owning company may guide them to create conservative sampling plans to address the probability of success and confidence levels. Depending on the product types and the level of risks that the company wants to accept, due diligence levels may change from company to company. Any reduction in these requirements and recommendations must be made under the supervision of a competent Process Authority.

As the last line of defense, determination of any potential risks by putting a majority of the burden on the producer prior to shipping the production batches, the qualification runs serve to demonstrate that all components of the system are under control and consistently producing safe aseptic products. Every aseptic production lot for a new process must be incubated and tested prior to release as per the recommendation of the PA. The incubation time depends on the risks and the nature of the products. The incubation time can be reduced based on the product nature, confidence levels, and the historic data from the production line. All sampling plans must be designed based on statistical principles (Stephens, 2001; Shmueli, 2016; Squeglia, 2008). This approach provides aseptic manufacturing companies with a chance to react and inspect the production lots prior to passing the risk on to consumers. This is a critical step toward "brand protection."

Once there is an organizational alignment on the risk-based approach with execution plans for F³DR analysis and development of positive release criteria, several methods of risk analyses can be applied to identify and communicate within the organization. One of the most used methodologies is Failure mode and effect analysis (FMEA) which helps quantification and prioritization of the potential risks at the equipment, process, or system levels.

14.5 FAILURE MODE AND EFFECT ANALYSIS

FMEA is an analytical technique that combines the technology and experience of people in identifying foreseeable failure modes of a product or process and planning for its elimination (Besterfield et al., 2003). In other words, FMEA can be explained as a group of activities intended to

1. Recognize and evaluate the potential failure of a product or process and its effects
2. Identify actions that could eliminate or reduce the chance of potential failures
3. Document the process

To quantify and prioritize the potential risks, Risk Priority Number (RPN) needs to be defined and explained for each risk. RPN is a numeric assessment of risk assigned to a process or steps in process as part of the FMEA.

$$RPN = S \times O \times D \tag{14.3}$$

where
S is the severity (1–6)
O is the occurrence (1–6)—the root-cause, not defect itself
D is the detectability (6 to 1)—reverse scale

The sum of RPNs can be considered as the overall RPN for the entire process. The minimum RPN can be determined as the system acceptance goals prior to starting the line or transferring into commercial production. *It is important to note that RPN is a measure for comparison within a single process; it should not be considered as a measure for comparing risk between processes or systems* (Tables 14.1 and 14.2).

TABLE 14.1   A TYPICAL FMEA RISK ANALYSIS TABLE TO CALCULATE AND PRIORITIZE RISKS ALONG WITH FURTHER ANALYSES OF THE RISKS AFTER THE PROPOSED MITIGATION ACTIONS

Process Step	Key Process Input		Potential Failure Effects	SEV (1–6)	Potential Causes	OCC (1–6)	Current Controls
What is the specific process step in the aseptic system	What is the Key Process Input?	In what ways does the Key Input go wrong?	What is the impact on the Key Output Variables (Customer Requirements) or internal requirements?	How Severe is the effect to the customer?	What causes the Key Input to go wrong?	How often does cause or Failure Mode occur?	What are the existing controls and procedures (inspection and test) that prevent either the cause or the Failure Mode?
Example: aseptic packaging	Bottle rinser	Bottle rinser effectiveness	Improperly rinsed bottles	6	Sterilization failure due to violation of the validation assumptions. Any particles carrying spores surviving through the rinser may create shadowing effect during the sterilization of the bottle surfaces and also add more spore load that might be higher than the initial validation assumptions	3	Existing inspection and controls of the process parameters

(*) EOC—ease of completion. In addition to EOC, "Impact" or environment health and safety—"EHS"—can also be added into RPN calculation.

"**Severity**" of the risk identifies and quantifies the consequences as a measure of impact on public health. Its scale can vary based on the granularity of the risk levels (e.g., 3, 6, 10, or more risk levels). The criticality is not the number of levels, it is the clear definition of each level for consistent application of the criteria. With this risk evaluation, we are trying to quantify the severity of the effect on the consumer (Table 14.3).

TABLE 14.2   A TYPICAL FMEA RISK ANALYSIS TABLE TO CALCULATE AND PRIORITIZE RISKS ALONG WITH FURTHER ANALYSES OF THE RISKS AFTER THE PROPOSED MITIGATION ACTIONS

DET (6-1)	RPN	EOC* (6-1)	Actions Recommended	Responsibility	Actions Taken	SEV	OCC	DET	RPN
How well can you detect cause or FM?	S × O × D		What are the actions for reducing the occurrence of the Cause, or improving detection? Should have actions only on high RPN's or easy fixes?	Who's responsible for the recommended action?	What are the completed actions taken with the recalculated RPN?				
6	108		Create a very conservative and realistic validation challenge protocol for the bottle rinser with all possible particle cross-contamination in terms of their sizes and types. Challenge the rinser with contaminants under the worst-case scenario parameters. Create routine monitoring schemes for the validated parameters	Quality and validation engineering team members	Rinser was challenged with contaminated bottles with predetermined parameters such as the air quality, flow, and distance	6	1	3	18

The **occurrence** component of the risk quantifies how often the cause and failure mode occur. This is where the risk analysis team considers the likelihood of this occurrence and all the existing test/inspection systems to prevent either the cause or the failure mode.

The third component of the risk quantification analysis is the **detectability** which deals with the company's capability to detect the cause or the failure mode. It must be noted that the quantification scale for detectability is in reverse order compared to severity and occurrence. In this

TABLE 14.3   QUANTIFICATION AND PRIORITIZATION OF THE HAZARDS BASED ON SEVERITY AND LIKELIHOOD OF THE POTENTIAL RISKS

Quantification and Prioritization of the Hazards							
Severity	Cause death	6						
	Cause serious illness	5			*Critical hazards*			
	Cause illness	4						
	Cause discomfort	3						
	Cause inconvenience	2	*Non-critical hazards*					
	No significant effect	1						
			1	2	3	4	5	6
			Very unlikely	Unlikely	Rarely possible	May occur	Likely occur	Very likely
			Occurrence—likelihood					

Scales can be changed based on the product/process types and applications.

scale, the harder it is to detect the defect, cause, or the failure mode, the higher the assigned number would be as a multiplier in the RPN calculations.

Once the RPN is calculated, the existing risks can be grouped or prioritized. This helps the second stage of the process which involves mitigation plans for each risk and recalculation of the RPN. With the mitigation plan, actions must be determined by reducing the occurrence of the cause or by improving the detection.

Based on this analysis and the list of RPN numbers, top risks can be identified, sorted, and validation actions can be determined based on these risk numbers. A predetermined range of RPNs can help make decisions for the validation and for the determination of the acceptable defect rates. For example, if the RPN for a specific risk is in the range of 1–6, a simple verification may be the only action that needs to be taken. If the RPN is between 16 and 48, the percent defect rate (PDR—consumer risk level—beta risk) can be relatively moderate. The PDR needs to be determined by thinking like a consumer who wants appropriate protection from any possible harm. If the RPN is between 64 and 125, a much smaller PDR (more conservative sampling may be required) must be selected for better consumer protection. Finally, any RPNs larger than 125 may indicate serious needs for improvement prior to any validation activities. Please note that these RPN numbers are arbitrarily selected for this example, and every team needs to develop a table for similar ranges and validation actions. Validation does always use consumer risk levels (PDR) as the criteria while organization picks an Acceptable Quality Level (AQL) that gives a good process the best chance of passing the validation actions (see Chapter 16 for definitions).

Since risk levels are different for various parts of the aseptic systems, different AQL and PDR may be selected based on the FMEA. For example, the risks for aseptic bottle sterilization and the application of the sleeves on the bottles are quite different and different AQL and PDR values are used to balance the consumer and manufacturer risk levels.

The type of data collection is also important to determine the defect rates. Depending on the data type (e.g., variable, attribute) and the stage of the validation (e.g., operational qualifications—OQ—or performance qualifications—PQ), different line/equipment reliability may be required. For example, if the team is looking for PDR at 95% Confidence Interval (CI) and the expected AQL is 0.01%, the system/equipment reliability would be 99.99% (Reliability = 100 – Percent Defective). According to the *American Heritage Dictionary*, the reliability is defined as the extent to which an experiment, test, or measurement procedure yields the same results on repeated trials. If 0.1% defect is considered for a specific equipment, the system should be 99.90% reliable, meaning 95% CI that the process is operating below 0.1% defective. This is the same as the statement that 95% CI that the process is at least 99.90% reliable. The selected statistical sampling plan must support these statements which will be driven by the consumer and manufacturer risks. Selecting the optimal sampling plan with the balance of alpha and beta risks can be achieved by developing and analyzing Operating Characteristic (OC) Curves (Shmueli, 2016; Stephens, 2001). In addition, collecting and reviewing the historic capability data is important to determine the reliability of the line/equipment. The cost of testing once and the cost of testing repeatedly must be considered in validation planning to avoid any unexpected financial burdens

14.6 ADDITIONAL RISK ANALYSIS METHODOLOGIES

In addition to FMEA as described above, there are many other primary methodologies that might be used in quality risk management by the food industry. This is not an exhaustive list of tools. It is important to note that no one methodology or set of tools is suitable for every situation that requires risk management. Some of these common tools (ICH, 2005. Quality Risk Management—Q9, Annex I) are Preliminary Hazard Analysis (PHA), Hazard Operability Analysis (HAZOP), Failure Mode Error and Criticality Analysis (FMECA), Fault Tree Analysis (FTA), and HACCP, as explained in Chapter 16.

Some of the common techniques are used to facilitate the execution of the risk management tools by organizing data and facilitating decision-making. These techniques include but are not limited to flowcharts, check sheets, process mapping, and cause-and-effect diagrams (Ishikawa diagram or fish bone diagram). Since these techniques are readily available in the literature, their details are not included in this chapter.

14.6.1 Statistical Methodologies for the Production Data Analyses and Better Decision-Making

Since risks are calculated based on probability, any risk-based analysis must use a sound statistical approach that supports risk management. Statistically sound methodologies can enable effective data assessment, help in determining the significance of the data set(s), and facilitate more reliable decision-making. A short list of the principal statistical tools commonly used in the food industry includes but is not limited to Control Charts (e.g., Acceptance Control Charts); Cumulative Sum Charts; Weighted Moving Average, Design of Experiments (DOE); Histograms; Pareto Charts; and Process Capability Analysis.

With technological advancements, there are numerous statistical software packages (e.g., Ignition [Inductive Automation], SAP, Infinity QS) readily available to capture the production data from the aseptic processing and packaging line and process them to display as reports or real-time analysis charts. Even though real-time control is still not very commonly used in the food industry, Statistical Process Control (SPC) software packages are also utilized to control the processes based on the production specification trends by some companies.

14.6.2 Preliminary Hazard Analysis (PHA)

PHA is a method of analysis that is based on applying prior experience or knowledge of a hazard, risk, or failure to identify future hazards, hazardous situations, and events that might cause harm, as well as to estimate their probability of occurrence for a given activity, facility, product, or system (International Council of Harmonization (ICH) Q9, 2005, https://database.ich.org/sites /default/files/Q9%20Guideline.pdf).

The tool consists of (1) the identification of the possibilities that the risk event happens, (2) the qualitative evaluation of the extent of possible injury or damage to health that could result, and (3) a relative ranking of the hazard using a combination of severity and likelihood of occurrence, and (4) the identification of possible remedial measures.

The main principles are very similar to the FMEA approach; however, there may be some differences in terms of their focus during the analysis. Although it is specifically written for medical devices, ISO 14971:2019 (and its guidance ISO/TR 24971:2020) is a great reference to explain the risk management tools and implementation. According to FDA with reference to ISO 14971, the focus of Hazard Analysis is on health- and safety-related issues. For the FMEA, equipment or product functionality can also be considered. This does not mean that PHA ignores functionality, especially if it impacts a process that may harm people.

PHA might be useful when analyzing existing systems or prioritizing hazards where situations prevent a more extensive technique from being used. It can be used for process, product, packaging, facility design, environment, and personal safety. Specificity of the analysis can be determined based on our desire to evaluate the types of hazards for the general product type, the product class, and finally the specific product.

PHA is commonly used at early stages in the development of a project when there is little information on design details or operating procedures. It is often considered as a precursor to further studies. Typically, hazards identified in the PHA are further assessed with other risk management tools such as those mentioned in this chapter.

Instead of using RPN as used in FMEA, PHA uses Risk Class (RC) to categorize the hazards. All the Severity, Occurrence (probability), and Detectability rankings are on the low, medium, and high scale with specific criteria for each one of them. RC can be determined by creating a matrix between Severity and Occurrence ranks. In this matrix, three Risk Classes are determined with RC1 through RC3. After this step, these RCs from the previous analysis are correlated with Detectability ranks for low, medium, or high-risk priority.

14.6.3 Hazard Operability Analysis (HAZOP)

HAZOP assumes that risk events are caused by deviations from the design or operational intentions. With this brainstorming tool, "guide-words" are used for identifying potential hazards. "Guidewords" (e.g., No, More, Less, Other Than Part of, etc.) are applied to relevant parameters (e.g., contamination, temperature) to help identify potential deviations from normal use or design intentions. It often uses a group of subject matter experts to evaluate the design of the processes or products and their application.

HAZOP tool can be used for the food and pharmaceutical industries to evaluate process safety hazards, including manufacturing processes, specific equipment, and facilities that produce a variety of products such as aseptically packaged food products. Like HACCP, the output of a HAZOP analysis is a list of critical operations that outlines the potential risks. Safeguards are determined for monitoring critical points in the manufacturing process.

There are three main steps: design, physical environment, and procedure. A good example of the use of this tool can be found at https://safetyculture.com/checklists/hazop/.

14.6.4 Hazard Analysis and Critical Control Point (HACCP)

HACCP is a systematic, proactive, and preventive tool for assuring product quality, reliability, and safety (WHO Technical Report Series No. 908, 2003 Annex 7). For the food industry, HACCP is mainly used for food safety plan development. It is a structured approach that applies technical and scientific principles to analyze, evaluate, prevent, and control the risk or adverse consequences of hazards at various stages of the process flow. There are so many online courses, books, and tools readily available to learn more about HACCP and its principles. The HACCP risk analysis tool is discussed in Chapter 16 in detail.

14.6.5 Failure Mode Effects and Criticality Analysis

Failure mode error analysis (FMEA) can be enhanced by incorporating an additional investigation for the "degree of severity" of the consequences, their respective probabilities of occurrence, and their detectability. With the additional severity focus, FMEA becomes FMECA (International Electrotechnical Commission—IEC; 60812:2018, FMEA and FMECA). With detailed product or process specifications, FMECA can identify areas where additional preventive actions might be appropriate to minimize severe risks.

FMECA application in the food/aseptic industry can be mostly for failures and risks associated with complexity of the technologies and continuous nature of the manufacturing processes; however, it is not limited to this application. The output of an FMECA is a relative risk "score" for each failure mode, which is used to rank the modes on a relative risk basis.

14.6.6 Fault Tree Analysis (FTA)

The FTA tool (IEC 61025:2006) is a graphical as well as mathematical tool with an approach that assumes failure of the functionality of a product or process. This tool evaluates the aseptic system or individual equipment group for failures one by one. Also, it is possible to combine multiple causes of failure by identifying a chain of causes. The results are presented in the form of a tree of fault modes. At each level in the tree, if there are combinations of fault modes, they are described with logical operators (e.g., AND, OR etc.). As the other risk analysis tools require, the FTA relies on the subject matter experts and their competencies to identify causal factors.

The FTA is used to create a pathway to the root cause of the failure. The FTA can be utilized for the investigations of complaints and/or deviations to understand their root cause. It is also useful to demonstrate assurance that suggested mitigations or enhancements will prevent the issue and not lead to other issues. The FTA is an excellent tool to show how multiple factors affect a given issue. The output of the FTA includes a visual representation of failure modes in a tree format. It can be used for both risk assessment and development of maintenance programs.

14.7 SUMMARY

Risk!—a four-letter word that has so much meaning for our industry. We sell quality and safety! Consumers' trust in our brands heavily relies on how we manage those risks by identifying them well in advance before they create concerns for the public and establish mechanisms to mitigate those risks for a sustainable process. For the aseptic processing and packaging lines, due to technological complexity and requirements of higher competency levels, the risk of failure is extremely detrimental to brand protection. Therefore, brand owners whom they manufacture aseptic products must think beyond the traditional and conventional food industry norms to overcome some of those risks.

Depending on the product category, without creating a huge burden on the organization, some of the risk management practices that are widely used in other industries such as pharmaceutical, medical devices, and nuclear can be adapted for the aseptic industry. Risk management is a cultural norm and should mutate the organization's DNA to be transferred from one generation to another for sustainability of the programs.

This chapter gives the reader an overall view of the risks, their identification, and mitigation methodologies for aseptic systems. The intent of this chapter is not to give the reader the details of each methodology, but rather highlighting what is available and pointing the reader in the right direction for further learning on this subject-critical topic.

Once the risks are identified and documented, the company has the obligation of addressing those risks by

1. Conducting risk analysis methodologies (e.g., FMEA) to quantify
2. Prioritizing based on RPN or similar tools
3. Considering other factors such as easy fixes (may be higher priority can be given to a risk if it does not take that much to complete the mitigation action), EHS concerns, and other serious impacts that may not be included in the analysis
4. Creating mitigation action plans and controls
5. Reprioritizing and communicating with leadership/business with potential consequences
6. Strategizing with the company's risk management program to create internal and external accountabilities
7. Identifying and justifying resources (e.g., people, capital) to complete actions and develop continuous monitoring programs

A systematic approach to the risk management within the company's risk-averse culture will create a sustainable brand protection to secure the company's future. Taking the risk-based approach from the efforts of individuals to the company's well-accepted cultural actions would minimize the occurrence and enhance the detectability; therefore, the consequences of the risks and severity would be minimized.

CASE STUDIES

The following real-life examples of system failures are shared to highlight the criticality of risk analysis at the early stages of the product life cycle.

CASE STUDY #1: SYSTEM OWNER OWNS THE RISK ANALYSIS AND MITIGATION

As discussed, the boundaries of validation depend on the worst-case scenarios based on the risk analyses. Even though the risks were determined by OEMs for each unit operation and hopefully mitigated, every installation is unique based on the company culture, utilities, operational variability, maintenance, and supporting programs. Therefore, additional risk analyses must be conducted to determine the possibility of any incremental risks at the owner's site.

In this example, Company X installed a full aseptic line to manufacture low-acid products in plastic containers. All the validation tests, as outlined in Chapter 15, passed successfully under the worst-case operational conditions with the current assumptions. When the performance criteria and test parameters are determined during the FSO analysis, initial bio-load assumptions are made as the bases of FSO calculations. Based on those assumptions, the initial load of spores on the package is assumed to be 100 per container.

Company X started experiencing sporadic spoilage in their products while the product was in the warehouse or in the distribution. After going through an extensive and stressful root-cause analysis, while the lines were shutdown, several suspects were identified. This type of large aseptic production line can easily cost six figures per day even if they are idle. Therefore, in addition to the opportunity cost, Company X is stuck with the burden of large operational costs. If there are several of these in your tenure with the company, they can be career-ending events!

In this case, all the batching, processing, and filling steps were investigated by multifunctional teams with suppliers' support. Each unit operation was found to be within the operational specifications and there were no suspects. Thousands of additional samples and microbiological testing took several weeks of investigation. The first thing was to determine the type of spoilage organisms and learn about the specific strain characteristics. Then, it was important to identify the same organisms somewhere in the aseptic system. All the aseptic zones of the line test results came back clean.

Sampling was expanded into incoming packaging and environment. Air samples from the areas that product and product contact surfaces have exposure to environment and atmosphere were systematically taken and analyzed. Air exchange rates and positive pressure differentials have been tested to understand the flow of the air in the critical aseptic areas. Microbiological tests of the air samples returned positive results as a match to the microorganisms found in the products. After this great finding, the most crucial step was to find the contamination mechanism by creating different hypotheses and scenarios.

"Sporadic" is an important keyword in this investigation. Since this was a potential cross-contamination case and does not affect every package, organisms were finding their way into the packages. Since the filler swabs came back clean in the aseptic zone, it was not the air sterilization of the filler. After checking the packages, the team did not find the same organisms.

In this type of investigation, it is always extremely helpful to have a timeline of the events and downtime analysis. Based on this analysis, the team found that there is a strong correlation between the downtimes (filler stops in this case) and the time stamps from the containers that showed spoilage. This is a great clue to bringing pieces of the puzzle together. We need to highlight the key clues: (1) The air in the filler room is contaminated with the same spoilage organisms. (2) The spoilage is sporadic and there is a strong correlation between the times of filling spoiling containers and the filler stoppages. (3) Packaging sampling did not show any positives.

It had been the plant practice to bring additional packaging material into the filler room for staging purposes. The surfaces of the staged containers were tested and found to have the same spoilage organisms. In addition, packages were tested right after the downtime (filler stoppage) and found some positive matches. This was the one before the last clue to determine the mechanism of the contamination. Of course, the last piece of information the team charged to find out was the main source of the contamination and the root cause of the failure.

After expanding the investigation into building, utilities, and the change management, it was found that a new procurement employee has decided to change the current Minimum Efficiency Reporting Value (MERV) ratings of the air-handling units to a smaller rating to save money. Although this was done with all the good intentions with the assumption that this is outside of the direct manufacturing area, the person(s) who made the decision did not have the background to do a risk analysis and did not run this change through the change management procedure. This person was not even on this site and did not have any exposure to aseptic systems. When running risk analyses and Operational Readiness scenarios, this kind of risk must be considered, and related mitigation plans should be implemented.

To summarize this case study, a change in an external filtration system outside of the equipment aseptic zones and significant changes in concentration of incoming particulates as the carriers of the spores significantly challenged the system beyond the assumed validation specifications and increased the risk of spoilage microorganisms reaching into filler room and finding their way into the product. This is the story to share with the equipment manufactures when next time they tell you that their equipment can be installed in the warehouse or in your garage and still deliver aseptic products!

This was a very costly learning experience for Company X and hopefully will help the readers of this chapter to prevent this food safety risk by understanding that risk analysis must be part of the entire product life cycle and not just one stage of the cycle.

CASE STUDY #2: EQUIPMENT MANUFACTURER (OEM) OWNS THE RISK ANALYSIS AND TRANSFER TO THE OWNER

As the previous case study was focusing on the additional risks which can be generated due to lack of training, gaps in preventive maintenance, and environmental changes, this case study will focus on the risks that can be identified and prevented at the design stage of the equipment. If one of the wings of an airplane falls off up in the air, what is the likelihood of landing this airplane safely? I believe the answer is a very slim chance that the airplane will land safely. Therefore, the focus must be on the failure mode that causes the wing to fall off. A proper risk analysis at the design stage with all the material failure predictions and possibilities should allow OEMs to put necessary mitigations in place to prevent this failure.

In this specific case, Company Y installed an aseptic filler as part of the aseptic production line for low-acid products in plastic containers. After several years of operation without failure of this part in the aseptic filler, a chemical sterilant that is used to sterilize the closure material was sensed in the filling area by chance—another advantage of installing the aseptic fillers in their own room with external controls such as air quality, exchange rates, positive pressure differential, and monitoring sensors. In this case, the location of the sensors also played a role since the chemical sterilant was leaking into the area at the top of the filler and sensors are not at eye level to protect the operators.

When the OEM designed the filler and the sterilant delivery mechanism for the closures, they did not consider this as a high-risk item and the corresponding mitigations and preventive maintenance were not implemented. Due to the failure of the material in the line, the sterilant delivery line started leaking a significant amount of the sterilant into the surrounding atmosphere. This failure may have two separate risks: (1) Personal hazard for mechanics if they climbed up on the filler and exposed to the sterilization chemical. (2) Food safety risk if not enough sterilant is delivered to the closures due to leakage.

Once the issue is identified, the solutions are straightforward! However, without proper warning sensors or notice by somebody, it is hard to detect this risky situation. If FMEA is applied to this risk, the highest contribution to RPN would come from the "detectability." Generally, it is very hard for the owner/user of the filler to determine this kind of risk since they will require technical knowledge of the operational principles of the filler. It is more suitable to cover this kind of failure mode analysis during the design stages by the OEM subject matter experts.

Many days of investigation and holding a substantial number of products during this period were very costly for Company Y and their economic loss multiplied by losing unbelievably valuable line availability for production (opportunity cost). Once the root cause was determined and documented, along with OEM collaboration, the risk was mitigated by placing additional sensors close to the application area for early detection of the failure, changing the preventive maintenance frequencies to match the material characteristics of the part, training the operators to inspect the areas for visual indicators, and add more programming alarms if there are any flow rate changes in the system.

As this case demonstrates, early detection of the failure modes throughout the life cycle of the product is extremely important to minimize the food and personal safety for the aseptic processing and packaging systems. The key is to consider the entire aseptic system with all its components, including the building and utilities. Conducting the risk analyses in silos may blindside us and may miss critical risks of failure.

GLOSSARY

Cause: Specific steps of the process can cause a failure mode to occur. For example, a material fatigue or worn (cause) on the peroxide sterilant delivery system may cause broken silicone connections (mode) in the assembly, which may cause the failure of delivering a satisfactory level of sterilant (effect) for proper sterilization.

Detection: An assessment to quantify based on the predetermined risk level and identify the likelihood that the current controls will detect the cause of the failure mode or the failure mode itself to prevent the failure effect from reaching your consumer/customer. The risk scale for detection is in the reverse order to indicate that the risk level is much higher if the process does not have all the detection tools for failure modes and/or causes. The customer in this case could be the next operation, subsequent operations, or the end user (consumer).

Effect: A product or process that does not perform according to design criteria may cause an adverse impact on the consumer or another process.

EOC: Ease of completion. This can be added as an assessment for the prioritization of the risks during the implementation of FMEA. The ranking is in reverse order to give a higher number to easy actions to address some of the risks. Giving higher numbers to an easy fix for relatively low hanging fruit will result in higher RPN with an elevated level of priority to address this risk.

Failure mode: The way in which a specific processing step or unit operation fails. If they are not detected, corrected, or removed, they may cause a negative effect to occur.

FMEA: Failure mode error analysis. This is a structured approach to estimate the risks associated with specific causes by identifying the possibilities that a product or process can fail.

Occurrence: An assessment to quantify based on a predetermined risk scale and identify the likelihood that a particular cause may occur and result in the failure mode.

pFMEA: Process FMEA. Capturing the entire process steps and unit operations to identify the risks and specific failure modes for each step in the process flow.

PNSU: Probability of non-sterile unit.

RPN: Risk Priority Number. This is a calculated number based on the quantification of severity, occurrence, and detectability and by multiplication of these values. RPN is not limited to these three factors. Other factors such as Environmental Health and Safety (EHS), Ease of Completion (EOC), and Impact Score (IS) can also be considered to prioritize the risks by creating a risk scale for each one of these additional factors.

Severity: An assessment to quantify based on a predetermined risk scale and identify how serious the failure effect (due to the failure mode) is to the consumer.

REFERENCES

Anderson, N., et al. 2011. Food Safety Objective Approach for *Controlling Clostridium botulinum* Growth and Toxin Production in Commercially Sterile Foods. *J. Food Prot.* 74(11):1956–1989.

Besterfield, D.H., Besterfield-Michna, C., Besterfield, G., and Besterfield-Sacre, M. 2003. Failure mode and effect analysis. In *Total Quality Management.* 3rd edition, Chapter 14. Upper Saddle, River, NJ: Prentice Hall.

Childs, Peter R.N. 2019. *Mechanical Design Engineering Handbook.* Edited by Das V Modell. 2nd edition. Amsterdam: Elsevier Ltd. https://doi.org/10.1016/C2016-0-05252-X

International Council of Harmonization (ICH). 2005. *Q9: Risk Management Methods and Tools.* https://database.ich.org/sites/default/files/Q9%20Guideline.pdf

Ocasio, W. 2019. Personal communications.

Shmueli, G. 2016. *Practical Acceptance Sampling,* 23. 2nd edition. Green Cove Springs, Florida, USA: Axelrod Schnall Publishers.

Squeglia, N.L. 2008. *Zero Acceptance Number Sampling Plans.* 5th edition. Milwaukee: ASQ, Quality Press.

Stephens, K.S. 2001. *The Handbook of Applied Acceptance Sampling.* Milwaukee: ASQ, Quality Press.

US FDA, 21 CFR Part 113. n.d. *Thermally Processed Low-acid Foods Packaged in Hermetically Sealed Containers.* https://www.accessdata.fda.gov/scripts/cdrh/cfdocs/cfcfr/CFRSearch.cfm?CFRPart=113

US FDA, 21 CFR Part 120. n.d. *Hazard Analysis and Critical Control Point (HACCP) Systems.* https://www.accessdata.fda.gov/scripts/cdrh/cfdocs/cfcfr/CFRSearch.cfm?CFRPart=120

Chapter 15

Establishing "Validated State" of Aseptic Processing and Packaging Systems

Dharmendra K. Mishra, Ferhan Ozadali, Patnarin Benyathiar, and Jairus R.D. David

CONTENTS

DOI: 10.1201/9781003158653-18

15.1 INTRODUCTION

This chapter deals with the validation concepts and their criticality for aseptic processing and packaging systems. Since extended shelf-life (ESL) systems are very similar to aseptic systems, a clear transfer function between the two systems will be obvious to the readers. Although the term *validation* is widely used, the true meaning is not very well understood in some of the industries, including the food manufacturing ones. Aseptic techniques are widely used in the pharmaceutical industry; therefore, some of the concepts covered in this chapter are also applicable to this industry.

The title of this chapter includes "*Validated State*" terminology which is much different than any other definitions in the manufacturing industries. What is the significance of this terminology? Why is Validation a "*state*"? Because validation is a gained status and not a time-dependent event. Due to its volatility, this status can be lost instantaneously without proper implementation of the required prerequisite programs. It is critical to clarify that validation is *not* a "*project*," it is a process! Projects have starting and ending dates, validation does not have the ending date until the retirement of a process or manufacturing line. The reader of this chapter will learn about the steps required to achieve this critical "*state*" of the process. Similarly, there are lots of information-misuse and definition-confusions in the industry regarding the terminologies used during the life cycle of the products. The reader will understand the differences and synergies among the qualifications, validation, and verification concepts.

"Validation is NOT a project!"

In previous chapters of this book, several concepts such as microbiology, risk-based analyses, food safety objective (FSO) determination, statistical sampling criteria, and some of the tools have been covered to prepare the reader with a foundation to build the framework for the overall validation concept. This chapter can be considered as a *"Roadmap"* to risk- and science-based implementation toward production of the safest and the highest-quality products with an appropriate level of confidence.

The next logical step is sharing the definition of validation. *"Validation is defined as the establishment of <u>documented evidence</u> which provides a <u>high degree of assurance</u> that a specific process will <u>consistently produce</u> a product meeting its <u>predetermined specifications and quality attributes</u>."* This is a powerful definition with several embedded valuable targets. We further discuss underlined words in this definition:

1. <u>Documented evidence</u>—indicates that objectives, tests protocols, data collection, and reports must be developed and executed.
2. <u>High degree of assurance</u>—asking the question if we have taken into account all factors that may affect the process/product and they have been effectively challenged.
3. <u>Consistently produce</u>—Is the process capable and stable with appropriate statistical process sustainability?
4. <u>Predetermined specifications</u>—What are our acceptance criteria and how do we know when we achieve them?

15.2 VALIDATION MASTER PLAN (VMP)

Due to commercial and quality benefits for certain product categories, aseptic processing and packaging systems have been very desirable. However, they are operationally and technically challenging compared to conventional technologies such as canning. In spite of the benefits, there are several business constraints such as high initial capital investment, operational complexity, requirement of the highly skilled employees, and development of "corporate aseptic culture." The type and the nature of aseptic processing and packaging equipment and facilities depend upon the market demand for product or product mixes, marketing research, caliber of research and development, innovativeness, entrepreneurship, regulatory constraints, and availability of capital funds for deployment.

A clearly written plan for the implementation of the *"Roadmap"* is the most crucial step that describes the details of processing and corresponding validation steps. Such a plan would address objective test parameters, product and process characteristics, predetermined specifications, resource needs, risks, and factors which will determine acceptable results. Although VMP is not a regulatory document, it is considered as a roadmap to success.

The objective of any validation process is to prove that a process works and can work routinely thereafter for routine production. The concept and process of validation pertaining to aseptic processing and packaging of food are complex. It is important to understand the scope, information content, and limitations of any validation process and its relationship to day-to-day process capability and deliverables. The VMP should include the overall commissioning approach for the aseptic processing and packaging systems and how it relates to the qualification of equipment. The qualification steps are design qualification (DQ), installation qualification (IQ), operational qualification (OQ), and performance qualification (PQ). These activities define the boundary of operations based on the equipment capabilities such as flow rate, pressure, viscosity, overpressure in aseptic filler, filler speed, etc. The maximum product runtime for the aseptic system should also be established to allow proper cleaning and sterilization. Risk and impact analysis as mentioned in the previous chapter should also be a part of the VMP. In addition to the technical

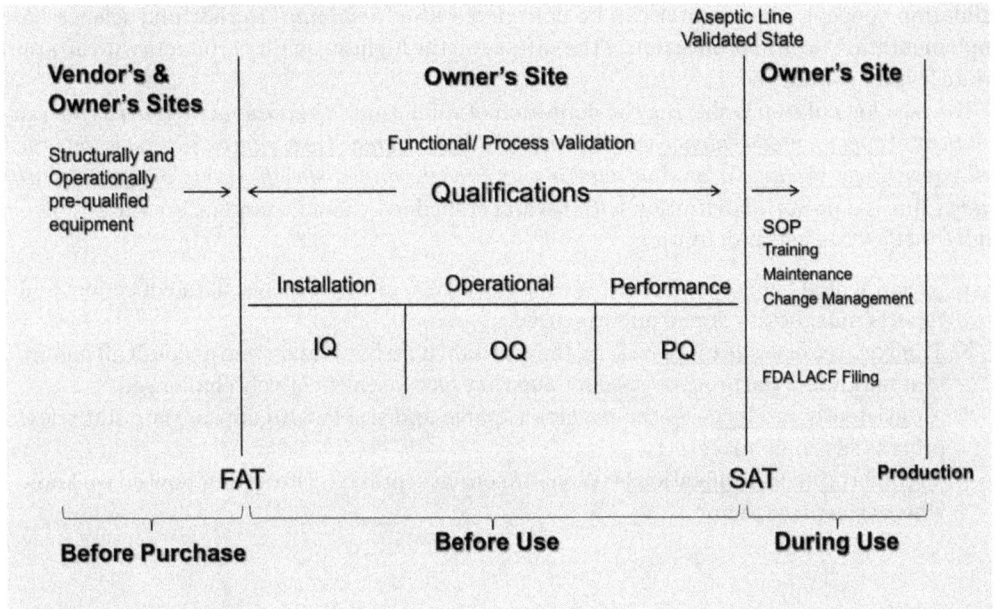

Figure 15.1 Who is doing what and where during the project timeline to satisfy the aseptic line's "validated state" and beyond. Since this sequence of events and their outcome would affect the success of the next event, the timely and systematically planned executions of these events are critical. Appropriate level of contractual obligations and accountabilities must support milestone deliverables such as the shipment of the equipment from the OEM location to owner's site after successful execution of the FAT protocols and required payments committed.

documentation, there are a few organizational structures such as team selection, change control, and project management that should be considered.

VMP outlines the qualification, testing requirements, reporting, organizational structures, roles, and responsibilities at various stages of the project timeline. The validated state of the aseptic line can only be achieved with a well-established series of activities and test protocols. As outlined in Figure 15.1, there are three stages of this timeline: before purchase of the equipment, before use, and during use. There are two main decision points that would qualify the line to be advanced to the next stage. The first one is the factory acceptance test (FAT) which is the decision step to allow shipping the equipment from the vendor site to the owner site. At this time, the equipment is considered prequalified for structural and operational specifications and significant financial obligations might be fulfilled.

Once the equipment arrives at the owner's site, all functional and process qualification protocols are executed before using the equipment for production. These stage activities will include installation qualifications, operational qualifications, and performance qualifications until the decision point after the Site Acceptance Tests (SAT). Upon successful execution of the SAT protocols, the line is a candidate to be at the "Validated State" (Mishra 2021a, b).

The last stage outlines the requirements to gain and maintain the validated state status for the aseptic line until the first commercial "salable" production. Requirements to maintain the "validated state" are, but not limited to:

1. Fully implemented Standard Operational Procedures (SOP)
2. Implemented Training program
3. Preventive Maintenance program
4. Implemented Change Management Program
5. Regulatory filing (e.g., USFDA LACF filing) of each product or product category based on the line validation documentation.

15.2.1 Microbiological Validation Plan

Microbiological validation of aseptic systems is one of the most important criteria for the safe production of acidified and low-acid food products. As compared to the traditional systems, the aseptic processing and packaging system typically has more critical control factors and requires proper commissioning and validation approach for the safe manufacturing of food products. The validation includes Ultra-high Temperature (UHT) inoculation tests for homogeneous and particulate food products, sterilization of aseptic tank and filters, sterilization of valve clusters, preproduction sterilization of aseptic filler, sterilization of packaging and lidding materials, and maintenance of aseptic zone sterility (Anderson et al., 2020). A suitable surrogate organism should be chosen for the microbiological validation work to prove the target log reduction and its correlation to the target pathogenic microorganism. In this chapter, a detailed outline is provided to accomplish the validation according to the regulatory frameworks and daily operations of the manufacturing facility.

There are multiple teams that are typically involved in the proper execution of the validation activities. The consequences of improper validation could be extremely costly as it may result in a food safety hazard. Hence, every aspect of the validation work must be considered properly as it may vary from one manufacturing facility to another, and it also depends on the types of products being manufactured. A comprehensive approach and overall validation framework should be developed at the onset. Multidisciplinary teams should include process authority, and representatives from product development, processing, packaging, quality, microbiology, chemistry, automation, and operations groups.

Aseptic processing and packaging equipment setup and operation is complex. Before one commercially produces aseptic products via a newly designed and installed aseptic system, it is necessary to perform a pre- and post-installation review, and to test and validate the equipment to facilitate process filing. The objective of any validation process is to prove that a process does what it says it does, or claims it does, and not anything else. It is interesting to note that most European regulations rely exclusively on "spoilage" data as a measure of how well an aseptic system works. The US Food and Drug Administration (FDA), however, requires microbiological challenges and chemical tests to document whether an aseptic system provides an adequate margin of safety. Based on the authors' extensive experience, it is desirable to have both challenging data and a comprehensive spoilage database for a complete understanding of process capability and deliverables.

It is desirable and necessary to follow the well-established validation V-model (Figure 15.2) for the validation of a new aseptic manufacturing process. The validation is not a project but a "state." Any inconsistencies and changes in the process, no matter how small, can easily jeopardize the validated state of the system and create a food safety risk. User Requirements Specification (URS) defines the requirements of a food manufacturer to produce specific food products. Based on the URS, a Functional Requirements Specification (FRS) is developed in collaboration with the equipment manufacturer and the food manufacturer. FRS describes the

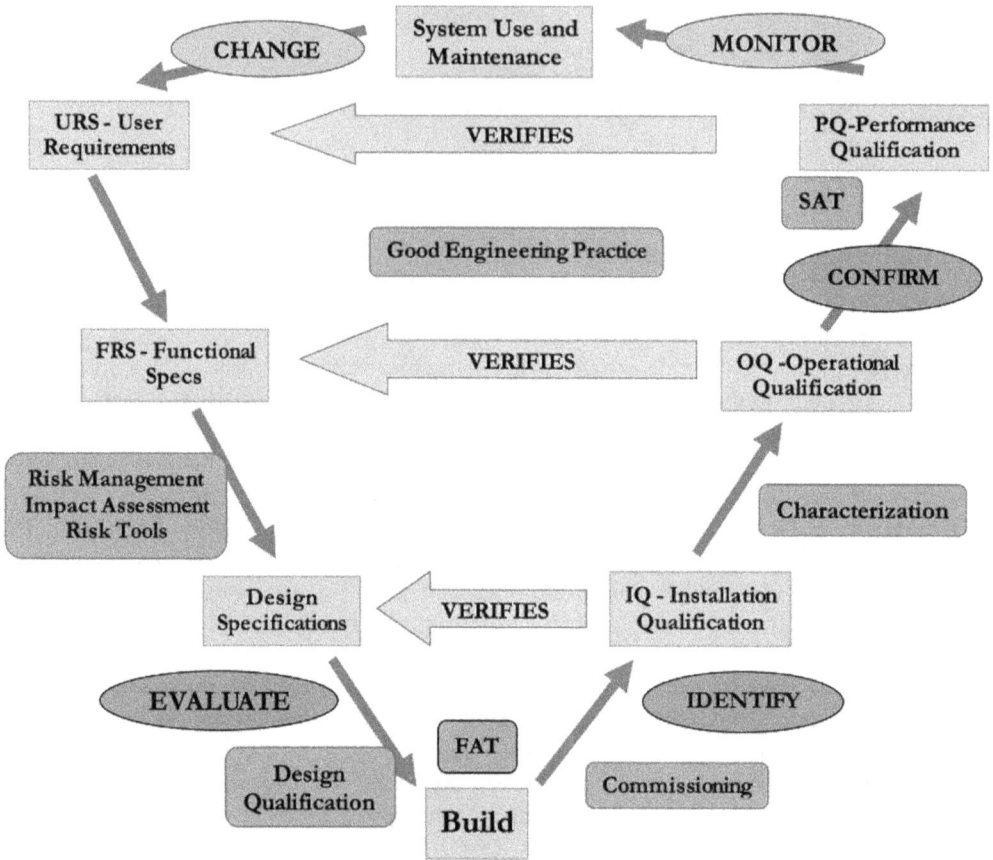

Figure 15.2 The V-diagram outlines the steps to determine the system specifications and design criteria along with the risk analyses on the left side of the diagram and the corresponding qualification steps on the right side of the diagram. At the top of the diagram, the aseptic system achieves the *"Validated State"* within the total life cycle for the equipment, production line, or the system.

functional characteristics of the equipment and sets the boundaries of operation such as maximum viscosity of a product that the system can handle. Using the information from URS and FRS, the Design Specifications (DS) are developed for the system. At this point, the system is ready to be built at the equipment manufacturer's site. Definitions of these document terminologies are provided in the Glossary section.

One of the commonly asked questions is about the level of qualification requirements for each unit operation in the aseptic system, including the facilities and utilities. The entire aseptic line must be evaluated in conjunction with the equipment vendors, integrators, and the engineering construction companies along with the internal functional represeners to execute an impact analysis for each unit operation on the entire system based on the risks. A typical impact analysis form (Table 15.1) can be constructed to identify the necessary protocols and qualification levels for each unit operation of the system. Different unit operations (e.g., aseptic filler, UHT, labeler, HVAC system) require different levels of qualifications based on their impact and risk analysis.

TABLE 15.1   A SAMPLE TEMPLATE FOR THE QUALIFICATION MATRIX FOR THE ASEPTIC SYSTEMS, INCLUDING THE SUPPORTING FACILITIES AND UTILITIES

Equipment/System	Impact			Qualification Requirements					
	D	I	N	COM	FAT	SAT	IQ	OQ	PQ

D	Direct Impact	SAT	Site Acceptance Test
I	Indirect Impact	IQ	Installation Qualification
N	No Impact	OQ	Operational Qualification
COM	Commissioning	PQ	Performance Qualification
FAT	Factory Acceptance Test	CSQ	Computer System Qualification

Once the list of the processes and unit operations are listed, key functions with their subject matter experts (SMEs) must have an alignment on the required test protocols, qualification levels, responsibilities, and acceptance criteria. It is critical to determine the acceptance criteria during the preliminary stages to minimize conflicting interpretations after the tests. For instance, while an aseptic filler will require a complete list of qualifications, a compressor may require just an operational qualification. Any part of the aseptic system that may impact the food safety and critical quality attributes should be subject to full qualification and validation steps. This table that is constructed with all unit operations and supporting utilities can be included in the Validation Master Plan as one of the key guidance documents.

15.2.2 Acceptance Criteria

The acceptance criteria should be established prior to conducting qualification and validation trials. Depending on the manufacturing operations, there could be a variety of acceptance criteria for various unit operations and equipment. Some examples of acceptance criteria are presented in this section that will help manufacturers to develop their own criteria for specific manufacturing lines.

15.2.2.1 Hydration of Dry Ingredients

The hydration parameters of dry ingredients are directly related to food safety and food spoilage risks. The dry ingredients can be either in powder or particle form. An example of powder dry ingredients is cocoa powder, and the proper hydration time and temperature conditions must be defined to avoid any potential failures during the final performance qualification trials. The mixing of cocoa powder can also be critical, and the final acceptable particle size can be defined for the batching operations. Another aspect of hydration is related to aseptic food products with discrete particles. An example of such a product is soup with pasta, and the pasta in this case is dry raw material. If the pasta is not hydrated sufficiently, there could be issues with the heat transfer and lower than minimum lethality at the center of the particle. Hence, acceptance criteria should be established to avoid potential problems during production.

15.2.2.2 Effectiveness of Particle Distribution

Uniform distribution of particles is critical for the aseptic processing of products containing particles. Non-uniform distribution of particles can pose a food safety risk due to the changing ratio of the carrying fluid to the number of particles. Significant variability in this ratio may affect the heat transfer characteristics, impact the particle clumping, and flow regimes. The blending tanks must be designed to operate in a way that the particles do not clump and that it provides a uniform distribution of particles. Acceptance criteria can include the blending temperature, mixer rpm, and sauce viscosity.

15.2.2.3 Batching Time and Temperature

For the upstream batching equipment, depending on the complexity of the product formulations, there could be several unit operations in series. The batch time and temperature must be designed to reduce the microbiological risk from mesophilic and thermophilic microorganism growth. Batching time is also important to allow enough time for uniform distribution of the particles and prevent any ingredient fallouts or precipitations. Selection of an appropriate batching temperature may help with creating suitable viscosities to keep the particles suspended throughout the batching process.

15.2.2.4 Aseptic Filler Operations

Acceptance criteria of aseptic fillers are based on the key performance deliverables that could include filler speed and defect rate. An example of defect rate is that filler is capable of producing a certain number of packages without any microbiological growth during incubation of the sample. The clean-in-place (CIP) and sterilization-in-place (SIP) time should also be established and tested during qualification trials. The microbiological log reductions for filler presterilization and packaging material sterilization should be listed in the acceptance criteria.

15.2.2.5 Aseptic Processing Operations

In order to achieve "*commercial sterility*" of the product, heat treatment is applied with a variety of heat exchangers to heat and cool the product under aseptic conditions. Depending on the type of the product and its characteristics such as pH, aseptic processing systems are designed to achieve target lethality for microorganisms of interest. Both high temperature short time (HTST) and UHT systems are used with different validation design criteria.

Throughout the production cycles, the same aseptic processing system can be utilized in different modes (CIP, SIP, production, and hibernation) for different purposes. Each mode may require different operational set points and validation requirements. Due to energy-saving purposes, the aseptic system can be designed to enter the "*hibernation*" mode when the line running on water and the flow are lowered and the required sterilization temperature is optimized to maintain the sterility. Validation acceptance criteria along with the operational parameters should be defined for all modes of operations. For instance, the maximum runtime without product fouling or to an acceptable level of fouling must be defined to prevent any process-related issues. Excessive fouling may change the flow and heating characteristics of the product in addition to issues related to effective cleaning.

15.3 ASEPTIC SYSTEM SPECIFICATIONS

As mentioned earlier, aseptic processing and packaging of food is a continuous process. Thus, the performance of various system components is interdependent (Bernard et al., 1987). Therefore, equipment design and selection must be viewed not as a selection of a number of individual components, but as a selection of a complete and compatible system. There are four major components of the aseptic system: (1) batching equipment, (2) UHT/HTST, (3) aseptic tank, and (4) aseptic filler. The interdependent connections are mainly clamped connections, welded connections, and sterile valve assembly. There are several design aspects to the sterile product-contact surfaces and protection from any contamination from the outside environment using a sterile barrier.

The first step in the selection of aseptic processing and packaging systems is to review the pertinent regulations, as described in Chapter 8. If the food product contains at least 3% red meat (raw basis) or 2% cooked poultry, the US Department of Agriculture (USDA) Food Safety and Inspection Service (FSIS) has regulatory oversight. These regulations are contained in "Guidelines for Aseptic Processing and Packaging Systems in Meat and Poultry Plants," published by the FSIS (USDA, 1984). All equipment needs approval or could be on the approved list already, prior to installation. Prior to producing product, FSIS requires that the processor submit an "acceptable" proposal for container and equipment testing, and an acceptable partial quality control (PQC) program covering the operation and maintenance of the aseptic system.

For products that do not contain meat or poultry as specified earlier, the FDA has regulatory oversight (with the exception of pet foods with meat). In addition, FDA has regulatory authority over all commercially sterile shelf-stable pet foods. FDA regulations can be found in Title 21, Parts 108, 117, 113 (foods with pH greater than 4.6 and water activity greater than 0.85), and Part 114 (foods with a pH less than 4.6 as a result of acidification and water activity greater than 0.85) of the Code of Federal Regulations. These regulations require that a process filing be submitted to the FDA prior to processing and packaging the product.

Finally, if the product to be produced is either a milk or milk-based product, the facility may additionally be required to comply with individual state-adopted Pasteurized Milk Ordinance promulgated by FDA's Milk Safety Branch (USPHS, 2019).

Due to the nature of the regulations involved, it is strongly recommended that a competent process authority having expert knowledge of regulations and thermal processing be consulted. Such consultation will help avoid costly mistakes and unnecessary delays in production startup. In addition to the pertinent regulations, a processor must take into consideration the type of product to be processed, the type of container to be used, the sterilant to be used, the production rate (i.e., product flow rate or container production rate), and a variety of other production parameters. The specifications are often interdependent, for example, product-related specifications could include viscosity, color, flavor, and overall quality. The determination of heating a heat exchanger should be based on the criteria established in the URS. As an example, if the product contains large particles, the type of filler to be used will be different than for products that are homogeneous and do not contain particles. The various factors involved in the selection process are largely interrelated and oftentimes dictate the choice of equipment available, especially for packaging.

15.3.1 Process Schematic

The first thing that should be done for any product—whether it be meat, poultry, or milk, or an acidified or acid product—is to prepare a process schematic. This should be in enough detail to determine that the product can be processed, the system can be sterilized, the system can be idled (i.e., stop, start, restart), the system can be cleaned, and the system can be switched from water to product or from product to water during production without losing sterility. The method(s) of heating and cooling are often determined by a process authority, and a cross-functional team consisting of process engineering, product development, sensory, packaging, microbiology and other disciplines.

The process schematic should be accompanied by a general written description, which will indicate how the various operations stated will be accomplished and agreed to by all personnel, including management, quality assurance and control, product development, and even sales. Then a more detailed process and instrument diagram (P&ID) should be prepared. In conjunction with the P&ID, a detailed description of all unit operations is required.

15.3.2 P&ID Schematic

The P&ID should be in much greater detail than the process schematic and must show every detail of valves, heat exchangers, holding tubes, product line sizes, product line changes, and length of line. On an accompanying sheet(s), because most commercial processes will be extensive, a listing of the various components should provide enough detail so that the engineers can know the important aspects required to design the system properly. For example, details given for a valve should include its size, the pressure of air that is required if it is air actuated, sanitary design, cleanability (CIP or SIP), hermeticity, and compliance to 3A standard.

The P&ID should be provided in numerous versions, that is, in stages and sequences indicating the flow of critical materials during various operations. Critical materials would be water during sterilization or product during operation, heating water or steam, cooling water (tower, well, or refrigerated), air, electrical, CIP solutions, chlorination, and so on. It should be determined by the engineer whether the pressures in the heat exchanger shells, or jackets are within the safe limit of the design of the heat exchangers.

The size of interconnecting tubing and the flow through the heat exchangers during sterilization, operation, and CIP should be determined and verified for their adequacy. For example, if the flow through the holding tube is not of the proper velocity, laminar flow may exist. If products such as eggs or chocolate toppings are processed, the holding tube may effectively decrease in size as the production run progresses through the day, due to product buildup or fouling, if not corrected.

15.3.3 Design Review

Once the equipment, processing, packaging, pumping, and controls are selected, the processor should work with a process authority in reviewing the overall design of the aseptic system. This review should include the processing and packaging systems and the interface, including the aseptic surge tank, if any. The process authority will review in detail the equipment presterilization procedures (processing systems, packaging systems, filler, aseptic zone, surge tanks, interfaces, and sterile gas lines), production procedures, procedures for maintaining sterility within the system at all times, postprocess cleanup procedures, record keeping, and quality control procedures.

During this review, a hazard analysis and critical control points (HACCP) system should be established. It is during this review that a hazard analysis (HA) is conducted, and critical control points (CCPs) are identified along with monitoring requirements and corrective actions necessary when they are out of specification. It is important to note that the low-acid canned food (LACF) regulations do all this.

The process authority should review the system in terms of critical factors that would have to be controlled, monitored, and recorded during all phases of the operations. Critical factors are those parameters that could, if out of specification, affect the eventual sterility and quality of the products. A preliminary list of critical factors is prepared at this time. A list of critical control parameters is generally a part of the process filing that will eventually be submitted to the appropriate regulatory agency for acceptance.

During this review, the process authority also reviews the instrumentation and controls on the aseptic processing and packaging systems to ensure that they meet the requirements of the regulations. As a part of this, the process control software and operation are reviewed to ensure that appropriate alarms and monitoring functions will be conducted during all phases of the operation. Any incompatibilities between the processing and packaging systems should be resolved at this point.

The last phase of the review involves the preliminary design of the system and challenge testing that would be considered necessary by the process authority to assure proper functioning of the equipment. A process authority should be cognizant of the type of data the regulatory agency will require during its review of the process filing. The tests should be designed to ultimately convince the regulatory agencies that the finished container of product would be commercially sterile and shelf-stable.

Any major modifications necessary should be implemented prior to the installation of the equipment. Modifications to equipment after installation are almost always more difficult and often expensive. A system may need to be revalidated after any modification, depending on the nature of the change and related impact analysis. All modifications should be handled through a dedicated change control management policy and procedure, and reviewed, endorsed, and documented by a cross-functional team. In some cases, the aseptic system may be partially revalidated if the impact of the change is well-studied with the appropriate risk analyses. For instance, some of the computer controls and automation changes may be tested with partial validation for the parts of the equipment affected.

15.3.4 Sterilization, Operation, Clean-in-Place, and Maintenance

It must be defined, in writing, how the system will be sterilized and cleaned. Sterilization can be readily checked through inoculated packs or inoculation of vulnerable areas, and if satisfactory, the filing can be made to the FDA. However, if the system operates only for 1–2 hours before fouling occurs, it must be stopped, cleaned, resterilized, and then switched to product for operation. This obviously is not a satisfactory procedure. The reason for fouling must be determined and corrections must be made.

The process system and packaging machine must be designed in such a manner that people normally available in food plants can operate the system with the appropriate aid provided. These aids include colored schematics and written instructions, along with periodic training and education and certification as appropriate.

Usually, process systems are fairly simple to operate compared to the packaging system machine. Even so, it is advisable to have the suppliers provide individuals to start the system, make adjustments if required, process product after initial sterilization, and CIP the system the first week or two of operation, and train the operators so that they can do the same thing with confidence.

If the system is one that requires a significant amount of maintenance, the maintenance department needs to know what makes up the system so that spare parts and critical items can be installed when required and checked at certain intervals. The maintenance crew should be trained additionally in preventive maintenance and aseptic techniques.

The same is true, and probably more so, with certain packaging systems that are used. Some packaging systems are fairly simple while others are very complex. The strength of the peroxide solution is generally checked in the quality control (QC) laboratory; hence, the lead operator must verify the information regarding the strength of the hydrogen peroxide on a regular basis by report. Anytime the source of hydrogen peroxide is changed (i.e., carboy or drum) or whenever the shipment is changed, the strength of the sterilizing solution must be recorded. Also, the operator must determine at the end of the day, and possibly throughout the day, the strength of the hydrogen peroxide and the consumption rate required to adequately perform the job required.

The cleaning crew, which may include individuals from the operation crew, should verify that cleaning has been performed properly and that the system is clean. This can be done with tubes by examining them with a flexible borescope on a routine basis. For other items of process or packaging equipment, a visual examination can indicate grossly soiled surfaces. Any surface that is suspect should be examined by a member of the QC laboratory, using swabbing or some other appropriate microbiological technique.

15.4 FACTORY ACCEPTANCE TEST

Aseptic operations are complex by nature and there are several key pieces of equipment that need to work flawlessly. It is critical to ensure that the purchased equipment functions properly and satisfies the contractual obligations. FAT helps assure both parties (equipment manufacturer and buyer) that the machine meets all the user-required specifications. The FAT is conducted at the equipment manufacturer's site before being shipped to the buyer (Figure 15.3).

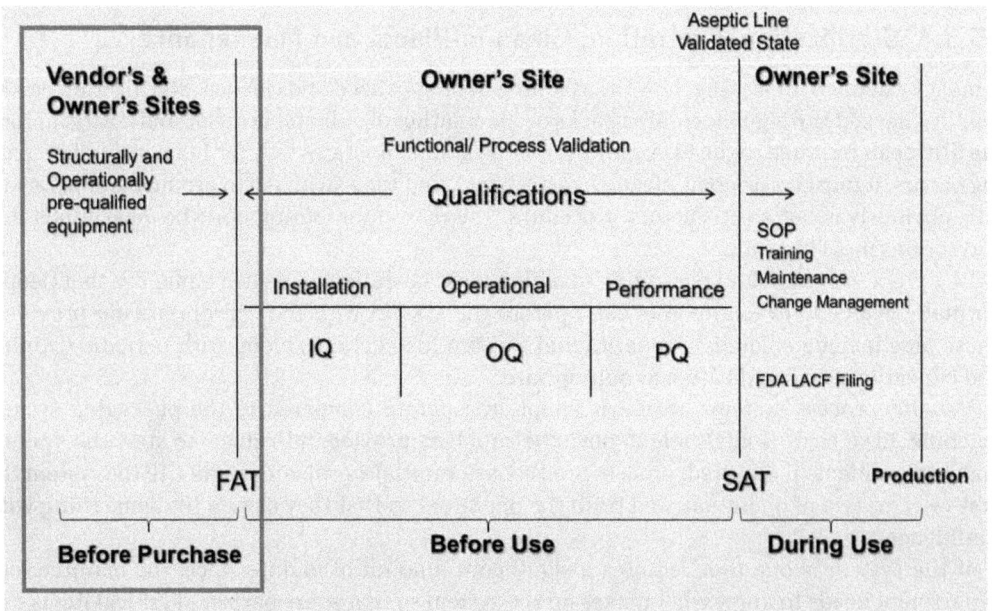

Figure 15.3 Repeat of Figure 15.1 highlighting the activities *before purchase* of the equipment.

A written plan should be developed that includes equipment specifications, drawings, functionality, and standards. This initial scope should be provided by the equipment manufacturer and updated with the customer's input. An exhaustive checklist needs to be developed that covers each of the following areas:

1. Mechanical specifications
2. Electrical specifications
3. Safety specifications
4. Equipment functional specifications
5. Regulatory specifications and compliance
6. Hygienic design
7. Materials of construction
8. Weld inspections
9. Equipment lubrications
10. Sterilization and process parameters
11. Clean-in-place (CIP)
12. Calibrations and certifications

The above list is only an example to demonstrate various requirements and testing that should be conducted during FAT. It is the author's recommendation that food manufacturers create an exhaustive checklist under each one of the major categories and work with the supplier to perform required tests during FAT. It establishes confidence in the system quality, functionality, and integrity for designed purposes. The tests should be designed to stress the system to get a better understanding of system operational flexibility. The tests should be conducted with extreme care and dedicated time. It should not be just a simple check-off list; instead, it should be conducted appropriately. The future of the line and the large multimillion-dollar investment depends on how well the equipment is going to run in future. An important result of the FAT test is a mutually developed punch list that both the supplier and buyer can agree upon. These items should be addressed before shipment of the equipment.

Software or PLC check should be an integral part of the FAT. The proper functionality of the machine is dependent on the PLC controls, and it should be a simple, intuitive machine interface for operators, and logical for overall system integration. FAT also provides a great opportunity to engage the operations team who will ultimately be responsible for daily production. Operators will get a chance to see and interact with the equipment for better familiarity. If it can be accommodated, training should also be a part of the FAT. The documentation during FAT provides a great template for commissioning and validation of the equipment.

The risk of not performing FAT is considered as "*kicking the can down the road*" and could be costly for the manufacturers and OEMs. It may impact the equipment integration within the overall line layout and create potential trouble for future operations. Hence, for a smooth start-up and commissioning of the line, FAT should be done to as much extent as possible. In some cases, a full product run could not be executed due to utility limitations at the equipment manufacturer site. However, all the functionality of the machine must be checked before accepting the machine for shipment to the buyer's site (Figure 15.1).

15.5 QUALIFICATION AND COMMISSIONING

At this point, we have completed the left-hand side of the validation "V" model (Figure 15.2) and we are going to delve into the right-hand side which deals with qualification and commissioning phases. While there are deadlines for commercial launch and the commissioning team may face a crunch and rush to finish the qualifications in a shorter timeline by reducing some activities, it is not advisable to do so. The cost of improper steps could be much higher in the long run or worst

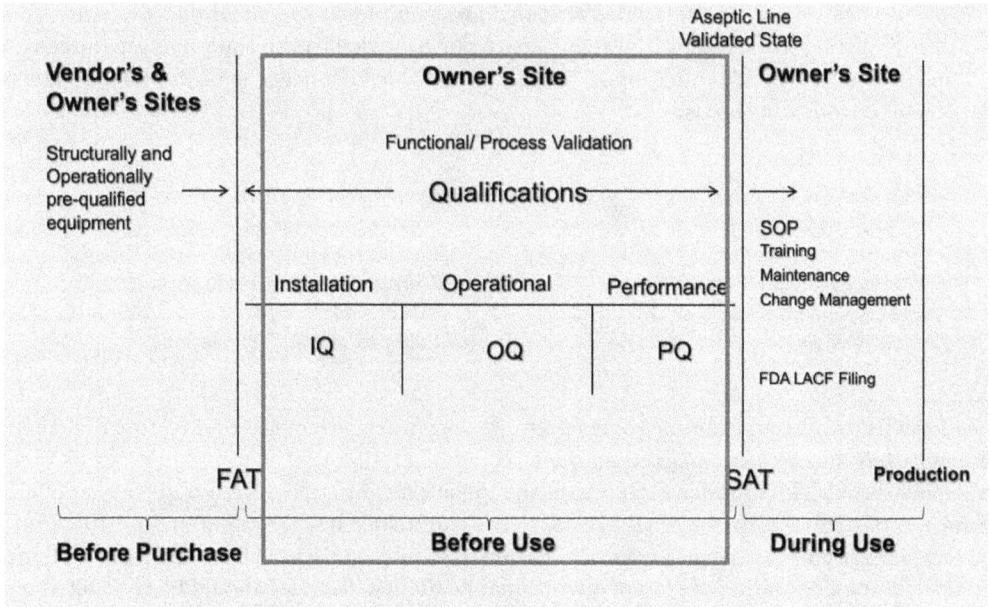

Figure 15.4   Repeat of Figure 15.1 highlighting the activities *before use* of the equipment.

case, a food safety risk for the company and consumers. A lack of proper qualification activity also jeopardizes the validation activities. Validation should not be a debugging activity for a lack of better qualification work. Commissioning activities include installation qualification (IQ), performance qualification (OQ), and performance qualification (PQ) as shown in Figure 15.4.

15.5.1 Installation Qualification (IQ)

IQ is the first step toward commissioning process. It verifies that the systems are installed properly and configured to the standards that were defined, reviewed, and approved in the design specifications. IQ is a document series of inspections and verifications to confirm the installation of equipment such as, UHT, aseptic tank, and aseptic filler, and that the interconnecting pieces have been installed and configured according to the specifications.

Prerequisite documents for IQ should be collected in advance and it can include a purchase agreement, FAT checklist, weld specifications, materials certificate, training documents, equipment cut sheets, P&ID, spare parts list, standard operating procedures, and operator manual. The first step in IQ is to verify the drawings to ensure that they reflect as-built conditions and confirm that all components and instruments are accurately labeled. It requires a walk down on the manufacturing floor to inspect layouts, P&IDs, wiring, cabling, and configurations. All the electrical installations should be checked for completion, and all process piping should be installed and ready for power-up. It is good practice to mark the drawings with a highlighter to ensure each section is checked and that all corrections are marked in red. Each item in the table should be initialed and dated by the person performing IQ.

Standard operating procedures should either be in an official final form or at least in a draft form which includes all the necessary steps for the proper functioning of the equipment. The preventative maintenance activities should be entered into the tracking system of the facility. The equipment logbook should also be created with appropriate information and located at an accessible controlled location. The main pieces of the system and subsystems should be field verified and documented for correct tag number, manufacturer, model, serial number, and other attributes. Compile a list of spare parts and consumables the manufacturer recommends to be maintained onsite for ease of replacement. Utility requirements such as electricity, compressed air, air exhaust, vacuum, water, and glycol should be verified for target values and range for specific equipment. Calibration of the instruments and sensors should be performed after installation to make sure they are accurate within the specified range.

The IQ activities should be closed out by completion of any deviation and discrepancy in accordance with the design. The final IQ report should have the initials and the date of all stakeholders.

15.5.2 Operational Qualification

Operational qualification of equipment provides documented evidence and verifies the functional specifications such as operations of sterilization, CIP, valve clusters, and process piping. It is important to establish that a piece of equipment or a subsystem is repeated operation within the specified range. Signal exchange between the UHT processor, aseptic tank, and filler should be verified as a first step. Since there are often multiple suppliers for this equipment, they must be present onsite to complete these activities. The signal exchange should be checked for CIP, SIP, production parameters, line flush, and product changeover. Another aspect of aseptic processing is to keep the system without any leaks, especially in the sterile zone.

15.5.2.1 Controls Validation

The control system used, in addition to providing interlocks for temperature–time during sterilization and temperature during operation, can provide interlocks on everything, including the amount of product that is available for processing, steam and cooling water temperatures/pressures, opening and closing of steam/water valves, starting off and stopping of pumps when the operation is complete, and so on. It must also be recognized that all systems, no matter how sophisticated, require an operator to be on the floor to interact with the machine and resolve alarms and other feedback requiring follow-up or corrective action. Due to the complexity of the aseptic systems, it is critical to have a good control system in place to control and monitor the entire operation. However, such a control system can only work perfectly if it has been properly challenged and validated. The purpose of controls validation is (1) to verify that the proper automated controls are working properly, and (2) to verify that whenever a corrective action is taken by the controls, it is documented. If the controls are not validated and it is not working properly, any further microbiological validation work might not be successful. Controls validation also helps to reduce downtime for the operations. Typical controls validation includes (1) CIP controls, (2) SIP controls, (3) production run mode controls, and (4) inspection system controls. It is important to examine the system design and programming logic and develop a detailed testing protocol. Some instruments and controls that monitor sterility for process and aseptic surge tanks are (1) RTDs, (2) pressure sensors, (3) flowmeter, and (4) PLC logic counters and timers. Examples of instruments that monitor sterility of aseptic fillers are (1) RTDs, (2) pressure sensors, (3) flowmeter, (4) level sensors, (5) temperature switches, (6) proximity switches, and (7) PLC logic counters and timers.

The very first time the test is executed, it is common to find inconsistencies in sequence of operations. Such issues must be fixed and checked again during the test. This should be continued until all the critical errors have been fixed. Then the second test should be conducted and fix any remaining issues that may still exist. The third test should be completed without any errors to pass the controls validation test. The document detailing the controls validation test should be reviewed by the personnel involved with the test, and the management of the change committee. It should be verified that all the critical points were tested adequately and that all records/charts are printed that reflect the test conditions. After the review of the document is completed, reviewers and the associated personnel should sign the appropriate forms stating that the validation test is completed and accepted.

15.5.2.2 CIP Validation

Cleaning of the aseptic system is critical for commercial production without any issues of spoilage and food safety risks. CIP prepares the lines for the next production batch by eliminating food that is stuck to the product-contact surfaces. If not cleaned properly, it could lead to bacterial contamination and formation of biofilms over time. There are four critical aspects of a successful CIP: (1) time, (2) temperature, (3) titration or concentration, and (4) turbulence. All of these factors are dependent on the type of product that is being produced and hygienic design of the equipment and interconnecting pipes. A successful CIP is possible when the fouled layer has a chance to swell and loosen, mechanical scrubbing action of the fouled layer, and velocity to remove the particles. A typical sequence of CIP consists of the following steps:

1. Initial water rinse to remove as much of the product as possible
2. Caustic wash to eliminate protein and fat residues
3. Water rinse to clean out the caustic
4. Acid wash to eliminate any mineral deposits
5. Final water rinse to eliminate CIP chemicals and prepare the line for SIP

The validation of the CIP cycle should be based on the worst-case scenario. The most difficult to clean product which could include allergen, high total solids, and high viscosity should be chosen in combination with the worst-case processing parameters such as high temperature. Also, the maximum production runtime should be considered for CIP validation. Some other aspects of the operations should also be considered that could impact CIP efficiency. For the aseptic surge tank, start and stop times for tank emptying can create drying of the product on the walls. The steam barriers can potentially provide a burn-on opportunity for the product and will be hard to clean. The criteria for successful validation should be established before conducting the CIP validation trials. A general recommendation is to pass three successive CIP tests on the worst-case product. After the trials, the identified hard to clean locations should be checked for no visible residue, ATP results of <200 RLU/swab, and aerobic plate count of <100 CFU/swab.

15.5.3 Performance Qualification

Performance qualification is generally done after the full line integration to check the performance of equipment based on the requirements outlined in the URS. The verification tests should be conducted and documented that the system is operating in the way it was designed to operate. The system functions and overall operation should be reliable and reproducible within a defined set of parameters. Also, the system must be in a state of control. All the critical parameters such as temperature, pressure, flow, overpressure, etc. must be stable for the duration of designed production time under normal and worst-case conditions. For an aseptic filler, some of the key performance deliverables are provided here as an example:

1. Hermetic seal of the package and the associated parameters must be checked against the set standards.
2. Spoilage defect rate that can be encountered during production. A proper sampling plan must be used to validate this key indicator.
3. Leak defect rate of the packages for a production cycle.
4. Foaming of the product and filling performance through the product filling nozzles.
5. Accuracy of the filling by measuring net weight of the packages.
6. The maximum filling rate of the packages. For example, if the filling rate of a particular filler is 600 bottles/minute, this must be verified using proper test methods.
7. Percent of oxygen in the headspace of the package. If a minimum, such as 5% headspace oxygen was defined in the URS, it must be checked at this stage and verified that the filler is capable of delivering against this set standard.
8. Rejection criteria based on the seal, fill volume, stoppages, or any other issue must be verified during performance qualification.

The key performance deliverables of the UHT, aseptic surge tank, and secondary packaging must also be verified during PQ trials.

15.6 STEPS TOWARD "VALIDATED STATE" OF THE ASEPTIC SYSTEM

Aseptic systems are extremely complex by their very nature. The major parts of the aseptic system include the UHT processing system, aseptic surge takes, the packaging system, interfaces and valve clusters, the cleaning method, sealing, splicing, start/stop, aborts, and extended runs. All parts and interdependent operations require separate validation tests to achieve the validated state of the aseptic system (Figure 15.5).

15.6.1 Commercial Sterility and System Operation

It is very essential to provide how the system of aseptic processing and packaging is cleaned and sterilized as a written document. Microbial challenge testing is the primary method to determine successful sterilization for food safety. Inoculation of the vulnerable area or inoculated packs can be used to check the sterilization of the aseptic system and the examination record can be used for the submission of the FDA filing.

Typically, the aseptic process system and packaging machine must be designed for a simple operation. The colored schematics of the system and written instructions can help the operators to perform in the same manner. Machine training provided by suppliers is needed for operators to learn about the system operation process, system adjustment, maintenance, and sterilization process such as COP and CIP after machine installation. Providing knowledge and education including periodic training and/or certificate training programs to factory operators is an important element of successful operation. With the appropriate training, the processing operation can be properly carried out with confidence and minor deviations.

Several concerns might come along such as frequent questions about training and other requirements for successful operation and safety. The first common question for the training might be who should be trained? One might think only floor operators. Actually, everyone who

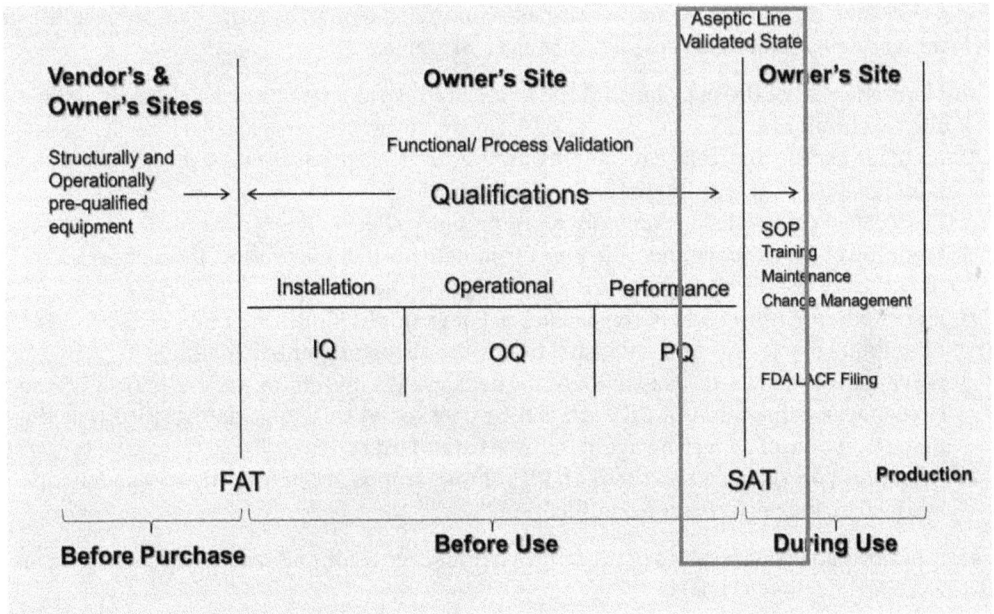

Figure 15.5   Repeat of Figure 15.1 highlighting the *validated state* of the equipment.

is related to aseptic production must understand and receive training. Factory operators, maintenance department, quality department, R&D department, management, and cleaning crew should be trained about any processing system.

It is important for the maintenance department to understand the details of the aseptic system. The maintenance crew should know not only the installation, preventive maintenance, and spare parts, but also other critical parts such as aseptic techniques and safety concerns, especially with high-pressure equipment and hazardous chemical use.

To provide confidence in aseptic operation and food quality, the quality control (QC) department is responsible for multiple analyses and safety. For example, the use or any changes of hazardous chemicals such as hydrogen peroxide and peracetic acid must be recorded. The sterilization of the packaging material which is essential for commercial sterility must follow regulatory requirements.

15.6.2 Selection of Surrogate Organism

To select the microorganisms for the validation test, it must be acceptable from the food manufacturing plant considering occupational safety, and public health perspectives. For microbiological validation of the processes, the surrogate or test microorganisms are used to correlate with target pathogenic microorganisms since pathogenic microorganisms cannot be used in manufacturing settings. The aim of microbiological validation is to verify the effectiveness of commercial sterility. The testing method must follow the safety and factory procedure, including work environment, bio-hazard handling, and disposal. The test organism to be used should be such that it has a higher resistance than spores of *Clostridium botulinum*. The highest lethality during the test is generally the time–temperature combination, concentration of chemical, or irradiation dose, resulting in commercial sterility to be used in the eventual filed process. This process is generally more severe than the process that would result in a 12-log reduction of *C. botulinum* for commercial sterility. For the UHT validation, the test organism most often used is *Clostridium sporogenes* (PA 3679) spores; however, other test organisms may be used. Denny et al. (1979) listed a number of test organisms suitable for various types of microbiological challenge tests. Surviving spores of *C. sporogenes* produce gas and a characteristic putrid odor, which is easy to detect, that confirm the presence of surviving test organisms. The common incubation temperature for this organism is 35°C. Sometimes detection, microbiological identification, and enumeration of surviving microorganisms may require more tedious and expensive procedures.

For steam sterilization, the spore of *Geobacillus stearothermophilus* is most typically used. The growth media contains an indicator dye called bromocresol purple, which turns yellow in the presence of spore growth (see Figure 15.7). Incubation temperature ranges from 45°C to 55°C. Spores of *Bacillus atrophaeus* can be used where the sterilization medium is either hydrogen peroxide or PAA. For irradiation sterilization, the choice usually is *Bacillus pumilus*. The surrogate microorganisms, which are recommended for the specific sterilization process, are listed in Table 15.2 (IFTPS, 2011).

15.6.3 UHT Validation

The ability of an aseptic processing system to deliver a calculated thermal process and produce commercially sterile products is generally tested by an inoculated pack test. An inoculated pack test consists of batch inoculating a product with an appropriate test organism and then processing and packaging the product at different levels of calculated lethality. The product to be used should be such that it will support the growth of any surviving test organisms, a nutritious media can also be used for the experiments.

TABLE 15.2   SURROGATE MICROORGANISMS USED FOR THE VALIDATION OF ASEPTIC FILLING MACHINE AND PACKAGING

Sterilization Method	Surrogate Microorganisms
Saturated steam Superheated steam	*Clostridium sporogenes* *Geobacillus stearothermophilus*
Superheated steam with dry heat	*Geobacillus stearothermophilus* *Bacillus polymyxa* *Bacillus atrophaeus*
Hydrogen peroxide with heat	*Bacillus atrophaeus* *Bacillus subtilis*
Hydrogen peroxide with UV-C	*Bacillus atrophaeus* *Bacillus subtilis*
Peroxyacetic acid	*Bacillus atrophaeus* *Bacillus subtilis* SA22
Heat from extrusion	*Geobacillus stearothermophilus*
Irradiation	*Bacillus pumilus*

The product processed during validation is packaged aseptically via the packaging machine that has yet to be validated with the assumption that all packages are sterile and hermetic. The containers are labeled to identify the process temperature used and incubated at a temperature suitable for growth of the test organism. If *C. sporogenes* is used, the incubation temperature is generally 35°C because the test organism is a mesophilic spore former. The product is incubated for a period long enough (3–4 weeks) to ensure that any surviving injured spores had an opportunity to grow and produce spoilage, that is, gas and odor.

The challenge level of the test organism used should be such that there are no survivors at the highest process temperature, and the product is commercially sterile. As an example, suppose that the test organism has a heat resistance of $D_{250°F} = 1.0$ minute in product, and that at least a 5-log reduction of the test organism is necessary for commercial sterility. If 100 containers of the product are to be incubated, the inoculum must be such that the product would contain at least 10,000 spores or surrogates per container. A successful test would result in no spoilage in product processed at the highest temperature, indicating that the time–temperature combination is adequate for commercial sterility and complete spoilage at the lowest temperature. Spoilage at the lowest temperature confirms that non-spoilage at the higher temperatures represents a kill rather than inability of the injured spores to grow in the product. Ideally, a test is designed to result in no spoilage at the two higher temperatures and spoilage at the two lower temperatures. This would provide an additional safety margin built into the calculated thermal process.

15.6.4 Aseptic Tank Validation

Aseptic surge tank is used to store the sterile product before filling in packages through the aseptic filler. One of the main aspects of the aseptic tank is that the product does not flow; however, positive pressure from sterile air or nitrogen is used to keep the product sterile. Tank is usually separated from the UHT and aseptic filler using automated valve clusters. Hence, all the interconnecting pipes, valves, filters, and the tank must be presterilized before receiving the sterile product. The surge tank, sterile air filters, valves, and piping are presterilized by saturated steam. The minimum sterilization requirement for a surge tank should be obtained from a competent process authority. Generally, 30 minutes at 121°C (250°F) or equivalent is sufficient for the SIP cycle.

Temperature distribution test is generally conducted for the aseptic tank. In addition to the reference RTD, several thermocouples are placed inside the tank to cover all the potential "cold spots." In Figure 15.6, there are 15 thermocouples for the 5k gallons tank capacity. If the tank is larger, one may want to place more thermocouples. These locations are determined by the process authority and can include the top of the tank, agitator, middle of the tank wall, bottom area, and the product outlet pipe. Temperatures at the locations are monitored to ensure that the sterilization cycle minimum time and temperature is attained and completed as designed. If any cold spots are identified in the system, a modified sterilization procedure is recommended by the process authority. Also, steam quality should be compliant according to PMO and USFDA low-acid canned food (LACF) requirements.

For the locations where a thermocouple could not be placed, a microbiological challenge test is recommended. These locations could be filter and filter housing and the valve clusters. The choice of the test organism is left to the process authority. As a general rule, heat-resistant spores of *G. stearothermophilus* are used. Because the strips recovered after the test are directly transferred into transparent tubes of an appropriate growth medium, detecting surviving organisms is relatively easy and quick (3–5 days at 45–55°C). However, other suitable organisms may be used with equal effectiveness. There are a number of ways of placing test organisms at selected

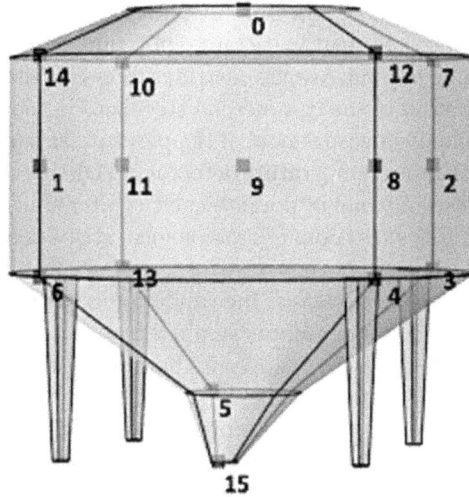

Figure 15.6 Thermocouple location for aseptic tank temperature distribution study. Locations and coding for the temperature probes and thermocouples:

0- Top entryway	7- Top – east
1- Middle – west	8- Middle – North
2- Middle – east	9- Middle of the tank (hanging)
3- Bottom – east	10- Top – west
4- Bottom – North	11- Middle – south
5- Bottom of the tank	12- Top – north
6- Bottom – west	13- Bottom – south
	14- Top – west
	15- Outlet.

locations. The most common method is to place a measured quantity of spore suspension on presterilized aluminum or stainless steel strips with a foil tape for attachment. Appropriate quantities of strips containing the test organism are placed at preselected locations in the surge tank, valves, and piping prior to running the surge tank sterilization cycle. After the test, the strips are aseptically recovered and placed in an appropriate growth medium and incubated at an appropriate temperature for recovery of the test organism. Based on the number of organisms on the strips (10^4, 10^5, or 10^6) and the heat resistance of the microorganism, the total lethality delivered can be estimated. The surge tank sterilization cycle recommended may be modified if the microbiological test results are not satisfactory. On occasion, instrumentation requirements and monitoring procedures may be changed.

If microbial filters are used for sterilizing air or nitrogen, the filters should be tested for integrity by the ASTM method prior to installation and to ensure that they are functioning properly. Depending on the design, a post-installation test may be necessary to ensure proper seating of the filter.

15.6.5 Pre-Production Sterilization of Aseptic Filler

Due to the complex nature of aseptic packaging machines, it is important to microbiologically validate preproduction sterilization to achieve commercial sterility of the aseptic zone, product, and package-contact surfaces. The product pathway is generally sterilized with steam, this includes the backside of the product valve on aseptic tank through the filler bowl and out through the filling nozzles and beyond. Aseptic cartridge filters are also sterilized with this cycle of SIP. General recommendation of the commercial sterilization temperature is 121°C (250°F) for at least 30 minutes. While the measurement of temperature at hard to heat areas (cold spot) can be monitored with a resistance temperature detector (RTD), there might be places where the placement of a thermocouple would not be possible such as filter housing. At those locations and depending on process authority's decision, microbiological strips can be placed to test the efficacy of steam sterilization. The microbiological strips for steam sterilization contain dried spots at 10^4 and 10^5 levels of *G. stearothermophilus*. The validation trials should be done in triplicates with the temperatures as close to the minimum steam temperature as possible.

Depending on the filler types, the sterile zone can be sterilized with radiation and/or chemicals. The most commonly used chemical sterilant is hydrogen peroxide, in either liquid or vapor form. The concentration of peroxide for regular production is 35%. However, during validation trials, the concentration is lowered to 32–34% to challenge the effectiveness of the sterilant. The temperature of the peroxide should be the minimum possible value recommended by the manufacturer of the filler. The lower concentration and temperature also help to clear any deviations that might be encountered during regular production of commercial products. Microbiological strips of *B. atrophaeus* at a concentration of 10^4 and 10^5 are used and placed at key locations within the sterile zone of the filler. Sufficient locations should be tested to provide a clear map and understanding of system behavior. The trials are performed in triplicates to provide statistically valid results. All critical records such as peroxide concentration, temperatures, and flow rate along with the location map should be kept. The validation records will be used to file the process with the regulatory agency.

15.6.6 Validation of Aseptic Packaging Material Sterilization

In aseptic food processing, packaging containers must be sterilized prior to entry into the sterile zone, and it is important to ensure that they are sterile. There are several sterilization methods which may be specific for packaging material and aseptic filling machine, designed by manufacturer, as discussed in Chapter 8. To ensure the sterility of packages, microbial tests are conducted to validate the sterilization process. Microbial inoculation on the packages including packaging containers and lidding materials is, therefore, required for the packaging sterilization validation.

Every aseptic filling machine has a specific sterilization method and the use of sterilization agent that is based on the design of the filling machine. The selection of surrogate microorganisms, which is used for microbial inoculation, can be varied depending on the sterilization method. Additionally, the sterilization process relies on the minimum conditions recommended by the equipment manufacturer such as minimum sterilant quantity, minimum temperature, minimum concentration of sterilant, and contact time. The position of inoculation on the packaging material is also very essential. It is based on packaging types, styles, and designs. It is important to understand the machine operation and system. For instance, the validation process of the aseptic packaging for fill-seal packaging system is generally different from the testing for form-fill-seal packaging machine. To achieve accuracy, mathematical modeling can provide a useful framework for visualization and understanding of the sterilization pattern on packaging containers. It can help to detect the cold spot areas of interior packaging container and minimize validation failure.

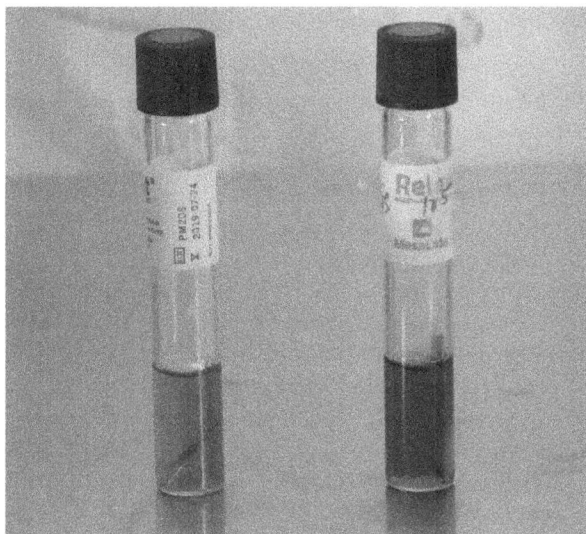

Figure 15.7   Growth/no-growth test results with spore-inoculated test strips negative (right: purplish color) and positive (left: yellowish brown color).

For the packaging validation testing, inoculated packaging containers are filled with sterile liquid food material or media, which is preferably sterilized through the aseptic UHT processing system. The packaging containers are sealed with closures or lidding materials, and then incubated under appropriate conditions for the test organism used. The results of incubation tests can be detected by the color changes of media or liquid food due to the growth of microorganisms (Figure 15.7).

For the form-fill-seal machines, the microbiological study is challenging since the method to inoculate the packaging container for the machine to form the containers is tricky. The inoculation process must be precise to ensure that the inoculation position on the packaging material eventually must be inside the container after forming. Like the container, the lidding material is inoculated and processed in the same manner. Both inoculated packaging containers and lidding materials are incubated to specific conditions as required for the surrogate organisms.

It is so important to ensure that the growth media filled inside the container is in contact with the microorganisms during the incubation period. A process authority should be consulted if any changes are made, because additional testing may be required. The data from the microbial validation studies are used for filing with the regulatory agency.

15.6.7 Conveyor Chain Sterilization Test

If a conveyor or conveyor chain is used to transfer the containers or container material through the aseptic zone, then it is necessary to presterilize the conveyor or conveyor chain prior to product filling and sealing. In addition, because the conveyor or chain leaves the aseptic zone and reenters, it can possibly recontaminate the aseptic zone. Thus, the conveyor or chain must be washed to remove any spilled product and continuously resterilized prior to entry into the aseptic zone. This is an extremely difficult task, and the procedure must be tested using microbiological tests described in earlier sections. As always, a test organism appropriate to the sterilant must be used.

15.6.8 Maintenance of Aseptic Zone Sterility

During production, the sterility of the aseptic zone is sustained by maintaining a positive pressure inside the aseptic zone, generally by sterile air. If this sterile air is produced by a flow of air through HEPA filters, the surface of the HEPA filter must be presterilized. The presterilization cycle for the HEPA filters must also be tested for adequacy in a manner similar to the procedure for the sterile zone. The most general way of testing the positive pressure is by particle count test of a specific size. An atomizer aerosol generator is used to create non-toxic, non-corrosive, non-soiling, and non-flammable smoke using oil. The particles should be generated within a range of values from 0.3 to 10 μm. Particle counters should be installed inside the sterile zone at different locations identified by the process authority. The values of particles that are >0.5 μm are reported and compared with the control sample. The control sample is the particle counts without the use of smoke. This test should be repeated for static and dynamic (filler in running mode with the conveyor) conditions to prove that the smoke did not get a chance to enter the sterile zone. The test is repeated three times to ensure proper operation of the filler in static and dynamic mode for the maintenance of sterility.

15.6.9 Commercial Sterility Test

Before releasing the line for commercial production, it is important to test the entire line operations, including UHT, aseptic surge tank, and aseptic filler. This test is commonly referred to as a commercial sterility test. The total number of packages to be tested is dependent on the acceptance sampling criteria. The sampling plan is based on exact upper confidence bounds. The question we try to answer here is: if x defectives are found in a sample of n, what is the upper 95% confidence bound for the proportion p of defectives in the lot? We are interested in the upper limit UCB which tell us [0, UCB] is a one-sided 95% CI for p. The confidence interval is calculated based on the F distribution. For example: for $x = 0$ defectives out of $n = 3,000$, UCB = 0.000998 ≈ 1 out of 1,000. This sampling plan is also called a zero-defect sampling plan. Hence, to say with confidence that a manufacturing plant is operating at a probability of 1:1,000 defect rate (AQL of 0.1%), we need to test 3,000 samples without any failures detected from the samples. Following the same principles, if one wants to operate at a probability of 1:10,000 defect rate (AQL of 0.01%), the total number of samples for testing would be 30,000 without any failures. Failure of sample in this case represents microbiological failures where an assignable cause cannot be established such as defective package. The trials are typically done in triplicates using media fill runs. The low-acid medium of choice in the food industry is skim milk with iron salt. The three consecutive runs are separated by deliberate stops. The subsequent runs are restarted without any intermittent presterilization step. The finished product is incubated for 14–21 days at 35°C, followed by microbiological and package integrity tests.

15.7 MANAGEMENT OF CHANGE (MOC) PROGRAM

There is an oft-forgotten program that its lack of implementation can jeopardize the "Validated State" of Aseptic manufacturing systems. This program is effectively implemented Management of Change (MOC) program. This is more than a simple paper exercise.

System cannot be called "Validated" without an implemented MOC program!

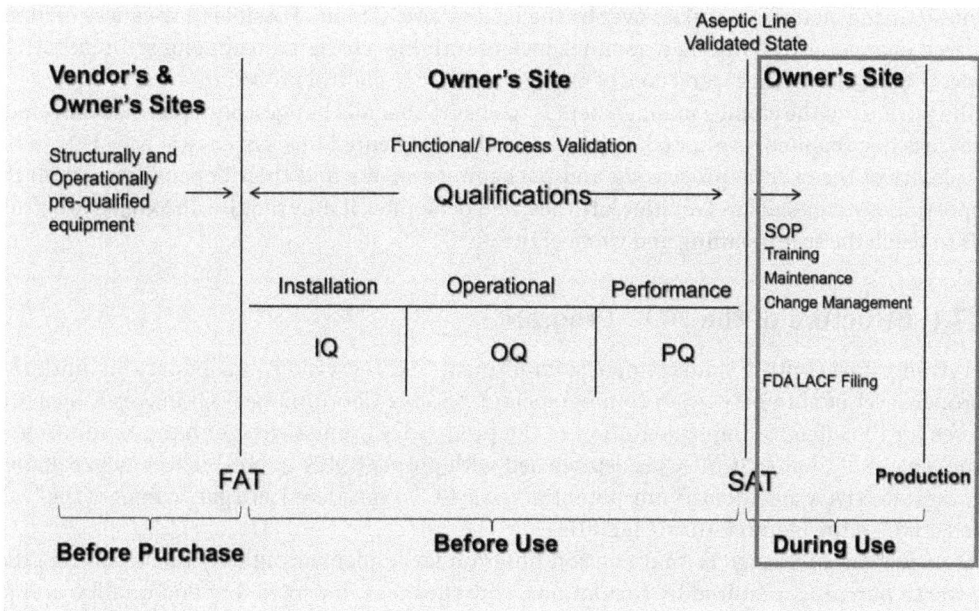

Figure 15.8   Repeat of Figure 15.1 highlighting the activities *during use of the equipment.*

All of the qualification (DQ, IQ, OQ, and PQ) are snapshot processes and the sustainability of their results are questionable if there are possibilities of changing operational and environmental conditions in addition to changes related to technologies, procedures, and personnel (Figure 15.8).

It is a very common question in the industry about the frequency of revalidation of the line. The quick answer is "depends"! Without knowing the goodness of the implementation of the MOC and any other supporting programs such as training, it is impossible to answer this question with a good level of confidence.

If the validation is seen as a "project" with start and end dates, it would not be possible to have a sustainable "validated state." Validation is not a project! There is no end to this process until the retirement of the manufacturing line.

Only with an implemented MOC, qualification activities can be turned into validation!

Changes are inevitable! Change will happen in the form of emergency, temporary, or planned. The question is if we are ready to deal with these changes? It is critical to understand the nature of the changes to prevent any unnecessary burden on the organization. The program can easily be turned into a cumbersome and lengthy document exercise that would discourage participation and acceptance within the organization. Therefore, the simplicity of the program is very important.

Due to the dynamic nature of project management, there may be numerous changes throughout the execution of the projects which may include some of the validation activities. During these stages, the manufacturing line is still managed by the project team and all the changes will be handled within the project change control team until the line is handed over to the manufacturing factory and taken over by the facility's MOC team. Possible root-causes of unauthorized changes are changing personnel, lack of training, changing technology, unauthorized scope, URS or FRS changes, and lack of emergency change control procedure.

The purpose of the change management is to ensure that all changes, both planned and emergency, are documented, evaluated, approved, and implemented in a systematic way. Due to the complexity of the aseptic processing and packaging systems and their dependency on all the supporting systems such as facilities, utilities, and personnel, it may require an extensive culture shift to teach the true meaning and value of the MOC.

15.7.1 Structure of the MOC Program

Everything starts with a management commitment and the vision to support this initiative. A proper level of authority given to a competent Change Coordinator (CC) makes a huge difference for the effective implementation of the program. CC must ensure that a Committee of Management of Change (CMOC) is established with proper SMEs to be able to analyze change requests effectively and identify any potential risks. CC is considered as a gatekeeper of the "validated state" of the manufacturing facilities.

Management of change is an execution program for implementing only those changes that are worth pursuing, required by regulations, consequences impact safety and quality, and for preventing unnecessary or overly costly changes.

It is important to note that most of the projects, if not a brand-new plant installation, are done in a host manufacturing facility with an existing MOC. Alignment between the existing MOC and the project team is critical to establish safe and sustainable processes and smooth transfer of the manufacturing line to the manufacturing facility. MOC acts as an agreement between the project and/or production team and the managers that are responsible for decision-making to evaluate the impact of a change before implementing it.

15.7.2 Structure of the Committee of MOC

Many changes that initially sound like good ideas will get thrown out once reviewed by the Committee of MOC. Not all changes qualify to go through the Change Control pathway (Figure 15.9). CMOC is made up of the decision-makers—project manager, subject matter experts, process authority, stakeholder or user representatives, vendors, and/or consultants. CC leads the CMOC team. The CMOC analyzes the impact of all requested changes to an established/validated system or projects and has the authority to approve or deny any change requests. The list of CMOC members should be written down in project charter or plant SOP and agreed upon, and each CMOC member should understand why the change control process is needed and what their role will be in the committee.

This process flow can be easily simplified to the steps as shown in Figure 15.10.

The evaluation and approval of the changes must be documented prior to implementation of the proposed change. To highlight and prioritize the proposed changes, every change should be categorized as a major, moderate, or minor based on their potential impact on product quality or safety. Without an approved protocol for testing the executed change by qualified personnel, the change process cannot be completed. The testing protocol must be well-thought-out and cover all the potential risks with related challenges.

"In God we trust, everyone else brings data" – USFDA – Gene Murano

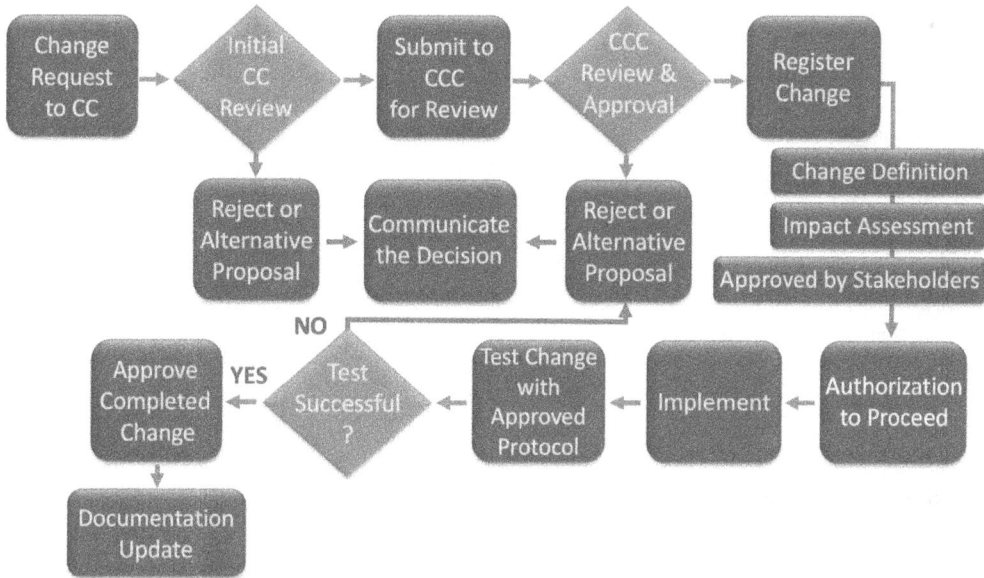

Figure 15.9   Management of change execution and decision flow diagram.

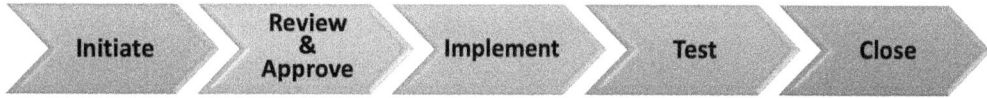

Figure 15.10   Simplified flow diagram to show the five basic steps of the MOC program.

Documentation of the test results and all related supporting data must be compiled with a final report and submitted for the closure of the case.

15.7.3 Impact Assessment of a Change

Once the change request for the aseptic system is submitted, before the approval of a change request, an impact assessment should be completed. There are more than 20 categories to consider when documenting the impact of a change. Some of those categories are, and not limited to, raw materials, specifications, facilities, support systems, equipment and computer hardware, processing steps, product quality, retesting, expiration dating, labeling and packaging materials, GMP compliance, products impacted, regulatory impact, validation impact, validated test methods, computer system changes, training impact, document changes, other sites, contractors, partners, or collaborators, and change categorization. These impacts are sometimes direct and sometimes indirect. Therefore, competent subject matter experts should be involved in the impact analysis (Rodriguez, 2006). The case studies for examples of failures and their consequences can be found in Chapter 14.

15.7.4 Change Request Process

Initial change requests can be made by anybody in the company. For an effective analysis and impact analysis, it is critical to include as much information as possible at the time of submission. However,

TABLE 15.3   COMMON FIELDS THAT SHOULD BE ADDED TO A CHANGE REQUEST FORMS

Description of the change	Reason for making the change	Equipment, systems, areas impacted
Products impacted	Assessment of any impact on product quality and safety	Assessment of impact on support documents such as engineering drawings, SOPs, etc.
Assessment of impact on regulatory status and filing (e.g., FDA filing)	Outline of testing required (or revalidation)	Approvals to implement
Implementation record (dates, timelines, testing results, remarks)	Record of testing review, acceptance, and approval to close	Links to the other programs such as CAPA, audit, maintenance, etc.

the completion of the information should not be an excuse to delay the submission of the change request. It is always possible to fill in the missing information later in the process. Recognition of the change request early in the process is much more important than the completeness of the information. There are many good practices in the industry and some of them are extremely simple and easy to implement based on the complexity of the systems and product categories. Initial request forms can be designed as simple as a handwritten form to an electronic spreadsheet (e.g., MS Excel) to a sophisticated software package (e.g., TrackWise, Honeywell Sparta Systems). Regardless of the format, the commonality of the forms are the key fields to be filled (Table 15.3).

15.7.5 Change and Urgency Classification

A proper classification of the change is important to adjust the level of attention and urgency for this request. Every change request will go through the same process; however, classifications can help effective and timely handling of the requests. Although it is possible to have more classifications, there are two common classification groups: type of change and level of urgency.

15.7.5.1 Types of Change

The types of change can be characterized as minor, moderate, or major based on its potential impacts on the aseptic processing packaging systems. If it is thought that the change would impact the specifications and processes fairly slightly or considered unlikely, it may be classified as "*Minor*." If the effect of the change on the aseptic systems should be evaluated to determine its potential impact, the change can be categorized as "*Moderate*." If the changes are significant and impacting the product quality and safety, the change must be categorized as "*Major*" and the products on that line must not be produced or shipped until closing the change control case.

15.7.5.2 Level of Urgency

After categorizing the type of change based on its potential impact on the aseptic systems, the second level of categorization is determination of its urgency. This is important to provide the change management team with resources, line time, and funds to execute the change effectively within expected timelines. There are four commonly used urgency levels: critical, high, medium, and low.

The highest degree of urgency is labeled as "*Critically Urgent*" and generally the system is down and unavailable to users. When a specific part or function is unavailable, and a serious business impediment is involved, the urgency is considered as *"Highly Urgent."* Once the change require-ment is recognized and the change involves fixing a problem with sufficient existing plans and resources, the urgency level is labeled as *"Medium Urgency."* And finally, if the change is an enhancement to current process, it is labeled as *"Low-Level Urgency."*

15.7.6 Implementation Plan, Final Disposition, and Closeout

An implementation plan should be clearly identified, generated, and revised. The plan should be used as a guide to follow up the execution of the changes, including the completion of SOPs, validation document/protocol, training records, and material.

Once all requirements specified in the change request form and implementation plan have been completed, the party responsible for verification/validation and the requestor will evaluate these documents for final approval and report. Any deviations from the plan must be evaluated by the subject matter experts and justification statement must be included

15.7.7 Conclusion and Recommendations

In this part of the validation chapter, the purpose is to highlight the criticality of the change management program for the aseptic processing and packaging systems and all the supporting facilities and utilities. As it is highlighted several times in this chapter, change management is the "keystone" for the "validated state" of the aseptic systems. Just one violation and bypass of the MOC can cause the loss of this status. Just to repeat again, what turns the qualification activities to validation is the implementation of the Change Management! Creating a MOC pro-cedure is not satisfactory until having a fully implemented change management program. The key to Change Management is to review and approve proposed changes before and after they are made. Unless the interaction is clearly identified in the protocol, multiple simultaneous changes should be avoided. Change management documents are the most valuable documents if there is an issue on the production line and need to find the root cause of the problem.

It is highly recommended to keep the MOC program simple to prevent the burden on the employees and discouragement due to excessive paperwork. Effective training is critical for proper implementation of this program since the change can be initiated by anybody in the com-pany. The audience is large and often the management is misguided to exclude employees who are not the first layer interfacing with the aseptic systems (e.g., a procurement person located in the headquarters of the company). Not all changes have to go through the change management program. Identifying like-to-like and routine maintenance changes would minimize the unnec-essary burden on the system and minimize any employee frustrations. One of the significant gaps in the food industry is the closing and follow-up steps for the changes introduced and exe-cuted. Even in well-designed and implemented MOC programs, the closure of the change request and checking the effectiveness of the execution may not be done effectively. There should be close attention to the closures of the changes initiated and executed. The MOC program should be assessed/audited periodically to understand the effectiveness of the program.

15.8 REGULATORY REQUIREMENTS, FILING, AND STANDARDS

Aseptic processing and packaging systems have been heavily regulated by the regulatory agen-cies (e.g., USFDA, Canadian Food Inspection Agency) all around the world. Some countries have a Memorandum of Understanding among them to harmonize some of the requirements and ease the export and import of affected products. USFDA is considered one of the anchor regulatory organizations in terms of creating and implementing aseptic requirements. Aseptic has a long history in the pharmaceutical industry due to the popularity of sterile product applications in the medical disciplines. However, there are significant differences between expectations of the USFDA's Drug/Medical Device decision versus the Food/LACF division.

The main difference comes from the nature of the products and their suitability for promoting microbiological growth. As most of the food products are great growth mediums for a variety of the microorganisms based on their characteristics and ingredients are originating from the nature with high bioburden levels, the probability of cross-contamination and microbial growth is much higher compared to a non-nutritive drug compound. However, aseptic concepts and validation requirements have lots of similarities between the two industries. In fact, some of the qualification principles used for aseptic systems are universal for many industries.

15.8.1 Regulations for Aseptic Systems

If a firm is located in the United States or located abroad and exporting to the United States, their facilities and products must be filed with the USFDA and include validation information as part of the supplemental information using a SUP-SID documentation.

Aseptic systems can fall under jurisdiction of either USFDA or USDA-FSIS. The determination is based on the product type and characteristics such as presence of meat-based ingredients in the products. As a starting point, a firm and its process authority must determine which requirements are pertinent for the specified product type.

Aseptically processed and packaged dairy products—as defined in Grade "A" Pasteurized Milk Ordinance (PMO)—must also comply with the applicable provisions of the PMO.

According to PMO (USPHS, 2019), the term "Aseptic Processing and Packaging," when used to describe a milk and/or milk product, means that the milk and/or milk product has been subjected to sufficient heat processing and packaged in a hermetically sealed container, to conform to the applicable requirements of 21 CFR Parts 108, 113, and 117 and to maintain the commercial sterility of the milk and/or milk product under normal non-refrigerated conditions. The PMO document can be assessed as https://www.fda.gov/media/140394/download.

In order to minimize the burden on the aseptic dairy industry due to double regulatory jurisdictions (USFDA and PMO) and potential conflicts, the Aseptic Processing and Packaging System (APPS) in a milk plant is comprised of the processes and equipment used to process and package aseptic Grade "A" low-acid milk and/or milk products. The APPS "shall" be regulated in accordance with the applicable requirements of 21 CFR Parts 108, 113, and 117. The APPS "shall" begin at the constant-level tank and end at the discharge of the packaging machine, provided that the Process Authority may provide written documentation which will clearly define additional processes and/or equipment that are considered critical to the commercial sterility of the product. The unofficial name for this definition in PMO is *"Aseptic Bubble."*

15.8.2 Regulatory Filing of Aseptic Lines and Products

With the technological developments (e.g., blow-fill-seal aseptic filler innovation) due to brand owners' demands and OEM competition, aseptic manufacturing systems are becoming more complex. This brings additional risks which need to be considered upfront. USFDA's LACF office and separate technical subject matter experts within the FDA review every new aseptic technology.

Expectations of the regulatory agency are similar to what the authors outlined in this section. A systematic science-based risk analysis with corresponding validation activities must be integrated into development and manufacturing systems. The time factor must be considered as the third dimension of the validation (sustainable process) and the "validated state" must be protected with proper Change Management program and other prerequisite programs such as HACCP, GMP, and implemented training programs.

For complex filing cases (e.g., aseptic products with particles or alternative processing technologies), it is advisable to contact the FDA SMEs well in advance, preferably at the design stage of the process and product development. When filing the process and product, attaching a brief explanation of the aseptic system and its components to help the reviewer is highly recommended.

It is important to highlight that the USFDA does *not* approve any filed product and/or process. The agency changes the status of the filing application to "filed" status by keeping the option of asking questions in the future and return the filing back to manufacturer. The applicant should anticipate potential questions and try to cover them in the comments section of the filing to prevent unnecessary returns. To prevent any returns due to mistakes, it is highly advisable to have colleagues and Process Authority review the filing documents before filing with the Agency.

"Do what you filed, file what you do!"

Filing of the aseptic manufacturing line brings legal obligations to manufacturing companies as declaration of the current practices at that facility. Therefore, it is important to match production facility's practices and supporting validation documentation with the filing documents in case the facility is audited by the FDA Auditors.

FDA regulations for aseptic processing and packaging are covered in 21 CFR 113.40(g). Regulations related to acidified products are covered under 21 CFR Part 114. According to Part 113, aseptic processing and packaging means the filling of a commercially sterilized cool product into presterilized containers, followed by aseptic hermetic sealing, with a presterilized closure, in an atmosphere free of microorganisms. 21 CFR Ch. 1, 113.3, the same Part of 21 CFR defines "Commercial Sterility of equipment" as the equipment free of viable microorganisms having public health significance, as well as microorganisms of non-health significance, capable of reproducing in the food under normal non-refrigerated conditions of storage and distribution (21 CFR Ch. 1, 113.3(e)(2))

The USFDA has improved the filing system by including smart forms and workflows to allow electronic filing while allowing filing with paper forms in mail. USFDA's filing system can be accessed at FDA's Acidified & Low-Acid Canned Foods Guidance Documents & Regulatory Information website

https://www.fda.gov/Food/GuidanceRegulation/GuidanceDocumentsRegulatoryInformation/AcidifiedLACF/default.htm

Every aseptic filing application is accompanied by detailed supporting documents, including validation documents. The USFDA does not prescribe every required document. The expected must-have documents are generally the microbiological challenge test results for the filler, aseptic processing, aseptic surge tank, and any sterility maintenance tests such as "smoke" test. In addition, any information related to the system and its operational principles, product information, and a good explanation of the packaging materials should be included.

15.8.3 Standards for Aseptic Equipment

When designing an aseptic processing and packaging line and all supporting utilities and facilities, knowledge of industry standards is overly critical. Any deviations from the standards during the design stage may cause very costly changes or lengthy spoilage investigations in the future. Unfortunately, there are no FDA or USDA approved standards available. Regulatory agencies will provide general guidance, but usually do not provide specifics (except for some items such as hold tube slope, RTD locations, etc.)

When the manufacturer is preparing the User Requirements Specifications (URS) for its vendors (OEMs), the background information about the industry standards would be very valuable. The authors of this chapter compiled a good list of these standards and issuing agencies as a reference to the reader. The intent is not to explain each standard here since they are self-explanatory and detailed information is available on their websites. Below is the list of the information that should be included in the URS document which is an important part of the product life cycle (Figure 15.2, V-Diagram) to guide the vendor when they are preparing Functional Requirement Specification (FRS). Any exceptions to this and additional standards required as part of the URS must be captured as deviations/exceptions by the vendor and the manufacturer and should be included in the risk analysis. Below is the list of the recommended documents and standards that should be included in the URS to be shared with the vendors to communicate the company's stand and expectations.

1. P&ID for processing system which detailed instrumentation and basic design
2. Specifications and KPIs for what each piece of equipment was to deliver
3. Codes and standards to reference (some of the standards may not be specifically for the aseptic products, however, they are still considered as good references)
   a. American Welding Society (AWS) https://www.aws.org
      i. AWS D1.1 Structural Welding Code https://www.aws.org/certification/detail/aws-d11-structural-welding-steel
   b. American Society for Testing and Materials (ASTM) https://www.astm.org/
      i. ASTM A36 Standard Specification for Carbon Structural Steel https://www.astm.org/a0036_a0036m-19.html
      ii. ASTM A194 Standard Specification for Carbon Steel Bolts https://www.astm.org/a0194_a0194m-20a.html
   c. American National Standards Institute (ANSI) https://www.ansi.org/
      i. ANSI S1.3 Specification for Sound Level Meters
      ii. NSF/ANSI/3-A 14159-1-2019 Hygiene requirements for design of Meat and Poultry Processing Equipment
   d. National Electrical Manufacturers Association (NEMA) https://www.nema.org/
      i. NEMA 250 Enclosures for Electrical Equipment https://www.nema.org/standards/view/nema-250-enclosure-types
   e. National Fire Protection Association (NFPA) https://www.nfpa.org/
      i. NFPA 70: 2020 National Electric Code https://www.nfpa.org/codes-and-standards/all-codes-and-standards/list-of-codes-and-standards/detail?code=70
      ii. NFPA 79 Electrical Standards for Industrial Machinery https://www.nfpa.org/codes-and-standards/all-codes-and-standards/list-of-codes-and-standards/detail?code=79
   f. Code Of Federal Regulations (CFR) https://www.accessdata.fda.gov/scripts/cdrh/cfdocs/cfCFR/CFRSearch.cfm

      i. **21-CFR-108** Emergency Permit Control
     ii. **21-CFR-110** Current Good Manufacturing Practice in Manufacturing, Packing, or Holding Human Food
   iii. **21-CFR-117** Current Good Manufacturing Practice, Hazard Analysis, And Risk-based Preventive Controls For Human Food
   iv. **21-CFR-113** Thermally Processed Low-Acid foods packaged in hermetically sealed containers
    v. **21-CFR-114** Acidified Foods
   vi. **21-CFR-11** Electronic Records, electronic signatures
   vii. **29-CFR-1910** Occupational Safety and Health Administration

g. Steel Structures Painting Council (SSPC) https://sspc.org/
      i. SSPC SP-2 Hand Tool Cleaning
     ii. SSPC SP-3 Power Tool Cleaning

h. 3A Sanitary Standards https://www.3-a.org/

i. European Hygienic Engineering and Design Group (EHEDG) https://www.ehedg.org/

j. Consumer Brands Association (CBA) – Formerly known as Grocery Manufacturers Association (GMA) https://consumerbrandsassociation.org/
      i. Validation Guidelines for Automated Control (bulletin 43-L) https://forms.consumerbrandsassociation.org//forms/store/ProductFormPublic/validation-guidelines-of-automated-control

k. American Society Of Mechanical Engineers (ASME) https://www.asme.org/
      i. Section VIII (latest edition) Unfired Pressure Vessels https://www.asme.org/certification-accreditation/boiler-and-pressure-vessel-certification
     ii. B31.3 (2018) Process Piping https://www.asme.org/codes-standards/find-codes-standards/b31-3-process-piping?productKey=A0371W:A0371W

l. Institute for Thermal Processing Specialists (IFTPS) https://iftps.org/
      i. Aseptic Validation Guidelines G.005.V1

m. Aseptic Processing Workshops at Purdue University https://ag.purdue.edu/foodsci/mishralab/workshops/

      i. Aseptic Processing and Packaging (Mishra, 2021a)
     ii. Validation workshop: Aseptic processing and filling (Mishra, 2021b)

15.9 SUMMARY AND RECOMMENDATIONS

With the risk of repeating some of the shared points throughout this chapter, the authors felt that key deliverables and take-home messages of this chapter should be highlighted in this summary. After the foundation of understanding our enemies, microorganisms, in Chapter 13, the risk analysis and mitigations methodologies are introduced to the reader. Those risks determined well in advance in the design stage allow subject matter experts to create validation protocols with acceptance criteria driven by the risks. After discussing a detailed approach to achieve "validated state" for the aseptic systems in this chapter, the following chapter will cover the organizational maintenance schemes such as Food Safety and Quality Management systems to maintain the aseptic culture and protect the brand.

Aseptic processing and packaging systems are technologically complex and require extensive work to validate and maintain the "validated state" of such systems. The aseptic system has many interconnected pieces of equipment that need to function flawlessly to avoid any issues of food safety and spoilage. In this chapter, the reader is presented with a holistic view of the planning and execution of the commissioning and validation of aseptic systems along with the regulatory aspects and supporting programs such as change management and preventive maintenance. The authors of this chapter have shared their views and recommendations based on their collective experiences of nearly 80 years in industry and academia.

The title of this chapter includes "*Validated State*" terminology which is much different than any other definitions in the manufacturing industries. What is the significance of this terminology? Why is Validation a "*state*"? The answer can be found in this chapter. Although Extended Shelf Life (ESL) systems are not specifically discussed, since the principles of the ESL systems are similar to aseptic systems, a clear transfer function between two systems will be obvious for the readers.

It is critical to reiterate that validation is *not* a "*project*," it is a process! Projects have starting and ending dates, validation does not have the ending date until the retirement of a process or manufacturing line. The reader of this chapter will learn about the steps required to achieve this critical "*state*" of the process.

Validation should not be considered in isolation and should be an integral part of business development. The afterthought validation approach has always been expensive and culturally not suitable to sustain the "validated state." Therefore, validation must be adapted and embedded in modern methodologies and business processes. Validation must be imbedded into several functions and programs such as quality systems, good engineering practices, and continuous improvement.

Before the implementation and creation of the validation culture within the company, it is important to understand the perceptions in the eyes of some decision-makers. Validation programs are often considered as follows:

1. Time-consuming
   a. Schedule delays expected
   b. Repeated activities

2. Costly
   a. Significant overruns

3. Not consistently applied
   a. Lack of standards
   b. Not following standards

4. Unpredictable
5. ROI not realized

To establish a sustainable aseptic validation culture, company leadership must deal with one big dilemma—balancing internal and regulatory requirements with the demands of the market. Without an established structure, business will pressure the company's unmature culture to take shortcuts for the purpose of speed-to-market. It is hard to blame them in the absence of company values and culture focusing on doing the right things. As a starting point, the following questions must be asked for self-assessment of the company culture:

1. What does compliance mean to the business?
2. What strategy is right to balance risk with compliance?

3. Who is responsible for compliance?
4. Where does validation fit in the organizational structure?
5. What is validation's role in compliance?
6. Why validate?

Validation should be included in the company culture as business and operational benefits, as well as regulatory requirements for compliance. It has been proven that validation increases usability and reliability resulting in decreased failures, fewer recalls and corrective actions, less risk to consumers, and reduced liability. It also reduces the long-term costs for sustaining the product (i.e., changes).

Companies must move validation from "discovery" to a repetitive, controlled, and properly managed business activity that strives for continuous improvement. Validation is not a development activity!

For securing the future of the company and protecting the company brands, validation must be incorporated into the company culture and imbedded into the business activities.

15.10 FREQUENTLY ASKED QUESTIONS

Q1. Do we have to revalidate? If so, when and why to revalidate?
A1. There is no set frequency for revalidation. The validated state of the aseptic systems is maintained with the change management and other supporting programs. For example, a major change in the packaging material or closure or ingredients may require full revalidation or partial validation.

Q2. Is swabbing performed after cleaning or sterilization of aseptic fillers?
A2. Yes, swabbing is performed after cleaning or CIP/COP, but not after sterilization. The system is in sterile condition after sterilization and hence no swabbing can be done.

Q3. What is the industry standard incubation time?
A3. The industry standard is 7–10 days. However, for the new lines, it can be 14–21 days at 35 °C for mesophilic and at 55°C for thermophilic microorganisms. It can be relaxed after gaining confidence in the system.

Q4. How is loss of sterility measured?
A4. The loss of sterility is dependent on the critical factors specific to the aseptic system, aseptic tank, and aseptic filler. For example, (1) loss of overpressure in aseptic tank, (2) loss of hold tube temperature for the product, (3) increase in the flow rate beyond the allowed maximum flow rate, and (4) loss of sterilant concentration or flow.

Q5. How do I decide on which method to employ between changing hold tube lengths versus changing my flow rate?
A5. This is a Process Authority and management of change control question. To achieve the target lethality, required time–temperature combinations can be determined to deliver the necessary quality and safety attributes. Depending on the product characteristics, the time–temperature combination can be adjusted. Some products may not handle very high temperatures and may need to make the necessary changes by increasing the residence time. Changing of the hold tube is not commonly done; however, if the new product demands a longer or shorter hold tube, it can be accomplished.

Q6. Why is F_0 temperature different for sterilization and pasteurization?
A6. Pasteurization process targets eliminating only vegetative pathogenic organisms with a limited shelf life for refrigerated products. On the other hand, sterilization, as the name indicates, targets eliminating pathogenic spore-forming bacteria for shelf-stable products. F_0 is used for sterilization purposes.

Q7. What is the best way to monitor H_2O_2 concentration?
A7. It can be measured using specific gravity at a given temperature. For vapor hydrogen peroxide (VHP), electronic sensors are available for different measurement ranges. Also, titration can be used to measure concentration.

Q8. If there is burnt substrate in the aseptic system, is it difficult to remove by CIP?
A8. If there is burn-on on the product-contact surfaces (e.g., aseptic valves), the possibility of cleaning failure is very high. White cloth tests and swabbing should determine the cleanliness and any burn-on in the system. CIP validation is critical to determine the effective cleaning parameters after long production campaigns. Some of the aseptic systems may have in-process cleaning (IPC) or aseptic intermediate cleaning (AIC) to remove excessive fouling by passing sterile cleaning chemicals without losing the sterility of the aseptic line.

Q9. What will our course of action be if there is a sudden power outage while in production?
A9. It usually ends up with a deviation if there are no battery backups or surge protectors installed for the aseptic line. The aseptic system may lose sterility. However, if the proper automation is validated, the product in the aseptic tank can be saved and released for production after a review by qualified personnel and process authority. On the batching side, the product can be saved if the backup refrigeration is working. Another situation which is similar to electricity failure is steam failure.

Q10. What is the most common gap or choke point you see in an aseptic operation? Personnel? Training?
A10. It is personnel training. Aseptic systems are complex and need constant training of the personnel to keep up with the technology. Lack of implemented change management closely follows the training gap.

Q11. Will steam barriers protect leaking valves from microbiological contamination?
A11. Steam barrier should not be relied on for a leaking valve. As soon as a leak is detected, corrective action must be taken to fix the problem.

Q12. How long does it take to get a decision on a FDA filing?
A12. Depends! The workload of the reviewers defines the filing time. Depending on the complexity and completeness of the filing dossier, FDA's response may be within 30–90 days (about three months). Products are considered "filed" as soon as the filing documents are submitted. However, for potentially questionable products, it is recommended to have prior discussions with the FDA LACF office or wait until receiving a communication from the FDA before shipping the product.

Q13. Is there a requirement to have feedback on whether the valve is opened or closed in an aseptic system to protect the aseptic zone?
A13. All the aseptic valve positions and feedback loops are tested during the controls validation. This is one of the best practices and highly advisable to document the functionality of valves and their responses to different events.

Q14. Are all of these validation activities (physical, temperature mapping) required by the FDA to be included in a filing?

A14. No, not everything has to be included in the filing dossier. Some of the container mapping and other validation support material can be kept in the facility and available if asked during the audits.

Q15. Is there a proper way to figure out where to place spore strips in a larger sterile zone, say a rotary or inline filler?

A15. Yes, aseptic fillers should be reviewed by the Process Authority and other subject matter experts such as vendor's technical teams and internal engineers. Airflow dynamics and patterns, temperature distribution, nozzle design characteristics, and machine surfaces in the aseptic zone must be evaluated to determine the worst-case locations for each location in the aseptic.

Q16. Is the FDA filing per product per package?

A16. Depends! FDA accepts filing applications for product families if the product characteristics are similar and critical factors are in the same filed ranges.

Q17. What is the purpose of blue milk in packaging validation?

A17. Blue milk is a nutritious microbial growth media used to support the growth of injured spores during validation. Since it contains bromocresol purple as an indicator of growth, hence the name blue milk.

Q18. Can all microbiological runs for the container be done in one day or do they need to be spread out over multiple days?

A18. Yes, depending on the size of the filler, packaging tests can be done in one day. However, it is unlikely to have enough time to run all of the packaging sterilization tests.

Disclaimer: Reasonable efforts have been made to answer FAQs based on our experience and knowledge, but we recommend that the readers work with a credible Process Authority and in full compliance to all applicable regulations, for implementation and results.

GLOSSARY

AC: Acceptance Criteria refers to a set of predefined requirements that must be met to mark the qualification and validation activity complete.

*Aseptic Bubble***:** PMO (2017) defines the aseptic processing and packaging system (APPS) as it begins at the constant-level tank and ends at the discharge of the packaging machine, provided that the Process Authority may provide written documentation which will clearly define additional processes and/or equipment that are considered critical to the commercial sterility of the product.

*CU***:**   Critical Utilities are the essential components of any aseptic processing and packaging systems with supporting facilities.

*DQ***:**   Design Qualification—a documented review of the design of new equipment and systems, at an appropriate stage in a project (preinstallation) for conformance to operational and regulatory expectations.

*FRS***:**   Functional Requirement Specifications—the initial definition of functional goals and objectives of the proposed equipment or system as compared with user requirements, management requirements, the operating environment, and the proposed design methodology. It also provides the basis for the design and qualification of the equipment or system. This document is generally created by the equipment vendor.

IQ: Installation Qualification is to establish documentary evidence that a subsystem or equipment is installed in compliance with technical specifications, standards, codes, and regulations. Another definition is to demons trate the performance of *documented* verification that equipment/system installation adheres to approved contract specifications and achieves design criteria.

The IQ is developed from P&ID's, electrical drawings, piping drawings, purchase specifications, purchase orders, instrument lists, engineering specifications, O&M manuals, and other necessary documentation. The IQ preceded the OQ.

OQ: Operation Qualification is establishing documentary evidence that a subsystem or equipment is capable of repeated operation within the limits defined in the specifications.

The documented verification that an equipment/system performs per design criteria over all defined operating ranges. The OQ includes the qualification of operating and maintenance records. The OQ precedes the PQ.

PQ: Performance Qualification is establishing documentary evidence that operating characteristics and product are in conformance with the limits defined in the specifications. Critical parameters (temperature, pressure, flow rates, etc.) must be stable over time and under both normal and worst-case conditions.

PQ is an organized event for the documented verification of equipment, systems, or processes that operate the way they are designed to do. The operation must be reliable and reproducible within a specified predetermined set of parameters, under normal production conditions, and must be in a state of control. This is the last qualification step before the SAT to bring the line to "Validated State."

PV: Process Validation is establishing *documented evidence* which provides a *high degree of assurance* that a specific process will *consistently produce* product meeting its *predetermined specifications and quality attributes*.

RA: Risk Assessment

URS: User Requirement Specifications—describes critical installation and operating parameters and performance standards that are required for the intended use of the equipment and provides the basis for qualification and maintenance of the equipment. It also describes the requirements of the systems or equipment in terms of the intended use or end product to be made. This document is generally created by the equipment or system owner.

SUP-SID: The Supplemental Submission Identifier is a code identifying the "Supplemental Submission" material to Form FDA 2541g for Aseptic product filing. Each Supplemental Submission is identified by a unique identifier, called a "SUP SID." The SUP-SID is a unique number associated with each Supplemental Submission for a facility. Filing company assign the SUP-SID. Each product filing would have a unique SID number with a SUP-SID for supporting documentation. Since it contains aseptic line specific information, the same SUP-SID can be assigned to many product SIDs.

VMP: Validation Master Plan—a written plan describing the process to validate, including production equipment and how validation will be conducted. Such a plan would address objective test parameters, product and process characteristics, predetermined specifications and factors which will determine acceptable results.

REFERENCES

Anderson, N.M, Benyathiar, P., and Mishra, D.K. 2020. Aseptic processing and packaging. In *Food Safety Engineering*, 661–692. New York: Springer.

Bernard, D.T., Gavin, A. Scott, V.N., Polvino, D.A. and Chandarana, D. 1987. *Establishing the Aseptic Processing and Packaging Operation, Chapter 8*, West Lafayette, Indiana, US: Purdue University Press. https://docs.lib.purdue.edu/aseptic/

Denny, C.B., Shafer, B., and Ito, K. 1979. Inactivation of bacterial spores in products and container surfaces. In International Conference on UHT Processing and Aseptic Packaging of Milk and Milk Products, November 27–29, 1979. Proceedings/Sponsored by Department of Food Science, North Carolina State University, Raleigh, NC, and Dairy Research, Inc., UDIA.

IFTPS. 2011. *Guidelines for Microbiological Validation of the Sterilization of Aseptic Filling Machines and Packages, Including Containers and Closures*. Institute for Thermal Processing Specialists.

Mishra, D.K. 2021a. *Aseptic Processing and Packaging*. West Lafayette, Indiana, USA: Purdue University

Mishra, D.K. 2021b. *Validation Workshop: Aseptic Processing and Filling*. West Lafayette, Indiana, USA: Purdue University.

Rodriguez, J. 2006. "The challenges of implementing a centralized change control management system." *J. GXP Compliance* 10(2):28–49.

USDA. 1984. *Guidelines for Aseptic Processing and Packaging System in Meat and Poultry Plants*. USDA.

USPHS. 2019. *Grade "A" Pasteurized Milk Ordinance (PMO)*. U.S. Department of Health and Human Services, Public Health Service, Food and Drug Administration, College Park, MD, USA https://www.fda.gov/media/140394/download

Chapter 16

Quality and Food Safety Management System (QFSMS) for Aseptic and ESL Manufacturing Companies

Tyler Dixon, Ferhan Ozadali, and Jairus R.D. David

CONTENTS

DOI: 10.1201/9781003158653-19

16.1 INTRODUCTION

This chapter is designed to give readers an outline of quality and food safety management systems (QFSMS) for aseptic and extended shelf-life (ESL) manufacturing companies. The scope and audience were determined as the manufacturing plant employees, industrial/academic/ regulatory quality, and food safety professionals, and of course students. As discussed in previous chapters of this section, continuous production and complex technologies require risk-based validation. However, most of the protocols mentioned and related executions are considered as snapshots of the systems at a given time. Therefore, well-established prerequisite quality and food safety programs are necessary to sustain the "validated state" (in Chapter 15) of the aseptic and ESL manufacturing systems. Although there are many transferrable quality and food safety schemes readily available, adaptability of these schemes in a cohesive way and creating a culture based on QFSMS are critical for sustainability of these programs.

The quality and food safety programs presented here are based upon pumpable fluid products with and without discrete particles which are sterilized, packaged in pretreated packages, and hermetically sealed. The finished product is targeted to be acid, acidified, or low-acid shelf stable under ambient or refrigerated conditions for aseptic or ESL products, respectively.

16.1.1 Concepts

Any food manufacturing operation consists of several value-added steps for transforming incoming raw materials—both food and packaging material—into finished product with acceptable safety, quality, and convenience. The *quality* of incoming raw materials and personnel/procedure/process/package *capabilities* of value-added steps will determine food safety, quality, and conformance of finished product. A quality assurance (QA) program will determine the success of any aseptic operation and is a true reflection of a company's commitment, mandate, and understanding of technologies, processes, and consumer needs.

Food safety and quality must be a top priority to the management, as important as return on investment (ROI), profitability, or market share. Without safety or quality, one cannot or does not have market share, profitability, or ROI. Companies in the food industry do not compete with each other based on food safety! A collective trust of consumers in the food industry is vital for the future of the specific product categories, including aseptic products. Therefore, minimum standards and regulations have been developed and enforced to keep the food industry safe and trustable. On the other hand, companies sell quality! The company's specific quality culture determines the profitability and longevity of that company. With the increased level of complexity (e.g., technologies, competency needs), aseptic processing and packaging systems require a unique aseptic culture to prevent any catastrophes.

The quality and food safety programs presented here are based upon pumpable fluid products with and without discrete particles, which are sterilized, packaged in pretreated packages, and hermetically sealed. The finished product is targeted to be acid, acidified, or low-acid shelf stable under ambient or refrigerated conditions for aseptic or ESL products, respectively.

16.1.2 Quality Control (QC)

It is imperative to start this discussion with a simple definition of quality, which is defined as the suitability or acceptability of a food product for its intended use by the targeted consumer base. Until the product reaches the consumer, the product life cycle has many risks to cause product failure against the intended uses of the product. Therefore, existence of validated quality systems is extremely important for the survival of a food company. Quality systems have QA programs, and QA programs have quality control procedures to minimize the risk for the consumer. Quality control is defined as the process through which one measures actual quality performance, compares it with a standard, and acts on the difference (Juran & Godfrey, 1999). In other words, quality control is a form of "gap analysis" followed by appropriate corrective action to close the gap. QC is very tactical and relies on standards and procedures. Therefore, organizations need quality programs at higher levels to determine the company culture and mission. Although QC is a very important part of the quality program and reduces the risk for the consumer, the purpose is not to create preventive strategies. While QC is focusing on the quality of the output in a very tactical way, there is a clear need to design, develop, implement, and inspect processes and programs for the company's quality culture. This is where QA comes into the picture.

16.1.3 Quality Assurance

QA is a subset of the quality systems which identifies the overall vision and mission of the company and outlines the quality culture of the organization with management commitments. Quality assurance is a form of insurance purchased at a relatively small expenditure for the purpose of securing protection against disasters. Quality assurance protection consists of information that the product is defect-free, the process is behaving normally, and the procedures are being followed (Juran & Godfrey, 1999). In a broad sense, in the food industry, QA represents a department with overarching responsibility for control of incoming food ingredients and packaging materials, batching, in-process checks, postprocess incubation, reports, inspections, regulatory compliance, and audits. There are general guidelines found in most textbooks, but they are usually far too broad to be of much use to anyone who plans to practice aseptic processing and packaging of foods.

One cannot instill quality and safety by inspection alone. Quality assurance is a competitive tool and should be implemented in a cost-effective manner. There is no upper limit to sampling and testing, and attendant cost. All testing and data gathering should have a clear purpose and utility.

16.2 QUALITY ASSURANCE FOR ASEPTICALLY PROCESSED AND PACKAGED FOODS

Compared to retort or hot-fill packaging in metal and glass containers, aseptic packaging of foods in flexible containers has shifted to the food processor increased responsibility for evolution of quality systems, identification of defects, quality control of products, and encouraging promulgation of appropriate regulations. This has given the food industry a strong incentive to develop a QA program in concert with suppliers of aseptic fillers, cans, lids, roll stocks, cups and lid stocks, and equipment for rapid testing of seal integrity and sterility (David, 1988, 1989).

A QA program must nurture a good interactive relationship with operators of sterilizers and aseptic fillers and QC line inspectors to assure manufacture of product in acceptable containers. Quality is not the responsibility of a department but is the joint responsibility of everyone involved in management, design, research, procurement, supply, testing, warehousing, and distribution. Some companies are currently integrating manufacturing and quality assurance through mandate, training, and at-line and online testing by line operators. Ultrahigh temperature (UHT) processing and aseptic packaging systems are sophisticated operations requiring skilled and trained individuals in production and quality assurance. This system leaves no room for errors.

For convenience, a quality assurance program for aseptically processed and packaged food is described as work outputs under (1) preprocess assurance, (2) in-process assurance, and (3) postprocess assurance functions.

16.2.1 Quality and Safety by Design

If the organization is relying on QC only to minimize the risks for the consumers, it is generally too late in the product life cycle and very expensive. Quality has to start from the "design" stage for the product development and manufacturing processes to ensure achievement of predetermined quality parameters and attributes for ultimate consumer satisfaction. Good Engineering Practices (GEP) are very important, however, not enough to achieve the quality levels with a sustainable process. It is critical to understand the process and variables that may affect the product quality and/or food safety. The industry history is full of examples of excellent engineering design without well-thought process design parameters to prevent catastrophic failures. Both Chapters 14 and 15 deal with the details of the risk identification and their use to set the validation parameters. The V-diagram of validation of aseptic systems in Chapter 15 clearly highlights the design steps on the left side of the "V" and their criticality prior to building (industry terms: cutting any metal) and equipment or systems. Every change after these steps would be more costly and the consequences of risks might be catastrophic if those risks are carried through the entire product life cycle without identifying and remediating. Therefore, a systematic quality and safety by design (QSbD) approach by understanding the quality and food safety requirements is required to prevent costly and risky consequences. This needs to be done at the equipment/process design stage based on proper risk analyses methodologies and an understanding of science and technologies.

QSbD must be applied to every product and process development. Industry and regulatory requirements, including hygienic, manufacturing, and engineering standards, must be used during the creation of the functional requirement specifications (FRS) and system/product design. For example, hygiene is a critical part of an aseptic design. Hygienic design of food processing facilities and equipment is crucial for eliminating the risk of food contamination. Considering the facility grounds, building layout, construction materials, and equipment design are vital for impeding contaminant ingression into the facility, equipment, and product. There is merit in careful consideration of building design and selection of materials for the successful operation of food production. The saying *"you get what you pay for"* is especially true in the aseptic industry. Taking the time to research, design, and maintain the facility grounds, building, and equipment is a worthwhile investment. Manufacturers and Original Equipment Manufacturers (OEMs) should take time to "build" hygiene into the operation and equipment designs.

16.2.2 Quality and Safety Risk Management

Quality and safety risk management is a scientific and systematic process for identifying and controlling risk throughout products' life cycle. An operation must consider all risks: those that may occur naturally, those introduced unintentionally, and those that may be intentionally introduced (either for economic gain or bodily harm). Procedures must be created for each scenario to adequately address the actual or perceived risk to the product, personnel, and the facility itself. All risks are identified for their severity and the likelihood of their occurrence. This can be achieved by using a risk assessment matrix, which can help identify hazards, and adequately determine risk. A risk assessment exercise can be conducted with an interdisciplinary group of individuals who understand the evaluated processes. A good risk assessment exercise will ask several questions: What might go wrong? If something is to go wrong, what is the probability that it will occur? Finally, if something does go wrong, what is the consequence of that failure if it is not appropriately addressed? The goal of this exercise should be to identify risk and also provide a way to reduce it. It is understood that risk is inherent to the system, and it can never be completely eliminated. However, having a proactive plan to identify and mitigate risk will provide a good framework for a Food Safety and Quality program. Several tools are available to adequately manage risk, including failure mode and effects analysis (FMEA), hazard operability analysis (HAZOP), and process hazard analysis (PHA). Utilizing appropriate tools and industry experts will significantly reduce the risk of producing an unsafe or poor-quality product.

For example, a risk analysis for an aseptic product may include an evaluation of the production process. Are there areas where mechanical failure may introduce foreign material (such as gaskets or metal) into the system? Is the CIP unit separated from the production process so that cleaning chemicals cannot be introduced? Can the thermal process introduce contaminants (such as furan or acrylamide), potentially leading to product safety or quality issues? Could someone purposefully introduce a harmful material at a step in the process undetected? For each possible scenario, the likelihood of occurrence and the severity of the hazard should it occur must be identified. Risks that have moderate to high levels of severity and likelihood must be addressed. If the addition of a harmful material to the process undetected is identified as moderate to high risk, cameras or other monitoring activities may be necessary to reduce the probability of occurrence.

Regardless of the scenario, risk must be handled proactively and monitored appropriately. Doing so will give consumers and producers a high level of confidence that the final product is safe for consumption. More information about the risks and analysis methodologies can be found in Chapter 13.

16.2.3 Corrective and Preventive Action (CAPA)

A critical tool in a proactive Food Quality and Food Safety plan is Corrective and Preventive Action (CAPA), especially considering the seventh step in a HACCP plan (verification). The principles of quality management must include this component for system improvement, which is comprised of corrective and preventive action taken in response to system or process deviation. A good plan should use this Root Cause Analysis tool for correction and future prevention of an existing deviation and as a proactive measure to ensure system health. A robust process in the hands of trained individuals can be an effective tool for maintenance and the improvement of quality, operations, and maintenance activities.

As defined by ISO 9001:2015 (ISO), a corrective action is "action to eliminate the cause of the non-conformity in order that it does not recur or occur elsewhere," and preventive action is "action to eliminate the cause of a potential non-conformity or other undesirable situation." Simply put, a corrective action is one taken to fix the existing problem, while preventive action inhibits the recurrence or existence of the problem itself.

Many models can be followed for the practical application of a CAPA process. Several common approaches, including FMEA and the DMAIC process, are discussed in a later chapter.

FDA training link provided: https://www.fda.gov/files/about%20fda/published/CDRH-Learn-Presenation--Corrective-and-Preventive-Action-Basics.pdf

16.2.3.1 Triggers for CAPA

An effective CAPA program should have defined triggers for action, which will initiate the CAPA process. A trigger is a non-conformity or deviation that when it occurs, "triggers" a CAPA investigation. Each process should have a set of triggers that will activate a CAPA process. Some companies will build these into their Key Performance Indicators (KPIs) to help define when action might be needed. For example, a trigger might be any safety incident that causes a recordable injury which keeps someone from returning to work. Perhaps a Loss of Sterility (LOS) event on a particular machine can cause downtime and reduced efficiency is defined as a trigger. It could even be a Quality or Food Safety deviation that may present itself and cause the product to be impounded.

Regardless of the process, it is advisable to establish triggers to help identify when and where a CAPA might be needed.

16.2.4 Change Control

Another essential tool in any food manufacturer's toolbox is change control. Change control is a systematic way of transforming a process or particular focus. Its goal should be to implement an effective strategy in which the particular change is affected or controlled. In addition to this, a productive change control process also helps those affected by the change adapt to it.

Change will happen. Often, it will happen quickly. Without a consistent and repeatable process for managing it, change will hinder progress. There are several critical factors in a successful change control program, which are discussed below.

16.2.4.1 Need for Change

One of the first items that must occur for a change control process to start is identifying a need for a process to change. This change could be as small as the wording in a Standard Operating Procedure to something as significant as introducing a new processing and filling line. It may be a change that is desired internally or one that is required due to regulatory updates. Regardless of the why, when a change is acknowledged as necessary, it is crucial to define the change, specify what needs to change, and identify the critical details of the change to be communicated to affected parties.

16.2.4.2 Leadership Support

As with any process, having the support of individuals who will determine the effect of the change is essential. Without leadership support for a change process, it will not sustain itself. It is essential to come prepared to any change control meeting and answer the who-what-when-why-how of the process. Having detailed drawings, schematics, process proposals will be essential to helping leadership understand the change and get behind it.

Once these questions have been answered, it is advisable to set up a change control meeting and include the appropriate individuals who might be affected by the change and the leadership supporting the change.

16.2.4.3 Expert Advice

In addition to leadership being present and supportive of the change control process, it is equally important to make sure that a diverse group of subject matter experts are present. This meeting should include members of Health and Safety, Quality, Operations, Regulatory, Finance, Maintenance, and any other department that might be affected by the change.

By doing this, each department can review a proposed change from their perspective and provide a high level of confidence in the change going forward. Each department should have an identified criteria for evaluation of a change that needs to be met before a change is approved. For example, a Quality department may require a HACCP process review of the proposed change to see if the Food Safety plan might be impacted. Perhaps additional training will be required for team members affected by this change. A Maintenance department may need to determine what might change in the preventive maintenance process. Does a new piece of equipment need to be added to the maintenance records? How often does the equipment need to be serviced? Are there new failure modes that need to be considered by bringing in this equipment? Perhaps a Process Authority needs to identify if the change might affect a process filing or critical parameter. Many considerations should be taken for the implementation of a change, especially for an aseptic process. Having each department review and approve of a change from their perspective is critical to change control success.

16.2.4.4 Executing Change Control

Once a change control process is underway and leadership and the affected departments have had time to review the change, a process must be established for following through with actionable items that might arise from the change control. Additional requested information should be gathered and provided. Actions needed for concurrence to the change should be completed before the change occurs unless otherwise agreed upon. All of this should be well documented and presentable for future reference, as well as for review.

Another critical piece in execution is ensuring that communication of the change is effective. Have the appropriate parties (often the individuals on the process floor that may be most affected) been apprised of the change? Are they in alignment? Perhaps new training or Operating Procedures will be required. These items must be considered in order for the change to be successful.

16.2.4.5 Monitoring and Verification

When a change has been implemented and success criteria for the change are defined, the process should be monitored to verify that the change is effective. Regular feedback and review of the change should happen until it is established that the change was successful, and the system can go back to a normal monitoring state. Asking questions such as "Have the success criteria been met?" and "Have actionable items from the change been implemented and communicated to all affected areas?" will help ensure the process is on track and healthy.

16.2.5 Process Authority Roles in the Food Industry

Process Authority (PA) title is widely used and recognized in the United States and in the countries exporting food products to the United States. Since this role is involved in risk analyses, validation, and maintaining the safety of the products after the validation through Food Safety and Quality management systems, it is critical to understand what this role brings to the table. There are clear regulatory definitions and expectations for the involvement of this role. However, with the evolving technological advancements, overall expectations from the PA role have also changed and the criticality of this role's involvement has become very clear, although aging regulatory requirements are focusing on the "thermal" processes, current PAs deal with a wide variety of sterilization technologies, including chemical (e.g., peroxide, PAA), irradiation (e.g., e-beam), and many other hurdle technologies. In addition, PAs have been at the center of the food safety discussions for the low-moisture food treatments. Therefore, Process Authorities deal with all kinds of processes, and not limited to thermal processing only. They look at the entire system with all of its components such as product, package, and process. This is generally expanded into storage and distribution in the supply chain throughout the product's shelf life.

The authors' definition of the Process Authority can be summarized as an individual or group of experts in the development, implementation, and evaluation of food processes throughout the products' life cycle for the protection of public health. Two important criteria for a PA credential: First, companies should be accountable to identify their PA to eliminate any Food Safety risks in their organizations. Official appointment letters by food companies must be issued to certify the authority, second, regulatory agencies (FDA, USDA, CFIA, etc.) should only accept filings or communications from recognized PAs.

The following are some of the required competencies for a PA:

- Knowledge of microbial risks, product and packaging characteristics, critical factors, commercial equipment, and manufacturing procedures, and their effects on the delivery of a process and maintenance of product sterility
- Knowledge of applicable regulations
- Knowledge of the underlying principles, process calculations, analysis tools, and evaluation techniques related to processing
- Experience in designing and conducting the appropriate studies relating to processing of food, such as heat penetration, temperature and heat transfer distribution studies, thermal-death-time experiments, process validation and verification studies, and applying other scientific methods related to aseptic and other processes, including thermal treatments
- Ability to analyze data generated by scientific studies and evaluate the effectiveness of processing (including thermal and other treatments) and packaging system to ensure safe and commercially sterile products.
- Experience and ability to identify and evaluate process deviations and spoilage incidents
- Ability to document process establishment methods and results, and communicate proper process requirements and recommendations

According to the FDA, a PA is a person who has "expert knowledge of thermal processing requirements for low-acid foods in hermetically sealed containers and having adequate facilities for making such determinations" (21 CFR 113.83).

According to CFIA (Chapter 13, Subject 1—Thermal Process Control Policy for Federally Registered Canneries), Process Authority means any person or organization that has been recognized by the Agency as being competent in developing and evaluating thermal processes. This definition continues with a list of the areas of competencies expected to deliver safe processes.

Therefore, a qualified PA will work as part of the manufacturing organization with proper credentials or work with a manufacturer through an external organization to determine the processing requirements for their systems to ensure safe food is being produced.

16.2.5.1 Thermal Process Review

A thermal process review can be conducted by a Process Authority, which relies upon experience, expert knowledge, and published scientific data to establish minimum criteria for safety. The process review, which will become the filed Scheduled Process, will identify operational processing limits that ensure product safety and protect quality.

A scheduled process, as defined by regulations, means the process selected by a processor as adequate for use under the conditions of manufacture for food in achieving and maintaining food that will not permit the growth of microorganisms having public health significance. It includes control of pH and other critical factors equivalent to the process established by a competent processing authority (21CFR 114.3e).

TABLE 16.1   RECOGNIZED LIST OF GFSI AUDITING SCHEMES

Recognized list of GFSI Auditing Schemes		
CanadaGAP	Global G.A.P.	JFSM: Japan Food Safety Management Association
Freshcare	IFS: International Featured Standards	Asia GAP
FSSC 22000: Food Safety System Certification	Global Seafood Alliance	SQF: Safe Quality Foods Institute
GRMS: Global Red Meat Standard	Primus GFS Standard	

16.2.6 External Certification Schemes (GFSI—BRC, SQF, FSSC2200)

Supplementation of internal verification and monitoring activities of a Food Quality and Food Safety program is advisable. Benefits of external auditing and certification include verification and continuous improvement of existing programs, and application of new or best in class processes. Companies that invest in additional oversight of their programs can deliver high levels of food safety and quality assurance to their stakeholders, investors, and clientele. Certification can be achieved in several ways. Multiple Food Safety certifications can be attained by companies seeking additional assurance for themselves, and their customers. Recently, collaboration with industry experts has begun to harmonize food safety standards, known as the Global Food Safety Initiative (GFSI).

GFSI is a food industry–driven mission for the development of food safety management systems to ensure food facilities are processing safe products (GFSI). Started in May of 2000, it is a private organization that oversees and recognizes auditing platforms that meet a set of criteria established by its steering committee. This committee involves retailers, manufacturers, suppliers, academia, and governments across the world to collaborate on best practices in food safety. The criterion established provides a universal standard of recognition in food safety auditing. The focus of this initiative is to apply a continuous improvement process in food safety management systems. Their vision is "Safe food for consumers everywhere" (GFSI) and is accomplished through benchmarking, collaboration, and capability-building. While GFSI itself does not provide food safety certification, it recognizes several certification programs that meet GFSI benchmarking requirements.

This means that a food processor or manufacturer can point to their GFSI certification and effectively show that their facility is operating with a structured, recognized, and implementable food safety program. This can provide consumer confidence in their products, and also foster a culture of food safety and pride within the company (Table 16.1).

16.3 THE QUALITY MANAGEMENT SYSTEMS MODEL

As with the Food Safety Management model, a good quality management system should take a similar approach. Utilizing HACCP methodology and process steps as described in Section 16.5.1, a comprehensive plan for food quality can be created. Considering this approach, identifying

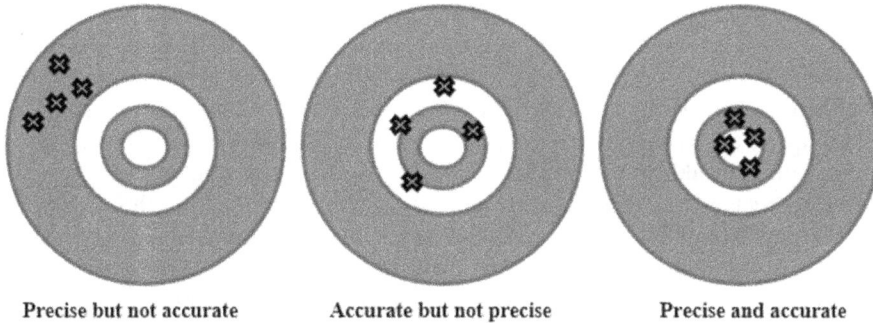

Precise but not accurate   Accurate but not precise   Precise and accurate

Figure 16.1   The importance of precision and accuracy.

quality control points (in place of control points) and critical quality control points (in place of critical control points), one can create a robust program for control of food quality systems. Where a Food Safety HACCP program looks at microbiological, chemical, and physical concerns, a QC program will look at specific indicators of process control (preprocess, in-process, and postprocess) that may be defined by the facility and/or by customers themselves. This can include considerations like nutritional or sensory quality, formulation conformity, inline checks, as well as controls that may also support food safety. Take for example a thermal process control point. There may be a minimum operating temperature that is defined by the process authority to make safe food. As long as the temperature exceeds that minimum value, food safety can be assured. However, if that value is significantly surpassed, the nutritional and sensory qualities of the product may be reduced. Therefore, it may be necessary to set upper operating limits for the particular point.

Other important aspects of a quality management program are repeatability and precision. How much variation is there in your QC points? How often do the control points exceed your established operating limits? Consider an archery target, with the ultimate goal of quality being the bull's-eye. As marks are made on the target, is there variation to the marks? Are they close to the target? How often do they hit the mark? These questions are some that should be asked (and addressed) in a good quality management system. Hallmarks of a good plan show repeatability (precision) and conformance to standards (accuracy) (Figure 16.1).

16.3.1 Preprocess Assurance

Preprocess assurance consists of vendor compliance of incoming raw material to company specifications through vendor certification programs, audits, and tests for verification. Quality programs must take into consideration high levels of reliability and conformity for the raw materials that will be used to manufacture a particular product. It is a well-known saying in the industry that "trash in" equals "trash out." A good quality program will have established standards for vendors of raw materials to comply with in order to ensure that the final product meets internal and customer expectations. Each material should have specific success criteria met for each lot that is brought in. This should be assured by the vendor (by COA or other means) and also verified by the manufacturing facility receiving the product. There are many factors of quality that must be considered for raw materials in aseptic and ESL processes. While some factors (such as microbiological) may be addressed by the process, many others cannot. Factors such as color, flavor, nutritional components, rancidity, toxins, and foreign material may not be able to be adequately addressed through processing and can be exacerbated by the process. Thus, exclusion of undesirable attributes is a critical part of successful aseptic and ESL processing.

It is important to be able to identify undesirable attributes of product prior to receiving them into the processing facility. Thus, qualified technicians and equipment for the detection of such attributes should be included in the quality program. These individuals need not be limited to laboratories only. Receivers and warehousers should have a good idea of the quality requirements of the process and be able to make determinations of the product before it goes to further analytical testing.

16.3.2 Raw Materials

The quality of any finished product depends to a great extent on the quality of incoming raw ingredients. All food ingredients for manufacture of shelf-stable products must meet physical, chemical, and microbiological specifications necessary for product safety, physicochemical stability, and quality. One relies on supplier certification and conformance with company specifications and incoming material testing. However, it is prudent to cross-check supplier certification using a skip lot sampling plan. Any shipment that does not meet specifications must be rejected. Accurate receiving records must be maintained showing vendor batch codes or lot numbers. Both dry and wet ingredients need to be stored at recommended temperature and humidity. Also, pest control, particularly rodent control, is very important.

All it takes is one leaker on a pallet and various pests will be attracted to whatever has spilled. The presence of fruit flies is a good indicator of a leaker. Even if one does not initially have a leaker, it takes very little effort for a mouse or rat to chew its way through a flexible or semirigid container. Therefore, proper pest control and sanitary conditions are essential in warehouses, as well as on trucks or on rail cars.

Incoming packaging materials and sterilants for incoming packaging (i.e., body and lid materials) are like food ingredients and must meet certain physical and functional specifications. These parameters are ultimately verified by the performance of packaging material during the forming, filling, and sealing steps via the aseptic packaging machine. It is common not to test packaging materials for microbiological load, because of the assumption that microorganisms cannot reproduce on the inert container and lid materials as it does in the foods. Research indicates very low levels of contamination with vegetative microbes rather than spores (Bockelmann, 1982). Packaging materials, though inert, must be stored properly wrapped to protect from dirt, dust, oil, debris, and moisture, and generally stacked not more than two pallets high to avoid stress. The procedures and programs applied to ingredients should also be applied to packaging materials to control and prevent contamination with dust, dirt, oil, and other debris.

For chemical sterilants such as H_2O_2, specific gravity measurement is common to assure strength and purity. As with all chemicals, caution should be exercised while handling chemical sterilants, as outlined in the safety data sheet (SDS).

16.3.3 In-Process Assurance

In-process assurance is a complex program involving control of batching, thermal processing, and aseptic packaging operations.

A scheduled thermal process is designed on the basis of an anticipated "average" initial load of microorganisms. Therefore, it is the responsibility of the manufacturer of shelf-stable food to keep the initial load to a minimum by following current good manufacturing practices (cGMP) during mixing, batching, and raw product storage. In addition to the initial load, the processor needs knowledge of the physiological types of bacteria such as spores of mesophiles, especially *Clostridium botulinum*, non-pathogenic mesophilic and thermophilic spore formers responsible for economic spoilage, and psychrotrophic bacteria present in raw product. A high initial number of bacteria, poor sanitary conditions, and storage at ambient conditions can lead to preprocess spoilage of raw product. Also, toxin production by heat-sensitive vegetative cells of *Staphylococcus aureus* in raw product must not be permitted, because this toxin is highly heat resistant, with a calculated $D_{280°F}$ of 4.6 minutes. Most UHT processes will not inactivate this toxin.

Refrigeration is capable of controlling the growth of most bacteria in food, except psychrotrophs, which can multiply and secrete extremely heat-resistant enzymes such as lipases and proteases. With UHT processing, there is little inactivation of such enzymes. Also, these enzymes may experience temporary inactivation and under certain conditions may regenerate to cause product bitterness and rancidity. According to the latest Pasteurized Milk Ordinance (PMO) (2019), raw milk must be chilled to less than 45°F. It is a good practice to process milk as quickly as possible. The total age of raw comingled milk should not exceed more than 72 hours.

The art of bringing together several ingredients in proper proportions via mixing, emulsifying, and homogenizing is complex. These steps do affect the quality and physicochemical properties of the finished product and, therefore, should be specified and fully documented as standard mix procedures.

16.3.4 Postprocess Assurance

Post-process assurance represents process analysis, data review, and tests that are performed after the production is completed. Often these tests are performed after storage at ambient temperature or incubation at an elevated temperature for prescribed time periods. The evaluation may include sensory, product stability, product thinning due to enzymes, and spoilage due to microbiological contamination.

It is well known that one cannot build quality and safety through extensive inspection of the finished product. However, these tests are performed on a small sample size as one last step in the assurance process for management, and for regulatory compliance.

16.3.5 Incubated Product Evaluation

Statistically sound sampling plan development is critical to have consistencies in industrial applications. Aseptic processing and packaging operations are not different when it comes to sampling requirements. A sampling strategy needs to be developed to clarify the risks at every stage of production.

Different statistical methodologies can be used at different levels of the production stages. During the qualification and validation activities, the sampling plans are designed based on strict criteria by considering the worst-case scenarios which would never happen under normal operational conditions. Once the aseptic line is qualified and validated, less strict but statistically sound sampling plans take over to minimize the risk of any catastrophic events and minimize the risks for the consumers.

The most common sampling plan follows military standards (MIL-STD-105E; ANSI/ASQ Z1.4). Similarly, ISO 2859-1 (BS 6001-1) also outlines sampling requirements with a given lot size and AQL. The original version of the standard (MIL STD 105A) was issued in 1950. The last revision (MIL STD 105E) was issued in 1989, but canceled in 1991. The standard was adopted by the International Standards Organization as ISO 2859 (Shmueli, 2016).

While determining the sampling plans, there are two additional factors to consider. One of them is the *"Inspection Level"* which determines the relation between the batch (number of aseptic products produced in a lot or a batch) and the sample size (the number of aseptic products that should be randomly chosen from a batch). There are three general inspection levels (Level I–III). Level II is considered as normal level where Level I requires half of the amount required for Level II and twice the amount when Level III is assigned to the sampling plan. Selection of these levels depends on the tolerance to discrimination and when the cost of sampling is critical. There are four additional special inspection levels (S-1 through S-4) where very small sample sizes are required due to sampling costs and the risks can be tolerated. These inspection levels are not considered in the case of food safety concerns.

The other important factor is the *"Type of Inspection"* which heavily depends on the history of the line performance. There are Normal, Tightened, and Reduced inspection types. As expected, sampling sizes for reduced inspection are smaller than the normal inspection type. If the line history has questionable events, due to increased probability of failure, tightened inspection should be utilized during the determination of the appropriate sampling plan.

After determining Inspection type and level, the size of the samples will heavily rely on the Acceptable Quality Level (AQL) selection. This is one of the most confusing topics in the food industry and relatively well-understood in some of the other industries such as automotive and pharmaceuticals. The complexity of this term for the food industry is due to the complexity of the food systems and root causes of the quality and food safety risks. Companies must predetermine the appropriate AQL values based on their product categories and the risks associated with the production of those products. For instance, FDA-regulated low-acid food manufacturing systems apply 0.01% as the AQL value for any food safety–related sampling plans and decisions. This AQL corresponds to a "probability" of 1 defect (spoilage) per 10,000 units of aseptic product. It must be noted that this value is used as a statistical probability to come up with a sampling plan and, by all means, does *not* reflect the acceptability of 1 defect for every 10,000 aseptic products. Zero-acceptance sampling plans are very common for low-acid aseptic manufacturing lines.

For some of the quality attributes and low risk defects, higher AQL (less strict) values can be used since the risk levels for the consumer are lower. When sampling plans are designed, the correlation between the risk levels of manufacturer and consumer needs to be understood and accepted company wide. This is critical to minimize the conflicts and confusion when the cost of sampling becomes a burden on the operations. Very conservative companies that serve to a very sensitive consumer base would generally elect to take most of the burden and cost on the manufacturer side to minimize the risks for the consumers.

Manufacturer risk is the probability that a batch form production line with the appropriate AQL is rejected. Very conservative and tight specs with rigorous inspection methodologies, false reject rates and probability of rejecting a good lot are very possible. On the other hand, a relaxed sampling plan and inspection systems can increase the probability that a batch with a non-conformity is accepted. The manufacturer risk is the risk that a batch of high quality/safety (based on the AQL) is rejected. The consumer risk is the risk that a batch of low quality/safety will be accepted.

In order to understand the risks for consumers and the manufacturer, the Operating Characteristic (OC) curve is used. The OC curve provides the sampling plan developers with a tool to observe how changing criteria (outlines required performance and risk levels) impact the amount of required testing. This type of curve can be used to balance the cost and potential risks. A typical curve is plotted as the percent defective versus the probability of acceptance of the lot. There are so many tools and software to help with the execution of this analysis. MS Excel, Minitab, and many other tools can be used to easily develop OC curves.

About 0.25–1% random samples (Department of Defense, 1989) are incubated for 7–14 days at 95–98°F (~35°C) to accelerate degradative reactions of rheologic and sensory characteristics, and any microbiological spoilage due to mesophiles. At the end of the incubation period, product samples are checked for all cosmetic and seal defects, microbiological growth, and other product characteristics such as fill weight, pH, color, flavor, taste, and viscosity.

If a lot contains between 35,000 and 150,000 containers, it is customary to sample and incubate 315 units. The lot should be accepted on zero defect and rejected on two defects. If there is one defect, the lot should be resampled at a more intensified rate. Whenever results are questionable, it is best that the lot be placed on hold and the tests repeated before the disposition of a lot.

16.3.6 Shelf-Life Determination and Stability of Aseptic and ESL Products

Once the product is made from various compounds, including fresh and preprocessed food ingredients from different sources at different conditions, the mixed product prior to any processing and/or packaging steps goes through initial ingredient-to-ingredient interactions and equilibration. Therefore, the formulation of the product with proper ingredients that are compatible with each other is extremely important for determination of the product shelf life. Before we discuss the approach to determine the shelf life of a product, it is important to define the true meaning of the "stability" of the product in its expected shelf life. The main misconception in the industry is that shelf life is defined based on "microbiological" stability. Although this may be true for some products such as ESL product, "physical" and "nutritional" stabilities are the main drivers to determine the shelf life to maintain a salable product on the shelf.

If we define the product as a system with several components, our approach is to analyze each component in terms of its impact on the shelf life of the product. The focus points for this system from a stability point of view are known as 3Ps—Product, Package, and Process. Each focus area will impact the shelf life of the product and heavily depend upon the external conditions. Therefore, it is also critical to determine the environmental (e.g., storage and distribution temperatures) and other conditions that would impact 3Ps.

It is important to understand the consumer/customer's desire for the quality expectations for the product(s) in question. If we define "quality" as the condition of the outcome which is the aseptic product in this case, then the expected outcome within the economically feasible time frame is the core of our discussion. Outlining the focus points for the process of determining the stability of the product will help identify the approach.

To better understand the overall concept, a simple equation can help us outline the approach:

$$SL_0 - \sum_{0}^{tb} PII\Delta t - \sum_{p1}^{p2} PI\Delta p - \sum_{tb}^{ts} PCI\Delta t - \sum_{tb}^{ts} ENI\Delta t \geq SLO$$

Here SL_0 is the current shelf-life expectation; tb is the product batching time; p is the processing condition, including time–temperature combinations; ts is the time on the shelf; PII is the product's ingredient interactions; PI is the processing impact; PCI is the packaging impact; ENI is the impact of environmental conditions; and SLO is the shelf-life objective.

The above equation needs to be considered to challenge the stability of the product for two categories: physical stability and nutritional (sometimes called "chemical") stability for desired SLO.

Product: Interaction between different ingredients is a well-known factor to impact the shelf life of food products. The impact may come from the oxidizing properties of the ingredients or creating a synergistic influence on the degradation of certain nutrients or attributes in the presence of some other ingredients or processing conditions. For instance, vitamin C is a desirable and extremely useful ingredient which can protect product from oxidation in the presence of dissolved or headspace oxygen content. However, in the presence of heat, vitamin C may create intermediate oxidizing compounds with a short life that may contribute to color and flavor degradations of the product during the shelf life.

Product developers and process engineers have to work together to run risk analyses and determine potential ingredient interactions that may accelerate the nutrient degradation or cause loss of quality attributes such as color or organoleptic characteristics.

Process: Depending on the severity of the processing conditions and desired food safety objectives, product quality and shelf life may be impacted due to exposure to processing conditions such as high temperatures for prolonged periods. For instance, certain nutrients are very sensitive to heat, and they can be significantly degraded during the processing. As a well-known advantage of aseptic processing systems over conventional retort processes in terms of retaining the higher percentages of the original nutrient contents, aseptic processing systems are considered gentler.

However, very high temperatures (close to 140°C) even for a very short time of exposure will cause some level of degradation. Product and process developers can compensate for the degraded amount by spiking the formula with degraded nutrients at the formulation stage. This is critical for certain products to meet the label claims for the specific nutrient content at the end of the shelf life of the product.

Package: Packages are considered as "the last-line-of-defense" for the product to protect the quality attributes and the nutritional value from the environment throughout the intended shelf life of the product. There are several factors that need to be taken into consideration during package design and packaging process known as "aseptic filling and sealing." Not every package was created equal. Depending on the packaging material, certain transmission rates through the material will impact the shelf life. To minimize the transmission of oxygen, light, and moisture, packaging material may need to be modified by adding layers of different material and/or "barriers." For instance, to minimize the oxygen transmission through the plastic containers, ethylene vinyl alcohol (EVOH) can be added by incorporating the material into either a multilayer coextruded roll stock structure or an extrusion laminated structure. Six to seven layers of packaging material with additional barrier layers for specific purposes are quite common in the food industry and the usage levels depend on the expected quality and stability at the end of the expected shelf-life objectives. Although adding more barriers to protect the product is useful, additional cost degrades the gross margins and impacts the financial feasibility of the products. Different plastic materials such as polypropylene (PP), polyethylene terephthalate (PET), and high-density polyethylene (HDPE) may be used for different aseptic and ESL products with additional barriers where applicable for specific purposes such as colorants for light protection and EVOH for oxygen transfer rate reduction.

Environment: As outlined with the above equation, shelf life of a product can be maximized by optimizing the product formulation, processing conditions, and packaging material; however, variability of the environmental conditions is so great that effective predictions are difficult to make. Therefore, accelerated shelf-life test conditions are created to mimic the worst-case scenarios for the stability of the products. Products' intended use and the climate conditions in the regions for the distribution are very critical for the determination of the accelerated stability test conditions.

Processed foods during storage may undergo numerous changes. It is well known that ingredients, conditions used for processing, packaging materials, and storage of foods may adversely affect the quality attributes of finished products. One or more of the quality attributes of a food may reach an undesirable state upon storage after a certain period, making the food unsuitable for human consumption, and the food is said to have reached the end of its shelf life. In other words, shelf life is producer-, retailer-, and ultimately consumer-driven through the understanding of how a product ages.

During storage and distribution, foods are exposed to a wide range of environmental conditions. Environmental factors such as temperature, humidity, oxygen, and light can initiate several reaction mechanisms that may lead to food degradation with attendant loss of quality attributes. Physical, chemical, nutritional, and microbiological changes are the major causes of food deterioration. Often these reaction(s) may occur independently or simultaneously.

The assignment of shelf life is very important. If it is too short, then manufacturing costs may be high and profit margins low; if it is too long, then there is the potential for age-related unacceptable quality, and/or spoilage, and/or the product that will not meet the nutritional value set by regulatory legislation and compliance. A longer shelf life gives producers more time to transport and to sell the product at retail, and to minimize food waste (CAST, 2016). A longer shelf life can be concerning to consumers as consumers are suspicious that a food product will look and taste just the same in two years as it does on the day the product was made (manufacturing date). Consumers may also believe that once opened, aseptic product will still be able to achieve its posted code date. It is important for producers of aseptic products to clearly identify unopened use by time frames and opened use by dates to bring clarity to this topic. Ideally producers would like to have their customers have a high-quality product at the time of consumption, to earn their trust, build customer loyalty as evidenced by repeat-purchase, and brand share.

Food processors must have a good understanding of different reactions that contribute to food deterioration, while developing specifications and specific test procedures for the evaluation of shelf life, and data-driven decisions on a product's code date, also known as a "best by" date. From a regulatory point of view, a product must meet certain requirements, such as maintaining the nutritional value printed (or differentiation claims made) on the label all the way through the stamped "best by" code date.

It is important to assign the shelf life in a systematic and scientific manner, taking all the relevant factors into consideration. Evaluation of shelf life is also important when products are reformulated, minor changes in product reformulation can have a major impact on quality attributes or growth of microorganisms. Thus, changing or reducing levels of salt, sugar, fat levels or types can all impact on shelf life. Therefore, any product formulation change should trigger a reevaluation of the product's shelf life.

In a broader sense, shelf-life assessment should be considered as a tool for Change Control Management to understand both intended and unintended consequences due to the following broad categories of changes:

- Ingredient changes for enhanced functionality or for cost reduction
- Optimization of severity of "kill step" to modulate process-induced insults or changes to product

- Change of packaging material for light weighting for lowering transportation fuel cost and reduction of planet carbon footprint. Also, any potential changes to barrier properties of packaging materials, seals, and closures
- Market expansion into new territories requiring long-distance shipment time and storage in different climates (temperature, humidity) and altitude

A change in one area may affect another unit of operation. For example, a change in ingredient may affect viscosity that may change the N_{Re} (Reynolds Number) impacting the nature of flow (turbulent vs. transient vs. laminar); and in turn affecting residence time in the hold tube, and process temperature at the end of the hold tube crucial for quality attributes (e.g., color) and assurance of food safety/spoilage avoidance.

Shelf life differs among different products. In this chapter, we will focus on aseptically processed and packaged ambient-stable food, and ESL chilled food.

A wide variety of factors may affect the shelf life of aseptic products. Non-exhaustive list of these factors:

- Enzyme activity: heat-resistant lipases and proteases in UHT milks and milk-based fluids
- Chemical reaction
- Oxidation: light or oxygen or trace metals-induced oxidations
- Non-enzymic browning or Maillard reaction
- Nutrient breakdown
- Physical breakdown of components
- Starch retrogradation
- Emulsion stability
- Microbiological activity: acceptable types and numbers for ESL chilled products with 90–100 days shelf life.

Some of the key parameters to measure shelf life quantitatively and qualitatively are color, flavor, and texture.

Changes in color and flavor are the major determinants of shelf life as an indicator of the impact of the process, ingredient interactions, and storage/distribution conditions. Therefore, they should be monitored and calibrated against the "best by" code date. Textural characteristics in the food during the shelf life may significantly impact the consumers' acceptance. These unwanted changes may be due to thermophilic spore growth during the shipping and storage or due to lack of textural stability which may cause settling or separation.

For liquid products, the measurement of viscosity can be used to monitor changes in texture. Color changes can be monitored subjectively with visual or instrumental measurements. Objective measurement of flavor change can be tested via HPLC and GC but are not suitable for routine use for off-line QC lab checks. A practical approach for the determination of "best by" code date is to rely on sensory and other analytical testing such as pH, viscosity, TSS, Brix, and in some cases, microbiological plating from preincubated chilled products produced via the ESL process.

16.3.6.1 Storage Protocol

Code dating for the shelf life should consider a standardized and research-based approach (e.g., stability studies) for aseptic and ESL products. Studies should be undertaken by considering the stability of the product over time in both normal and stressed storage conditions such as temperature, humidity, and light. For the ESL products, that may mean holding product under normal refrigeration conditions (4°C) to the code date and sampling the product for specific quality attributes (taste, smell, separation, oxidation, color, pH, titratable acidity) over this period. Additionally, products should be held at stressed non-refrigerated conditions to code date (7°C) to help determine product stability under non-optimal storage conditions. This will help identify maximum shelf life at optimal quality attributes.

There are several critical factors that can be considered while setting up a shelf-life study. Some of these factors are the storage time, storage temperature, light, humidity, and barrier properties of the packaging.

Storage time: Shelf-stable aseptic products have a normative shelf life of 12–24 months (12 months for finished products packaged in flexible and semirigid containers; and 18–24 months for finished products packaged in rigid containers as per canned goods). Market velocity at retail and consumption may be much earlier than 12–24 months stamped shelf life. Storage studies are usually done for a period of 12 months and sample pulls at 0, 1, 2, 3, 6, and 12 months, or test in between in case of premature quality failure. ESL products are dated for about 90–100 days (about three and a half months) at a chilled storage temperature of 4°C.

Storage temperature: 25°C is considered as the normative temperature, and 32–35°C as the abuse temperature. For different geographic locations, different distribution zones can be determined to identify the boundaries of the stability studies. For tropical climates and for military rations that may be distributed in arid climates, 30°C average and 40–45°C temperature abuse conditions are recommended. The ESL products are studied at normative 7°C, 7–10 days, and an abuse temperature of 21–24°C, 2 days. Sometimes it is important to control the humidity of storage chambers if the products are going to be marketed in areas with a humid climate.

16.3.6.2 Sensory Analysis

Sensory analysis is done to assess the failure mode of a product, i.e., "threshold" as to when it becomes unacceptable for human consumption. Different sensory tools can be used to assess threshold failure:

- Discrimination test using untrained panel for discerning differences
- Triangular test
- Paired comparison test
- Descriptive test using trained panel to measure rate of change of quality attributes
- Quantitative Descriptive Analysis (QDA)
- Nine-Point Hedonic Rating to assess preference between aged and new/freshly made product for degree of preference. The panel is untrained to reflect real-world customers/consumers.

All of the above sensory testing data will provide directional/guidance results rather than absolute results. If necessary and appropriate, absolute results can be generated via full Consumer Research (Carpenter et al., 2000).

16.3.7 Microbiological Testing for Sterility and Sample Size Consideration

It is important to note that chemical composition characteristics, nutritional value, and sensory characteristics can be established for a batch or lot by random testing of end products with a high level of confidence due to uniformity throughout the production batch. The same, however, is not true for assuring microbiological safety of foods because spoilage events are very rare and could be sporadic, random, or systemic.

TABLE 16.2   STATISTICAL NUMBER OF SAMPLES BASED ON ACCEPTABLE QUALITY LEVELS, DIFFERENT CONFIDENCE INTERVALS, AND ZERO DEFECT WITHOUT ANY UNASSIGNABLE ROOT CAUSES

Overall Maximum Acceptable Defect Rate (%AQL)	Probability*		
	99%	90%	95%
	Number of Samples to be Tested		
1:100 (1%)	225	300	460
1:1,000 (0.1%)	2,250	3,000	4,600
1:10,000 (0.01%)	22,500	30,000	46,000
1:100,000	225,000	300,000	460,000
1:1,000,000	2,250,000	3,000,000	4,600,000

* Probability of finding at least one defective package in the sample if the total unsterility lies at the maximum acceptable defect rate.

Microorganisms are inactivated (dead) if they cannot reproduce to appear as colonies on a microbiological solid medium plate or make liquid medium turbid or redox color change. Sterility is the absence of all life-forms, and it is a concept of negative state! One cannot measure microorganisms when they are not present and therefore negative test results do not prove anything conclusively.

During troubleshooting to identify cause-and-effect relationships, extensive sampling is followed by approved bacteriological validation tests. However, lack of sterility is a rare event in a large population and can only be identified by a very large sample size and destructive testing. For example, to verify an overall defect rate of 1:10,000, one needs to sample 22,500, 30,000, and 46,000 units at 90%, 95%, and 99% confidence limits, respectively, as shown in Table 16.2. It is important to note that no sampling inspection plan for finished products will catch all defective units. Therefore, routine inspection is intended to detect serious errors.

It is important to clarify that sampling requirements are determined based on the risks identified (see Chapter 14). Depending on the history of the aseptic line and the risks involved with the product type and process, the sampling plan for validation may be as conservative as 3X for 30,000 samples per run to demonstrate the sterility reliability at AQL level 0.01%, 95% confidence interval, and delivering "zero" defects (spoilage) after destructively testing of 30,000 samples per lot and demonstrating sustainable commercial sterility by repeating three times. These validation requirements can be relaxed with the Process Authority's involvement and support from different functions such as QA, food safety, and engineering. Analyses are based on the historic safety data specific to the line and other product/package/process characteristics that may bring additional safety cautions or further restrictions.

Because the sample size is large and testing is expensive, it is better to control and check all preproduction and process variables via QA and hazard analysis critical control point (HACCP) programs, rather than extensive examination, destructive testing, and rework of finished products. Rework is expensive and dishonorable to employees (Crosby, 1984).

In reality, commercial sterility in the food industry is assured by design, validation, and control, and by testing a small but statistically random sample, followed by either visual testing or by subculturing and bacteriological analysis of preincubated product (Food and Drug Administration, 1992). This protocol is adequate for regulatory compliance and for providing assurance to management.

16.3.8 Distribution, Handling, and Storage

The role of quality assurance does not end with the release of product to distribution. Products can get severely abused and damaged in distribution. Semirigid or flexible containers are far less resistant to abuse than metal cans from a Dole aseptic canner and conventional canning. All it takes is a crack or fracture in one layer of packaging material and the container may lose its sterility. Pallets of such products must be treated with care. They should not be stacked more than two high. Rack stacking gives additional protection, and slip sheets can provide stability. Many pallets are shrink wrapped for additional support during transit. Storage temperatures will play a key role in keeping quality of a product.

16.3.9 ASTM Drop Test

With the popularity of the e-commerce marketing, shipping of the case quantity and individual containers may pose additional structural damage risks due to rough handling. A wide variety of transportation modes and package handling by many untrained delivery personnel increased the product abuses. To understand the products' tolerance levels to these kinds of abuses, testing the product under worst-case conditions would help companies to design robust packaging. One of the critical qualifying tests employed by any processor or copacker is the drop test as outlined in ASTM-D-775-80 (ASTM, 1987) for finished product in secondary packaging. This test verifies the ability of the secondary package (i.e., cardboard case containing primary containers) to withstand damage after being dropped from a predetermined height. Since a hermetical seal is required for commercial sterility of the aseptic products, dropping the products from a certain height may create damages to the body of the package or more importantly on the hermetically sealed areas. These seal damages may pose food safety risks or consumer dissatisfaction due to spoilage. Most of the aseptic packages are designed with lightweight material to reduce the cost and can easily be damaged by any external physical impact. Extra attention to the design of the packaging material and handling of the products during the shipping is highly desirable.

16.3.10 Cumulative Assurance and Product Release

There is no "one" single "magic" test that can unequivocally prove the microbiological safety or sterility of the production lot. In the food industry, it is common to release non-USDA products into commerce using "cumulative assurance," which is a form of "parametric release" practiced in the pharmaceutical industry for release of parenteral solution terminally sterilized by moist heat (K.S. Purohit, Process Tek, personal communication, 2008). The principle of parametric release consists of review of all QA data—especially sterilizer control, aseptic filler control, and routine QC monitoring programs—to judge the sterility and safety assurance level, without the need of finished product sterility testing. USDA meat-containing products must be incubated for about 10 days at 35°C, followed by appropriate tests, prior to release, according to USDA regulations (9 CFR, Parts 318 and 381, 1986).

16.4 SYSTEM REQUIREMENTS FOR PREREQUISITE PROGRAMS

This section is dedicated to programs that support the Food Quality and Food Safety systems. These vital programs will help determine the sustainability of the process. Significant investment must be made early on in the process to have long-term success in the aseptic industry. This starts with the facility and equipment and includes investment in personnel resources for maintenance, continuous improvement, and troubleshooting capabilities.

16.4.1 Facilities and Equipment

16.4.1.1 Grounds and Environment

Facility grounds should take into consideration several important factors. Is there anything nearby that might serve as a pollutant to the facility? Is the area prone to flooding or other natural disasters? Facility design should include considerations for excluding pests and pest harborage areas, which is a regulatory requirement (21 CFR 117.35c). The area should be properly graded for drainage and exclude standing water. Landscaping should be kept to a minimum around the facility to reduce harborage points for pests and allow for pest control to manage the building perimeter effectively.

16.4.1.2 Facility Layout

The design and layout of the facility itself should abide by good hygiene practices to protect against contamination before, during, and after production. The facility must be cleanable to a hygienic level—constructed in a like manner and meticulously maintained. All operations should be carried out in such a way that the risk of contamination is minimal. Building construction should be made of impermeable material that can withstand harsh wet down, temperature, and chemical conditions and be adequately cleaned, inspected, and maintained. The design of the facility should protect against the accumulation of dirt, dust, and other contaminants.

Higher risk areas (such as finished product areas) require segregation from lower risk areas (preprocessing areas). In many cases, the general design of the facility must abide by the regulatory body's requirements. Additional considerations for appropriate lighting, airflow, and traffic patterns should occur.

Cleaning of the facility must consider proper drainage of residuals and be appropriately designed to handle the flow of waste. Is the drain system able to handle large amounts of cleaning material? Can all areas of the facility be easily reached and cleaned effectively? While an aseptic operation involves liquids, keeping the building and equipment as dry as possible will help reduce the chance of microbial harborage. Equipment, utilities, and support structures should be constructed to be easy to clean and work around. Sloped panels and coverings help highlight areas that need to be cleaned and allow for liquid runoff.

Traffic patterns should be such that access of personnel and material can be controlled. Designated walkways and traffic routes should be established. Clean rooms for preparation prior to entry into critical hygiene zones allow for cleaning, gowning, and transfer of individuals. The flow of the facility should be such that contaminant ingression is minimized. Is an area where raw materials are received and stored near the next step in the process, such as batching? Or are raw materials stored near-finished products and packaging? How far is one processing step from another? Lunchrooms, restrooms, and other areas of human interaction should be designed away from operation areas, and reentry into processing areas should require a "re-dedication" step for personnel and material.

16.4.1.3 Production Equipment

Finally, careful consideration for appropriately designed and installed equipment should occur. Equipment must be constructed of corrosion-resistant, non-toxic, easily cleanable material and be designed to eliminate harborage and accumulation. Where welding is required, hygienic welding is required, and inspection for weld spatter, gaps, and incomplete welds should occur. Crevices, hollow areas, and accumulation points must be eliminated to ensure proper cleaning of equipment. Rounded, seamless equipment mounted off the floor (pedestal mounted) will reduce harborage areas. As a general rule, equipment kept 3 feet from walls, support structures, and other equipment will allow enough room for proper cleaning and maintenance activities.

A facility should consider the levels of humidity required for the products being made. Higher levels of humidity can lead to a growth environment for microorganisms. Positive pressure air within a higher hygienic zone will reduce the inward flow of undesirable airborne contaminants. CIP systems should utilize best practices, such as separation from product lines (block and bleed valves, manifolds, mix proof valves), and validated against the production process. All equipment and piping should be designed to be drainable, generally with a 1° downward slope to the drain. Piping should avoid abrupt sizing changes, which might allow for standing residual product or chemical, and be free of "dead ends" and "T-junctions." If a T-junction is necessary, it should not extend above more than 1.5 times the primary pipe diameter so that CIP flow can still be adequate for those areas.

Appropriate hygienic connections (self-aligning with appropriate gaskets) will help aid in the cleaning process. A regular gasket management inspection and replacement program will be essential for a successful aseptic food safety plan. This should be built into the facility preventive maintenance program. A record of all gaskets used in the system, their material, and the date of last installation should be known, at a minimum. Gasket management should consider the type of gasket material (Viton, EPDM, silicon) in relation to the process type and the chemical compatibility of the material with the CIP system.

16.4.1.4 Thermal Processing Operations

Usually, the thermal processing operation is monitored, controlled, and documented by operators, that is, personnel in production. However, QA monitors interact with operators to ascertain information on quality parameters such as color, viscosity, total soluble solids (TSS), pH, and product burn-on due to fouling, and to provide information for putting products "on hold" should there be a process deviation due to temperature, flow or pressure drop, power outage, or any other reason.

Thermal processing records reside in the QA department. These records are checked within 24 hours for process delivery and integrity by a QA supervisor or process authority and a state inspector periodically. Most process charts and records are mandated to be retained for three years; however, the food industry typically retains its records for about five years.

Based on experience, it is correct to assume that thermal processing operations are usually conservative, robust, and, therefore, less prone to deviations than the downstream operation of aseptic filling and packaging.

Components for heating and cooling required for UHT processing include a product tank, timing pump, heater, holding tube, cooler, homogenizer, pump and filter for sterile additives, back-pressure valve, temperature-measuring device, and interface to aseptic filler. Aseptic fillers are discussed in the next section. Some operations do include aseptic surge tanks because filler speeds do not correlate with processing flow rates. Many food processors are cautious in employing surge tanks as an interface between upstream processing and downstream packaging. Surge tanks on the cold side may be vulnerable to recontamination and therefore represent a weak link in the "aseptic chain."

A commercial UHT process consists of sterilizing the product in a hold tube, in which every particle of food is held for at least a specific minimum residence time. No portion of the holding tube can be heated from an external source, and it must be sloped upward at least 0.25 in/ft (US CFR 21 Part 113 requirements). A temperature-recording device must be installed in the product at the holding tube outlet. The thermal process delivered is based on the flow rate and length of the holding time, taking into account the residence time of the fastest-moving particle and the lowest temperature indicated at the hold tube outlet. Both direct and indirect methods of heating and cooling are employed for sterilizing food. Compliance with cGMP and the PMO (for dairy products) is essential for a successful operation.

Direct heating: The direct steam injection and infusion methods of heating consist of direct contact between culinary steam (ingredient quality) and product. After the hold tube, cooling may be accomplished by vacuum flashing or regeneration. Indirect heat exchangers may be employed for preheating by regeneration of heat. Monitoring of the pressure differential during regeneration between raw cold products and sterile hot products is required by the Pasteurized Milk Ordinances (PMO), US Food and Drug Administration (FDA) (21 CFR, 1987), and US Department of Agriculture (USDA) (9 CFR, 1986).

The following are some of the characteristics of direct heating:

1. Rapid "square wave" heating and cooling
2. No burn-on or fouling of heat exchanger surfaces, with most products
3. Better for low-viscosity products
4. Fewer moving parts compared to a scraped surface indirect heat exchanger
5. Vacuum on the cooling side is susceptible to "gasping" and recontamination
6. Need specialized clean-in-place (CIP) and sanitation program
7. Careful moisture stripping is needed during flashing for proper control of solids and fat in the finished product

Indirect heating and cooling: Plate, tubular, and scraped surface heat exchangers are used for indirect heating and cooling. This heating and cooling method is used for products that may require gradual heating and cooling compared to direct methods. Systems may need intermittent flushing to reduce fouling or burn-on. By the use of regeneration, indirect heating and cooling is more energy-efficient than direct heating and cooling. Because of slower heating and cooling than direct steam heating, the total heat treatment may be considered more impactful on the quality attributes for the indirect heating systems, resulting in greater intensity of desirable or undesirable flavors.

Most equipment suppliers will provide a detailed operations manual showing all critical control points (established during validation and establishment of aseptic processing and packaging operations, as outlined in Chapter 5, and Appendix 3) and ways to monitor these and make the necessary adjustments if needed. Many equipment suppliers will train employees either on location or at their factory and will provide them with the necessary operational programs, including quality procedures.

16.4.1.5 Aseptic Filling and Packaging Operations

Aseptic filling and packaging functions represent the crux of the entire aseptic operation. The aseptic filler is where several operations are choreographed and synchronized: maintenance of sterility, forming, sterilization of container and lid, filling, hermetic sealing, and labeling.

Successful operation of the aseptic filler depends on a well-thought-out QA program, preventive maintenance, proper CIP and sanitation, the incoming quality of packaging materials, dedicated personnel, and continuous training and education. The output from the aseptic filler serves as a focal point for the QA and food safety program and a primary index of machine performance.

There are many vendors of aseptic filling and packaging systems. Vendors can sell a whole production line together, or the process can be pieced together from several different vendors. A manufacturing facility should consider multiple vendors and identify the ideal filling and packaging equipment. Different types of packaging will have advantages and disadvantages that need to be considered. For instance, an HDPE plastic bottle will have different barrier capabilities than a PET bottle, which will differ from a pack-formed multilayer container. In addition to packaging, many different fillers will have alternate methods of container sterilization, different run speeds, and in many cases, differing varieties of package sizes available to a manufacturer. Filling systems and vendors have evolved at a rapid pace over the last 20 years. Today, there are some systems capable of filling packaging at a rate of 800 bottles per minute! Each system will come with its challenges, including requirements for maintenance, operator competency, and process efficiency (see Appendix 3 for more detailed information about the aseptic fillers).

Additionally, filling systems with different technologies may handle a wider variety of package sizes and types, with minimal to moderate changeover requirements. However, it is important to note that the more changes introduced into a filling system, the more control and verification required. Large amounts of variation and changeover will also affect operational efficiency, which is something that leadership should carefully consider. In recent years, filler OEMs (e.g., Krones, Sidel) introduced blow-fill-seal (BFS) aseptic fillers that included a bottle-blowing equipment as an integrated part of the fillers. Some of these BFS technologies rely on the sterilization of preform (packaging material prior to blowing step with a final design of finish) with hydrogen peroxide vapor prior to bottle blowers. In these systems, bottles are considered sterile when they are blown and enter the drying and filling areas. As one can imagine, these additional bottle-blowing technologies and complexities add more controls and training requirements to minimize the associated risks for quality and food safety. Table 16.3 lists several equipment vendors, with links to their sites for more information.

Type of the sterilant (e.g., hydrogen peroxide—H_2O_2) used and its form directly affects the efficacy against certain target microorganisms, including the ones with public health concerns. Proper and consistent application of the sterilant to sterilize the aseptic filler and packaging material is generally unique for each aseptic filler design. Each method has its advantages and disadvantages toward the levels of assurance and confidence determined based on the Food Safety Objectives (FSO) for a particular aseptic line. For instance, H_2O_2 is a very common sterilant used in the aseptic industry. Every filler company that uses this chemical as a sterilant has its way to generate and apply based on their machine design. As a very potent chemical, H_2O_2 (liquid or vapor forms) may be unstable, and the generation/distribution

TABLE 16.3   ESL AND ASEPTIC FILLING AND PACKAGING VENDORS

ESL and Aseptic Filling and Packaging Vendors			
Shibuya	Scholle	Tetra Pak	GEA
JBTC	Hassia	Bausch	Sidel
Sig-Combibloc	FOGG	Krones	Evergreen

system may pose an additional risk if the H_2O_2 quality is not stable throughout the production cycles. The inactivation efficacy of this chemical depends on the surface exposure and effective distribution within the container. If the vapor form is used, maintenance of the vapor state until the application point in the container, proper flow (with air velocity profiles), effective condensation on the packaging and filler (sterile zone) surfaces, and ideal time/temperature combination to evaporate from the surfaces are the most important deliverables of the aseptic sterilization design. These are the core of the qualification and validation tests to prove that the system consistently delivers acceptable sterilization assurance under predetermined worst-case parameters. All the set points determined during these tests are filed with the regulatory agencies (e.g., the US FDA) and used as the foundation for the ongoing production critical control points. There are additional requirements to monitor the effectiveness of the sterilant and its removal. One of those requirements is related to removal of the H_2O_2 from the surfaces. According to the FDA's requirements, the H_2O_2 residual amount after the drying process shall be equal to or less than 0.5 ppm (measured in distilled water). H_2O_2 is removed from the containers during the drying process right after injecting the H_2O_2 into the containers. Generally, hot air streams through several nozzles are used to evaporate peroxide down to acceptable concentrations. This process is not as simple as it sounds since the drying process is also part of the sterilization efficacy and plays a huge role in accumulative sterility value and the effective distribution of the sterilant in the container.

With the wide variety of sterilants, including peracetic acid (PAA), irradiation, steam, and some additional hurdle technologies, the proposed FSO for each system has to be determined by the process authority and a validation plan must be prepared to challenge the system to deliver sustainable assurances for food safety. Regardless of the system weighing of the benefits of each process compared to the budget, the product being made, and overall machine capability should be carefully considered.

16.4.2 Utilities

Knowledge of legal requirements and best practices regarding utilities in aseptic and ESL processing is needed for optimal operation. Considerations for main utilities such as steam, air, and nitrogen are a must. Processing systems with steam coming into food and food-contact areas are required to be of culinary quality. Chemical additives to steam must meet regulatory requirements and must not contaminate the product or product surfaces. The use of appropriate filtration can help achieve this process. If possible, boilers providing steam should be dedicated to the aseptic system and sized appropriately to handle load requirements.

Air and/or nitrogen that is used to maintain sterile zones or used in drying must also be sterile. Filters for sterile air that provide positive air pressure during processing operations must remove contaminants such as bacteria, dust, pollen, aerosols, chemical vapors, and hydrocarbons, which can compromise the sterile condition and spoil the product. These filters must be changed out regularly and validated upon replacement. Air filter housings in aseptic filtration installation must meet hygienic design standards and have sanitary connections. Redundant filter setups may be required by regulatory agencies to provide additional confidence in case of filter failure. This can reduce the risk of losing the system's sterile condition and improve overall quality. Filter integrity testing is an important step to confirm the maintenance of the aseptic conditions. Depending on the filter type used in aseptic installations, different integrity tests may be applied. There are two main types of filters used in the aseptic systems—membrane and depth filters. Membrane filters are in cartridge format and the maximum pore size is 0.2 μm. These are also called sterilization membrane filters. The depth filters rely on the tortuous path structure to capture any contaminants in the path of air and gasses. Most of the high-efficiency particulate air (HEPA) filters belong to this category and are designed to remove at least 99.97% of bacteria and airborne particles with a size of 0.3 μm. This type of filter can be in a cartridge or an in-box format. The depth filters are widely used in aseptic fillers to provide high-volume sterile air into the sterile zones to create positive pressure and prevent any cross-contamination from the environment. Since their structure may depend on creating a tortuous path, it requires a certain thickness which makes the sterilization process limited to certain chemicals such as hydrogen peroxide. Institute for Thermal Process Specialists (IFTPS) has been working on several guidance documents for the aseptic systems (https://www.iftps.org/guidelines/). In addition to Aseptic Validation Guidelines (G.005.V1—IFTPS Guidelines), a very detailed guideline on Aseptic Filtration is in press for publication during the preparations of this chapter.

The criticality of the filters used in the aseptic systems should be obvious. However, there are other filters in the manufacturing facility that may pose similar risks for the aseptic line. The most important among them is the air intake filters into the building to provide positive pressure in hygienic zones. These filters are not sterile filters; however, they filter out the airborne particles which may carry harmful organisms into the manufacturing area and cross-contaminate the equipment or packaging material. These filters are generally categorized based on their removal rates against the posing risks in the area. Minimum Efficiency Reporting Values (MERV) report these filters' ability to capture particles between 0.3 and 10 μm. The rating is derived from a test method developed by the American Society of Heating, Refrigerating, and Air Conditioning Engineers (ASHRAE); https://www.ashrae.org/. The higher the MERV rating, the better the filter at trapping specific types of particles. For example, MERV 1–4 correspond to average particle size efficiency of 3–10 μm at or below 20%. The efficiency jumps to 80-89.9% with 1–3-μm particles for filters with MERV 12 rating. The same filter would remove 3–10-μm size particles at 90% or greater efficiency. When MERV ratings are in the 14–16 range, filter can start removing as small as 0.3–1.0 μm with the 75–84% efficiencies.

Aseptic systems may demand specialized and dedicated electrical equipment as well. Where it is feasible, elimination of electrical equipment from production rooms and wet down areas should occur. Moving electrical equipment to a dry area can help prevent failure due to getting soaked or corroded. If critical electronics are susceptible to moisture or other environmental exposures, their installations must follow the Ingress (sometimes "I" stands for International) Protection (IP) Code which is defined in International Electrical Commission (IEC) Standard EN 60529 (British BS EN 60529:1992, European IEC 60509:1989). This standard classifies and rates the degree of protection provided by mechanical casings and electrical enclosures against intrusion, dust, accidental contact, and water.

In the nomenclature of the IP ratings, every digit identifies different levels of protection. Ratings start with letters IP (Ingress Protection) followed by the first digit for "Solid Protection" and the second digit for "Liquid Protection." The range of the first digit is from zero (or "X") for no protection to 6 which indicates dust-tight and full protection. The range of the second digit is from zero for no protection to 9 (or "K"), indicating protection against water through direct jets of high temperature and pressure, wash downs, or steam-cleaning practices. For example, IP69K indicates complete protection against ingress of dust and water even under very harsh conditions such as high pressure and temperature. During the design stage, based on Quality and Safety by Design principles, proper protection of the electronics and instrumentation must be specified with proper references to the standards.

16.4.3 Inline Inspection Requirements

Inline inspections are an essential part of both a Food Quality and a Food Safety program. Every 15–30 minutes, line inspectors should check for container cosmetic defects and record weights, vacuum, pH, total solids or Brix degree, viscosity, flavor, and filling temperatures. This helps ensure conformity to the process and also consistency within the quality program. In some cases, such as those for weight, inline check weighers can perform a complete inspection of each package and track any inconsistencies in weight. This helps meet legal requirements for label content and can also help facilities manage product loss due to overfilling effectively. Multiple other online inspection systems allow for 100% inspection of products in a non-destructive way. This can include vision systems that can inspect cap placement and closure, squeezing, or voltage systems that can detect leaks and verify proper hermetic sealing.

Metal detection is common in the industry and can be used at multiple points in the process, both in-process and post-process. Metal detection can be used on finished product to aid in foreign material investigation. In some cases, application of inline magnets and/or filters for catching metal shavings is extremely useful, especially in batching processes and when installed past pumps throughout the process. These tools can be highly effective for a manufacturer and improve the operation efficiency of the system. One of the limitations in QC checks of semirigid or flexible packages compared to metal containers is the inability to record simple vacuum readings as a reliable index of a hermetic seal because these packages have ambient pressure or have nitrogen-flushed headspace. This limitation has been overcome by evolving package-specific QC test procedures based upon defect criteria and levels. The line inspectors serve as direct feedback to avoid mass production of defective products or containers.

In some cases, an inline inspection may require destructive testing. An example of this can be leak detection for Tetra Brik pack-style packages. This process requires an operator to take a package, apply a dye to the package interior, and then cut the package open to inspect for dye seepage past the seal junctions.

Other inline checks may be required during the batching process or through the thermal process. Checks for macronutrient content or other formula requirements may be necessary. In many cases, this is done manually by taking a sample from a designated point in the process. There are, however, established, and emerging technologies that can provide an inline analysis of the critical components. A FOSS inline analyzer, for instance, can determine the fat or protein content of a product as it is flowing through a line by utilizing near-infrared (NIR) technology. These systems can be installed new or retrofitted into existing pipe and allow real-time analysis of product flowing from one area to the next.

16.4.4 Internal Assessments and Auditing

Conducting validation activities around aseptic and ESL food quality and food safety management systems is a regulatory requirement for operation. Additionally, verification activities will be required to show that the validated state can be maintained and that the process is being followed. One important aspect of this is internal assessments. An excellent internal assessment program will include follow-up and inspections of the management system. This will consist of "mini-audits" led by qualified individuals. Ideally, these inspections should be monthly to quarterly to verify that the validated process is being followed. This may include checks of employee adherence to cGMPs or residual allergen inspection of a CIP process. Ideally, each supporting program for the Food Quality and Food Safety systems should have a regular internal inspection schedule so that manufacturing leadership has confidence in the process. This will also aid in keeping the facility in an "audit ready" state at all times. Failure to plan and practice for audits will greatly increase the chance of a failed audit or significant findings that could delay production or force a product recall. Because this process can be resource-heavy, it is good to have qualified leadership and operators participate in these. This will help spread the assessment process across multiple resource groups and help keep engagement in Food Quality and Food Safety high.

In addition to regular assessments of programs, an annual review and audit of the Food Safety and Food Quality plan should be conducted. Review of the Food Safety plan is required at least every year, or when the program changes, such as a new line addition or a new formulation. (21CFR 123.8a) This process should look at the documentation associated with the programs and how the written program is being executed on the production floor. Is the process that is written in the plan accurate when compared to activities on the floor? If there are differences, review the validated process, determine if it is correct, and align the program with the floor activities. In some cases, a review will uncover that a revalidation effort is necessary. Detailed documentation of the activities, reviews, and process changes is required as evidence of the verification of the process.

16.4.5 Human Resources and Personnel Development

Training is a critical area of focus for a successful aseptic, retort, or ESL operation that often lacks investment. Aseptic systems are complex and require a degree of knowledge that should be continuously reinforced through experience and supplemented by regular and recurring training. These systems may have unique operations, equipment, and methods that require a higher degree of skill to operate, maintain, and troubleshoot than conventional systems. Significant resources should be invested in the facility for maintaining and enhancing competency and capability. Operators will be working with pressurized vessels and steam, sterile zones, packaging, and product. Multiple critical control points must be monitored and verified. Monitoring activities for an aseptic, retort, or ESL systems are often much more complex and labor-intensive, and scrutiny from regulatory agencies will be the same. In light of this, training must be at the forefront of a successful manufacturing operation.

Training should not be focused on operations only but holistically through the manufacturing system. There should be specific and recurring training for senior management, floor leadership, operations, maintenance, laboratory, and quality assurance disciplines. Technical instruction on the system, such as operations and maintenance, will always be needed. However, additional areas of study in aseptic theory, regulatory requirements, risk management, food safety, and quality systems must also be considered.

It is important to note that in some cases, training is a regulatory requirement (21CFR 113.10 and 21CFR 117.3-4, 21CFR 121.4), which state that the training provided must be acceptable by the regulatory body and supplied and supported by management. Additionally, training should be provided to all individuals engaged in manufacturing, processing, packing, or holding food and be appropriately recorded. One should note that the regulations differentiate levels of training for individuals in operations and those in supervisory or management roles.

Therefore, a comprehensive, tailored training plan is necessary for regulatory compliance and the safe, efficient, and quality-assured operation of the aseptic system. There are multiple resources that one can consider for effective training programs to supplement one designed and given by the facility's leadership. There are also several levels of training required for different roles. Some level of training should be considered for all individuals at the establishment (Table 16.4).

TABLE 16.4   INDUSTRY AND REGULATORY TRAININGS

Industry and Regulatory Training for Aseptic Processing				
Aseptic processing and packaging	Overpressure and retort thermal processing	Preventive control for qualified individuals	Food fraud mitigation	Critical thinking
Aseptic validation	Better process control school	Aseptic operation	Vendor operator training	Risk management for food operations
Thermal processing workshops	HACCP	Food safety and defense	Process controls (Rockwell, Yokogawa, Beckhoff, etc.)	Canned foods: principles of thermal process control

It is recommended and required by many third-party certification schemes that training take place and be annually refreshed, at a minimum, for all individuals within the facility. The amount of training should vary by role and be regularly reviewed for supplementation. Leadership should have a good understanding of the aseptic operation at a higher level (theory) and the regulatory requirements, risk analysis, and mitigation strategies. Operators and maintenance should be considered for some of these areas but should have more training focused on technical operations, good manufacturing practices, and preventive control. Floor supervisors and leaders should have a good mix of both and be heavily invested in training. Operation Managers, Quality Assurance, and QC managers will require the highest levels of training, being both parts of the upper management team and heavily involved in the operations process.

16.4.5.1 Management Responsibility

Management responsibility for the creation, implementation, execution, and review of the Food Safety and Food Quality plans, and the development of a sustainable culture within the food manufacturing facility cannot be understated. The plan should start with the management team identifying the mission and values which will be at the core of the process.

While the focuses of these two plans are different, the requirements and expectations of management within the scope of an overall food program are remarkably similar. Whereas food safety is federally mandated for compliance, food quality is often governed by the facility (or parent company) and more so, their customers. It is important to note, however, that both plans are intertwined and cannot be separated. Management must be aligned in solidarity in these aspects, and the plan for execution must be well-defined and communicated.

16.4.5.2 Plan Creation

Each plan must be created by the management team, in conjunction with the process and regulatory experts, and a food safety and quality manager. Each member of the management team should be familiar with the programs, and in agreement with their implementation. This will require a level of training and competency necessary to be able to make recommendations and enforce application of the program. Specific and recurring training in the areas of food safety, quality management, and relevant training within the scope of the facility operations should occur. There must be at least one individual in a facility that possesses the required credentials for a Food Safety Program oversight (21CFR 120.13; 21CFR113.10).

16.4.5.3 Implementation

Once a plan is created, and agreed upon by the team, it must be clearly communicated to the operating facility. Care should be taken to make sure that the process is understood by all members of the facility. This may include having the plan available in multiple languages. Members that have been trained and understand should acknowledge with a commitment signature. Each member of the team should have a basic understanding of the plan, and how to successfully achieve it. Steps should be taken to not only drive awareness of the Food Safety and Food Quality process, but also to build upon it so that all members of the facility are able to contribute and improve the process. There are two major phases within the implementation step—validation of the plan, and, once complete, verification of the plan. Validation means "that element of verification focused on collecting and evaluating scientific and technical information to determine whether the HACCP plan, when properly implemented, will effectively control the identified food hazards" (21 CFR 120.3p). In other words, validation asks the question: "Is the right process being effectively implemented?" Whereas verification means "those activities, other than monitoring, that establish the validity of the HACCP plan and that the system is operating according to the plan." (21 CFR 120.3q) Think of verification as a step to ensure that the plan is actually being followed. Once a plan has been shown to be effective (validation), it must be followed (verification) for it to work. A good plan is only as effective as it is executed.

16.4.5.4 Review

As mentioned previously, review and reassessment of the plan is necessary. In order to keep up with an ever-changing environment (internal and external), management and process experts should be critical of their own programs. At a minimum, Food Safety plans should be reviewed on an annual basis, or whenever a change to the plan has occurred. This could be something as simple as a new ingredient addition, or as complex as the installation of a new processing line. Regardless of the change, the process should be examined for additional hazards, and for any impact that the new change may bring to other areas and processes. In addition to internal review, it is always beneficial to consider review by an independent body, such as those mentioned in earlier sections. Review must be conducted with not only the management team, but also with on floor experts, and, as applicable, vendors or suppliers of the new change. By including a diverse team in the process, additional clarity, depth, and impact will be added to the plan.

16.4.5.5 Sustainability

The final step in the responsibility and oversight of a Food Safety or Food Quality program is sustainability. Is the process able to hold itself up once it is implemented? Do all members of leadership acknowledge and embrace the process? Have team members been trained to follow the process, and teach others what to do? Most importantly, is the process creating a culture of Food Safety and Food Quality, which places value in the assurance that the product of the process is what was intended to be made, and meets all applicable regulatory and customer expectations? Often, and especially with Food Quality, resources dedicated to maintenance of this area are seen as profit-consuming, rather than profit-generating. While this may appear true on the surface, one must realize that the true benefit of a Food Safety and Quality plan is not in the number of units produced or sold, but in the growing confidence of the consumer purchasing the product. A well-executed program will keep customers loyal and bring more into the fold. This should be the focus of any food manufacturing industry facility. When choosing to do the right thing over short-term profit and gains or "getting the product out the door," the system will flourish.

16.4.5.6 Summary

A holistic approach to the Food Safety and Food Quality for the aseptic systems is critical for the sustainability of the "validated state" of the system. Proactive and risk-based aseptic culture development with the implementation of proven preventative methods such as HACCP and all prerequisite supporting programs are the main elements to secure the future of the business. They need to work in harmony. Failure of one in this structure may impact the entire program and the quality/food safety objectives of the organization. Therefore, starting the quality and food safety at the design stage is extremely important to avoid any unwanted costly consequences. A good program will not only have minimum requirements from a regulatory standpoint, but rather supplement programs on a risk-based basis, dependent on the product and processing within the facility. Aseptic culture starts with the leadership/management support, and it is sustained only with the implemented standards, policies, procedures, and focused training at every level of the organization. People are the most important assets of the aseptic systems!

16.5 THE FOOD SAFETY MANAGEMENT SYSTEMS MODELS

Food safety management systems (FSMS) are a preventive approach toward identifying and controlling (reduction or elimination) foodborne hazards. This is to ensure that food is safe for consumption. A well-designed program can help food manufacturers comply with regulations, meet customer expectations, and deliver a safe product every time. The FSMS is a controlled process for the management of food safety. Each step in the production chain must be defined, and risks for each step identified. An appropriate program will follow the established HACCP process and other supporting programs to make an impactful management system. A detailed list of requirements for FSMS is found at 21 CFR 117.126.

The USDA has defined *food safety* as "the handling, processing and storage of food in order to prevent food-borne illnesses" (USDA). The Federal Government has recently (2011) passed sweeping regulatory changes to the Food Safety world through the Food Safety Modernization Act. This act, also known as FMSA, is intended to have commercial food handlers and their leadership teams take a proactive approach to food safety rather than a reactive approach.

One of the critical requirements of a Food Safety Plan is system review. In today's constantly changing environment (both industry and regulatory), the plan will and must be fluid for progress to occur. For this to happen, management and employees must be trained and aware of the process, science, and regulatory changes that will happen in their particular industry.

Therefore, emphasis and execution of a risk-based approach model is not only a best practice; it is also a legal requirement (21CFR 117.126). As such, the Federal Government has defined acceptability for different approaches within the food sector. There are two major food safety management models recognized by the Federal Government, which will be discussed below.

16.5.1 Hazard Analysis Critical Control Point (HACCP) Program

When used in conjunction with other supporting food safety policies, HACCP is a preventive control tool that makes up a Food Safety Program. This program aims to eliminate, reduce, or control to an acceptable level identified food safety risks. Regulatory agencies and industry organizations have promoted adoption by industry of the hazard analysis critical control points program. HACCP is mandated in the food industries listed in Table 16.5.

TABLE 16.5   HACCP MANDATED INDUSTRIES AND FSMA

HACCP Mandated Industries and FSMA		
Juice (21 CFR Part 120)	Meat and Poultry (9 CFR Part 417)	The Food Safety Modernization Act requires a Food Safety Plan for all other food processing facilities (21 U.S.C.350g)
Seafood (21 CFR Part 123)	Thermally Processed Low-Acid and Acidified Foods (21 CFR Parts 113 and 114)	

16.5.1.1 Principles of HACCP

HACCP is a comprehensive tracking of an entire process, which happens on the plant floor. It is a systematic and quantitative risk assessment tool. HACCP is distinctly different from a QA program and should not be misconstrued as a mere listing of all control points and critical control points. HACCP programs emphasize prevention rather than inspection for defectives. The evolution of a HACCP program depends upon careful gleaning of historical QA programs encompassing incoming food and packaging ingredients, sanitation and CIP, maintenance, batching, operation of processing systems and packaging equipment, downstream handling, warehousing, distribution, retailing of finished products, and consumer handling. HACCP is compatible with most existing QC and QA programs (Schothorst & Jorgeneel, 1994). It should be noted that a HACCP plan must be built into the overall Food Safety program and is required by the FDA. The HACCP plan itself is not a stand-alone program but must be used in conjunction with other food safety plans for an effective safety program. HACCP principles include:

1. Identify potential hazards
2. Establish critical control points (CCPs)
3. Establish specifications for CCPs
4. Establish monitoring procedures for CCPs
5. Develop procedure/policy for corrective actions for deviations
6. Keep records
7. Develop a procedure for verification

16.5.1.2 Categories of Hazards

The sole objective of HACCP is food safety, and safety only. It considers microbiological, physical and chemical hazards. These three hazards may naturally occur in the food, exist within the environment, or be generated by a mistake in the manufacturing procedure, or process, or package. Physical and chemical hazards are non-living; they do not grow and multiply. Their incoming concentration may remain constant or may get diluted during batching, thus reducing the associated risk. On the other hand, microbiological hazards are living; they can grow and multiply under ideal growth conditions, leading to toxin production, pH changes, gas formation, and other chemical degradation, thereby increasing the associated risk. Although chemical hazards are the most feared by consumers and physical hazards are the most commonly identified by consumers, microbiological hazards are the most serious from a public health perspective. For example, a piece of metal (physical hazard) in food may result in a chipped tooth for one consumer, but contamination of a batch of milk with *Salmonella* or meat with *E. coli* O157:H7 may affect hundreds or even thousands of consumers. Because every production line is different, the HACCP program must be tailored to be product-process-package-, line-, and factory-specific.

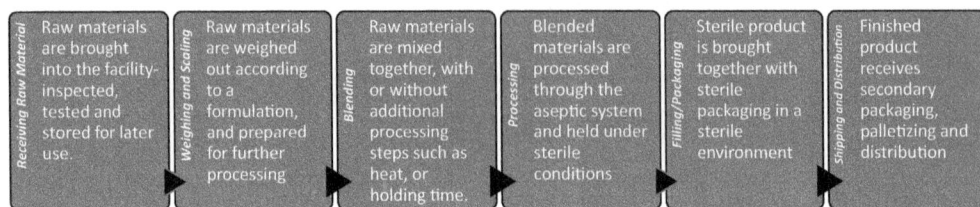

Figure 16.2   HACCP flow diagram.

Further work in the areas of calibration of monitoring methods and alternate methods for CCPs is desired.

A HACCP program is required for aseptically processed and packaged foods because the technology and associated operations are highly complex (Figure 16.2).

Figure 16.2 shows a basic HACCP flow diagram. Within each step of the flow diagram, hazards (biological, chemical, or physical) must be determined, and those hazards must be controlled. Hazards may be controlled within that process step or at a subsequent step along the processing route.

For each hazard, control points or critical control points (depending on the level of severity and risk) must be established. Operating limits surrounding those control points can then be developed, and monitoring activities to ensure that the process stays within those limits can proceed.

16.5.2 Hazard Analysis and Risk-Based Preventive Control

In most instances, domestic and foreign food manufacturers required to register with Section 415 of the FD&CA must also comply with the requirements for risk-based preventive controls and cGMPs. This requires food facilities to have a food safety plan with several supporting preventive control programs identified (FSMA).

Risk-based preventive control takes the same process steps in a HACCP program and identifies hazards that are known or reasonably able to be identified. In addition to biological, chemical, and physical hazards, it also takes into consideration radiological hazards.

Manufacturing facilities can identify their hazards and controls, provided that they meet established scientific standards for control. Several required controls will be discussed in more detail below.

- Process controls
- Food allergen controls
- Sanitation controls
- Recall controls
- Supply chain controls
- Other controls (if necessary for the production facility and process)

When a preventive control for a hazard is identified, the facility must ensure that corresponding controls are in place, monitored by qualified personnel, and consistently satisfied the critical parameters. Monitoring procedures should be designed to ensure that control of a process is maintained. While an aseptic process is running, process control points must be continually

TABLE 16.6   SUPPORTING PROGRAMS IN FSMS

Supporting Programs in Food Safety Management Systems				
cGMP	Training	Calibration	Pest control	Maintenance
Sanitation	Water monitoring	Foreign material	Supplier approval	Shipping and receiving
Waste management	Allergen	Recall	Environmental monitoring	Food defense

monitored to ensure the system stays within the critical limits and maintains its commercial sterility. On the other hand, some of the potential hazards can be checked with triggering activities such as start-up, shutdown, or downtime. For instance, allergen cross-contamination verification may be a monitoring activity that takes place only at the beginning and end of a production run. It is important to note that all monitoring activities must be clearly defined (who, what, when, how, and how often).

Additionally, corrective and preventive actions must be considered at this step as well and be well-defined. Verification activities, which give confidence to the validated process, must be conducted at regular intervals (typically annually or as defined by the process needs).

16.5.2.1 Supporting Programs for Food Safety Management

There are multiple programs that should be utilized in addition to a HACCP or other process preventive control program that will enhance the overall capabilities of the system. Table 16.6 highlights several common yet crucial supporting programs.

Some FSMS will require more supporting programs, depending on the product and process utilized. While there are some differences, each FSMS will have several required supporting programs. Regardless of which specific programs need to be administered, it is essential to note that HACCP or Preventive Control is not a stand-alone program. It must have these supporting programs to make a successful FSMS. A preventive control scheme will have these built-in as a requirement of FSMA (Figure 16.3).

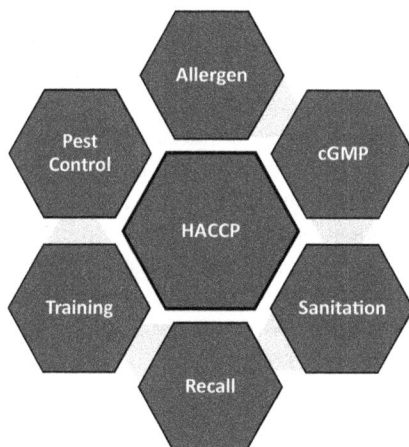

Figure 16.3   HACCP support, showing the correlation between the food safety program and the supporting programs that assist in the execution of the overall food safety program.

16.5.2.2 Process Controls

Process controls, also called critical control points in a HACCP program (and process preventive controls in a preventive control program), establish how identified hazards are controlled or eliminated. This can include steps like UHT processing, acidification, refrigeration controls, and others. It is important to remember that process controls cannot be arbitrarily established but must have scientific evidence showing their effectiveness. The FDA and other entities provide guidance for establishing process controls within a particular food process. The Pasteurized Milk Ordinance (PMO) is an example of a regulatory-approved process for pasteurization and sterilization of dairy products and gives specific guidance for process control.

16.5.2.3 Current Good Manufacturing Practices

Current good manufacturing practices, known in short as cGMPs, are a required supplemental program in an FSMS. cGMPs are found under 21 CFR 117, whose purpose is to ensure that food has been manufactured under conditions that ensure it is fit for consumption and has not been adulterated. cGMPs cover a wide variety of considerations in order to prevent food from becoming adulterated. *Food adulteration* can be defined as any food that is unsafe, unwholesome, or impure and may be deleterious to health. (21 USC 342). Specific considerations within the code include personnel, grounds and facilities, operations, sanitation, and several other areas of manufacturing that require specific conditions to be met for food to be deemed as "fit for consumption."

There are several updates to cGMPs from the FMSA regulations, such as education and training, which are required competencies in an FSMS. Management must ensure employees who work in the facility are qualified and have a training record that reflects those qualifications. In some cases, a combination of training and experience will suffice for the definition of a "qualified individual." There must also be an individual(s) that has gone through specific training to be qualified as a "Preventive Control Qualified Individual" present on-site at all times while manufacturing is underway. A link to certifications and training is below:

https://www.ifsh.iit.edu/fspca/fspca-preventive-controls-human-food

Purdue University offers Aseptic Packaging and Processing workshops, as well as Validation workshops:

https://ag.purdue.edu/foodsci/mishralab/workshops/aseptic-processing/
https://ag.purdue.edu/foodsci/mishralab/workshops/validation-workshop/

16.5.2.4 Allergen Program

The United States has identified nine allergens that make up roughly 90% of all known food allergies, which require specific labeling on all foods and plans to prevent the cross-contamination of foods with other allergens. These allergens are milk, eggs, fish, shellfish, tree nuts, peanuts, wheat, soy, and sesame. Proper identification, labeling, and prevention of cross-contamination of allergens is a critical part of an FSMS. As food allergens are capable of causing great harm to individuals within the population who are susceptible to specific food proteins, care must be taken to prevent the adulteration of food with an undeclared allergen.

In 2004, Congress passed FALCPA to help combat and identify food allergens. This law identified eight major food allergens: milk, eggs, fish, shellfish, tree nuts, peanuts, wheat, and soy, and was required to be explicitly declared on all food labels. More recently (2021), the Food Allergy Safety, Treatment, Education, and Research (FASTER) Act was signed into law. Under part of the provision of this law, sesame was declared as a major food allergen, in addition to the ones listed above (Allergies).

A food allergen program must take into consideration what allergens are present within the manufacturing facility, their use and distribution, and how they must be controlled. Additionally, the program should have verification steps in place to prevent the adulteration of food with an allergenic ingredient that is not listed on the label of that particular product. Care must be taken to ensure that residual product from previous runs is not left in an aseptic system, thereby contaminating the following batch of product.

In addition, the FSMA text specifically calls out allergen cross-contact as a consideration point and must be addressed adequately in the cGMP program. A resource, which details this is listed below:

https://www.fda.gov/food/food-safety-modernization-act-fsma/fsma-final-rule-preventive -controls-human-food

16.5.2.5 Sanitation Program

The FDA has described cleaning and sanitation in detail in 21 CFR 117.35 and describes cleaning as steps taken to ensure that food-contact and non-food-contact areas are protected against allergen cross-contact and against contamination of food. It describes sanitation as "Sanitize means to adequately treat cleaned surfaces by a process that is effective in destroying vegetative cells of pathogens, and in substantially reducing numbers of other undesirable microorganisms, but without adversely affecting the product or its safety for the consumer" (21 CFR 117.3).

A sanitation program should include a structured, orderly approach to cleaning and sanitizing all areas in the facility to prevent contamination of food. A Master Sanitation Schedule (MSS) should be developed to identify the kind of cleaning, frequency, and verification steps necessary to provide evidence that a room, equipment, or utensil has been adequately cleaned. This should not be limited to the equipment only, but should be a holistic approach that encompasses receiving, storage, process, and finished product–holding areas.

Cleaning is a critical component of all food manufacturing processes, especially aseptic and ESL operations. If aseptic and ESL systems are not adequately cleaned, they cannot be properly sterilized, putting product at risk. It is important to highlight that 99% clean is still 1% dirty! A validated cleaning plan is essential to the successful operations of any aseptic line. Many systems rely on a Clean-in-place (CIP) methodology for effective cleaning without disassembling the equipment and pipelines. CIP systems will depend on several characteristics such as flowing water, caustic solution, and acid solution at predetermined concentrations. These solutions have to go through the system at an established temperature, flow rate, and certain level of turbulence for a given time in order to remove soils and ensure proper cleaning of the aseptic system. The time of the CIP will be dependent on several factors, including the size of the system, the flow rate capability, and the level of soil accumulated in the system. However, recirculation times between 20 and 30 minutes for chemicals are common in commercial size processing lines. Flow rates have been generally recognized as adequate if they are capable of exceeding 5ft/s (1.52 m/s) of flow through the entire system and/or are flowing faster than the product (generally 125% of product flow). Temperature will also be dependent on factors, including the recommended cleaning temperatures of the chemicals used. Regardless, heat will be required to help break down stubborn soils, including fat, at temperatures in excess of 170°F or higher.

CIP cleanings for aseptic systems often follow a common flow sequence:

An initial rinse is conducted to remove loose soil and debris. This will help prepare the system for the following steps. Introduction of caustic solution, usually at a concentration of 1–2% will follow the initial rinse. This will help dig into protein and fat residues, as well as helping remove some of the "fouling" (burnt on product) from the system. The caustic solution will then be rinsed out of the system, so that an acid-based solution can be added. This step will help remove remaining mineral residues from the system. A final, freshwater rinse of the system should be conducted to eliminate any acid residual in the system. A final step that can occur based on the product and process type is sanitization. Some systems will add a final sanitizer rinse to help reduce any remaining microbiological populations that may still remain prior to a sterilization cycle.

16.5.2.6 CIP System Validation

All systems and components for aseptic processing must be validated to ensure a proper clean is achievable with a high degree of reproducibility. CIP and line validations should take into consideration several factors for success. Criteria for these systems should include adherence to the CIP recipe (run order, time, temperature, flow), the identification of appropriate cleaning chemical concentrations, and the ability to inspect the process once the CIP has completed. Validation activities should challenge the line in triplicate, under worst-case scenarios in order to have assurance that the CIP system is capable of performing according to expectations. Worst-case scenarios will typically include running a product that is normally difficult to clean (high solids, protein, fat, viscous, allergenic, strong flavored) for maximum operation time followed by a CIP cleaning operating at the minimum level of acceptability. Challenge criteria will often include visually inspecting the equipment and supplemented by adenosine triphosphate (ATP) swabbing, aerobic plate count investigations, and allergenic residual verification.

16.5.2.7 Aseptic Intermediate Cleanings (AIC)

Since the aseptic lines are designed for high-temperature treatment, even for short exposure times, some of the heat-sensitive products are prone to form fouling in the line due to their characteristics such as protein content. The formation of fouling inside of the equipment may change the heating characteristics and even the flow regimes if there are interventions. The line fouling may shorten the runtimes and eventually the line efficiencies. Therefore, in some systems, aseptic intermediate cleaning steps have been implemented with separate controls and programming to extend the runtimes without losing the sterility of the line. Some companies use in-process CIP (IPC) terminology. This process will allow an operator to maintain the system sterility while running a special cleaning step with a sterile cleaning solution. This will extend the operating time of a unit in between full CIP cleanings. The intention of an AIC is to reduce some of the fouling that may start to build up during normal operation. This cleaning will not be as effective as a full CIP and should only be utilized in support of a validated cleaning process.

16.5.2.8 Recall

A recall or withdrawal plan should be prepared in advance, describing actions to be taken when a defective product leaves the supply chain and gets into the store or consumers' hands. The plan must identify a process to notify receivers of the product (including consumers), a way to verify the plan (mock recall exercises) and define how the recalled product will be handled.

It should also define who is responsible for performing actions in product recovery, investigation, correction, and future prevention of non-compliance. These procedures and policies do not have to accompany a process filing; however, a firm doing any commercial sterilization, whether it is aseptic or canning, should have these. An annual mock recall is recommended to test the reflexes and efficiency of the plan and system.

The Food Safety Modernization Act (FSMA) regulation requires a written recall plan if preventive control is required. Additionally, certain products regulated by the USDA are required to develop and maintain recall plans as well. The Bioterrorism Act of 2002, which exists to enhance security in the United States, requires traceability for incoming materials and the recipients of FDA-regulated products. It also provides FDA with the power to access records under certain emergency conditions (Martino et al., 2020).

All food manufacturers must have a recall plan that fully considers risks present within the facility and any affected facilities related to production, inventory, distribution procedures, existing personnel, and other considerations. Should the need for a recall occur, this plan will be the standard operating procedure. The plan should be reviewed by leadership and those responsible for verification activities at least annually. In addition, a reassessment should occur anytime there are significant changes in the company organizational structure, personnel, product line, or distribution areas, or when a "mock" or genuine recall indicates a need.

An accessible and updated list of supplier contacts should be maintained. The recall plan should also include responsible parties and their contact information for all essential personnel within the company and for external subject matter experts used for product analyses, scientific experts (e.g., toxicologists, microbiologists, chemists, physicians, food allergy specialists), trade associations, public relations firms, and other outside support elements required (https://link .springer.com/chapter/10.1007%2F978-3-030-42660-6_11).

A strong recall program, in addition to excellent document control and product tracking, is necessary for aseptic operations for several reasons. First, due to the complexities of the aseptic process, many things can go wrong. Processes must be reviewed, and products pass all operating criteria successfully, or non-conforming products may enter the retail space. This will include programs that can perform accurate traceback and forward-trace exercises on the raw, in-process, and finished product and critical operational factors. For instance, in an aseptic process, a failure of the sterile air filters will be a critical deviation and result in product impoundment, withdrawal, or recall. In conventional processing systems, the failure of a system or room air filter may have minor quality implications and not prohibit the product from entering the marketplace.

It is also worth noting that due to the longer shelf life for aseptic and ESL products, there is more time for the product to have quality issues that may negatively impact the consumer experience. Many companies will have finished products under third-party control for storage and distribution. They may have to set up customized plans for holding and releasing a product, not in their direct control. Due to this, manufacturers will need to keep much more detailed information for more extended periods than conventional processes.

In conclusion, recalls are the last line of defense for the food safety programs. Although the consequences are not good for the company, a company recalling or withdrawing their products may give the perception of their commitment to their food safety program in case they caught the issue quickly and successfully prevented the product from reaching consumers. A successful recall cannot be executed without proper document control and record-keeping systems. Therefore, record-keeping is extremely important for aseptic systems since it is much harder to isolate and minimize the financial risks with continuous processing systems. Good manufacturing practices allow the recall team to identify the root cause(s) effectively and in a timely manner and determine the affected batched/lots to start the withdrawal/recall process. Implemented recall program for aseptic and ESL companies is a regulatory requirement. It is important to highlight that regulatory agency has the authority to enforce a requirement of a recall if they deem it necessary. Due to the unique nature of the product, process, and packages for the aseptic systems, much more investment in capability and competencies may be required.

16.5.2.9 Consumer Complaints

Close contact between the quality assurance and consumer relations departments needs to be maintained so that any consumer complaint is promptly brought to the attention of QA. Consumer complaints often provide a valuable early warning of a potentially big problem.

Also, consumer complaints can be used as a proactive QA tool. Proper databasing on consumer complaints is necessary to evaluate the performance of a product out in the field and to compute complaint or defect rates and analyze trends.

When a consumer complaint is brought to the attention of a manufacturer, care should be taken to carefully analyze the complaint, and compare it to existing retained samples that the company has on hand. Generally, keeping retainer samples timestamped 15–30 minutes apart will help in the investigation process. These retainer samples can also be shelf-life study samples, which are discussed elsewhere in this chapter.

If a customer complains about a particular quality or packaging defect, an investigation should be started. Having a strong CAPA resolution to a consumer complaint can often identify a potential issue early, and prevent repeated issues, or even a recall.

16.5.2.10 Supply Chain

Manufacturers are also required to implement a risk-based supply chain program if the hazard analysis identifies a significant hazard that requires control and if that control is to be applied in the supply chain. A manufacturer would not need a supply chain program if they, or a subsequent processor, will control the hazard through further processing. Manufacturers are responsible for verifying that ingredients and materials are coming in from approved vendors (as approved by a company risk analysis, audit, or other acceptable methods). Procedures should be in place for approving all vendors providing materials to the manufacturing site. Hazard control and verification at this step mean that once a supplier of material has been identified and approved for use, the vendor and supplier both should have verification activities for the product. This may mean that the vendor has provided a Certificate of Assurance (COA) or Certificate of Compliance (COC) and that the manufacturer who has received this product also conducts verification activities (inspection of the product to ensure that it conforms to the COA).

16.6 CONCLUSIONS AND FINAL RECOMMENDATIONS

The most critical factor for the aseptic industry is brand protection. The risks for the aseptic processing and packaging systems are at a much higher level due to complexities of the technologies, continuous nature of the processes, and the competency requirements. Therefore, in this section, identification and mitigation of the risks, validation of the systems with consideration of those risks, and finally, maintaining the "validated state" of the aseptic systems with well-designed and implemented quality and food safety management systems are necessary. The purpose of this chapter is to share the basic requirements of such systems to have a sustainable and safe aseptic manufacturing line. The intent of the authors was not to be prescriptive with details of the programs which can easily be found in the literature and market. The intent was to give highlights of the important aspects of these programs that would allow manufacturers to have sustainable safety tracking and establish "public trust" within the aseptic industry. When it comes to food safety, there are no secrets in the industry and companies do not compete with food safety. One food safety concern in the industry would damage the image of overall industry and public trust and would affect every company in the industry.

To maintain the validated state of the aseptic lines, the most important prerequisite programs are training, preventive maintenance, change management, and well-implemented QFSMS. For these programs, companies need to have competent resources to be able to analyze the risks and develop mitigation and maintenance plans to achieve organizational quality and safety visions. Since every vision would require strong leadership support and sponsorship, the organization's quality and safety culture need to be the starting point for the development of such programs. Aseptic culture is different than many other conventional food manufacturing systems and the requirements to sustain the acceptable quality levels.

The authors highly recommend using internal and external competencies to design, evaluate, validate, and maintain the quality and food safety programs for the aseptic manufacturing systems. It may require political and managing-up type strategies to influence the decisions and initiatives of the company executives. Such programs are expensive and may require creative approaches to justify the related costs at the beginning of the initiatives, during the validation and start-up of the system and throughout the process life cycle.

GLOSSARY

Acceptable quality level (AQL)
The maximum percentage of non-conforming products, which is considered, for inspection purposes. Different AQLs may be designated for different types of defects. It is common to use an AQL of 0.01% for FDA-regulated industries.

Acceptance limit
The upper limit on the number of non-conforming items in a sample, which would still lead to the acceptance of the entire lot. If the number of non-conforming items in the sample exceeds this number, the entire lot must not be accepted.

Accuracy
Data gathered from repeated analysis are close to the established ranges (targets).

Acid foods
Acid foods means foods that have a natural pH of 4.6 or below (21 CFR 114.3).

Acidified food
Acidified foods means low-acid foods to which acid(s) or acid food(s) are added; these foods include, but are not limited to, beans, cucumbers, cabbage, artichokes, cauliflower, puddings, peppers, tropical fruits, and fish, singly or in any combination. They have a water activity (aw) greater than 0.85 and have a finished equilibrium pH of 4.6 or below (21 CFR 114.3).

Aseptic processing and packaging of foods
The filling of a commercially sterilized-cooled product into presterilized containers, followed by hermetic sealing with a presterilized closure in an atmosphere free of microorganisms (Canned Food, 2007).

Batch
The batch size is the number of items in a lot or a batch.

Commercial sterility of thermally processed food
Refers to absence of disease-causing microorganisms, absence of toxic substances, and absence of spoilage-causing microorganisms capable of multiplication under normal non-refrigerated conditions of storage and distribution (APHA, 2001).

Corrective and Preventive Action (CAPA)

A corrective action is "action to eliminate the cause of the non-conformity in order that it does not recur or occur elsewhere", and preventive action is "action to eliminate the cause of a potential non-conformity or other undesirable situation" (ISO).

Current Good Manufacturing Practices (cGMP)

Program whose purpose is to ensure that food has been manufactured under conditions that ensure it is fit for consumption and has not been adulterated.

DMAIC process

An acronym (define, measure, analyze, improve, control) is a quality based, data-driven strategy used in continuous improvement processes.

Extended Shelf-Life (ESL)

Extended shelf-life (ESL) product is produced by utilizing aspects of aseptic manufacturing, such as thermal processing and aseptic storage, but fills product into clean containers and must be stored under refrigerated conditions.

Failure mode and effects analysis (FMEA)

Failure mode and effects analysis (FMEA) is an analytical technique that combines the technology and experience of people in identifying foreseeable failure modes of a product or process and planning for its elimination (Besterfield et al., 2003).

Food adulteration

Food adulteration can be defined as any food that is unsafe, unwholesome, or impure and may be deleterious to health (21 USC 342).

Hazard analysis and critical control points (HACCP)

HACCP is a preventive control tool that makes up a food safety program. This program aims to eliminate, reduce, or control to an acceptable level identified food safety risks.

Hermetic seal

A type of container sealing that makes an airtight seal.

Inspection levels for Military Standard 105E

The inspection level determines the relation between the batch size and sample size.

The four special inspection levels S-1, S-2, S-3, and S-4 use very small samples, and should be employed when small sample sizes are necessary, and when large sampling risks can be tolerated.

Levels I, II, and III are general inspection levels:

- Level II is designated as normal.
- Level I requires about half the amount of inspection as Level II, and is used when reduced sampling cost is required and a lower level of discrimination (or power) can be tolerated.
- Level III requires about twice the amount of inspection as Level II, and is used when more discrimination (or power) is needed.

Key performance indicators (KPI)

A measurable value which demonstrates the effectiveness of a company in achieving its business goals.

Low-acid food

Low-acid foods means any foods, other than alcoholic beverages, with a finished equilibrium pH greater than 4.6 and a water activity (aw) greater than 0.85. Tomatoes and tomato products having a finished equilibrium pH less than 4.7 are not classed as low-acid foods (21 CFR 114.3).

Manufacturer's risk

The chance that a batch from a process with AQL is rejected. This is generally an additional costly risk.

Master sanitation schedule (MSS)

A master cleaning schedule is a program which includes the items, process, frequency, and responsible individuals for the acceptable management of the overall sanitation program within a manufacturing facility.

Percent non-conforming

The percent or proportion of non-conforming items in a batch or in a process. In many cases this is unknown, but it is used to learn about scenarios for different values of p.

Precision

Data gathered from repeated analysis are close to each other.

Process Authority (PA)

An individual or group of experts in the development, implementation, and evaluation of food processes throughout the products' life cycle for the protection of public health.

Quality assurance

QA is a form of insurance purchased at a relatively small expenditure for the purpose of securing protection against disasters.

Quality control

QC is defined as the process through which one measures actual quality performance, compares it with a standard, and acts on the difference (Juran & Godfrey, 1999).

Reynolds' Number

This number can be described as the ratio of inertial forces to viscous forces within fluid systems. Low numbers indicate laminar flow, while higher numbers may indicate turbulent flow.

Safety data sheet (SDS)

A safety data sheet is a document that lists information relating to occupational safety and health for the safe use of chemicals.

Sample size

The number of products that should be randomly chosen from a batch.

Scheduled process

Scheduled process means the process selected by a processor as adequate for use under the conditions of manufacture for a food in achieving and maintaining a food that will not permit the growth of microorganisms having public health significance. It includes control of pH and other critical factors equivalent to the process established by a competent processing authority (21 CFR 114).

Shelf-stable food

Refers to food capable of being stored without refrigeration at ambient environmental conditions for 1–2 years.

Type of inspection for military standards

There are three types of inspection:
- Normal inspection is used at the start of the inspection activity.
- Tightened inspection is used when the vendor's recent quality history has deteriorated (acceptance criteria are more stringent than under normal inspection).

Reduced inspection is used when the vendor's recent quality history has been exceptionally good (sample sizes are usually smaller than under normal inspection).

Validation

Validation means "that element of verification focused on collecting and evaluating scientific and technical information to determine whether the HACCP plan, when properly implemented, will effectively control the identified food hazards" (21 CFR 120.3p).

Verification

Verification means "those activities, other than monitoring, that establish the validity of the HACCP plan and that the system is operating according to the plan" (21 CFR 120.3q).

REFERENCES

9 CFR Parts 308, 318, 320, 327 and 381. 1986. *Canning of Meat and Poultry Products Final Rule Part II.* Washington, DC: Department of Agriculture.

21 CFR 113 Thermally Processed Low-acid Foods Packaged in Hermetically Sealed Containers. U.S. Government Publishing Office, Washington, D.C. 44 FR 16215, Mar. 16, 1979, 76 FR 81363. 2011, December 28. 21 CFR 113.83. https://www.accessdata.fda.gov/scripts/cdrh/cfdocs/cfcfr/CFRSearch.cfm?fr=113.83

21 CFR 114 Acidified Foods. U.S. Government Publishing Office, Washington, D.C. 44 FR 16235. 1979, March 16. 21 CFR 114.3. https://www.accessdata.fda.gov/scripts/cdrh/cfdocs/cfcfr/CFRSearch.cfm?fr=114.3

21 CFR 117 Current Good Manufacturing Practice, Hazard Analysis, and Risk-Based Preventive Controls for Human Food. U.S. Government Publishing Office, Washington, D.C. 80 FR 56145. 2015, September 17. 21 CFR 117.26. https://www.accessdata.fda.gov/scripts/cdrh/cfdocs/cfcfr/CFRSearch.cfm?fr=117.126

21 CFR Parts 110, 113, and 114. 1987. Washington, DC: U.S. Government Printing Office.

ASTM. 1987. *Annual Book of ASTM Standards Paper; Packaging; Flexible Barrier Materials; Business Copy Products*, Vol. 15.09. American Society for testing and Materials.

Besterfield, D.H., Besterfield-Michna, C., Besterfield, G., and Besterfield-Sacre, M. 2003. Failure mode and effect analysis. In *Total Quality Management*. 3rd edition, Chapter 14. Upper Saddle, River, NJ: Prentice Hall.

Bockelmann, V.B. 1982. *Aseptic Packaging and Processing: A Collection of Lectures from Tetra Pak Seminars.* Chapters 3 and 7. Lund, Sweden: Tetra Pak.

Carpenter, R.P., Lyon, D.H., and Hasdell, T.A. 2000. *Guidelines for Sensory Analysis in Food Product Development and Quality Control.* Gaithersburg, MD: Aspen Publishers, Inc.

CAST (Center for Agriculture and Technology). 2016. *Food Waste Across the Supply Chain: US Perspective on a Global Problem.* Ames, IA: CAST.

Crosby, P.B. 1984. *Quality Without Tears.* New York: McGraw-Hill.

David, J.R.D. 1988. Realization of a commercial low acid thermoform-fill-seal (TF-F-S) machine in the U.S. Part B: Quality assurance and regulatory aspects. In Proceedings of the Fifth International Conference on Aseptic Packaging, "ASEPTIPAK 1988," June 8–10, Bloomingdale, IL.

David, J.R.D. 1989. *Quality Assurance for UHT-sterilized and Aseptically Packaged Food at Real Fresh, Inc. Aseptic Processing and Packaging Course. Food Engineering and Technology Series.* Madison, WI: University of Wisconsin.

Department of Defense. 1989. *Military Standard Sampling Procedures and Tables for Inspection by Attributes, MIL-STD-105E.* Washington, DC: Department of Defense.

Food and Drug Administration. 1992. *Bacteriological Analytical Manual.* 7th edition, Chapters 23 and 24. Arlington, VA: Association of Official Analytical Chemists.

GFSI. 2021. GFSI overview. https://mygfsi.com/who-we-are/overview/

Grade "A" Pasteurized Milk Ordinance (PMO). 2019 Revision. U.S. Department of Health and Human Services, Public Health Service, Food and Drug Administration (FDA), Washington, DC.

International Organization for Standardization. 2015. ISO 9000:2015: Quality management systems: Requirements. http://www.iso.org

Juran, J.M., and Godfrey, A.B. 1999. *Quality Planning and Analysis*. 2nd edition. New York: McGraw-Hill.

Martino, K., Stone, W., and Ozadali, F. 2020. Product recalls as part of the last line of food safety defense. In *Food Safety Engineering*, Chapter 11, 247–263. New York: Springer International Publishing. https://link.springer.com/chapter/10.1007%2F978-3-030-42660-6_11

National Food Processors Association. 1989. *Flexible Package Integrity Bulletin. Flexible Packaging Integrity Committee*. Washington, DC: National Food Processors Association.

National Research Council (U.S.) Food Protection Committee Subcommittee on Microbiological Criteria. 1985. *An Evaluation of the Role of Microbiological Criteria for Foods and Food Ingredients*. Washington, DC: National Academy Press.

Schothorst, V.M., and Jorgeneel, S. 1994. Line monitoring, HACCP, and food safety. *Food Control* 5(2):107–110.

Schmueli, G. (2016). *Practical Acceptance Sampling*, 2nd Edition. Axelrod Schnall Publishers, ISBN-13:978-0-9915766-7-8]

Shafer, B.D., Balestrini, C.G., and Weddig, L.M. 2007. *Canned Food: Principles of Thermal Process Control, Acidification and Container Closure Evaluation*. 7th edition. Washington, DC: GMA Science & Education Foundation.

United States Code. 2021, October 11. USC 342. http://uscode.house.gov/view.xhtml?req=21+USC+342&f=treesort&fq=true&num=126&hl=true&edition=prelim&granuleId=USC-prelim-title21-section342

United States Department of Agriculture. 2021. USDA. https://nifa.usda.gov/glossary#F

United States FDA. 2021a, October 1. 21 CFR 113.10. https://www.accessdata.fda.gov/scripts/cdrh/cfdocs/cfcfr/CFRSearch.cfm?fr=113.10

United States FDA. 2021b, October 1. 21 CFR 117. https://www.accessdata.fda.gov/scripts/cdrh/cfdocs/cfcfr/CFRSearch.cfm?CFRPart=117

United States FDA. 2021c, October 1. 21 CFR 117.3. https://www.accessdata.fda.gov/scripts/cdrh/cfdocs/cfcfr/CFRSearch.cfm?fr=117.3

United States FDA. 2021d, October 1. 21 CFR 117.4. https://www.accessdata.fda.gov/scripts/cdrh/cfdocs/cfcfr/CFRSearch.cfm?fr=117.4

United States FDA. 2021e, October 1. 21 CFR 120.3p, q. https://www.accessdata.fda.gov/scripts/cdrh/cfdocs/cfcfr/CFRSearch.cfm?fr=120.3

United States FDA. 2021f, October 1. 21 CFR 120.13. https://www.accessdata.fda.gov/scripts/cdrh/cfdocs/cfcfr/CFRSearch.cfm?fr=120.13

United States FDA. 2021g, October 1. 21 CFR 121.4. https://www.accessdata.fda.gov/scripts/cdrh/cfdocs/cfcfr/CFRSearch.cfm?fr=121.4

United States FDA. 2021h, October 1. 21 CFR 123.8. https://www.accessdata.fda.gov/scripts/cdrh/cfdocs/cfcfr/CFRSearch.cfm?fr=123.8

United States FDA. 2021i. Allergies. https://www.fda.gov/food/food-safety-modernization-act-fsma/fsma-final-rule-preventive-controls-human-food

United States FDA. 2021j. FMSA. https://www.fda.gov/food/food-safety-modernization-act-fsma/fsma-final-rule-preventive-controls-human-food

United States FDA. 2021k. Key Requirements-Supply Chain. https://www.fda.gov/food/food-labeling-nutrition/food-allergies

US Department of Health and Human Services. Public Health Services FDA. 2021, December 31. Grade "A" Pasteurized Milk Ordinance 2019 revision. https://www.fda.gov/media/140394/download

Part IV

Frontiers and R&D Opportunities and Challenges

Chapter 17

Computational and Numerical Models and Simulations for Aseptic Processing

George N. Stoforos, Deepti Salvi, Tunc Koray Palazoglu,
Nihat Yavuz, Pablo M. Coronel, and Josip Simunovic

CONTENTS

17.1 INTRODUCTION

Computational models and simulations are a broad and continuously expanding set of basic and advanced tools for simulation, evaluation, optimization, development, and decision-making in a variety of processes, industries, and activities. The knowledge base and techniques developed using these tools have often been further developed and implemented as the basis for real-time recognition, alarms, decision-making, and optimization through the application of artificial intelligence tools and techniques.

For aseptic processing and packaging, these tools are invaluable in making operational, tactical, and strategic decisions, especially in development and implementation of novel and emerging technologies and optimization and problem-solving in analyses and modifications of existing and established technologies and installations.

Computational models range from very simple series of interconnected equations and calculations implemented in a spreadsheet format, to stand-alone, dedicated computer programs with variable input values, to very complex applications developed using one of the several established and emerging Multiphysics packages of software systems. Simulations are performed by repeated application of models to generate relevant calculated data and results which can be used to make and implement decisions in equipment and process construction and modifications, process and plant design as well as strategic decisions about the selection of particular equipment elements and capacities and identifying the need for research and development of new technologies or novel implementations of existing ones.

In this chapter, we present five examples of applications of computer models and simulations using the available and self-developed software tools and simulations to address the identified and emerging issues in aseptic processing of more complex and difficult foods and biomaterials, such as viscous, poorly conductive, thermosensitive and complex, multiphase foods containing varying sizes and loads of solid discrete particles.

Until recently, aseptic processing and packaging has been limited in its application range primarily to thin, fluid, uniform, and mostly dairy-based products. Development and commercialization of novel, more powerful technologies for continuous-flow heating and sterilization of aseptically packaged products have opened new possibilities in the product range, quality optimization, and nutrient retention maximization of numerous materials, which have previously been excluded from aseptic processing due to a variety of technological limitations.

Computational modeling and simulations have proven to be valuable and necessary tools in all aspects of development, implementation, and eventual industrial success of such novel tools and technologies. The following examples illustrate the breadth and flexibility of these tools when applied to several critical and emerging application areas in the food and beverage processing using aseptics.

17.2 COMPUTATIONAL FLUID DYNAMICS AND HEAT TRANSFER MODELING: GENERAL

Aseptic food processing includes the steps of heating, holding, and cooling the product, aiming for the commercial sterility and safety of the finished product while extending the product's shelf life and preserving most of the food's quality attributes. The applied thermal processing requires an accurate design of the employed process and processing parameters and the equipment to achieve the desired food safety and product quality results. The application of thermal processing is conducted via conventional or advanced process technologies. Conventional methods that involve heating and cooling via indirect systems, namely, plate, tubular, and scraped surface heat exchangers, are still the most popular process methods used in aseptic processing. Optimization of thermal processing and equipment design requires understanding and solving complex partial differential equations based on fluid motion and heat transfer. The design and efficiency of conventional heating and cooling system have undergone great development in the last decades due to the availability of advanced modeling and computational techniques such as computational fluid dynamics (CFD) (Scott & Richardson, 1997; Tomas & Sun, 2006; Yanniotis & Stoforos, 2014; Park & Yoon, 2018).

CFD has become the most common method for the initial design, investigation, and test-ing of new processes and equipment or ways to improve the used processing methods (Scott & Richardson, 1997; Tomas & Sun, 2006; Yanniotis & Stoforos, 2014; Park & Yoon, 2018). Computer simulation is proven to be an alternative to experimentation for gaining a quantitative and qual-itative understanding of food process applications. CFD is a simulation modeling tool that can analyze and solve aseptic processing systems' complex fluid flow and heat transfer problems and associated phenomena. The governing equations of fluid flow and heat transfer, known as Navier–Stokes equations, can be considered mathematical formulations of fluid mechanics' con-servation laws. CFD comprises the fundamental governing equations of fluid dynamics, involv-ing the solution of the equations of continuity (Equation 17.1), momentum (Equation 17.2), and energy (Equation 17.4), which are solved numerically to estimate parameters such as velocity, shear, temperature, and pressure profiles inside the system being examined (Scott & Richardson, 1997; Tomas & Sun, 2006; Yanniotis & Stoforos, 2014; Park & Yoon, 2018).

Continuity or conservation of mass equation describes the overall mass balance within the tube flow:

$$\nabla(\vec{v}\cdot\rho)+\frac{\partial\rho}{\partial t}=0 \tag{17.1}$$

where $\vec{v}$ is the fluid velocity vector (m/s), ρ is fluid's density (kg/m³), and ∇ is the del operator.

The momentum equation describes the motion of fluids. The terms in Equation 17.2 corre-spond to the *inertial forces* (1), *pressure forces* (p [Pa]) (2), *viscous forces* (σ is the stress component [Pa]) (3), and the *external forces* (F) applied to the fluid (4).

$$\underbrace{\rho\cdot(\vec{v}\cdot\nabla)\cdot\vec{v}+\frac{\partial(\rho\cdot\vec{v})}{\partial t}}_{1}=\underbrace{-\nabla p}_{2}+\underbrace{(\nabla\sigma)}_{3}+\underbrace{F}_{4} \tag{17.2}$$

In CFD models in the momentum equation, usually the Boussinesq approximation (Equation 17.3) is adopted to model density variation caused by buoyancy, based on a reference density (ρ_{ref} [kg/m³]) value at a given reference temperature (T_{ref} [K]). However, the Boussinesq approxima-tion is not valid when the temperature difference between the evaluated temperature and the reference temperature is high. In this case, density can be modeled as follows:

$$\rho = \rho_{ref}\cdot\left[1-a\cdot(T-T_{ref})\right] \tag{17.3}$$

Finally, one of the governing equations applied in aseptic thermal processing CFD modeling is the energy equation (Equation 17.4). The energy equation (the first law of thermodynamics) describes the rate of energy change of a fluid element that equals the work done and the heat generation on the element.

$$\frac{\partial(\rho\cdot c_p\cdot\vec{v})}{\partial t}+\nabla(\vec{v}\cdot T)=\nabla\cdot(k\cdot\nabla T)+Q \tag{17.4}$$

Equation 17.4 describes the heat transfer in fluids, where c_p is the fluid's specific heat (J/[kg·K]), T is the temperature component (K), and Q describes the external heat sources (W/ m²).

The solution to the above mathematical set of equations gives the velocity, pressure, and tem-perature of the fluid at any given point of the modeled domain. The accuracy of the CFD heat transfer problem's solution depends on the input initial and boundary conditions of the problem and the selected structure of the examined domain (Scott & Richardson, 1997; Tomas & Sun,

2006; Park & Yoon, 2018). An example of boundary conditions in a heat exchanger CFD problem can be the non-slip (zero velocity) and the external natural convention applied in the modeled heat exchanger's boundary pipe walls (Stoforos & Simunovic, 2018). Selection of the examined structure (geometry) of the modeled domain directly affects the required time for the solution. The three most common modeled structures used in a CFD heat transfer problem are the axisymmetric approach, two-dimensional or three-dimensional domain, depending on the complexity and the requirements of the given problem. The more complex the examined geometry, the longer the simulation time required for the problem to be solved.

Furthermore, the accuracy of the solution of the CFD heat transfer problem depends on the meshing method and time-step (for unsteady-state problems) selected to solve the problem. Time-stepping expresses the frequency of computations in the model (Scott & Richardson, 1997; Tomas & Sun, 2006; Park & Yoon, 2018). Time-stepping is a critical parameter in time-dependent, unsteady-state problems, where the flow properties are functions of time. In this type of problem, the selection of time-step(s) is critical, and it should be short enough to avoid numerical error(s) and instability of the solution. The selection of the proper time-step applied to the problem depends on the fluid velocity field and the size of the mesh cells applied to the examined domain/ geometry (Scott & Richardson, 1997; Tomas & Sun, 2006; Park & Yoon, 2018; Ansys, 2021b). The proper meshing of the examined subdomains is critical to ensure valid solutions across the common interfaces. For a three-dimensional model, tetrahedron mesh is more common, while meshes such as quadrilaterals and triangular are typical options for two-dimensional structures. The governing equations are then discretized and solved inside each of these subdomains. One of the three following methods is typically used to numerically solve the partial differential equations of the CFD heat transfer problem: finite volumes, finite elements, and finite differences (Scott & Richardson, 1997; Tomas & Sun, 2006; Park & Yoon, 2018; Ansys, 2021b).

17.2.1 Turbulent Flow Modeling

The Navier–Stokes equations described above govern the velocity and pressure of a fluid flow and are applicable to simulate problems for both laminar and turbulent flow conditions. Fluid flow is classified into two main categories, laminar or turbulent flow, and depends on the driven forces such as viscous forces. Reynolds number is used to classify which flow regime fits the examined fluid flow. Reynolds number is a dimensional number estimated based on static and dynamic fluid properties such as velocity, density, and a characteristic linear dimension of the flow (hydraulic diameter is used for pipe flow). In general, for a Reynolds number at or below 2,300, the flow is considered laminar, while for a Reynolds number above 4,000, the flow is characterized as turbulent (Ansys, 2021b; Comsol, 2021b; Simscale, 2021). The intermediate Reynolds number range, 2,300–4,000, is considered as the transitional flow regime. Note that for CFD models in aseptic processing, evaluating food safety, and product sterilization, the critical Reynolds number used to identify laminar or turbulent flow should be properly selected, based on literature data and industrial practice, as it may vary based on food product formulation, for example, for products with high starch content.

Under laminar flow conditions, the fluid elements flow within streamlines, through a smooth path, with no disruption between adjacent paths. Laminar flow assumption is typically applicable to simulate CFD and heat transfer problems associated with viscous foods such as fruit and vegetable puree. In contrast, under turbulent flow conditions, fluid flows through a chaotic path that comprises eddies, swirls, and flow instabilities (Figure 17.1). Turbulent flow conditions are usually used for low-viscous foods such as fruit juices and so on. The equations described

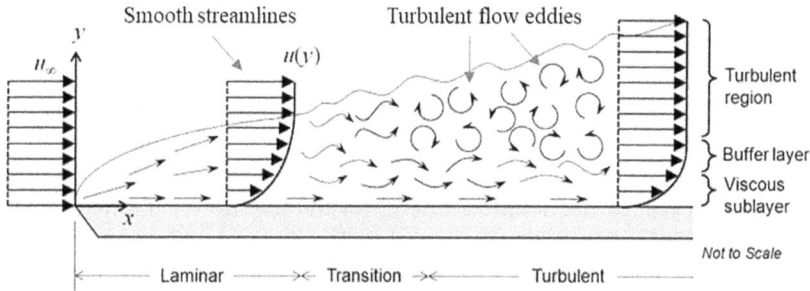

Figure 17.1   Schematic presentation of laminar, turbulent, and transitional flow regimes; showing the smooth streamlines of the fully developed laminar flow and the eddy formation during turbulent flow conditions (Comsol, 2021b).

above can give a good approximation of laminar flow CFD heat transfer problems. On the other hand, tubular flow occurs in minimal distance and timescales and is characterized by fluctuating velocity fields and unsteady aperiodic motion, which requires a more complex solution and additional equation(s) to be solved. As a result of that, supplementary to the Navier–Stokes equations, turbulent flow models have been developed.

Selection of the correct turbulence model requires a good understanding of the physical parameters applied to the problem, a literature review to identify suitable models for the examined case, and finally applying a "trial and error" approach, concurrently testing different models to identify the model that fits best and delivers the more accurate prediction (Ansys, 2021b; Comsol, 2021b; Simscale, 2021). The turbulent model used can make a big difference in the CFD simulation results. Although a universal turbulent flow model does not exist, several different turbulent models have been developed and are available in all CFD platforms to solve the turbulent phenomena more accurately. The main purpose of turbulence models is to prompt equations to estimate the time-averaged velocity, pressure, and temperature fields, without calculating the complete turbulent flow pattern. The turbulence models are classified based on the governing equation(s) and the numerical method used to calculate turbulent viscosity. The following are the three main categories of turbulence models used in CFD: (i) direct numerical simulation (DNS), (ii) large eddy simulation (LES), and (iii) Reynolds-averaged Navier–Stokes (RANS) turbulence models (Ansys, 2021b; Comsol, 2021b; Simscale, 2021). The DNS model uses a direct implementation of fluctuated values to Navier–Stokes' equations to compute turbulent flow. DNS model requires precise fine grid mesh resolution and large computer capacity for the simulation, making it not practically suitable for any CFD aseptic processing simulations. LES turbulent models are based on the macroscopic evaluation of the turbulent flow. LES models use modified Navier–Stokes equations examining only the large eddies formed in a turbulent flow; thus, using the LES models in simulations is rarer (Ansys, 2021b; Comsol, 2021b; Simscale, 2021; Park and Yoon, 2018).

Finally, RANS turbulent models are algebraic models based on mean values of flow parameters and variables used to describe the turbulent flow. RANS simulate turbulence flow by calculating the average of both steady-state and dynamic flow variables. Furthermore, RANS turbulence models can be subdivided into two broad types: eddy viscosity models and Reynolds stress models, which differ on how the flow close to boundaries (wall) is modeled, the number of additional variables solved, and what these variables represent. All RANS models augment the Navier–Stokes equations with an additional turbulence eddy viscosity term, but they differ on how it is computed. One of the most popular and widely used RANS turbulent models is the k-ε model and its variations (Park & Yoon, 2018; Ansys, 2021b; Comsol, 2021b; Simscale, 2021). The k-ε model is based on the eddy viscosity hypothesis, expressing eddy viscosity (the turbulent transfer of momentum by eddies) through turbulence production and destruction. The k-ε model uses two partial differential equations to estimate local eddy viscosity in a turbulent flow using the two variables of turbulent kinetic energy (k) and the turbulent kinetic energy dissipation (ε) (Park & Yoon, 2018; Ansys, 2021b; Comsol, 2021b; Simscale, 2021). The model gives good results when the problem involves small pressure gradients but is not very accurate in computing flow fields that exhibit adverse pressure gradients, strong curvature to the flow, or jet flow. Consequently, a combination of different turbulence models can be applied to overcome the weak points of each model and hence to get a more accurate simulation of the turbulent fluid flow (Comsol, 2021b; Simscale, 2021).

17.2.2 Non-Newtonian Flow Behavior

Aseptic processing, by definition, is a continuous-flow process, and hence accurately modeling and solving CFD heat transfer problems require taking into consideration the examined food product's rheological characteristics. Food product's rheological characteristics play a critical role in determining the flow behavior, such as the type of flow, laminar or turbulent, and the velocity profile. The majority of the aseptically processed food products present a non-Newtonian behavior. For CFD models, any type of independent incompressible and compressible flow model can accurately predict non-Newtonian laminar flow behavior for single-phase and multiphase products. CFD models have available all different rheological models, such as the Newtonian model, the power-law model, the Bingham model, the Herschel Bulkley model, and so on, to express the rheological properties as best fitted to each examined case. Furthermore, CFD models provide the option to express product viscosity as a temperature-dependent or time-dependent function as applicable (Ansys, 2021a; Comsol, 2021a).

For non-Newtonian turbulent flow problems, an extra equation may be required to be solved for viscosity, in addition to the turbulent models described above. However, it should be noted that modeling turbulent flow on non-Newtonian fluids is a complex case and still an area of research. Non-Newtonian turbulent flow modeling requires selection of the correct geometry, mesh, and time-step to provide accurate results and convergence of the solution (Comsol, 2021a).

17.2.3 CFD Heat Transfer Models in Aseptic Processing

Aseptic processing is a continuous-flow process that results in a commercially sterile product using minimal treatment, high temperatures for a short period of time, minimizing the product's quality degradation. To optimize the equipment and the processing parameters (temperature, flow rate, pressure, and so on) and improve aseptic processing operations and finished product food quality, computer simulation modeling, such as CFD and heat transfer models, provides an excellent tool. Computer simulation is proven to be an alternative to experimentation for gaining a quantitative and qualitative understanding of food process operations and has been used for initial conceptual studies of new designs, detailed product development, and scale-up. The computational models has been used to optimize the design and efficiency of equipment such as heat exchangers and static mixers (Kumar, 1995; Grijspeerdt et al., 2003; Fourcade et al., 2001; Stoforos & Simunovic, 2018), validate the sterility of single- and multiphase (with particulates) products, and study and optimize the application of advanced heating technologies such as microwave heating (Meher et al., 2017; Park & Yoon, 2018). The next sections of this chapter will provide examples of computational models used to study and optimize aseptic processing applications.

17.3 EXAMPLES

Five examples of the applications of modeling and simulations have been selected for inclusion in this chapter. The first four are examples of the utility of simulations to provide the data to be used as a basis for decision-making for equipment and processing systems design as well as the tools for research and study of advanced thermal processing methods and the need for modifications and optimization. The final example can be implemented for both conventional and established processing technologies and advanced and emerging technologies, and their comparisons. These examples are as follows:

1. Conventional tube-in-tube cooling
2. Continuous flow microwave heating: first generation
3. Continuous-flow microwave heating: second generation
4. Continuous-flow heating of products containing particles
5. Using spreadsheet-based models for evaluating thermal processes

All of the presented examples have been used to justify decision-making in R&D, equipment design, system design and processing facility installations.

Specifically, in the case of continuous-flow microwave processing technology and its applications, simulation results from these models have also been used to identify the limiting aspects in individual designs and applications of each generation of modeled technology, and a justification and tool for development of better optimized next generations of technology.

17.3.1 Example 1: Conventional Tube-in-Tube Cooling

Computer simulation studies have been used to optimize the efficiency of the cooling stage of aseptic processing. In aseptic processing, a rapid cooling process is required to minimize product degradation. Computer simulation studies have been conducted to examine different designs and process parameters to optimize conventional tube-in-tube heat exchangers used in the cooling stage of aseptic processing (Stoforos & Simunovic, 2018). In the example described below, computer simulations using CFD and heat transfer models were implemented to study the cooling of sweet potato puree, examining different flow configurations and process parameters.

In Stoforos and Simunovic (2018) example study, Multiphysics software Comsol 5.2 (Comsol Inc., Burlington, MA, USA) was used to simulate the cooling of sweet potato puree in a three-dimensional model of a tube-in-tube heat exchanger. Simulation studies were used to compare the cooling efficiency under two different flow regimes within the same stainless steel (300 series) tube-in-tube heat exchanger, with sweet potato puree: (i) flowing within the inner tube of the heat exchanger and (ii) flowing within the annulus. The model comprises differential equations for CFD and heat transfer, using incompressible steady-state conditions under laminar flow for sweet potato puree. The sweet potato puree's viscosity was modeled using the power-law model, with the temperature-dependent rheological parameters of sweet potato puree obtained from experimental data. The examined flow rate for sweet potato puree was low (<5 gpm), due to the equipment limitations of the experimental study used to validate the model. For the coolant (water), RANS Algebraic yPlus turbulent model was applied under the same conditions of inlet temperature (4°C) and flow rate (~10 gpm). Boundary conditions of the CFD model were based on non-slip wall conditions. Natural convection heat transfer on a long horizontal cylinder was used as the boundary condition of the heat exchanger's outer surface. The boundary conditions at the pipe wall of the heat exchanger were modeled based on the heat flux across the stainless steel pipe wall. A fine mesh grid, with a total of 150,296 mesh cells, was used for the discretization of heat exchanger geometry and to obtain the data of velocity and temperature during cooling.

Using computer simulation models, sweet potato puree's temperature and velocity profile were compared between the above two different flow configurations. The different flow configurations were examined using different inlet product temperatures (120–140°C), comparing counter-current and co-current regimes, and using different inner tube diameters (I.D.) The most efficient cooling of sweet potato puree was observed with the product flowing within the annular passage of a heat exchanger, with the equal cross-sectional area (the outside diameter of the inner tube at 0.0558 m [2.2 in]) for both inner tube and annular product flow. The simulation results indicate that under an identical set of operating conditions, cooling of sweet potato puree was significantly improved for the annular product flow compared to the other case, resulting in a bulk temperature at the exit of the heat exchanger (T_{bulk}) lower by 12 ± 4°C, improving cooling efficiency by 25% for sweet potato puree (Stoforos & Simunovic, 2018). The improvement in cooling observed with the annular flow resulted from the heat transfer between the product and the coolant and the additional heat transfer to the environment. Performing energy balance calculations, the additional thermal energy lost from the product flowing in the annulus to the environment was in the range of 88–400 W. While for the other case, the heat transferred between the environment and the coolant was in the range of 22–326 W, having a negative impact on cooling efficiency (heat transfer, cost). The energy balance results indicate that the heat transfer to the environment increased as the temperature and the velocity of the product were higher (Stoforos & Simunovic, 2018).

17.3.1.1 Comparison of Cooling Between Annular and Inner Tube Product Flow for Counter-Current and Co-Current Heat Exchangers

CFD and heat transfer modeling on the cooling of sweet potato puree for the two examined heat exchanger configurations were studied, using a constant uniform product inlet temperature (T_{inlet}), at 120°C, comparing the counter-current and co-current flow heat exchangers. For all the examined cases, the same heat exchanger design with an identical hydraulic diameter (I.D. of the inner tube at 0.038 m) for both the inner tube and the annulus was used. The result of this simulation showed that cooling of sweet potato puree was more efficient with the food flowing in the annulus, with practically no difference between co-current and counter-current flow. Cooling of sweet potato puree for the annular product flow resulted in a T_{bulk} lower by 8–10°C compared to the inner tube product flow (Figure 17.2) (Stoforos & Simunovic, 2018).

17.3.1.2 Comparison of the Fastest-Moving and the Least-Cooled Fluid Particles

Using a T_{inlet} of 140°C, the temperature of the fastest-moving (at the center of the flow cross-sectional area) and the least-cooled fluid particles (hot spot) across the length of the heat exchanger ("hot spot") were compared for the two examined flow configurations, product flowing in (i) the annulus versus (ii) the inner tube. For the annular flow, the "hot spot" was located at the center

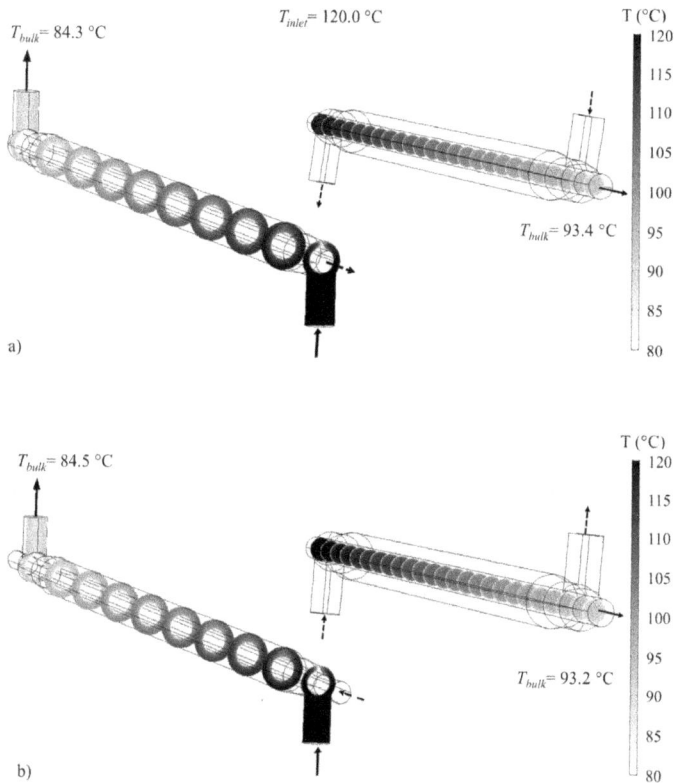

Figure 17.2 Comparison of temperature distribution results for (a) counter-current and (b) co-current flow modes, during cooling of sweet potato puree (T_{inlet} =120°C) flowing within the annulus (left side) and the inner tube (right side). Note the black arrows indicate the flow path for the product, while the blue arrows indicate the flow path for the coolant.

of the inner tube and close to the outer tube's inside wall, while the fastest-moving particle was located in the middle of the annular cross section. For the case of the inner tube flow, the "hot spot" and the fastest-moving fluid particle was the same located at the center of the tube. The simulation results showed that both the "hot spot" and the fastest-moving fluid particles were cooled down faster in the case of annular flow. Cooling of sweet potato puree was more efficient for the annular flow, with a T_{bulk} at 97.5°C compared with 105.3°C for the inner tube product flow (Figure 17.3) (Stoforos & Simunovic, 2018).

Finally, based on the simulation results, it is important to mention that the velocity of the product in the transition area to/from the annulus passage is significantly lower compared to the rest of the field. These "static" zones located in the transition area of the annular flow generate concern regarding the efficiency of cleaning in place procedures (Figure 17.4) (Stoforos & Simunovic, 2018). The example of the issue mentioned above shows the importance of implementing computer simulation modeling in the initial design of new equipment or new parameters in aseptic food processing. Using computer simulation studies helps observe, understand, study, and solve initial ideas and designs prior to experimental testing, validation, or industrial scale-up, improving the final design and decisions, while eliminating any costly mistakes.

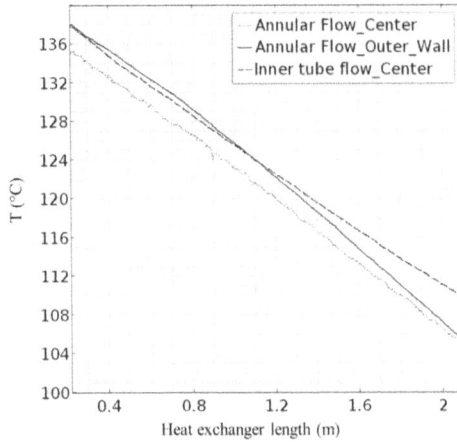

Figure 17.3 Simulation results during cooling of sweet potato puree ($T_{inlet} = 140°C$), comparing the temperature field across the heat exchanger length for the fastest-moving (dotted line for annular flow, dashed line for inner tube flow) and the least-cooled food particle (black line for annular flow). Note the "hot spot" and the fastest-moving fluid particle are the same for the case of the inner tube flow located at the center of the tube.

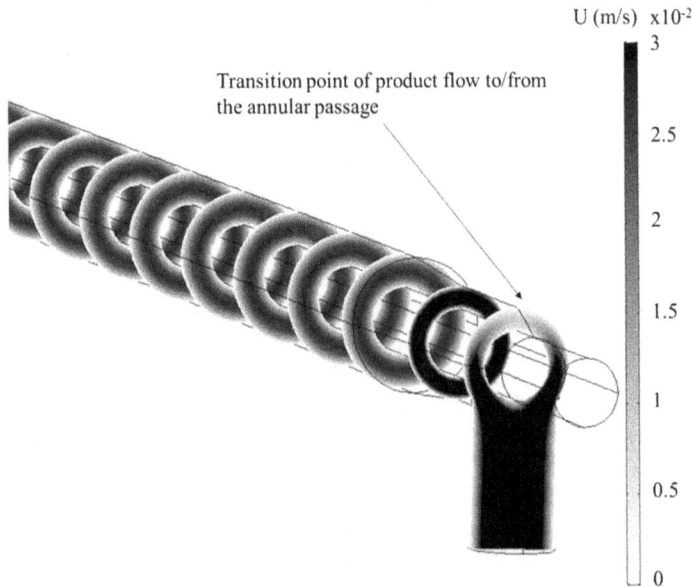

Figure 17.4 Simulation results of velocity distribution during cooling of sweet potato puree flowing in the annulus. Note the "static" zones located at the transition area of product flow to/from the annular passage.

17.3.2 Example 2: Continuous-Flow Microwave Heating—First Generation

Microwave heating is a viable alternative for high-temperature short-time thermal processing as volumetric heating of products due to microwave leads to faster heat transfer rates and shorter processing time than conventional thermal processes. The use of continuous-flow focused microwave systems can provide uniform heating as systems are optimized for suitable electric field distribution to the liquid food products flowing through the microwave cavity (Falqui-Cao et al., 2001). For optimization of the process, the knowledge of the product's three-dimensional temperature profile distribution by numerical or experimental means is essential (Knoerzer et al., 2005). As traditional temperature measurement devices can interfere with electromagnetic field and therefore heating, numerical models can be used to predict temperature profiles and optimize the process of microwave heating. Modeling of continuous-flow focused microwave involves solving Maxwell's equations, Fourier's energy balance equation, and Navier–Stokes equation which describe the process. Instead of solving these equations, simplified approaches, including plug flow and homogeneous heat dissipation (Mudgett, 1986; Le Bail et al., 2000), or exponential decay of the microwave energy (Datta et al., 1992), have been used by researchers. The numerical models involving coupling of high-frequency electromagnetism, heat transfer, and fluid flow are very recent developments in the field and are critical for developing a fundamental understanding of microwave heating in a continuous-flow focused microwave system. Recent studies have coupled electromagnetism with fluid flow and convective heat transfer to get temperature profiles in continuous-flow focused microwave systems for model or real food products (Zhang et al., 2000; Ratanadecho et al., 2002; Zhu et al., 2007a–c, Sabliov et al., 2007; Salvi et al., 2008; 2010, 2011; Cuccurullo et al., 2013; Tuta & Palazoğlu, 2017).

As an example of numerical modeling and validation, a study on heating CMC solution in a 915 MHz, 5 kW continuous-flow microwave by Salvi et al. (2008, 2010) is presented here. A COMSOL Multiphysics model was developed by coupling electromagnetism, fluid flow, and heat transport, including phase change and non-Newtonian flow. The results from model were validated by comparison of outcomes against previously developed ANSYS model data as well as experimental data.

Maxwell's equations were solved to determine electric field distribution in a microwave cavity as follows:

$$\nabla \times \left(\frac{1}{\mu'} \nabla \times \vec{E} \right) - \frac{\omega^2}{c} \left(\varepsilon' - i\varepsilon'' \right) \vec{E} = 0 \qquad (17.5)$$

where $\vec{E}$ is the electric field intensity (V/m), ε' is the relative permittivity or dielectric constant of a material, ε'' is the relative dielectric loss of a material, ω is the angular wave frequency ($2\pi f$, rad/s), μ' is the relative permeability of the material, and c is the speed of light in free space (3×10^8, m/s).

Using the electric field intensity obtained from the above equation (Figure 17.5) and material properties, the volumetric power generation (W/m³) due to microwave exposure were calculated by the following equation:

$$Q = \sigma |\vec{E}|^2 = 2\pi \varepsilon_0 \varepsilon'' f |\vec{E}|^2 \qquad (17.6)$$

where σ is the electrical conductivity of the material (S/m), ε_0 is the free space permittivity (8.854 $\times 10^{-12}$, F/m), and f is the frequency (Hz).

This volumetric power generation term calculated above is used as the energy source term in solving convective heat transfer using Fourier's energy balance equation and Navier–Stokes equations to calculate the temperature distribution. Fluid flow was assumed to be incompressible

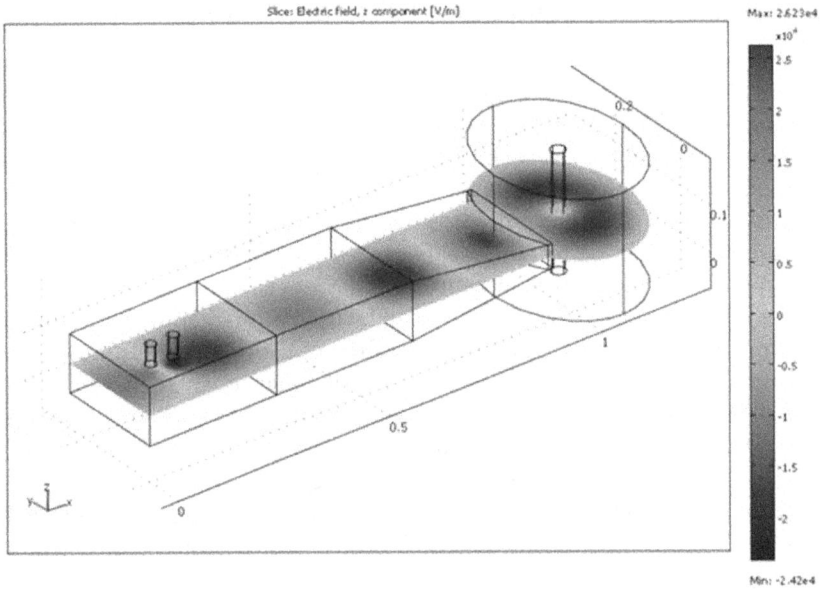

Figure 17.5   Electric field intensity (V/m) distribution inside waveguide and microwave cavity.

and the PTFE tube was considered completely transparent to microwaves. The model was these were coupled iteratively by using temperature-dependent dielectric and physical properties. First Maxwell's equations were solved to calculate heat generation, which was then used to solve Fourier's energy balance equation and Navier–Stokes equation. The temperature profiles obtained were then used to recalculate the new dielectric properties and, thus, a new heat generation term. The process was repeated until it reached convergence.

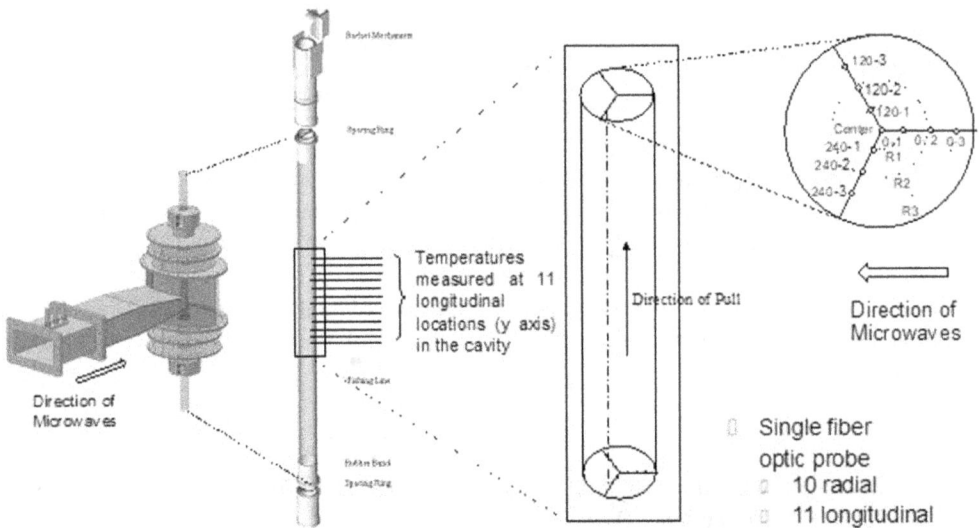

Figure 17.6   Continuous-flow microwave system showing locations of temperature measurement using a fiber-optics probe (Salvi et. al., 2009).

For temperature measurement experiments, a fiber-optic probe system (Salvi et al., 2009) was used to measure temperature at 110 different radial and longitudinal locations in the dielectric undergoing continuous-flow microwave heating, as shown in Figure 17.6. Non-Newtonian fluid, i.e., 0.5% CMC solution, was heated through the continuous-flow microwave system (Industrial Microwave Systems, NC, USA) at 4 kW of power and a flow rate of 1 liter/m. Within the PTFE applicator tube (3.81 cm diameter and 25.4 cm height), temperature was measured at ten radial

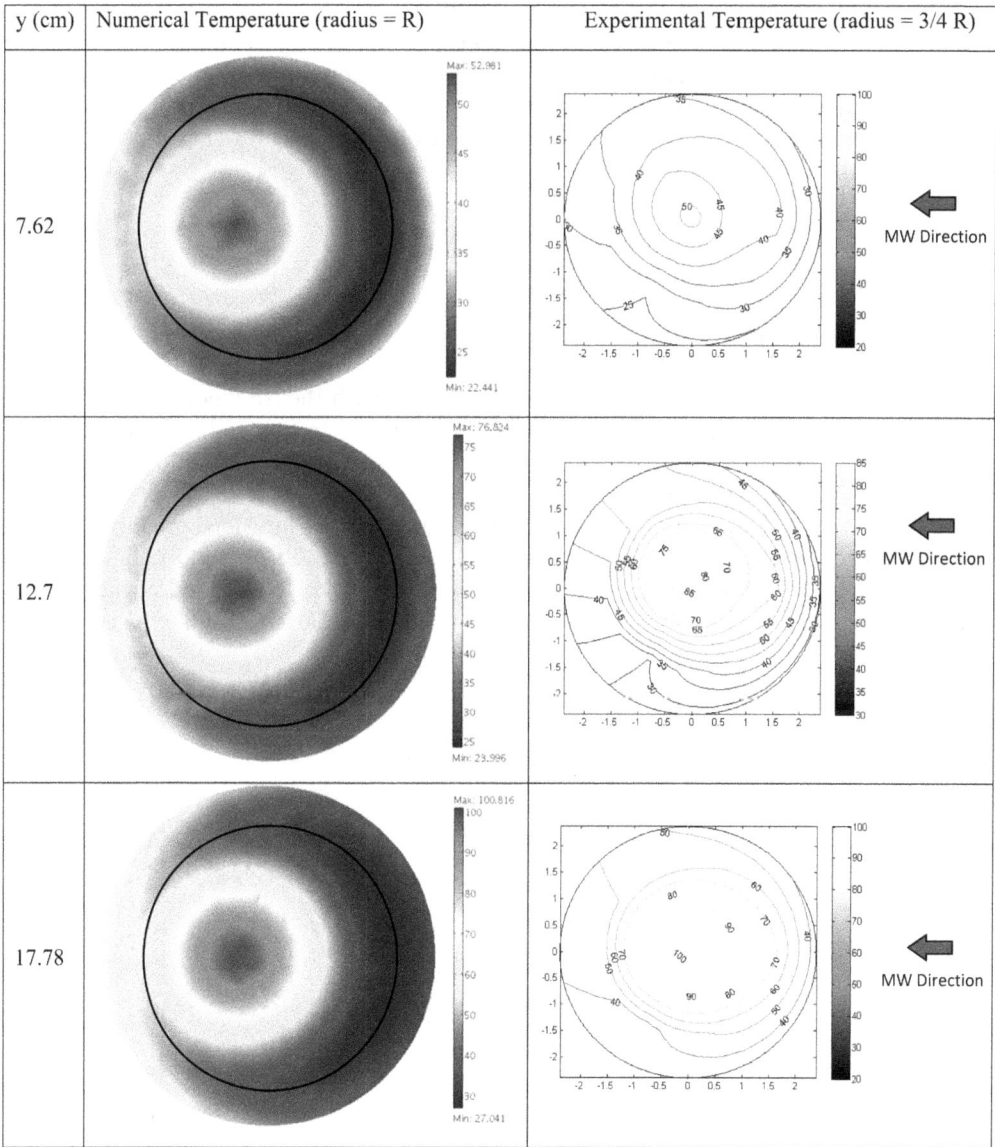

Figure 17.7 Numerical and experimental temperatures in cross-sectional spatial plane at different longitudinal distances (y) for CMC solution at 1 liter/m. *Note:* Black circle inside numerical profile shows ¾ R for easier comparison to experimental data.

points in a cross section of liquid; the tube was then divided into 11 cross sections at different longitudinal distances (Figure 17.6). This provided comprehensive three-dimensional temperature profile distribution in the heated products. While most of the methods used around the time provided a few temperature points within the product, the 3D temperature profile provided an important tool for optimization of microwave heating and for extensive validation of the numerical model.

A comparison between average experimental and average simulated temperature at each cross section at 11 longitudinal locations shown for CMC solution showed a fairly good agreement (R^2 value 0.89). Figure 17.7 shows experimental and simulated cross section spatial temperature profiles at different longitudinal locations ($y = 7.62$ cm, 12.7 cm, and 17.78 cm). Both experiments and simulations showed a high-temperature region near the center and lower temperature ones near the edge of the tube, suggesting good agreement between both. The results from the COMSOL model were also comparable with a previously developed ANSYS model (Sabliov et al, 2007, 2010).

17.3.3 Example 3: Continuous-Flow Microwave Heating—Second Generation

The following is another modeling example of continuous-flow thermal processing where liquid whole egg (LWE) is being heated in a continuous-flow microwave applicator. The finite element model was developed using COMSOL Multiphysics along with the RF module (v4.4, COMSOL Inc., Burlington, MA, USA), and the equations governing electromagnetics (Maxwell's equations), heat transfer (Fourier's equation), and fluid flow (Navier–Stokes equations) were solved simultaneously to obtain the temperature field.

The system was designed for controlled heating of heat-sensitive fluid food products, and consisted of two components: a rectangular waveguide assembly (WR975) and a tubing (0.038 m I.D.), as shown in Figure 17.8. In this system, microwaves (915 MHz, 2 kW) and LWE (20°C) enter from the same end and move in the same direction. Liquid whole egg absorbs the energy as it

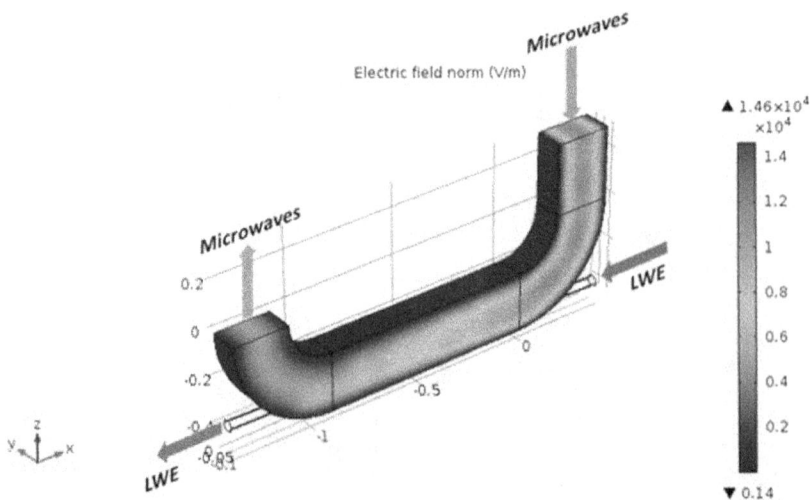

Figure 17.8   Electric field in waveguide.

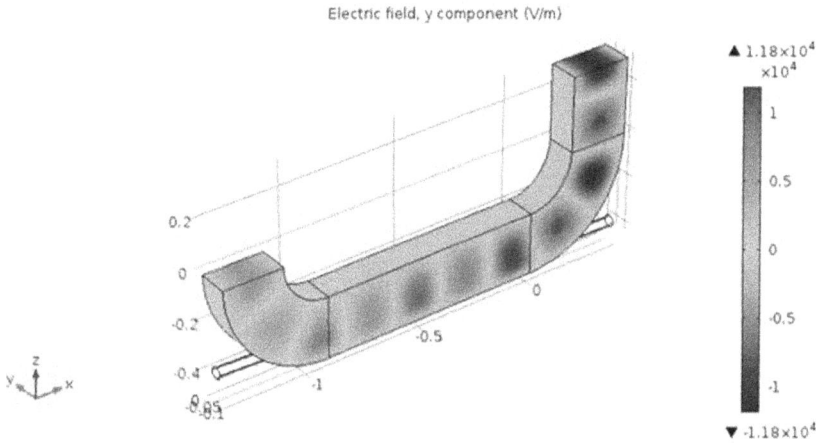

Figure 17.9 *y*-Component of electric field.

moves toward the exit causing attenuation of the microwaves propagating in the same direction, which in turn results in its receiving progressively lower energy as it gets heated. This is clearly shown by the electric field strength plot presented in Figure 17.8.

Attenuation of microwaves toward the waveguide exit can also be seen from the *y*-component of the electric field presented in Figure 17.9. Total absorbed microwave power under steady-state conditions is 1.595 kW (79.8% of the incident power).

The electric field strength in the tubing is depicted in Figure 17.10. The figure shows that the electric field is concentrated in the lower part of the tubing presumably due to that part being closer to the waveguide surface. It can also be seen that the strength of the electric field decreases as LWE moves away from the waveguide entrance where the intensity of microwave energy is the highest.

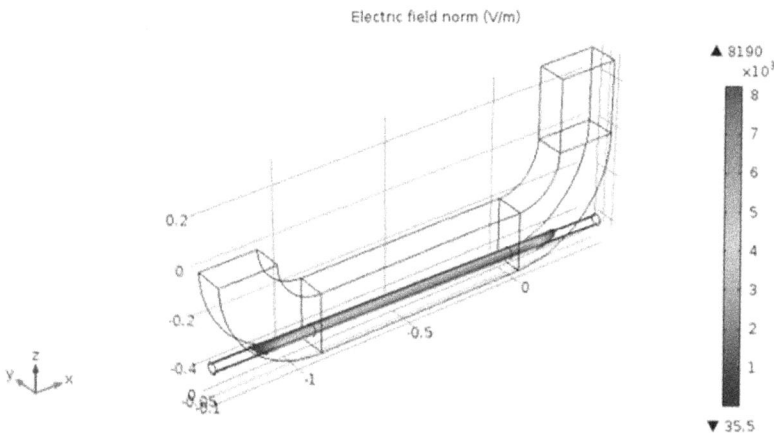

Figure 17.10 Electric field in tubing.

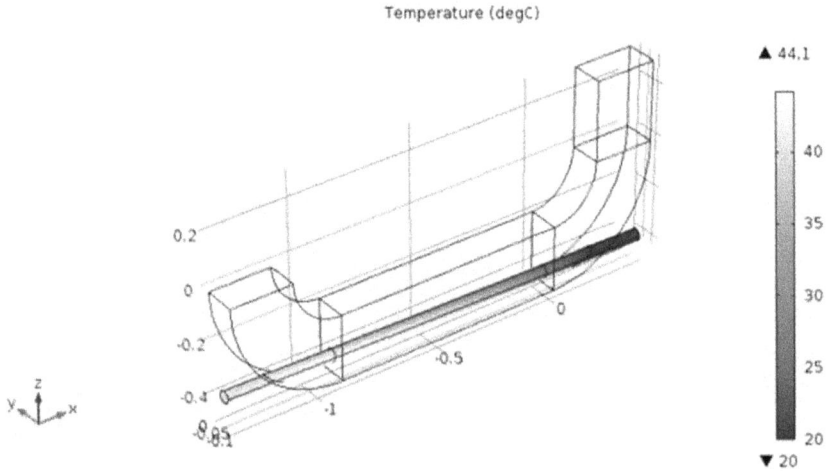

Figure 17.11　Temperature contour plot for LWE.

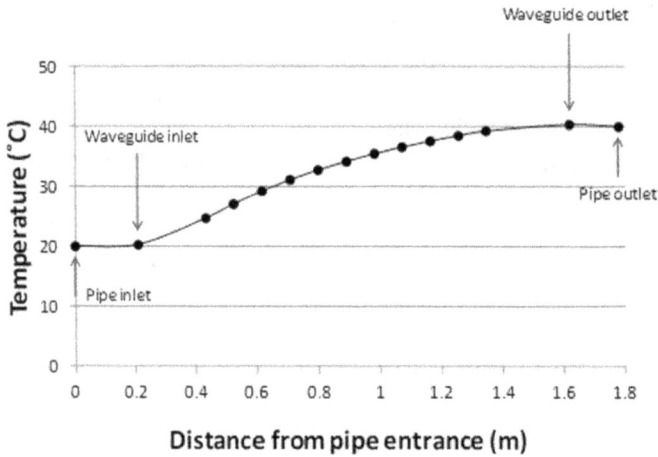

Figure 17.12　Temperature profile of LWE.

Figures 17.11 and 17.12 show the average temperature contour plot and temperature change with distance from the pipe entrance. It is obvious that the target temperature at the outlet can be easily achieved by adjusting the process parameters such as incident microwave power and flow rate of LWE in the developed model.

As can be seen, mathematical modeling provides great insight into the physics of continuous-flow microwave heating, a novel and emerging technology that is considered an alternative to using conventional heat exchangers. Through the use of such models, equipment and process design can be optimized to enhance product quality, reduce nutrient losses, and improve utilization of microwave power.

17.3.4 Example 4: Continuous-Flow Heating of Products Containing Particles

Aseptic processes for food products including particle–liquid mixtures can be optimized and investigated in greater detail using mathematical modeling, thanks to advancements in computational technology. Critical parameters such as lethality value at the cold spot of the fastest-moving particle, temperature distribution within the products, and overall nutritional quality can be predicted to reduce the number of experiments to finalize the process design. However, care must be taken to develop models yielding results comparable to actual experiments. The important points such as estimating thermal properties of foods, fluid-to-particle heat transfer coefficient, residence time distributions of solid particles, and other heat transfer–related problems have been extensively studied in the literature and summarized by Ramaswamy and others (1997).

Due to the complex nature of aseptic processing of particulate–liquid mixture type foods, some simplifications are required to develop a mathematical model which does not take too long time to converge. It is also a common engineering practice to build a model based on the worst-case scenario. Since the fluid-to-particle heat transfer coefficient is one of the most significant variables affecting the modeling results, the worst-case scenario is the one where the fluid-to-particle heat transfer coefficient is the lowest. Sastry and others (1989) concluded that when the particle and liquid velocities are the same within the processing system, the Nusselt number is equal to 2, which gives the lowest fluid-to-particle heat transfer coefficient for a spherical particle. The particle to be considered for the lethality calculation must be the largest one with the least residence time in the system. Although a continuous aseptic processing system includes heating, holding, and cooling sections, only the time in the holding section is counted for the lethality accumulation. While the particles move in the holding tube, the liquid temperature will drop. Therefore, the exit temperature of the holding tube should be used in the mathematical model setup (Ramaswamy et al., 1997).

Considering the aforementioned points, a mathematical model can be developed using a conjugate heat transfer analysis. Such model examples exist in the literature. Palazoğlu and Sandeep (2002) studied the effects of fluid-to-particle heat transfer coefficient on the lethality for *Clostridium botulinum* and the nutritional degradation of thiamin and lysine. The assumption of $N_{Nu} = 2$ for particles with various radii resulted in different fluid-to-particle heat transfer coefficient values. Constant thermal properties for the food particles and uniform initial temperature distribution were assumed. The numerical model was run for either experimentally determined residence times of the particles or the time which results in an F_o value of 3 minutes. The results of the analytical solution for the same heat transfer problem agreed with the numerical model's predictions. The disadvantage of designing an aseptic process based on the worst-case scenario is the overheating of the product with a reduced quality. The numerical model was able to show the expected trend for the nutritional loss of the products. Using a commercially available software (Comsol Multiphysics version 5.3), we calculated the temperature distribution for the particles of different sizes studied by Palazoğlu and Sandeep (2002) (Figure 17.13). The temperature distributions are shown for the end of heating time which resulted in F_o values of 3 minutes for each particle. The Comsol model was based on the same initial and boundary conditions selected by Palazoğlu and Sandeep (2002). A 3D time-dependent model was set up and the following continuity, momentum, heat equations for fluids and solids were solved:

$$\frac{\partial \rho}{\partial t} + \nabla \cdot (\rho u) = 0$$

$$\rho \frac{\partial u}{\partial t} + \rho u \cdot \nabla u = -\nabla p + \nabla \cdot \left(\mu \left(\nabla u + (\nabla u)^T \right) - \frac{2}{3} \mu (\nabla \cdot u) I \right) + F$$

$$\left(\rho C_p \left(\frac{\partial T}{\partial t} + (u \cdot \nabla) T \right) = -(\nabla \cdot q) + \tau : S - \frac{T}{\rho} \frac{\partial \rho}{\partial T} \Big|_p \left(\frac{\partial p}{\partial t} + (u \cdot \nabla) p \right) + Q \right)$$

$$\rho C_p \frac{\partial T}{\partial t} = -(\nabla \cdot q) - T \frac{\partial E}{\partial t} + Q$$

Here,

t is the time (s)
u is the velocity vector (m/s)
p is the pressure (Pa)
F is the body force vector (-)
C_p is specific heat capacity at constant pressure (J/[kg·K])
T is the absolute temperature (K)
q is the conduction heat flux (J/[s·m²])
S is the strain-rate tensor (s¹)
Q is the heat source (W/m³)
E is the entropy of elastic contribution (J/[m³·K])
ρ is the density (kg/m³)
μ is the dynamic viscosity (kg/[m·s])
τ is the viscous stress tensor (Pa)

Aside from the worst-case scenario, Palazoğlu and Sandeep (2002) also investigated the case in which the fluid-to-particle heat transfer coefficient was infinite. To do so, the Biot number was set to 40, and the required values of the fluid-to-particle heat transfer coefficient were calculated. The bottom row in Figure 17.13 shows the temperature distributions at the central plane of each particle when $N_{Bi} = 40$. The differences between the temperature values achieved at the end of heating time for two different scenarios can be observed on the temperature scales given. The time–temperature history at the cold spot of each particle in both scenarios is also calculated and given in Figures 17.14 and 17.15.

One of the major drawbacks of mathematical modeling of continuous-flow aseptic processing of particulate–liquid mixtures is the comparison of the model with experimentally obtained data. Jasrotia and others (2008) developed an experimental method to validate aseptic processing of different solid particles in liquids. A design approach allowed constructing polymeric simulated particles which had the abilities of slow heating and fast moving compared to real food particles. The designed particles were half-inch cubes and composed of a wall material, an inner cavity, and a two-part (a cap and a body) configuration.

The polymeric particles and the real food particles (carrot and potato cubes of the same size) were heated under the same experimental conditions, and F_o values of the particles were lower than those of food particles. In order to further advance the design of such simulated particles mathematical models with the options allowing different material properties assigned to the cap and body, temperature-dependent thermal properties, etc. may be set up.

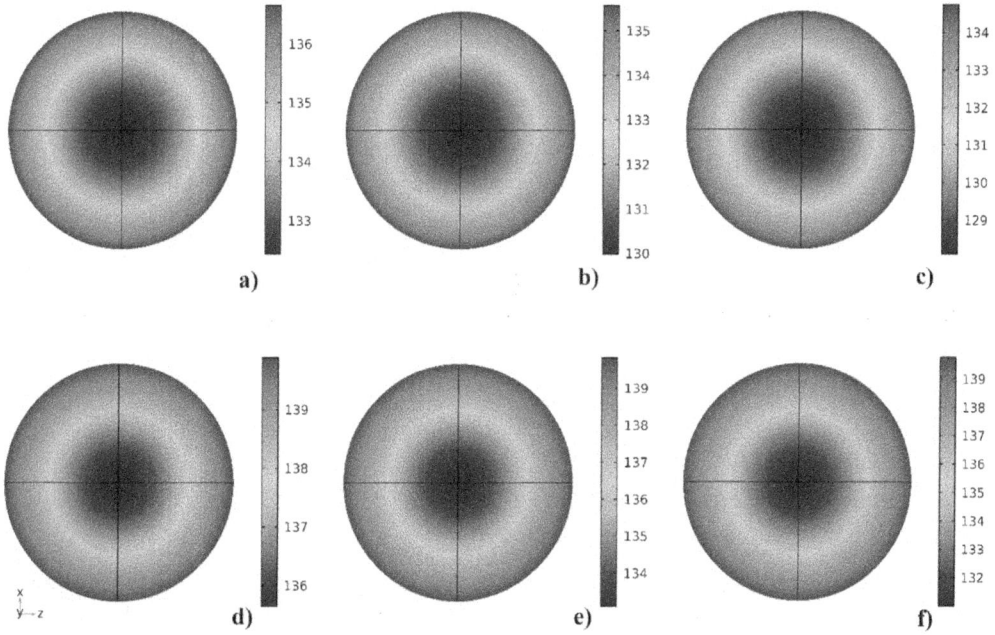

Figure 17.13 Temperature distribution at the central plane of (a) 0.005-m particle radius, (b) 0.0075-m particle radius, and (c) 0.01-m particle radius. The top row is for the worst-case scenario. The bottom row shows the temperature distribution at the central plane of (d) 0.005-m particle radius, (e) 0.0075-m particle radius, and (f) 0.01-m particle radius when $N_{Bi} = 40$.

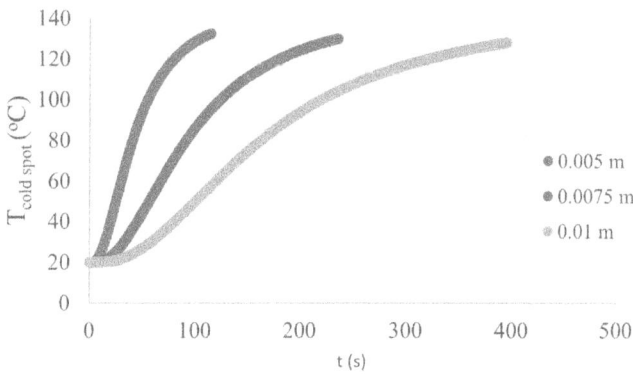

Figure 17.14 Comsol model prediction of time–temperature data for the cold spot of each particle studied by Palazoğlu and Sandeep (2002) based on the worst-case scenario.

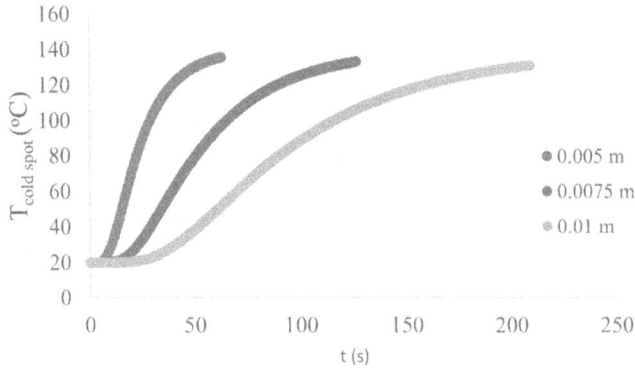

Figure 17.15   Comsol model prediction of time–temperature data for the cold spot of each particle studied by Palazoğlu and Sandeep (2002) when $N_{Bi} = 40$.

17.3.5 Example 5: Using Spreadsheet-Based Models for Evaluating Thermal Processes

Spreadsheets are one of the most used tools since the inception of personal computers. Spreadsheets are useful for making tables, and performing basic operations, and while they were initially developed for accounting, there is a wide space for them in the scientific and engineering use (Singh, 1996). Spreadsheets can perform most mathematical calculations, matrix operations, statistical analysis, and have excellent search within table functions. The ability to expand their functionality with macros and incorporate programs with more complex equations makes this tool a very simple and efficient way to find the first estimates for process establishment and evaluation of processes. This example centers around generating data needed for safety calculations and provides all the elements needed for FDA filings (form 2451g), centering around the critical parameters in the hold tube.

In most aseptic processes, the thermal process can only be credited in the hold tube: at the temperature at the exit of the hold tube, which must be a CCP together with the flow rate. It is necessary to calculate average residence time, process residence time (laminar or turbulent determination), and minimum temperature at the end of the hold tube. SIP is also a CCP and depending on the temperature that can be attained at the coldest return point defined during the design of the process line, the target microbial reduction can be achieved in a shorter time.

Bigelow's general method is used for calculations of sterility (Equation 17.7), where the lethality (F) is calculated as a function of temperature (T) and thermal resistance of the microorganisms, product characteristics, etc. (T_{ref} and z) as defined in Chapter 4, making the equations straightforward and easy to use with a judicious use of logarithms and exponentials. Some of the assumptions are that the hold tube is a long straight tube, with a uniform flow profile, and that temperature at the exit of the hold tube is uniform throughout the whole holding section, thus providing a conservative estimate since there will be temperature losses between the entry and exit and due to the presence of bends, the residence time and flow profile will not be as extreme as the ideal. Residence time (t_{res}) in the hold tube can be calculated by knowing the volumetric flow rate and tube dimensions, which are standardized and is easy to make a searchable table of dimensions which includes nominal D, ID, area, and volume per unit length.

$$F = 10^{\frac{T - T_{ref}}{z}} \frac{t}{D} \tag{17.7}$$

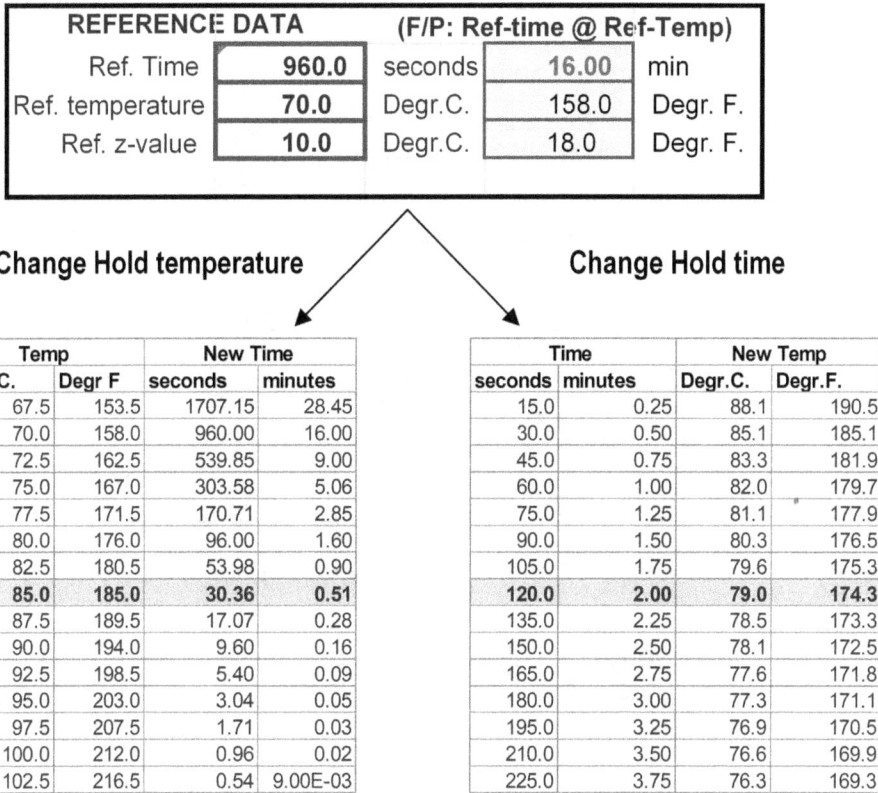

REFERENCE DATA			(F/P: Ref-time @ Ref-Temp)	
Ref. Time	960.0	seconds	16.00	min
Ref. temperature	70.0	Degr.C.	158.0	Degr. F.
Ref. z-value	10.0	Degr.C.	18.0	Degr. F.

Change Hold temperature **Change Hold time**

Temp		New Time		Time		New Temp	
Degr.C.	Degr F	seconds	minutes	seconds	minutes	Degr.C.	Degr.F.
67.5	153.5	1707.15	28.45	15.0	0.25	88.1	190.5
70.0	158.0	960.00	16.00	30.0	0.50	85.1	185.1
72.5	162.5	539.85	9.00	45.0	0.75	83.3	181.9
75.0	167.0	303.58	5.06	60.0	1.00	82.0	179.7
77.5	171.5	170.71	2.85	75.0	1.25	81.1	177.9
80.0	176.0	96.00	1.60	90.0	1.50	80.3	176.5
82.5	180.5	53.98	0.90	105.0	1.75	79.6	175.3
85.0	185.0	30.36	0.51	120.0	2.00	79.0	174.3
87.5	189.5	17.07	0.28	135.0	2.25	78.5	173.3
90.0	194.0	9.60	0.16	150.0	2.50	78.1	172.5
92.5	198.5	5.40	0.09	165.0	2.75	77.6	171.8
95.0	203.0	3.04	0.05	180.0	3.00	77.3	171.1
97.5	207.5	1.71	0.03	195.0	3.25	76.9	170.5
100.0	212.0	0.96	0.02	210.0	3.50	76.6	169.9
102.5	216.5	0.54	9.00E-03	225.0	3.75	76.3	169.3

Figure 17.16 Use of tables to find equivalent processes.

First, we can use the ability to create tables for finding equivalent processes that match the lethality of a reference thermal process (Figure 17.16). The equivalency is found by either changing the temperature and finding a new process time (Equation 17.8) or by changing the process time to find a new temperature.

$$t = 10^{\frac{T_{ref}-T}{z}} DF \tag{17.8}$$

Such table of equivalency has proven useful during R&D and process development to explore different combinations of time–temperature and the effects they might have on product quality and safety.

Residence time and flow rate can be calculated; when process is being developed, it allows for exploring the length and diameter of the hold tube. The diameter of the hold tube is standardized and the search future (VLOOKUP function) on a table of sizes of tubing helps in making the process faster. Once the residence time is defined, there is a need to account for thermal expansion and for the flow regime to provide sufficient data for establishing the process.

Thermal expansion can be calculated using correlations for water, known as steam tables, such as the ones by Popiel and Wojtowiak (1998) or by simply looking at the specific volume at the inlet and maximum temperature. With the assumption that food products will follow the same expansion rates as water, in indirect heat exchangers the expansion is between 0% and 6% depending on the difference between inlet and outlet temperatures. When a direct steam

VOLUMETRIC FLOW CHANGE DURING STEAM INJECTION			
	Injection 1	Injection 2	Injection 3
Inlet temperature [C]	70	75	60
Outlet Temperature [C]	129	129	129
Steam Temperature	**160**	**160**	**165**
Steam Pressure [bar]	5.10	5.10	5.91
RESULTS			
Specific Volume			
Inlet [m3/kg]	0.001023	0.001026	0.001017
Outlet [m3/kg]	0.001065	0.001065	0.001065
Steam Enthalpy [kJ/kg]	2758.1	2758.1	2763.5
cp [kJ kg-1 °C-1]	4.22	4.22	4.21
δm (% increase in mass flow rate)	11.23	10.29	13.08
δv (% increase in volumetric flow rate)	11.7	10.7	13.7

Figure 17.17 Calculation of expansion of the flow rate due to steam injection.

PRODUCT	Pudding	Milk	Tomato	Banana
Flow Rate [gpm]	36	50	36	36
Hold tube Nominal D [in]	1.5	2	1.5	1.5
I.D. [in]	1.37	1.87	1.37	1.37
Hold tube length [in]	**950.00**	**334.00**	**950.00**	**480**
Expansion Coefficient	1	1.2	1	1
Target F-value [min]	8.4	5	5	0.5
Tref [F]	250	250	205	250
z-value [F]	18	18	10	18
RESULTS				
Hold Time -average [s]	10.1	4.0	10.1	5.1
Hold Time -average [min]	0.17	0.07	0.17	0.09
Divert Temperature				
Laminar Time [s]	5.05	1.99	5.05	2.55
Laminar Temp [F]	285.98	289.23	222.74	269.26
Turbulent Time[s]	8.39	3.30	8.39	4.24
Turbulent Temp [F]	282.02	285.26	220.54	265.30
Reynolds - Water	**38,477**	**39,634**	**29,357**	**36,005**
SET POINT	291	294	228	274
Back Pressure [psig]	**67**	**71**	**24**	**52**

Figure 17.18 Spreadsheet used for calculation of the values used in Appendix 5.

injection is used as the heating method (Figure 17.17), expansion becomes critical, steam is added directly to the product adding mass to the flow, and thus increasing the mass flow rate by the added volume of steam, which must then be added to the expansion by temperature. The amount of steam added to the product depends on the steam pressure (temperature) and difference between inlet and required temperature, and this calculation must be performed together with the supplier of the steam injection equipment.

Adding all this together, provides all the information required for filling form 2451g into a table, as shown in Figure 17.18. This tool was used to calculate the examples in Appendix 5.

This tool can be expanded as needed and several other functionality can be added by using programming and macros functions, such as calculating microbial reduction, analysis of deviations, and optimization of existing processes. The limits of the processes must, however, be set by a competent process authority in order to ensure the safety of the products.

17.4 CONCLUSION

Computational and numerical models and simulations have been a necessary tool for research, development, process, and systems design and implementation for aseptic processing for several decades now. The more recent advances in development of sophisticated Multiphysics software packages like Comsol and Ansys, further catalyzed by the rapid advances in hardware and processing power of computer processors, memory, video, and communication signals processors, have enabled the capability to handle ever more advanced and complex models and simulations. These new and emerging capabilities have proven to be invaluable in discovery, development, and implementation of a broader range of technologies, leading to improved processes, more varied, higher quality, and improved nutrition products previously difficult or impossible to achieve using conventional aseptic processing technologies. The examples presented in this chapter represent a window into a small segment of these new capabilities and their potential to bring these positive changes to commercial reality in aseptics. Currently pending and future developments will further build upon this potential to enable the implementation of real-time optimization and decision-making in aseptic processing and packaging, where similar models and simulations will be performed as building blocks of artificial intelligence applications. These AI applications will contribute to additional improvements of quality, safety, variety, and nutritional benefits of shelf-stable aseptically processed product lines.

LIST OF SYMBOLS

$\vec{v}$:   fluid velocity vector (m/s)
ρ: fluid's density (kg/m^3)
∇:   the del operator
p: pressure forces (Pa)
σ: stress component (Pa)
F: external forces applied to the fluid
ρ_{ref}:   reference density (kg/m^3)
T_{ref}:   reference temperature (K)
c_p: specific heat (J/[kg·K])
T: temperature component (K)
Q: external heat sources (W/ m^2)
$\bar{E}$:   electric field intensity (V/m)

ε':   relative permittivity or dielectric constant of a material

ε: relative dielectric loss of a material

ω: angular wave frequency ($2\pi f$, rad/s)

μ':   relative permeability of the material

c: speed of light in free space (3×10^8, m/s)

Q: volumetric power generation (W/m^3) due to microwave exposure

σ: electrical conductivity of the material (S/m)

ε_0:   free space permittivity (8.854×10^{12}, F/m)

f: frequency (Hz)

t: time (s)

t_{res}:   residence time on the hold tube (s)

u: velocity vector (m/s)

p: pressure (Pa)

F: body force vector (-)

C_p: specific heat capacity at constant pressure (J/[kg·K])

T: absolute temperature (K)

q: conduction heat flux (J/[s·m^2])

S: strain-rate tensor (s^1)

Q: heat source (W/m^3)

E: entropy of elastic contribution (J/[m^3·K])

ρ: density (kg/m^3)

μ: dynamic viscosity (kg/m·s])

τ: viscous stress tensor (Pa)

D: decimal reduction time (minute)

z: z-value (C)

$F_{Tref/z}$:   Equivalent time at a reference temperature T_{ref} and a known z-value

δ_v: Change of volumetric flow rate

δ_v: Change of mass flow rate

REFERENCES

Ansys. 2021a. 8.4.5 Viscosity for non-Newtonian fluids. https://www.afs.enea.it/project/neptunius/docs/fluent/html/ug/node297.htm#sec-viscosity-non-newtonian. Accessed on April 16, 2021.

Ansys. 2021b. ANSYS fluent tutorial guide. http://users.abo.fi/rzevenho/ansys%20fluent%2018%20tutorial%20guide.pdf. Accessed on April 16, 2021.

Comsol. 2021a. Simulate fluid flow applications with the CFD module. https://www.comsol.com/cfd-module. Accessed on April 16, 2021.

Comsol. 2021b. Turbulence modeling. https://www.comsol.com/blogs/which-turbulence-model-should-choose-cfd-application/. Accessed on April 16, 2021.

Cuccurullo, G., Giordano, L., and Viccione, G. 2013. An analytical approximation for continuous flow microwave heating of liquids. *Adv. Mech. Eng.* 5:929236.

Datta, A., Prosetya, H., and Hu, W. 1992. Mathematical modeling of batch heating of liquids in a microwave cavity. *J. Microwave Power Electromagn. Energy* 27(1):38–48.

Falqui-Cao, C., Wang, Z., Urruty, L., Pommier, J., and Montury, M. 2001. Focused microwave assistance for extracting some pesticide residues from strawberries into water before their determination by SPME/HPLC/DAD. *J. Agric. Food Chem.* 49:5092–5097.

Fourcade, E., Wadley, R., Hoefsloot, H. C. J., Green, A., and Iedema, P. D. 2001. CFD calculation of laminar striation thinning in static mixer reactors. *Chem. Eng. Sci.* 56:6729e6741.

Grijspeerdt, K., Hazarika, B., and Vucinic, D. 2003. Application of computational fluid dynamics to model the hydrodynamics of plate heat exchangers for milk processing. *J. Food Eng.* 57:237–242.

Jasrotia, A.K.S., Simunovic, J., Sandeep, K.P., Palazoglu, T.K., and Swartzel, K.R. 2008. Design of conservative simulated particles for validation of a multiphase aseptic process. *J. Food Sci.* 73:E193–E201.

Knoerzer, K., Regier, M., and Schubert, H. 2005. Measuring temperature distributions during microwave processing. In *The Microwave Processing of Foods*, edited by Schubert, H., and Regier, M., 243–263. Boca Raton, FL: CRC Press.

Kumar, A. 1995. Numerical investigation of secondary flows in helical heat exchangers. In *Institute of Food Technologists Annual Meeting, 148*, Anaheim, CA.

Le Bail, A., Koutchma, T., and Ramaswamy, H.S. 2000. Modeling of temperature profiles under continuous tube-flow microwave and steam heating conditions. *J. Food Process Eng.* 23:1–24.

Meher, J.M., Keshav, A., and Dixit, N. 2017. CFD study of non-Newtonian fluid food particle used in sterilization processing. *Int. J. Eng. Technol. Manage. Appl. Sci.* 5(4):46–54, ISSN 2349-4476.

Mudgett, R.E. 1986. Microwave properties and heating characteristics of foods. *Food Technol.* 40(6):84–93.

Palazoglu, T.K., and Sandeep, K.P. 2002. Assessment of the effect of fluid-to-particle heat transfer coefficient on microbial and nutrient destruction during aseptic processing of particulate foods. *J. Food Sci.* 67(9):3359–3364.

Park, H.W., and Yoon, W.B. 2018. Computational fluid dynamics (CFD) modelling and application for sterilization of foods: A review. *Processes* 6(6):62. https://www.mdpi.com/2227-9717/6/6/62

Popiel, C.O., and Wojtkowiak, J. 1998. Simple formulas for thermophysical properties of liquid water for heat transfer calculations (from 0 C to 150 C). *Heat Transfer Eng.* 19(3):87–101.

Ramaswamy, H. S., Awuah, G. B., and Simpson, B. K. 1997. Heat transfer and lethality considerations in aseptic processing of liquid/particle mixtures: A review. *Crit. Rev. Food Sci. Nutr.* 37(3):253–286.

Ratanadecho, P., Auki, K., and Akahori, M. 2002. A numerical and experimental investigation of the modeling of microwave heating for liquid layers using a rectangular wave guide (effects of natural convection and dielectric properties). *Appl. Math. Model.* 26:449–472.

Sabliov, C.M., Salvi, D.A., and Boldor, D. 2007. High frequency electromagnetism, heat transfer, and fluid flow coupling in ANSYS multiphysics. *J. Microwave Power Electromagn. Energy* 41(4):4–16.

Salvi, D.A., Boldor, D., Sabliov, C.M., and Rusch, K.A. 2008. Numerical and experimental analysis of continuous microwave heating of ballast water as preventive treatment for introduction of invasive species. *J. Marine Environ. Eng.* 9(1):45–64.

Salvi, D.A., Boldor, D., Sabliov, C.M., Ortego, J., and Arauz, C. 2009. Experimental temperature measurement of liquids during continuous flow microwave heating to study effect of different dielectric and physical properties on temperature distribution. *J. Food Eng.* 93(2):149–157.

Salvi, D.A., Boldor, D., Ortego, J., Aita, G.M., and Sabliov, C.M. 2010. Numerical modeling of continuous flow microwave heating: A critical comparison of COMSOL and ANSYS. *J. Microwave Power Electromagn. Energy* 44(4): 187–197.

Salvi, D.A., Boldor, D., Aita, G.M., and Sabliov, C.M. 2011. COMSOL multiphysics model for continuous flow microwave heating of liquids. *J. Food Eng.* 104:422–429.

Sastry, S.K., Heskitt, B.F., and Blaisdell, J.L. 1989. Experimental and modeling studies on convective heat transfer at the particle-liquid interface in aseptic processing systems. *Food Technol.* 43(3):132–136.

Scott, G., and Richardson, P. 1997. The application of computational fluid dynamics in the food Industry. *Trends Food Sci. Technol.* 81:119–124.

Simscale. 2021. https://www.simscale.com/blog/2017/12/turbulence-cfd-analysis/. Accessed on April 16, 2021.

Singh, R.P. 1996. *Computer Applications in Food Technology: Use of Spreadsheets in Graphical, Statistical, and Process Analysis.* Amsterdam: Elsevier.

Stoforos, G.N., and Simunovic, J. 2018. Computer-aided design and experimental testing of continuous flow cooling of viscous foods. Submitted for publication. *J. Food Process Eng.* 41(8): e12913. https://doi.org/10.1111/jfpe.12913.

Tomas, N., and Sun, D.W. 2006. Computational fluid dynamics (CFD): An effective and efficient design and analysis tool for the food industry: A review. *Trends Food Sci. Technol.* 17(11, November 2006):600–620. https://doi.org/10.1016/j.tifs.2006.05.004

Tuta, S., and Palazoğlu, T.K. 2017. Finite element modeling of continuous-flow microwave heating of fluid foods and experimental validation. *J. Food Eng.* 192:79–92.

Yanniotis, S., and Stoforos, N.G. 2014. Modeling food processing operations with computational fluid dynamics: A review. *Sci. Agric. Bohemica* 45:1–10. https://doi.org/10.7160/sab.2014.450101

Zhang, Q., Jackson, T.H., and Ungan, A. 2000. Numerical modeling of microwave induced natural convection. *Int. J. Heat Mass Transfer* 43:2141–2154.

Zhu J., Kuznetsov, A.V., and Sandeep, K.P. 2007a. Mathematical modeling of continuous flow microwave heating of liquids (effects of dielectric properties and design parameters). *Int. J. Therm. Sci.* 46(4):328–341.

Zhu J., Kuznetsov, A.V., and Sandeep, K.P. 2007b. Numerical modeling of a moving particle in a continuous flow Subjected to microwave heating. *Numer. Heat Transfer, Part A* 52:417–439.

Zhu J., Kuznetsov, A.V., and Sandeep, K.P. 2007c. Numerical simulation of forced convection in a duct subjected to microwave heating. *Heat Mass Transfer* 43(3):255–264.

Chapter 18

Frontiers and Research and Development

Challenges and Opportunities

Jairus R.D. David, Pablo M. Coronel, and Josip Simunovic

CONTENTS

DOI: 10.1201/9781003158653-22

18.1 INTRODUCTION

Aseptic processing is neither new nor old, but is a commercially successful and robust technology. It is approximately 70 years old and began with the invention of the first aseptic line of the hot–cool–fill (HCF) system by Dr. C. Olin Ball (1936). In a true sense, aseptic technology became a commercial reality with installation of Dr. W.M. Martin's Dole aseptic canner in 1951 (Appendices 1 and 2). In 1981, approval from the US Food and Drug Administration (FDA) of the food additive petition for use of hydrogen peroxide as a sterilant for food-contact surfaces provided the impetus for introduction of various aseptic filling and packaging systems into the US market (Figure 18.1). It is interesting to note, using 1981 as an origin, that aseptic processing technology is only 70 years old in the United States. Basic and applied research in the areas of sterilization microbiology, food engineering design, thermal processing technology, packaging technology, and chemistry were the original driving forces for the successful commercialization of aseptic processing and packaging technology in the United States and Europe. Also, the discipline and manufacturing mechanics for aseptic processing of foods is based upon theory, practice, and regulations originating from conventional canning, continuous pasteurization, clean room technology, and the biopharmaceutical industry.

It is axiomatic to consider aseptic processing and packaging as the benchmark in optimization, for manufacture of sterile, shelf-stable low-acid food. Contrasted with retorting or in-container sterilization, in aseptic, the product and package are independently sterilized by optimal processes wherein microbial inactivation and quality factor degradation are also co-optimized. Aseptic systems permit sterilizing the product and container separately and appropriately, without the rate-limiting heat transfer modes, or the attendant thermal and pressure stress to the container closure and seal integrity, and enabling high-temperature short-time (HTST) or ultra-high temperature (UHT) processing of heat-labile products, without excessive quality factor degradation, while achieving the requisite commercial sterility (see Figure 7.1). The pre-1981 market drivers were better quality low-acid heat-sensitive products, compared to retorted versions. The

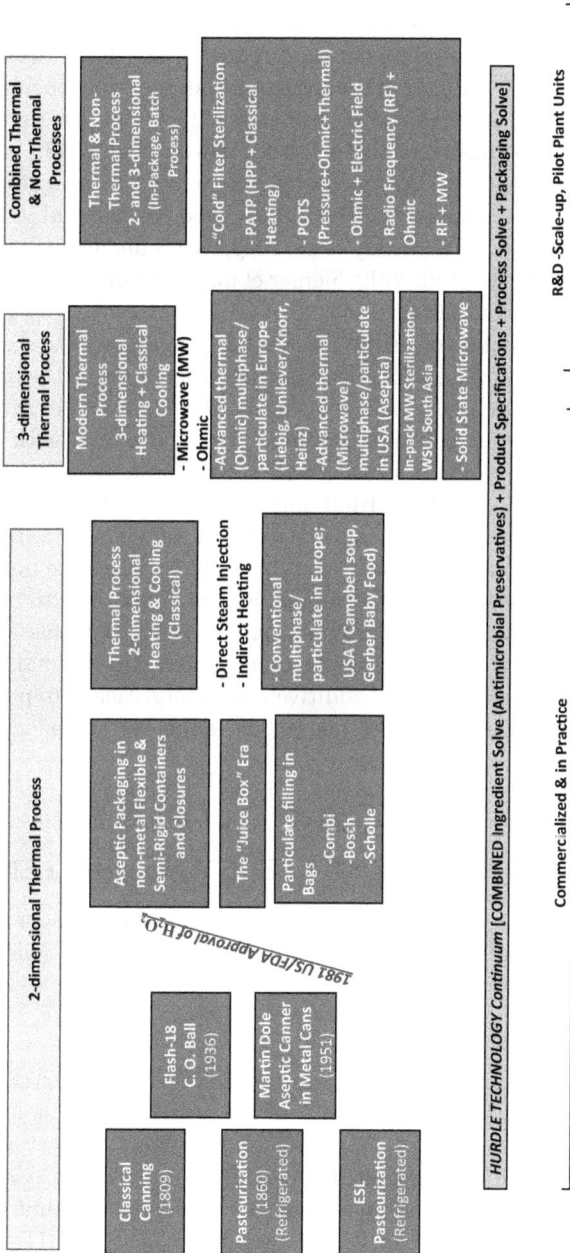

2-dimensional Thermal Process

Classical Canning (1809)

Pasteurization (1860) (Refrigerated)

ESL Pasteurization (Refrigerated)

Flash-18 C. O. Ball (1936)

Martin Dole Aseptic Canner in Metal Cans (1951)

1981 US/FDA Approval of H₂O₂

Aseptic Packaging in non-metal Flexible & Semi-Rigid Containers and Closures

The "Juice Box" Era

Particulate filling in Bags
-Combi
-Bosch
-Scholle

Thermal Process 2-dimensional Heating & Cooling (Classical)

- Direct Steam Injection
- Indirect Heating

- Conventional multiphase/particulate in Europe;
USA (Campbell soup, Gerber Baby Food)

3-dimensional Thermal Process

Modern Thermal Process 3-dimensional Heating + Classical Cooling

- **Microwave (MW)**
- **Ohmic**

-Advanced thermal (Ohmic) multiphase/ particulate in Europe (Liebig, Unilever/Knorr, Heinz)

-Advanced thermal (Microwave) multiphase/particulate in USA (Aseptia)

In-pack MW Sterilization- WSU, South Asia

- Solid State Microwave

Combined Thermal & Non-Thermal Processes

Thermal & Non-Thermal Process 2- and 3-dimensional (In-Package, Batch Process)

- "Cold" Filter Sterilization
- PATP (HPP + Classical Heating)
- POTS (Pressure+Ohmic+Thermal)
- Ohmic + Electric Field
- Radio Frequency (RF) + Ohmic
- RF + MW

HURDLE TECHNOLOGY Continuum [COMBINED Ingredient Solve (Antimicrobial Preservatives) + Product Specifications + Process Solve + Packaging Solve]

Commercialized & in Practice

R&D -Scale-up, Pilot Plant Units

Figure 18.1   Optimization continuum for aseptically processed products, filled and packaged in hermetically sealed containers for non-refrigerated ambient distribution. Shown here are classical two-dimensional and three-dimensional thermal processing technologies, non-thermal processing technologies, combination technologies, and hurdle technology.

post-1981 basic drivers are the use of alternate packages, consumer convenience, lightweighting, cost reduction, superior quality, and package sterilization by optimal methods.

Technology has been diversified in recent years and concepts of aseptic processing and packaging that are somewhat removed from the core of the original technologies are being adapted to liquid and solid foods—ambient shelf stable or refrigerated.

The future of aseptic processing will depend upon reducing the departures from process optima, as discussed in Chapters 1 and 7, and capturing findings from ongoing research in both basic and applied research leading to use of both classical and novel thermal sterilization methods coupled with innovative non-thermal processes. Shown in Figure 18.1 is a technology/ innovation timeline depicting aseptic processing technology relative to emerging technologies (Beckerich, 2012; Cho et al., 1999; Coronel et al., 2003, 2005, 2008; Kumar et al., 2007; Loghavi et al., 2009; Municino, 2018; Park et al., 2013; Reddy et al., 2003; Sastry and Cornelius, 2002; Sastry, 2014; Sandeep et al., 2004; Shynkaryk et al., 2010; Siemer et al., 2014; Somavat et al., 2012; Sun et al., 2011; Tahir et al., 2009; Zhang et al. 2011).

It is important to underscore that no one single technology can replace the shelf-stable capabilities of either classical retorting or aseptic processing. However, many of the innovative thermal and non-thermal processing technologies can be used either additively or synergistically to build "hurdles" in tandem with an objective to produce superior products with minimal heat-induced damages. The future of aseptic processing will depend upon reducing the departures from process optima, and capturing findings from ongoing research in both basic and applied research leading to the use of both novel thermal sterilization methods (two-dimensional and three-dimensional such as ohmic and microwave) coupled with innovative non-thermal processes, sensors, validation methods, and nanotechnologies. It is important to underscore that no one single technology can replace the shelf-stable capabilities of either classical retorting or aseptic processing. However, many of the innovative thermal and non-thermal processes, sensors, and nanotechnologies can be used either additively or synergistically to produce superior products with minimal heat-induced damage and at an affordable price (see Figure 18.1, David, 2016, 2020; Legan & David, 2021).

18.2 RESEARCH AND DEVELOPMENT NEEDS AND CHALLENGES

For ease, research and development needs and challenges may be grouped under four categories: raw product, processing, aseptic filling and packaging, and finished product (David, 2013a, b).

18.2.1 Raw Product

Because the roots of aseptic processing and packaging are in the dairy industry, this chapter is written with milk and dairy products as examples, but the principles can be applied easily to other food products.

Consistent quality of raw ingredients and established mix procedures are essential for reproducible finished products. Globally sourced ingredients need to have the same level of quality and microbiological control as locally sourced. It is an established fact that UHT-processed products have less thermal degradation compared to that of conventionally canned products. But this can very easily be masked by our inability to control the quality and uniformity of incoming raw ingredients. Also, affordable quality control screening tests for checking microbiological, chemical, and physical quality are needed.

18.2.1.1 Raw Food Quality

It is a well-established fact that raw milk quality determines the sensory and quality characteristics of finished milk. Raw milk quality is determined by herd/udder hygiene, barn sanitation, feeding method, milking method, collecting tank storage temperature and time, refrigerated transportation, and elapsed time between barn to plant. According to the Pasteurized Milk Ordinance (PMO), 40°F is the recommended storage temperature. Refrigeration is an effective hurdle for control of heat-resistant toxin secretion by coagulase-positive *Staphylococcus aureus*. However, psychrotrophs can survive refrigeration temperature, and grow and secrete heat-resistant enzymes that can survive the UHT sterilization process, leading to long-term flavor problems due to proteolysis and lipolysis. Refrigerated storage is a short-term insurance and is not a panacea to cover a multitude of handling problems.

18.2.1.2 Thermization

Often, fluid milk and probably other dairy products such as cream are heat treated, not for elimination of pathogens, but to reduce the number of psychrotrophic bacteria capable of secreting heat-resistant proteases and lipases. Thermization is a less severe heat treatment process than pasteurization or ultra-pasteurization; heat is applied for about 15 seconds at temperatures in the range of 65–70°C (149–158°F), followed by cooling to below 6°C (43°F). Chilled milk is distributed to manufacturing plants for further formulation and heat processing and packaging.

18.2.1.3 Enzyme Blockers and Biotechnology

Plasmin is the major native protease in milk, which causes gelation and flavor problems. Plasminogen is the precursor for plasmin and is present in a larger amount and is heat resistant. Unfortunately, the heat treatment required for inactivation of heat-stable plasminogen often exceeds that required for control of microorganisms and their spores, and inactivation of the plasmin system.

Research by Kohlman et al. (1991) points to the presence of heat-resistant protease activators (PAs) and heat-sensitive protease inhibitors (PIs) controlling the kinetics of expression of plasmin from plasminogen in plasmin-free UHT sterilized milk. Unfortunately, PIs are susceptible to sterilization processes and the reaction is overwhelmingly driven by PAs. Research on enzyme blockers as proposed by S.S. Nielsen of Purdue University (personal communication, 1994) consists of removing PIs prior to UHT processing or bulk producing them as a biotechnology product, filter sterilizing, and aseptically adding them back to milk after UHT treatment. PIs can improve product quality by suppressing the expression of plasmin from plasminogen, necessary for gelatin and off-flavor.

Other heat-stable enzymes in foods that affect product quality are pectin methyl esterase in citrus juices causing cloud loss, and amylase in aseptically processed pudding causing thinning.

18.2.1.4 Economic Spoilage and Control

Economic spoilage due to thermophilic spores in certain sensitive ingredients was presented in Chapters 1, 13, and 16. Ingredients of concern are cocoa, non-fat dry milk (NFDM), carboxymethylcellulose (CMC), starches, sugar, corn, mushrooms, and spices. It is a good practice to monitor, measure, and track and control the mesophilic and thermophilic spore loads of each batch of raw product based on ingredients in a formulation. In addition, these ingredients must be completely hydrated prior to batching to ensure that any spores present are fully exposed to a designed and delivered thermal process via a direct or indirect method of heating. Thermophilic spores are of concern in known-to-world and new-to-world ingredients that are sourced both globally and locally, as species that were not considered in the initial thermal process design might be present in such ingredients. Since these new species might lead to potential spoilage, reevaluation of the thermal process and methods to detect such spores should be incorporated in the supply chain verification before import.

Some pretreatment procedures such as preheating, irradiation, tyndallization, proper hydration of cocoa, and H_2O_2 and nisin treatments are common. Research is needed in the area of validation and approval of heat-stable and encapsulated natural antimicrobials or biopreservatives for efficacy at UHT and HTST sterilization regimes.

18.2.2 Processing

Agitated retorts, "Flash 18" process, and aseptic processing represent technologies that optimize thermal sterilization and product quality by reducing heat-induced damages prevalent in still or static retorts. However, in aseptic processing, continuous process delivery is complex and needs to be carefully controlled and automated.

18.2.2.1 12D "Bot Cook" for Milk

Milk is one of the most fragile food systems known. A food industry goal has been to reduce the heat-induced insults to quality degradation during pre- and post-sterilization phases of unit operations. Currently, shelf-stable UHT milk is regulated by the PMO and the US Department of Agriculture (USDA) and is treated as a low-acid canned food (LACF) by the FDA, with a required F_0 of 3 minutes. In reality, the $F_{process}$ that is actually delivered without any lethality credit is on the order of 6–7 minutes in direct systems and 12–15 minutes in indirect systems. This is an incredible assault on a very fragile food system leading to quality defects, including unacceptable "cooked flavor." This is one reason that UHT milk has had difficulty competing with refrigerated pasteurized milks in the dairy case in the United States.

Many have questioned the relevance of 12D reduction or delivery of F_0 of 5 minutes for shelf-stable milks. Europeans produce shelf-stable UHT milk using an F_0 of about 1.22–3.87 minutes, which is equivalent to a required heat process for ultra-pasteurized milks, retailed chilled, in the United States, as shown in Table 18.1. Research is needed in the area of thermobacteriology pertaining to heat resistance of *Clostridium botulinum* in milk menstruum and other foods containing plant-based novel ingredients at HTST and UHT regimes for factual design of safe thermal processes, rather than empirical and ultra-conservative design (CCFRA, 2008; Casolari, 1994).

18.2.2.2 Lethality Credit for Come-Up Time

In classical canning, it is common but sometimes erroneous to claim lethality credit for come-up time (CUT) and cool-down time because transient or unsteady-state heat transfer is lengthy. In UHT processing, sterilization is done at very high temperatures with short CUT and exposure times. However, very short transient or CUT does carry a significant amount of lethality relative to brief exposure periods on the order of a few seconds. In the United States, traditionally one only accrues lethality due to isothermal temperatures and residence times in the hold tube, and cannot claim any credit for either CUT or cool-down time. Deleting thermal lethal contributions in heaters and coolers, while consistent with safety, is a departure from process optimum. More recently, regulatory authorities in the United States have become open to the lethality credit claim on a case-by-case basis.

TABLE 18.1   ULTRA-PASTEURIZATION HEAT PROCESS, APPROXIMATIONS OF HEAT PROCESSES FOR DESTRUCTION OF *CLOSTRIDIUM BOTULINUM*, AND COMMERCIAL STERILITY AS COMPARED TO THE EUROPEAN RANGE OF UHT PROCESSES

Process	Storage	Heat/Hold	Sterilizing Value ($z = 18\,F°$)
Ultra-pasteurization	Refrigerated	280°F, 2 seconds	F_o 1.5 minutes
Minimum to destroy			
Clostridium botulinum spores	Nonrefrigerated	280°F, 4 seconds	F_o 3.0 minutes[a]
Commercial sterility	Nonrefrigerated	280°F, 8 seconds	F_o 6.0 minutes
European UHT range	Nonrefrigerated	275°F, 3 seconds	F_o 1.22 minutes
		284°F, 3 seconds	F_o 3.87 minutes

Source: McGarrahan, E.T., 1982, Considerations Necessary to Provide for Sterilized Milk and Milk Products in Hermetically Sealed Non-Refrigerated Containers, *Journal Dairy Science* 65:2023–2034.

[a] Minimum for public health safety varies based on product and process.

18.2.2.3 Control of Hold Time and Temperature

Precise control of flow and temperature is essential for proper design and delivery of an approved thermal process, followed by prompt cooling. Inability to control temperatures within 5°F (2.5°C) in the UHT regime can lead to irreversible damage, because this may correspond to a doubling or tripling of process lethality (F_o) (see Figure 18.2). Figure 18.2 depicts the effect of a

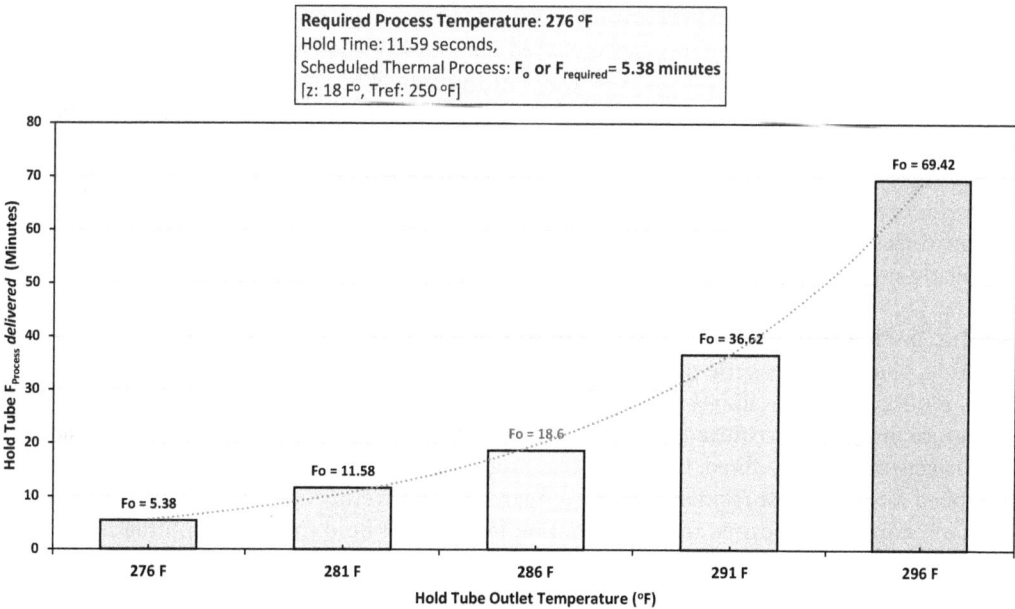

Required Process Temperature: 276 °F
Hold Time: 11.59 seconds,
Scheduled Thermal Process: F_o or $F_{required}$= 5.38 minutes
[z: 18 F°, Tref: 250 °F]

Figure 18.2   Damaging effect of process outlet temperature in hold tube on $F_{Process}$ delivered.

step temperature increase of 5°F on F_0 or lethality delivered. This illustration is based upon a scheduled process for milk of F_0 of 5.38 minutes and hold time of 11.59 seconds at 276°F. In reality, one can notice very significant differences between flavors of milk sterilized for 4 and 4.7 seconds at 280°F. The small difference of 0.7 seconds can cause a big flavor difference in UHT regimes. Food industry should strive for defining the safety and quality and bio-functionality limits in a scheduled process to prevent departure from process and product quality optima.

It is common in the pharmaceutical industry to specify both an F_0 minimum and an F_0 maximum in a scheduled process. The food industry should strive for defining the safety and quality and bio-functionality limits in a scheduled thermal process to prevent departure from process optimum.

18.2.2.4 Heat Exchangers and Product Quality

Certain types or combinations of heat exchangers are best suited for processing of specific products. For instant heating and cooling without excessive transient heat for low-viscosity products such as milk, it is more appropriate to use direct steam heating. It is important to note that during cooling, there may be flashing due to the presence of vacuum, and caution needs to be exercised in order not to allow in any microbial contaminants, or lose volatiles, too. Most attempts to heat a viscous product like cheese sauce yields a wonderful product but risks fouling the vacuum chamber, which may be impervious to clean-in-place (CIP) and routine cleaning. This situation leads to anaerobic pockets within 6–8 hours prior to cleanup and can promote 100% spoilage in subsequent production runs. These issues provide an impetus and motivation for developers of new products and processes in the industrial aseptic applications to pursue development and optimization of advanced volumetric methods of heating as well as novel and improved methods of cooling under continuous flow conditions. Some of these efforts are already underway, and are expected to increase in frequency and intensity as more challenging and thermosensitive foods and beverages are introduced into the commercial marketplace.

Shell-and-tube heat exchangers are better suited for chocolate drinks to promote caramelization. Tubular heat exchangers are better suited for soups. Viscous dairy products can be produced in coiled heat exchangers, as the heat transfer rate is increased and axial mixing helps achieve uniform temperatures. Tubular and scraped surface heat exchangers (SSHE) may be best suited for cheese sauces and puddings. Additionally, novel and emerging technologies of thermal processing optimized for sterilization of viscous, poorly conductive, and thermosensitive products prior to aseptic packaging, such as continuous flow microwave (MW) processing, have been demonstrated in both larger scale pilot plant trials and commercial applications as capable of producing superior quality products with maximized nutrient retention levels.

18.2.2.5 Holding Tubes

A critical phase of an aseptic process that requires proper technical consideration is the holding of product at the sterilization temperature. Liquids and purees need a hold tube that is long enough to provide the required thermal treatment approved by a process authority. Residence time is generally well-defined, by rheological properties and flow rate. Generally, a factor of 0.5 is applied for the fastest residence time compared to the average residence time, this is based upon a theoretical residence time distribution for an ideal fluid (Newtonian) in laminar flow, most food products are non-Newtonian and the ratio of average to maximum velocity needs to be estimated, as shown in Chapter 6. However, careful design is needed in case products can deposit or foul in the walls of the hold tube, creating a tunnel through which product would flow at a shorter hold time than initially designed.

When the product contains particles, the residence time and temperature of each individual particle during continuous processing cannot be practically measured, thus leaving room for potential cold (underprocessed) spots. Moreover, temperatures might not be constant from inlet to discharge of conventional holding tubes due to transfer of heat from liquid to particle or vice versa, and heat losses from the external surface of the tube. Mathematical modeling of the process, including hold tube, can alleviate these unknowns and needs to be validated for accuracy and monitored as CCP during processing. Sensors like the ones used for the validation in Appendix 6 are a way to determine the average and minimum residence times of particles, and the use of an engineered simulated worst-case particle ensures the safety of the product.

18.2.2.5.1 Flow in Conventional Holding Tubes

A conventional holding tube consists of straight sections connected by 45°, 90°, and 180° bends. Flow in the straight sections typically exhibits a range of velocities from the tube wall to the center; this is significant in laminar flow, which is likely to prevail in viscous products. In the bends, a swirling motion called secondary flow, known as the Dean effect, caused by centrifugal forces, is initiated. This secondary flow, depending upon its intensity, acts much like turbulence in reducing the velocity spread in the tube. However, this advantage is quickly lost in the next straight section where the original flow pattern resumes. Coiled tubular heat exchangers help to solve this issue by continually changing directions, and thus using the Dean effect to a maximum degree. The positive influence of the Dean effect is well-established for homogeneous fluid products; however, its role in equalizing the residence times and therefore cumulative thermal exposure for heterogeneous (particle-containing) products still needs to be properly researched, even as there are clear indications of its positive potential. This is especially important for products containing multiple particle components, differing in shapes, sizes, loads, and densities. The Dean effect may be utilized to even out some of these variabilities, but should not be relied upon without detailed studies of real products and real processing installations.

A condition that may develop with products containing particles flowing in straight tubes is partial adhesion of some of the particles to the tube wall, effectively reducing the area available for flow. While these adhering particles will be held for longer than twice the designed holding time, some other particles and portions of the liquid may be going through the tube in less than one half the calculated holding time. This has not been recognized by the FDA as a problem with homogeneous liquid products in that the overall product is "safe." However, significant holding time reduction has been experienced in processing chocolate syrup, which does not contain discrete particles. At the start of syrup production, the specified hold is achieved. As time elapses, the chocolate coats the tube wall, gradually reducing the actual holding time. At the end of the day, the effective diameter of the holding tube may be only half of the tube ID, which would cut the holding time to a quarter of the specified time for which the holding tube was designed. Bacteriological analyses through the day correlate well with the reduction of time of hold; that is, as the holding time decreases, the bacterial count increases. This phenomenon is also evident from the work of certain investigators with more viscous products, such as eggs, which tend to coagulate or burn at the walls of the holding tube (Carlson, 1994, 1996). One product-related, active processing element in the multiphase/particulate products that helps reduce the appearance of similar tube wall coating issues and the related reductions in the process line diameters is the continuous flow and therefore presence of subsequently arriving particles causing the collisions with the temporarily stationary/stuck particles which tends to result in their removal from the wall as a result of these continuous impingements. This is especially important for novel volumetric heating technologies such as continuous flow microwave heating where the presence of a stationary particle within the high energy density application area can result in its rapid overheating, causing occasional catastrophic breaches in the tube wall and process failures.

Regarding the product components getting temporarily or permanently immobilized at the heat exchange surfaces and tube walls in general, there have been efforts to minimize or eliminate this issue by the introduction of hydrophobic or superhydrophobic coatings to the surfaces in contact with the product. This has been particularly investigated in order to increase the efficiency of continuous flow cooling of thick viscous products thermally sterilized by the application of advanced thermal technologies like microwave heating. Typically, the viscous product will subsequently coat the heat exchange surface during the cooling stage of the process and the rate of the heat exchange in this stage will be minimized, occasionally resulting in the hot and sometimes boiling product reaching the filler. This issue has been addressed by the introduction of superhydrophobic layers to the product-contact surfaces—some temporary and some permanent. Generally, imparting the superhydrophobic characteristics to the product-contact surfaces throughout the processing system will have very positive effects on the speed of both heating and cooling and consequently maintaining the quality of the finished products, especially when they are viscous, poorly conductive, and thermosensitive in nature. In order to achieve this, efforts should continue to implement the high-performance engineering polymers, glasses, and ceramics throughout the processing systems. High-precision laser etching of stainless steel processing equipment is also known to impart the superhydrophobic characteristics to such surfaces. Considering some of the questions that have come about regarding the conventional holding tube, a holding tube should be designed in such a manner which will render a product commercially sterile.

18.2.2.5.2 Proper Holding Tube Design
The holding tube should be designed as one continuous helical coil with an adequate coil radius-to-tube diameter ratio. This satisfies the legal requirements of a 2% slope and takes advantage of the velocity-leveling effect of the secondary flow in curved tubes. The secondary flow is effective in mixing the phases and keeping the particles in uniform suspension without adhesion to the tube wall. The fluid agitation also helps considerably during CIP. Furthermore, the holding tube should be insulated from the environment to minimize heat loss and temperature drop, but it can't be heat traced in any way to comply with regulation. This approach eliminates the need to oversize the holding tube to account for the wide variation in velocities in straight tubes. A properly designed holding tube will act significantly toward providing the best-quality aseptic product. HydroCoil® holding tubes are designed based on the aforementioned principles.

18.2.2.6 Cooling Cycle and Leak Detection
Prompt cooling of UHT heated product to 50–60°F is an important part of optimization. Thermal process equipment design should incorporate adequate cooling. Inadequate cooling or slow rate of cooling can promote economic spoilage due to non-pathogenic thermophilic spores. Failure modes and effect pertaining to the cooling cycle were discussed in Chapters 13, 14, and 15. Proper equipment preparation and setup (EPSU) and leak detection tests are essential for control and prevention. There is a need to develop a reliable leak test to check system integrity.

18.2.2.7 Surge Tank
Certain viscous products, such as puddings and cheese sauces, when processed using conventional technologies, foul the processing equipment and must be cleaned at frequent intervals. To contend with this requirement, aseptic surge tanks are used. The surge tank is filled at the same time fouling is anticipated. The surge tank is large enough so that the processing system can be cleaned, resterilized, and put back on stream before the surge tank is empty. Also, as mentioned

in Chapters 4 and 5, most operations include aseptic surge tanks, because filler speeds do not correlate with upstream processing flow rates.

Surge tanks are expensive. They are usually one of the first places considered as contamination points if contamination is detected, and they are complicated to operate, which is somewhat simplified through the use of programmable logic controllers (PLCs) and proper programming. Agitators and steam seals must be monitored and equipped with alarms, as well as the SIP process which must include the connection piping to the filler. Therefore, someone knowledgeable in electronics and programming is required during design, construction, commissioning, and validation. This resource must be available when changes are needed. CIP and SIP of tanks have improved with the use of rotating spray heads, but large-volume tanks are more complex, especially when products with a high-yield stress are produced; these materials can solidify and require high-energy shear for removal. It would be advantageous if processing equipment could be developed that could be cleaned easily or inexpensively so that the operation and delivery of the product to the packaging system are not interrupted. This is another potential benefit of treating the equipment surfaces so that they have superhydrophobic properties, as well as the use of novel high-performance engineered materials for contact surfaces.

18.2.2.8 Aseptic Processing of Low-Acid Particulate Foods

To aid processors in developing a validated process schedule for low-acid particulate foods, a series of industry, university, and government workshops on aseptic processing of multiphase foods were conducted by the National Center for Food Safety and Technology in Chicago, University of California at Davis, and the Center for Aseptic Processing and Packaging Studies at North Carolina State University in Raleigh. This resulted in the publication of the "Case Study for Condensed Cream of Potato Soup from the Aseptic Processing of Multiphase Foods Workshop" (Anon., 1996). Other results of the workshop were summarized in a series of articles that appeared in *Food Technology* magazine (Damiano et al., 1997). Based on the recommendations of the workshops, Tetra Pak, Inc. (with the assistance of the National Food Processors Association [NFPA]) developed the necessary data required for filing a scheduled process for aseptic processing of a low-acid product (cream of potato soup). The filing resulted in a "no objection" letter from the FDA in May 1997. It was thus demonstrated that it was indeed possible for processors to adopt the recommendations of the workshop to gain the acceptance of the FDA for aseptically processing low-acid foods containing large particulates. A detailed description of process filing using the FDA form 2541c (for aseptic processing of low-acid foods) has been described by Sastry and Cornelius (2002).

Researchers at North Carolina State University and Aseptia, Inc. developed, commercialized, and implemented a series of validation methods and tools which were used to validate foods with large particles and presented to FDA (see Figure 18.4). These validations used simulated particles to determine the residence time distribution, worst possible case scenario for heat transfer and spore survival at the same time. The use of such particles eliminates many constraints that were identified in the 1997 workshop, such as the need for a large number of test particles to statistically support the worst-case scenario. Mathematical modeling and simulations are needed as a prerequisite for a successful validation (Chapter 17). Aseptia, Inc. used continuous flow microwave heating for rapid heat transfer and uniformity of temperature. This validation methodology is summarized for the first time in Appendices 6 and 7. Aseptia, Inc. received Letters of No Question (LONQ) which are shown in Appendix 6 for a population of similar particles, and in Appendix 7 for a mixed population of particles. The validation system described in Appendix 6 had already been used for large particles in Europe and is being tested by several companies in the United States.

The system is based on establishing the design criteria for conservative (fast) flow characteristics and conservative (slow) heat penetration characteristics, thus enabling the fabrication of a population of "worst case" simulated particles to carry the bacterial spore loaded implants through the aseptic process sterilization. These particles are designed to flow faster and heat slower than any of the real particulate food ingredients in the product. Sastry and Cornelius (2002) described the statistical approach (Table 7.1), which demonstrates that 299 sample particles are the minimum required number to ensure that at least one of such particles is among the fastest 1% within the sampled population. By engineering the properties of every simulated particle to be the "worst case," further statistical studies are needed to justify reduction of the minimal and recommended number of samples to achieve the equal level of confidence. This will further simplify the implementation of this validation technology, bringing more aseptically processed particulate products to the consumer marketplace.

Prior to producing aseptically processed and packaged foods containing particles, a processor is required to demonstrate that the food is commercially sterile by the regulatory agencies (FDA or USDA, depending on the type of food) with the involvement of a properly trained Process Authority. FDA regulations are contained in Title 21 CFR 108 and 113, and USDA regulations are contained in Title 9 CFR 318 and 381 (Chapter 2). General FDA requirements for establishment registration, thermal process filing, and good manufacturing practices for low-acid canned foods are covered in 21 CFR 108, 21 CFR 117, 21 CFR 113, and 21 CFR 114. These and other listed regulations and forms are also accessible through contact with FDA directly or from website (www.cfsan.fda.gov).

18.2.2.8.1 Thermal Process Design

A typical thermal process design basically consists of a combination of process temperature and a process time. For aseptically processed foods containing particles, the temperature of the interior of the particles cannot be monitored and must be estimated mathematically. As shown in Appendix 6, simulated food particles with implants and corresponding non-invasive sensors and bio-loads can be used to determine the residence time distribution and reduction of microbial loads.

Mathematical modeling of the system, which takes into account the three-dimensional nature of the food particles and the heat transfer method (conventional or advanced), is needed to perform an initial estimation of the thermal process, and several software packages are available to do this. Verification of the model prediction is needed as part of the validation process. Ultimately, thermal process should be confirmed with a biological validation test, either by introducing spores into products or using simulated food particles with thermoresistant spores as a worst-case scenario, as discussed in Chapter 17 and Appendices 5, 6, and 7.

18.2.2.8.2 Biological Validation Test

A biological validation test is a confirmation test meant to ensure that the process developed by calculation and implemented operationally is effective in commercially sterilizing food. The test should be designed such that the test organism selected is appropriate for the type of food under testing and the technology and temperature levels used to process the food. In addition, the food may need to be supplemented with nutrients such that any surviving test organisms, which may be in an injured state, would be able to recover and grow out. This would ensure that the absence of growth of the test organisms represents the actual destruction of the organism (Chapter 13).

Any procedure used to establish an aseptic process for a low-acid food product containing particulates should include a biological validation step. Biological validation is complicated by the need to document the particle residence time distribution throughout the system and the enumeration of control mechanisms for the critical heating rate factors. Traditional biological enumeration methods used for canned food products are not directly applicable to continuous aseptic particulate systems (Larkin, 1993). Appendix 6 shows an approach to biological validation in which the residence time of the particles is documented thoroughly, as well as monitoring of the reduction of microbial spores which overcame many of the hurdles which were encountered in previous experiences. This approach simplifies the validation of products with particulates while ensuring the safety of such products.

It is strongly recommended that any processor pursuing low-acid aseptic particulate processing should maintain a continuous dialogue with the appropriate regulatory agency and a competent process authority. A continuous dialogue will help resolve any confusion or issues that may occur before the final process submission is prepared.

18.2.2.9 Ohmic Heating

Ohmic or electrical resistance heating of foods to sterilization temperatures is under intensive study with a prototype system capable of sterilizing large particulates. As with microwave heating, there is, in principle, no upper limit to the temperatures that can be achieved. The resulting sterile product can be as high in quality as minimally processed foods (Figures 18.3).

Shown in Figure 18.3 are time–temperature profiles for low-acid particulated Pistou soup thermally processed in ohmic, tubular, and static retort. For a $F_{required (Designed)} > 6$ min, area under

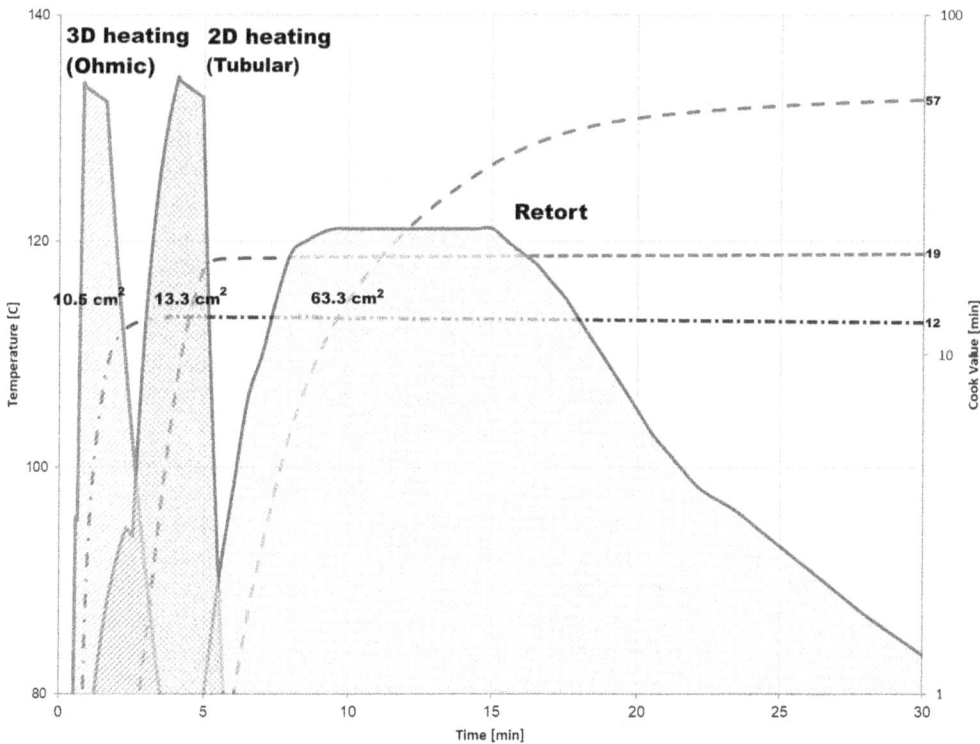

Figure 18.3 Time–temperature profiles for low-acid particulated Pistou soup thermally processed in ohmic, tubular, and static retort. For a $F_{required (Designed)} > 6$ min, area under the curve is 10.5, 13.3, and 63.3 cm^2, for ohmic, tubular, and still retort, respectively. Also shown are Cook Value (C_o) as an index of heat-induced quality degradation—12, 19, and 57 minutes, respectively, for ohmic, tubular, and still retort (see Figure 7.1).

the curve is 10.5, 13.3, and 63.3 cm^2 for ohmic, tubular, and still retort, respectively. Ideal "square wave" $T(t)$ profiles are realized for both ohmic and tubular heating, compared to transient $T(t)$ profile demonstrated for unsteady state heating in still retort. Also shown are Cook Value (C_o) as an index of heat-induced quality degradation—12, 19, and 57 minutes, respectively, for ohmic, tubular and still retort.

The technology is a commercial success in Europe and Japan for human and pet foods. In the United Kingdom, Sous Chef (H.J. Heinz) is commercially producing shelf-stable meat and vegetable entrées using ohmic heating, followed by scraped surface cooling and aseptic filling into Bosch trays. Around 2010, Liebig (Campbell Soup) and Knorr (Unilever) concurrently introduced soups with large particles, filled in CombiBloc cartons which are processed using ohmic heating. Ohmic heating has been coupled to bulk aseptic packaging of acid foods, tomato products, and low-acid foods with particulates.

Further research is needed to ensure food safety in ohmic heating, in particular the effects of changing particle composition and concentration, as well as balancing conductivity between liquid and solid phases in a liquid particulate slurry (Municino, 2018; Shynkaryk et al., 2010; Somavat et al., 2012; Sun et al., 2011).

18.2.2.10 Microwave Heating

Microwave (MW) processing of low-acid foods is known and recognized by FDA. The first process that received regulatory approval was the sweet potato puree by Yamco in Snow Hill, North Carolina, in 2007 based on microwave heating technologies developed at North Carolina State University. This plant has expanded its capacity several times, as well as the variety of products offered to include purees of pumpkin, cauliflower, squash, carrot, etc. Aseptia, Inc. opened a factory in 2012, and successfully introduced several lines of products in the market such as high-acid fruit purees, beverages, and low-acid soups and drinks, and was recognized with the 2015 IFT Food Technology Industrial Achievement Award. Recently, Sinnovatek in Raleigh, NC, has also started offering smaller scale continuous flow microwave sterilization units integrated with aseptic packaging for aseptic R&D and small product series production (Coronel et al., 2003, 2005, 2008; Tahir et al., 2009).

The need to replace the contact heating of viscous, poorly conductive, and particulate foods was and is at least an order of magnitude greater than the need for advanced cooling (Figures 18.3 and 18.4). The quality and nutrient degradation caused by inefficient heating is far greater than the one caused by slow cooling in any scenario. This is due to the fact that more of the processed food mass is in greater surface contact for much longer periods of time with the extreme temperature during conventional heating, especially toward the final stages of sterilization, than it is during cooling, regardless of the method. So, if temperatures over 100°C is the "danger zone" for quality and nutrient losses, for the bulk of the food product this danger zone exposure happens almost exclusively during heating. Therefore, there is no need to wait for 3D cooling technology to be developed in order to implement the 3D heating as it is justified on its own merit. To recap, the quality improvement advantage of implementing volumetric heating such as continuous flow microwave heating or ohmic heating for sensitive and particulate foods far outweighs the potential quality improvement of the implementation of 3D cooling, whether alone or in combination with 3D heating (Stoforos et al., 2021). $T(t)$ profiles from microwave processes are similar to the "square wave" $T(t)$ profile shown for ohmic heating illustrated in Figure 18.3.

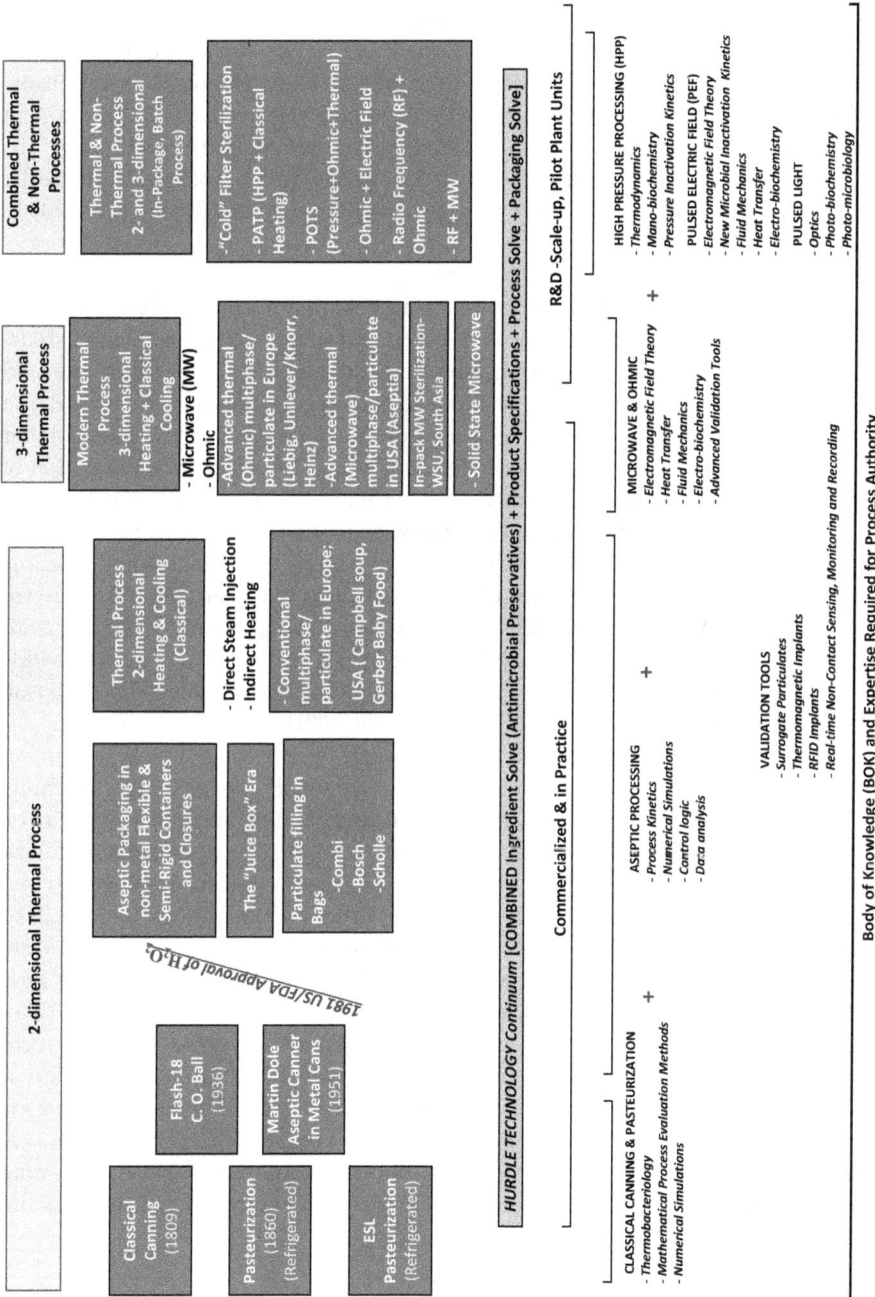

Figure 18.4 Optimization continuum for aseptically processed products, filled and packaged in hermetically sealed containers for non-refrigerated ambient distribution. Also shown is Body of Knowledge (BOK) and expertise required for process authority for the management of classical two-dimensional and three-dimensional thermal processing technologies, non-thermal processing technologies, combination technologies, and hurdle technology.

18.2.2.11 Other Non-Thermal Processes

The ultrahigh pressure (UHP) pasteurization process may also have potential applications in the aseptic processing of some high-acid foods. Pressure can inactivate vegetative microorganisms, but not spores or enzymes, and therefore may be used to commercially sterilize acid foods. Because no heat is applied, this method is sometimes known as cold pasteurization (Figure 18.1).

The Institute of Food Safety and Health (IFSH) at the Moffett Center, Illinois, led the consortium work on pressure-assisted thermal sterilization (PATS) and was accepted by the FDA in 2009 as a new alternative to commercially sterilize low-acid food products. The technology utilizes high pressure to produce temperatures that ensure the production of commercially sterile low-acid food products while significantly improving food quality attributes (Reddy et al., 2003).

18.2.2.12 Additive and Synergistic Processes

Non-thermal processes capable of cold pasteurization, including microfiltration of clear thin liquids, can be combined with classical heat treatments such as pasteurization or ultra-pasteurization and other novel methods of thermal treatment to yield superior high-acid or low-acid refrigerated premium products with minimal heat-induced damages. Food safety is achieved by a combination of milder hurdle technologies rather than by delivery of a single severe one. Further progress in this area is dependent upon a better understanding of the mechanism of action of these novel processes (National Advisory Committee on Microbiological Criteria for Foods, 2006).

Shown in Figure 18.4 is the *continuum* of Body of Knowledge (BOK) and technical expertise required for a contemporary Process Authority (PA) for managing two-dimensional heating: classical canning, aseptic processing; three-dimensional or volumetric heating (ohmic, microwave), non-thermal processing (HPP, pulsed electric, pulsed light, UV light or processing by radiation, etc.), combined thermal and non-thermal processing (PATP, POTS, PEF + Heat, HPP + PEF, etc., Sastry, 2014), and Hurdle Technologies (Process solve + Ingredient solve + Formulation solve + Packaging solve, etc.) (David, 2016, 2020; Legan & David, 2021).

The role and responsibilities of PA in the context of aseptic processing are paramount in designing and delivery of processes from "front-door to back-door"—ingredients, batching, schedule processes for two-dimensional thermal processing and cooling, barriers, sterile tanks, aseptic filler sterile work zone, package integrity, and commercial sterility of finished products. The PA must have a rigorous BOK and background in process kinetics, residence time distribution, numerical simulations, control logic, data analysis, and thermobacteriology (Figure 18.4). The PA managing three-dimensional heating or a combination of two-dimensional and three-dimensional heating must also have a good understanding of electromagnetic field theory, heat transfer, fluid mechanics, electro-biochemistry, and advanced validation tools. In addition, there is the additional complexity, skill set, and BOK required of the PA managing non-thermal technology, or combined thermal and non-thermal technologies, and multifactorial preservation or "Hurdle Technology" (Sastry, 2014; David, 2016, 2020; Legan & David, 2021). Furthermore, there is a need for standardizing BOK and for developing Better Process Control School (BPCS) program and curriculum for training and continuous education for PAs managing three-dimensional thermal processing technologies, non-thermal processing technologies, combined thermal and non-thermal processing technologies, and hurdle technology.

18.2.3 Aseptic Filling and Packaging

18.2.3.1 Line Speed

It is usually desired to increase the speed of equipment. What the optimum speed of aseptic packaging equipment should be is not exactly known. With small containers (e.g., coffee creamers), a speed of 2,000 containers per minute is now possible. This is accomplished on a reliable basis using preformed cups. An analysis must be made by a process/industrial engineer considering costs, demographics, and so forth to determine the ideal machine speed.

There is a definite need for increasing the line speed of aseptic fillers to be competitive and comparable to dairy processing and conventional canning. It is worthwhile to note that multiple operations such as labeling and packaging are performed at the time of processing, justifying slower speeds. In contrast, bright cans or jars from still retorts are staged in cages for subsequent handling; hot water washing and air-drying is used to remove food debris from outside of glass jars prior to labeling and casing. Higher-speed Dole canner, Tetra Pak, and Bosch cup machines have been introduced to meet the current industry demand for higher speeds, comparable to dairy products and ESL products in paperboard cartons, but do not fare well with conventional canning.

18.2.3.2 At-Line and Online Measurements

The current food industry trend is for integration of manufacturing with quality assurance, thereby shifting and diffusing quality responsibility to people involved in making of the product. To facilitate this difficult transition, several infrared and nuclear magnetic resonance (NMR) instruments at-line and online are available to measure solids, sugars, fat, and so forth to validate a batch while flowing continuously. There is a definite need for reeducating production personnel as well as designing and implementing the training programs for new generations of production operators and supervisors in quality control, instrumentation, and continuous refinement.

18.2.3.3 Packaging Issues

The consumer, to a degree, will dictate what packages are convenient and satisfactory and which packages may be desirable. Their wishes must, obviously, be tempered with practical realities and economics.

One of the major problems that exist today is that the package does not adequately protect the product, the product takes on flavors from the packaging material, or the product loses desirable flavors to the packaging material; hence, the strength of the flavor may depend upon the age of the product. Other factors can affect the packaged product quality, including the average temperature the product is exposed to between the time it was packed and the time it was consumed, and what competing flavors or odors were present during the life of the product (e.g., was it placed near a bin of onions and the onion flavor penetrated the packaging material and gained entrance to the product). In other words, it would be desirable to develop coatings or packaging materials that are more impermeable, for example, glass or some of the coating systems used with metal cans.

High-quality, glasslike transparent packaging has been achieved by using recyclable polyethylene terephthalate (PET) bottles for water and fruit beverages, using either preformed or blow-molded in-filler bottles. Please refer to Chapters 8 and 9 and Appendices 3 and 4 (Szemplenski, 2013).

18.2.3.4 Bulk Packaging (Chapter 10)

The greatest growth in aseptic package has been in the realm of bulk packaging into barrier flexible pouch packages. Those most commonly produced are from preformed pouches or bags incorporating valves through which the product is filled. Empty pouches are presterilized by ionizing radiation—1.5 Mrad for high-acid foods and 3 Mrad for low-acid foods (Chapter 10).

With industrial containers, a movement appears to be underway that calls for the centralization of the majority of the producers with one or maybe two large plants in populated areas, for example, the corridor from Boston to Washington, the southeast area from Atlanta to Florida, Texas, and some of the Southern states, and California and the rest of the West Coast. The main plants appear to be in the Midwest. Obviously, there are exceptions, but this appears to be the trend. The plants being built are designed to package large quantities of product. Hence, processing and packaging equipment must be able to cope with such needs. At the present time, this is possible.

While this high level of centralization of operating facilities may be an optimal arrangement for the United States, it may not represent the best solution for less-developed countries and the rest of the world. More recently, there has been discussion and in some cases implementation of more distributed, smaller capacity aseptic processing plants in closer proximity to the sources of ingredients such as farms and orchards. For sensitive, perishable products like ripe fruits and vegetables, this approach to minimize transportation and potential injury of recently harvested produce seems to have merit for certain products and geographic locations and also includes some industrial installations in the United States (Yamco, LLC., Snow Hill, North Carolina, USA).

Also, there appears to be a need for packaging products with particulates, which may be low acid, at fairly high rates into 2–300-gallon bags. These products would be aseptically packaged during the season, which may be quite short, and then repacked in the off-season when they would be aseptically reprocessed, retort processed, frozen, prepared as a deli item, and so forth.

The concept of packaging large-volume commodity products, for example, tomato paste and orange juice or their concentrates, has proven its economic value, while providing quality products to the consumer. The same concept could be used if processing facilities existed for processing low-acid products with particulates. These products could then be repackaged using techniques that were legal and performed satisfactorily.

Another machine developed and in use is a pouch or bag-in-box system to handle institutional quantities of 1–2 gallons economically. This packaging equipment is capable of packaging products containing particles, for example, soups, stews, entrées such as macaroni and cheese, baked beans, and nacho cheese dip. Cheese dip products are now used in convenience stores using custom-made hand-operated dispensing equipment. Please refer to Chapters 8, 9, and 10 on aseptic bulk packaging.

18.2.3.5 Pulsed Light Technology

Yet, another possible technology for aseptic packaging is pulsed light to sterilize surfaces of materials used to form packages. The source is capable of surface effects in eliminating microbial load with virtually no energy penetration. The lethal effect is a combination of ultraviolet (UV), visible light–based thermal cycle, and photodynamic effect requiring oxygen and a photochrome substrate.

A European company has produced low-acid product in a vertical form–fill–seal machine using a pulsed light energy source in place of H_2O_2 for sterilization of food-contact surfaces. The energy is suitable for sterilization of clear polypropylene but is not suitable for recyclable PET bottles. Pulsed light technology needs to be evaluated in regard to bulb longevity, reliability, and online PLC monitoring of treatment delivery and efficacy.

18.2.3.6 Seal Integrity

The terminal retort process in canning is a de facto seal integrity tester, and marginal seals usually do not survive the time, temperature, and pressure of the retort scheduled process. In aseptic, not only are the product and package decoupled, but also the final seal, made in a sterile work zone, is not trauma tested. In aseptic, the reliability of seal inspections, mandatory incubation holds, and sterility tests take on additional significance. Thus, seal integrity and maintenance of hermeticity in aseptic is a key determinant of sterility assurance.

18.2.3.7 Aseptic Filler or Sterile Work Zone Integrity and Validation

Aseptic filler is undoubtedly the heart of the aseptic system, and is crucial to the delivery of a sterile product. Although it can be presterilized and validated to be sterile to a sterility assurance level of 10^{-6}, procedures and practices to maintain or monitor its sterility are unavailable. It is assumed to be sterile by virtue of presterilization and the use of sterile, HEPA, or laminar flow air. Thus, the work zone is at best, sterile or passive, and not sterilizing or active, and thus not able to overcome recontamination. The vectors of recontamination include filter failures, grow through, and non-sterile air aspiration by the work zone due to loss of laminarity, eddy currents at interfaces or leaks, and the discharge of finished units from the zone. In the pharmaceutical industry, media fills have been used to validate the overall sterility assurance of an aseptic system to a 10^{-3} level, but this requires three successive simulated runs of a broad-spectrum growth media of 3,000 units each, with no failures. As long as the work zone remains passive or sterile (opposed to active or sterilizing), the media fill "lack of recontamination" is the only way to validate aseptic systems.

As discussed in Chapters 15 and 16, $3 \times 3,000$ media fill runs are performed for validation of the aseptic zone. These runs are of short duration of about 5–30 minutes, and are, therefore, not reflective of long production runs (8–16 hours) or extended runs (72–120 hours). The other limitation is that these runs are done immediately after a "severe" presterilization process, and, therefore, may not be reflective of the subsequent "gentler" and dynamic decontamination prevailing in "sterile" work zone in aseptic filler. It is recommended that media fill runs be done at production start-up (Beginning of run [BOR]), middle of run (MOR), and end of run (EOR) to be meaningful. It is good to evaluate both validation and incubated spoilage data, and field complaints, for proper evaluation of system capability and deliverables.

It should be reemphasized that despite subsystem validation to 10^{-6}, overall sterility assurance is governed by the aseptic work zone, wherein maintenance of sterility cannot be assured and monitored, and any recontamination cannot be detected, removed, or inactivated. An exception is the active or sterilizing work zone (using superheated steam), which is an integral part of the very first aseptic system, that is, Dole (K.S. Purohit, Process Tek, personal communication, 2008). However, it is important to note that the Dole aseptic system is somewhat vulnerable, when superheated cans are "tempered down" prior to filling to avoid product-contact burn-on.

Reliable, robust aseptic systems require that all product pathways, critical and direct, as well as incidental and indirect, be initially sterilized with appropriate sterilants and maintain sterility throughout production. While direct product (HTST/UHT, filler) and package (formers, cabinet, sealer) pathways tend to get the required attention, sometimes the air and its pathways that contact packages and other critical equipment surfaces are not rigorously presterilized. The work zone, where the sterile filling and sealing of presterilized product into presterilized containers occurs, is also presterilized.

The methods for presterilization of critical air pathways and air-contact surfaces and pathways, as well as those for product and package, vary by manufacturer but should be validated rigorously. The subtle differences between HEPA, microfiltration, and incineration must be kept in mind during presterilization, production monitoring, and validation. In the absence of online monitoring of the integrity of the sterilizing filters, product release criteria are needed to document all pathways and surfaces—direct or indirect—were sterile and maintained sterility throughout production.

Quality, convenience, and economy are the drivers of activity in aseptic processing and packaging. Although this is valid, as we move toward low-acid particulates, we should work to enhance the reliability of aseptic systems, such that they can be designed, monitored, controlled, and validated to comparable or higher sterility assurance levels as canning.

There is the need to develop either artificial intelligence (AI) or fuzzy logic for real-time monitoring and control of the integrity of aseptic systems during production runs (Chapter 17). Computing power of hardware and modeling and simulation capabilities of software solutions have increased by orders of magnitude over the last decade. Applications of artificial intelligence enabled by these developments have become the default methods of navigation systems, facial and mechanical recognition and investment, and real-time decision-making systems. Block chain and the internet environments have become the basis for their respective applications to food safety, traceability and industrial automation, and record-keeping. Emerging developments taking advantage of these capabilities and information management environments, integrating them into artificial intelligence applications, are expected to continue to improve and displace both human and electronic monitoring and decision-making systems in the aseptic processing industry and in the food industry in general.

18.2.3.8 Computational and Numerical Models and Simulations

Computational and numerical models and simulations have been a necessary tool for research, development, process and systems design, and implementation for aseptic processing for several decades now. The more recent advances in development of sophisticated multi-physics software packages like Comsol and Ansys, further catalyzed by the rapid advances in hardware and processing power of computer processors, memory, video and communication signals processors, have enabled the capability to handle ever more advanced and complex models and simulations (Chapter 17). These new and emerging capabilities have proven to be invaluable in discovery, development, and implementation of a broader range of technologies, leading to improved processes, more varied, higher quality and improved nutrition products previously difficult or impossible to achieve using conventional aseptic processing technologies. Currently pending and future developments will further build upon this potential to enable the implementation of real-time optimization and decision-making in aseptic processing and packaging, where similar models and simulations will be performed as building blocks of artificial intelligence applications. These AI applications will contribute to additional improvements of quality, safety, variety, and nutritional benefits of shelf-stable aseptically processed product lines.

18.2.3.9 Cleanup and Extended Run

Using the same line of reasoning, packaging equipment should be designed so that it can be intermittently cleaned—at least to the point where operation can continue after intermittent cleaning for some period of time. This argument is equally valid for microbiological filters used for postprocess sterile additive or HEPA filters that provide the work zone with sterile air.

The limiting factor should be how long it takes organisms to grow from the contaminated air, or item, or material to the sterile enclosed area, thereby causing contamination. What the length of time should be is not known exactly; however, it should be longer than the 1 day that is presently used. Perhaps 2 days, 3 days, or 1 week of continuous operation should be possible. One of the considerations with aseptic packaging equipment is that the equipment requires considerable lengths of time for cleaning and sterilization. Hence, once these operations are completed, and the packaging equipment put back online, if the equipment could operate continuously for an extended period of time, it would be advantageous and beneficial to the manufacturers.

TABLE 18.2   SUMMARY OF DEFECT RATE FOR CANNING AND ASEPTIC PROCESSES

Thermal Processing Operation	P_{defect} of System or Sterility Assurance Level (SAL)	
	Calculated or Standard	Industry
Canning (de Morgan Theorem)	1.2×10^{-4}	1×10^{-4} to 1×10^{-5} or better
Aseptic processes (de Morgan Theorem)	1.2×10^{-3}	1×10^{-3} to 1×10^{-4}?
Literature values		
United Kingdom		
Hersom, 1985	2×10^{-4}	
Burton, 1988	1×10^{-3}	
The United States		1×10^{-3} or better
Pflug, 1987	1×10^{-3a}	
Denny, 1988[b]	1×10^{-4}	
Vendors, 1982	1×10^{-3}	

Source:
David, J.R.D., 1992. Aseptic Processing of Foods: Market Advantages and Microbiological Risks. In
   "Advances in Aseptic Processing Technologies." Singh, R.K., and Nelson, P, E. (Editors),
   Chapter 8, Pages 189–216. Elsevier Applied Science, London & New York.
David, J.R.D., 1995, Research Needs to Ensure Safety of Shelf-Stable Low-Acid Aseptically Packaged
   Products. Symposium on Current Status of Aseptic Processing and Packaging: An Industry
   Perspective, Paper Nos. 4–5, Annual IFT Meeting, Anaheim, California, June 4.
[a]   For aseptically assembled products in the pharmaceutical industry.
[b]   C. Denny, National Food Processors Association, personal communication, 1988.

Research is needed in validation and measurement of "repeatability" of CIP and sanitation cycle as a function of soil, fouling, and biofilm loads (Chapter 12).

18.2.3.10 Defect Rate or Sterility Assurance Level (SAL)

It is not correct to compare canning and aseptic processing with respect to defect rates. However, it is not practical and profitable to run lines that are prone to too many breakdowns, too many defectives, and higher defect rates. It is possible and very common to run retorts obtaining a defect rate of 1 in 100,000 units, as shown in Table 18.2 (David, 1992, 1994, 1995). A defect rate of 1 in 10,000 holds good for all low-acid foods aseptically packaged in rigid, semirigid, or flexible containers (C. Denny, National Food Processors Association, personal communication 1988). However, only 1 in 1,000 or 1 in 3,000 is guaranteed by vendors of aseptic filler machines.

The safety of low-acid canned foods has been well-studied, and 12D Bot-Cook is an accepted design objective for thermal processes. In-container sterilization is a terminal process, and it is valid to extrapolate the probability of a non-sterile unit being one in a million or lower, if the designed process is delivered and postprocess contamination does not occur. It is important to note that sterility assurance of aseptically processed and packaged foods, while similarly computed, cannot be comparably validated. In aseptic, the overall sterility assurance is governed primarily by the lack of recontamination, and even though scheduled processes are delivered to each subsystem (food, package, closure, air, machine, etc.), the sterile work zone for filling and sealing and the maintenance of its sterility (with seal integrity) are key determinants of the probability of non-sterile units.

Numerous safety and regulatory hurdles have had to be scaled, and additional, increasingly difficult ones still need to be overcome if aseptic processing and packaging is to live up to its potential. Safety, reliability, and robustness as well as assurance of sterility (at least in comparison to in-container sterilization) have all had to be nominally and directionally compromised to avail of the quality and convenience that aseptic offers. This is especially true when one compares an overall assurance of sterility of 10^{-3}, or perhaps 10^{-4} to which aseptic systems can and have been validated, against at least 10^{-6} to which terminal processes have been routinely validated. As the locus of research and development and commercialization of aseptic moves toward large particulates, sterile formulation, and higher throughputs, demands for control and monitoring will no doubt increase. Maintenance of sterility and prevention of recontamination are the aseptic equivalents of postprocess contamination (historically the canning industry's Achilles' heel), except that in aseptic they are so intertwined with the system's design and operation that they will require constant, significant, and specialized attention to keep aseptic systems acceptable from public safety concerns.

18.2.4 Finished Product and Package

18.2.4.1 Flavor Problems

In addition to understanding bitter and rancid flavors caused by heat-resistant enzymes in milk, there is immediate research needed in the area of chemistry and control of flavor defects: cooked flavor, including origins and reactions involved; other heated flavors, especially "UHT-milk" flavor; stale flavor; and influence of storage conditions on flavor.

18.2.4.2 Gelation and Other Physical Defects

Further work is needed in understanding conditions that influence gelation of milk, concentrated milk, and milk-based products. It will be important to understand the relationships between gelation and the following: sedimentation, protease inhibitors and activators, fat separation, and variation in milk supply.

18.2.4.3 Rapid Microbiological Methods

Reliable and affordable rapid methods are needed for assessing commercial sterility, and to shorten the lengthy product incubation period and the associated inventory buildup in warehouses.

During the test time of 10–14 days, a great deal of production can be packed before the suspect problem is discovered. Also, there is a considerable time lapse before product can be shipped into commerce. The current food industry trend is to prevent inventory buildup and to reduce warehouse space and associated costs. A typical food industry with $1 billion in gross sales could potentially realize a one-time "working capital" savings of about $1 million to $2 million through inventory reduction and interest monies. There are several techniques used in the food industry to optimize preincubation of intact product from 10–14 days to 3–5 days followed by testing on instrumentation systems that measure impedance, conductivity, or microbial ATP increase as a function of microbial growth in suspect preincubated products. There are also new instruments that are capable of measuring the actual growth of microorganisms (instead of end products as a result of growth or ATP) in preincubated product samples by subculturing in "gourmet media" for 24–48 hours. Applied research is needed to compare reliability and veracity between instruments based on end products of microbial growth or actual growth versus plate counts to minimize the occurrence of false-positives and false-negatives (comparative instrument output vs. spread or streak plate data) while testing preincubated production samples, for final product release to distribution centers (DCs) and into commerce.

Proper identification and speciation via modern molecular methods is needed for proper root cause(s) analysis (RCA) and definitive corrective action and preventive action (CAPA) (Chapter 13). Tracking sources of microbial contaminants have been a concern, given the uncertainties associated with the integrity and maintenance decontamination of "sterile working zone" or the aseptic tunnel during normative and extended production runs (24–72 hours). However, recent advances in the development of molecular subtyping methods have provided tools that allow more rapid and highly accurate determinations of these sources. Only trained individuals with a good understanding of the molecular subtyping methods and plant floor experience can evaluate the reliability of a link between plant ecology, food, and the incidence of spoilage. Some of the more commonly used subtyping methods are ribotyping, polymerase chain reaction (PCR) methods (RPAD and REP-PCR), DNA sequencing-based subtyping, and other characterization methods (Kornacki, 2010).

18.2.4.4 Consumer Education

The food industry has the responsibility to educate consumers about the product rather than the technological intricacies. The reason for many premature failures of outstanding aseptic products is due to the lack of educating the consumer and providing adequate information concerning product capabilities and deliverables. For example, consumers need information in regard to UHT milk and its advantages over refrigerated milk. Consumers are confused as to how a carton of milk can stay fresh at ambient temperatures without refrigeration. Caution should be exercised when using terms such as pasteurized, ultra-pasteurized, UHT, long life, extended shelf-life (ESL), and the like for marketing and advertising.

18.2.4.5 "Aseptic" versus Quality Fresh

Also, consumers are confused by the word *aseptic*, which to them connotes a food preserved with lots of chemicals similar to an "antiseptic" mouthwash! Some in the industry, based on quantitative consumer surveys, have proposed alternative names for aseptic, such as "quality fresh," "shelf fresh," and "fresh system."

18.2.4.6 Product Development

Spore-sensitive ingredients of concern are cocoa, tapioca granules, non-fat dry milk (NFDM), carboxymethylcellulose (CMC), starches, sugar, corn, mushrooms, and spices. It is a good practice to monitor, measure, track, and control the mesophilic and thermophilic spore loads of each batch of raw product based on ingredients in a formulation. In addition, these ingredients must be completely hydrated prior to batching to ensure that any spores (if and when present) are fully exposed to a designed and delivered thermal process via a direct or indirect method of heating for prevention of economic spoilage (Chapters 13 and 16). It is important to note that novel plant-based ingredients that are used to formulate premium clean label natural beverages and food have little history or track record in regard to their interaction with other ingredients. There is the need to develop a good working knowledge on the use of novel plant-based ingredients (both globally sourced and locally sourced) and their impact on viscosity which in turn can determine flow and residence time distribution (RTD) characteristics (Chapter 6) and lethality in the hold tube; thermal degradation and fouling behavior; shelf-life stability, and heat resistance and survivability of thermoduric and thermophilic spores via designed thermal process leading to potential microbial spoilage.

Further work is needed in introducing products specific to medical, dietary, athletic, geriatric, and ethnic needs. Another market that needs attention is development of mesophilic cultures for production of shelf-stable cultured and aseptically processed fermented products such as yogurts and yogurt drinks.

Ingredients, especially functional ones such as flavors, sweeteners, fat replacers, stabilizers, and emulsifiers, need to be thermostable at UHT processing regimes. Research is needed in the areas of thermostable artificial sweeteners, fat replacers, and processing aids for formulation of low-acid and high-acid drinks, beverages, and solid foods.

18.2.5 Process Controls and Electronic Records

Aseptic processing involves continuous and rapid heating of food product that is then held at a predetermined sterilization temperature for a specific minimum amount of time, followed by the rapid cooling and filling of the product in presterilized containers under sterile conditions. Product and packaging sterilization, besides the aseptic working zone sterile conditions, are critical parameters, and hence they should be controlled, monitored, and recorded. Everything (pumping, temperature, and pressure) within the aseptic process that will jeopardize the finished product's commercial sterility is critical and requires to be adequately controlled and monitored. Various control systems are typically used to interface with various temperature-monitoring equipment, timers, pressure sensors, and all of aseptic processing system's critical monitoring instruments.

A control system may be relatively basic and simple or sophisticated, with each operation being accomplished automatically. However, considering the number of complicated controls required in aseptic processing, almost every new aseptic processing system is controlled by a sophisticated programmable logic controller (PLC). Control system used in an aseptic system provides interlocks (alarms) for temperature, flow rate, pressure during presterilization, production, and other operations (CIP, water recirculation, etc.), also including interlocks for level sensors (amount of available product), steam/cooling water temperature and pressure, differential pressure in the cooling system (regeneration), opening and closing of steam/water valves, starting and stopping of pumps, and others germane to the processing configuration.

Recent regulations do allow aseptic processing and packaging sophisticated control systems to adapt complete automation. However, this adaptation involves regulatory requirements and necessitates operations properly functioned and controlled. The control software and system performance should be verified at the installation of the new aseptic system, and routinely thereafter. Changes to the control software should be made only by authorized personnel, and any software change that is made should be reevaluated and documented. Verification of the automatic control systems software is required to ensure that adequate controls and interlocks are in place and systems response functions as intended; validation of the automated control system includes installation qualification, operational qualification, critical instrument calibration, etc. Bulletin 43-L of the National Food Processors Association (NFPA, 2002) provides an excellent guide to developing, operating, and validating automated control systems.

Although aseptic processing control systems are automated and sophisticated, they require operators' interaction with the control system (resolve alarms and other feedback requiring follow-up or corrective action) and record-keeping of the system's critical parameters. The USFDA requires the observation and measurements of operating conditions to be made and recorded at intervals of sufficient frequency to ensure that the food product is commercially sterile. Documentation of the critical parameters can be done either with handwritten records or paperless electronic recording (21 CFR Part 113, 2021). Electronic records may be necessary for complex aseptic packaging systems with over 100 critical parameters to control, monitor, and record. Electronic record-keeping may improve efficiency, simplify manufacturing, and help to standardize the process. If an aseptic processing plant keeps electronic records, its control systems *shall* comply with the 21 CFR Part 11 (1997).

18.2.5.1 21 CFR Part 11

21 CFR Part 11 applies to records in electronic form, created, modified, maintained, archived, retrieved, or transmitted under any records requirements set by the FDA. Records generated by an aseptic processing facility that are not required or regulated by FDA do not need to comply with this regulation. If an aseptic processing facility can prove that its electronic records comply with 21 CFR Part 11, then FDA will accept electronic records instead of paper records. This mandatory regulation provides criteria for using electronic records and signatures instead of their handwritten equivalents. According to 21 CFR Part 11, a computer control system (including hardware and software) closed (controlled system access) or open (system access is not controlled) requires having proper controls to protect data within the system and ensure that all electronic records are authentic, incorruptible, and reliable. The main pillars that a control system with electronic records and signatures should have to comply with 21 CFR Part 11 are as follows (Ignition, 2021; Onfi, 2021):

- *Validation:* 21 CFR Part 11 requires verification to be conducted for a computer system (including a complex PLC program, an electronic chart recorder, or a simple spreadsheet) used in an aseptic system that provides electronic records regulated and reviewed by FDA. Validation and documentation of a computer system should be performed to ensure the system's proper performance, verifying systems' accuracy, reliability, and the ability to discern invalid or altered records. Validation requires systematic documentation of computer system requirements, combined with documented testing (requirement specifications and testing protocols), demonstrating that the computer system functions as intended.
- *Electronic signatures:* In a computer system regulated by FDA, the aseptic processing facility should have written policies and procedures (SOPs) to define practices for using electronic signatures. The written SOPs should clearly state that the electronic signing is the same as a person's handwritten signature. All electronic signatures must include the signer's printed name and the date/time the signature was applied. Electronic signatures must not be separable from their record and must be unique to a single user.
- *Accurate generation of records:* A computer system should be validated to verify that it can generate, store, and retrieve accurate and complete copies of records in the form of both electronic copies (export to file capabilities) and paper copies/printouts. Audit trail information and any associated electronic signature information must also be available. Electronic records must not be corrupted and must be readily accessible throughout the record retention period.
- *Limited system access:* System security: FDA-regulated computer systems must have controls to ensure that only authorized personnel can operate the system, meaning the system and records generated and stored in the system should be accessible only by authorized users. The computer system should be validated to ensure that the security, authenticity, integrity of electronic records are accessed only by authorized users. System security typically is enforced in two primary ways: (i) user passwords and (ii) programmed time-outs to put the system into a locked state.
- *Protection of records (data storage):* Computer systems should ensure that electronic records regulated by FDA are securely retained throughout the required retention period of the

data. The data stored should be protected throughout the entire record retention period; electronic records should be accessed or retrieved within a reasonable time.

- *Audit trail:* Audit trails are required by 21 CFR Part 11 for every document. Audit trails should be generated independent of the operator and include User ID, sequence of events, the local date, and the actions that alter the record. Record changes shall not obscure previously recorded information/data and must be stored as long as the record itself is held and be available for review.

In an aseptic processing system using electronic records, the computer system used should be validated at the installation and routinely thereafter to ensure that it meets all 21 CFR Part 11 requirements. Validation study should involve:

i) Installation qualification phase to ensure that all system components are correctly installed (review of hardware and software connections, input/output (I/O) checks, calibration of all instruments, etc.)

ii) Operational qualification phase to verify via challenge tests that all SOPs are in place, and confirm all functions of the computer system work as expected (data collection and storage, data security, data access, data retention, audit trail, electronic signature, etc.)

iii) Performance qualification phase to confirm that the computer system performs as intended. Moreover, computer control personnel, users responsible for operating the system, entering, editing, or retrieving data in the system, and responsible for using an electronic signature, should be adequately trained to be aware of all systems requirements and functionalities.

The Institute for Thermal Processing Specialists (IFTPS) has published a draft guidance document that presents a good overview of all steps and conditions for proper validation study of an aseptic processing system that uses electronic records and signatures to comply with 21 CFR Part 11 (IFTPS, 2018).

Finally, it is important to note that aseptic processing systems using electronic records and signatures regulated by the United States Department of Agriculture (USDA) do not have similar specific regulation, i.e., 21 CFR Part 11 does not apply to these systems. However, similar validation practices for the computer control system have to be met (G. Stoforos, personal communication, 2022).

18.3 MANAGEMENT AND ADMINISTRATIVE CHALLENGES AND OPPORTUNITIES

18.3.1 Capital Cost

Current aseptic packaging equipment is expensive. Depending upon the type of container to be packed, rate, manufacturer, and so on, the prices can range from the equivalent of $500,000 to $7,000,000 for machines to sterilize–fill–seal consumer-sized packages. Furthermore, the equipment is not readily available, and most of it is custom-made to a given order. Hence, it may be anywhere from 6–8 months to 1–2 years before the equipment can be obtained.

When equipment for packing in institutional- or industrial-sized packages (bag-in-box) is considered, the prices may be reduced to $150,000 from $1,000,000. This is still expensive. In addition, there is cost associated with start-up validation and operations.

18.3.2 Complexity

Large aseptic packaging equipment is complicated. Granted, more operations are required than with conventional equipment; however, the number of automatic operations, critical controls, and interlocks used seems excessive. It is difficult for an ordinary plant-operating person to acquire the skills to operate a system without considerable education or training. Because the equipment is complicated, many users are reluctant to progress into fields that require aseptic packaging. They do not want to undertake a packaging system that involves operations they do not understand. Also, there is the lead time and expenses related to validation and commissioning and start-up, and training and education.

18.3.3 Reliability

Poor aseptic packaging equipment reliability may be due to several factors. The first is improper design. One of the problems with designing equipment is it may be initially sterilized with heat, meaning clearances must be provided during this phase. In a later phase, that is, during operation or CIP, the equipment will be exposed to temperatures that are considerably cooler; hence, shrinkage and contraction must be considered. These dimensional differences can cause problems of proper fit during operation and potentially impact safety.

18.3.4 Repair and Maintenance

The other major problem with aseptic packaging machines that exist today is the intervals and difficulty of repair and maintenance. The machines are operating in conditions that are difficult to begin with (high temperatures, corrosive atmospheric conditions, cleaning and sanitizing agents are used at times, etc.). Designs call for precision fits so that leakage of air or contaminants into the sterile section(s) does not occur, and everything possible is done to operate the machines so that problems will not result. The machines are first sterilized at high temperatures and then operated at room temperature. Aseptic machinery also is expected to operate at very high production rates. Oftentimes the rate is a function of how fast the food product can be placed in the container and the container can be transferred through the machine without slopping or spilling—particularly if the machine must be stopped because of some failure or if it is an intermittent motion machine. Due to all this complexity, preventive maintenance must be carried out at shorter intervals than in other industries and it requires specialized maintenance personnel.

Furthermore, today's machines are sophisticated from a mechanical standpoint, and they employ state-of-the-art electronics. This makes it difficult for any one individual to be skilled in both areas and be able to correct a problem. The problem may be electronic or may be caused by a mechanical operation that is not being properly performed. Likewise, an electrical or electronic failure may be causing mechanical problems; hence, the first assumption of someone investigating a problem is that it is due to mechanical failure.

These types of problems are accentuated in the United States because many plants are eliminating or reducing the number of engineers and maintenance personnel that are on staff. They depend more on vendors who may or may not have qualified people or special service organizations who may or may not be qualified on a given machine. There is also a time delay in addressing issues caused by the absence of qualified personnel on-site which can further exacerbate these problems.

Many times, the best approach is to have the vendor provide highly qualified technical people who can be supported by the local personnel. Also, it is not uncommon to require at least two different individuals (e.g., an electronic and a mechanical engineer) on a given machine to resolve a problem. The maintenance program for any given operation involving a specific type of machinery must be resolved by each company-plant. However, it would not be unreasonable to see plants with individuals on their staff skilled both in electrical, mechanical, computer programing, and operation of the equipment.

The needs of the industry, obviously, are to have designs available that are simpler and equipment that is more reliable. Usually, increased reliability is a direct function of how simple or complicated the equipment design. Certain equipment may be simplified if this has been stated as a primary objective of the development group. However, if the primary objective is to make a machine that has high speeds or develop a machine that can handle a wide range of containers, then the design may be more complicated. The basic reason is that there are many more tools that the design engineer has to work with, and these are often used. Unfortunately, many management people do not understand the engineering details and must make decisions based upon limited experience. It is still a requirement that the packaging equipment be simple, reliable, and safe for both the quality of the food being packed and the operators and maintenance personnel who are working with the machines.

18.3.5 Education and Continual Learning, and Funding for Research

Simplification and streamlining of industrial controls and means of human interaction also needs to become a goal to be pursued by upstream suppliers of industrial controls, and vendors of highly automated aseptic filler machines. It is important to recognize the widening "technical chasm" between automation with all bells and whistles and human/operators on the floor. People on the floor are central to successful production of safe and high-quality products, shift-on-shift. However, the technical skills needed to confidently, i.e., without intimidation, operate highly complex "human–machine interface" (HMI) may not be readily available in today's workforce. Craft/trade schools and local technical community colleges have a crucial role to play in training, continuing education, and preparing the next generation of technical personnel who can comfortably work with automation, PLCs, use of HMI, touch screens, and software. Humans must leverage automation for scale, for maximum efficiency and for productivity.

The role and responsibilities of PA in the context of aseptic processing is paramount in designing and delivery of processes from "front-door to back-door"—ingredients, batching, schedule processes for two-dimensional thermal processing and cooling, barriers, sterile tanks, package integrity, and commercial sterility of finished products. The PA must possess rigorous BOK and background in process kinetics, residence time distribution, numerical simulations, control logic, data analysis, and thermobacteriology (Figure 18.4). The PA managing three-dimensional heating or a combination of two-dimensional and three-dimensional heating must also have a good understanding of electromagnetic field theory, heat transfer, fluid mechanics, electro-biochemistry and advanced validation tools. In addition, there is the additional complexity, skill set, and BOK required of the PA managing non-thermal technology, or combined thermal and non-thermal technologies, and multifactorial preservation or "Hurdle Technology" (Sastry, 2014; David, 2020; Legan & David, 2021). Finally, there is the need for standardizing BOK and for developing Better Process Control School (BPCS) program and curriculum for training and continuous education for PAs managing three-dimensional thermal technologies, non-thermal technologies, combined thermal and non-thermal technologies, and hurdle technology.

There is the need for funding for supporting applied research on aseptic processing conducted at the universities and other centers of excellence. Cooperative research between large CPG (consumer packaged goods) companies, vendors, academic centers, and regulatory agencies is a must to address voids in both fundamental and applied research identified in this book.

18.4 FUTURE

The trends for the future of aseptic processing and packaging continue to look bright. Improved processing equipment and controls will upgrade flavor profiles, which will provide greater consumer acceptance. These new methods will include the processing of particulates, such as meat, rice, beans, and other vegetables. Barrier properties of container materials will continue to improve, reducing costs and extending shelf life. Speeds of processing and packaging, along with more automated controls, will be another important factor in cost reduction.

UHT processing is energy-intensive and can impart objectionable cooked flavors. The amount of cooked flavor is often dependent on the type of process used and the consumer's sensitivity to it. Direct steam injection or infusion is considered to give an improved flavor profile in some products, such as milk. Three-dimensional or volumetric heating systems such as ohmic and continuous flow microwave processing can further improve flavor and nutrient retention with their speed of heating and lack of hot walls. Indirect systems, however, have improved immeasurably with the advent of hot-water sets or the use of hot water under pressure in lieu of steam to reduce heat shock to the product. The cooked flavor and odors tend to dissipate during storage with age.

In order to be competitive with high-speed retort canning speeds, aseptic canners and form–fill–seal units must operate at higher speeds in the future. In order to accomplish this, engineers need to come up with alternatives to heat and hydrogen peroxide as the methods of sterilizing containers prior to filling. The exposure time required for heat and chemicals makes it difficult to achieve any degree of speed. The "Flash 18" process used by Nestlé Carnation is a unique hybrid alternative in that the product is filled in the can at 255°F in a room under 18 psi pressure. This prevents flashing, and the product and containers are sterilized simultaneously.

Cans as preferred containers will continue to get a boost because of their strength and barrier properties; however, they are not sustainable and use space very poorly, besides opening cans is responsible for a large number of nicks and cuts. Lighter-weight cartons help with sustainability due to their lower transportation cost and the ability to recycle some of the materials. Multipolymer film-layered aseptic bags and pouches are slowly replacing aluminum-lined bags which also increases the ability to recycle the materials.

Form–fill–seal laminated materials may be sterilized during their coextrusion process when temperatures routinely surpass sterilization temperatures. The protective outer layer is then peeled off when entering the filler sterile zone. This is the method employed on the ERCA aseptic units and may be expanded to other manufacturers as the patents expire.

The development of products that take full advantage of aseptic technology has often come in conjunction with refinement or improvement of two-dimensional and three-dimensional thermal processing equipment, packaging material, and packaging machines that provide enhanced convenience, health, and wellness for the consumers. Concerns for filling low-acid particulates and implications in seal integrity are not trivial.

18.5 SUMMARY

Aseptic processing and packaging is an attractive and a challenging alternative compared to conventional methods of canning of foods. Continuous sterilization of heat-sensitive foods at ultrahigh temperatures followed by prompt cooling results in a superior finished product, which can be filled into containers of varying compositions, of different shapes, and with many consumer-attractive features. Compared to classical canning, the definitive market advantage of aseptically processed and packaged foods originates from the ability to incorporate several value-added features such as substantially increased sensory and nutritional qualities, microwaveability, several user-friendly conveniences, and cost-saving from use of plastics.

The future success of aseptic processing and packaging in the United States and throughout the world will depend on breakthroughs in processing, packaging, and aseptic fillers; maximizing optimization potential and minimizing departures from optima; well thought-out and top-down mandated quality assurance and hazard analysis critical control point (HACCP) programs; ability to validate and measure "repeatability" of CIP and sanitation cycles; ability to validate, maintain, and measure "dynamic sterility" of aseptic zones; development of reliable online instruments for assuring package seal integrity; concerns for filling of legally approved particulates and implications in seal integrity are not trivial; innovative containers convenient to consumers; consumer education; capturing findings from ongoing research in both basic and applied research leading to the use of both novel thermal sterilization methods (two-dimensional and three-dimensional such as ohmic and microwave) coupled with innovative non-thermal processes, sensors, validation methods, and nanotechnologies. It is important to underscore that no one single technology can replace the shelf-stable capabilities of either classical retorting or aseptic processing. However, many of the innovative thermal and non-thermal processes, sensors, and nanotechnologies can be used either additively or synergistically to build "hurdles" in tandem with an objective to produce superior products with minimal heat-induced damage and at an affordable price (see Figures 18.1 and 18.4) (David, 2016, 2020).

Sustainable packaging and longer shelf life are of great importance for the food and beverage industry, along with cost of environmental benefits in terms of ambient shipping and storage. The cost of refrigeration is higher and is a key driver for companies to invest in aseptically processed and packaged products that are ambient shelf stable for 12–24 months. This advantage must be fully leveraged to realize savings on energy and carbon footprint. Also, the global refrigerated supply chain or cold chain is rapidly expanding in developing countries. The sustainability impacts of many of these changes are unknown, given the complexity of interacting social, economic, and technical factors (Heard & Miller, 2016).

The future of aseptic processing and packaging of foods and beverages will be driven by customer-facing convenience and taste, use of current and new premium clean label natural ingredients, use of multifactorial preservation or hurdle technology for maximizing quality, and sustainable packaging with claims and messaging (David, 2016, 2020).

REFERENCES

21 CFR Part 113. 2021. *Code of Federal Regulations, Title 21, Part 113: Thermally Processed Low-Acid Foods Packaged in Hermetically Sealed Containers.* http://www.accessdata.fda.gov/scripts/cdrh/cfdocs/ cfcfr/CFRSearch.cfm?CFRPart=113. Accessed on March 30, 2021.

Anonymous. 1996. *Case Study for Condensed Cream of Potato Soup from the Aseptic Processing of Multiphase Foods Workshop. Sponsored by the National Center for Food Safety and Technology, Summit, Argo, IL, and the Center for Aseptic Processing and Packaging Studies.* Raleigh, NC: North Carolina State University and Davis, CA: University of California at Davis.

Ball, C.O. 1936. Apparatus for and a method of canning. U.S. Patent 2,029,303, issued February 4, 1936.

Beckerich, I. 2012. Use of ohmic heating in aseptic systems: Performance advantages: Qualification steps. In IFTPS 31st Annual Conference, San Antonio, TX, March 7–8, 2012.

Burton, H. 1988. *Ultra-High-Temperature Processing of Milk and Milk Products.* London/New York: Elsevier Applied Science.

Carlson, V.R. March 1994. *Holding TUBES.* The ASTEC Report 14(1).

Carlson, V.R. 1996. Food processing equipment: Historical and modern designs. In *Aseptic Processing and Packaging of Food: A Food Industry Perspective,* edited by David, J.R.D, Graves, R.H., and Carlson, V.R., Chapter 6, 95–127. Boca Raton, FL: CRC Press, Taylor and Francis Group.

Casolari, A. 1994. About basic parameters of food sterilization technology. *Food Microbiol.* 11:75–84.

CCFRA. 2008. *History of the Minimum Botulinum Cook for Low-acid Canned Foods.* R&D report 260.

Cho, H.Y., Yousef, A.E., and Sastry, S.K. 1999. Kinetics of inactivation of *Bacillus subtilis* spores by continuous or intermittent ohmic and conventional heating. *Biotechnol. Bioeng.* 62(3) 368–372.

Coronel, P., Simunovic, J., and Sandeep, K.P. 2003. Temperature profiles within milk after heating in a continuous-flow tubular microwave system operating at 915 MHz. *J. Food Sci.* 68(6):1976–1981.

Coronel, P., Truong, V.D., Simunovic, J., Sandeep, K.P., and Cartwright, G.D. 2005. Aseptic processing o sweet potato purees using a continuous flow microwave system. *J. Food Sci.*70(9):E531–E536.

Coronel, P., Simunovic, J., Sandeep, K.P., Cartwright, G.D., and Kumar, P. 2008. Sterilization solutions for aseptic processing using a continuous flow microwave system. *J. Food Eng.* 85:528–536.

Damiano, D., Digeronimo, M., Garthwright, W., Marcy, J., and Sastry, S.K. 1997. Workshop targets continuous multiphase aseptic processing of foods. *Food Technol.* 51(10):43–62.

David, J.R.D. 1992. Aseptic processing of foods: Market advantages and microbiological risks. In *Advances in Aseptic Processing Technologies,* edited by Singh, R.K., and Nelson, P.E, Chapter 8, 189–216. London and New York: Elsevier Applied Science.

David, J.R.D. 1994. Aseptic processing and packaging of foods: Food industry perspective. Part I. Presented at Canners International Permanent Committee (CIPC) of France—Symposium, NFPA 87th Annual Convention, November 2–5, Los Angeles.

David, J.R.D. 1995. Research needs to ensure safety of shelf stable low acid aseptically packaged products. In Symposium on Current Status of Aseptic Processing and Packaging: An Industry Perspective. Paper Nos. 4–5, Annual IFT Meeting, June 4, Anaheim, CA.

David, J.R.D. 2013a. Industry research and development, and management needs and challenges. In *Handbook of Aseptic Processing and Packaging,* edited by David, J.R.D, Graves, R.H., and Szemplenski, T., Chapter 15, 263–288. 2nd edition. Boca Raton, FL: CRC Press, Taylor and Francis Group.

David, J.R.D. 2013b. Thermal processing and optimization. In *Handbook of Aseptic Processing and Packaging,* edited by David, J.R.D, Graves, R.H., and Szemplenski, T. Chapter 11, 167–186. 2nd edition. Boca Raton, FL: CRC Press, Taylor and Francis Group.

David, J.R.D. 2016. What is clean label? A food industry perspective. Presented at the Institute of Food Technologists (IFT) Annual Meeting, Chicago, IL.

David, J.R.D. 2020. Hurdle technology: Multifactorial food preservation for high quality foods. Presented at the 7th Clean Label Conference, 14–16, March 26, 2020, Westin Hotel, Itasca, IL. 2020 Clean Label Conference Proceedings. Global Food Forum, Inc.

Heard, B.R., and Miller, S.A. 2016. Critical research needed to examine the environmental impacts of expanded refrigeration on the food system. *Environ. Sci. Technol.* 50(22):12060–12071.

Hersom, A.C. 1985. Aseptic processing and packaging of food. *Food Rev. Int.* 1(2):215–270.

IFTPS. 2018. Institute for thermal processing specialists (IFTPS) guidelines for validating electronic chart recorders. Retrieved from: http://www.iftps.org/wp-content/uploads/2018/06/Electronic-Chart-Recorders-06-01-2018.pdf. Accessed on March 30, 2021.

Ignition. 2021. 21 CFR ignition compliance. Retrieved from: http://pages.inductiveautomation.com/rs/inductiveautomation/images/21cfr11_ignition_compliance.pdf. Accessed on March 30, 2021.

Kohlman, K.L., Neilsen, S.S., and Ladisch, M.R. 1991. Effects of a low concentration of added plasmin on ultra-high temperature processed milk. *J. Dairy Sci.* 74:1151–1156.

Kornacki, J.L. 2010. *Principles of Microbiological Troubleshooting in the Industrial Food Processing Environment.* New York: Springer.

Kumar, P., Coronel, P., Simunovic, J., Truong, V.D., Sandeep, K.P. 2007. Measurement of dielectric properties of pumpable food materials under static and continuous flow conditions. *J. Food Sci.* 72(4) E117–E183

Larkin, J.W. 1993. Considerations for biological validation of aseptically processed particulate food products. Presented at Symposium on Aseptic Processing and Packaging Technology, FIRDI, May 19–21, Taiwan, R.O.C.

Legan, J.D., and David, J.R.D. 2021. Hurdle technology: Or is it? Multifactorial food preservation for the 21st century. In *Antimicrobials in Foods*, edited by Davidson, P.M., Taylor, M.T., and David, J.R.D., Chapter 21, 695–714. 4th edition. Boca Raton, FL: CRC Press, Taylor and Francis Group.

Loghavi, L., Sastry, S.K., and Yousef, A.E. 2009. Effect of moderate electric field frequency and growth stage on the cell membrane permeability of *Lactobacillus acidophilus*. *Biotechnol. Prog.* 25(1):85–94.

Martin, W.M. 1951. Apparatus and method for preserving products in sealed containers. U.S. Patent 2,549,216, issued April 17, 1951.

McGarrahan, E.T. 1982. Considerations necessary to provide for sterilized milk and milk products in hermetically sealed non-refrigerated containers. *J. Dairy Sci.* 65:2023–2034.

Municino, F. 2018. Advances in ohmic heating. In International Thermal Processing Conference, Campden BRI, June 7–8, 2018.

National Advisory Committee on Microbiological Criteria for Foods (NACMCF). 2006. Requisite scientific parameters for establishing the equivalence of alternative methods of pasteurization. *J. Food Prot.* 69(5, Suppl.):1190–1216.

NFPA. 2002. *NFPA Bulletin 43-L Validation Guidelines for Automated Control of Food Processing Systems Used for the Processing and Packaging of Preserved Foods.* 2nd edition. Washington, DC: National Food Processors Association.

Onfi. 2021. Introduction to 21 CFR Part 11. Retrieved from: http://www.ofnisystems.com/information/resources/introduction-to-21-cfr-11/. Accessed on March 30, 2021.

Park, S.H., Balasubramaniam, V.M., Sastry, S.K., and Lee, J. 2013. Pressure–ohmic–thermal sterilization: A feasible approach for the inactivation of *Bacillus amyloliquefaciens* and *Geobacillus stearothermophilus* spores. *Innovative Food Sci. Emerg. Technol.* 19(2013):115–123.

Pflug, I.J. 1987. Endpoint of a preservation process. *J. Food Prot.* 50(4):347–351.

Reddy, N.R., H.M. Solomon, Tetzloff, R.C., and Rhodehamel, E.J. 2003. Inactivation of *Clostridium botulinum* type A spores by high-pressure processing at elevated temperatures. *J. Food Prot.* 66(8): 1402–1407.

Sandeep, K.P., Simunovic, J., and Swartzel, K.R. 2004. Developments in aseptic processing. In *Improving the Thermal Processing of Foods*, edited by Richardson, P. Boca Raton, FL: CRC Press.

Sastry, S.K. 2014. Advanced thermal and nonthermal food safety technologies: Academic perspective and future research. Presented at the International Nonthermal Conference, September, 2014. Columbus, OH: Ohio State University.

Sastry, S.K., and Cornelius, B.D. 2002. *Aseptic Processing of Foods Containing Solid Particulates*. New York: Wiley-Interscience.

Shynkaryk, M.V., Ji, T., Alvarez, T.J., and Sastry, S.K. 2010. Ohmic heating of peaches in the wide range of frequencies (50 Jz to 1 MHz). *J. Food Sci.* 75(7):E493–E500.

Siemer, C., Toepfl, S., and Heinz, V. 2014. Inactivation of *Bacillus subtilis* spores by pulsed electric fields (PEF) in combination with thermal energy: I. Influence of process and product parameters. *Food Control* 39:163–171.

Somavat, R.A., Hussein M.H., Mohamed, B., Yoon-Kyung Chung, C., Ahmed E. Yousef, A.E., and Sastry, S.K. 2012. Accelerated inactivation of *Geobacillus stearothermophilus* spores by ohmic heating. *J. Food Eng.* 108(2012):69–76.

Stoforos, G.N., Rezaei, F., Simunovic, J., and Sandeep, K.P. 2021. Enhancement of continuous flow cooling using hydrophobic surface treatment. *J. Food Eng.* 300(July 2021):110524. https://doi.org/10.1016/j.jfoodeng.2021.110524

Sun, H., Masuda, F, Kawamura, S, Himoto, J.I., Asano, K., and Kimura, T. 2011. Effect of electric current of ohmic heating on nonthermal injury to Streptococcus thermophilus in milk. *J. Food Process Eng.* 34(2011): 878–892.

Tahir, A., Mateen, B., Univerdi, A., KaraGoban, O., and Zengin, M. 2009. Simple method to study the mechanism of thermal and non-thermal bactericidal action of microwave radiations on different bacterial species. *J. Bacteriol. Res.* 1(5):058–063.

Zhang, H.Q., Barbosa-Canovas, G.V., Balasubramaniam, V.M., Dunne, P.C., Farkas, D.F., and Yuan, J.T.C., eds. 2011. *Nonthermal Processing Technologies for Food*. New York: IFT Press, Wiley Blackwell.

Part V

Appendices

Appendix 1: United States History & Evolution

Ralph H. Graves

Adapted from-

Graves, R.H. 2013. United States History & Evolution. In *Aseptic Processing and Packaging of Food*, edited by David, J.R.D, Graves, R.H., and Szemplenski, T., Chapter 2, 9–17. 2nd edition. Boca Raton, FL: CRC Press, Taylor and Francis Group.

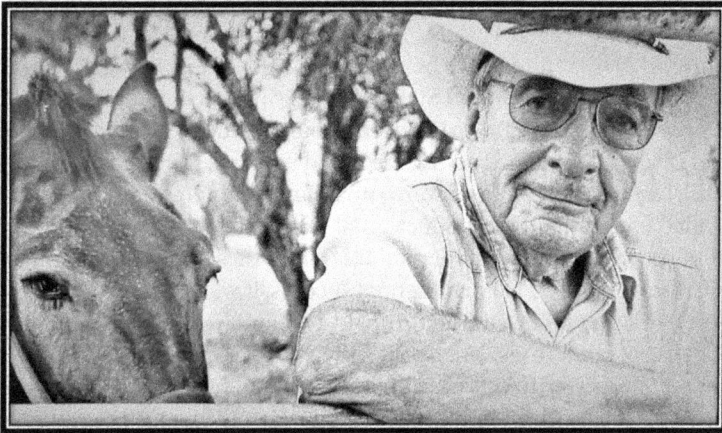

Ralph H. Graves

(1926 – 2019)

Plant Manager, Real Fresh Inc., Aseptic Processing & Packaging Operations, Visalia, California
(Naturalist, Back Country Horseman, and "The Muleskinner")

Figure A1.1   Olin Ball.

1. **Early pioneers**

C. Olin Ball (Figure A1.1) is credited with being the earliest pioneer in the development of aseptic processing and packaging in the United States. His research in conjunction with the American Can Research Department in 1927 led to the development of the Heat/hold–Fill/hold–Cool (HCF) process (Ball and Olson, 1957). The processes received its name from the first letters of the words heat, cool, and fill. Numerous pilot tests were run on many different products. In 1938, two HCF units were installed for commercial production of a chocolate milk beverage. The HCF process did not expand beyond these two commercial lines, and these lines are no longer in operation. Although the HCF process was not a commercial success, it was a great success in that it was the initiator of all the work that followed (Goldblith, Joslyn, and Nickerson, 1961).

In 1942, at the Avoset plant in Gustine, California, George Grindrod developed the Avoset process. Whipping cream was sterilized by steam injection and packaged in cans, and later in glass bottles. Containers were sterilized in retorts using saturated steam. The retort method of sterilization was eventually abandoned and replaced by a continuous hot-air system and utilized ultraviolet lamps to protect the filling and closing area. Like the HCF process, the Avoset process is no longer in operation, but it served as a stepping stone in the evolution of aseptic processing and packaging technology.

In 1948, William McKinley Martin (Figure A1.2) entered the arena with the Dole aseptic process. The Dole aseptic process involved four separate operations: (1) sterilization of the product in a tubular heat exchanger system; (2) sterilization of the containers and covers with superheated steam; (3) aseptic filling of the cooled sterile product into the sterile containers; and (4) sealing the lids in an atmosphere of superheated steam. Martin's aseptic canner overcame many of the obstacles that prevented wider application of the HCF unit. The use of superheated steam at atmospheric pressure eliminated the need for rotary valves for passing empty cans into the system and finished cans out of the system. Also, the use of atmospheric pressure negated the need for construction of high-pressure equipment. This early unit was the forerunner of the Dole aseptic canner. Dole aseptic canners are still in use today. The first commercial units were installed in 1951 for Andersen Pea Soup and sterilized milk at the Med-O-Milk plant in East Stanwood, Washington.

Figure A1.2   William McKinley Martin.

Figure A1.3   "Pea Soup" Andersen in front of a Martin aseptic filler, which made perfect soup possible.

2. **The Graves era**

Roy Graves (Figure A1.3) had a restless mind. He was a creative genius that could have used a full-time staff to follow him around and pick up on his pearls of wisdom. A brief look at his accomplishments over a span of 70 years in the dairy industry is impressive. After completing graduate work at the University of Missouri in 1912, he became the head of the fledgling Dairy Department at Oregon State College at Corvallis, Oregon. After World War II, he became part of the US Department of

Agriculture's (USDA) Food for Peace Program, which encouraged new food sources. Graves joined the Bureau of Dairy Services, USDA, rising to become chief of the Division of Cattle Breeding, Feeding, and Management. His duties included overseeing the extensive research programs at its Beltsville, Maryland, facility. He was the author of over 100 scientific papers and bulletins, and acted in an advisory capacity to several university and college dairy departments. In 1931, he served as the US delegate to the World's Dairy Congress in Copenhagen. From 1935 to 1939, he was secretary-treasurer of the American Dairy Science Association and served as its president in 1937.

He patented the combine-pipeline milking machine along with the merry-go-round milking platform called the Rotolactor. The first unit was installed at the Walker Gordon Dairy in New Jersey and later became a live display at the 1939 World's Fair in New York. Subsequently, these patents were acquired by De Laval Ltd., which continues to manufacture and market adaptations of the original combine milking machine.

In the 1940s, Graves began work with other scientists on sterilizing milk. They reasoned that fresh milk could be drawn from the cow without it being exposed to the air, sterilized and packaged aseptically, and retain its natural quality and nutrition. His patent-owning company, Graves–Stambaugh, Inc., obtained the patents that were put to commercial test by Real Fresh, Inc., and several other licensees.

The original patents described a method whereby milk was transferred from the cow via a pipeline to a vacuum tank mounted on wheels. The vacuum tank was then transported to the dairy plant; the milk was removed by pump, heat treated to 285°F, cooled immediately to room temperature, and packaged aseptically into metal cans. The product found immediate acceptance by the US military, diplomatic, and commercial personnel living around the world with no access to fresh milk.

Graves actively continued improving his original ideas up to his death in 1976 at the age of 90. The patents have expired, and the vacuum tanks retired, but the understanding of the need for high-quality raw milk remains a top priority for all ultrahigh temperature (UHT) processing and aseptic packaging plants.

3. Jack Stambaugh

Jack Stambaugh, Graves' patent coholder, was the owner of Wood Jon Farms and the dairy in Valparaiso, Indiana, where the original research work was performed from 1946 to 1950. The friendship between Stambaugh and Graves began in Washington, DC, during World War II. Stambaugh, an army officer, was assigned to the USDA to oversee a program to increase dairy foods production. Motivated by an interest in purebred Holstein dairy cattle, he looked to Graves for advice. Graves in turn found a willing and eager ear for his ideas on the genetics of dairy cattle, methods for preserving green roughage, loose housing, milking parlors, and UHT processing. This dialogue included discussions on a new type of heat exchanger Graves had designed for UHT processing and a method for aseptically packaging milk so that it did not need refrigeration to extend its shelf life.

When Graves retired from the USDA in 1945, he moved his own herd of purebred Holsteins to Indiana and joined in a partnership with Stambaugh. They enlisted help from the Continental Can Company, which provided financial and technical expertise from its Chicago research center. Together, they shared a vision for the future of the dairy industry.

4. The first commercial aseptic plant

The first commercial plant to use the Graves–Stambaugh process was built in 1951 at East Stanwood, Washington. Their brand name was Med-O-Milk. One of the first Martin (later became

Dole) aseptic canning machines was purchased and installed there to carry out the aseptic packaging part of the process. The initial market was Alaska, because most of its milk supplies had to be imported from the lower 48 states.

Two other licensees quickly followed: one in Visalia, California, using the brand name Real Fresh, and the other, the International Milk Processors, Inc., plant in Ridgeland, Wisconsin.

5. The Real Fresh, Inc., company

In 1952, utilizing the Graves–Stambaugh patents, Robert Graves, Roy Graves' son, opened the new Real Fresh Milk, Inc. in Visalia, California. Though largely unnoticed by the rest of the dairy industry, it was a truly unique facility as it was the second American dairy plant, in one year, dedicated solely to sterilizing milk at ultrahigh temperatures and packaging aseptically.

6. The first aseptic form–fill–seal packages

In 1959, work was underway at Real Fresh to move beyond metal cans into paper–foil–plastic laminated packages. It began with the conversion of the Tetra Pak tetrahedron filler of pasteurized dairy products into an aseptic unit. In 1962, Roy Graves' patent-owning company received patent numbers 3,063,845 and 3,063,211 on this chemically sterilized aseptic packaging system using chlorine as the web sterilant. The tetrahedron package, however, was never widely accepted by the American consumer, but large quantities were produced for the US Navy during the Korean conflict. In 1981, Tetra Pak returned to the United States and introduced the Brik Pak carton. The system used hydrogen peroxide as a sterilant on the form-fill container and was approved by the US Food and Drug Administration (FDA) in January 1981. Real Fresh commissioned and operated the first commercial Brik Pak filler in the United States, in partnership with the National Food Processors Association (NFPA) and Tetra Pak Company.

The Graves–Stambaugh heat exchangers were used at Real Fresh for over 20 years, when they were replaced by steam injection-direct heating systems and a new state-of-the-art tubular/indirect heating processor. Graves believed that enzyme reactivation and oxidative deterioration due to exposure to air, along with the reaction from certain strains of bacteria, were the prime culprits in preventing preservation of unrefrigerated milk for extended periods of time. His patents and subsequent research were devoted to correcting those problems and were instrumental in bringing many new aseptically packaged products to market. To name a few: infant formula, flavored milks, sour cream, cheese spreads and sauces, hollandaise sauce, a dairy spread (butter substitute), eggnog, ice cream and ice milk mixes, meal replacement drinks, soups, and puddings.

7. Early aseptic packers

Although Graves is credited with being the pioneer in producing commercially viable sterilized milk in the United States, concurrently Nestle with its Bear Brand was an early aseptic canner of milk in Switzerland. Other early entries in the United States were Tom Conley at Amboy Sterile Packaging in Amboy, Illinois; Dr. Robert Stewart's plant in Corning, Iowa; the Land O'Lakes plant in Clear Lake, Wisconsin; Sol Zausner in New Holland, Pennsylvania; the Maryland–Virginia Milk Producers Association's manufacturing plant in Laurel, Maryland; the Avoset Company in Gustine, California; the Borden company at its Galloway West plant in Fond Du Lac, Wisconsin; and Foremost Dairies in Newman, California.

Many of these companies produced a single product. A good example is Andersen Pea Soup. Andersen's Restaurant in Buellton, California, was famous for its split pea soup. Tom Andersen

Figure A1.4   Roy R. Graves.

wanted to expand his sales and market his soup through grocery stores. He tried retort canning and was disappointed with the results, so he turned to aseptic canning in 1952 (Figure A1.4). This provided the flavor profile he wanted. After a succession of copackers came and went, Real Fresh bought the name and rights to can and market Andersen Soups, which now covers a span of over 40 years. Others of note were Avoset with its aseptically canned creams, and Foremost Dairies, which produced Fresh Tasting Evap, an aseptically canned evaporated milk, in Newman, California. Foremost also purchased and operated the original Med-O-Milk plant in East Stanwood, Washington, and the Ridgeland, Wisconsin, plant originally owned and operated by International Milk Processors in Chicago, Illinois. Real Fresh, Foremost, General Mills, Hunt Wesson, and the Del Monte were pioneers in aseptically packaged puddings. Borden was also an early entry with its aseptically canned eggnog from its plant in Fond Du Lac, Wisconsin. In 1982, Dairymen, Inc., a Georgia-based cooperative, began the production and marketing of sterilized milk in Tetra Brik containers. This continues to the present under the Farm Fresh brand name. Early producers of cheese sauce and puddings for the institutional market in #10 cans were the AMPI plant in Dawson, Minnesota; Michigan Fruit in Benton Harbor, Michigan; John Gehl in Germantown, Wisconsin; the Land O'Lakes plant in Clear Lake, Wisconsin; Dean's plant in Dixon, Illinois; Carnation with its Flash 18 process; and Real Fresh, Inc., in Visalia, California. Many of these plants have shifted from cans to aseptically filled pouches and bags for economic reasons.

Real Fresh, Inc. was the only firm completely committed to aseptic food packaging. This required the flexibility of several different processing and packaging lines to pack the diverse number of products it marketed and contract packed for other store labels.

There are now over 30 plants in the United States producing one or more low-acid, shelf-stable, aseptically packaged products. They range from milks (including soymilk), flavored milks, and drinks to cheese sauces and dips, puddings, soups, sauces (Aunt Penny's Hollandaise Sauce, Atwater Canning Company, Atwater, California), diet drinks, infant formulas, creams, and milkshake mixes.

8. Restrictions for growth

There are many factors that have restricted growth of UHT processing and aseptic packaging. The following list is not necessarily in order of dominance. First, due to lack of controls and technical specifications, much has been left up to the expertise and skill of the individual operators. Second, regulatory control of low-acid food products marketed unrefrigerated in the United States is onerous. The major players are the Food Safety Branch of the FDA and the Milk Safety Branch in the FDA, which oversee states' inspection of milk under the Pasteurized Milk Ordinance. The USDA also plays two roles in that it is responsible for the inspection of dairy products sold to the government and for the inspection of meat products under the Meat Inspection Branch (FSIS) for all of the US market. So, we not only have federal and state involvement, but also county and municipal health departments having to be dealt with and educated. Third, the cost of aseptic packaging is generally higher. Milk, for example, has a higher energy cost due to the higher processing temperatures, and the packaging costs are more because of the more robust barrier properties needed to maintain asepsis and oxygen barrier. The speeds of aseptic fillers are generally slower than pasteurized milk fillers, which also add to higher costs. The offsetting costs will be unrefrigerated storage and distribution along with reduced outdated returns, which plague pasteurized and ultra-pasteurized milk distributors. Thus, a product–by-product evaluation must be carried out to determine if the advantages of superior value outweigh the additional cost of production and packaging.

9. Trends for the future

The trends for the future continue to look bright. Improved processing equipment and controls will upgrade flavor profiles, which will provide greater consumer acceptance. These new methods will include the processing of particulates, such as meat and rice, beans, and other vegetables. Barrier properties of container materials will continue to improve, reducing cost and extending shelf life. Line speeds of processing and packaging, along with more automated controls, will be another important factor in cost reduction.

UHT processing is energy intensive and can impart objectionable cooked flavors. The amount of cooked flavor is often dependent on the type of process used and the product itself. Direct steam injection or infusion is considered to give an improved flavor profile in some products, such as milk. Indirect systems, however, have improved immeasurably with the advent of hot water sets or the use of hot water under 33 psig in lieu of steam to reduce heat shock to the product. The cooked flavor and odor tend to dissipate with age.

Look for more emphasis in the future on "cold sterilization" methods. Radiation, electron beams, pulsed light energy, and radio frequency (RF) are some methods that will receive considerable attention, and, hopefully, improved consumer acceptance. Other systems may include microfiltration and even a genetically modified bacteria that secretes a bacteriostatic substance such as bacteriocins. This is already in use in cheese-making here and abroad.

In order to be competitive with high-speed retort canning speeds, aseptic canners and form–fill–seal units must operate at higher speeds in the future. To accomplish this, engineers need to come up with alternatives to heat and hydrogen peroxide as the method of sterilizing containers prior to filling. The exposure time required for heat and chemicals to achieve sterilization adds additional constraints, increasing line speeds. The "Flash 18" process used by Carnation is a unique hybrid alternative in that the product is filled in the can at 255°F in a room less than 18 psi pressure. This prevents flashing, and the product and container are sterilized simultaneously.

Cans as a preferred container will continue to get a boost because of their strength and barrier properties. Lighter weight plate will reduce cost, and strength can be retained by injecting nitrogen or carbon dioxide in the headspace prior to sealing.

Fillers have been developed for pouches that can fill particulates up to ¾ inch, in pouch sizes ranging from 8 fluid ounces to 4 liters. This technology will undoubtedly be made available to other types of fillers.

REFERENCES

Ball, C.O., and Olson, F.C.W. 1957. *Sterilization in Food Technology: Theory, Practice, and Calculations.* 2nd ed. New York: McGraw-Hill.

David, J.R.D., Graves, R.H., and Szemplenski, T. E. 2013. *Aseptic Processing and Packaging of Food: A Food Industry Perspective*, Chapter 2. 2nd Edition. Boca Raton, FL: CRC Press.

Goldblith, S.A., Joslyn, M.A., and Nickerson, J.T.R. 1961. *Introduction to Thermal Processing of Foods*, Vol. 1. Westport, CT: AVI Publishing.

Appendix 2: Dr. William McKinley Martin—Father of Aseptic Canning

Compiled by Jairus R.D. David

Complete equipment system for high-temperature-short-time (HTST) processing and sterile-packaging foods — particulate as well as liquiform products — in all types and sizes of containers.

W. McKinley Martin

QUICK HIGH-TEMPERATURE CANNING PROCESS
for better quality canned foods

500 ALMER RD., BURLINGAME, CA. 94010

~~437 Virginia Avenue~~   ~~San Mateo, California 94402~~

343-1716 (Area Code 415)

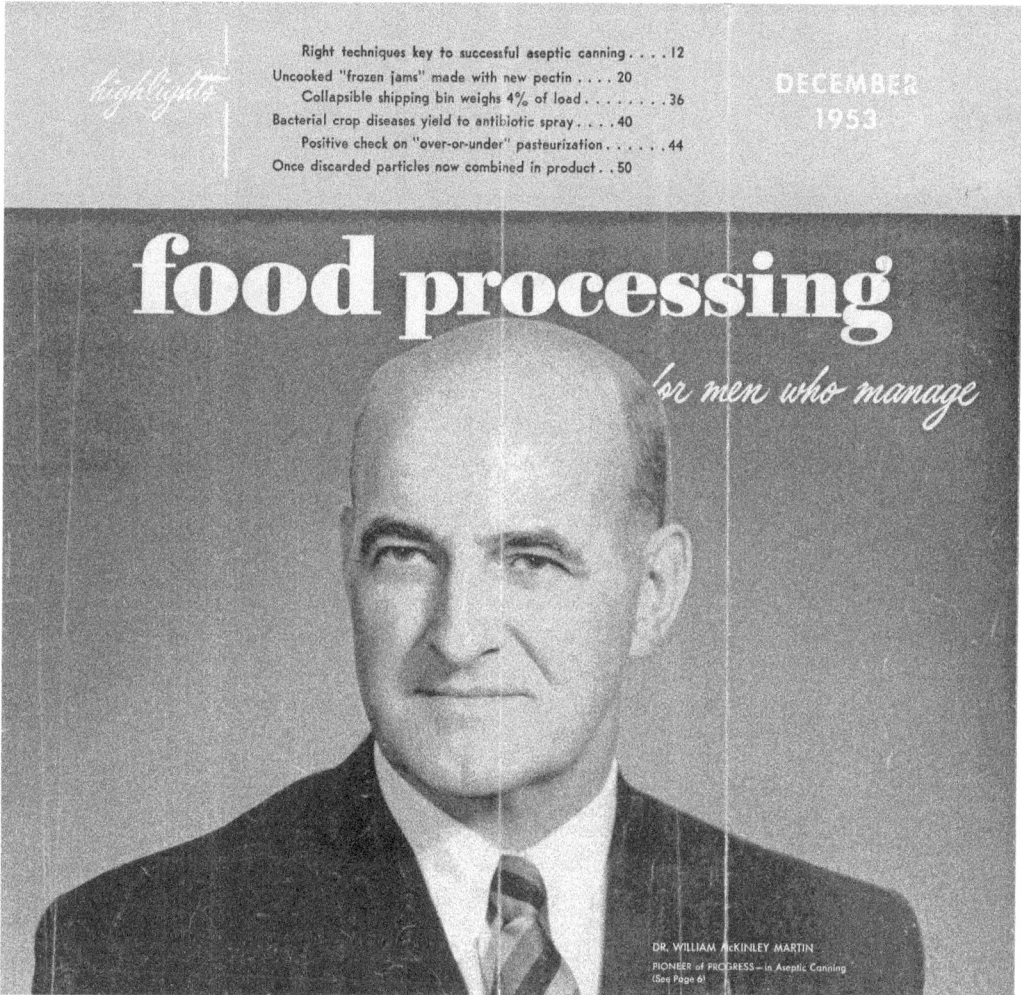

DECEMBER
1953

food processing

for men who manage

DR. WILLIAM McKINLEY MARTIN
PIONEER of PROGRESS—in Aseptic Canning
(See Page 6)

2006-2

PATENT SPECIFICATION

DRAWINGS ATTACHED

978,807

978,807

Date of Application and filing Complete Specification May 9, 1962.

No. 19288/63.

(Divided out of No. 978,806).

Complete Specification Published Dec. 23, 1964.

© Crown Copyright 1964.

Index at acceptance: —A2 D(2E2, 3A); B1 C(1, 4, 10, 11, 12, 14, 19CX, 19F1, 19F4A, 19F4D, 25); G1 H(2A, 2B3, 5)

International Classification: —A 23 b (B 01 f, G 01 f, G 05 d)

COMPLETE SPECIFICATION

Continuous Aseptic Canning Process and Apparatus

I, WILLIAM McKINLEY MARTIN, a citizen of the United States of America, of 457 Virginia Avenue, San Mateo, California, United States of America, do hereby declare the
5 invention, for which I pray that a patent may be granted to me, and the method by which it is to be performed, to be particularly described in and by the following statement: —
10 This invention relates to a continuous heating method and apparatus for sterilising liquids and flowable liquids containing solid particles for use in aseptic canning systems and in other food and chemical processing sys-
15 tems. It relates especially to the aseptic caning of foods containing suspended solids, for example such foods as vegetable soup, beef stew, and the like. The invention comprises a substantially complete process in which
20 presized solid pieces of food are blanched and fed in metered amounts into a liquid phase of the food product and are mixed with it. Short-time, high temperature sterilisation is employed in a novel manner. The sterilised
25 food is then cooled and dispensed into cans by a novel filler. The invention includes unusual coaction between several different parts of the complete system and also incorporates novel features in many of the elements them-
30 selves, including an improved method and improved apparatus for metering and blanching sizable particles of food; and an improved apparatus for mixing the solid components with metered amounts of a liquid
35 foodstuff.

The present invention has the aim of preventing disintegration, attrition, or mushing of the solid components in the food product while assuring their complete sterilisation,
40 and their accurate and rapid filling into the presterilised containers. It also enables the continuous production of canned fluid or semi-fluid food products containing solid

[Price 4s. 6d.]

pieces and having better flavour, colour, texture, and uniformity than are produced by 45 some known canning methods. The invention can also be used to produce homogeneous liquid and semi-liquid canned products of improved quality.

According to the invention the continuous 50 aseptic canning process for a product consisting of food containing large particles, comprises continuously pumping a said food; spreading said pumped food into a moving film; contacting, without intermixture, the 55 surface only of said film with superheated steam under pressure sufficient to prevent its condensation so as to raise the temperature thereof in the range of 275° F. to 300° F. while maintaining all other surfaces of said 60 film in contact with cooler temperature bodies; sterilising said food at said temperature; cooling said food below its flash temperature at atmospheric pressure; and measuring said food and dispensing it under 65 sterile conditions at atmospheric pressure into sterile containers while holding the back pressure between said contacting step and said measuring step at the point of said measuring step. 70

The invention also includes an apparatus for carrying out the process comprising the combination of metering means for both solid and liquid components of the food, mixing and feeding means, a product heater having a 75 supporting surface for distributing a product to be heated over the surface in a thin, gently-moving stream, means for continuously pumping the food, means for continuously contacting the surface only of the 80 stream with hot gas, means for keeping the supporting surface at all places cooler than the product in contact with it, and a closed housing having a product inlet, a product outlet near the bottom of the housing, pump- 85 ing means for pumping the mixture of the

United States Patent Office

3,138,178
Patented June 23, 1964

1

3,138,178
ASEPTIC CANNING SYSTEM
William McK. Martin, 457 Virginia Ave.,
San Mateo, Calif.
Original application Oct. 12, 1959, Ser. No. 845,774.
Divided and this application Mar. 30, 1961, Ser. No.
108,695

16 Claims. (Cl. 141—82)

This invention relates to method and apparatus improvements for use in aseptic canning systems. It relates especially to the aseptic canning of foods containing suspended solids, such as vegetable soup, beef stew, and the like. The invention comprises a substantially complete process in which presized solid pieces of food are blanched and fed in metered amounts into a liquid phase of the food product and are mixed with it. Short-time, high-temperature sterilization is employed in a novel manner. The sterilized food is then cooled and dispensed into cans by a novel filler. The invention includes unusual coaction between several different parts of the complete system.

The invention also incorporates novel features in many of the elements themselves. Thus, it relates, as well, to an improved method and improved apparatus for metering and blanching sizable particles of food; an improved apparatus for mixing the solid components with metered amounts of a liquid foodstuff, an improved method and apparatus for sterilizing them in a short time at elevated temperatures, and an improved method and apparatus for filling the products into presterilized containers under aseptic conditions.

This application is a division of my application Serial Number 845,744, now Patent No. 3,041,185, filed October 12, 1959, which was a continuation-in-part of my application Serial Number 759,098, filed September 4, 1958, now abandoned, which was a continuation-in-part of my application Serial Number 546,306, filed November 14, 1955, now abandoned.

A very important object of the present invention is to prevent disintegration, attrition, or mushing of the solid components in the food product while assuring their accurate measurement, their blanching, their complete sterilization, and their accurate and rapid filling into the presterilized containers.

Another important object of the invention is to provide for the continuous production of canned fluid or semifluid food products containing solid pieces and having better flavor, color, texture, and uniformity than can be produced by conventional canning methods. The invention can also be used to produce homogeneous liquid and semiliquid canned products of improved quality.

Although the apparatus and methods of this invention will be described in connection with an aseptic canning system, many features are useful elsewhere in other food and chemical processing systems; so the invention is not to be interpreted as confined too narrowly.

THE ASEPTIC CANNING PROCESS CONSIDERED GENERALLY

The aseptic canning process differs from conventional canning methods in that the product to be canned is sterilized before it is sealed in the containers, or even put into them, whereas in the conventional methods, the product is first put into the containers and sealed, and then the sealed containers are heated in a pressure cooker or retort to sterilize the product. In aseptic canning, the product is quickly heated to an elevated temperature in the range of 275–300° F., is maintained at that temperature for sufficient time to effect sterilization, and is then rapidly cooled to 90–110° F.; the cooled sterile product is filled into presterilized containers in a sterile

2

atmosphere, and the containers are sealed with sterile covers while still in the sterile atmosphere.

The heat-treatment received by the product in the sterilization step of the aseptic canning method is a matter of seconds, as compared with minutes in the conventional canning methods. For example, the conventional in-can sterilization process for green split pea soup in 303 x 406 cans (16-oz. size) comprises heating the sealed can of soup for 55 minutes at a temperature of 250° F. In comparison, the aseptic canning method achieves sterilization of the same product before filling by holding it for only 8.8 seconds at 286° F. In the process of this invention, it takes only one or two seconds to heat the soup to 286° F., for a total heating time of about ten or eleven seconds to effect sterilization.

While the short-time, high-temperature sterilization process of this invention provides for continuous high-speed aseptic canning with more precise automatic control and consequent savings in labor and heat-energy, these savings and this speed are not its only advantages. Equally important is the fact that the finished canned product has better flavor, color, texture and vitamin content than the product resulting from lower temperature sterilization.

This outstanding improvement in quality is due to the fact that the lethal effect of heat upon bacterial spores increases at a very much higher exponential rate with increasing temperature than do the chemical changes that cause the degradation of flavor, color, texture and vitamin constituents of the product. In fact, the sterilizing effect or lethality, time being constant, increases tenfold while the chemical reactions responsible for degradation of food quality increase only twofold, with each increase in 18° F. in process temperature. Some idea of the importance of this interesting relationship can be grasped by remembering that 2^4 is 16, while 10^4 is 10,000.

QUANTITATIVE EVALUATION OF LETHALITY

The quantitative evaluation of lethality, or the sterilizing effect, of short-time, high-temperature processes for canned foods is expressed in the formula (given in the National Canners Association Laboratory Manual for the canning industry, 2nd Edition, Chapter 12, page 37):

$$F_0 = \frac{S}{60} 10^x$$

where

$$X = \frac{T - 250}{z}$$

S = the time in seconds during which the product is held at a process temperature T.

T = the temperature in ° F. of the product during the process time S, and

z = the slope of the thermal death-time curve in ° F., which for most of the common low-acid food products has been found to be 18° F.

In the above formula, 250° F. is taken as a standard reference temperature, and the sterilization value F_0 is expressed as time in minutes at this temperature. The sterilization values (F_0) are thus expressed on a comparable basis, regardless of the actual process temperature.

To illustrate the practical significance of the quick, high-temperature sterilization process used in the present invention, let us see how temperature affects the "minimum botulinus cook," i.e. what it takes to kill the dangerous bacterium, *Clostridium botulinum*. For the destruction of its heat-resistant spores $F_0 = 4$. A 100 percent margin of safety would be given by using $F_0 = 8$. Now compare the times necessary for equivalent sterilization processes at various temperatures, shown in the following table:

1831

No. 737,144

ISSUED June 28, 1966
CLASS   21-28

CANADIAN PATENT

ASEPTIC CANNING PROCESS AND APPARATUS

William McK. Martin, San Mateo, California, U.S.A.

APPLICATION No.   840,006
FILED   Jan. 15, 1962
PRIORITY DATE

No. OF CLAIMS   26

DISTRIBUTED BY THE PATENT OFFICE, OTTAWA.
SOS-400-8-1 (REV. 6/63)

Memorandum MP76-32-10

UNITED STATES PATENTS
ISSUED TO W. McK. MARTIN

I. PATENTS ASSIGNED TO AMERICAN CAN COMPANY - 1935-45

1,996,336   Mechanical display device for circulating liquids in
tubes. (With J. R. Green). November 6, 1934.

2,227,226   The Tenderometer.
Method and Apparatus for Grading Peas and The Like.
A machine for evaluating the maturity of peas and other
products in grading for quality. December 31, 1940.
Assigned to American Can Co. and subsequently to
Canning Industry Research, Inc. (NCA) Washington, D. C.

2,239,726   The Consistometer.
Device for Measuring Consistency of Product.
A mechanical apparatus for measuring the consistency
and index of Gelation (Thixotropy) of semi-fluid and
semi-solid products in the "undisturbed" state in cans.
April 29, 1941. Assigned to American Can Co.

2,340,336   Filling Machine.
Baffle device for milk filling machine to prevent foam
from being drawn into container in filling homogenized
milk. February 1, 1944. Assigned to American Can Co.

2,345-617   Filling Machine.
A mechanical device for disintegrating foam in filling
liquids into containers. April 4, 1944. Assigned to
American Can Co.

2,351,059   Filling Machine.
An apparatus for dispersing foam on the surface of liquids
by impact of centrifugally projected droplets of the liquid.
June 13, 1944. Assigned to American Can Co.

2,354,693   Liquid Control System. A float-actuated butterfly valve
and accessory apparatus of sanitary design for controlling
the flow of milk, fruit and vegetable juices, and other
free-foaming liquids with especial reference to preventing
foaming and aeration. August 1, 1944. Assigned to
American Can Co.

2,375,806   Float Valves.
A float-actuated valve for automatically regulating flow of
liquids in intermittent operations. May 15, 1945.
Assigned to American Can Co.

-2-

2,383,507   Container Filling System.
Flow control and filling system for packaging fluid
products. August 23, 1945. Assigned to American
Can Co.

2,455,938   Method of making tubular container with solderless
six-layer side seams. December 14, 1948. Assigned
to American Can Co.

2,461,559   Apparatus for washing the side-seams of can bodies.
Removal of solder flux from black iron cans on high-
speed bodymaker. February 15, 1949. (With W. F.
Pillnik and D. E. Wobbe). Assigned to American Can
Co.

2,650,178   Method of Washing the Side-Seams of Can Bodies.
Apparatus designed for use on high-speed bodymaker
for removing flux residue. (With W. F. Pillnik and
D. E. Wobbe). August 25, 1953. Assigned to American
Can Co.

II.   PATENTS ASSIGNED TO JAMES DOLE ENGINEERING COMPANY -
1945-53.

2,507,797   Apparatus and Method for Deaeration of Liquids. A
centrifugal deaerator for viscous or heavy consistency
products. May 16, 1950. Assigned to James Dole
Engineering Co.

*2,549,216   Apparatus and Method for Preserving Products in
Sealed Containers. Method embodying principles of
short-time high-temperature sterilization of food products and equipment for aseptically canning same.
April 17, 1951. Assigned to James Dole Engineering Co.

2,607,698   Product Sterilizing Apparatus and Method. Product
sealed in can and processed in atmosphere of superheated
steam or hot gases with high rate of agitation.
August 19, 1952. Assigned to James Dole Engineering
Co.

2,631,628   Apparatus for Peeling Fruit or Vegetable Products.
Superheated steam and hot gases used in conjunction with
high-pressure water sprays for peeling fruits and vegetables. March 17, 1953. Assigned to James Dole Engineering Co.

*2,631,768   Filling Apparatus and Method. A straightline slit-type
filler with internal rotating helix delivering multiple
streams of product into moving containers. (With A. E.
Post). March 17, 1953. Assigned to James Dole
Engineering Co.

-3-

*2,667,424	Apparatus and Method for Filling Products in Containers. January 26, 1954. Assigned to James Dole Engineering Co.
*2,672,270	Apparatus and Method for Filling Containers with Products. March 16, 1954. Assigned to James Dole Engineering Co.
*2,685,520	Apparatus and Method for Preserving Products in Sealed Containers. Embodies means for reducing head-space vacuum in cans. August 3, 1954. Assigned to James Dole Engineering Co.
*2,771,644	Apparatus for Sterilizing Container Covers, November 27, 1956. Assigned to James Dole Engineering Co.
*2,771,645	Apparatus for Sterilizing Food Containers. November 27, 1956. Assigned to James Dole Engineering Co.
*2,774,531	Container Washing Apparatus. May 8, 1956. Assigned to James Dole Engineering Co.
*2,855,314	Method and Apparatus for Preserving Products in Sealed Containers. October 7, 1958. Assigned to James Dole Engineering Co.
*2,870,024	Apparatus and Method for Preserving Products in Sealed Containers. Aseptic canning method as applied to glass containers. January 2, 1959. Assigned to James Dole Engineering Co.

NOTE * - ASEPTIC CANNING PATENTS

III.   PATENTS NOT ASSIGNED (Owned by William McKinley Martin).

2,992,778	Liquid Spray Apparatus. Centrifugal spray lining method for application of enamel linings in 55-gallon drums, 5-gallon pails and cans. July 18, 1961.
3,018,184	Aseptic Canning Process and Apparatus, January 23, 1962.
3,041,185	Aseptic Canning, issued June 26, 1962.
3,101,752	Aseptic Filling Machine, issued August 27, 1963.
3,138,178	Aseptic Canning System, issued June 23, 1964.
3,146,691	Heating Device, issued September 1, 1964.
3,178,066	Solids Metering and Feeding Device, April 13, 1965.

-4-

3,180,740   Process for Sterile Packaging, issued April 27, 1965.

3,212,674   Liquid Metering and Mixing Device, issued October 19, 1965.

3,291,563   Apparatus for Sterile Packaging, issued December 13, 1966.

FOREIGN PATENTS

Foreign patents are being issued in Australia, Belgium, Canada, Denmark, France, Germany, Great Britain, Holland, South Africa and Switzerland.

W. McK. Martin
January 23, 1967

Source:
Mrs. Jacquelyn M. Trankle (Daughter of Dr. William McKinley Martin), Hillsborough, California. Personal communication, 1995.

Appendix 3: Aseptic Filler Profiles

Thomas Szemplenski

Contract manufacturers play a very crucial role in the introduction of innovative and consumer-convenient new products to the market in a timely and cost-effective manner. The high capital investment and instinct required to run complex aseptic operations tend to preclude many start-up companies entering aseptic operations. Contract packers fill this void by facilitating prototyping, product sensory and specification development and launch, and go/no-go business decisions. Shown are aseptic filler profiles in this Appendix 3, and a list of contract packers in Appendix 4.

AMPACK AMMANN

Ampack Ammann & Co. KG
Lechfeldgraben 7
D-86343 Konigsbrunn
Germany
Tel: (49) 8231/6005-0
Fax: (49) 8231/6005-11

Ampack has no representation in the United States.

Products

Rotary indexing and linear aseptic fillers for plastic bottles and linear cups fillers using preformed cups.

Products Currently Being Filled Using Ampack Fillers

Fruit juices, milk, cream, soups, sauces, puddings, fruit conserve with whole fruit pieces, cheese, quark, stirred and layered yogurt, infant formula, sport drinks.

Equipment Models

Figures A3.1 and A3.2.

Figure A3.1   Ampack linear bottle filler. (Photograph courtesy of Ampack & Evergreen Packaging.)

Figure A3.2   Ampack aseptic preformed cup filler. (Photograph courtesy of Ampack & Evergreen Packaging.)

Type filler	Models equipment	Production rate
Aseptic	Linear preformed bottle filler	36,000 bph
Aseptic	Linear preformed cup filler	65,000 cph
ESL	Rotary preformed cup filler	16,000 cph
ESL	Linear preformed cup filler	65,000 cph
ESL	Linear preformed bottle filler	36,000 bph

Sterilization of filler—Vaporized H_2O_2 used in the sterile zone. Pressurized steam in the fill system.

Packaging—PP, HDPE, and PET with foil lidding.

Sterilization of packaging—High-temperature steam mixed with H_2O_2 followed by a hot sterile air overpressure.

Ampack aseptic fillers are currently not FDA validated for filling low-acid foods.

Additional Information

- Ampack Ammann has been manufacturing aseptic filling equipment since 1978.
- Ampack fillers are capable of high production rates.
- Ampack has more than 130 aseptic installations.
- Ampack manufactures both rotary and linear aseptic fillers.
- The time required to change bottle sizes is very short.

ASTEPO S.P.A.

Via Pilastrello, 13
43044 Collecchio-Parma (Italy)
Ph: (39) 0521 800054
Fax: (39) 0521 802064
Web: www.astepo.com
E-mail: info@astepo.com

Product

Manufacturer of aseptic filling equipment for bag-in-box, bag-in-drum, bag-in-bin.

Products Currently Being Filled Using Astepo Fillers

High- and low-acid food and beverage products, including but not limited to: fruit purees and concentrates, tomato products, vegetables, and dairy products.

Equipment Models

See Figures A3.3 and A3.4.

Type	Model	Filling head	Package sizes
T.A.F. Thousand liter	1H	Single head	300 gallons
	Up to 2"	100 gallons	55 gallons
T.A.F. Thousand liter	2H	Dual head	300 gallons
	Up to 2"	100 gallons	55 gallons
C.A.F. Compact Aseptic Filler	1H	Single head	55 gallons
	Up to 2"		
C.A.F.	2H	Dual head	55 gallons
	Up to 2"		

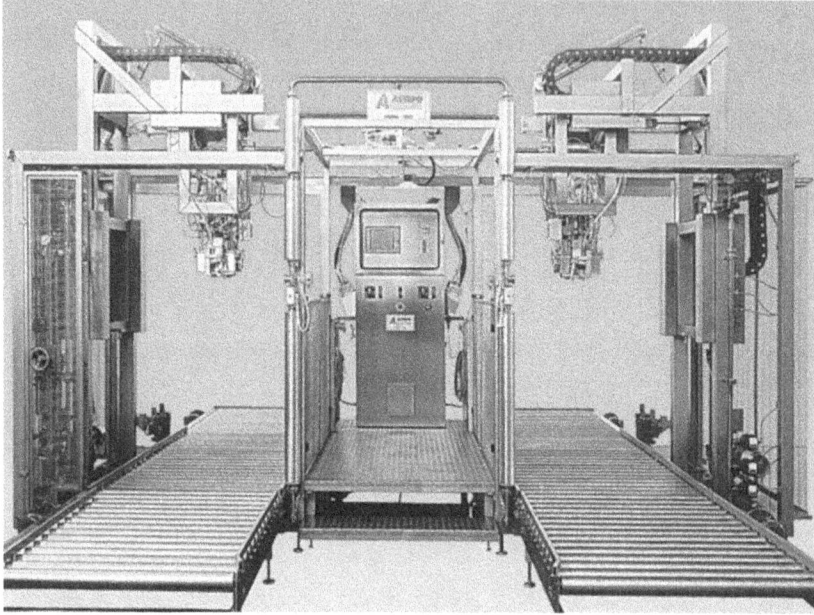

Figure A3.3   Astepo T.A.F. 2H 1,000-liter aseptic filler. (Photograph from Astepo literature, courtesy of VR Food Equipment.)

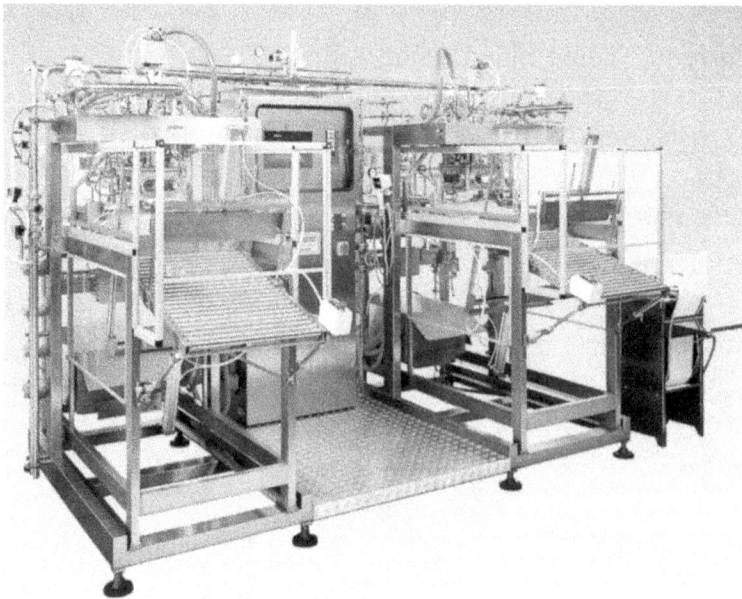

Figure A3.4   Astepo C.A.F. web-fed aseptic filler. (Photograph from Astepo literature, courtesy of VR Food Equipment.)

Filler—Fill is controlled using a flowmeter.

Filler operation—The T.A.F. filler must be loaded with the bag and manually started. The C.A.F. filler is a continuous, automatic, web-fed filler. After starting, the filler automatically resterilizes the spout and cap, removes the cap, and volumetrically meters the filling of the bag. The filler then automatically replaces the cap and ejects the filled bag.

Sterilization of filler—The presterilization of the filler is completely automatic and is accomplished with either superheated water or steam. During sterilization, the automatic cycling of the valves ensures that all product contact surfaces receive enough high-temperature water or steam to effect sterilization. Sterilization time of the filler is approximately 1 hour.

Packaging—Flexible bags of composite polymers with fitments. Astepo fillers can fill bag sizes up to 1,000 liters.

Sterilization of packaging—Preformed and sealed bags are sterilized by gamma radiation. The cap and spout are resterilized just before filling by steam or hydrogen peroxide.

Astepo fillers are FDA validated for aseptically filling low-acid foods.

Additional Information

- The Astepo aseptic filler is one of the most popular bag-in-box fillers for filling low-acid food and beverages in the United States.
- The Astepo filler is manufactured in all 304 stainless steel, except the filling heads, which are constructed of 316 stainless steel.
- Control panel contains PLC with interface video for the operator and electrical components for the filler operation.
- Control panel also includes a diagnostics printer.
- The fillers have automatic presterilization and CIP circuits.

ROBERT BOSCH GMBH

PA-PH/PJM3
Postfach 11 27
D 71301 Waibingen
Germany
Ph: 49 711 811-57133
Fax: 49 711 811-57230

Sales and service office in the United States:

Robert Bosch Packaging Technology, Inc.
8700 Wyoming Avenue North
Brooklyn Park, MN 55445
Ph: (941) 373-9130
Fax: (941) 373-9130
Web: www.boschpackaging.com

Product

Manufacturer of aseptic filling equipment for form–fill–seal plastic cups, bottles, coffee creamers, and multilaminar pouches.

Products Currently Being Aseptically Filled Using Bosch Fillers

Infant formulas, pureed baby food, fruit juices, milk, cream dishes, coffee creamers, puddings, cheese sauces, fruit gels, tomato products, convenience foods, and pet food.

Equipment Models

See Figures A3.5–A3.8.

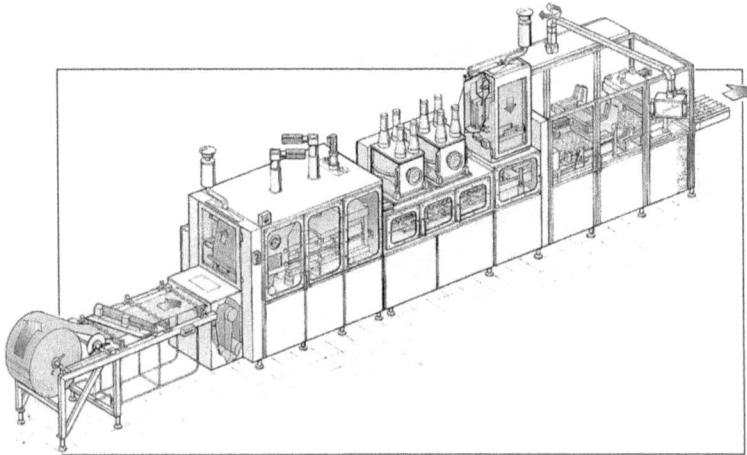

Figure A3.5   Bosch TFA 4818/TFA 4830 aseptic cup filler. (Diagram from Bosch literature.)

Figure A3.6   Some of the products aseptically packaged on a Bosch aseptic cup filler. (Photograph from Bosch literature.)

Figure A3.7   Aseptically operating thermoform fill and seal machine TFA 4940 for safe packaging of coffee cream. (Photograph from Bosch literature.)

Figure A3.8   Bosch SVA 2000/3000 aseptic pouch filler. (Photograph from Bosch literature.)

Type filler	Model	Production rate
Form-fill-seal cups	TFA 4818	350 cpm
	TFA 4830	720 cpm
Form-fill-seal coffee creamers	TFA 2520	800 cpm
	TFA 4940	1,650 cpm
Form-fill-seal pouches	SVA 2000	Up to 40 ppm
	SVA 3000	Up to 25 ppm

Sterilization of filler—Accomplished by steam and hydrogen peroxide. After sterilization the filler is maintained sterile by positive filtered, sterile air pressure.

Sterilization time—Up to 1.5 hours.

Packaging—Form–fill–seal cups of various polymers (including polypropylene) from 5 to 500 mL and form–fill–seal pouches up to 5,000 mL.

Sterilization of packaging—All packaging material is sterilized by heated hydrogen peroxide (35%).

All four Bosch fillers are FDA validated for aseptically filling low-acid food and beverages.

DOLE CANNING SYSTEMS

Division of Graham Engineering Corporation
1420 Sixth Avenue
York, PA 17403
Ph: (717) 849-4095
Fax: (717) 854-1931

Product

Manufacturer of aseptic filling equipment for cans.

Products Currently Being Filled Using Dole Canners

Puddings, cheddar cheese sauces, gravies, soups, ketchup, alfredo sauce, ice cream mix, eggnog, pumpkin, sandwich spreads, pizza sauce, dietetic drinks, applesauce, tomato paste, banana puree, toppings and syrups.

Equipment Models

See Figure A3.9.

Figure A3.9   Dole aseptic canner. (Photograph courtesy of Dole.)

Models	Can sizes	Production rate
330	18–96 fl. oz	70–100 cpm
520	4.5–46 fl. oz	124–225 cpm
530	4.5–46 fl. oz	145–338 cpm
540	4.5–46 fl. oz	155–450 cpm
1210 S	4.5–46 fl. oz	23–60 cpm
1220 L	18–96 fl. oz	30–50 cpm

Filler—The basic model is a slit-type filler. With the slit filler, the cans pass underneath a slit opening in a tube-type filler. The slit is approximately ¼-inch wide by 6 inches long, although the size is variable depending upon the type of product and speed of filling.

The slit filler consists of a thick-walled, stainless steel pipe that is slit along the lower side. This pipe has a second inner pipe that has multiple parts and provides consistent flow to the filling system. The sterile product is fed into the inner pipe and flows out to the slit in the outside pipe.

Sterilization of the filler—By superheated steam using electric heaters.

Sterilization time of filler—30 minutes.

Packaging—Metal cans with seam-on ends are used in all Dole systems. Can sizes range from 4.5 ounces (202 × 214) to the #10 can (603 × 700), and all are compatible with the system. Smaller cans are generally aluminum. The larger cans are two- or three-piece steel. The ends/lids can be standard or easy-open types. The cans are manufactured with heat-resistant coatings, common to the can industry. The can is either left undecorated (or "bright") or has temperature-resistant lithography.

Sterilization of packaging—Cans and lids are sterilized by using superheated (500°F) steam with electric heaters.

The Dole canner is FDA validated for aseptically filling low-acid foods.

Additional Information

- The Dole canner was the original aseptic packaging source and used in the basis for aseptic canning.
- The Dole canner is only leased.
- A royalty based on volume is charged for the Dole canner's use.

DUPONT/LIQUI-BOX

6950 Worthington-Galena Rd.
Worthington, Ohio 43085
Ph: (800) 260-4376
Fax: (614) 888-0982

Product: Bag-in-Box

Manufacturer of aseptic filling equipment for bag-in-box, bag-in-drum, bag-in-bin, and associated preformed packaging of various flexible packaging laminates.

Products Currently Being Filled Using DuPont/Liqui-Box Bag-in-Box Filling Equipment

Processed fruits, juices, purees and concentrates, citrus, vegetable purees, cheese and other sauces, tofu products, sour cream, and dairy products, including condensed milk.

Equipment Models

See Figures A3.10 and A3.11.

Type/model	Operation	Package sizes	Production rate
StarAsept™ Model 1307	Semiautomatic	6–370 gallons	Up to 100 gpm
Model 2000 C1T-0-A	Automatic	1–5 gallons	Up to 15 bpm

Filler—Programmable logic-controlled filling; by flowmeter; and by weight cells depending upon the filler.

With Model 2000 C1T-0-A, the bags are automatically fed through a sterilization tunnel where the bag spout is sterilized, the cap is removed, and the bag is filled to a predetermined volume. After filling, the bag is recapped and ejected. The next bag then indexes to be filled.

Sterilization of filler—Steam.

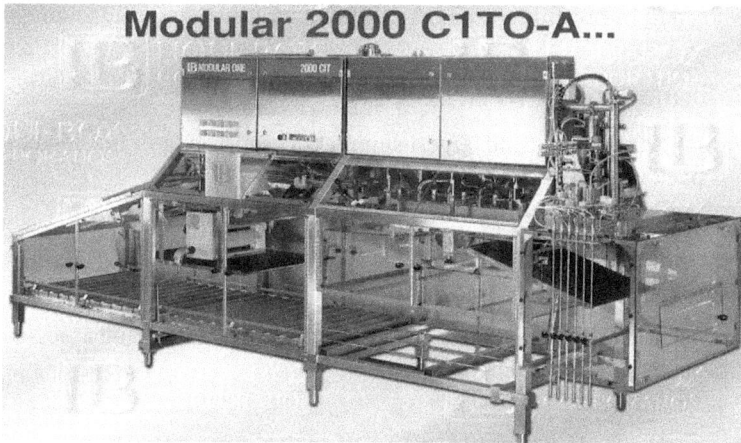

Figure A3.10   DuPont/Liqui-Box web-fed, low-acid aseptic filler. (Photograph from DuPont/Liqui-Box literature.)

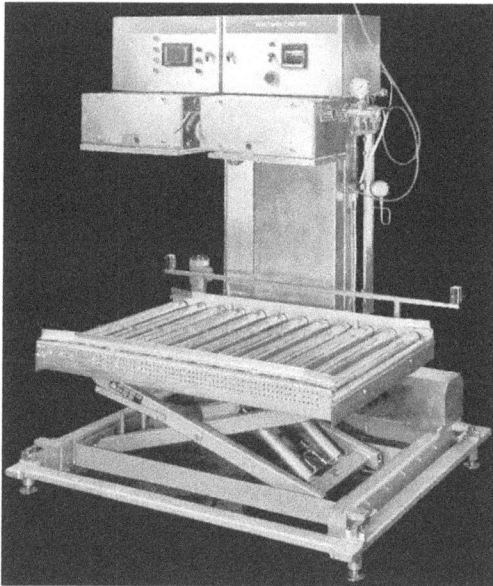

Figure A3.11   DuPont/Liqui-Box StarAsept™ Model 1307 bulk aseptic filler. (Photograph from DuPont/ Liqui-Box literature.)

Sterilization time—Approximately 45 minutes.

Package types—Premade, presterilized bags or various flexible packaging laminates with fitments.

Sterilization of packaging—Preformed, sealed bags are sterilized by gamma radiation. The cap and spout is resterilized using either steam (StarAsept) or hydrogen peroxide.

DuPont/Liqui-Box bag-in-box fillers are FDA validated for filling low-acid foods.

Additional Information

- DuPont/Liqui-Box is the second largest supplier of aseptic bag-in-box fillers and aseptic bag-in-box packaging.
- There are more than 220 StarAsept fillers installed worldwide.

Product: Pouches

Manufacturer of aseptic filling equipment for form–fill–seal pouches and associated packaging of various flexible packaging laminates.

Products Currently Being Aseptically Filled Using DuPont/Liqui-Box Pouch Filling Equipment

Milk, cream, cheese and other sauces, sour cream, nutritional food supplements specialty beverages, ice cream mix, and fruit juices.

Equipment Models

Model	Operation	Package sizes	Production rate
A-6	Automatic	200 mL to 1 liter	Up to 120 ppm
DA-4000	Automatic	200 mL to 2 liters	Up to 100 ppm

Filler—Double-head vertical form–fill–seal continuous filling with programmable logic controller.

Sterilization of filler—Steam.

Sterilization time—Approximately 45 minutes.

Package types—Form–fill–seal packaging with low, medium, and high barrier films.

Sterilization of packaging—Sterilization of packaging is with 35% hydrogen peroxide.

DuPont/Liqui-Box aseptic pouch fillers are FDA validated for filling low-acid foods.

Additional information

- Nitrogen injection can be supplied as an option.
- DuPont Liqui-Box is the second largest supplier of aseptic pouch fillers in the United States.

ELPO

FBR-Elpo Sp.A.
Via Arnaldo de Brescia, 12/A

43100 Parma (Italy)
Ph: (39) 0521 267511
Fax: (39) 0521 267676
Web: www.fbr-elpo.it

Product

Manufacturer of aseptic filling equipment for bag-in-boxes, bag-in-drums, and bag-in-cartons.

Products Currently Being Filled Using Elpo Fillers

Tomato products, fruits, juices, purees, concentrates.

Equipment Models

See Figure A3.12.

Figure A3.12   Elpo single-head aseptic filler. (Photograph courtesy of Elpo.)

Model	Operation	Package sizes	Production rate
Bulk filler	Semiautomatic	2–1,000 liters	Manual
Web-fed	Automatic	5–30 liters	10–12 bpm

Filler—The filler must be loaded with the bag and manually started. Then the filler automatically resterilizes the spout and cap with steam, removes the cap and volumetrically meters the filling of the bag, replaces the cap, and ejects the filled bag.

Sterilization of filler—Steam.

Sterilization time—Approximately 1 hour.

Package types—Premade, presterilized, multilayer bags with "plug"-type fitments.

Sterilization of the packaging—Preformed, sealed bags are sterilized by gamma radiation.

Elpo fillers are currently not FDA validated for aseptically filling low-acid food products.

Additional Information

- Elpo has been very successful in marketing its fillers in Europe, Latin America, the Middle East, and Asia.
- Elpo fillers are used in these regions to fill large and small bags with high- and low-acid products.
- Elpo has no representation in the United States.

FENCO

Fenco S.p.A.
Via Prampolini
40-43044 Lemignano
Parma (Italy)
Ph: (39) 0521 303429
Fax: (39) 0521 303428
Web: www.fenco.it

Product

Manufacturer of aseptic filling equipment for bag-in-drum and bag-in-bin.

Products Currently Being Filled Using Fenco Fillers

Tomato products, processed fruits, juices, purees, concentrates.

Equipment Models

See Figures A3.13 and A3.14.

Figure A3.13   Fenco Model 154 aseptic bag filler. (Photograph from Fenco literature.)

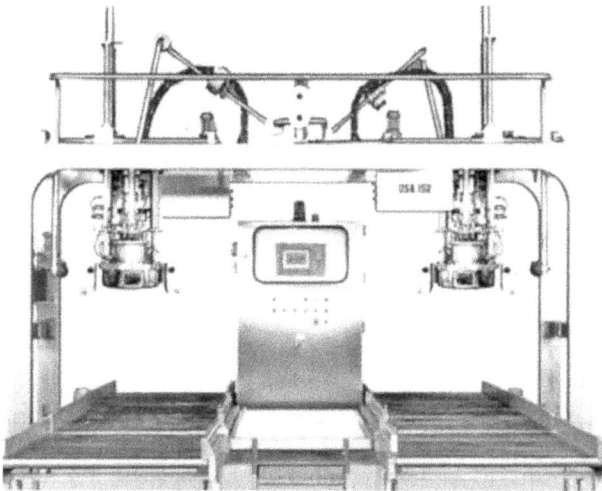

Figure A3.14   Fenco Model 152 aseptic bag filler. (Photograph from Fenco literature.)

Model	Package sizes	Production rate
154	Up to 20 liters	Manual
152 bulk filler	25–1,000 liters	Semimanual

Filler—Fill is controlled by volumetric flowmeter. The filler must be loaded with the bag and manually started. The filler then automatically resterilizes the spout and can with steam, removes the cap, and volumetrically meters the filling of the bag. The cap is then replaced and the bag is ejected.

Sterilization of filler—Steam.

Sterilization time—Approximately 45 minutes.

Packaging—Premade, multilayer bags with fitments. Bag sizes from 25–1,000 liters.

Sterilization of bags—By gamma radiation. The spout and cap are resterilized with steam.

Fenco fillers are currently not FDA validated for aseptically filling low-acid foods.

Additional Information

- Fenco was established in 1986.
- Fenco can also supply the aseptic processing equipment to sterilize the products prior to filling.
- Fenco does not have representation in the United States.
- The bags used for Fenco fillers use the "plug"-style fitment design.

FRES-CO SYSTEM

3005 State Road
Telford, PA 18969-1033
Ph: (215) 721-4600
Fax: (215) 721-4414
Web: www.fresco.com

Product

Manufacturer of aseptic filling equipment and related packaging for form–fill–seal pouches.

Products Currently Being Filled Using Fres-co Fillers

Cheese sauces, puddings, and other dairy products. The Fres-co aseptic fillers are capable of filling foods with particulates.

Equipment Models

See Figure A3.15.

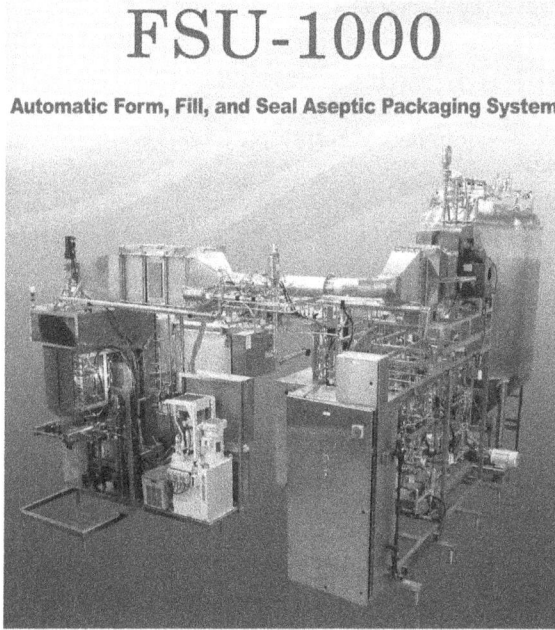

Figure A3.15   FSU-1000. (Photo courtesy of Fres-co Systems, USA.)

Type/model	Package sizes	Production rate
FSU 1000A	10 mL to 10 liters	Up to 30 one-gallon ppm Up to 500 half ounce ppm
FSU 800 high acid[a]	50 mL to 10 liters	Up to 30 pouches[b] ppm

[a] For products with a pH <4.6, aseptic cold fill, ESL low acid or hot fill of high acid.
[b] Either flat or stand-up pouches. Fitment attachment is optional.

Filler—Options: Time/pressure/orifice filler, positive displacement piston filler, rotary pump filler, and magnetic or mass flowmetering fillers.

Sterilization of filler—By steam and H_2O_2.

Sterilization time—Approximately 1 hour.

Packaging—Form–fill–seal, multiply high-barrier laminates with and without fitments. Pouches can be printed or unprinted.

Sterilization of packaging—Packaging material is sterilized with H_2O_2.

The Fres-co aseptic pouch filler is FDA validated for aseptically filling low-acid foods and beverages.

Additional Information

- Fres-co engineering has many years of experience developing and improving aseptic form–fill–seal equipment.

- The current aseptic filling equipment being supplied by Fres-co incorporates new technology that will surely secure Fres-co as a leading supplier of aseptic filling equipment for pouches.
- Fres-co is already one of the leading suppliers in the world for packaging material.
- Fres-co is one of the few suppliers of aseptic packaging equipment that is manufactured in the United States.
- Fres-co can supply aseptic pouches with and without fitments.
- Fres-co also manufactures high-speed equipment for hot filling form–fill–seal pouches (acid products) and equipment for retort pouches.

HRS PROCESS TECHNOLOGY

5035 N. 55th Ave., Suite 6
Glendale, AZ 85301
Ph: (623) 915-4328
Fax: (623) 939-6168
Web: www.hrs-america.com

Product

Manufacturer of aseptic filling equipment for bag-in-box, bag-in-drum, bag-in-bin.

Products Currently Being Aseptically Filled Using HRS Fillers

Processed diced fruits, fruit concentrates, tomato products, citrus products, vegetables, coffee concentrates, and soups.

Equipment Models

See Figure A3.16.

Model	Package sizes	Type filler
HRS AF-1	Up to 6 gallons	R & D Portable
HRS AF-1-A	Up to 6 gallons	Automatic
HRS AF-2	Up to 6 gallons and 55 gallons	Automatic
HRS AF-3	55 and 300 gallons	Manual

Filler—Fill controlled by flowmeter or by weight.
Sterilization of filler—By steam and hot water.
Sterilization time of filler—Approximately 45 minutes to 1 hour.
Package types—Premade, presterilized bags with fitments. HRS does not supply the bags.
    Scholle, Aran, Rapak, and Liqui-Box bags can be aseptically filled on the HRS filler.
Sterilization of the bags—Preformed, sealed bags are sterilized by gamma radiation.

Figure A3.16   HRS aseptic filler for bag-in-drum. (Courtesy of HRS.)

HRS fillers are not FDA validated for filling low-acid foods.

Additional Information

- HRS aseptic fillers are manufactured in Argentina and Spain.
- HRS has nearly 100 aseptic bag fillers in operation.
- HRS aseptic fillers can fill food products containing particulates.
- HRS has a testing facility in Spain where interested processors can aseptically fill products into bag-in-box.

JBT FOODTECH

2300 Industrial Avenue, Box A
Madera, CA 93639
Ph: (559) 661 3200
Fax: (559) 661 3222
Web: www.jbtfoodtech.com

European office:

John Bean Technologies S.p.A.
Via Mantova 63/A
43100 Parma, Italy
Ph: 39 0521 908411

Fax: 39 0521 460897
Web: www.jbtfoodtech.com

Product

Manufacturer of aseptic filling equipment for bag-in-drum and bag-in-bin packaging. JBT FoodTech also manufactures and supplies the complementary aseptic processing systems. JBT also owns the Metal Box aseptic cup filler.

Products Currently Being Filled Using JBT FoodTech Fillers

Diced tomatoes and paste, fruit purees and concentrates, yogurt fruit sauces, citrus juices, pulp and concentrates, cheese sauce (low acid), and sauces into bag-in-box.

Equipment Models

See Figures A3.17 and A3.18.

Model	Operation	Package sizes	Production rate
ABF-0.200	Semiautomatic	1–230 liters	Up to thirty 230-liter bags/hour
ABF-1200	Semiautomatic	5–1,200 liters	Up to fifteen 1,200-liter bags/hour

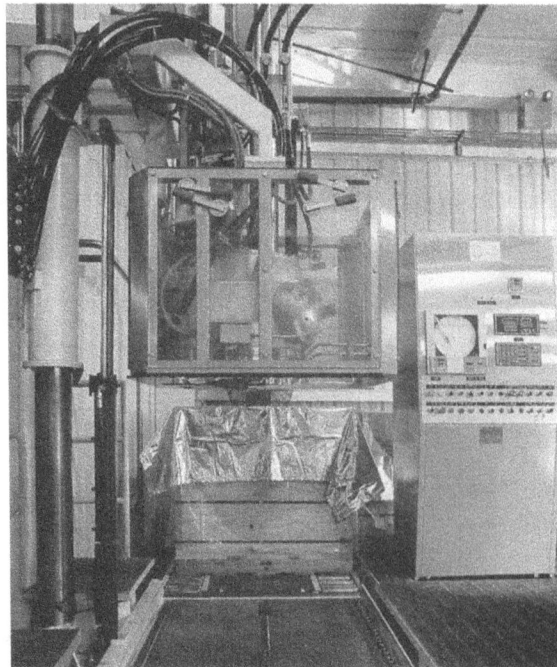

Figure A3.17   JBT FoodTech aseptic bag filler. (From Fran Rica literature.)

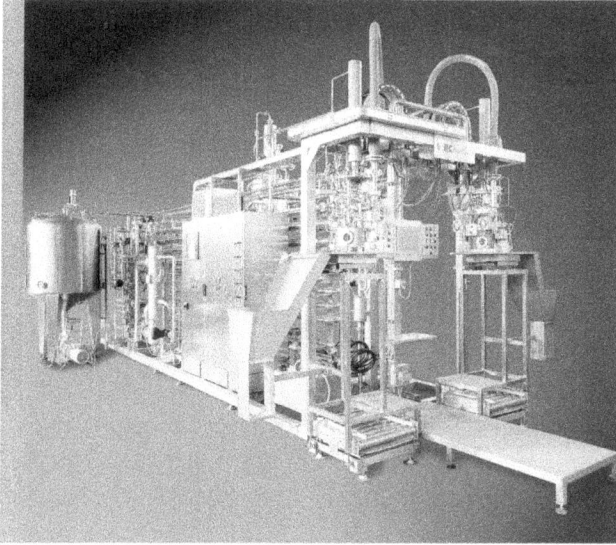

Figure A3.18 JBT FoodTech "compact" aseptic plant skidded aseptic processing system and filler. (From JBT FoodTech literature.)

Filler—Bag fillers are controlled load cells (primary) and by flowmeters (optionally).

Sterilization of fillers—Steam and hot water.

Sterilization time—Approximately 1 hour for the various fillers.

Package types—Bag filler utilizes premade, presterilized bags with fitments, including Fran Rica 63-mm and 3-inch membrane fitments, 1- and 2-inch snap caps, and 1- and 2-inch plug fitments.

Sterilization of packaging—Bags are presterilized by gamma radiation. Cups are sterilized by hydrogen peroxide.

The ABF-0.200 bag filler is FDA validated for aseptically filling low-acid foods.

Additional Information

- JBT FoodTech has many years of experience in aseptic processing and packaging food products.
- JBT FoodTech is one of the few companies that can supply both the aseptic packaging and processing system.
- JBT FoodTech does not supply the aseptic packaging material for their fillers.

KHS AG

Juchostrasse 20
D-44143 Dortmund
Germany

Ph: (49) 231 569 0
Fax: (49) 231 569 141
Web: www.khs-ag.com

Sales and service office in the United States:

KHS USA
880 Bahcall Ct.
Waukesha, WI 53186
Ph: (262) 787-7200
Fax: (262) 787-0025

Product

Rotary aseptic bottle filler for beverages and related aseptic processing systems.

Products Currently Being Filled Using KHS Fillers

Fruit juices, juice drinks, teas, and milk and other dairy products.

Equipment Models

KHS Alfill rotary capable of aseptically filling up to 800 bpm of high-acid (<pH 4.6) and 600 bpm of low-acid (>pH 4.6) beverages.

Filler—Electronic controlled volumetric filling by weight cells.
Sterilization of filler—Either dry vapor and H_2O_2 or wet peracetic acid.
Sterilization time of filler—Approximately 90 minutes.
Packaging—HDPE and PET bottles.
Sterilization of packaging—Dry bottle sterilization with H_2O_2 and warm air; or wet bottle sterilization using peracetic acid.

The KHS bottle filler has received FDA validation for aseptically filling low-acid beverages.

Additional Information

- KHS has a manufacturing, sales, and service organization located in Waukesha, Wisconsin.
- KHS can sterilize the packaging with either peracetic acid or hydrogen peroxide.
- KHS can supply the mutually dependent processing equipment for a turnkey aseptic system.

KRONES AG

Bohmerwaldstrabe 5
D-93068 Nuestraubling
Germany
Ph: (40) 94 01/70-0
Fax: (40) 94 01/70 24 88

Sales and service office in the United States:

Krones USA
9600 S. 58th St.
Franklin, WI 53132
Ph: (414) 409-4000
Fax: (414) 409-4100

Product

Manufacturer of rotary aseptic and extended shelf-life (ESL) bottle fillers for beverages and related aseptic processing systems.

Products Currently Being Filled Using Krones Fillers

Water, teas, fruit juices, and extended shelf-life dairy products.

Equipment Models

Model	Bottle sizes	Production rate
VODM-PET		36,000 bph
CAF	Up to 2 liters	Up to 46,500 bph

Approximate delivery—6 months.
Filler—Volumetric filling valve with inductive flowmeter.
Sterilization of filler—Presterilized by steam and water at up to 135°C. The rinser, filler, and closer are in a clean room designed to comply with Class 100 sterile room.
Sterilization time of filler—Approximately 90 minutes.
Packaging—PET bottles up to 2 liters with screw cap closure, including sports caps.
Sterilization of bottles—By means of either peracetic acid or a gaseous hydrogen peroxide achieving a log 5 reduction.

Krones aseptic bottle fillers have not been validated by the FDA for filling low-acid beverages.

Additional Information

- Krones installed their first aseptic filler in Switzerland in April 1999 aseptically filling iced tea.
- Krones can supply the entire aseptic processing system, including premixing (formulating), processing, filling, labeling, and palletizing.
- Krones aseptic fillers cannot fill beverages with discrete particulates.
- Krones has a testing facility in Germany.
- Krones has no low-acid aseptic fillers installed in the United States.

OYSTAR HASSIA

Verpackungsmaschien GmbH
P.O. Box 1120
63689 Ranstadt, Germany
Ph: 49 6041 810
Fax: 49 6041 81213
Web: www.OYSTAR.hassia.de

Sales and service office in the United States:

Hassia USA Inc.
1210 Campus Drive West
Morganville, NJ 07751
Ph: (732) 536-8770
Fax: (732) 536-8850
E-mail: sales@oystarusa.com

Product

Manufacturer of aseptic filling equipment for form–fill–seal plastic cups and StickPack sachets.

Products Currently Being Filled Using Hassia Fillers

Baby foods, fruit juices, desserts, soups, milk, cream dishes, coffee creamers, puddings, cheese sauces, fruit gels, and sour cream.

Equipment Models

See Figures A3.19–A3.22.

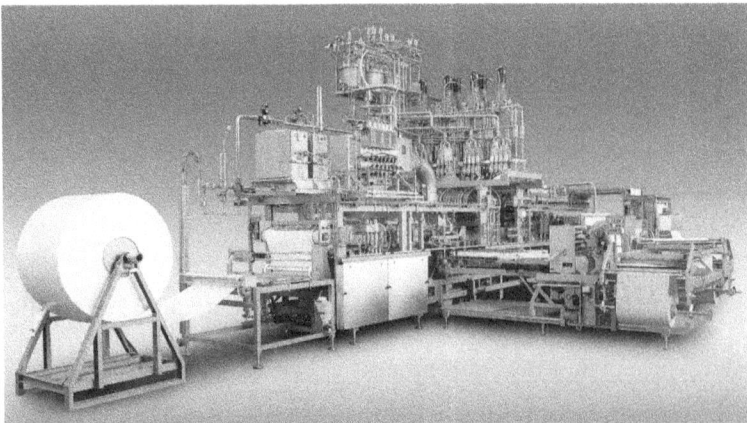

Figure A3.19   OYSTAR Hassia form–fill–seal aseptic cup filler. (Photograph courtesy of OYSTAR Hassia, USA.)

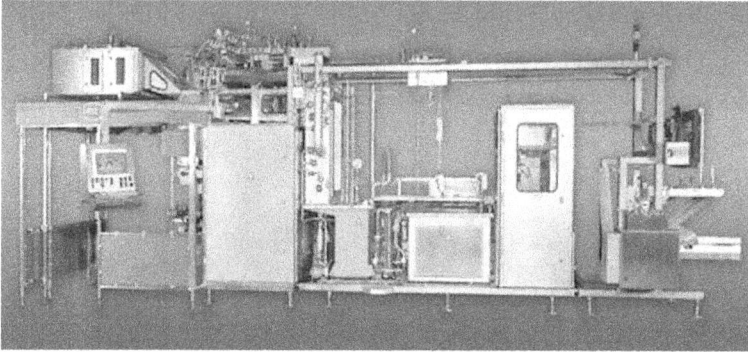

Figure A3.20 Hassia aseptic stick pack filler Model SVP 20/30. (Photograph from OYSTAR Hassia literature.)

Figure A3.21 Aseptically filled products using Hassia cup filler. (Photograph from OYSTAR Hassia literature.)

Type	Model	Production rate
Form–fill–seal cups	TAS 8/48	280 cups per minute
	TAS 16/48	420 cups per minute
	TAS 32/48	720 cups per minute
	TAS16/80	840 cups per minute
	TAS 32/80	1680 cups per minute
Form–fill–seal stick packs	SAS 20/60	480; 2.2 oz sticks per minute

Figure A3.22   Products aseptically filled on the Hassia aseptic stick filler. (Photograph from OYSTAR Hassia literature.)

Filler—Up to three stage fillers are possible, controlled by computer. Accurate up to ±2 grams. Fillers can fill particulates up to 12 mm.

Production Rate of Aseptic Cup Fillers

TAS 8/48	TAS 16/48	TAS 32/48	TAS 16/80	TAS 32/80
6 up	12 up	24 up	24 up	48 up
35 strokes	35 strokes	30 strokes	35 strokes	40 strokes
16,800 cph	25,200 cph	43,200 cph	50,400 cph	100,800 cph

Sterilization of filler—Accomplished by H_2O_2 followed by drying. After sterilization, the filler is maintained sterile by positive filtered, sterile air over pressure.

Sterilization time of filler—1 hour.

Packaging—Form–fill–seal of various barrier polymers (including polypropylene) from 3 to 8 oz.

Sterilization of packaging—Sterilization of the packaging depends upon the material. Hassia can sterilize the packaging by:

- Dry heat
- Radiation
- Moist heat (steam)
- Chemical (hydrogen peroxide)

Hassia fillers are FDA validated for aseptic filling low-acid food products.

Additional Information

- OYSTAR Hassia is the leading supplier of aseptic filling equipment for plastic cups in the United States.
- OYSTAR Hassia has a major sales, service, and spare parts company located in New Jersey.
- OYSTAR Hassia also owns these other aseptic and extended shelf-life fillers:
  - OYSTAR Gasti
  - OYSTAR Hamba
  - OYSTAR Erca
  - OYSTAR Hassia also manufactures aseptic coffee creamer filling equipment.
  - Hassia has many years of experience manufacturing aseptic filling equipment.

PROCOMAC

GEA Procomac S.p.A.
Via Fedolfi, 29
43038 Sala Baganza (Parma) Italy
Ph: 39 0521 839411
Fax: 39 0521 833879
Web: www.procomac.it

Sales and service office in the United States:

GEA Process Engineering Inc.
1600 O'Keefe Road
Hudson, WI 54016
Ph: (715) 386-9371
Fax: (715) 386-9376

Product

Manufacturer of rotary aseptic bottle fillers for beverages. Procomac also manufactures and supplies the product preparation equipment and mutually dependent aseptic processing system.

Products Currently Being Filled Using Procomac Fillers

Fruit juices, sport drinks, dairy products, soy milk, iced tea, and nutriceutical products.

Equipment Models

See Figures A3.23–A3.26.

Model	Packaging size	Production rate
Fillstar Fx	250 mL to 2 liters	Up to 800 bpm

Current delivery—5–6 months.

Filler—Counter-pressure volumetric electronic filling head with magnetic flowmeters on each filling valve. The Procomac filler is capable of filling round, square, and rectangular bottles.

Sterilization of filler—Steam and peracetic acid: product path is sterilized by steam.

Sterilization time of filler—Time for clean-in-place (CIP) followed by sterilization-in-place (SIP) is approximately 4.5 hours for automated cycle.

Packaging—HDPE and PET bottles.

Sterilization of packaging—By the use of peracetic acid obtaining a log 6 reduction of reference target microorganism and with peroxide sterilization.

Bottle sealing—Either with screw cap or aluminum foil lined fitment put on in sterile zone.

Procomac aseptic fillers have been FDA validated to fill low-acid beverages.

Additional Information

- Procomac can normally change bottle sizes in 30 minutes.
- Procomac has been supplying aseptic fillers since 1996.

GEA Procomac Aseptic Filling Line

Figure A3.23   GEA Procomac aseptic filling line. (Photograph from Procomac literature.)

GEA Procomac PET Bottle Filling

Figure A3.24   GEA Procomac PET bottle filling. (Photograph from Procomac literature.)

Figure A3.25   Procomac aseptic rotary bottle filler. (Photograph from Procomac literature.)

- Procomac has installed 95 aseptic fillers worldwide, one-third are filling low-acid products; 35 of these fillers are aseptically filling low-acid beverages.
- Procomac aseptic fillers can fill particulates up to 5 mm × 5 mm × 5 mm.
- Procomac has a test facility in Parma, Italy.
- Procomac can engineer, manufacture, and supply the mutually dependent aseptic processing system.

Figure A3.26   Products aseptically filled with Procomac filler.

- Procomac trains operators, either at its own processing plant or in Italy.
- Procomac has Internet modem-based diagnostic services.

PURITY

Genpak
68 Warren Street
Glens Falls, NY 12801-0727
Ph: (724) 457-3326
Fax: (724) 457-3328

Product

Manufacturer of aseptic filling equipment for preformed coffee creamers.

Equipment Models

Model	Package size	Production rate
SC 2104A	10–20 mL	Up to 1,500 creamers/minute

Filler—Purity manufactures a linear indexing filler that fills volumetrically by the use of 3A approved diaphragm pumps.
Approximate delivery—1 year.
Packaging—Multiple materials, including HIPS and high-impact polystyrene.
Sterilization of filler—Accomplished by steam and hydrogen peroxide.
Sterilization time—30 minutes.
Sterilization of packaging—Accomplished by H_2O_2 spray, dried with filtered hot air.

The Purity filler is FDA validated for aseptically filling low-acid foods.

Additional Information

- Purity fillers are manufactured in Toronto, Canada.
- The aluminum lidding material is also produced in Canada.
- The cups are manufactured in Longview, Texas.

RAPAK

D S Smith Plc
Beech House
Whitebrook Park
68 Lower Cookham Road
Maidenhead
Berkshire SL6 8XY
Ph: 44 1628 583 400

Offices in the United States:

Rapak USA, Division of D S Smith Plastics
1201 Windham Parkway
Romeoville, IL 60446
Ph: (630) 296-2000
Fax: (630) 296-2195
Web: www.rapak.com

and

Rapak USA
299959 Ahern Ave.
Union City, CA 94587
Ph: (510) 324-0170
Fax: (510) 324-0180

Product

Manufacturer of aseptic bags and aseptic filling equipment for bag-in-box, bag-in-drum, bag-in-bin, and supply of packaging.

Products Currently Being Aseptically Filled Using Rapak/Intasept Fillers

Fruits and vegetable purees, dairy products, sauces, juices, processed fruit, milkshake base, and ice cream mix.

Equipment Models

See Figure A3.27.

Figure A3.27   Intasept aseptic fully automatic bag-in-box filler. (Photograph from Rapak literature.)

Model	Packaging sizes	Production rate
Intasept™ 2400	5–10 liters	Up to 4 bpm
Intasept 2600	5–20 liters	3–5 bpm
Intasept 2800 Manual	Up to 1,000 liters	4.5–18 bph

Filler—The manual filler must be loaded with the bag and manually started. The filler then automatically resterilizes the fitment with steam, punctures the film over the fitment, and volumetrically meters the filling of the bag. The filler then reseals the film to the backside of the fitment and ejects the filled bag.

The automatic filler feeds the preperforated webbed bags to a guillotine that separates the bag. It then inserts the bag into the filling chamber where the filler resterilizes the fitment with steam, punctures the film over the fitment, volumetrically meters the filling of the bag, reseals the film to the backside of the fitment, and then ejects the filled bag into an optional carton-loading system.

Sterilization of filler—By steam and maintained by sterile water and air.

Sterilization time of filler—Approximately 1 hour.

Package types—Premade, presterilized bags with fitments. Rapak manufactures bags up to 100 liters.

Sterilization of packaging—Bags are presterilized by gamma radiation.

Rapak/Intasept fillers are FDA validated for aseptically filling low-acid food and beverages.

ROSSI CATELLI S.P.A.

Via Traversetolo, 2/A
43100 Parma
Italy

Ph: (39) 0521 463284
Fax: (39) 0521 463284
Web: www.rossicatelli.com

In the United States:

Process Resource, Inc.
P.O. Box 1620
Oakdale, CA 95361
Ph: (209) 499-1974
Fax: (209) 847-4821

Product

Manufacturer of aseptic filling equipment for bag-in-box, bag-in-drum, and bag-in-bin.

Products Currently Being Filled Using Rossi Catelli Fillers

Tomato paste and diced, processed fruits, vegetables, and dairy products.

Equipment Models

See Figure A3.28.

Figure A3.28   Rossi Catelli Macropak™ 2000/2 aseptic filler. (Photograph from Rossi Catelli literature.)

Model	Filling heads	Package sizes
Macropak RVL/2T	Dual	Up to 230 liters
Macropak TM 2000/2	Single and dual	230–1,500 liters

Filler—Fill is controlled by electromagnetic flowmeter or weight cells. Up to 3-inch opening.

Sterilization of filler—Steam.

Sterilization time—Approximately 45 minutes.

Package types—Premade, presterilized bags with and without fitments. Rossi Catelli does not supply the packaging.

Sterilization of packaging—By gamma radiation. The cap and spout are resterilized using steam.

Rossi Catelli fillers are currently not FDA validated for aseptic filling of low-acid foods.

Additional Information

- Rossi Catelli has been manufacturing aseptic fillers for many years.
- Rossi Catelli purchased Manzini, an Italian manufacturer of aseptic bag-in-box fillers.
- Rossi Catelli can supply the aseptic processing system to sterilize the product prior to filling.
- Rossi Catelli has over 265 aseptic fillers currently in operation worldwide.

SCHOLLE

Scholle Corp.
19520 Jamboree Road
Irvine, CA 92612
Ph: (949) 955-1750
Fax: (949) 250-1462

Regional and sales office:

Scholle Corp.
200 W. North Ave.
Northlake, IL 60164
Ph: (708) 562-7290
Fax: (708) 562-6569
Web: www.scholle.com

Product

Manufacturer of aseptic filling equipment for bag-in-box, bag-in-drum, and bag-in-bin. Also, Scholle is the world's leading supplier of presterilized, aseptic bags.

Products Currently Being Filled Using Scholle Fillers

Processed fruits, juices, purees and concentrates, sauces, dairy products, liquid eggs, tomato products, vegetable products, pumpkin puree, citrus products, and coffee creamer.

Equipment Models

See Figures A3.29 and A3.30.

Model	Package sizes	Production rate
AF10-2E (high acid)	5–200 liters	Manual
AF10-2LA (low acid)	5–200 liters	Manual
AF14 (high acid)	200–1,150 liters	Manual
AF19-A (high acid)	5–20 liters	5–7 bags/minute
Surefill 22 (high acid)	5–20 liters	Web fed; up to 15 bpm
Surefill 30 LA	.5–5 gallons	Web fed; up to 15 bpm

Filler—Fill is controlled by flowmeter.
Sterilization of filler—By steam, chlorine, and hot water.

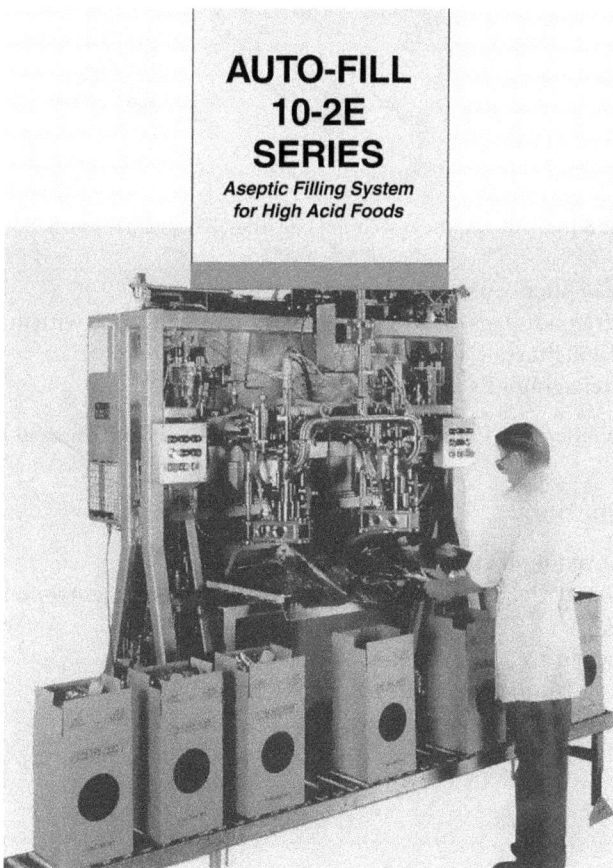

Figure A3.29   Scholle aseptic bag-in-box filler. (Photograph from Scholle literature.)

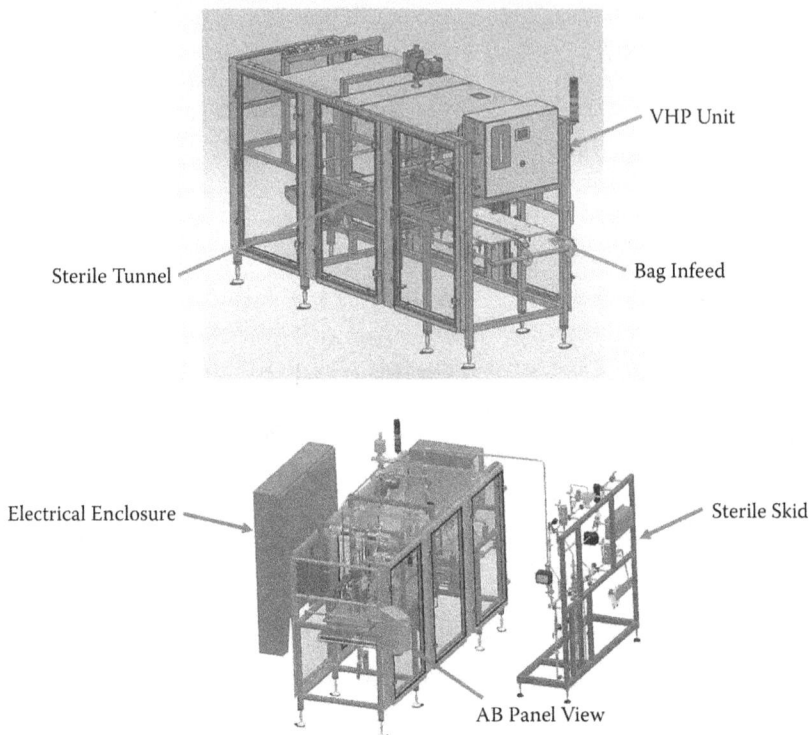

Figure A3.30   Scholle Surefill 30 LA linear web-fed bag filler. (Diagrams from Scholle literature.)

Sterilization time of filler—Approximately 45 minutes.
Package types—Premade bags or various polymers, manufactured with fitments. Bag sizes
   are from 5 to 1,150 liters.
Sterilization of packaging—By gamma radiation.

Some Scholle aseptic fillers are FDA validated for aseptically filling low-acid foods.

Additional Information

- Scholle is the inventor of aseptic bag-in-box packaging.
- Scholle is the world's leading supplier of aseptic filling equipment for bag-in-box/drum/
  bin packaging.
- Scholle is the world's leading supplier of aseptic bags.
- Scholle has manufacturing facilities on five continents.
- Scholle fillers can be purchased or leased.

SERAC

Serac Group
12 route de Mamers BP 46

72402 La Ferte Bernard Cedex
France
Ph: (33) 2 43 60 28 28
Fax: (33) 2 43 60 28 39
www.serac-group.com

Manufacturing, sales, and service office in the United States:

Serac, Inc.
300 S. Westgate Dr.
Carol Stream, IL 60188
Ph: (630) 510-9343
Fax: (630) 510-9357
Web: www.serac-usa.com

Product

Manufacturer of rotary aseptic fillers for beverages into plastic bottles (see Figures A3.31 and A3.32).

Products Currently Being Filled Using Serac Fillers

Fruit juices, milk, milkshakes, yogurt drinks, coffees, non-carbonated drinks, eggnog, dairy beverages, teas, sports drinks, soups, and mineral water.

Figure A3.31   Serac rotary bottle filler. (Photograph courtesy of Serac.)

Figure A3.32   Some products filled using Serac fillers. (Photograph from Serac literature.)

Equipment Models

Model	Packaging sizes	Production rate
Lab filler	75 mL to 3 liters	Approx. 20 bpm
SAS 4 TF	75 mL to 3 liters	Up to 600 bpm
SAS 4 TF	75 mL to 3 liters	Up to 800 bpm

Approximate delivery—7–8 months.

Filler—Fill is by net weight.

Sterilization of filler—By peracetic acid and superheated water at 285°F.

Sterilization time of filler—Approximately 2 hours.

Packaging—PET, HDPE, Barex, and polyethylene bottles from 75 mL to 3 liters. Serac filler can also fill steel and aluminum cans for aerosol whipped cream.

Sterilization of packaging—Preformed bottles are sterilized by the use of peracetic acid obtaining a log 6 reduction.

Serac bottle fillers are not currently FDA validated for aseptically filling low-acid beverages.

Additional Information

- Serac has a complete manufacturing, sales, and service organization, including spare parts located in Carol Stream, Illinois.
- Serac is a leading supplier of aseptic fillers for plastic bottles with more than 80 installations worldwide of which 50 are filling low-acid beverages.

- Serac fillers can fill particulates up to 10 mm in diameter.
- Serac also manufactures a filler (model STAS) for aseptic and ESL filling of HDPE blow-molded (sterilized and sealed in the blow-molding process) bottles.
- Serac has a fully equipped testing facility located in France.
- Comparative floor space requirement for Serac bottle fillers is small.

SHIBUYA KOGYO

Shibuya Kogyo Co., LTD.
Mameda-Honmachi
Kanazawa 920
Japan
Ph: 0762-62-1200
Fax: 0762-23-1921
Web: www.shibuya-int.com

Sales and service office in the United States:

Shibuya Hoppmann Corp.
13129 Airpark Drive, Suite 120
Elkwood, VA 22718
Ph: (800) 368-3582
Fax: (540) 829-1724

Product

Manufacturer of rotary aseptic filling equipment for PET, HDPE, and other plastic bottles.

Products Currently Being Aseptically Filled Using Shibuya Fillers

Flavored milk beverages, apple juice, tomato juice, milk, milk tea, milk coffee, and Japanese, Chinese, English, and barley teas.

Equipment Models

Model	Production rate	Package size
NWF36-120	1200 bottles per minute	Up to 16 oz
NWF32-108	900 bottles per minute	Up to 16 oz
NWF32-81	600 bottles per minute	Up to 32 oz
NWF4390F/SR	400 bottles per minute	Up to 64 oz

Current delivery—9 months.
Filler—Rotary volumetric aseptic filler for plastic bottles.
Sterilization of filler—The Shibuya filler and capping system is based on a closed chamber principle. All the equipment for sterilizing the incoming components, filling, and

capping is contained in stainless steel chambers that have been presterilized with H_2O_2 with heat. The fully automatic presterilization is carefully controlled, making it possible to fill sterile product into sterile containers in a sterile environment. Once the filler and chamber are sterile, they are maintained in a sterile state by the use of ultra-filtered air overpressure. See Figure A3.33.

Sterilization time of filler—Approximately 2 hours.

Packaging—Standard 500–2,000 mL PET bottles and other suitable plastic containers (i.e., HDPE). The system can handle round, square, or rectangular bottles with minimal change parts. Other container geometries could be handled with extra change parts. The closure is an aseptic type screw cap without under cap foil seal.

Sterilization of packaging—H_2O_2 with heated air and electron beam.

The Shibuya filler is FDA validated for aseptically filling low-acid beverages.

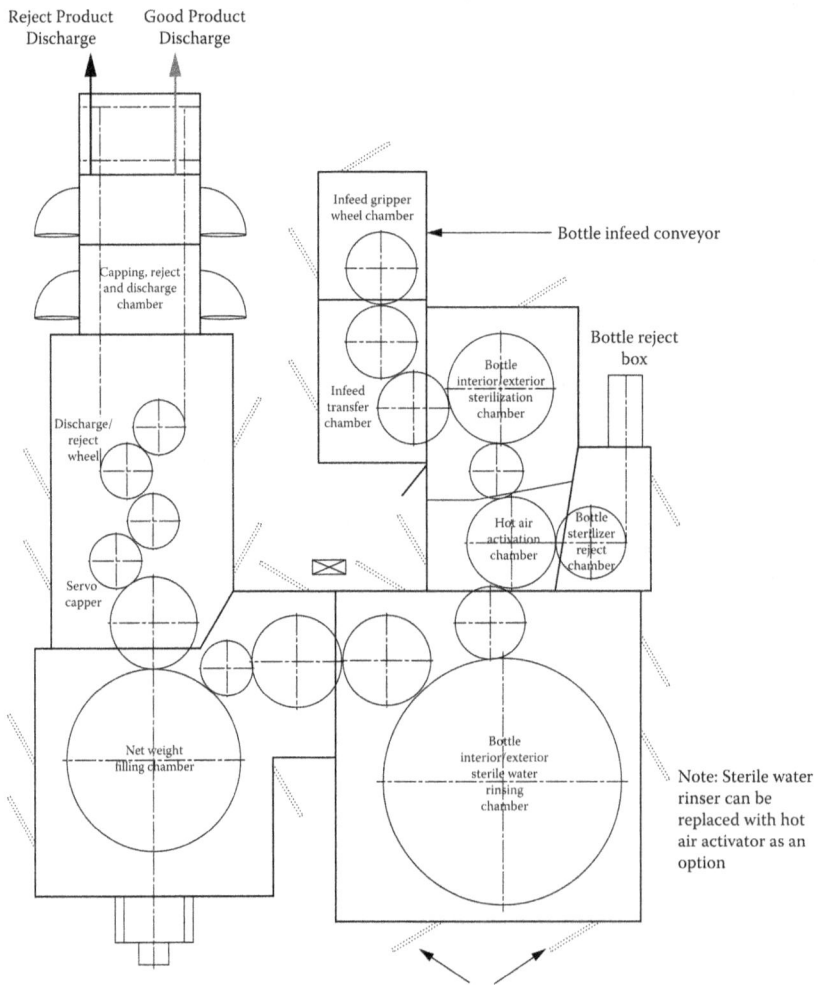

Figure A3.33   Aseptic system chamber layout.

Additional Information

- At 1,200 bottles per minute, the Shibuya aseptic filler is the highest production speed filler for low-acid beverages on the market.
- Shibuya has years of experience at installations in Asia and the Far East.
- Shibuya is one of the world's leading suppliers of aseptic filling equipment for plastic bottles and has over 100 aseptic filler installations.
- Shibuya has recently introduced electron beam sterilization of PET bottles and can obtain a log 6 reduction.
- Shibuya has a test facility in Japan where it can aseptically process and package high- and low-acid beverages.
- Shibuya cannot aseptically fill products with particulates.
- Shibuya aseptic fillers utilize more floor space than any other aseptic beverage filler.
- It was directly reported by a processor at an installation in the United States that there is 4 miles of 4-inch stainless steel piping installed just for the use of the Shibuya filler and not associated with the processing of the product.

The Aseptic System Chamber Layout

The aseptic filling system consists of a number of stainless steel chambers joined together. Normally the chambers are arranged in a horseshoe formation. The chambers are further subdivided by stainless steel partitions as shown below.

Note: The heavy solid lines denote the chamber walls and partitions, the light broken lines denote equipment inside the chambers.

SIDEL

Sidel S.p.A.
Via La Spezia 241/A
43100 Parma, Italy
Ph: +39 0521 9991
Fax: +39 0521 959009
Web: www.sidel.com

Sales and service office in the United States:

Sidel
5600 Sun Court
Norcross, GA 30092
Ph: (678) 221-3000
Fax: (678) 221-3266

Product

Manufacturer of rotary aseptic and extended shelf-life fillers for filling beverages into plastic bottles. Sidel also manufactures associated sterile blow molders. Sidel USA is additionally responsible for LFA-20 (originally Tetra Pak) linear aseptic bottle fillers installed in the United States.

Products Currently Being Filled Using Sidel Fillers

Fruit juices, Powerade, energy drinks, teas, soy milk, milk and other dairy products, and coffee creamers.

Equipment Models

See Figures A3.34–A3.36.

Type filler	Package sizes	Production rate
Combi Predis™ FMa	100–2,000 mL	Up to 36,000 bph
Sensofill™ FMa	100–2,000 mL	Up to 60,000 bph
LFA-20	100 mL–1 liter	

Figure A3.34   Sensofill FMa. (Diagram from Sidel literature.)

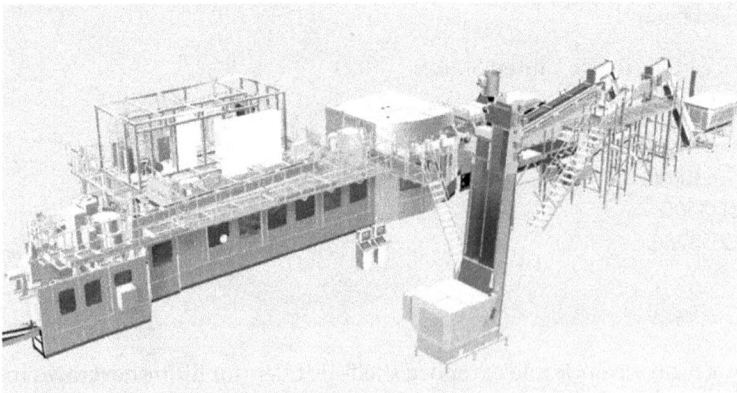

Figure A3.35   Combi Sensofill FMa. (Diagram from Sidel literature.)

Figure A3.36   Some products filled using Sidel fillers. (Photograph from Sidel literature.)

Current delivery—6 months.

Filler—Magnetic gravity-level filler with flowmeters using no membrane that is capable of filling pulp.

Sterilization of filler—Presterilization is accomplished by 280°F steam. Sterility is maintained by positive filtered, sterile air pressure.

Sterilization time of filler—70 minutes.

Packaging—PET, PP, and HDPE bottles with screw caps, heat-sealed foil caps, and sports caps.

Sterilization of packaging—Predis™, accomplished by H_2O_2 vapor; sterilization up to 4 log; Sensofill™, accomplished with peracetic acid.

Sidel Combi and Sensofill fillers are currently not FDA validated for aseptic filling of low-acid beverages. The Tetra Pak designed linear fillers (LFA-20) are FDA validated for aseptically filling low-acid beverages.

Additional Information

- Sidel has approximately 200 Combi and 100 aseptic fillers installed around the world.
- Sidel has testing facilities in Italy and France.
- Sidel fillers are capable of aseptically filling small particulates.
- Sidel has strong service and spare parts in the United States.
- Combi Predis™ FMa and Sensofill™ FMa are suitable for high-acid aseptic filling.
- Sidel employs approximately 5,500 people around the world and is a division of Tetra Laval.

SIG COMBIBLOC GMBH

Rurstrasse 58
D-52441 Linnich
Germany

Ph: 49 2462 79-0
Fax: 49 2462 79-2519

Sales and service office in the United States:

SIG Combibloc, Inc
2501 Seaport Drive, Suite 100
Chester, PA 19013
Ph: (610) 546-4140
Fax: (610) 546-4340
www.sig.biz

Product

Aseptic carton packaging equipment for liquid products into preformed paperboard cartons.

Products Currently Being Filled Using Combibloc Fillers

Juice and juice concentrates, nectars, milk, cream, condensed milk, rice milk, cream soups, soups with particulates, coffee, sauces, tomato products, baby food, fruit toppings and purees, syrups, and teas.

Equipment Models

Note: SIG Combibloc manufactures many different models of fillers. Contact Combibloc for exact model desired. The following is the production speeds of the fillers.

Model	Production rate	Package sizes
Small size package	Up to 24,000 pph	125–150 to 200–250 mL
Medium size package	Up to 12,000 pph	250–350 to 400–500 mL
Large size package	Up to 9000 pph	1000–2000 mL

Approximate delivery—High-acid filler, 9 months; low-acid filler, 12 months.
Filler—Available in either single- or double-lane filler configurations. Capable of filling particulates up to 15 mm. Fill is above the product facilitating the filling of foods containing particulates.
Sterilization of filler—By steam and hot water.
Sterilization time of the filler—1 hour.
Packaging—Preformed, flat, folded sleeves, which are printed, die cut, and flame sealed. The sleeve is fed into the Combibloc filler where it is opened, the bottom is sealed, and the formed carton is then filled with sterile product and sealed above the liquid contents. See Figures A3.37–A3.41.

Combibloc fillers are FDA validated for aseptically filling low-acid foods.

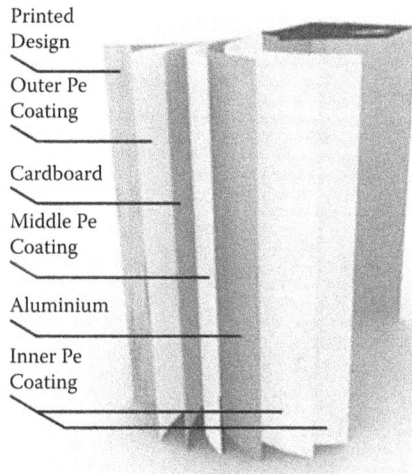

Figure A3.37   Combibloc packaging structure. (Diagram from SIG Combibloc literature.)

Figure A3.38   Combibloc and Combifit packages. (Photograph from SIG Combibloc literature.)

Additional Information

- Unlike Tetra Pak, there is no royalty payment assessed on packaging volume.
- Offers 20 different size packages (from 150 mL to 2 liters).
- Combibloc fillers are capable of food and beverages containing discrete particulates.
- Packaging provides headspace in the package to allow filling of particulates, shakeability, and to avoid spilling when opened.

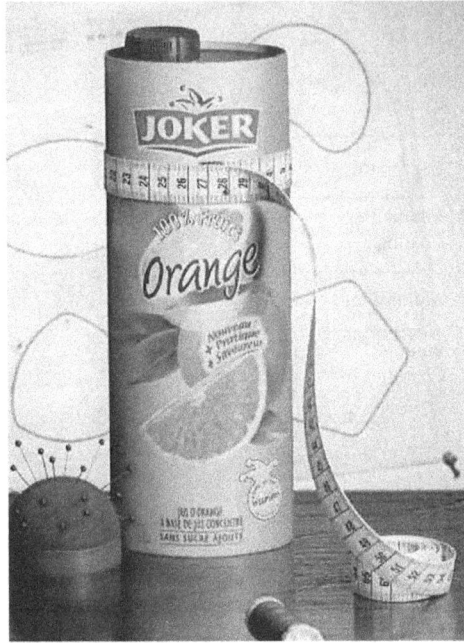

Figure A3.39   New package: Combibloc Combishape. (Photograph from SIG Combibloc literature.)

Figure A3.40   Types of Combibloc opening options.

- Combibloc has a testing laboratory in Germany.
- Changeover time for different size packages is approximately 2 minutes without changing machine parts.

- There is no loss of sterility during changeover of carton size.
- Unique patented Pour 'n Seal CombiTop fitment available for recloseable carton.

STORK

Stork Food and Dairy Systems B.V
Deccaweg 32
1042 AD Amsterdam
Netherlands
Ph: 31 20 634 89 11
Fax: 31 20 636 97 54

Figure A3.41   Combibloc 724 Food Filler. (Photograph from SIG Combibloc literature.)

Sales and service office in the United States:

Stork Food & Dairy Systems
1024 Airport Parkway
P.O. Box 1258
Gainesville, GA
Ph: (770) 535-1875
Fax: (770) 532-9867
Web: www.stork-usa.com

Product

Manufacturer of linear aseptic fillers for plastic bottles and related aseptic processing systems.

Products Currently Being Filled Using Stork Fillers

Fruit juices, milk and other dairy products, stirred yogurt, soy milk, and coffee drinks.

Equipment Models

See Figures A3.42 and A3.43.

Model	Package size	Production rate
Asep-Tec	20 mL to 2 liters	Up to 400 bpm

Lanes	Filling Stages	Nominal capacity	
		1 liter	0.5 liter
8	1	9,400	12,000
8	2	12,000	12,000
12	1	14,000	18,000
12	2	18,000	18,000
16	1		24,000

Figure A3.42   Some products aseptically filled using Stork fillers.

Figure A3.43   Stork linear aseptic filler for plastic bottles. (Photograph from Stork literature.)

Approximate delivery—6–8 months.

Filler—Linear aseptic filler for various plastic bottles for high- and low-acid beverages capable of log >6 sterility. A flowmeter principle is used and can fill particulates up to a 9-mm cube. Bottles can be nitrogen purged after sterilization to remove air and again after filling to enhance shelf life.

Sterilization of filler—Accomplished by steam and H_2O_2 followed by drying and cooling. Sterility is then maintained by positive filtered, sterile air pressure.

Sterilization time of filler—Approximately 90 minutes.

Packaging—PET, HDPE, PP, and other plastic bottles from 20 mL to 2 liters. Bottles can be monolayer, three layers with light barrier, or up to six layers with both light and oxygen barriers. Bottles can be closed with either seals or screw cap.

Sterilization of packaging—Preformed bottles are sterilized with hydrogen peroxide.

Stork aseptic fillers have been FDA validated for aseptically filling low-acid beverages.

Additional Information

- Stork is a leading supplier of aseptic fillers for bottles with many installations throughout the world with more than 60 aseptic fillers in operation.
- Stork manufactures the blow-molding equipment for bottles that can be directly coupled to the aseptic filler.
- Stork can change the container size or the product without an intermediate sterilization cycle.
- The Stork linear, aseptic bottle filler has substantially reduced maintenance compared to rotary fillers.
- Stork can supply the mutually dependent aseptic processing systems.
- Stork has a testing facility in Holland where it can aseptically process and package beverages with direct and indirect heat exchange systems.
- The Stork filler is capable of filling beverages with discrete particulates up to 9 mm × 9 mm × 9 mm cubes.
- Stork was the first aseptic bottle fillers that received FDA validation for aseptically filling low-acid dairy beverages.
- Stork also manufactures the conveying system, secondary packaging, and palletizing equipment.

TETRA PAK

Tetra Pak International S.A.
70 Avenue General-Guisan
CH-1009 Pully/Lausanne
Switzerland
Ph: 41 21 729 21 11
Fax: 41 21 729 22 88
www.tetrapak.com

Sales and service office in the United States:

Tetra Pak, Inc.
101 Vernon Hills Parkway
Vernon Hills, IL 60061
Ph: (847) 955-6000
Fax: (847) 955-6500
www.tetrapakusa.com

Product

Tetra Pak is the world's leading manufacturer of aseptic filling equipment for form–fill–seal paperboard cartons, and supplier of aseptic packaging material and associated materials-handling equipment. Tetra Pak is also one of the world's leading suppliers of aseptic processing systems.

Products Currently Being Filled Using Tetra Pak Fillers

Fruit juices, vegetable juices, milk, cream, pudding, ice cream mix, custard, cheese sauce, yogurt, wine, coffee and tea drinks, soups, milkshake mix, broths, rice and soy beverages, and specialty sauces.

Equipment Models

See Figures A3.44–A3.46.

Figure A3.44   Tetra Pak A3/Flex. (Photograph from Tetra Pak literature.)

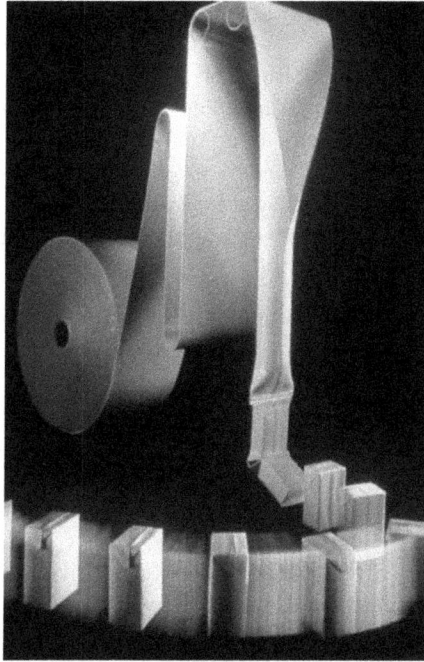

Figure A3.45   All Tetra Pak aseptic fillers generate packaging based on form–fill–seal. (Diagram courtesy of Tetra Pak.)

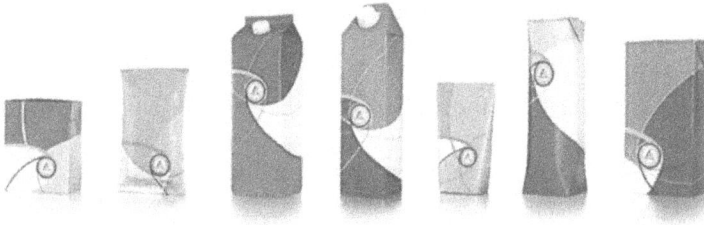

Figure A3.46   Several configurations of Tetra Pak packaging. (From Tetra Pak literature.)

Model	Package size range	Production rate
A3/Flex	200–2,000 mL	Up to 8,000 p/h
A3/Flex Speed	See following note	Up to 24,000 p/h

Filler—Tetra Pak has many models of Tetra Pak form–fill–seal continuous fillers capable of producing a myriad of different shapes and sizes of packages with different types of openings. See Tetra Pak website www.tetrapak.com for further details.

Note: The A3/Flex can produce 22 different Tetra Pak Aseptic and Tetra Prisma Aseptic packages. The A3/Flex Speed can produce Tetra Brik, Aseptic 1000 mL Baseline, Slimline, and Square Line packages.

Sterilization of filler—Accomplished by hot air and hydrogen peroxide.

Sterilization time—45 minutes.

Tetra Pak packaging—Form–fill–seal from roll stock consisting of paper, plastic, and aluminum foil in a variety of combinations. Package sizes range from 125 mL to 2 liters. Cartons are printed by offset method in up to four colors.

Sterilization of packaging—Accomplished by 30% hydrogen peroxide at 70°C for 6 seconds. Hydrogen peroxide is then removed by either rollers or hot air.

Packaging options—Pull tab, CIP unit, cap applicator, case packer, accumulator, straw applicator, multishrink.

All Tetra Pak aseptic fillers are FDA validated for filling low-acid food and beverages.

Additional Information

- Tetra Pak is the world's largest supplier of aseptic packaging and fillers.
- In 2009, Tetra Pak had 9115 Tetra Pak fillers in operation.
- Tetra Pak's new A/3 Flex filler can convert from Brik to Prism and can convert 2 volumes in 10 minutes.
- The A3 Flex filler can also produce 22 different Tetra Brik packages.
- Tetra Pak's A3 Speed can produce Tetra Brik, Aseptic Baseline, Slimline, and Squareline packages.
- Some Tetra Pak fillers can fill products containing discrete particulate matter.
- Tetra Pak has offices in 165 markets and employs more than 21,000 employees.
- Tetra Pak has 59 services centers and 19 research and development centers worldwide.
- Tetra Pak is one of the few companies in the world that also manufactures and supplies mutually dependent aseptic processing systems.
- Tetra Pak has a fully equipped and staffed testing laboratory in Denton, Texas. This testing laboratory:
  - Is validated by the FDA and conforms to the PMO for aseptically processing and packaging product.
- Has a number of different processing options, including direct and indirect heat exchange systems.

Appendix 4: Aseptic Contract Manufacturers in the United States

Thomas Szemplenski

BAG-IN-BOX

- Associated Milk Producers, Inc. (AMPI)
- Bay Valley Foods
- Beverage House
- California Natural Products
- CASP
- Cutrale Citrus
- Fruit Crown
- Gehl Foods
- Gossner Foods
- Island Oasis
- Langer Juice
- Lyon's Magnus
- Pacific Fruit
- Pacific Natural Foods
- Sabroso
- Stahlbush Farms
- Steuben Foods
- Wild Aseptics

CANS AND CUPS

- Advanced Foods (California)
- Advanced Foods (Pennsylvania)
- Advanced Foods (Wisconsin)
- AMPI
- Bay Valley Foods
- Gehl Foods
- IFP/Leahy
- Steuben Foods

SIG COMBIBLOC AND TETRA PAK

- Advanced Foods (California)
- American Soy
- Beverage Concepts
- California Natural Foods
- Coastlog
- Cutrale Citrus
- Foods Swing
- Gossner Foods
- IFP/Leahy
- Indulac
- Island Oasis
- Jasper Foods
- Johanna Foods
- Kerry Ingredients
- Kiko Foods
- Lyon's Magnus
- Morningstar
- NorCal
- Ocean Spray
- Pacific Natural Foods
- Pacific Nutritional Foods
- Schroeder (Michigan)
- Schroeder (Minnesota)
- Steuben Foods
- SunOpta (California)
- SunOpta (Minnesota)
- Whitlock Packaging (New Jersey)
- Whitlock Packaging (Oklahoma)

PLASTIC BOTTLES

- Aseptic Solutions
- Flavors Inc.
- Gehl Foods
- HP Hood
- Jasper Foods
- Kan Pak
- Lyon's Magnus
- Steuben Foods

POUCHES

- Advanced Foods (California)
- Advanced Foods (Pennsylvania)

- Advanced Foods (Wisconsin)
- AMPI
- Bay Valley Foods
- Gehl Foods
- Kan Pak
- Morningstar
- Steuben Foods

Processor	Location	Plastic bottles	Cups and cans	Bag-in-box	Pouches	Tetra Pak and Combibloc
Advanced Foods	402 S. Custer Ave. New Holland, PA 17557 Ph. (717) 355-8667		L		L	
	600 First Ave. West Clear Lake, WI 54005 Ph. (715) 263-2956		L		L	
	1211 E. Nobel Ave. Visalia, CA 93277 Ph. (559) 627-2070		L		L	L
American Soy	1474 N. Woodland Dr. Saline, MI 48176 Ph. (734) 736-9230					L
AMPI	E. Highway 212 Dawson, MN 56232 Ph. (320) 769-2994		L	L	L	
Aseptic Solutions	4848 Alcoa Circle Corona, CA 92880 Ph. (951) 736-9230	H				
Bay Valley Foods	820 Palmyra Ave. Dixon, IL 61021 Ph. (815) 288-4097		L	L	L	
Beverage Concepts	30322 Esperanza Rancho Santa Margarita, CA 92688 Ph. (949) 459–2922					H
Beverage House	107 North Ave. Cartersville, GA 30120 Ph. (770) 387-0451			H		

Processor	Location	Plastic bottles	Cups and cans	Bag-in-box	Pouches	Tetra Pak and Combibloc
California Natural Products	1250 Lathrop Road Lathrop, CA 95330 Ph. (209) 858-2525			L		L
CASP	105 Horizon Park Dr. Penn Yan, NY 14527 Ph. (315) 531-8080			L		
Coastlog	209 Theodore Rice Rd. New Bedford, MA 02745 Ph. (248) 344-9556					L
Cutrale Citrus	602 S. McKean Street Auburndale, FL 33823 Ph. (863) 965-5000			H		H
Flavors, Inc.	575 Alcoa Circle Corona, CA 92880 Ph. (949) 459-2660	H				
Food Swing	904 Woods Road Cambridge, MD 21613 Ph. (410) 228-1644					L
Fruit Crown	250 Adams Blvd. Farmingdale, NY 11735 Ph. (518) 694-5800			H		
Gehl Foods	N116 W15970 Main St. Germantown, WI 53022 Ph. (262) 251-8572	L	L	L	L	
Gossner Foods	1105 N. 1000 West Logan, UT 84321 Ph. (435) 752-9365			L		L
HP Hood	160 Hood Way Winchester, VA 22602 Ph. (540) 969-0045	L				

Processor	Location	Plastic bottles	Cups and cans	Bag-in-box	Pouches	Tetra Pak and Combibloc
IFP/Leahy	401 N. Main Street Rosendale, WI 54974 Ph. (920) 872-2181		H			H
Indulac	198 Chardon Ave. Hato Rey, Puerto Rico 00918 Ph. (787) 753-0974					L
Island Aseptic	100 Hope Road Byesville, OH 43723 Ph. (740) 685-2548			H		L
Island Oasis	141 Norfolk St. Walpole, MA 02081 Ph. (800) 999-5674			H		L
Jasper Foods	3877 E. 27th St. Joplin, MO 64804 Ph. (417) 206-3333	L				L
Johanna Foods	Johanna Farms Rd. Flemington, NJ 08822 Ph. (908) 788-2200					H
Kan Pak	1016 Summitt St. Arkansas City, KS 67005 Ph. (800) 378-1265	L			L	
Kerry Ingredients	11 Artley Road Savannah, GA 31408 Ph. (912) 330-7955					L
Kiko Foods	5510 Jefferson Hwy. Jefferson, LA 70123 Ph. (504) 736-0220					L
Langer Juice Co.	16195 Stephens St. City of Industry, CA 91745 Ph. (626) 336-1666			H		
Lyon's Magnus	1636 S. Second St. Fresno, CA 93702 Ph. (559) 268-5966	H		H		H

Processor	Location	Plastic bottles	Cups and cans	Bag-in-box	Pouches	Tetra Pak and Combibloc
Morningstar	6364 Valley Park Mt. Crawford, VA 22841 Ph. (540) 434-1948					L
	500 Jackson St. N Sulphur Springs, TX 75482 Ph. (903) 885-7573				L	
NorCal	2286 Stone Blvd. Sacramento, CA 95691 Ph. (916) 372-0660					H
Ocean Spray	7800 S. 60th Ave. Kenosha, WI 53142 Ph. (262) 942-5351					H
Pacific Fruit	121 Center Street South Gate, CA 90280 Ph. (562) 531-1770			H		
Pacific Natural	19480 SW 97th Ave. Tualatin, OR 97062 Ph. (503) 692-9666			L		L
Pacifie Nutritional	9960 SW Potano Tualatin, OR 97062 Ph. (503) 692-3498					L
Rio Bravo	36889 Hwy. 58 Buttonwillow, CA 93206 Ph. (661) 764-9000			H		
Sabroso	690 S. Grape St. Medford, OR 97502 Ph. (541) 772-5653			H		
Schroeder	2080 Rice St. Maplewood, MN 55113 Ph. (561) 855-6418					L
	5252 Clay Ave. Grand Rapids, MI 49548 Ph. (616) 538-3822					L

Processor	Location	Plastic bottles	Cups and cans	Bag-in-box	Pouches	Tetra Pak and Combibloc
Stahlbush Farms	3122 Stahlbush Island Rd. Corvalis, OR 97333 Ph. (541) 757-1497			L		
Steuben Foods	150 Maple Rd. Elma, NY 14059 Ph. (716) 291-9484	L	L	L	L	L
SunOpta	3915 Minnesota St. Alexandria, MN 5630 Ph. (320)763-9822					L
SunOpta	555 Mariposa Rd. Modesto, CA 95354 Ph. (209) 818-0032					L
Whitlock Packaging	1701 S. Lee Ft. Gibson, OK 74434 Ph. (918) 478-4300 92 Main Street Wharton, NJ 07885 Ph. (973) 361-9794					H H
Wild Aseptics	2924 Wyetta Drive Beloit, WI 53511 Ph. (608) 362-5012			H		

Note: H = high acid; L = low acid.

Appendix 5: Examples of Typical Thermal Process Design for Aseptically Processed Fluids and Purees

Pablo M. Coronel and Jairus R.D. David

CONTENTS

A5.1: WHITE MILK

Typical thermal process design for aseptically processed *liquid food* (*)

Heating method: direct steam injection system (two-dimensional heating)

Cooling: vacuum flash cooling followed by indirect tubular series of heat regen/heat exchangers

Product: **White milk**—shelf-stable (non-refrigerated) with an ambient shelf life of 12–18 months

- pH: >4.6 (low acid)
- Target pathogenic microorganism: mesophilic/proteolytic spores of *Clostridium botulinum*, requiring "Bot Cook."
- Specific gravity: 1.023 (whole)—1.035 (skim)
- Density (ρ): -0.0341 to 0.0345 lb/in^3 943.8–955 kg/m^3
- Viscosity (μ): 2.0–2.5 cps at 77°F
- 2.19×10^4 lb/in-s (3.94×10^4 kg/m-S) at 290°F—estimated using viscosity of water
- Flow rate: 50 gpm—3.15×10^3 m^3/s or 11 m^3/h
- Corrected flow rate (steam condensation and volume expansion):
  expansion factor 1.2
  60 gpm (231 in^3/s)
  3.78×10^3 m^3/s or 13.6 ton/h
- Hold tube ID: 1.87 inches or 4.75×10^3 m (nominal ID 2.0 inches)
- Hold tube length: 334.0 inches—8.48 m
- Least required F_o: 5 minutes

Compute:

1. Hold time
2. Reynolds number (N_{Re}; dimensionless) to determine flow profile
3. Process temperature for milk in hold tube based on flow profile

1. **Hold time (s)**

Volume of hold tube (in³)/(flow rate in cubic inches per second	Volume of hold tube (m³)/flow rate in cubic meters per second
917.3173 in³/(60 gpm × 3.85 in³/s-gpm)	1.503×10^2 m³/3.78×10^3 m³/s

= **3.97 seconds** (average hold time)

2. **Determination of flow profile based on NRe**
   N_{Re} values for flow patterns:
   Laminar: 2,000 (correction factor 0.5)
   Turbulent: >6,000 (correction factor 0.83)
   Transitional: 2,000–6,000 (use 0.5 correction)
   $$N_{Re} = \frac{4 \, \text{Flow rate} \, \rho}{\pi D \mu} = \frac{42310 \cdot 03413}{\pi 1.872 \cdot 1910^{-4}} = 239,971 \ \textbf{Flow is turbulent}$$

   Correction for laminar flow (0.5) = 1.986 s
   Correction for turbulent flow (0.83) = 3.296 s
3. **Process temperature in hold tube with turbulent flow:**

Log (F_o minutes/hold time in minutes) × 18°F + 250°F	Log (F_o minutes/hold time in minutes) × 10°C + 121.1°F
Log [5 / (3.296/60)] × 18 + 250	Log [5 / (3.296 / 60)] × 10 + 121.1
= [1.959] × 18 + 250	= [1.959] × 10 + 121.1
= 35.204 + 250	= 19.59 + 121.1
= 285.264°F	= 140.69°C
Turbulent flow:	**Turbulent flow:**
285.2°F = 286°F	**140.69°C = 141°C**
(283.8°F for the average, 289.2°F for laminar flow)	(139.9°C for the average, 142.9°C for laminar flow)

Answers:

White milk

1. **Hold time (seconds):** 3.97 s "Average Hold Time"
2. **Determination of flow profile based on N_{Re}:**
   N_{Re} of 239,971
   Flow is turbulent
   Correction for turbulent flow (0.83 × 3.97 s) = 3.296 s
3. **Process temperature in hold tube with turbulent flow:**
   285.2°F (140.7°C) = 286°F (141°C)
4. **Process temperature set point and alarms:**
   Minimum process temperature: 286°F (141°C) measured at the end of the hold tube
   Set point: 294 °F (145.5°C)—6–8°F (3–4°C) higher than process temperature
   Low-temperature alarm: 288°F (142.2°C)—2°F (1°C) higher than process temperature)

Process deviation: any process temperature lower than 286°F (141°C) is a process deviation, and therefore will require shutdown and complete CIP/SIP of the process system and restart.

*Disclaimer:*Thermal process design calculations must be completed by a Process Authority. Process Authority should be aware of any validation made on the process system to understand the environment in which the day-to-day process will be carried out and the needed temperature and flow allowances.Need to consider variations in incoming heat-resistant microbial spores loads in sensitive ingredients; hydration of difficult-to-dissolve raw ingredients such as cocoa powder, starches, milk powder, gums, stabilizers, etc.; and supply reliability of plant utilities such as steam, water, compressed air, cooling, and electricity during peak demands.

(*) Using Thermal Process Calculator developed by Dr. Pablo Coronel, CRB Consulting Engineers, Raleigh, North Carolina

A5.2: CHOCOLATE PUDDING

Typical thermal process design for aseptically processed *VISCOUS Homogeneous* food (*)
    Heating method: indirect series of tubular heat exchangers (two-dimensional heating)
    Cooling: indirect tubular series of heat regen/exchangers
    Product: **Chocolate pudding**—shelf-stable (non-refrigerated) with a shelf life of 12 to
8 months

- pH: >4.6 (low acid)
- Target pathogenic microorganism: mesophilic/proteolytic spores of *C. botulinum.*
- Specific gravity: 0.7–0.8
- Viscosity: 1,000 – 2,500 cps
- Flow rate: 36 gpm – 2.27×10^3 m^3/s or 8.2 m^3/h
- Expansion coefficient = 1 (no water added)
- Hold tube ID: 1.37 inches or 3.48×10^3 m (nominal 1.5 inches)
- Hold tube length: 950.0 inches—24.13 m
- Least F_0: 8.4 minutes based on mesophilic pathogenic microorganism—mesophilic/proteolytic spores of *C. botulinum.* Non-pathogenic mesophilic and thermophilic aerobic and anaerobic spores—*Geobacillus stearothermophilus* (flat-sour spoilage in low-acid food) and *Clostridium thermosaccharolyticum* with higher heat resistance with a D_{250F} of 1–4 minutes are responsible for economic spoilage. F_0 for "Thermophilic Cook" can range from 5 to 20 minutes and process of this severity must be very damaging to product quality.
- In lieu of Thermophilic Cook, in the food industry, economic spoilage is controlled by vendor-specific HACCP program for incoming ingredients known to contain high numbers of non-pathogenic mesophilic and thermophilic aerobic and anaerobic spores. The sole intention here is to "minimize" the initial spore load so as to not to "overwhelm" a designed thermal process for "Bot Cook." Also, some ingredients such as cocoa powder and other dried ingredients must be properly hydrated and dissolved during batching. It is acceptable to selectively pretreat sensitive or contaminated ingredients by appropriate physical or chemical methods, prior to delivering an optimal thermal process (see Sections 2.1, 2.6, and 4.2).

Compute:

1. Hold time
2. Reynolds number (N_{Re}; dimensionless) to determine flow profile
3. Process temperature in hold tube based on flow profile

1. **Hold time (s)**

Volume of hold tube (in^3])/flow rate in cubic inches per second	Volume of hold tube (m^3)/(flow rate in cubic meters per second)
1400.4 in^3 / (36 gpm × 3.85 in^3/s-gpm)	2.295×10^2 m^3 / 2.27×10^3 m^3/s

    = 10.1 seconds—average hold time in hold tube

2. **Determination of flow profile based on N_{Re}**
    N_{Re} values for flow patterns:

Laminar: 2,000 (correction factor 0.5)
Turbulent: >6,000 (correction factor 0.83)
Transitional: 2000–6000 (use 0.5 correction)

$$N_{Re} = \frac{4 \, Flow \, Rate \, \rho}{\pi D \mu} = \frac{4138.60 \cdot 0257}{\pi 1.441 \cdot 1110^{-1}} = 293 \; \textbf{Flow is laminar}$$

Correction for laminar flow (0.5) = 5.05 s
Correction for turbulent flow (0.83) = 8.39 s

3. **Process temperature in hold tube with laminar flow:**

Log (F_o minutes/hold time in minutes) × 18°F + 250°F	Log (F_o minutes/hold time in minutes) × 10°C + 121.1°F
Log [8.4 / (5.05 / 60)] × 18 + 250 = [1.999] × 18 + 250 = 35.98 + 250 = 285.98°F	Log [8.4 / (5.05 / 60)] × 10 + 121.1 = [1.999] × 10 + 121.1 = 19.99 + 121.1 = 140.69°C
Laminar flow: 285.98°F = 286°F (280.6 F for the average, 282.0°F for turbulent flow)	**Laminar flow: 141.09°C = 141.5°C** (138.09°C for the average, 138.9°C for turbulent flow)

Answers:

Chocolate pudding

1. **Hold time (seconds):** 10.1 seconds—average hold time in hold tube
2. **Determination of flow profile based on N_{Re}:**
   N_{Re} of 293
   Flow is laminar
   Correction for laminar flow (0.5 × 10.1 seconds) = 5.05 seconds
3. **Process temperature in hold tube with laminar flow:**
   285.98°F (141.1°C) = 286°F (141.5°C)
4. **Process temperature set point and alarms:**
   Minimum process temperature: 286°F (141.5°C) measured at the end of the hold tube
   Set point: 291°F (144°C)—5°F (2–3°C)—higher than process temperature
   Low-temperature alarm: 288°F (142°C)—2°F higher than process temperature
   Process deviation: any process temperature lower than 286 °F (141.5°C) is a process deviation, and therefore will require shutdown and complete CIP/SIP of the process system and restart.

*Disclaimer:*Thermal process design calculations must be completed by a Process Authority. Process Authority should be aware of any validation made on the process system to understand the environment in which the day-to-day process will be carried out and the needed temperature and flow allowances.Need to consider variations in incoming heat-resistant microbial spores loads in sensitive ingredients; hydration of difficult-to-dissolve raw ingredients such as cocoa powder, starches, milk powder, gums, stabilizers, etc.; and supply reliability of plant utilities such as steam, water, compressed air, cooling, and electricity during peak demands.

(*) Using Thermal Process Calculator developed by Dr. Pablo Coronel, CRB Consulting Engineers, Raleigh, North Carolina

A5.3: TOMATO PUREE

Typical thermal process design for aseptically processed *Viscous homogeneous food* (*)
Heating method: indirect tubular heat exchanger (two-dimensional heating)
Cooling: tubular heat regen/exchangers

- Product: **Tomato puree**
- pH: ~4.5
  - Border line pH in each batch must be tested and controlled carefully via acidification
- Specific gravity: 1.0
- Viscosity power law: $K = 868$ $n = 0.36$ (mPa-s) at 77°F
- Flow rate: 36 gpm—2.27×10^3 m³/s or 8.2 m³/h
- Hold tube ID: 1.37 inches or 3.48×10^3 m (nominal 1.5 inches)
- Hold tube length: 950.0 inches—24.13 m
- Least $\mathbf{F_{205/18F}}$ ($F_{96/10}$): 5 minutes based *on Bacillus coagulans, Bacillus macerans,* and *Bacillus polymyxa; Bacillus thermoacidurans* (flat-sour spoilage in acid products especially tomato) will require $\mathbf{F_{235/18}}$ ($F_{112/10}$) of 5 minutes

Compute:

1. Hold time
2. Reynolds number (N_{Re}; dimensionless) to determine flow profile
3. Process temperature in hold tube based on flow profile

1. **Hold time (s)**

Volume of hold tube (in³)/flow rate in cubic inches per second	Volume of hold tube (m³)/flow rate in cubic meters per second
1400.4 in³/(36 gpm × 3.85 in³/s-gpm)	2.295×10^2 m³ / 2.27×10^3 m³/s

= **10.1 seconds**—average hold time in hold cell

2. **Determination of flow profile based on N_{Re}**
   N_{Re} values for flow patterns:
   Laminar: 2,000 (correction factor 0.5)
   Turbulent: >6000 (correction factor 0.83)
   Transitional: 2,000–6,000 (use 0.5 correction)
   $$N_{Re} = \frac{4\,\text{Flow rate}\,\rho}{\pi D \mu} = \frac{4138.60 \cdot 0345}{\pi 1.374 \cdot 5310^{-3}} = 980 \text{ Flow is laminar}$$

   Correction for laminar flow (0.5) = 5.05 s
   Correction for turbulent flow (0.83) = 8.39 s
3. **Process temperature in hold tube for laminar flow:**

Log ($F_{205/18}$ minutes/hold time in minutes) × 18°F + 205°F	Log ($F_{96/10}$ minutes/hold time in minutes) × 10°C + 96°C
Log [5/(5.05/60)] × 18 + 205	Log [5/(5.05 / 60)] × 10 + 96
= [1.999] × 18 + 205	= [1.999] × 10 + 96
= 35.98 + 205	= 19.99 + 96
= 236.9°F	= 113.79°C

Laminar flow: 236.9°F = 237°F	Laminar flow: 113.79°C = 114°C
(231.5°F for the average, 233.0°F for turbulent flow)	(110.8°C for the average, 111.6°C for turbulent flow)

Process for Spoilage:

Log ($F_{235/18}$ minutes/hold time in minutes) × 18°F + 235°F	Log ($F_{112/10}$ minutes/hold time in minutes) × 10°C + 112°C
Log [5 / (5.05 / 60)] × 18 + 235 = [1.7738] × 18 + 235 = 31.93 + 235 = 266.92°F	Log [5 / (5.05 / 60)] × 10 + 112.8 = [1.7738] × 10 + 112.8 = 17.74 + 112.8 = 130.55°C
Laminar flow: 266.9°F = 267°F (261.5°F for the average, 263.0°F for turbulent flow)	**Laminar flow: 130.54°C = 131°C** (127.5°C for the average, 128.3°C for turbulent flow)

Answers:

Tomato puree

1. **Hold time (seconds):** 10.1 seconds—average hold time in hold tube
2. **Determination of flow profile based on N_{Re}:**
   N_{Re} of 980
   Flow is laminar
   Correction for laminar flow (0.5 × 10.1 seconds) = 5.05 seconds
3. **Process temperature in hold tube with laminar flow:**
   • For safety 236.5°F = 237°F (114°C)
   • For spoilage 267°F (131°C)
4. **Process temperature set point and alarms:**
   Minimum process temperature: 237°F (114°C) measured at the end of the hold tube
   Set point: 242°F (117°C)—5°F higher than process temperature
   Low-temperature alarm: 239°F (115°C)—2°F higher than process temperature
   Process deviation: any process temperature lower than 237°F (114°C) is a process deviation, and therefore will require shutdown and complete CIP/SIP of the process system and restart.

*Disclaimer:*Thermal process design calculations must be completed by a Process Authority. Process Authority should be aware of any validation made on the process system to understand the environment in which the day-to-day process will be carried out and the needed temperature and flow allowances.Need to consider variations in incoming heat-resistant microbial spores loads in sensitive ingredients; hydration of difficult-to-dissolve raw ingredients such as cocoa powder, starches, milk powder, gums, stabilizers, etc.; and supply reliability of plant utilities such as steam, water, compressed air, cooling, and electricity during peak demands.

(*) Using Thermal Process Calculator developed by Dr. Pablo Coronel, CRB Consulting Engineers, Raleigh, North Carolina

A5.4: BANANA PUREE

Typical thermal process design for aseptically processed *FLUID Homogeneous* Food (*)
Indirect heating and cooling: tubular heat exchanger

- Product: **Banana puree**—pH: ~5.2 (low acid)
- Specific gravity: 1.06
- Viscosity power law: 1,500 cps at 77°F
- Flow rate: 36 gpm—2.27×10^3 m^3/s or 8.2 m^3/h
- Hold tube ID: 1.37 inches or 3.48×10^3 m (nominal 1.5 inches)
- Hold tube length: 950 inches—24.13 m
- Least F_0: 0.5 minutes for low acid
- Target microorganisms: spore formers such as *Bacillus licheniformis* and *B. coagulans*

Compute:

1. Hold time
2. Reynolds number (N_{Re}; dimensionless) to determine flow profile
3. Process temperature in hold tube based on flow profile

1. **Hold time (s)**

Volume of hold tube (in^3)/flow rate in cubic inches per second	Volume of hold tube (m^3)/Flow rate in cubic meters per second
1400.4 in^3/(36 gpm × 3.85 in^3/s-gpm)	2.295×10^2 m^3/2.27×10^3 m^3/s

=1400.41 in^3/(36 gpm × 3.85 in^3/s-gpm)
= **10.1 seconds**—average hold time in hold cell

2. **Determination of flow profile based on N_{Re}**
   N_{Re} values for flow patterns:
   Laminar: 2,000 (correction factor 0.5)
   Turbulent: >6,000 (correction factor 0.83)
   Transitional: 2,000–6,000 (use 0.5 correction)
   $$N_{Re} = \frac{4\,\text{Flow rate}\,\rho}{\pi D \mu} = \frac{4138.60 \cdot 0341}{\pi 1.371 \cdot 8510^{-2}} = 237 \; \textbf{Flow is laminar}$$

   Correction for laminar flow (0.5) = 5.05 s
   Correction for turbulent flow (0.83) = 8.39 s
3. **Process temperature in hold tube with laminar flow**

Log (F_0 minutes/hold time in minutes) × 18°F + 250°F	Log (F_0 minutes/hold time in minutes) × 10°C + 121.1°F

Log [0.5 / (5.05 / 60)] × 18 + 250
= [0.774] × 18 + 250
= 13.93 + 250
= 263.93°F

Log [0.5 / (5.05 / 60)] × 10 + 121.1
= [0.774] × 10 + 121.1
= 7.74 + 121.1
= 128.8°C

= Log [0.5 / (10.1 / 60)] × 18 + 250
= [0.473] × 18 + 250
= 8.51 + 250
= 258.51 F for the average. Laminar = **263.9°F = 264 F**

Laminar flow: 263.93°F = 264°F
(258.5°F for the average, 260.0°F for turbulent flow)

Laminar flow: 128.8°C = 129°C
(125.8°C for the average, 126.6°C for turbulent flow)

Note:

- Often times aseptically processed low-acid banana puree packaged in bulk containers using thermal process shown above are *reprocessed* into retail or food service packages using the same above process. The final product in retail and food service containers is aseptically processed twice.
- Sometimes, aseptically processed and packaged banana puree in bulk containers is either retorted or reprocessed for packaging in retail and food service containers using an acidification process (pH of about $\geq$4.2; lower pH tends to promote tartness and aggravate *"pinking or reddening of banana."* From a food safety point of view, pH must not exceed 4.5 maximum).
  - Acidified banana process is based on an F200F/16F = 20 minutes. This process is designed to destroy spoilage microorganisms, such as non-spore formers, yeast and molds, including a small number of heat-resistant molds. It will also destroy spore formers such as *B. licheniformis* and butyric acid anaerobes which could grow in this acidified product. It will not however, destroy *C. botulinum* spores (they will not grow in acidified bananas), not the more heat-resistant bacillus such as *G. stearothermophilus* (which will not grow in acidified bananas), or *Bacillus coagulans* (small numbers will be destroyed but large numbers will not).
    (Keith Ito, National Food Processors Association, Dublin, California, Personal Communication, 1996).

Answers:

Banana puree

1. **Hold time (seconds):** 10.1 seconds—average hold time in hold tube
2. **Determination of flow profile based on N_{Re}:**
   N_{Re} of 237
   Flow is laminar
   Correction for laminar flow (0.5 × 10.1 seconds) = 5.05 seconds
3. **Process temperature in hold tube with laminar flow:**
   263.93.5°F (128.8°C) = 264°F (129°C)
4. **Process temperature set point and alarms:**
   Minimum process temperature: 264°F (129°C) measured at the end of the hold tube

Set point: 269°F (131.5°C)—5°F (2–3°C) higher than process temperature

Low-temperature alarm: 266°F (130°C)—2°F (1°C) higher than process temperature)

Process deviation: any process temperature lower than 264°F is a process deviation, and therefore will require shutdown and complete CIP/SIP of the process system and restart.

(*) Using Thermal Process Calculator developed by Dr. Pablo Coronel, CRB Consulting Engineers, Raleigh, North Carolina

Disclaimer: Thermal process design calculations must be completed by a Process Authority. Process Authority should be aware of any validation made on the processing system to understand the environment in which the day-to-day process will be carried out and the needed temperature and flow rate allowances. Need to consider variations in incoming heat-resistant microbial spores loads in raw ingredients; hydration of difficult-to-dissolve raw ingredients such as cocoa powder, starches, milk powder, gums, stabilizers, etc.; and supply reliability of plant utilities such as steam, water, compressed air, cooling, and electricity during peak demands.

Appendix 6: Process Design and Microbial Validation of a Product with Large Particulates

Pablo M. Coronel and Josip Simunovic

A6.1 INTRODUCTION

Since its inception in 2006, the team of scientists and engineers at Aseptia worked on building a scientific base and an integrated approach for advancing the continuous flow sterilization and aseptic packaging of shelf-stable complex foods and biomaterials. These efforts were taken in order to enable production, distribution, and commercialization of previously unavailable, convenient, high-quality and nutritionally superior products to the US and international markets.

The following is a case study for a tomato soup with whole-corn pieces up to ½" (12.77 mm) size for aseptic processing. The method for production and validation was the result of many years of R&D and was built upon the success of previous case studies. While using technology available at the time, it required the development of methods and apparatuses to support the end goal, resulting in several patents that are listed in the references section. The process and product were tested extensively in order to ensure their safety and the ability to be produced at industrial scale. This case study was presented to FDA for evaluation of the process and validation methodology, and eventually received a LONQ (Letter of No Questions).

Aseptically processed and packaged heterogeneous foods, such as soups and stews, have been long recognized as one of the most challenging product categories to process and preserve to the state of shelf stability under continuous flow conditions, and this pursuit was the central focus of Aseptia's concentration.

The critical missing elements in the processing of this category of products were identified early in the process and in collaboration with other researchers, and appropriate technologies to achieve this goal have been studied, developed, and implemented at various R&D and production scales.

The two key groups of science and technology components have been independently commercialized and tested in production prior to their integration for processing of complex particulate products:

A) A new sterilization technology for difficult, viscous, poorly conductive and particulate materials (continuous-flow microwave-assisted thermal processing) has been invented, developed, and successfully implemented to a wide variety of high-acid and low-acid homogeneous viscous materials as well as several acidified multiphase products.

B) A system of real-time and post-process particle flow monitoring, recording, and validation methods, devices, and tools has been developed, commercialized, and implemented for process and product safety validation by third-party industrial partners for both conventional (tube-in-tube heat exchangers) and advanced (ohmic and microwave) continuous-flow particulate sterilization systems.

This system is based on the design and fabrication of customized, conservative surrogate particles to appropriately (conservatively) emulate the flow and heating behavior of various food components of selected complex products to be sterilized. Conservative emulation means that the surrogate particles are designed and fabricated to heat slower and travel faster within the processing system than any real food product to be processed. These surrogate particles are used as carriers for residence time and cumulative thermal lethality implant carriers and passed through the system intermittently within the flow of product under representative processing conditions.

The particle residence times are monitored using the proprietary sensing, recording, and real-time and postprocess analyses. The validation of the safety of the applied processes and produced products is based on successful inactivation of bacterial spore (bio-load) implants contained within these conservative carriers—providing the proof of adequate and sufficient thermal exposure for any component subjected to a higher level of thermal and temporal lethality accumulation—i.e. all real food particles and components contained within the targeted product.

Additionally and concurrently, a process simulation and design software package was developed and used in establishment of appropriate conservative models for each targeted product and product range. The model was based on over two decades of basic and applied research of food and biomaterial product properties, flow dynamics and continuous flow heat exchange simulation models using both conventional and advanced thermal processing methods of food, beverage, and biomaterial sterilization (ohmic and continuous-flow microwave processing).

Integrating those major missing elements into the process development, design, implementation, and validation matrix, Aseptia's approach to processing and packaging of complex multiphase foods was implemented in several stages. These stages have been performed consecutively and/or concurrently as appropriate.

As a first case study, Aseptia developed a process for a tomato soup with corn pieces up to 12.77 mm (0.5") in size, which would be packaged in aseptic bag in box filler. For the purpose of this report, only the process establishment and validation of such process will be discussed. Aseptic filler and container integrity will be discussed in the separate filing documents.

The biological challenges and validations were carried out using surrogate simulated particles. These particles consist of spheres made of a combination of polymers which contain magnetic implants for residence time measurements, as well as biological loads for microbial validation. These simulated particles are built in such a way that they are conservative for both residence time (flow conservative) and thermal penetration (thermal conservative), thus becoming the worst-case particle in themselves, and allowing for a precise determination of microbial inactivation for such case.

In order to establish the process, the following steps were taken:

- PREREQUISITES
    1. Develop a model of the process
        - Define a mathematical model of the system
        - Measure thermophysical and dielectric properties of carrier fluid and vegetable particles
    2. Establish a conservative surrogate simulated particle
        - Determine and compare the heat penetration rates of vegetable particles and surrogate particles

-   Establish the conservative flow characteristics (critical density) for the surrogate particle
-   Establish the biovalidation surrogate

3. Proper design of the process line
    -   Design of process line based on model
        -   Proper operation of the equipment
        -   Design of pilot plant based on modeling
        -   Sterilization capability
        -   Maintenance of back pressure
        -   Control of flow rate
        -   Control of temperature
    -   Construction of process line
    -   Installation and commissioning
4. Proper formulation of the product
    -   Viscosity control
    -   Particulate concentration
    -   Particulate size

-   PROCESS ESTABLISHMENT
    1. Establish process based on the model of the system
        -   Model system, and product
        -   Determine the worst-case particle for modeling
        -   Verify model with measurements from preliminary test runs
    2. Biological validation of the product using the conservatively designed surrogate particles
        -   Verify the model-established process based on the data from previous experiments
        -   Challenge the system with surrogate particles loaded with spores and verify the predictions of the model

These steps were followed through to establish and verify that the product will be safe for consumers, and that the established process can be transferred to other production lines in the future.

A6.2 PREREQUISITES

A6.2.1 Model of the System

A mathematical model of thermal processing has been developed and called thermal process simulator (TPS). TPS models thermal process only, and does not include batching/mixing or packaging. This model is based on the finite difference method to calculate the heat transfer to liquid, bulk particle and a "worst-case" particle. It solves the differential energy equations per element, including convection, conduction, and generative (volumetric) heat transfer per finite element, thus calculating the temperature per element of liquid and particles (Figure A6.1).

This software can be used to model several types of heat exchangers and products with particulates. Particles need to be approximated to simpler geometries, and average sizes and properties.

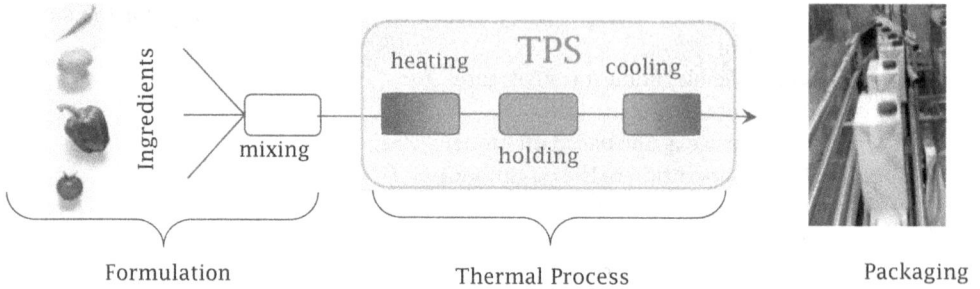

Figure A6.1   Scope of thermal process simulator (TPS),

The fluid, bulk, and the worst case can have different properties and can also have different residence time distributions which are accounted for in the calculations. Formulation is based on real measurements and is part of the input.

The process line is modeled as units, which can be of several types such as hold tube (no heating), tubular heat exchanger, scraped surface, multi-tube, ohmic, or microwave and each unit has a series of parameters which are part of the input.

Post-calculations for safety and quality parameters include the standard F_0 value, which is hardcoded in the system, as well as several other quality and safety parameters which can be detailed in the input file. The input file is in the format of an MS Excel spreadsheet and the outputs are both a summary file (ASCII text) and a complete report with all the finite element data in the form of an Excel Spreadsheet (Figure A6.2).

The model was validated by comparing the results of biological tests to the predictions of the model. The model could only be valid if it was conservative, i.e., predicted lethalities that were equal or lesser than those observed in testing.

In order to produce a valid process, the model was used to define the steps needed to design and build a process line, and to have a conservative approach to soups with particles. To do so, the input data for the model was investigated in the form of thermophysical properties of liquid carrier and particles, and the worst-case (conservative) simulated particle was defined.

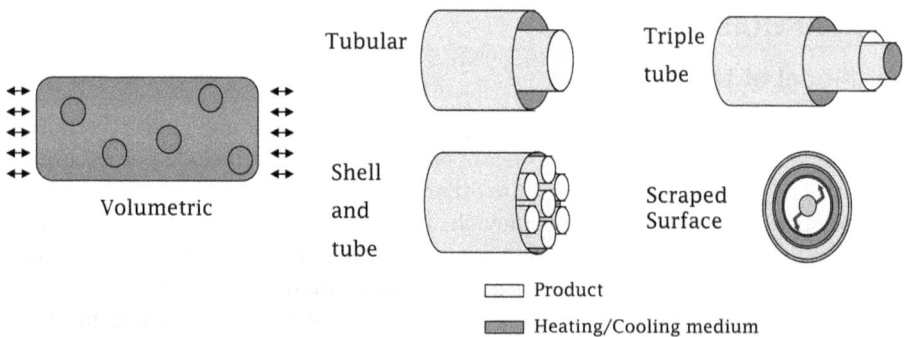

Figure A6.2   TPS modeling software. (Continued)

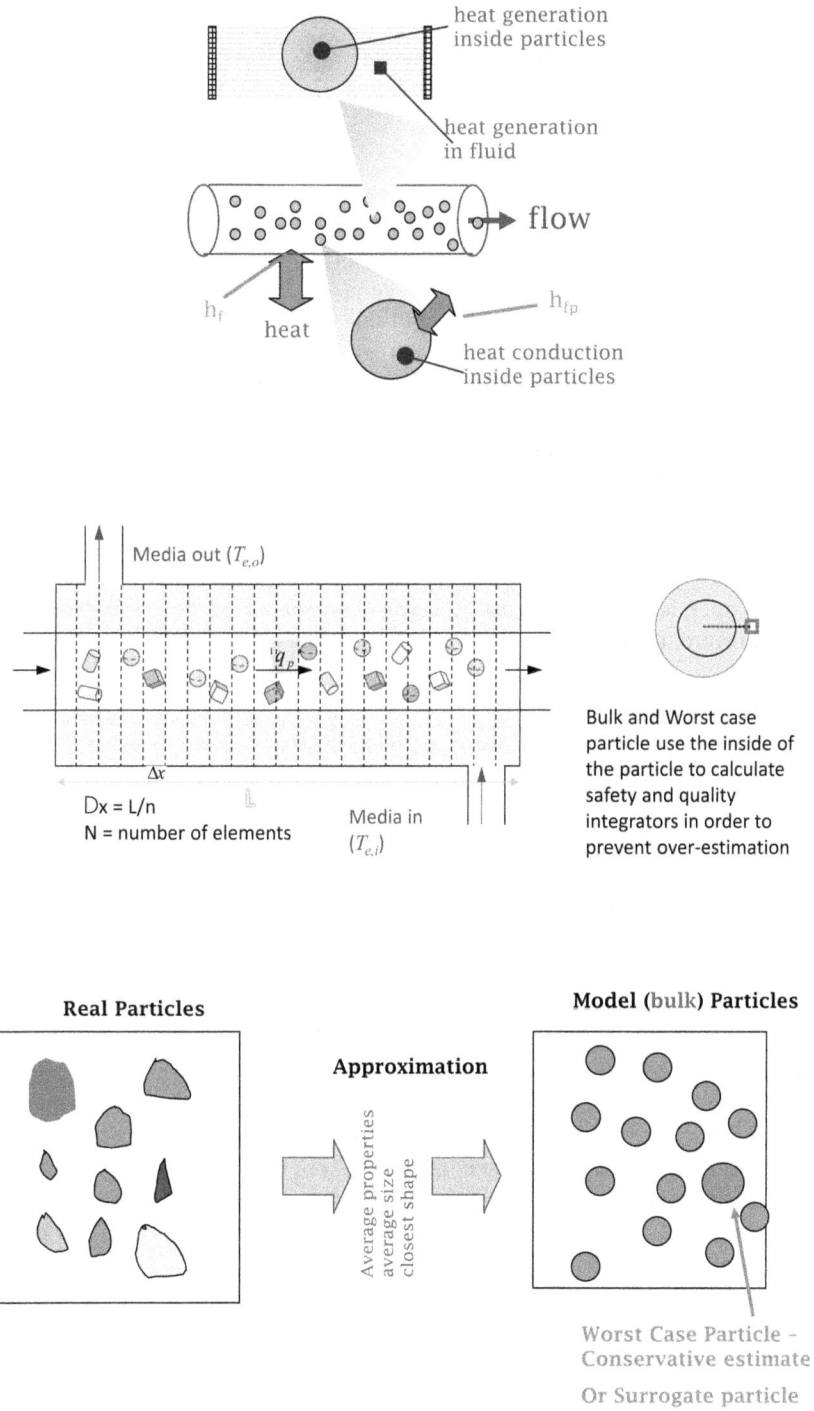

Figure A6.2 (Continued)   TPS modeling software.

A6.2.2 Thermophysical Properties of Particles and Carrier Fluid

Thermophysical properties of real food particles were measured using a Thermal Analyzer (KD2 Pro, Decagon). This analyzer allows the measurement of thermal conductivity, volumetric specific heat (C_v), and thermal diffusivity.

Particles that were not large enough to be tested by the analyzer were first homogenized using a blender and then placed in a test cell where the temperature could be adjusted from 20 to 140°C. Particles that could be directly read by the analyzer have been used as such.

Properties were also estimated using COSTHERM (EU Cost90 project) based on the analysis of the different particles in the temperature range of 20–130°C. The most conservative of both values was recommended for further modeling (Table A6.1).

Readings from the K2D were very unreliable at temperatures above 50°C, thus readings were performed only at room temperature and compared to Costherm results. In most cases, Costherm results were more conservative and thus recommended.

Thermal penetration studies were carried out concurrently in the same group of vegetables, comparing each product to surrogate particle materials, which resulted in a correspondence between surrogate particulates and the different vegetables that were identified as candidates for the study. Detailed results are found in Thermal Penetration report, which is added as an attachment.

Dielectric properties were measured using the open end coaxial probe method, as described by Coronel et al. (2003) for the same families of vegetables and the results are given in Figure A6.3.

All of the above data was used to create models of a process line, which will be built afterward in order to be able to properly process fluids with particles. In order to have a proper validation, a conservative surrogate particle needed to be designed and constructed.

A6.2.3 Determination of a Conservative Surrogate Simulated Particle

Validation of the thermal process given to the particulates was carried out using simulated particles. Simulated particles consist of spheres, cylinders, or cubes made in two parts of different polymers (TPX, Ultem, polysulfone, polypropylene), and sized to match the worst-case product particle (1/4", ½", 5/8"). The particles are built in such a way that an area in which implanted loads can be placed is in the center of the simulated particle (Figure A6.4). Such particles can be fabricated combining polymers, in order to match the desired engineering properties, such as density, thermal conductivity, or electrical properties. These properties for the individual polymers are listed in Table A6.2.

The goal of using simulated surrogate particles is to engineer a particle which is conservative in both flow and thermal properties, such that it will move faster and heat slower than any food particle in the system. To do so, residence time studies and thermal penetration studies needed to be carried out.

During the assembly of the surrogate particles, ballast loads in the way of glass beads or steel bearings can be added to match the required effective density of the particle. The unadjusted simulated particles have air in the central cavity, which makes the effective density of the particles much lower than the density of the constituent polymers. Residence time studies which are submitted in a separate document (RTD-Studies) showed that the density of the simulated particles is critical in obtaining flow characteristics of these particles which will result in a flow conservative particle.

TABLE A6.1   THERMOPHYSICAL PROPERTIES OF VEGETABLE PARTICLES

Product/ingredient	Thermal Conductivity	Specific Heat	Thermal Diffusivity
Units	(W/m-K)	(kJ/kg-K)	(m^2/s) ´ 10e-6
Beans			
Kidney—raw	0.5	3.28	0.15
Kidney—cooked	0.5	3.3	0.15
Black—raw	0.5	3.28	0.15
Black—cooked	0.5	3.28	0.15
Lima beans—raw	0.46	2.78	0.175
Lentils—raw	0.45	2.8	0.165
Lentils—cooked	0.5	3.3	0.155
Sweet peas	0.47	2.8	0.175
Sweet peas—cooked	0.5	3.7	0.17
Black eye peas	0.5	3.2	0.15
Garbanzo beans	0.5	3.2	0.15
Green beans	0.59	3.8	0.16
Potatoes			
Russett potatoes raw	0.55	3.49	0.18
Russett Potatoes—cooked	0.5	3.55	0.175
Corn			
Sweet Corn	0.51	3.55	0.160
Sweet corn—cooked	0.56	3.58	0.164
Vegetables			
Celery	0.55	3.99	0.161
Onion raw	0.42	3.9	0.12
Bell pepper	0.59	4	0.16
Carrot	0.6	3.9	0.16
Mushroom			
Cremini raw			
Stems	0.45	2.55	0.185
Caps	0.38	2	0.182
Cremini—cooked (>60°C)	0.55	2.85	0.16

Critical density of the particles has to be such that it matches closely the density of the fluid at sterilization temperatures. In that way particles are neutrally buoyant and travel close to the center of the flow, thus being the fastest moving particles at velocities similar or equal to those of the fluid.

Surrogate simulated particles can also be engineered to be thermally conservative. As it can be observed in Table A6.2, the polymers used to fabricate the particles have lower thermal

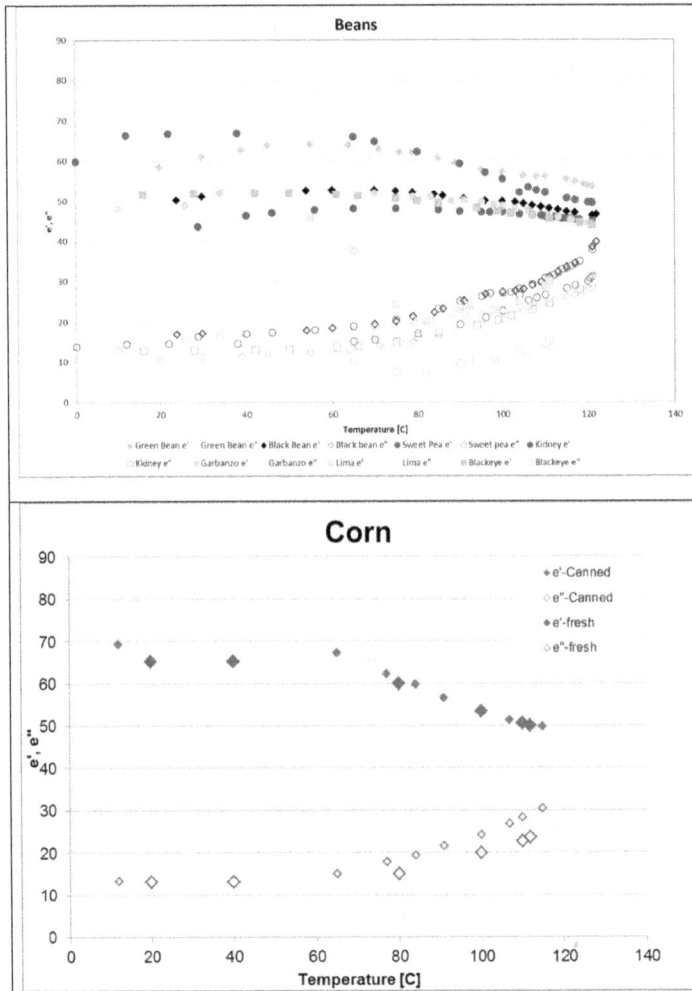

Figure A6.3   Dielectric properties of selected vegetables. (Continued)

conductivity than most food materials, however this needed to be proven experimentally. The results of this can be seen in the attached Thermal Penetration Studies.

The surrogate particles also carried a postprocess implant, which consists of a 6 mm (1/4") sphere with a load of a bioindicator (see Figure A6.4). The postprocess implant is to be incubated and analyzed.

For these studies, *Geobacillus stearothermophilus* (ATCC 7953) spores were used. These spores were acquired from MesaLabs (ProSpore) and are suspended in a color indicator broth that changes color in the presence of survivors during the incubation at elevated temperatures (55–60°C). Kinetic values for these spores are provided by MesaLabs and are shown in Figure A6.5 for the lot used in these experiments (Lot 619).

The suspension also contains a biological indicator, which turns yellow if the pH goes below 5.2 (bromocresol purple). In those cases, when the surrogate particles got stained by the carrier

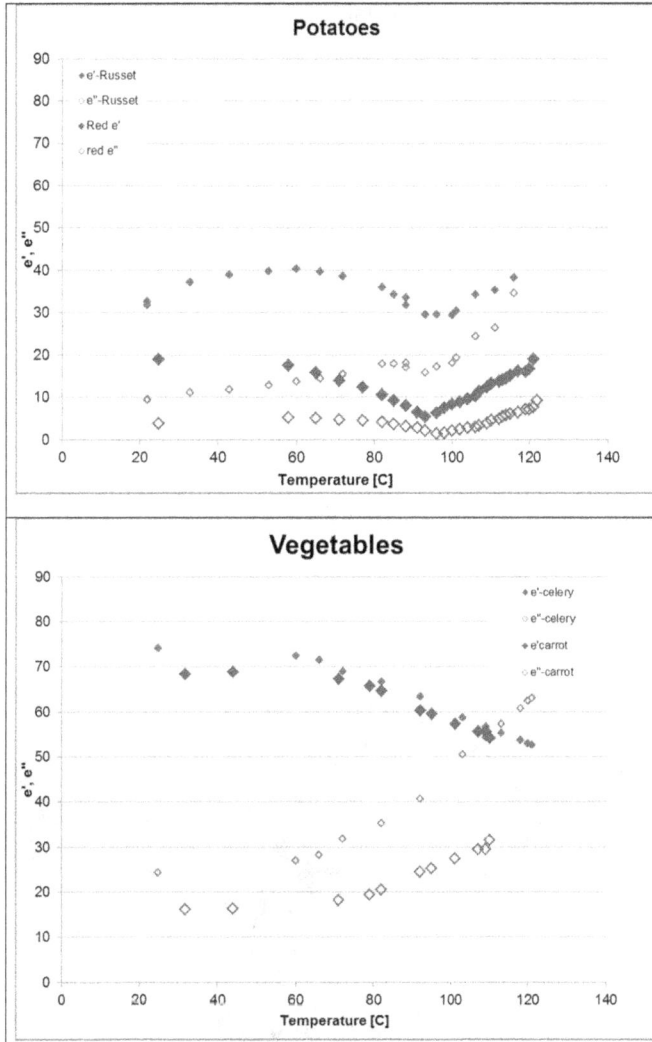

Figure A6.3 (Continued)   Dielectric properties of selected vegetables. (Continued)

fluid used, their pH value could be used as an indicator of the presence or absence of survivors. In Figure A6.6, a comparison is presented of the pH of spore suspension before incubation, after incubation, spore suspension that was inactivated and incubated, and spore suspension that was incubated.

In order to ensure that the spore solution was also conservative while exposed to the micro-wave field, the dielectric properties of the suspension were measured and are shown in Figure A6.7. It can be observed that properties of the spore suspension are also modeled.

After the initial modeling work was performed, a pilot line was designed, using the process synthesis methodology. This methodology takes into account the desired outcomes and inputs of the process, and generates a network of possible processes. These processes are then limited by constraints, which lead to the designs that are both suited for the purpose and efficient.

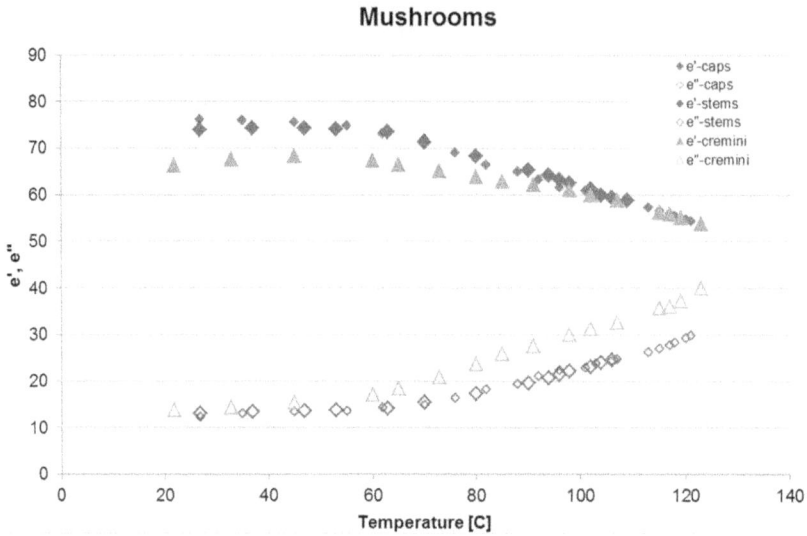

Figure A6.3 (Continued)   Dielectric properties of selected vegetables.

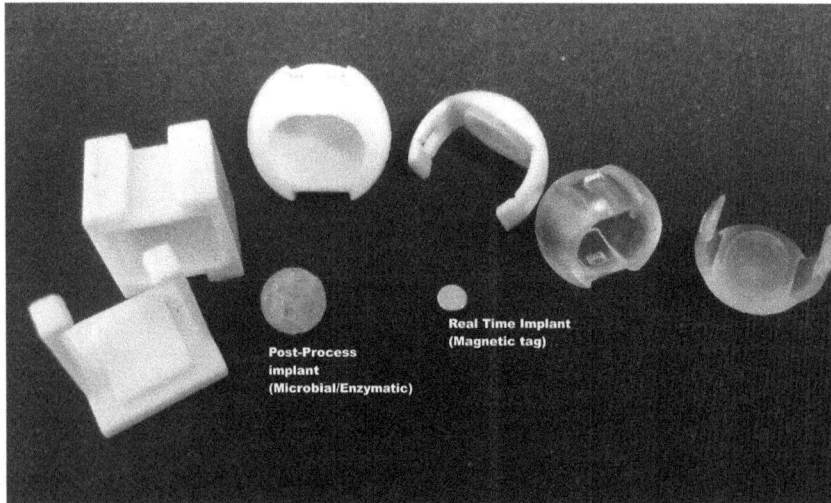

Figure A6.4   Design of simulated particles.

TABLE A6.2   ENGINEERING PROPERTIES OF THE POLYMERS USED IN THE FABRICATION OF SIMULATED SURROGATE PARTICLES

	TPX (Polymethylpentene)	Polypropylene	Ultem (Polyetherimide)	Polysulfone
Density (g/cm³)	0.83	0.91	1.34	1.24
Water absorption, 24 hours (%)	0.01	0.12	0.21	0.3
Thermal conductivity (W/m-K)	0.167	0.117	0.123	0.194
Dielectric strength (short time, 1/8" thick) (V/mil)	20	500	830	380
Dielectric constant at 1 MHz	2.12	2.30	3.15	3.50
Dissipation factor at 1 MHz	0.000025	0.0017	0.0013	0.0022
Volume resistivity at 50% RH (ohm-cm)	$>10^{16}$	$>10^{16}$	$>10^{13}$	1.7×10^{15}

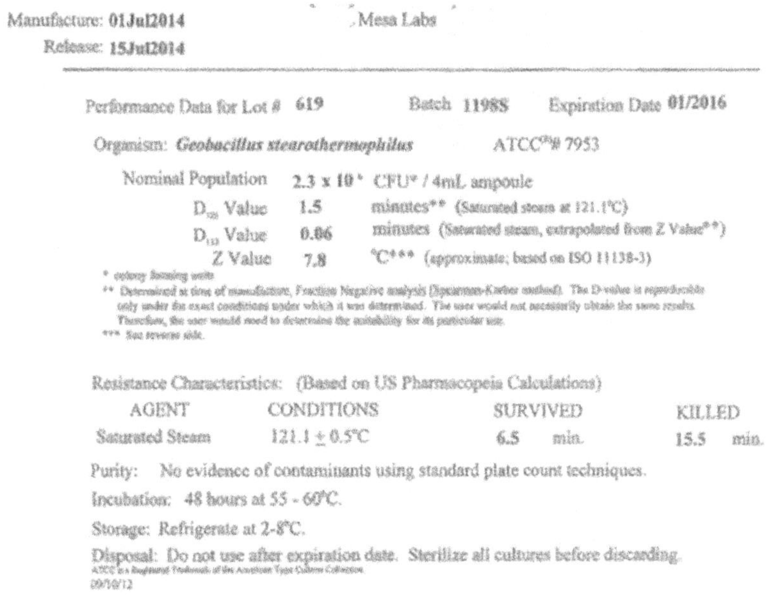

Manufacture: 01Jul2014   Mesa Labs

Release: 15Jul2014

Performance Data for Lot # 619   Batch 11988   Expiration Date 01/2016

Organism: *Geobacillus stearothermophilus*   ATCC®# 7953

Nominal Population   2.3 x 10⁵ CFU* / 4mL ampoule

D_{95} Value   1.5   minutes** (Saturated steam at 121.1°C)

D_{10} Value   0.86   minutes (Saturated steam, extrapolated from Z Value**)

Z Value   7.8   °C*** (approximate; based on ISO 11138-3)

* colony forming units

** Determined at time of manufacture, Fraction Negative analysis (Spearman-Karber method). The D-value is reproducible only under the exact conditions under which it was determined. The user would not necessarily obtain the same results. Therefore, the user would need to determine the suitability for its particular use.

*** See reverse side.

Resistance Characteristics: (Based on US Pharmacopeia Calculations)

AGENT	CONDITIONS	SURVIVED	KILLED
Saturated Steam	121.1 ± 0.5°C	6.5 min.	15.5 min.

Purity: No evidence of contaminants using standard plate count techniques.

Incubation: 48 hours at 55 - 60°C.

Storage: Refrigerate at 2-8°C.

Disposal: Do not use after expiration date. Sterilize all cultures before discarding.

ATCC is a Registered Trademark of the American Type Culture Collection.

09/30/12

Figure A6.5   Kinetic parameters of *Geobacillus stearothermophilus* used as surrogate microorganism for these trials.

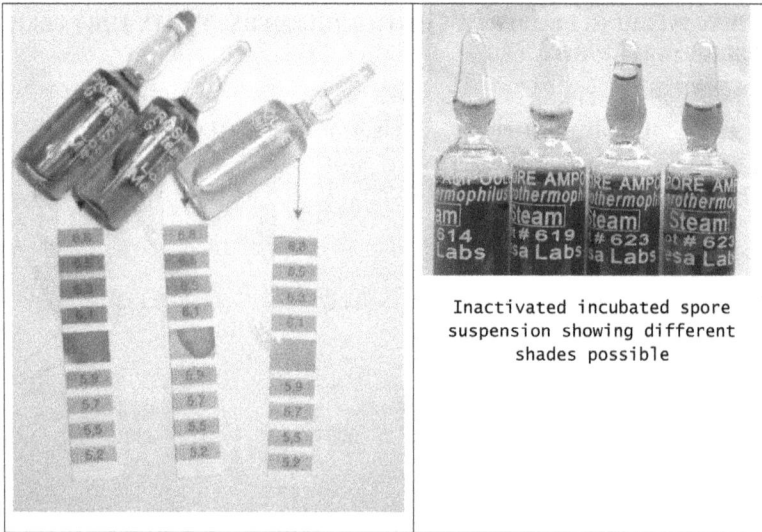

Figure A6.6   Spore solution pH—non-incubated—non-treated, incubated inactivated, incubated active.

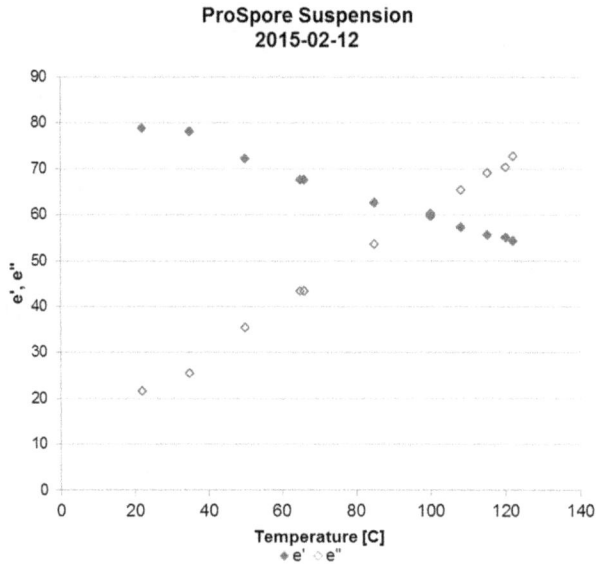

Figure A6.7   Dielectric properties of spore suspension.

A6.3 PROCESSING LINE

Based on the results of modeling, and taking into account some flexibility, a pilot processing line was designed and constructed, for which a schematic depiction can be seen in Figure A6.8 that consists of:

Blend tank
Piston Pump (Emmepiemme SPM80B)

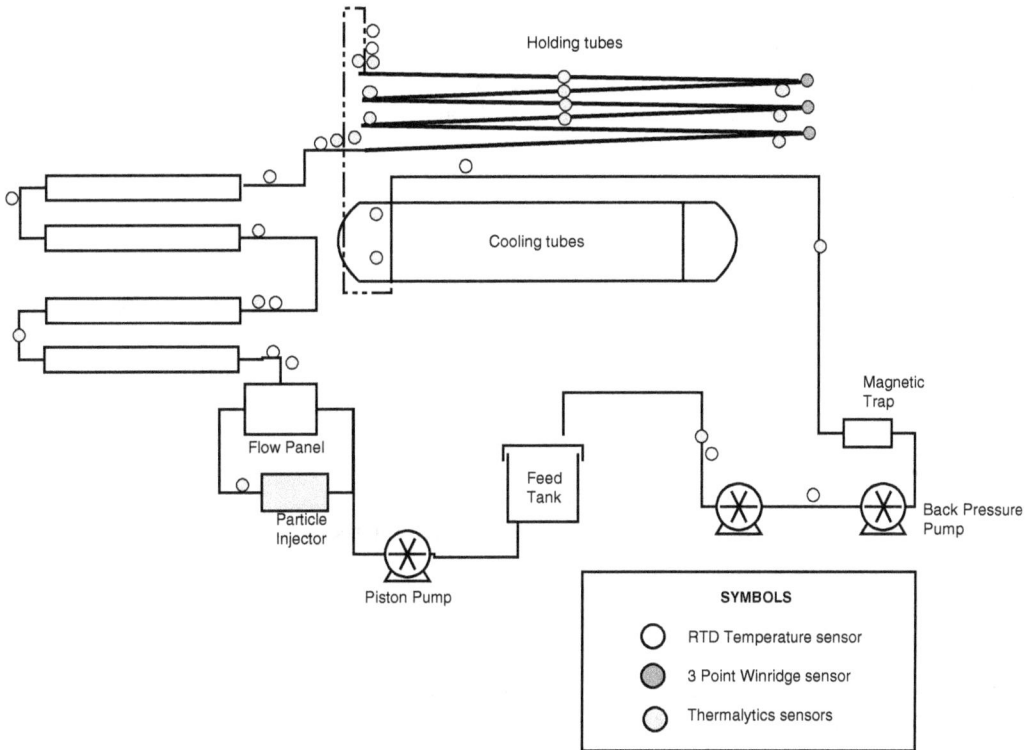

Figure A6.8   Diagram of the processing system.

Microwave applicators and connecting tubes
Hold tube divided into eight 20¢ segments
Cooler, tubular
Magnetic trap to recover surrogate particles
Back pressure system of positive displacement pumps

Instrumentation was added to the system in order to have a full description of the process as seen in the diagram. It includes:

RTD Temperature sensors used for control and data logging 3 point temperature sensors (WindRidge sensors) at the exit of sections 1,3,5 of the hold tube
Flowmeter (E+H magnetic type)
Pressure sensors after the pump, before hold tube, after cooling tubes
Thermalytics sensor summation amplifiers for detection of magnetic implants. Thirty-two total sensor amplifiers were located in the system, with 16 amplifiers covering the hold tube. Each amplifier processed the detection signals from 4 to 8 GMR (Giant Magneto-Resistive sensors) in real time, and a Windows computer was used for the real-time signal displays, recording, and postprocess playback of recorded detection events for each recorded channel concurrently.

Particle launching and recovery system were also developed in order to insert particles individually and enable the tracing of the passage of each at individual detection locations, thus enabling the measurement of residence times of each tagged particle throughout the processing system.

Recovery of the particles was carried out in a magnetic trap, since all the surrogate particles contained magnetic implants.

The plant was installed and commissioned at Aseptia's facility by August 2014. Trials were then carried out to determine the flow profiles (RTD) and to find the most suitable test products for verification of the model.

A6.4 PRODUCT FORMULATION

The product used in these experiments was tomato vegetable soup with 10% corn kernels; the composition is shown in Table A6.3.

Proper formulation was critical for the proper suspension and flow of the product.

Viscosity was measured using a Brookfield DV-II at 77°F, 60 RPM, and it ranged between 1,500 and 2,000 cps.

pH was designed at 5.2, with a maximum of 6.0.

Density: 1.025–1.030

Size of the corn pieces measured, taking samples from the received IQF corn and the results are as follows:

Length 10.2 mm ± 3.2 mm
Thickness 4.6 mm ± 1.3 mm

Therefore, using the equivalent diameter concept introduced by Jennings and Parslow (1988), a sphere of 12.77 mm (1/2") was considered a conservative model for corn, and used for modeling. The 12.77-mm sphere had a thermal critical path larger than any corn kernel but was chosen taken into account the possibility of kernels clumping (Figure A6.9).

TABLE A6.3   FORMULATION OF PRODUCT

Ingredient	Percent
Corn whole kernel IQF	10.0%
Water	42.5%
Tomato puree	15.3%
Carrot puree	13.8%
Supersweet corn puree, frozen	7.2%
Celery puree	5.4%
Red bell pepper puree	2.0%
Sugar	1.4%
Onion puree	1.4%
Thyme, IQF	0.23%
Garlic puree	0.27%
Mirepoix base concentrate	0.15%
Salt, Kosher	0.40%
Shallot Aquaresin	0.01%

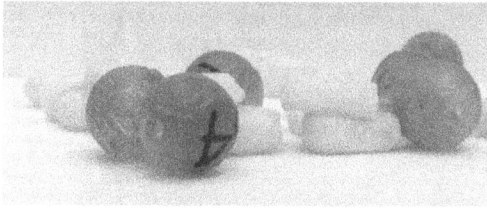

½" simulated spheres
compared to actual corn kernels

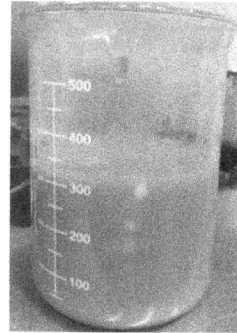

Figure A6.9   Product formulation and particles.

TABLE A6.4   THERMOPHYSICAL PROPERTIES USED IN MODELING

	Units	Bulk Particle	Carrier Fluid
Thermal conductivity	W/m-K	0.5	0.6
Density	kg/m³	1020	1025
Heat capacity	J/kg-K	3550	3800
Thermal diffusivity	m²/s	1.60E-07	1.55E-07
Dielectric properties T_{ref}	C	70	70
Loss tangent T_{ref}	[]	0.9	1.3
Temperature factor	1/C	0.011	0.011

Thermal properties of the carrier fluid and corn kernels were measured and the most conservative values were selected for input to the model, as shown in Table A6.4.

All the above data was fed into the Model of the System for verification. This model was verified during the preliminary runs by comparing the simulated temperature profile with the temperatures observed in these runs. Adjustments to the model parameters were performed in order to match both predictions and the real data. Once the model predictions matched the real temperatures observed, it was used to calculate the lethality in the hold tube only.

In order to validate the process, the model can simulate a worst-case particle, which in this case was a simulated particle. The worst-case particle was to be a surrogate-simulated particle, and that one needed to be engineered based on the definition of the conservative particle outlined above.

A6.4.1 Construction of Particles for Microbial Challenge

Based on the previous observations, the following characteristics were used to assemble conservative simulated particles for microbial challenges:

Diameter½" (12.77 mm)
Materials of construction:TPX (polymethylpentene) body

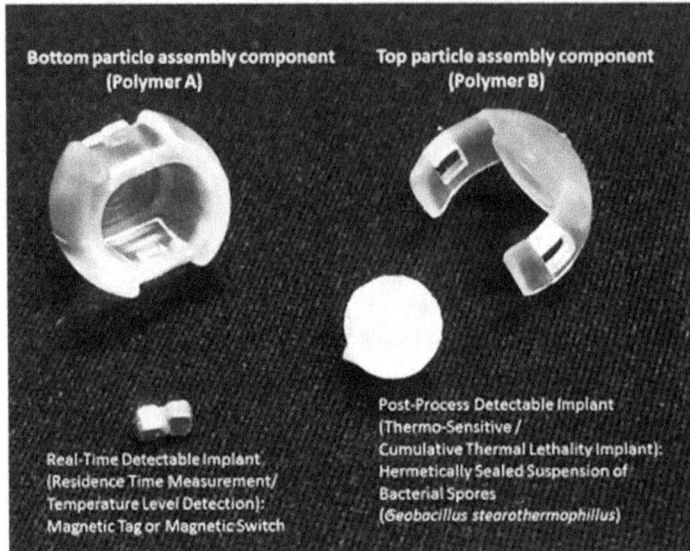

Figure A6.10   Final design of surrogate simulated particle.

Polysulfone cap

Biological Implant¼" Polysulfone sphere filled with 40 sL of spore suspension (2.3 x 10^4 spores)

Magnetic Implant1 rare earth magnet

Final density:0.83 ± 0.05 g/cm^3

Several challenges were encountered during the development of the surrogate particle, mostly with achieving a proper sealing so that the product fluid would not leak into the particle cavity, or the spore suspension would not leak out of the particle during processing. The final solution was to add a layer of acrylic adhesive (3M VHB 4910) in the sealing section of the particles, and then cover the particle with a silicone gel. The drawback of this solution was that density control was more difficult, since even such small variation in the coating layer added weight to the particle (Figure A6.10).

A6.5 THERMAL PROCESS DESIGN

A6.5.1 Thermal Process Design Confirmation

In order to confirm the process design and validate the model, several experiments were carried out. These experiments involved heating the product to the intended process temperatures and observing the temperature profiles in the hold tube, as well as the residence time of the surrogate simulated particles in each section of the hold tube. Residence time studies are presented in the RTD_Studies document.

The parameters used for the model of the thermal process are shown in Table A6.5.

Temperature expected from modeling was 131 at the exit of the last heater and 124.7 at the exit of the hold tube (Table A6.6). However, reality showed that while the proportion was correct,

TABLE A6.5   PARAMETERS FOR MODEL THERMAL PROCESS VALIDATION

Hold tube length	1,590" (39.8 m)	6 straights ´ 6 m + 5 turns ´ 0.21 m + 5 straight ´ 0.15 for sensors + connections
Hold tube ID	1.91" (0.0485 m)	
Flow rate	3.0 gpm [680 l/h]	
Residence time		
Fluid average	390 seconds	
Correction factor	0.62	For liquid
Particle fastest (seconds) From observation	291 0.67	For particle
Min temperature at exit of hold tube	252°F (122.5°C)	
Min temperature inlet of hold tube	260°F (126.5°C)	

TABLE A6.6   EXPECTED TEMPERATURES ACCORDING TO THE MODEL

No.	Unit	Type	x(m)	T_{fluid} (°C)	T_{bulk} (°C)
0	Inlet	--	0	35	35
1	Connection	Holding tube	2	34.9	35
2	Heater 1	Volumetric	5	66.2	58.9
3	Connection	Holding tube	6	66	60.2
4	Heater 2	Volumetric	9	91.9	83
5	Connection	Holding tube	11	91.2	86.6
6	Heater 3	Volumetric	14	111.5	104.9
7	Connection	Holding tube	15	111.1	106.1
8	Heater 4	Volumetric	18	130.8	124.1
9	Connection	Holding tube	20	129.4	126
10	Hold Tube 1	Holding tube	28	128.2	128.4
11	Hold Tube 2	Holding tube	34	127.5	128
12	Hold Tube 3	Holding tube	40	126.8	127.3
13	Hold Tube 4	Holding tube	47	126.1	126.6
14	Hold Tube 5	Holding tube	53	125.4	125.9
15	Hold Tube 6	Holding tube	60	124.7	125.2
16	Connection	Holding tube	63	124	124.8
17	Cooler	Tubular	88	62.5	72
18	Cooler	Tubular	112	34.6	39

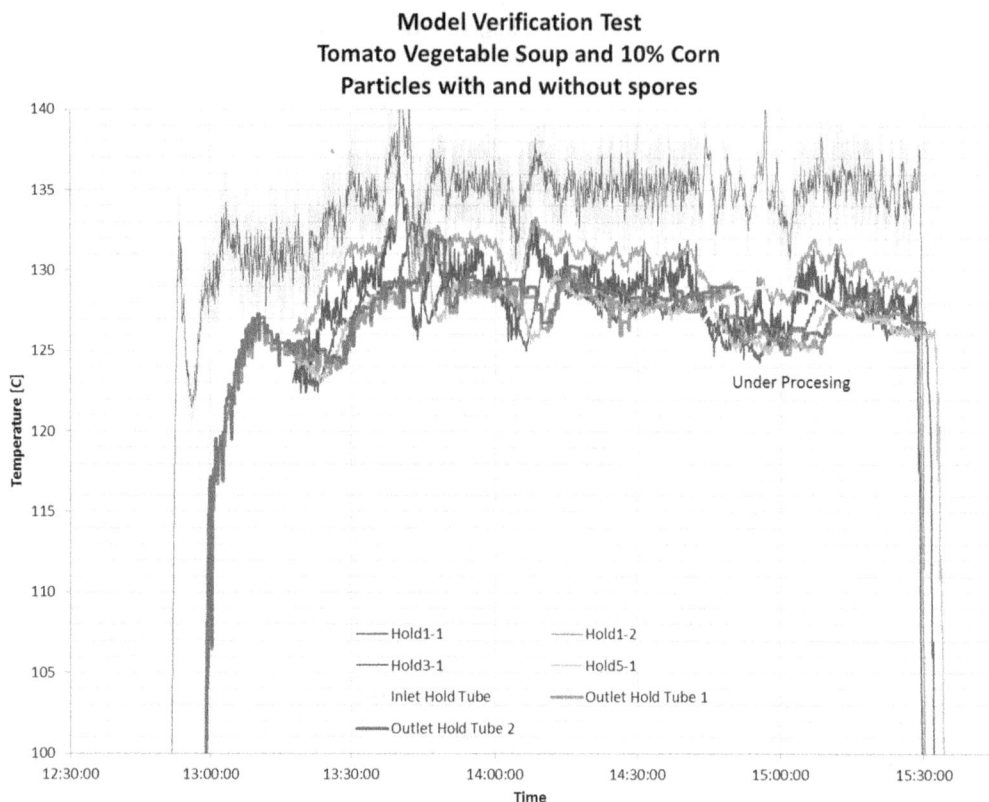

Figure A6.11   Model verification test temperatures.

the system was less stable than expected, and thus it was overshooting the temperature setpoint (Figure A6.11). During this run, toward the end, and in order to verify the conservative nature of the model, the setpoint was lowered to achieve underprocessing of the simulated particles (end point $F_0 < 6$), as shown in Figure A6.11. The expectation was that the particles inserted during this time should show unsterility (i.e., growth) after incubation. A subset of ten particles were inserted during this time.

A total of 106 simulated particles were inserted, 96 of them carrying spores. During the process, temperature was lowered to a point in which the model predicted underprocessing (14:50 to

TABLE A6.7 MODEL VERIFICATION MICROBIAL RESULTS	
Sterile	67
Non-sterile	3
Leak	26
Empty	10
Total	**96**

15:10 on the timeline) and during this time, a subset of ten particles was inserted into the system. Results after 48 h of incubation (Table A6.7 Model verification microbial results) showed that three of the particles which were part of the deliberately underprocessed subset showed some growth, and two were unreadable (leak), while the remaining five did not show any growth (sterile).

This experiment therefore confirmed that the predictions of the model were conservative.

For the final process establishment and validation, an additional section of hold tube was added, which increased the length of the hold tube by 40 feet (480"). It was also observed that the first hold tube should be considered as a stabilization section, in order to smooth out temperature differences between the fluid and the particles, and allow credit for lethality only in the sections in which temperature is uniform.

A6.5.2 Thermal Process Design for Tomato Soup with Corn

Thermal process was designed to achieve an F_0 of 8 minutes in the center of the corn particles, as recommended by Pflug (1988). The process design was performed by reconstructing the initial testing using the modeling software and the maximum flow rate possible in the process line (3 gpm). The temperature profiles observed during the validation runs can be seen in Figure A6.12. Each unit mentioned previously is separated by a gray line, and for purposes of the process setting, and because there was a large temperature difference at the end of the heaters, the first section of the hold tube was not considered for accumulation of lethality and it is called the "equilibration" tube. It can be observed in Figure A6.12 that the center of the simulated particle should have received a thermal process equivalent to F_0 7.2 minutes in the hold tube only, while the real food particle should have received at least twice that (Table A6.8).

The process setting, based on the modeling was set as shown in Table A6.9.

These parameters were used for subsequent microbiological challenges in which simulated surrogate particles loaded with the above-mentioned *G. Stearothermophilus* spores were used for testing.

A6.6 MICROBIOLOGICAL CHALLENGE STUDIES

Microbiological challenge studies were performed in order to verify the process as it was set. The operating conditions were as follows:

Flow rate:3.0 gpm
Temperature at the exit of the heating section:135°C
Temperature at the exit of the hold tube:>122.5°C

First, a challenge was performed with 100 particles to verify the residence time distribution and that the process was sufficient to achieve the required lethality. Product was prepared according to the formulation, and simulated particles were inserted one at a time, at 90 second intervals while recording the times of their passage at individual locations, their residence times, and temperatures of the carrier fluid at different locations throughout the processing system.

A6.6.1 100 Particle Challenge

Product was prepared according to the formulation shown in Table A6.3 in batches of 1,500 lb. System was brought up to temperature and the insertion of simulated particles began. A total of 110 particles were inserted one at a time at 30–90-second intervals.

(a)

(b)

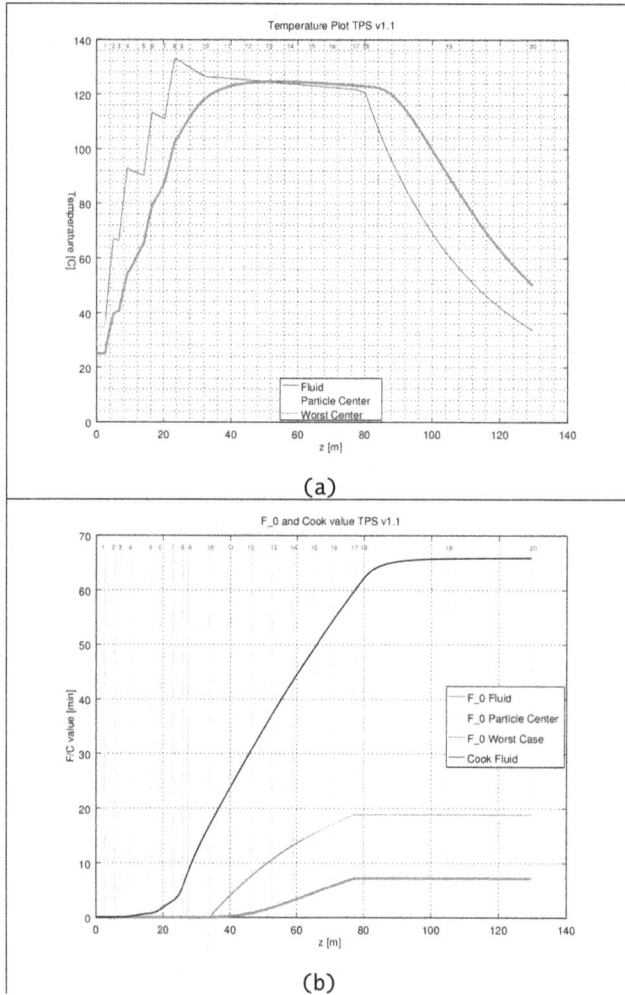

Figure A6.12   Temperature profiles (a) and lethality accumulation (b) for process setting.

Times of passage by individual sensors, residence times, and temperatures were recorded and the results are listed in Table A6.10 and Figure A6.13.

It can be observed that the residence time has a minimum of 340.7 seconds, which is 2.1σ of the average, with a maxima of 378.19 which is also 2σ of the average. The average coincided with the design times, but due to the flow conservatism of the simulated surrogate particles, these moved faster through the system than expected.

Temperatures during this experiment were recorded and can be observed in Figure A6.14. Exit of the heater temperatures and inlet of the hold tube were in the same order as the modeled temperatures; however, the end of the hold tube was 1°C lower than expected. Thus, underprocessing was a possibility, based on the model results.

Simulated particles were recovered and placed in incubation at 55°C for 48 hours in accordance to the instructions of the manufacturer to observe any potential growth from spores.

TABLE A6.8   MODEL PARAMETERS FOR PROCESS ESTABLISHMENT

Unit No.	Unit name	Temperature (°C)	Temperature (°F)
0	Inlet	35	95
1	Connection	34.9	94.82
2	Heater 1	67	152.6
3	Connection	66.3	151.34
4	Heater 2	92.7	198.86
5	Connection	89.9	193.82
6	Heater 3	112.9	235.22
7	Connection	110.8	231.44
8	Heater 4	132.9	271.22
9	Connection	131.9	269.42
10	Estabilizer	127.1	260.78
11	Hold Tube 2	126.5	259.7
12	Hold Tube 3	125.8	258.44
13	Hold Tube 4	125.2	257.36
14	Hold Tube 5	124.5	256.1
15	Hold Tube 6	123.8	254.84
16	Hold Tube 7	123.2	253.76
17	Hold Tube 8	**122.5**	252.5
18	Connection	121.3	250.34
19	Cooler	61.2	142.16
20	Cooler	34	93.2

TABLE A6.9   PARAMETERS FOR PROCESS SETTING FOR TOMATO SOUP WITH CORN

Hold tube length	1,822" (46.1 m)	(7 straights ´ 6 m + 6 turns ´ 0.21 m + 5 straight ´ 0.15 for sensors + connections)
Hold tube ID	1.91" (0.0485 m)	
Flow rate	3.0 gpm []	
Residence time		
Fluid average	440 seconds	
Fluid correction factor	0.5	
Particle fastest observed	294 seconds 0.67	Based on 100 particle run
Min temperature at exit of hold tube	252 °F (122.5°C)	
Min temperature inlet of hold tube	260°F (126.5°C)	

TABLE A6.10	RESIDENCE TIME DISTRIBUTION
Min (s)	340.71
Max (s)	378.19
Average (s)	360.89
σ (s)	9.124

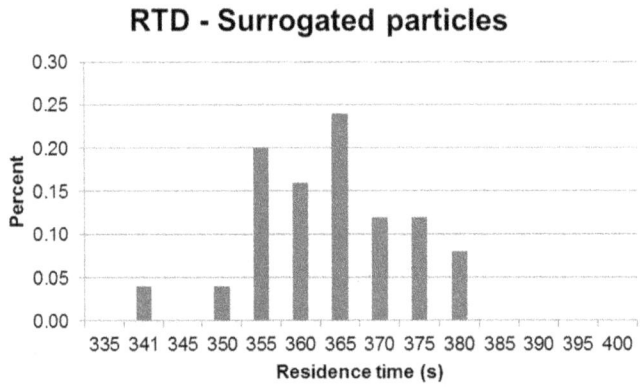

Figure A6.13   Residence time for pretrial.

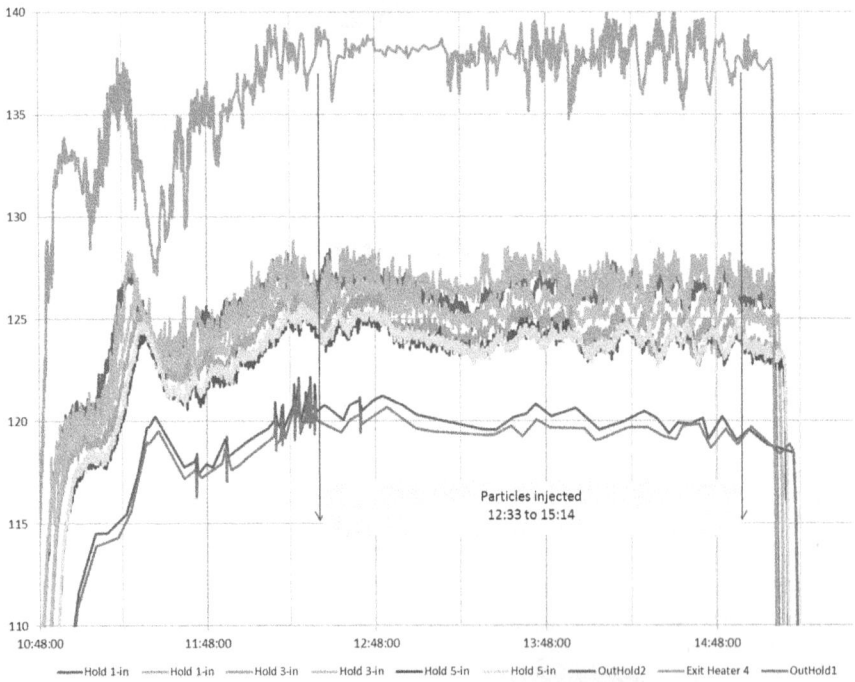

Figure A6.14   Temperature record.

TABLE A6.11   RESULTS OF SURROGATE SIMULATED PARTICLES IN PRETRIAL

	After Processing	48-Hour Incubation	Notes
Sterile	87	66	
Non-sterile	0	0	
Leak	15	8	pH tested
Empty	0	28	
Lost	8	8	
Total	**110**	**110**	

Particles were characterized as Negative (purple color), Positive (yellow color), Leak (soup had a leak inside the cavity), or Empty (implant had lost the liquid).

Results from the implants before and after the 48-hour incubation are shown in Table A6.11. No positives have been observed; thus, the conservative character of the model was verified. However, many implants had lost the liquid and thus the number of empties increased significantly.

pH of the incubated particles that were considered leaks was also tested to check for positives using the pH-specific testing strips (peHanon 904-16) and it was found to be above 5.6 pH in all cases.

Statistically, F_0 at the center of the surrogate particles can be estimated using the endpoint method calculation (Equation 1)

$$\text{Average log reduction} = \log(\text{innoculum}) - \log\left[\ln\left(\frac{\text{Total No. particles}}{\text{No. Sterile particles}}\right)\right]$$

Equation 1: Endpoint log reduction calculation

With the existing data, the result is that the average log reduction was equivalent to 5.15 log, which is equivalent to the F_0 of 7.725 minutes.

With the results of this test, 300+ particle test was planned, and in order to compensate for the 20% of unusable particles, 390 particles were prepared for insertion.

A6.6.2 300+ Particle Challenge

Product was prepared according to the formulation shown in Table A6.3 in batches of 1,500 lb. System was brought up to temperature and the insertion of simulated particles commenced. A total of 390 particles were inserted into the system at 30-90-second intervals.

For this challenge, a modification was made to the particle launching system which allowed for a faster insertion cycle.

Residence times and temperatures were recorded and these results are listed in Table A6.12.

It can be observed that the minimum residence time was lower than in the previous trial, and the standard deviation was larger. During this trial, it was observed that a few particles got stuck in the temperature sensors in the hold tube, and thus skewed the data. However, since each particle is individually tagged and followed, the fastest ones could be isolated (Figure A6.15).

A segment-to-segment analysis (Table A6.13) shows that particles were stuck in section 1 and 11 marked in bold. This coincides with the locations of temperature sensors and that extended the residence time of some particles.

TABLE A6.12   RESIDENCE TIME DISTRIBUTION RESULTS MICROBIAL CHALLENGE

	Time	Flow Factor
Average fluid (s)	505.0	
Min (s)	337.0	0.66
Max (s)	483.6	0.95
Average (s)	374.0	0.74
σ (s)	21.593	

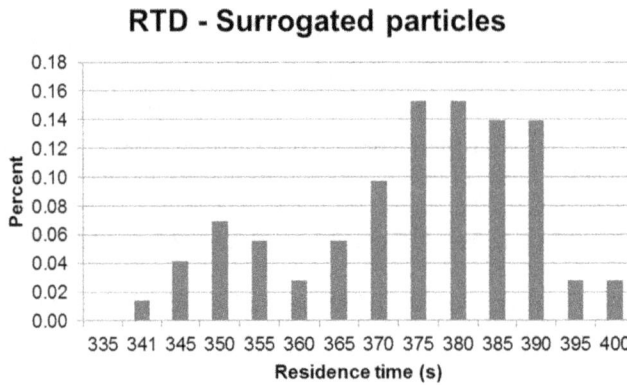

Figure A6.15   Residence time distribution.

Temperatures during this experiment were recorded and can be observed in Figure A6.16. Exit of heater temperatures and inlet of hold tube followed trends similar to the 100-particle challenge, and temperature at the end of the hold tube fell below the expectation for a short time. During this time, particles 90–120 were inserted and these were expected to be minimally processed. However, after incubation, they were also sterile (Table A6.14).

A total of 390 particles were inserted, and 385 were recovered (the remaining 5 were subsequently recovered during CIP). From those, 295 were easily recognizable as sterile, as can be seen in Figure A6.17 marked as GOOD. After 48-hour incubation, the ones marked as unsure or leak were confirmed as sterile, by testing the pH of the solution, as seen in Figure A6.18.

With these results, the average log reduction was 5.89 in the surrogate-simulated particles, which is an equivalent of F_0 8.3 minutes.

A6.7 CONCLUSION

The results from all the tests show that process setting was implemented correctly by applying a process-driven model of the line. All the processed particles and surrogates were sterile.

Validation of the model settings proved that the model was conservative, and so were the surrogate-simulated particles.

Thus, these particles can be used for further validation of food particles that are smaller and more thermally conductive than the simulated particles.

TABLE A6.13   RTD SEGMENT ANALYSIS

Segment	1	2	3	4	5	6	7	8	9	10	11	12	13	14	Total
Min (s)	5.0	39.3	3.4	17.8	21.3	5.1	17.9	20.8	3.4	20.7	21.7	5.0	20.6	122.9	337
Max (s)	**80.7**	48.2	5.7	23.2	24.7	15.1	23.9	26.3	15.2	26.4	**132.8**	19.8	26.1	148.5	483
Average (s)	7.5	43.7	4.4	21.3	23.0	6.8	21.6	23.2	4.9	23.1	26.1	7.5	23.1	138.2	374
Std Dev (s)	8.783	1.956	0.571	0.980	0.903	1.388	1.017	1.165	1.539	1.274	13.015	2.317	1.361	6.690	21.593

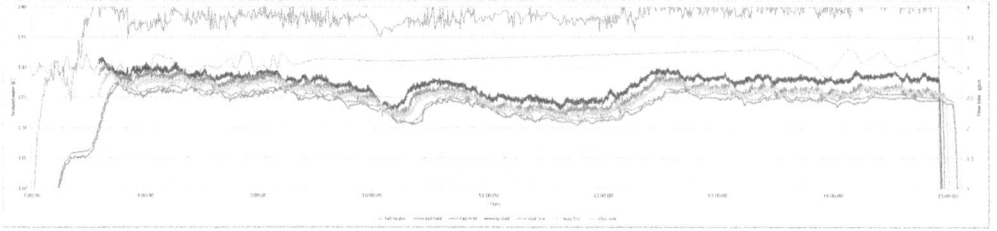

Figure A6.16   Temperature trace during 300 particle microbial challenge.

TABLE A6.14   RESULTS OF PARTICLES BEFORE AND AFTER INCUBATION

	After Processing	48-Hour Incubation	Notes
Sterile	278	305	
Non-sterile	0	0	
Unsure	39	0	pH tested
Leak	23	0	pH tested
Empty	43	80	
Lost	5	5	
Total	**390**	**390**	

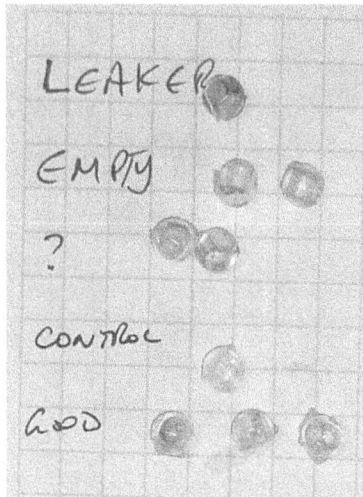

Figure A6.17   Example of each type of implant result.

Residence time distribution needs to be further analyzed statistically in order to determine the required number of particles that will guarantee that a significant population will be in the fastest possible moving fraction for validation.

Sterility of liquid and food particles was confirmed by the experiments carried out, and thus production under the specified conditions will yield a safe product.

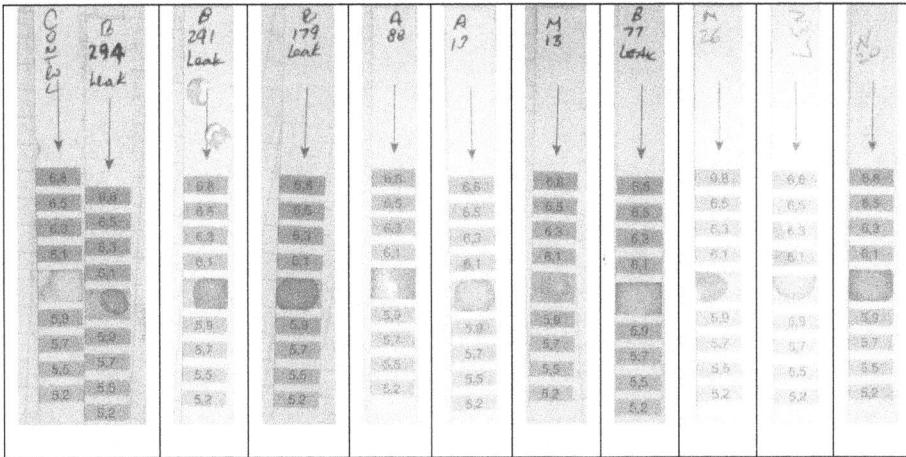

Figure A6.18   Record of pH testing of questionable particles.

A6.8 THERMAL PENETRATION IN SIMULATED FOOD PARTICLES COMPARED TO REAL FOOD PARTICLES: DETERMINATION OF THERMALLY CONSERVATIVE SIMULATED FOOD PARTICLES

A6.8.1 Introduction

Aseptic processing of foods with particles requires the experimental determination of proper processing inside of the particles (geometric center/cold spot). Aseptia proposes the use of conservative simulated particles that can carry real-time (magnetic) and postprocessing (microbial/enzymatic) implants for validation as well as inert loads to provide the means of density adjustment to achieve the conservative flow properties if needed. These simulated particles need to be representative of the worst possible case for thermal processing, i.e., thermally conservative, flow conservative, and microwave conservative for installations using microwave heating for sterilization.

The purpose of these experiments was to compare the heat penetration rates into existing simulated particles versus the penetration into the real food particles and determine whether the available fabricated particles could be used as conservative surrogates.

Simulated particles consist of spheres, cylinders, or cubes made of two components of different polymers (TPX, Ultem, Polysulfone, Polypropylene), and sized to match the worst-case product particle (1/4", ½", ¾"). The particles are constructed in such a way that a space in which the implanted loads can be placed is in the center of the simulated particle (cold spot) (Figure A6.19).

In order to determine the particle that could be used as surrogate, the following measurements were performed:

1. Thermal properties of food particles
2. Dielectric properties of food particles
3. Thermal properties of simulated particles
4. Thermal penetration studies on selected food and surrogate particles

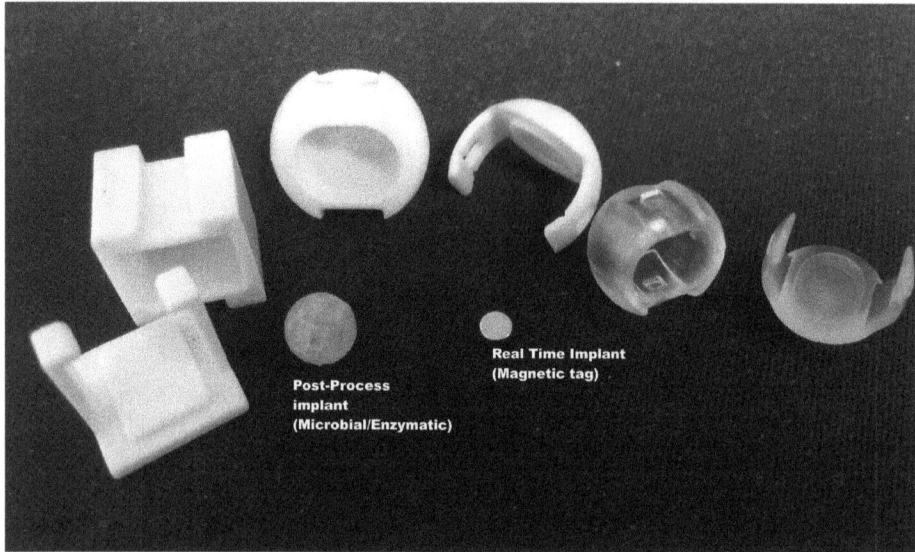

Figure A6.19   Simulated particles. PS: polysulfone; ULT: Ultem 1000/polyetherimide; TPX/PMP: poly-methylpentene; PP: polypropylene.

A6.8.2 Materials and Methods

The products to be tested were defined as a mix of vegetables. These products were further separated into the following six categories:

	Vegetable	Category
1	Corn	Corn
2	Green beans	Beans
3	Green peas	Beans
4	Black beans	Beans
5	Lentils	Beans
6	Red kidney beans	Beans
7	Lima beans	Beans
8	Garbanzo beans	Beans
9	Whole Cremini mushrooms	Mushrooms
10	Button mushrooms (stems and pieces)	Mushrooms
11	Diced carrots	Vegetables
12	Celery bits	Vegetables
13	Sliced onions	Vegetables
14	Red bell pepper dices and/or pieces	Vegetables
15	Diced potatoes	Vegetables

	Vegetable	Category
16	Rice pieces	Vegetables
17	Barley	Vegetables
18	Diced chicken	Meat
19	Spaghetti pasta	Pasta
20	Fettuccine pasta	Pasta
21	Rotini pasta	Pasta
22	Penne pasta	Pasta

A6.8.3 Thermal Properties of Particles

Thermal properties of particles were measured using a Thermal Analyzer (KD2 Pro, Decagon Devices Inc., Pullman, WA). This analyzer allows the measurement of thermal conductivity, volumetric specific heat, and thermal diffusivity.

Particles that were not large enough to be tested by the analyzer were first homogenized using a blender and then placed in a test cell where the temperature could be adjusted to 20–140°C. Particles that could be directly read by the analyzer were used as such.

Properties were also estimated using COSTHERM (EU Cost90 project) software in the temperature range of 20–130°C.

The more conservative of these two values was recommended for use in further modeling.

Product/ingredient	Thermal Conductivity	Specific Heat	Thermal Diffusivity
Units	W/m-K	kJ/kg-K	(m²/s) ´ 10e-6
Beans			
Kidney—raw	0.5	3.28	0.15
Kidney—cooked	0.5	3.3	0.15
Black—raw	0.5	3.28	0.15
Black—cooked	0.5	3.28	0.15
Lima beans—raw	0.46	2.78	0.175
Lentils—raw	0.45	2.8	0.165
Lentils—cooked	0.5	3.3	0.155
Sweet peas	0.47	2.8	0.175
Sweet peas—cooked	0.5	3.7	0.17
Blackeye peas	0.5	3.2	0.15
Garbanzo beans	0.5	3.2	0.15
Green beans	0.59	3.8	0.16
Potatoes			
Russett potatoes—raw	0.55	3.6	0.18
Russett potatoes—cooked	0.5	3.55	0.175

Product/ingredient	Thermal Conductivity	Specific Heat	Thermal Diffusivity
Corn			
Sweet corn	0.55	3.55	0.17
Sweet corn—cooked	0.59	3.58	0.171
Vegetables			
Celery	0.55	3.99	0.161
Onion raw	0.42	3.9	0.12
Bell pepper	0.59	4	0.16
Carrot	0.6	3.9	0.16
Mushroom			
Cremini raw			
Stems	0.45	2.55	0.185
Caps	0.38	2	0.182
Cremini cooked (>60°C)	0.55	2.85	0.16

It is visible that most of the beans have very similar thermal properties, so red kidney beans were chosen to represent black, navy, black-eye, and red kidney beans, while lima beans were tested separately.

Sweet peas, garbanzo, corn, green beans, potatoes, carrots, and mushrooms were also tested for thermal penetration.

Readings from the KD2 Pro were very unreliable at temperatures above 50°C. Thus, these reading were performed only at a few temperature points and compared to Costherm results. In most cases, Costherm was the more conservative and thus recommended.

A6.8.4 Thermal Penetration Tests

In order to confirm the thermally conservative properties of the simulated particles, thermal penetration of the particles was compared to those of real food particles.

These tests consisted of immersing the particles at room temperature into boiling water (100°C), while a data logger (Tempscan 1100—Iotech) was used to record the temperatures of both simulated and real particles at intervals of less than 1 second.

A special testing setup was developed, in which the heat penetration into both real and simulated particles could be tested concurrently. On a 3" blanking plate, TriClamp 12 Type T thin-wire thermocouples (Omega Engineering Inc., Stamford, CN) were inserted using stainless sanitary fittings (Swagelok, Solon, OH) in a rectangular grid arrangement (Figure A6.20).

Simulated particles of different materials, sizes, and shapes were affixed at the end of some of these thermocouples using temperature-resistant epoxy. The temperature sensors were connected to a data logger capable of reading every 0.1 second with a precision of 0.1°C.

Once the temperature inside all the particles was uniform, the food particles were cooled and discarded.

Experiments were repeated at least five times for each type of particles selected.

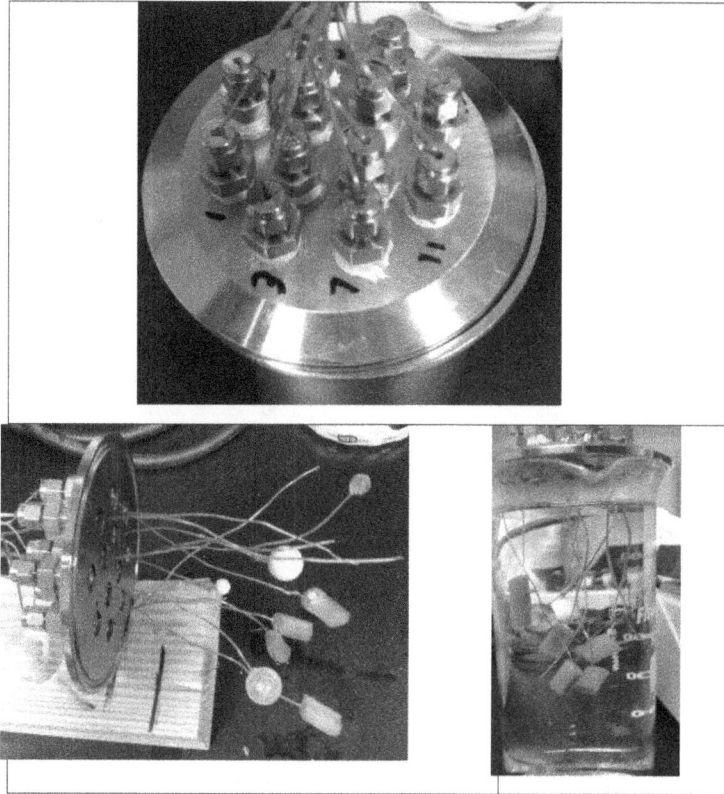

Figure A6.20   Thermal testing apparatus.

Data Analysis

The data obtained in the previous experiments was analyzed in two ways:

1. Temperature curves

Direct measurements at the approximate centers of the particles were compared to observe the rate at which the centers of the particles heated. Both real and simulated particles can be compared directly this way.

These experiments took into account the worst-case scenario in which a cold particle is introduced into the very hot liquid. This provides a large $T_{\infty} - T$ and would be representative of a particle which did not receive any thermal treatment while traveling through the heating stages, but is exposed to thermal treatment as soon as it enters the hold tube (Figure A6.21).

2. Modeling of particles

In order to model the behavior of the particles in the processing line, a comparison was made to yield the numeric values which could be entered into the TPS (Thermal Process Simulator) software.

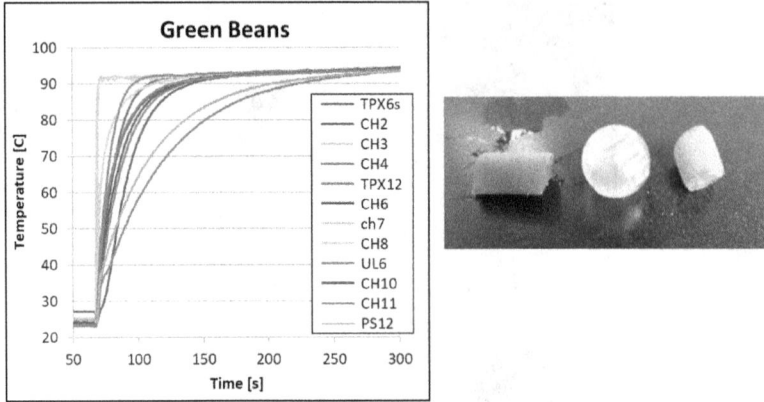

Figure A6.21   Example of temperature penetration study.

In order to do so, the B_i number needs to be calculated, which is performed by comparing a normalized temperature curve to models.

Then the B_i and F_0 numbers for the particles are calculated using a model based on a solution of the heat transfer equations.

By having this equivalence we can in the future predict the way the particles will heat and thus provide us with a tool for process design and prevalidation.

The temperature can be normalized as follows:

$$\Theta = \frac{T_i - T_\infty}{T_0 - T_\infty}$$

Here T_i is the temperature of the particle at any given time, T_0 is the initial temperature, and T_∞ is the temperature of the carrier fluid surrounding the particles.

This normalization eliminates the noise that can be created by the differences of initial temperatures of the vegetables and simulated particles, resulting in a graphical representation very similar to a Heisler Chart used to determine the heat transfer in heating or cooling (Kern, 1965).

By comparing the F_0 and B_i numbers in both cases, we can estimate the heat transfer and internal temperature for different inlet temperatures and carrier fluid temperatures.

A6.8.5 Temperature Penetration into the Simulated Particles

Surrogate particles chosen for the study were spheres of ¼" (6 mm), ½" (12.7 mm), and 5/8" (15 mm), and cubes of the respectively identical dimensions. The surrogate particles were made of polymethylpentene (TPX), polysulfone, Ultem, and polypropylene for testing. The properties of these materials, including the thermal properties, are presented in Appendix 1, based on the manufacturer's specifications.

The fabricated polymer particles were first tested for penetration of heat into the center of each particle, with and without the implants contained in the carrier particle cavity.

The heating curves confirmed that increasing the size of the particles results in the reduction of the rate of heat penetration into their center, as well as the influence of material properties.

It must also be noted that two sizes of particles were tested with bioload carrying implants (6 mm particle) inside of them. It can be observed that these particles were much more thermally insulating than the single particles without the implant subparticles contained within the cavity. The polysulfone particles (PS12s) have a very low temperature resistance by themselves, being the fastest heating among the 12 mm spheres. However, when the implant is added, they heat slower than the slowest heating 12 mm sphere without the subimplants.

These thermal characteristics needed to be compared to real food particles, as listed above.

A6.8.6 Temperature Penetration into the Real Food Particles

The same procedure was repeated for real food particles, for which the samples were purchased fresh from a local supermarket and also frozen for comparison and the worst cases used.

From each batch, the largest vegetables were selected by visual inspection and measured. Dimensions of the beans and corn were measured, and they were below ½" with the exception of garbanzo beans, which were between 1/2" and 5/8".

Different vegetable particles tested are compared in Figure A6.22 to the simulated food particles to ensure the conservative heat penetration behavior. The slowest heating vegetable particles from every experiment (five replicates) were used.

It is worth mentioning that the simulated particles carry a smaller subimplant inside containing the microbial spore suspension, and this is used for comparison (Figure A6.23).

Based on the above comparisons, the following simulated particles are to be used for surrogate of vegetable particles in order to ensure their conservative (slowest heating) behavior (Table A6.15)

The properties of the polymers used to manufacture simulated particles are provided in Table A6.16.

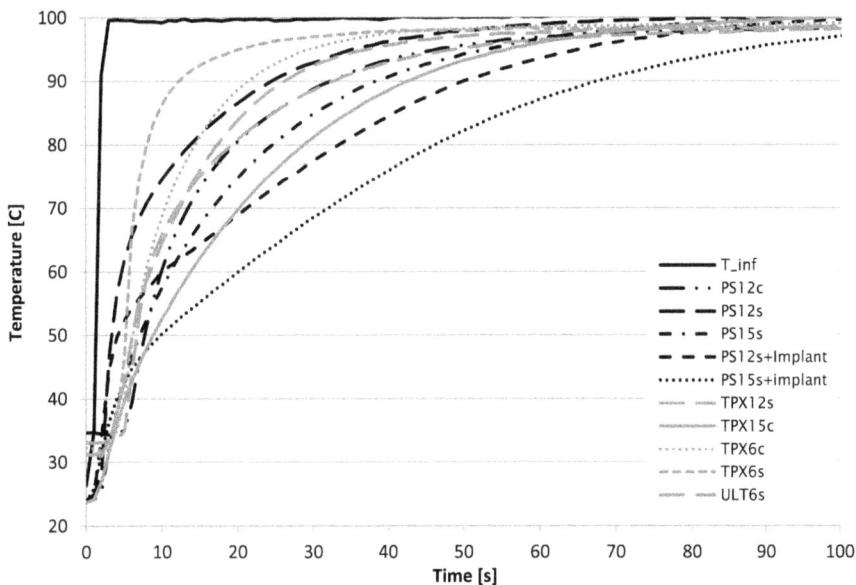

Figure A6.22   Temperature penetration of simulated particles.

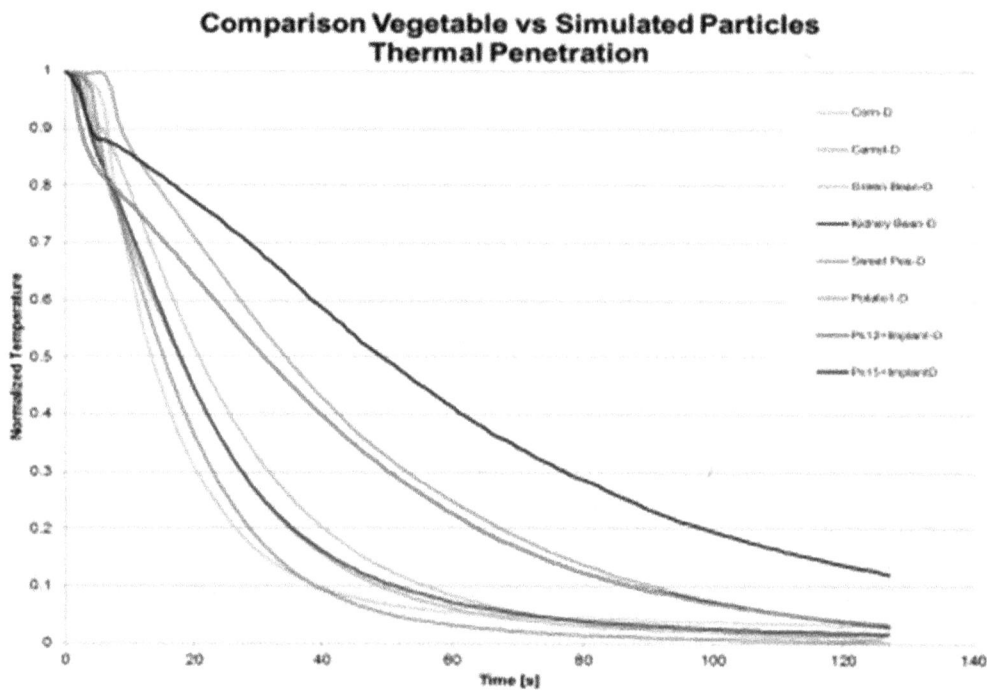

Figure A6.23   Comparison of vegetables to simulated particles.

TABLE A6.15   CORRELATION OF INGREDIENTS AND SIMULATED PARTICLES

Product/Ingredient	Dimensions	Simulated Particle
Beans		
Kidney, black, navy, blackeye peas, lima	Length ½-5/8" Thickness ¼-3/8"	12 mm TPX or PS sphere
Sweet peas	½" max	12 mm TPX or PS sphere
Lentils	½" max	12 mm TPX or PS sphere
Chickpeas / garbanzo	5/8" max	15 mm TPX or PS sphere
Green beans	15 mm max length, max diameter 3/8"	12 mm TPX or PS sphere
Potatoes		
Russett potatoes raw	½" cubes	Depending on the cut – needs more specification
Corn		
Sweet corn	3/8"– ½"	PS 6 mm sphere TPX 12 mm sphere
Vegetables		
Carrot (1/2" cubes)	½" cubes	12 mm TPX or PS sphere

TABLE A6.16 PROPERTIES OF THE POLYMERS USED TO MANUFACTURE SIMULATED PARTICLES

	TPX (Polymethylpentene)	Polypropylene	Ultem (Polyetherimide)	Polysulfone
Density (g/cm³)	0.83	0.91	1.34	1.24
Water absorption, 24 hours (%)	0.01	0.12	0.21	0.3
Thermal conductivity (W/m-K)	0.167	0.117	0.123	0.194
Dielectric strength (short time, 1/8" thick) (V/mil)	20	500	830	380
Dielectric constant at 1 MHz	2.12	2.30	3.15	3.50
Dissipation factor at 1 MHz	0.000025	0.0017	0.0013	0.0022
Volume resistivity at 50% RH (ohm-cm)	$>10^{16}$	$>10^{16}$	$>10^{13}$	1.7×10^{15}

A6.9 RESIDENCE TIME DISTRIBUTION STUDIES

A6.9.1 Introduction

Surrogate simulated particles fabricated from microwave-transparent polymers have been used in the validation of heterogeneous products by Aseptia Inc. These particles can be engineered to be conservative in both flow and thermal characteristics when compared to real food particles.

In this study, the effect of density on the residence time is examined, and a method to design the most flow conservative particle is presented.

Residence time is important to determine the fastest moving particle in a heterogeneous product. It is well-known that the presence of particles modifies the behavior of flows, to the point where an approximation to non-Newtonian flow can be used (Inoue et al., 2013). It has been recognized that a turbulent flow approximation is sufficient to model particles moving through a system, and that the density of the particulates is the critical parameter for their residence time (Kechichian et al., 2012; Ramaswamy et al., 1995). It is however important to experimentally determine the fastest moving particle which is performed by using simulated particles.

Residence time distribution (RTD) is determined by inserting and passing magnetically tagged particles through a processing system using the representative product environment and underrepresentative flow rate, pressure, and temperature conditions, in which sensors have been installed to observe and record the time periods it takes the particles to travel through different sections of the system. One advantage of this approach is that each particle is individually tagged and weighed, so that the characteristics are known for each. By inserting and following the flow history of each particle, the influence of particle shape, weight, and density on residence time under representative conditions can be determined.

A6.9.2 Materials and Methods

Aseptia's R&D Line was used for these studies, and the schematic depiction of that line is shown in Figure A6.24.

Thermalytics sensor networks (multisensor arrays of giant magnetoresistive detectors and summation amplifiers) were installed in the process line used in the validation of the product. The system was fitted with a particle insertion device for the magnetically tagged particles and a magnetic trap to collect such particles at the end of the line. A total of 32 amplified sensor arrays were located as follows and shown in Figures A6.24 and A6.25:

Eight (A1–A8) at the insertion point/launch, and across the heating stages (yellow)
Sixteen (B1–B16) on the hold tube (green)
Eight (C1–C8) after holding tube (yellow)

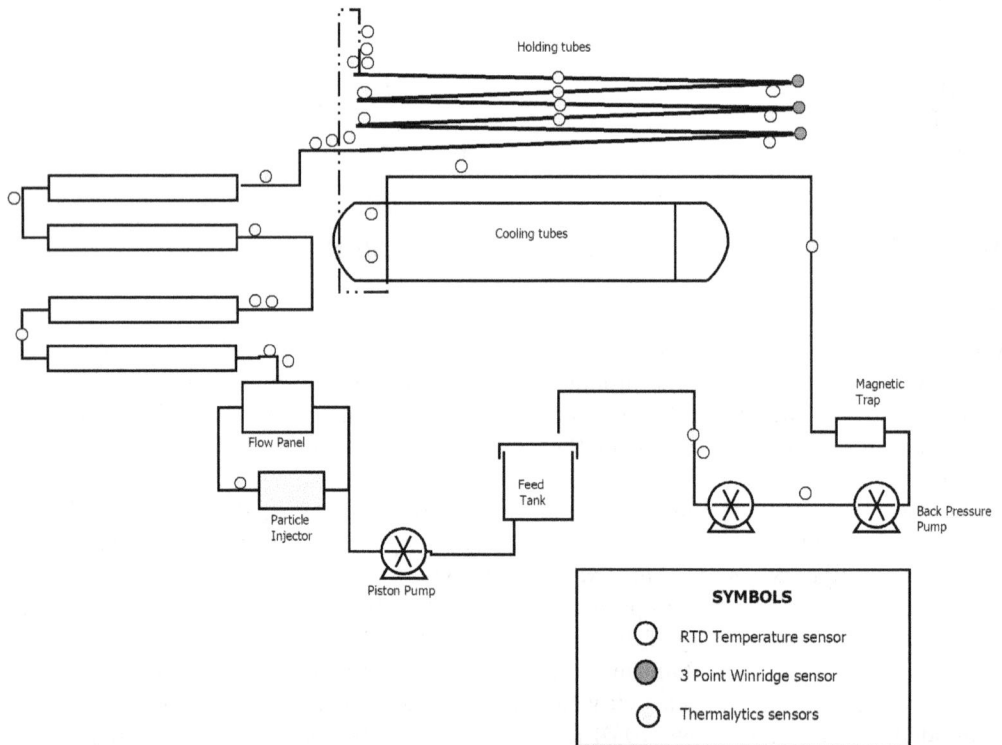

Figure A6.24 Diagram of the process line used for validation studies.

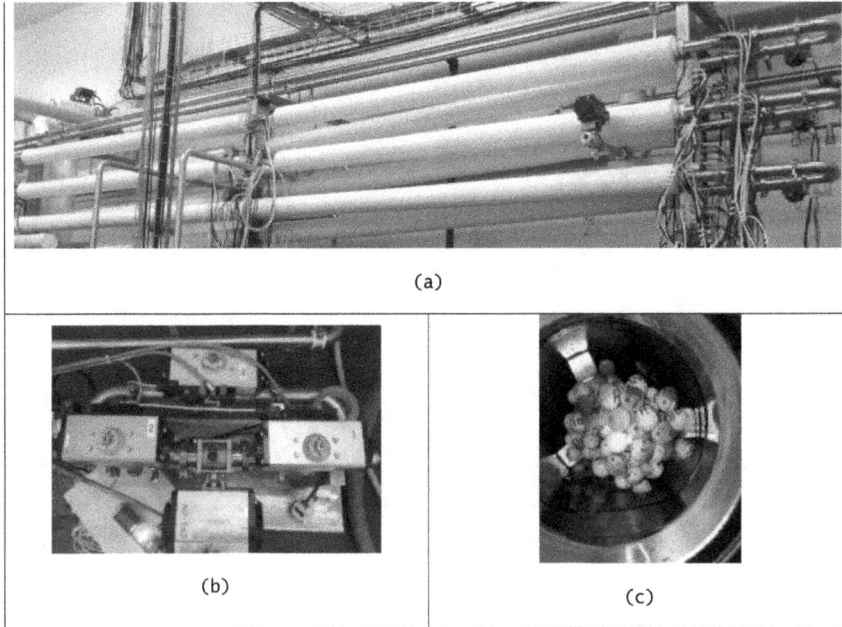

Figure A6.25 System used to trace the magnetic particles. (a) View of hold tubes with sensors. (b) Particle insertion assembly. (c) Magnetic trap.

A6.9.3 Simulated Particles

Simulated particles consist of spheres, cylinders, or cubes assembled of two parts of different polymers (TPX, Ultem, polysulfone, polypropylene), and sized to match the worst-case product particles (1/4", ½", 5/8"). The particles are constructed in such a way that space in which implanted loads can be placed is a volume contained within the center of the simulated particle (Figure A6.26).

The particles can be fabricated by combining different polymers, in order to match desired engineering properties, such as density, thermal conductivity, or electrical properties.

A6.9.4 Critical Density Determination

In order to determine the properties that will result in the most conservative particles for flow (fastest travel), the critical parameter of density was tested. In order to do so, particles of ½" and 5/8" were assembled using 16 combinations of polymers and added weight, in order to cover a wide range of densities. These particles were then inserted in the system at representative flow rates, and tracked in the processing line to observe the individual residence times. Effective densities ranged from 0.75 to 1.10 g/cm³ and the particles were numbered from 1 to 16 (Table A6.17).

Four of each of these particles were inserted into a 12% suspension of corn in CMC, monitored and recorded during the flow through the system using the Thermalytics Particle Flow Monitoring System and real-time recording and analysis software, and recovered to determine and analyze the residence time distributions. These experiments were repeated in triplicate and the results are presented in Figure A6.27.

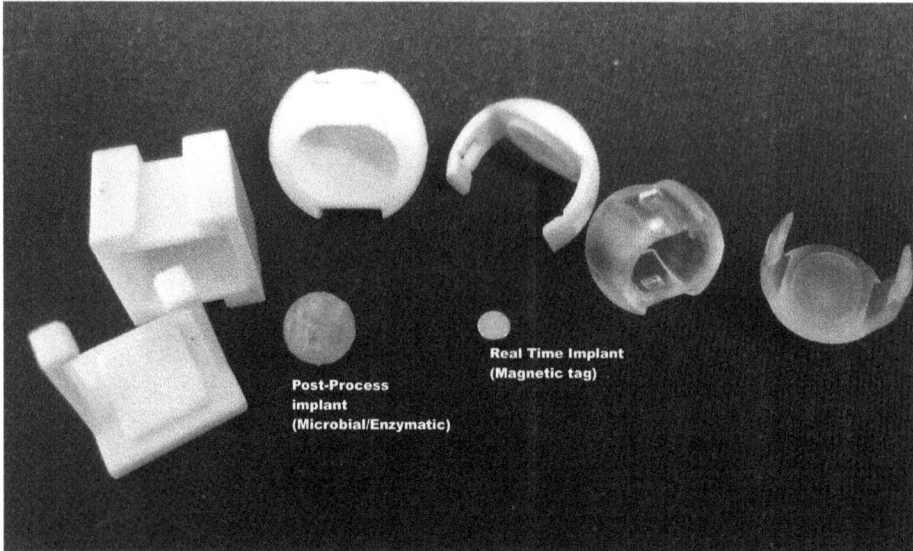

Figure A6.26   Examples of simulated particles used in these studies.

TABLE A6.17   SIMULATED PARTICLES USED IN THE DENSITY SCAN.

Code	Cap	Body	Density (g/cm^3)
1	TPX	TPX	0.75
2	PP	TPX	0.77
3	TPX	PP	0.79
4	PP	PP	0.82
5	PS	TPX	0.84
6	PP	PS	0.86
7	TPX	ULT	0.88
8	PP	ULT	0.90
9	PS	PS	0.93
10	ULT	TPX	0.95
11	PS	PP	0.97
12	ULT	PP	0.99
13	TPX	PS	1.01
14	ULT	PS	1.03
15	PS	ULT	1.06
16	ULT	ULT	1.10

PS: polysulfone; ULT: Ultem 1000/polyetherimide; TPX / PMP: polymethyl-pentene; PP: polypropylene

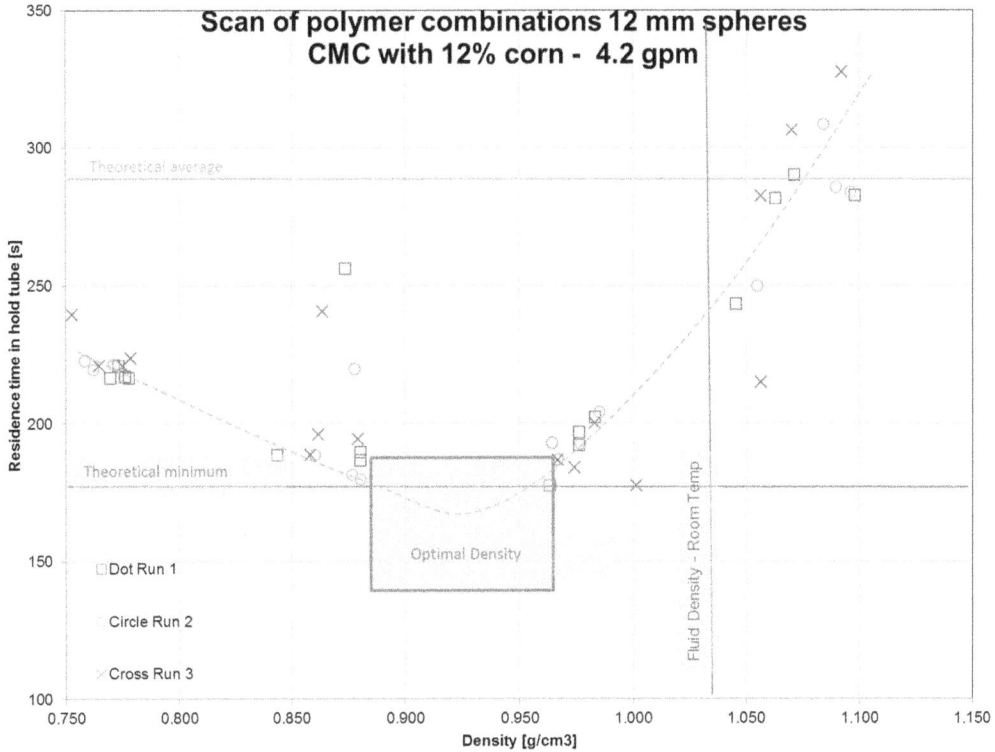

Figure A6.27 Residence time as function of density for determination of critical density for simulated particles.

With the results of this screening, the heavier particles were not used in a scan of residence times for the final product (10% corn in tomato soup). The results of this scan are presented in Figure A6.28 and it can be seen that a density of 0.85 ± 0.05 is the one that results in fastest possible particles (flow factor 0.75). Thus, the simulated particles to be used in the validation were engineered to be within that density range (i.e., the critical density).

Confirmation of the critical density was carried out by an experiment in which 50 particles of the four fastest populations were inserted and tracked at a representative flow rate (3 gpm) with temperature at the sterilization levels (120°C). The residence time of each population was measured and can be seen in Figure A6.29. It can be observed that the population with the shortest residence time has an initial/ambient temperature density of 0.83 ± 0.02 and a flow correction factor of 0.72, thus confirming the previous observation.

The particles with a spore load and a density of 0.83 showed the following residence times:

	Time	Factor
Flow average	424.9	1
Minimum	288.9	0.67
Average	301.2	0.72
Maximum	369.3	0.86

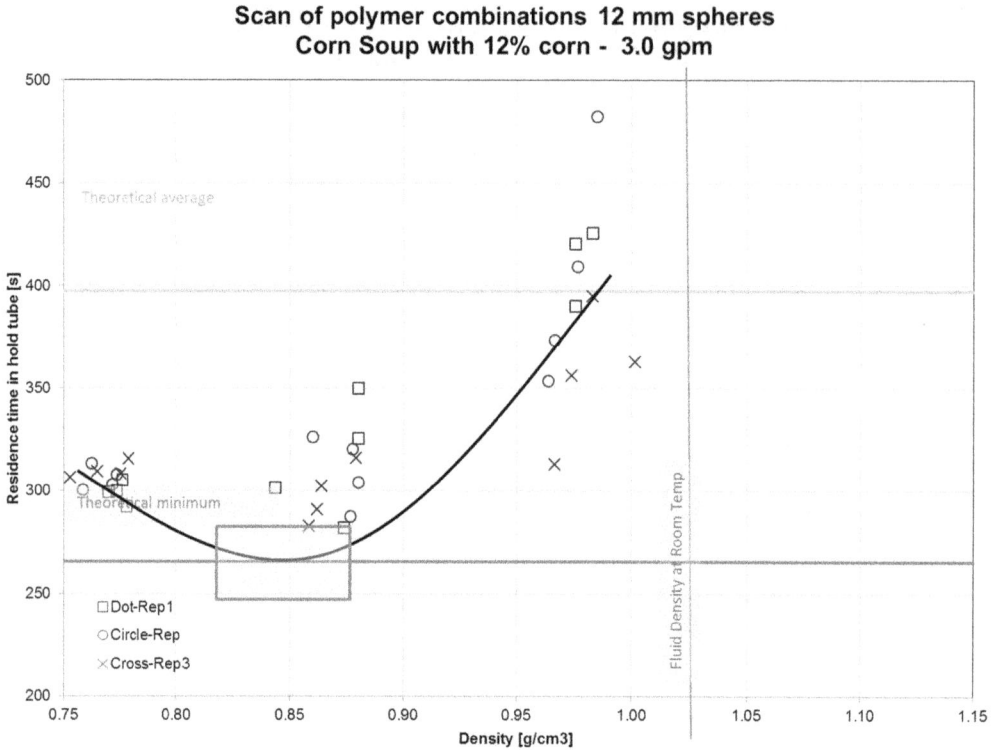

Figure A6.28   Residence time as function of density for the product to be validated.

It was interesting also to observe that in bends, the flow factor for the magnetically tagged particles was very close to 1, which reflects the mixing that bends are known to have due to the Taylor flow effect.

A6.9.5 Conclusion

Based on these observations, a particle with an initial (ambient temperature) density of 0.83 ± 0.03 was recommended for the conservative flow behavior. Flow factor observed was 0.72 with a minimum of 0.67 and this is recommended for the setting of the process (Figure A6.30).

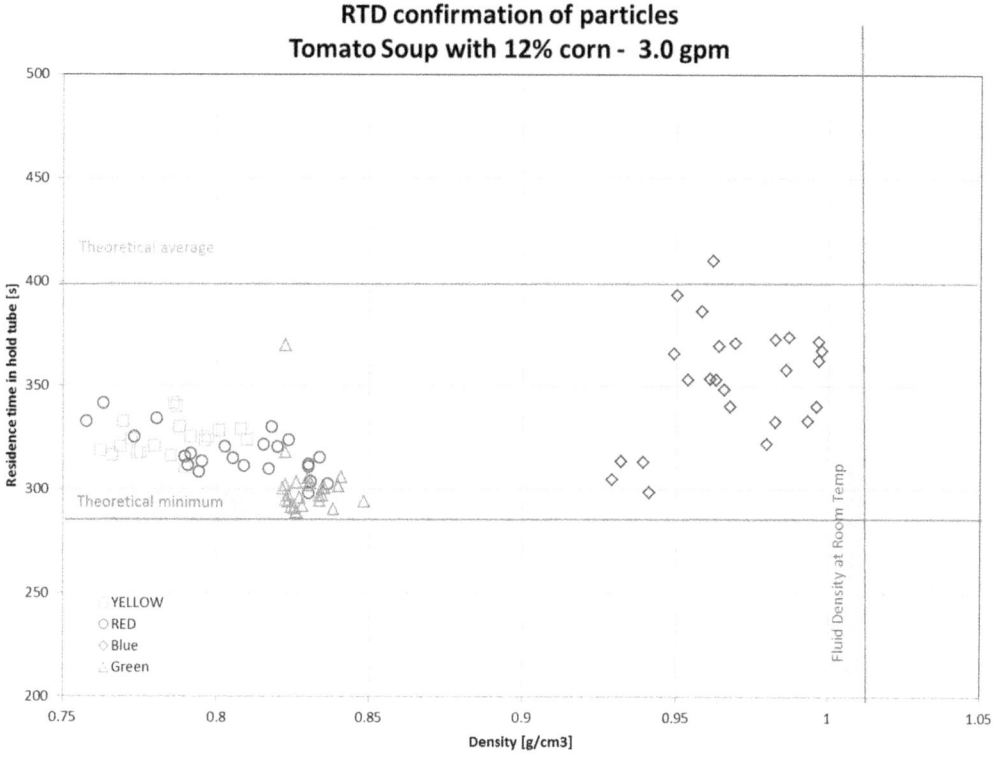

Figure A6.29   RTD confirmation of critical density.

DEPARTMENT OF HEALTH AND HUMAN SERVICES   Public Health Service

Food and Drug Administration
College Park, MD 20740

July 31, 2015

Pablo Coronel
Process Authority
Aseptia Technologies, Inc./Wright Foods

Dear Dr. Coronel:

We are responding to the information that your firm voluntarily submitted on April 7th 2015 regarding the processing and validation of flows with large particulates (up to 5/8") applied to a single-particle product, corn kernels in tomato soup.

We have no additional questions at this time about the information you submitted. The Wright Foods Food Canning Establishment (FCE) may, as appropriate, refer to the information you submitted when filing scheduled processes for a single-particle product, corn kernels in tomato soup. As always, it is the continuing responsibility of the FCE to ensure that their food products are in compliance with all applicable statutory and regulatory requirements.

If you have any additional questions, feel free to contact us.

Sincerely,

Nathan Anderson, PhD
Food and Drug Administration
Center for Food Safety and Applied Nutrition
Process Engineering Branch

Figure A6.30   LONQ received from FDA.

REFERENCES

Coronel, P., Simunovic, J, and Sandeep, K.P. 2003. Temperature profiles within milk after heating in a continuous-flow tubular microwave system operating at 915 MHz. *J. Food Sci.* 68(6):1976–1981.

Inoue, C., Versluis, P., Coronel, P., and Elberse, J.M.M. 2013. Effect of particle phase volume, shape and liquid phase concentrations on rheological properties of large particulate-liquid model food systems by using ball measuring system. *J. Chem.* 7:643–652.

Jennings, B.R., and Parslow, K. 1988. Particle size measurement: The equivalent spherical diameter. *Proc. R. Soc. London. Ser. A*, 419:137–149.

Kechichian, V., Crivellari, G.P., Gut, J.A., and Tadini, C.C. 2012. Modeling of continuous thermal processing of a non-Newtonian liquid food under diffusive laminar flow in a tubular system. *Int. J. Heat Mass Transfer* 55(21):5783–5792.

Kern, D.Q. 1965. *Process Heat Transfer*. New York: McGraw Hill.

Pflug I.J. 1988. *Selected Papers on Microbiology and Engineering of Sterilization Processes*. 5th edition. Fowler IN: Environmental Laboratories Inc.

Ramaswamy, H.S., Abdelrahim, K.A., Simpson, B.K., and Smith, J.P. 1995. Residence time distribution (RTD) in aseptic processing of particulate foods: A review. *Food Res. Int.* 28(3):291–310.

Appendix 7: Process Design and Microbial Validation of a Product with Large Particulates of Multiple Types

Pablo M. Coronel and Josip Simunovic

A7.1 INTRODUCTION

The following is a case study for a tomato soup with carrot cubes, navy beans, black-eyed peas, and diced onion with pieces up to 5/8" (15 mm) for aseptic processing. The method for production and validation was the result of many years of R&D and was built upon the success of previous case studies. While using technology available at the time, it required the development of methods and apparatuses to support the end goal, resulting in several patents that are listed in the references section. The process and product were tested extensively in order to ensure its safety and the ability to be produced at industrial scale. This case study was presented to FDA for the evaluation of the process and validation methodology, and eventually received a LONQ (Letter of No Questions).

This is a follow-up of the case study with a single particulate type shown in Appendix 6 and has been carried out at Aseptia's R&D facility. We ask the reader to refer to Appendix 6 for prerequisites.

Since its inception in 2006, the team of scientists and engineers at Aseptia worked on building a scientific base and an integrated approach for advancing the continuous-flow sterilization and aseptic packaging of shelf-stable complex foods and biomaterials. These efforts were taken in order to enable production, distribution, and commercialization of previously unavailable, convenient, high-quality, and nutritionally superior products to the US and international markets.

Aseptia developed a process for a tomato soup with vegetable pieces, which will be packaged in an aseptic pouch. For the purpose of this report only the process establishment and validation of such process will be discussed. Aseptic filler and container integrity will be discussed in separate filing documents.

The biological challenges and validation were carried out using surrogate-simulated particles. These particles consist of spheres made of a combination polymers which also contain magnetic implants for residence time measurements, as well as biological loads for microbial validation. These simulated particles are built in such a way that they are conservative for both the residence time (flow conservative) and thermal penetration (thermally conservative), thus becoming the worst-case particle in themselves, and allowing for a precise determination of microbial inactivation.

In order to establish the process, the following steps were taken:

- Prerequisites
    1. Develop a model of the process
        - Define a mathematical model of the system

       – Measure thermophysical and dielectric properties of the carrier fluid and vegetable particles

  2. Establish a conservative surrogate-simulated particle
- Determine the thermal penetration properties of vegetable particles and surrogate particles
- Establish conservative flow characteristics (critical density) of surrogate particles
- Establish the surrogate carrier particle construction to be used for biovalidation

  3. Proper design of the process line
- Design of process line based on the model
- Construction of process line based on the established design
- Installation and commissioning

  4. Proper operation of the equipment
- Design of pilot plant based on modeling
- Sterilization capability
- Maintenance of back pressure
- Control of flow rate
- Control of temperature

  5. Proper formulation of the product
- Viscosity control
- Particulate concentration
- Particulate size

- Process establishment
  1. Establish the process based on the model of the system
     - Model system and product
     - Determine the worst-case particle for modeling
     - Verify model with measurements from preliminary test runs
  2. Biological validation of the product using the conservative surrogate
     - Verify model-established process based on the data from previous experiments
     - Challenge the system with surrogate particles loaded with spores and verify the predictions of the model

These steps were followed through to establish and verify that the product will be safe for consumers, and that the established process can be transferred to other production lines in the future.

A7.2 PREREQUISITES

A7.2.1 Model of the System

A mathematical model of thermal processing has been developed and designated as thermal process simulator (TPS). A more complete description can be found in Appendix 6.

The model was validated by comparing the results of biological tests to the predictions of the model. The model could only be valid if it was conservative, i.e., if it predicted lethalities that were equal to or lower than the lethalities observed in testing.

In order to produce a valid process, the model was used to define the steps needed to design and build a process line, and to have a conservative approach to soups with particles. To do so,

TABLE A7.1   THERMOPHYSICAL PROPERTIES OF THE INGREDIENTS

Product/Ingredient	Thermal Conductivity	Specific Heat	Thermal Diffusivity (m²/s)
Units	W/m-K	kJ/kg-K	x 10e-6
Carrot	0.6	3.9	0.16
Navy beans	0.5	3.3	0.15
Black-eyed peas	0.5	3.2	0.15
Onion	0.42	3.9	0.12

the input data for the model was investigated in the form of thermophysical properties of liquid carrier and fluid, and a worst-case (conservative) simulated particle was defined.

A7.2.2 Thermophysical Properties of Particles and Carrier Fluid

Thermophysical properties of real food particles and dielectric properties were analyzed following the same procedures outlined in Appendix 6 and the results are summarized in Table A7.1 and Figure A7.16.

Acquired data was used to create models of a process line, which will be constructed subsequently in order to be able to properly process fluids with particles. In order to ensure that a conservative validation was achieved, a conservative surrogate particle needed to be designed and fabricated,

A7.2.3 Design of a Conservative Surrogate-Simulated Particle

Validation of the thermal process given to the particulates was carried out using simulated particles. Simulated particles consist of spheres, cylinders, or cubes made in two parts of different polymers (TPX, Ultem, polysulfone, polypropylene), and sized to match the worst-case product particle (1/4", 1/2", 5/8"). The particles are constructed in such a way that a hollow cavity volume in which implanted loads can be placed is in the center of the simulated particle (Figure A7.1). The particles can be fabricated combining polymers, in order to match desired engineering properties, such as density, thermal conductivity, or electrical properties.

During the assembly of the surrogate particles, ballast loads made of glass beads or steel bearings can be added to match the required effective density of the particle. The simulated particles have air in the cavity, which makes the effective density of the particles much lower than the density of the polymers. Residence time studies which are in a separate document (RTD-Studies) showed that density of the simulated particles is critical in obtaining flow characteristics of these particles that will result in a flow conservative particle.

Critical density of the particles has to be such that it matches closely the density of the fluid at sterilization temperatures. In that way particles are neutrally buoyant and travel close to the center of the flow, thus being the fastest-moving particles at velocities similar or equal to those of the fluid.

The goal of using simulated surrogate particles is to engineer a particle which is conservative in both flow and thermal properties, such that it will move faster and heat slower than any food particle in the system. To do so, residence time studies and thermal penetration studies needed to be carried out, as detailed in Appendix 6.

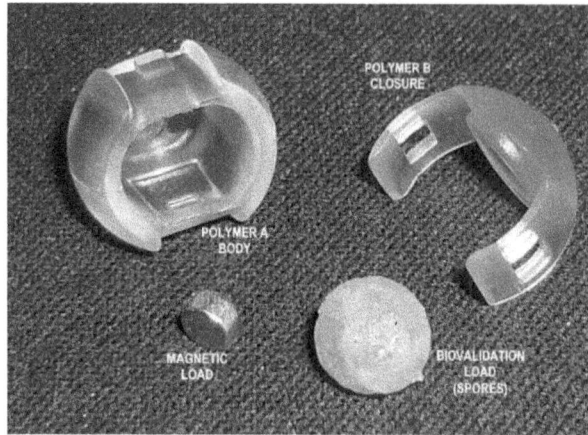

Figure A7.1   Surrogate particle construction.

Figure A7.2   Kinetic parameters of *Geobacillus stearothermophilus* used as surrogate microorganism for these trials.

For these studies, *Geobacillus Stearothermophilus* (ATCC 7953) spores were used. These spores were acquired from Mesa Labs (ProSpore) and are suspended in a color indicator broth that changes color in the presence of survivor growth during incubation at elevated temperatures (55–60°C). Thermal inactivation kinetic values for this lot of spores (619) are shown in Figure A7.2.

After the initial modeling work was performed, a pilot line was designed, using process synthesis methodology. This methodology takes into account the desired outcomes and inputs of the process, and generates a network of possible processes. These processes are then limited by constraints which lead to designs that are both suited for this purpose and efficient. The process line was used for the experiments carried out for single and multiple populations of particles and was validated as shown in Appendix 6.

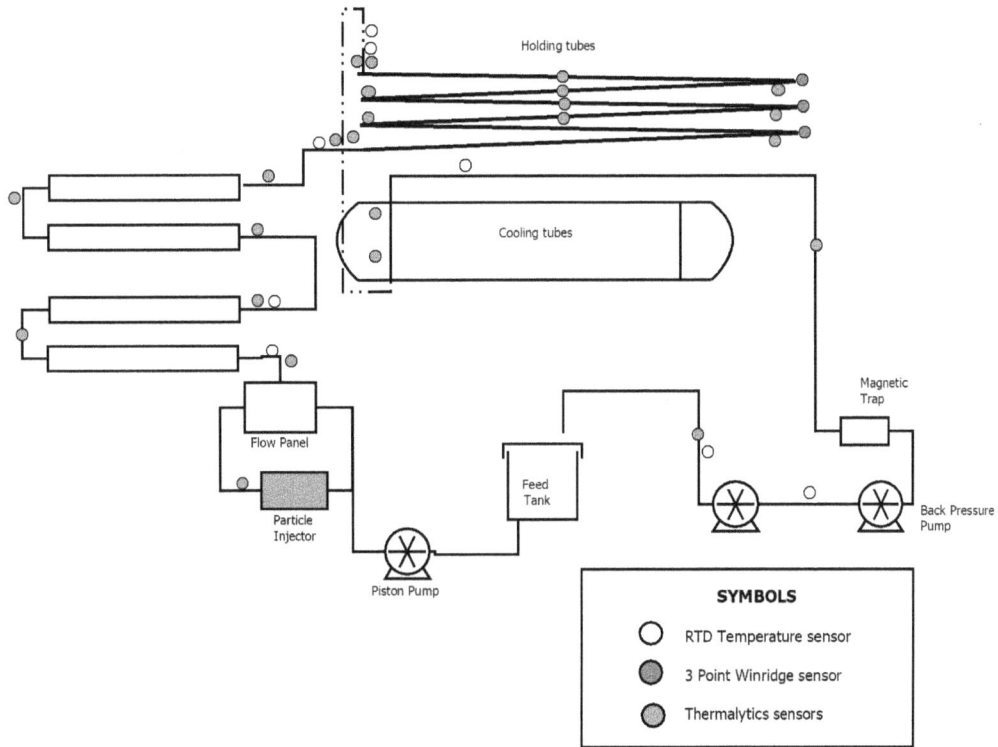

Figure A7.3 Diagram of the processing system.

A7.3 PROCESSING LINE

Based on the results of modeling, and taking into account some flexibility, a processing line suitable to process viscous fluids with large particulates was designed and built—a schematic can be seen in Figure A7.3 and consists of:

Blend tank
Piston Pump (Emmepiemme SPM80B)
Four microwave applicators and connecting tubes
Hold tube divided into eight 20¢ segments
Cooler, tubular
Back pressure system consisting of positive displacement pumpsControl system

Hold tube length is a critical factor in the thermal process, and it can be broken into segments, which yield the following total length:

Quantity	Description	Length (m)	Total (m)	Total (in)
8	Straight sections	6.10	48.8	1,920
7	180° turns	0.31	2.2	85

6	Extensions for temperature sensors (exit of 1-3-5)	0.15	0.9	35
	Inlet connection		1.5	59
	Outlet connection		1	39
	Total hold tube		**54.4**	**2140**

Instrumentation was added to the system in order to have a full description of the process, as seen in the diagram. It includes:

- RTD temperature sensors used for control and data logging at the inlet and exit of each microwave unit, inlet of hold tube, exit of hold tube (double), and exit of cooling section.
- Three-point temperature sensors (WindRidge sensors) at the exit of sections 1,3,5 of the hold tube.
- Flowmeter (magnetic type).
- Pressure sensors after the pump, before the hold tube and after the cooling tubes.
- Thermalytics sensor summation amplifiers for detection of magnetic implants. A total of 32 total sensor amplifiers were located in the system, with 16 amplifiers covering the hold tube. Each amplifier processed the detection signals from 4 to 8 GMR (giant magneto-resistive sensors) in real time, and a Windows computer was used for the real-time signal displays, recording and postprocess playback of recorded detection events for each recorded channel concurrently.
- Particle launching and recovery system were also developed in order to insert particles individually and enable the tracing of the passage of each at individual detection locations, thus enabling the measurement of residence times of each tagged particle throughout the processing system. Recovery of the particles was carried out in a magnetic trap, since all the surrogate particles contained magnetic implants.

Recovery of the particles was carried out using a custom-designed multiregion magnetic trap, since all the surrogate particles contained magnetic implants. Recovery system was located before the back pressure pumps, to prevent damage to particles or pump if one or more happen to get trapped in the lobes of the pump.

This plant was also used for the particulate product case for single particle, described in Appendix 6, and extensive testing was carried out to extend the validation to real food products with a varied population of particulates up to 15 mm (5/8") in size.

A7.4 PRODUCT FORMULATION

The product used in these experiments was a tomato vegetable soup as carrier fluid with ~15% vegetable pieces of four different varieties and the composition is shown later in Figure A7.16.

Proper formulation was critical for the proper suspension and flow of the product. Carrier fluid properties:

- Viscosity of the carrier fluid was measured using a Brookfield DV-II at 77°F, 60 RPM, and it was 1,500-2,000 cps
- pH was designed at 5.2, with a maximum of 6.0
- Density 1.025–1.030 at room temperature

TABLE A7.2   FORMULATION OF PRODUCT

	Quantity	Max Dimensions (L ´ W ´ H)
Tomato vegetable carrier	85%	
IQF carrot dices (3/8" max)	3%	0.45 ´ 0.38 ´ 0.38 in (11.4 ´ 9.65 ´ 9.65 mm)
IQF navy beans	5%	0.83 ´ 0.35 ´ 0.35 in (21.08 ´ 8.89 ´ 8.89) mm
IQF black-eyed peas	5%	0.51 ´ 0.26 ´ 0.35 in (12.95 ´ 6 ´ 6 ´ 8.89 mm)
IQF onion pieces (3/8" max)	2%	0.3 ´ 0.38 ´ 0.38 in (7.62 ´ 9.65 ´ 9.65 mm)

Vegetable pieces used where IQF carrot dices, IQF navy beans, IQF black-eyed peas, and IQF onion pieces. Hundred pieces of each were measured, in order to determine the largest ones visually and the maximum dimensions are listed in the table of ingredients.

It is noticeable that the beans were large in one dimension but their diameter was below 3/8" (15 mm). The onion pieces were very fragile, as they tended to "flake" into the layers of the onion and make very thin and flat pieces.

Therefore, using the equivalent diameter concept introduced by Jennings and Parslow (1988), a sphere of 10.97 mm (0.432") could be used as conservative model for the kidney beans and black-eyed peas, while 9.87 mm (0.389") spheres could be used for the carrot pieces and onion pieces.

Based on the dimensions, and on our previous thermal penetration studies, two simulated particle sizes were chosen: ½" and 5/8" as both of these will be more conservative thermally than any of the chosen ingredients.

Thermal properties of the carrier fluid and different pieces were measured and the most conservative values selected for input to the model, as shown in Table A7.2

All the above data was used as input into the Model of the System for verification. This model was verified during the preliminary runs by comparing the simulated temperature profiles with the temperatures observed in these runs. Adjustments to the model parameters were performed in order to match both prediction and real data. Once the model predictions matched the real temperatures observed, it was used to calculate the lethality in the hold tube only.

In order to validate the process, the model can simulate a worst-case particle, which in our case is initially a simulated particle. The worst-case particle was to be a surrogate-simulated particle, and that one needed to be engineered based on the definition of conservative particle above.

A7.4.1 Construction of the Conservative Surrogate Particles for Microbial Challenge/Biovalidation

Based on the previous observations for microbial challenges, conservative simulated particles were assembled, with characteristics shown in Table A7.3. Density was a critical parameter for RTD and each particle was individually tagged, and its density was tabulated (Figure A7.4).

TABLE A7.3   THERMOPHYSICAL PROPERTIES USED IN MODELING

	Units	Kidney Bean	Black-Eyed Pea	Carrot	Carrier fluid
Thermal conductivity	W/m-K	0.5	0.5	0.6	0.6
Density	kg/m³	990	1010	995	1025
Heat capacity	J/kg-K	3300	3200	3800	3800
Thermal diffusivity	m²/s	1.48E-07	1.48E-07	1.49E-07	1.55E-07
Dielectric properties T_{ref}	°C	70	70	70	70
Loss tangent T_{ref}	[]	0.880	0.883	0.987	1.3
Temperature factor	1/°C	0.011	0.011	0.011	0.011

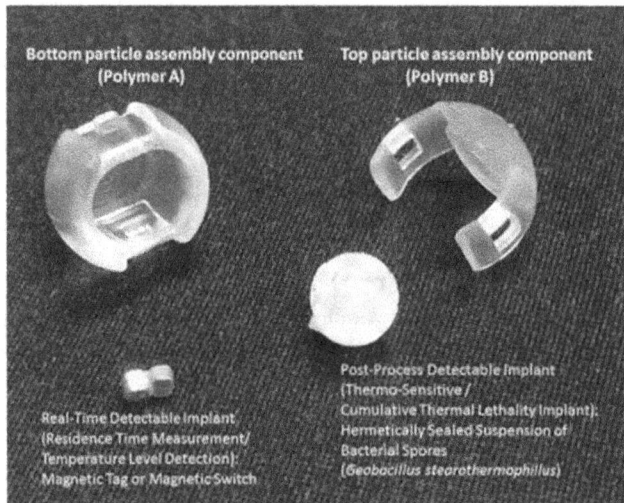

Figure A7.4   Final design of surrogate-simulated particle.

A7.5 THERMAL PROCESS DESIGN

Thermal process was designed to achieve an F_0 of 6 minutes in the center of the simulated 5/8" particles, as this would match the value recommended by Pflug (1988) for navy beans and black-eyed peas. The process design was performed by reconstructing the initial testing using the modeling software and the maximum flow possible in the process line (3 gpm). The modeled temperature profiles observed during the validation runs can be seen in Figure A7.5. Each unit mentioned before is separated by a gray line, and for purposes of the process setting, and because there was a large temperature difference at the end of the heaters, the first section of the hold tube was not considered for accumulation of lethality and it is called "equilibration" tube. It can be observed in the figure that the center of the simulated particle should have received a thermal

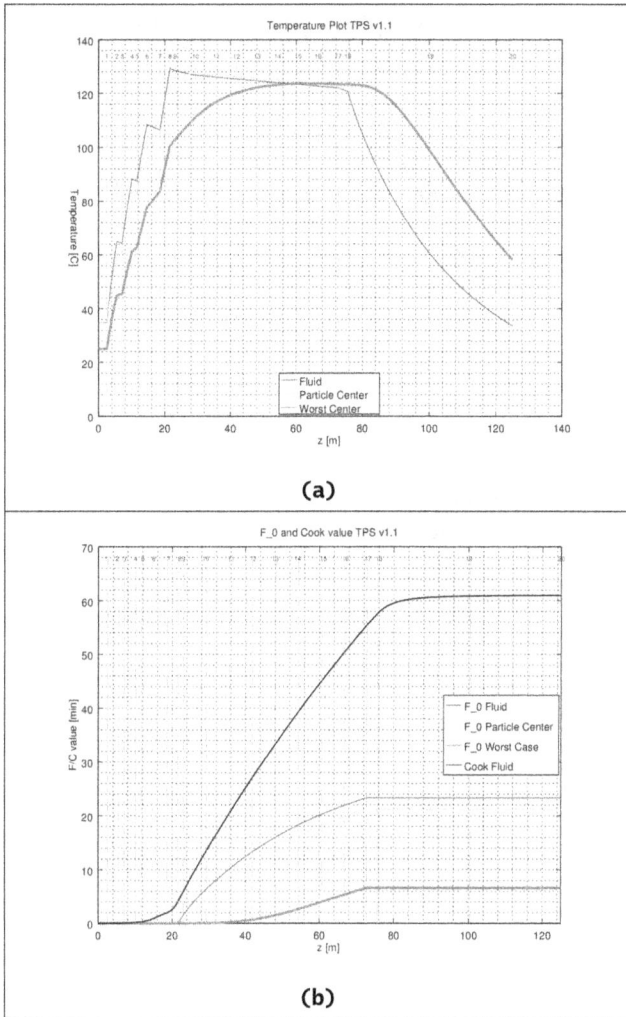

Figure A7.5   Temperature profiles (a) and lethality accumulation (b) for process setting.

process equivalent to F_0 of 6.7 minutes in the hold tube only, while the real food particle would have received at least twice that level of cumulative lethality.

The process setting parameters were set, based on the model, and the parameters are listed in Table A7.4. These parameters were used for subsequent microbiological challenges in which simulated surrogate particles loaded with the above-mentioned *G. Stearothermophilus* spores were used for testing.

A7.5.1 Design of Thermal Process for Underprocessing

In order to verify that the surrogate particles were being processed correctly, a process in which the thermal reduction would be in the order of the inoculum (5 log) was designed. This treatment

TABLE A7.4   DENSITY OF PARTICLES (G/CM³) USED IN MICROBIAL CHALLENGE.

Diameter	½" (12.77 mm)	5/8" (15.88 mm)
Materials of construction	TPX (polymethylpentene) body—polysulfone cap	
Biological implant	¼" TPX sphere filled with 40 sL of spore suspension (2.3 ´ 10⁴ spores)	
Magnetic implant (rare earth magnets)	1	2
Non-magnetic implant	Steel balls to have the same weight as the magnet	
Density at room temperature (g/cm³)		
Average	0.92 ± 0.023	0.90 ± 0.033
Maximum	0.98	1.00
Minimum	0.85	0.83

would render surviving spores which would later grow and thus produce positive test implants. The design of this process was produced in the model so that F_0 should be below 4 minutes.

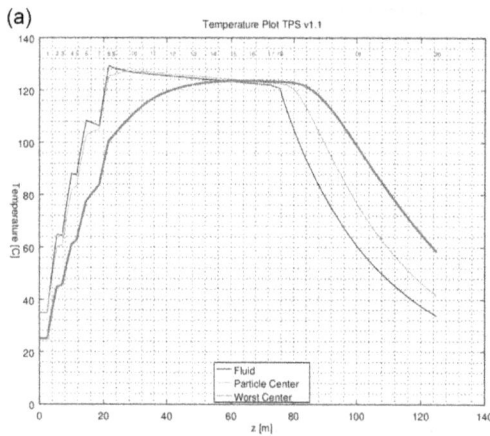

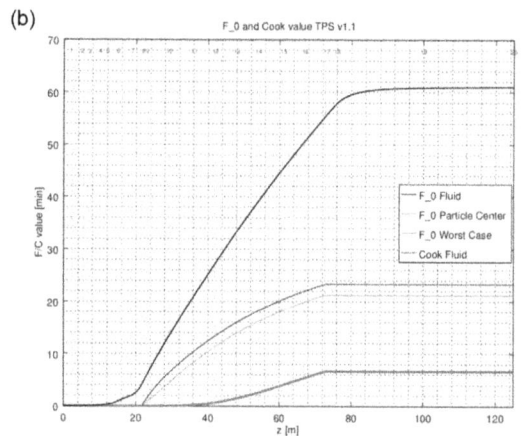

The main change was the temperature at the exit of the hold tube, which for this experiment was to be kept at 115°C or below (see Table A7.8). The expected F_0 was 3.89 for the hold tube, and 5.5 total. With these parameters and the initial concentration of spores in the particles, all of the particles should have survivors, and thus present growth.

A7.6 MICROBIOLOGICAL CHALLENGE STUDIES

Microbial challenge studies were made to verify the process as described in Table A7.5. The operating conditions were:

Flow rate:3.0 gpm
Temperature at the exit of heating section:132°C
Temperature at the exit of hold tube:>120°C

TABLE A7.5   MODEL PARAMETERS FOR PROCESS ESTABLISHMENT

Unit No.	Unit Name	Temperature (°C)	Temperature (°F)
0	Inlet	35	95.0
1	Connection	34.9	94.8
2	Heater 1	65	148.8
3	Connection	64.4	147.9
4	Heater 2	88.3	190.9
5	Connection	87.5	189.5
6	Heater 3	108.4	227.1
7	Connection	107.5	225.5
8	Heater 4	131.3	268.3
9	Connection	129.4	264.9
10	Stabilizer	126.0	258.8
11	Hold tube 2	125.0	257.0
12	Hold tube 3	124.3	255.7
13	Hold tube 4	123.6	254.5
14	Hold tube 5	123.0	253.4
15	Hold tube 6	122.3	252.1
16	Hold tube 7	121.6	250.9
17	Hold tube 8	**120.0**	248.0
18	Connection	121.6	249.3
19	Cooler	60.3	140.5
20	Cooler	33.6	92.5

Product was prepared according to the formulation, and magnetically loaded simulated particles were inserted at 90-second intervals while recording residence time and temperature. Non-magnetically loaded particles were inserted at 60-second intervals.

Statistically, F_0 at the center of the surrogate particles can be estimated using the endpoint method calculation (Equation 1):

$$\text{Average log reduction} = \log(\text{innoculum}) - \log\left[\ln\left(\frac{\text{Total No. particles}}{\text{No. sterile particles}}\right)\right]$$

Equation 1 End point log reduction calculation

With the existing innoculum if there were no survivors, the result is that the average log reduction was equivalent to 5.15 log which is equivalent to F_0 7.725.

A7.6.1 Particle Challenge

In order to verify the process setting, a test with more than 300 surrogate particles was performed. Since two populations of particles were used concurrently, the target was to insert

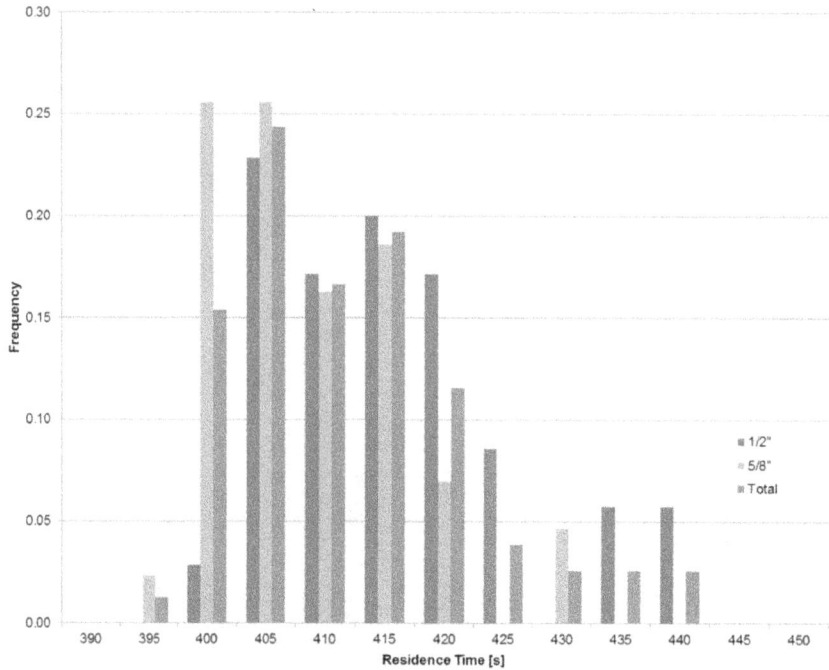

Figure A7.6   Residence time distribution during microbial challenge for 1/2" and 5/8" surrogate particles with microbial implants.

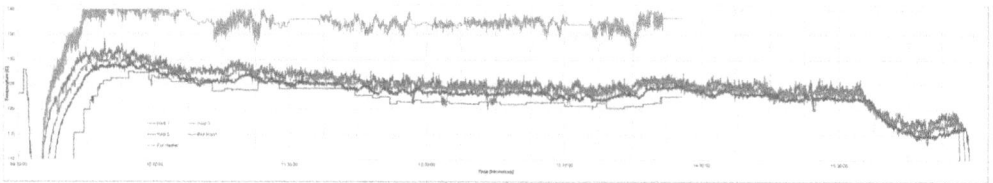

Figure A7.7   Temperature trace during particle microbial challenge.

440 particles. However, this count also included the particles used for the last part of the experiment in which the operating temperatures of the system were lowered.

Product was prepared according to the formulation shown in Table A7.2 in batches of 1,500 lb. System was brought up to temperature and the insertion of simulated particles began. A total of 390 particles were inserted at 30-90-second intervals. Particles had both magnetic tags and microbial loads to confirm both their residence times and thermal processing. During this experiment, temperatures were recorded at the inlet of the system, at the exit of each heating unit, at the beginning and exit of hold tube, and at the exit of cooling. The liquid temperature serves as an indication of the temperature of the whole product if it is uniform, and thus the additional temperature sensors at the exit of holding section 1 (stabilization section), 3, and 5 which were used to observe uniformity of temperatures. The results are shown in Figures A7.6 and A7.7 and it can be observed that the traces marked Hold 1, Hold 3, and Hold 5 are within a narrow

TABLE A7.6   PARAMETERS FOR PROCESS SETTING DESIGNED TO UNDERPROCESS THE SURROGATE PARTICLES

Hold tube length	1822" (46.1 m)	7 straights ´ 6 m + 6 turns ´ 0.21 m + 5 straight ´ 0.15 for sensors + connections
Hold tube ID	1.91" (0.0485 m)	
Flow rate	3.0 gpm [680 l/h]	
Residence time		
Fluid average	440 s	
Fluid correction factor	0.5	
Particle fastest observed	0.72	Based on RTD studies
Max temperature at exit of hold tube	2°F (116°C)	
Max temperature inlet of hold tube	20°F (122.0°C)	

temperature range, with differences that can be attributed to environmental losses. This confirms the process setting in which the first section of the hold tube was considered for stabilization only.

Residence time of the magnetically tagged particles was recorded and the results are listed in Table A7.6. Please note that the RTD was measured along the stabilization and Hold section of the hold tube, thus adding time to the proposed hold time. This, however, can be compared to the theoretical residence time in the full eight sections.

The residence time per population correlated with the expected for the RTD studies, with a minimum observed flow correction of 0.74.

Temperatures during this experiment were recorded and can be observed in Figure A7.7. Temperatures at the exit of heater and the inlet of hold tube showed a 13°C difference, which coincided with the model prediction, as the first hold tube segment was used for stabilization of temperatures. At the exit of the first hold tube, the temperature was 10°C below the inlet of the hold tube, and from then on most of the losses were environmental. At the exit of the hold tube, the temperature was maintained above 120°C, with the minimum of 120.3°C at 13:50:00

At 15:35, the control-input exit temperature of the heater was lowered in order to produce a non-sterile product to prove the validity of the implemented sterilization and particle construction, tagging, and biovalidation implants. The control set point was lowered by 5°C, which resulted in a temperature at the exit of hold tube to 114.5°C.

To verify the validity of the process modeling and the final sterility, 394 surrogate particles were inserted into the product flow stream, and 394 were recovered. The results of the microbial incubation are tabulated in Table A7.7. The first inspection was carried out immediately after recovery of the particles, and the particles were marked as shown in Figure A7.8. The particles were then incubated at 55°C for 48 hours and reinspected.

During the initial visual inspection, 284 were easily recognizable as sterile as can be seen in Figure A7.8 marked as GOOD. The ones in which the color was not easy to observe were marked unsure. If carrier fluid had leaked into the particle, they were marked as leaks, and if the spore

TABLE A7.7   RESULTS OF PARTICLES BEFORE AND AFTER INCUBATION

	After Processing		48-Hour Incubation			Notes
	½"	5/8"	½"	5/8"	Total	
Sterile	172	112	176	111	289	
Non-sterile	0	0	0	0	0	
Unsure	5	3	0	0	0	pH tested
Leak	10	49	10	49	59	pH tested
Empty	16	27	17	29	46	
Lost	0	0	0	0	0	
Total	**203**	**191**	**203,203**	**191**	**394**	

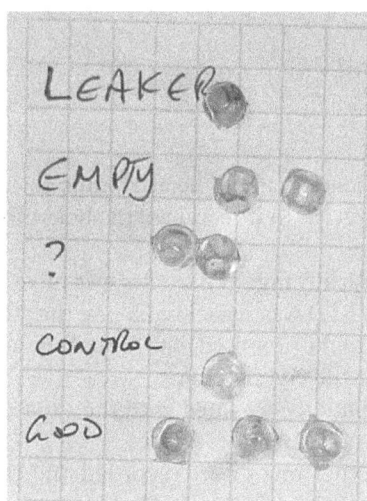

Figure A7.8   Examples of each type of implant result.

suspension had leaked out of the implant, they were marked as empty. The ones marked as unsure or leak were tested by reading the pH of the spore suspension using Pehanon pH paper, as seen in Figure A7.9.

It can be observed that the main issue was the leaking of the carrier fluid into the larger particles (5/8"). The results of the ½" particles match the previously presented data.

With these results, the average log reduction was of 5.15 in the surrogate-simulated particles, which is an equivalent to F_0 of 7.2 minutes.

A7.6.2 Underprocessed Particulates

The set of underprocessed particles, with process conditions as shown in Table A7.8, was also analyzed following the same principles as before, and the results are shown in Table A7.2. All of the recovered particles from this set showed growth, marked by a change of color of suspension from purple to yellow after incubation and are shown in Table A7.9.

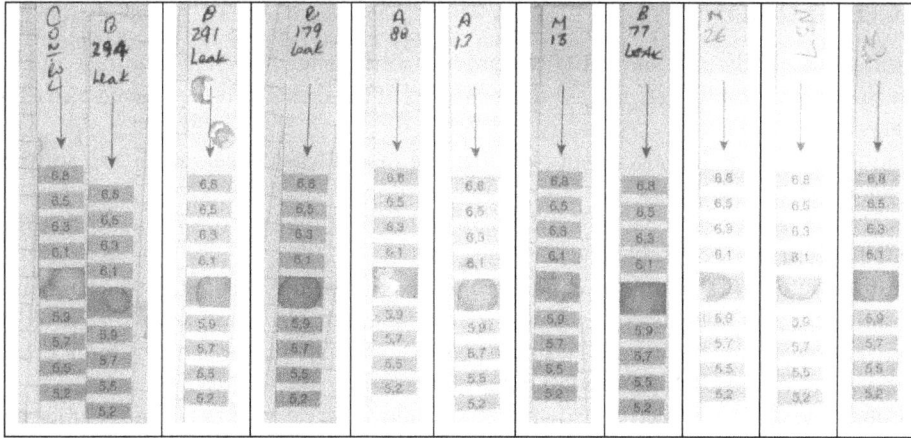

Figure A7.9   Record of pH testing of questionable particles.

TABLE A7.8   PARAMETERS FOR PROCESS SETTING FOR TOMATO SOUP WITH CORN

Hold tube length	1822" (46.1 m)	7 straights ´ 6 m + 6 turns ´ 0.21 m + 5 straight ´ 0.15 for sensors + connections
Hold tube ID	1.91" (0.0485 m)	
Flow rate	3.0 gpm [680 l/h]	
Residence time		
Fluid average	440 s	
Fluid correction factor	0.5	
Particle fastest observed	0.72	Based on RTD studies
Min temperature at exit of hold tube	251.5°F (120.0°C)	
Min temperature inlet of hold tube	256.0°F (124.5°C)	

TABLE A7.9   RESULTS OF BIOVALIDATION FOR UNDERPROCESSED PARTICLES BEFORE AND AFTER INCUBATION.

	After Processing		48-Hour Incubation			Notes
	½"	5/8"	½"	5/8"	Total	
Sterile	21	18	0	0	0	
Non-sterile	0	0	20	18	38	
Unsure	0	0	0	0	0	pH tested
Leak	4	6	4	6	10	
Empty	0	1	1	1	2	
Lost	0	0	0	0	0	
Total	**25**	**25**	**25**	**25**	**50**	

A7.7 CONCLUSION

The results from all the test show that the process settings have been determined and implemented correctly by applying a process-driven model of the line. All the processed particles and surrogates were sterile.

Validation of the model settings proved that the model was conservative, and so were the surrogate-simulated particles.

Thus, these particles can be used for further validation of food particles that are smaller and more conductive than the established surrogate particles.

For future validation studies, since the surrogate particles have been established as conservative, both in flow and in heat penetration characteristics (i.e., each properly constructed simulated particle represents the worst case or the process cold spot carrier), and the reduction of the number of particles required for validation may be justified.

In order to do this, residence time distribution and conservative flow characteristics need to be further analyzed statistically in order to determine the required number of particles that will guarantee that a significant population of surrogate particles will be representative of the fastest possible moving fraction of the real multiphase product for safety validation.

Sterility of the carrier fluid and the food particles was confirmed by the experiments carried out, and thus it has been proven that production of this product under the specified conditions would yield a safe shelf-stable product.

This case study was awarded a LONQ (Letter of No Questions) by the FDA as a viable methodology for validation of safety for heterogenous foods with a population of several particulates as shown later in Figure A7.15.

REFERENCES

Coronel, P., Simunovic, J., and Sandeep, K.P. 2003. Temperature profiles within milk after heating in a continuous-flow tubular microwave system operating at 915 MHz. *J. Food Sci.* 68(6):1976–1981.

Jennings, B.R., and Parslow, K. 1988. Particle size measurement: The equivalent spherical diameter. *Proc. R. Soc. London. Ser. A.* 419:137–149.

Pflug, I.J. 1988. *Selected Papers on Microbiology and Engineering of Sterilization Processes.* 5th edition. Fowler IN: Environmental Sterilization Laboratories, LLC.

A7.8 RESIDENCE TIME DISTRIBUTION STUDIES: CRITICAL PARTICLE DETERMINATION AND VERIFICATION

A7.8.1 Introduction

Surrogate-simulated particles fabricated from microwave-transparent polymers have been used in the validation of heterogeneous particles containing products by Aseptia Inc. These particles can be engineered to be conservative in both flow and thermal characteristics when compared to real food particles.

In this study, the effect of density on the residence time is examined, and a method to design the most flow conservative particle is presented.

Residence time measurements are important to determine the fastest-moving particle in a heterogeneous product thermally processed under continuous-flow conditions. It is well-known that the presence of particles modifies the flow behavior of fluids and homogeneous carrier media, to the point where an approximation to non-Newtonian flow can be used (Inoue et al., 2013). It has been recognized that turbulent flow approximation is sufficient to model particles moving through a system, and that the density of the particulates is the critical parameter for their residence time (Kechichian et al., 2012; Ramaswamy et al., 1995, Simunovic, 1998). Density is the most critical factor when it comes to residence time, particles with density slightly lower than the liquid at processing temperatures will travel closer to the center of the flow where the velocity is higher (Simunovic, 1998).

Depending on the degree of inclination of the hold tube segment of the processing system, the critical density (density range with the highest likelihood of containing the fastest-particle population) will be slightly lower than the carrier fluid density at the prevailing hold tube temperature level. It was observed that particles that were outside the critical range of densities would have big dispersion of RT, while particles in the critical range would travel close to the center of the flow.

Due to the complexity of multiphase product environments and the dynamic changes transpiring at sterilization temperature levels (denaturation of proteins, interaction of macromolecules with water, lipid melting, escape and/or uptake of fluids from the particulate product phase etc.), it is impossible to predict the critical density of the product at hold tube temperatures under representative production conditions, so the fastest-particle density range needs to be determined experimentally under those conditions. In order to experimentally determine the fastest-moving particle population, simulated particles with a preadjusted and incrementally distributed range of densities are used.

Residence time distribution (RTD) is measured by inserting and passing these magnetically tagged particles through a processing system using a representative product environment and underrepresentative flow rate, pressure, and temperature conditions.

Non-contact magnetic sensors installed to monitor the magnetically tagged particle progress through the system are used to observe and record the time the simulated and tagged particles take to travel through different sections of the system.

One advantage of this approach is that each particle is individually tagged and weighed, so the effective density at room temperature for each simulated particle is known. By inserting and following the flow history of each particle, the influence of particle shape, weight, and density on residence time under representative conditions can be determined.

A7.8.2 Materials and Methods

Aseptia's R&D Line was used for these studies, with a schematic depiction of the line that can be seen on Figure A7.10. Heating section consists of four microwave applicator sections, which are interconnected with sections of tube in which mixing is promoted by changes in direction (elbows). Hold tube consists of eight straight sections, joined by 180 degree elbows and also having some extension sections to locate the additional temperature sensors at the end of straight sections 1, 3, and 5. Total length and the lengths of each individual segment can be found in Table A7.10.

Aseptia/Thermalytics sensor networks (multisensor arrays of giant magnetoresistive detectors and summation amplifiers) were installed in the process line used for the validation of the

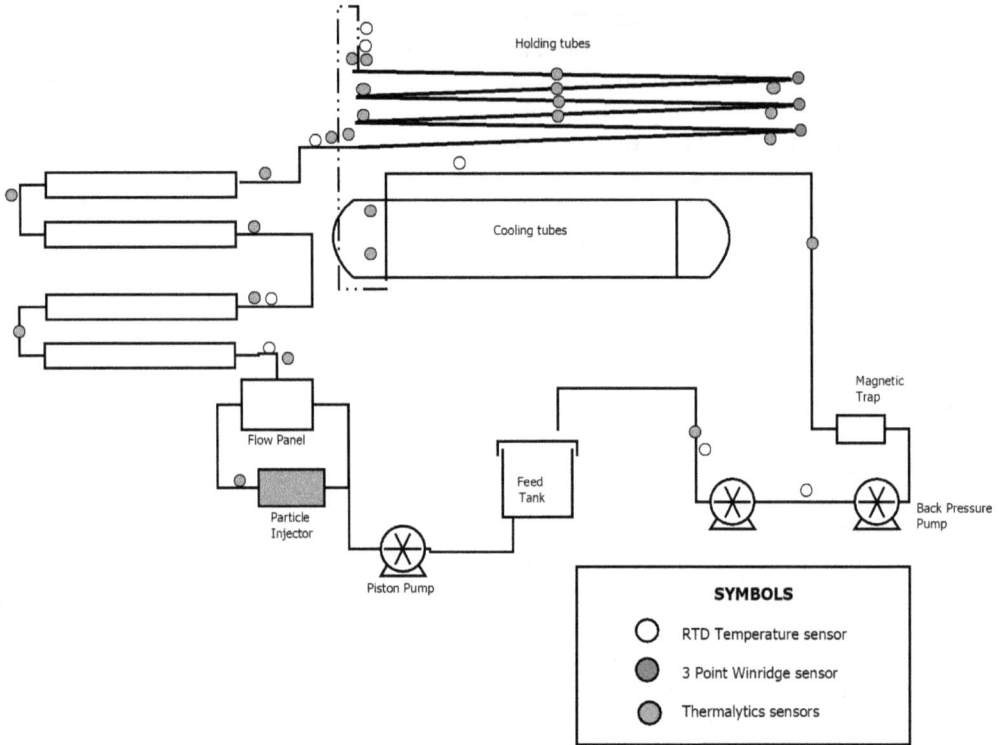

Figure A7.10 Diagram of the process line used for validation studies.

TABLE A7.10   DIMENSIONS OF HOLD TUBE

Quantity	Description	Length (m)	Total (m)	Total (in)
8	Straight sections	6.10	48.8	1920
7	180° turns	0.31	2.2	85
6	Extensions for temperature sensors (exit of 1-3-5)	0.15	0.9	35
	Inlet connection		1.5	59
	Outlet connection		1	39
	Total hold tube		**54.4**	**2,140**

product. The system was fitted with a particle insertion device for the magnetically tagged particles and a magnetic trap to collect such particles at the end of the line. A total of 32 amplified sensor arrays were located as follows and as shown in Figures A7.10 and A7.11:

8 (A1–A8) at the insertion point/launch, and across the heating stages (yellow)
16 (B1–B16) on the hold tube (green)
8 (C1–C8) after holding tube (yellow)

Figure A7.11   System used to trace the magnetic particles. (a) View of hold tubes with sensors. (b) Particle insertion system. (c) Magnetic trap for recovery.

Soup was formulated as follows:

	Quantity	Max dimensions (L x W x H)
Tomato vegetable carrier	80%	
IQF Carrot dices (3/8" max)	5%	0.45 ´ 0.38 ´ 0.38
IQF navy beans	5%	0.83 ´ 0.35 ´ 0.35
IQF black-eyed peas	5%	0.51 ´ 0.26 ´ 0.35
IQF onion pieces (3/8" max)	5%	0.3 ´ 0.38 ´ 0.38

Vegetable pieces used were IQF carrot dices, IQF navy beans, IQF black-eyed peas, and IQF onion pieces. Hundred pieces of each were measured, trying to determine the largest ones visually and the maximum dimensions are listed in the table of ingredients.

It is noticeable that the beans were large in one dimension but their diameter was below 3/8".

The onion pieces were very fragile, as they tended to "flake" into the layers of the onion and make very thin and flat pieces.

Based on the dimensions, and our previous thermal penetration studies, two test particle sizes were chosen: ½" and 5/8" as those have been experimentally shown to be more conservative thermally (slower heating) than any of the chosen particulate food ingredients.

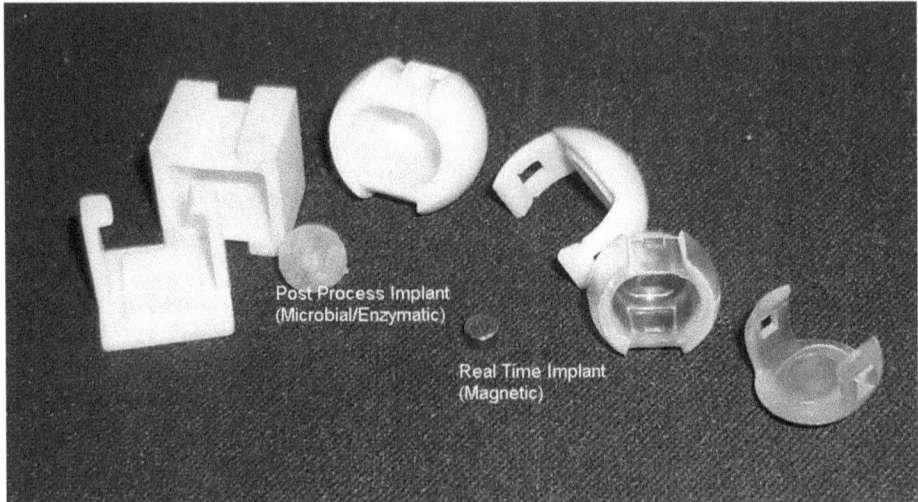

Figure A7.12   Example of simulated particles used in these studies.

A7.8.3 Simulated Particles

Simulated particles consist of spheres assembled of two parts of different polymers (TPX, Ultem, polysulfone, polypropylene), and sized to match the worst-case product particle (½", and 5/8"). The particles are built in such a way that space in which implanted loads can be located is a cavity contained in the center of the simulated particle (Figure A7.12).

The particles can be fabricated by combining different polymers, in order to match desired engineering properties, such as density, thermal conductivity, dielectric, and/or electrical properties. Ballast can be added to the particles in the form of microwave transparent glass beads to modify density to the desired level or to match the density of the food particles. Density is the most critical factor when it comes to residence time, particles with density similar to that of the liquid at processing temperatures will travel closer to the center of the flow where the velocity is higher (Simunovic, 1998).

A7.8.4 Critical Density Determinant

In order to determine the particle density that will result in the most conservative particles for flow (fastest travel), a set of density-adjusted particles was built. Particles of ½" and 5/8" were assembled using magnets for detection, and combinations of polymers and added weight (ballast), in order to cover a wide range of densities. Effective densities determined at room temperature are shown in Table A7.11, the range observed was from 0.77 to 1.10 g/cm^3 and the related particle density classes were numbered from 1 to 12 for 12.77-mm diameter (1/2") and L1 to L12 for 15.88-mm diameter (5/8").

Theoretically, the critical density of simulated particles has to be slightly lower than that of the carrier fluid at process temperatures (Simunovic, 1998). Since it was not feasible to measure the density of the carrier fluid at sterilization temperatures, or to construct simulated particles for which the temperature/density dependence would closely mimic that of

TABLE A7.11   SIMULATED PARTICLES USED IN THE DENSITY SCAN

Code	Diameter (mm)	Density (g/cm³)		
1	12.77	0.777	±	0.007
2	12.77	0.821	±	0.008
3	12.77	0.878	±	0.012
4	12.77	0.905	±	0.011
5	12.77	0.884	±	0.014
6	12.77	0.923	±	0.007
7	12.77	0.991	±	0.007
8	12.77	1.008	±	0.015
9	12.77	0.978	±	0.010
10	12.77	1.014	±	0.015
11	12.77	1.058	±	0.010
12	12.77	1.109	±	0.004
L1	15.88	0.779	±	0.023
L2	15.88	0.809	±	0.010
L3	15.88	0.816	±	0.014
L4	15.88	0.844	±	0.005
L5	15.88	0.880	±	0.006
L6	15.88	0.905	±	0.002
L7	15.88	0.934	±	0.007
L8	15.88	0.952	±	0.008
L9	15.88	0.980	±	0.004
L10	15.88	0.992	±	0.009
L11	15.88	1.037	±	0.008
L12	15.88	1.064	±	0.010

the carrier fluid, an initial approximation was applied, assuming the temperature dependence of density of the carrier fluid would follow the same trend as water; thus, at 125°C, it would be ~0.939 g/cm³. This value was used as the referent midpoint for the population of density-adjusted fabricated particles, assuming a minimal temperature-related change in the effective density of polymer particles.

These particles were inserted into the continuously flowing soup formulation, which was subjected to processing to representative sterilization-level temperatures (125°C target within hold tube) and flow rate (3 gpm) to monitor and record their residence times in the system and determine the critical density.

Residence time was monitored and recorded during the flow through the system using the Aseptia/Thermalytics Particle Flow Monitoring System and real-time recording and analysis software, and recovered to determine and analyze the residence time distributions.

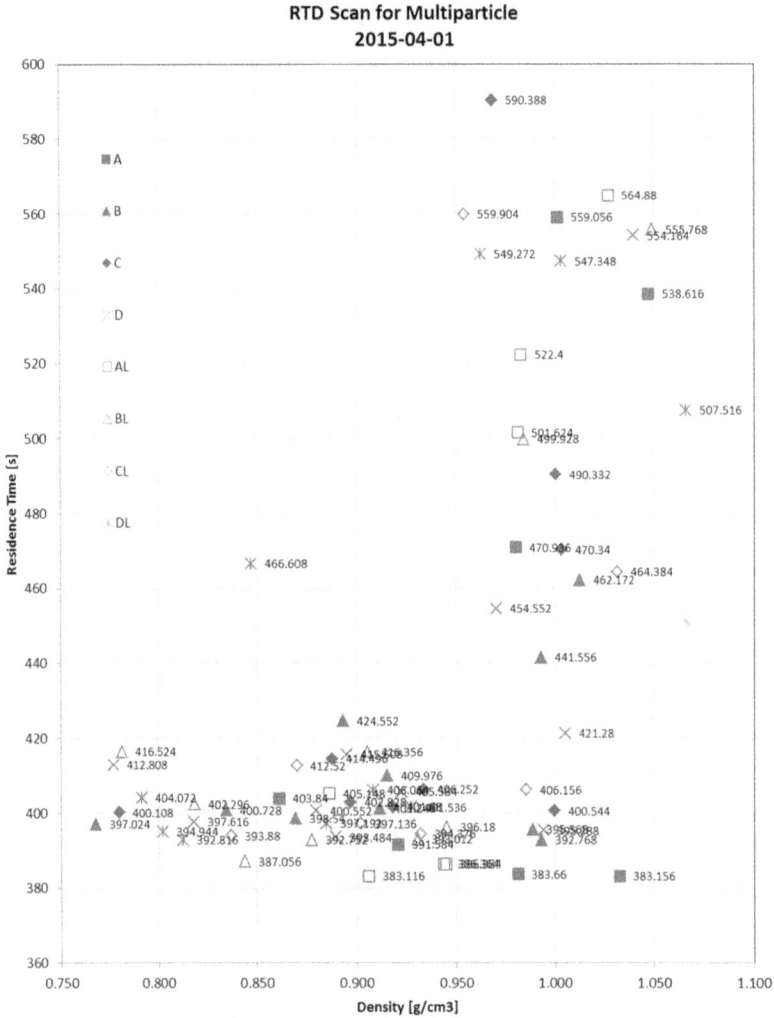

Figure A7.13 Residence time as function of density for product to be validated—legend corresponds to the nomenclature in Table A7.11.

This experiment was performed using four particles of each adjusted density category (A,B,C,D for each number) and the experiment was carried out in duplicate and the results are presented in Figure A7.13.

The residence time of the density-adjusted particles for ½″ and 5/8″ is shown in Table A7.12, with the 5/8″ particles data *in red italics font*. Data shows a trend in which particles with density above 0.95 have a very large residence time distribution, with a minimum of 383 and a maximum of 595 seconds (average 543 ± 131 seconds). In contrast, particles with density between 0.85 and 0.95 have a residence time with a minimum of 383 and a maximum of 424 (average 408 ± 32 seconds). It is also very interesting to observe that particles of different sizes, but within the same density class, have very similar RTD. Thus, the critical density was determined to be 0.92 ± 0.05 g/cm³, which results in a residence time of 410 seconds ± 35 seconds with a minimum

TABLE A7.12   RESIDENCE TIME DISTRIBUTION PER DENSITY CATEGORY FOR 1/2" (12.77 MM) PARTICLES AND 5/8" (15 MM) PARTICLES

Density range (g/cm³)	Minimum (s)	Maximum (s)	Average (s)	σ (s)
0.74-0.85	397.02	412.81	401.66	6.43
	387.063	*466.61*	*407.27*	*25.61*
0.85-0.90	393.48	424.55	406.07	9.96
	387.06	*466.61*	*408.34*	*26.98*
0.90-0.95	391.58	409.98	402.72	6.33
	383.12	*559.90*	*410.95*	*50.32*
0.95-1.00	383.66	590.39	433.15	62.71
	406.16	*747.19*	*550.63*	*99.04*
1.00-1.15	383.16	834.83	543.70	131.88
	464.38	*774.05*	*584.86*	*95.97*

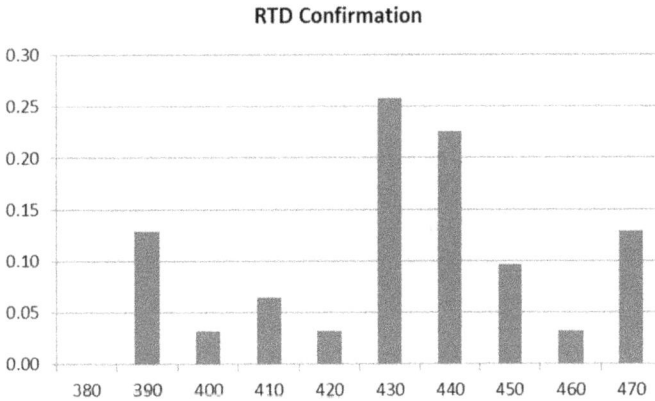

Figure A7.14   RTD confirmation of critical density.

of 383 seconds. This converted into flow correction factor is a minimum of 0.72 and average 0.77. Based on this data, a confirmation experiment was carried out that will confirm the distribution observed.

Confirmation of the critical density was carried out by an experiment in which 50 particles of ½" with initial density of 0.94 g/cm³ were inserted and tracked at the same conditions as used for the critical density determination. The residence time distribution of this population was measured and can be seen in Figure A7.14 and Table A7.13. From the observations in Table A7.12, it was expected to have a residence time of 402 ± 6 seconds with a minimum of 391 seconds.

The average residence time of the simulated particles was 432.5 seconds (correction factor 0.81) and the fastest observed had a RT of 382.9 seconds, which is within the expected range. It

TABLE A7.13   RTD FOR CONFIRMATION OF CRITICAL DENSITY

	Time	Factor
Flow average	530.8	1.00
Minimum (s)	382.9	0.72
Average (s)	432.5	0.81
Maximum (s)	527.1	0.99
σ (s)	23.8	

DEPARTMENT OF HEALTH AND HUMAN SERVICES   Public Health Service

Food and Drug Administration
College Park, MD 20740

October 15, 2015

Pablo Coronel
Process Authority
Aseptia Technologies, Inc./Wright Foods
723 West Johnson Street,
Raleigh, NC 27603

Dear Dr. Coronel:

We are responding to the information that your firm voluntarily submitted on June 25th 2015 regarding the processing and validation of flows with large particulates (up to 5/8") applied to a multiple-particle product, tomato soup with beans, carrots and onions.

We have no additional questions at this time about the information you submitted. The Wright Foods Food Canning Establishment (FCE) may, as appropriate, refer to the information you submitted when filing scheduled processes for a multiple-particle product, tomato soup with beans, carrots and onions. As always, it is the continuing responsibility of the FCE to ensure that their food products are in compliance with all applicable statutory and regulatory requirements.

If you have any additional questions, feel free to contact us.

Sincerely,

Nathan Anderson, PhD
Food and Drug Administration
Center for Food Safety and Applied Nutrition
Process Engineering Branch

cc:
HFS-302 (Mignogna, Brecher, Mallory, Geffin)
HFC-450 (Anderson)

Figure A7.15   LONQ awarded by FDA.

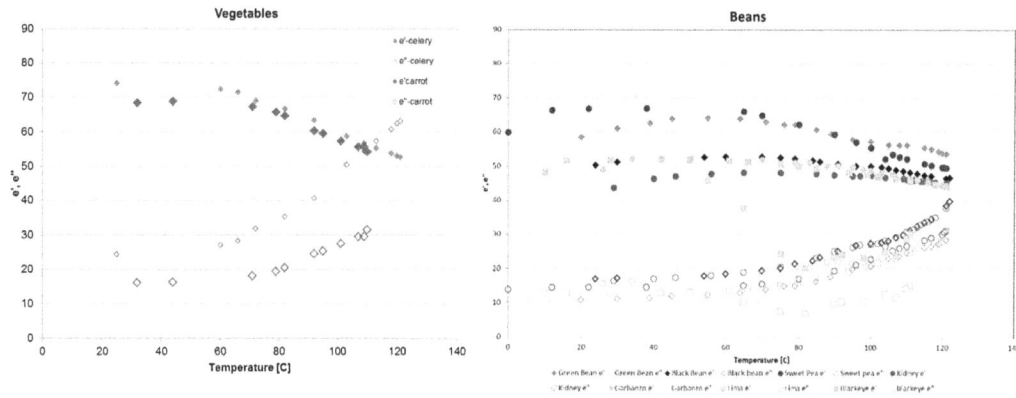

Figure A7.16 Dielectric properties of the ingredients.

TABLE A7.14 RESIDENCE TIME DISTRIBUTION RESULTS—MICROBIAL CHALLENGE

	½"		5/8"		Total	
	Time	**Flow Factor**	**Time**	**Flow Factor**	**Time**	**Flow Factor**
Average fluid (s)	530.87					
Min (s)	396.1	0.75	394.3	0.74	394.3	0.74
Max (s)	437.0	0.82	428.3	0.81	437.0	0.82
Average (s)	413.3	0.78	406.1	0.76	409.3	0.77
σ (s)	10.33		8.47		9.95	

was observed that a few of the particles were slowed down in the temperature sensor locations, and thus some slow particles observed. However, the fastest particles observed in this test were even faster than the expected fastest ones, but within the expected range for the critical density (Figures A7.15 and A7.16; Tables A7.2–A7.8 and A7.14).

7.8.5 Conclusion

Based on these observations, a particle with an initial (ambient temperature) density of 0.92 ± 0.05 was recommended for conservative flow behavior. Flow factor observed had a minimum of 0.72 and this is recommended for simulations of the process and the setting of the process.

REFERENCES

Inoue, C., Versluis, P., Coronel, P., and Elberse, J.M.M. 2013. Effect of particle phase volume, shape and liquid phase concentrations on rheological properties of large particulate-liquid model food systems by using ball measuring system. *J. Chem.* 7:643–652.

Kechichian, V., Crivellari, G.P., Gut, J.A., and Tadini, C.C. 2012. Modeling of continuous thermal processing of a non-Newtonian liquid food under diffusive laminar flow in a tubular system. *Int. J. Heat Mass Transfer.* 55(21):5783–5792.

Ramaswamy, H.S., Abdelrahim, K.A., Simpson, B.K., and Smith, J.P. 1995. Residence time distribution (RTD) in aseptic processing of particulate foods: A review. *Food Res. Int.* 28(3):291–310.

Simunovic, J. 1998. *Particle flow monitoring in multiphase aseptic systems.* Doctoral Dissertation. North Carolina State University.

Appendix 8: Thermal Processing Methods

Jairus R. D. David

[Excerpted from *"Handbook of Aseptic Processing and Packaging,"* 2nd Edition. (2013).

Chapter 11, Thermal Processing and Optimization. J.R.D. David. Pages 167–186. CRC Press, Boca Raton, Florida.]

A8.1 CONVENTIONAL CANNING

Compared to pasteurization, canning is a severe form of heat treatment to specifically target and inactivate mesophilic spores of *Clostridium botulinum* and prevent economic loss in low-acid foods. The heat resistance of *C. botulinum* is characterized by a $D_{250°F}$ of 0.21 minutes and a z-value of 18 $F°$. The food must experience at least a 12D cycle reduction at the "cold point" to be made commercially sterile. This is also referred to as F_o of 3 minutes. In practice, a thermal process of equal to or greater than F_o of 3 minutes is delivered in closed pressure cookers or retorts above atmospheric pressure. Canning is sufficient to inactivate heat-resistant enzymes, microorganisms, and mesophilic spores that cause spoilage under normal room temperature storage conditions.

The food to be canned is filled in metal cans, glass jars, retortable semirigid plastic containers, or pouches, followed by double seaming or heat sealing with a cap or lid. Such containers are heated and cooled in a pressurized batch or continuous retort.

This type of conventional canning is also referred to as in-container terminal sterilization because the thermal process is delivered after filling and sealing operations. Both container and product experience an identical method of sterilization. Heterogeneous high- and low-acid particulated foods such as chunky fruits and soups with meats and vegetables can be processed by this method (9 CFR 1986). The technology of containers and seaming and capping are very well-established. Canned products are regulated by the Code of Federal Regulations (21 CFR 1987).

A8.2 PASTEURIZATION

Pasteurization is a mild heat treatment process for milk and fluid foods to specifically inactivate certain pathogenic vegetative microorganisms with low heat resistance. The usual minimum time–temperature combinations are low-temperature, long-time (LTLT) 145°F for 30 minutes or high-temperature, short-time (HTST) 162°F for 15 seconds. The process can be delivered either by batch or continuous heating. It is important to note that the heat treatment is not intended or sufficient to inactivate all spoilage-causing vegetative cells or any heat-resistant spores, if present. This fact determines a short keeping-quality period up to about 2–3 weeks under refrigerated conditions (less than 40°F). In other words, the finished product (low acid) is not commercially sterile. However, pasteurized high-acid fluid packaged via the hot–fill–hold method in a hermetically sealed container may yield a commercially sterile shelf-stable product.

Pasteurization is warranted for high-acid foods such as juices and beverages, and low-acid refrigerated foods such as milk and dairy products that have a rapid turnover in commerce. Pasteurized milk and dairy products are regulated primarily by Grade "A" Pasteurized Milk Ordinance (2009).

A8.3 THE "FLASH 18" PROCESS

The "Flash 18" process represents an interim milestone in optimization, between agitated retorts and UHT processing and aseptic packaging of foods. The "Flash 18" process is similar to the hot–fill–hold–cool process routinely used for processing of high-acid or acidified foods but is done for low-acid foods at very high temperatures.

In this procedure, low-acid foods such as soups containing particulates are brought to high temperature of 265°F in bulk, for example, through steam injection heating, and then pumped while at sterilizing temperature to a hot-fill operation, carried out under pressure to accomplish sterilization of the metal container and lid. Conventional filling equipment and steam-flow can sealers are housed in a pressurized room or vessel maintained at 18 psig of air pressure. Hot product enters the vessel at a temperature up to 265°F, followed by flash cooling to 255°F—boiling point at 18 psig of pressure. The filled and sealed cans are then processed through a continuous horizontal retort for a controlled hold time at 255°F to sterilize the inside surfaces of can and lid, and deliver the scheduled process time before final cooling and release to the outside through pressure-sealed valves. This system is used primarily for large institutional-size cans that would otherwise require long retort processes, yielding unacceptable product quality.

It is interesting to note that benefits of UHT sterilization have been achieved by partitioning bulk sterilization of low-acid food, followed by heat–hold–fill–hold–cool for sterilization of container and lid. This represents a definite departure from heat–hold–cool–aseptic fill Dole aseptic canner, in which UHT-sterilized low-acid food is promptly cooled prior to filling and double seaming. Superheated steam is used to presterilize the container, lid, and aseptic zone at ambient pressure. Also, in the "Flash 18" process, food safety constraints related to residence time distribution in hold tube for aseptic processing of low-acid foods containing particulates is overcome by "scheduled" hold time in individual hot-filled and hermetically sealed containers, in a continuous horizontal retort.

A8.4 ULTRAHIGH TEMPERATURE (UHT) PROCESSING AND ASEPTIC PACKAGING

Aseptic processing and packaging encompass the filling of sterilized and cooled food into presterilized containers, followed by hermetic sealing with a presterilized closure in a presterilized tunnel or aseptic zone. The principles and manufacturing mechanics for aseptic processing of foods is based upon theory, practice, and regulations originating from pharmaceutical, continuous dairy pasteurization, and classical canning industries.

Raw product sterilization is accomplished by continuous methods of heating at HTST (up to 265°F) or UHT (270–300°F) regimes. During UHT sterilization, a pumpable fluid food product is exposed to brief but intense heating, normally temperatures in the range of 130–145°C (265–295°F). Common holding times for fluid foods range from 2 to 45 seconds. The principles of thermal processing are similar to conventional canning; in this case, an equivalent process or $F_{process}$ is delivered at much higher temperatures allowing for shorter exposure times (David 1985).

UHT processing has been concentrated in the dairy industry. However, other products such as low-acid viscous liquids and high-acid particulate foods are also similarly processed. The benefits of aseptic processing stem from UHT sterilization of foods. These benefits depend on a high temperature being maintained for only a few seconds, as time–temperature response for microbial inactivation and for chemical reactions are different. In other words, heat treatment of products at much higher temperatures for only a few seconds can achieve sterilization with greatly reduced product sensory and nutritional damage. The benefits of UHT sterilization cannot be realized in conventional canning or terminal sterilization due to long process times needed to achieve high temperatures. Come-up times (CUT) associated with transient or unsteady-state heat transfer can account for significant thermal degradation yielding overcooked or mushy product. The other overriding mechanical constraint is the inability of containers to withstand high internal pressure in retorts corresponding to temperatures in the UHT regimes. In aseptic processing, this is overcome by partitioning UHT sterilization of raw food product from that of container and lid. Compared to raw product, container and lid may experience different and milder methods of sterilization. The containers and lids are sterilized using steam, hydrogen peroxide, heat of coextrusion, or irradiation, where approved. Also, other sterilants have been approved and are used (National Food Processors Association 2004). The resistance to heat transfer in UHT applications is not determined by container shape or dimension as it is in canning. This results in more uniform product quality over hours of production run, independent of container shape and dimension.

Finally, the cooled sterile product is filled into sterile containers and hermetically sealed in a presterilized and continuously decontaminated tunnel or aseptic zone. Similar to canning, the products are shelf stable with a shelf life of about 1–2 years at ambient temperatures. The shelf life is a function of the barrier characteristics of the container in terms of loss of moisture and oxygen transmission, which may possibly cause physicochemical changes, and not necessarily due to contaminating microorganisms. Storage at high temperatures may dramatically reduce the shelf life of heat-sensitive products or promote economic spoilage due to the presence of thermophilic anaerobes.

A8.5 ULTRA-PASTEURIZATION

Ultra-pasteurization refers to pasteurization at very high temperatures of 280°F or above for 2 seconds or longer. The objective is similar to a pasteurization process and further extends the shelf life of the product. However, this high-temperature process is sufficient to destroy a greater proportion of spoilage microorganisms leading to extended shelf-life (ESL) of about 6–8 weeks under refrigeration, compared to 2–3 weeks for traditionally pasteurized products. This process is also known as ESL technology. Ultra-pasteurization is not UHT sterilization as the processed product is not commercially sterile and shelf stable (refer to Chapter 18, Table 18.1). This process has been effectively used for half-and-half, chocolate and flavored milks, other dairy products,

TABLE A8.1   COMPARISON OF CONVENTIONAL CANNING AND ASEPTIC PROCESSING AND PACKAGING OF FOODS

Criteria	Retorting	Aseptic Processing and Packaging of Foods
I. Sterilization		
A. Product		
1. Temperature regime	220–250°F	HTST (180–220°F) and UHT (260–290°F)
2. Delivery	Unsteady state	Precise—square wave
3. Heat/cool lethality credit	Possible	Isothermal temperature in hold tube only
4. Process calculation		
Fluid	Routine–convection	Routine
Particulate	Routine–conduction or broken heating	Complex
B. Other sterilization (process equipment, container, lid, and aseptic tunnel)	None	Complex; many
C. Energy efficiency	Lower	30% saving or more
II. Quality		
A. Psychophysical or sensory	Mushy; not suitable for heat-sensitive and nutritional products	Superior; suitable for homogeneous heat-sensitive and nutritional products
B. Nutrient loss	High	Minimum
C. Value-added	Lower	Higher
Convenience	Shelf stable	Shelf stable and other features
Microwaveability	Semirigid containers (bowls and trays)	All non-foil flexible and rigid containers
D. Product quality	Dependent on container size and shape	Independent of container size and shape
III. Production aspects		
A. Container speed	High; 600–1,000+ containers per minute	Medium; 500–700 containers per minute is common; higher speeds via multiple lanes
B. Handling/labor	High	Low
C. Downtime	Minimal, typically at seamer and labeler	Resterilization due to sterility loss in sterilizer or filler
D. Versatility/flexibility for manufacture of a product in different container sizes	Different size containers need different process delivery and/or retorts	Need one or two aseptic fillers to fill different size containers

(Continued)

TABLE A8.1 (CONTINUED)   COMPARISON OF CONVENTIONAL CANNING AND ASEPTIC PROCESSING AND PACKAGING OF FOODS

Criteria	Retorting	Aseptic Processing and Packaging of Foods
IV. **Process deviation**		
A. Under processing	Reprocess intact container-in-product after removal of labels if possible	Reprocess product after destructive opening of containers
B. Survival of heat-resistant enzymes	Rare	Common in certain foods, especially in fluid milk; problem could be overcome, if system is designed properly (refer to Chapter 18)
V. **Spoilage analysis**		
A. Troubleshooting	Simple; preprocess, inadequate process and postprocess recontamination	Complex; need to deal with compromises in CIP and sanitation, and aseptic zone and its elements (refer to Chapters 12 and 13)
B. Traceability : container code versus case code	Usually not identical; may differ by 1/2 to 6 hours	Usually, identical
VI. **Low-acid particulate processing**	In practice	Work in progress by industry consortium and regulatory agencies; data from numerous high-acid particulate systems may be used for design basis of low-acid systems with additional validation and modeling
VII. **Postprocess prepackage sterile additives**	Not possible	Possible and in practice to add filter-sterilized enzymes (lactase), bioactive compounds, therapeutic agents, and probiotics

and non-dairy creamers in portion pack cups and tabletop containers. Ultra-pasteurized dairy products are defined and regulated by the Grade A PMO. Ultra-pasteurization gives producers great latitude to regulate the stock rotation, product velocity, or turnover rate, and reduction of spoilage returns due to postdated conventionally pasteurized products.

A8.6 REFRIGERATED ASEPTIC PRODUCTS

Several food companies retail their aseptically processed products such as puddings, refrigerated or chilled for diversification, volume saturation, market velocity, and repositioning. Placing shelf-stable products in the dairy case or in chilled cabinets in supermarkets may be one way of meeting the current consumer demand for "fresh-like" premium foods. Interestingly, refrigeration circumvents the regulatory requirements that govern thermally processed low-acid, shelf-stable foods. Low-acid vegetable and acidified products filled using aseptic fillers and processes that are not approved by the US Food and Drug Administration (FDA) or US Department of Agriculture (USDA) must be refrigerated throughout the supply chain.

A8.7 DEFINITIONS

The following are definitions of terms used in this Appendix.

Acid food is food with a natural equilibrium pH less than or equal to 4.6 and a water activity (a_w) equal to or greater than 0.85.

Acidified foods refer to low-acid foods to which acid(s) or acid food(s) are added to lower the finished equilibrium pH to 4.6 or below. Usually, acidified foods have a water activity (a_w) greater than 0.85.

Aseptic processing and packaging of foods is the filling of a commercially sterilized-cooled product into presterilized containers, followed by hermetic sealing with a presterilized closure in an atmosphere free of microorganisms (Canned Food 2007).

Aseptic processing, when used to describe a milk product, means that the product has been subjected to sufficient heat processing, and packaged in a hermetically sealed container, to conform to the applicable requirements of 21 CFR 113 and the provisions of Section 7, Item 16p of Pasteurized Milk Ordinance (PMO), and to maintain the commercial sterility of the product under normal non-refrigerated conditions (PMO 2009).

Commercial sterility of equipment and containers used for aseptic processing and packaging of food means the condition achieved by application of heat, chemical sterilant(s), or other appropriate treatment that renders the equipment and containers free of viable microorganisms having public health significance, as well as microorganisms of non-health significance, capable of reproducing in the food under normal non-refrigerated conditions of storage and distribution (21 CFR 113.3 2001).

Commercial sterility of thermally processed food refers to absence of disease-causing microorganisms, absence of toxic substances, and absence of spoilage-causing microorganisms capable of multiplication under normal non-refrigerated conditions of storage and distribution (APHA 2001).

F or F value is time required to destroy a given percentage of microorganisms at a reference temperature and z-value.

F^z_{Tref} is a more specific notation for F or F-value and is defined as the time in minutes at reference temperature of T_{ref} required to destroy a given percentage of microorganisms whose thermal resistance is characterized by z.

F_o is F value when T_{ref} is 250°F and z is 18 $F°$. ($F_o = F_{250}{}^{18}$).

$F_{required}$ is time in minutes required in a thermal process to achieve a desired degree of commercial sterility or "end point" or probability of a non-sterile unit (PNSU); it is defined by the equation

$$\left(F^Z_{Trerf}\right)_{required} = F_{required} = D_T\left(\log N_i - \log N_i\right)$$

$F_{process}$ is the time–temperature effect of a thermal process expressed as time in minutes at reference temperature T_{ref} for microorganisms with a given z-value; it is defined by the equation

$$\left(F_{T_{ref}}^{Z}\right)_{process} = F_{process} = \int_{t_i}^{t_f} \frac{dt}{10^{(T_{ref}-T(t))/z}}$$

Hermetically sealed container means a container that is designed and intended to be secure against the entry of microorganisms and thereby maintain the commercial sterility of its contents after processing (21 CFR 113.3 1994).

Low-acid food is food with an equilibrium pH greater than 4.6 and a water activity (a_w) equal to or greater than 0.85; excludes alcoholic beverages. pH 4.5 serves as a benchmark in the state of California. Tomatoes and tomato products having a finished equilibrium pH less than 4.7 are not classified as low-acid foods.

Pasteurization, pasteurized, and similar terms mean the process of heating every particle of milk or milk product in properly designed and operated equipment, to one of the seven temperatures given, and held continuously at or above the specified temperature and for specified hold time. The process can range from 145°F for 30 minutes to 212°F for 0.01 second (PMO 2009).

Shelf-stable food refers to food capable of being stored without refrigeration at ambient environmental conditions for 1–2 years.

Ultra-pasteurized, when used to describe dairy products, means that such products shall have been thermally processed at or above 280°F for at least 2 seconds, either before or after packaging, so as to produce a product that has an extended shelf-life under refrigerated conditions (PMO 2009).

A8.8 NOMENCLATURE

a_w: Water activity; a measure of available water in foods to support the growth of microorganisms. It is the ratio of vapor pressures of a food product to that of pure water at a specified temperature.

pH: A measure of the intensity of acidity or alkalinity. The negative logarithm of the hydrogen ion concentration.

T_{ref}: Reference temperature (°F); common reference temperatures are 250°F for low-acid foods, 212°F and 200°F for high-acid and acidified foods, and 145°F and 161°F for pasteurization of milk.

z: Temperature difference that causes a tenfold change in the rate of microbial destruction, in Fahrenheit (F°).

REFERENCES

21 CFR, Parts 110–113 and 114. 1994. Washington, DC: U.S. Government Printing Office.

9 CFR, Parts 308,318,320,327 and 381. 1986. Canning of Meat and Poultry Products, U.S. Department of Agriculture.

Canned Food: Principles of Thermal Process Control, Acidification and Container Closure Evaluation, 7th ed. 2007. Washington, DC: GMA Science & Education Foundation.

David, J.R.D. 1985. Kinetics of inactivation of bacterial spores at high temperature in a computer-controlled reactor. Ph.D. dissertation, University of California at Davis.

David, J.R.D., Graves, R.H., and Szemplenski, T. E. 2013. *Aseptic Processing and Packaging of Food: A Food Industry Perspective*, Chapter 2. 2nd Edition. Boca Raton, FL: CRC Press

Downes, F.P., and Ito, K., eds. 2001. *Compendium of Methods for the Microbiological Examination of Foods*, 4th ed. Washington, DC: American Public Health Association.

Grade "A" Pasteurized Milk Ordinance. 2009 Revision. U.S. Department of Health and Human Services, Public Health Service, Food and Drug Administration (FDA).

National Food Processors Association (NFPA). 2004. Technical Overview of Alternate Sterilants for Use in Aseptic Processing. National Food Processors Association Workshop, June, Arlington, Virginia

Index

A

For Product Safety Concerns and Information please contact our EU
representative GPSR@taylorandfrancis.com
Taylor & Francis Verlag GmbH, Kaufingerstraße 24, 80331 München, Germany

www.ingramcontent.com/pod-product-compliance
Lightning Source LLC
Chambersburg PA
CBHW080342220326
41598CB00030B/4582